Biochemistry
Molecules, Cells
and the Body

Biochemistry
Molecules, Cells and the Body

JOCELYN DOW
GORDON LINDSAY
JIM MORRISON
Division of Biochemistry and Molecular Biology
Institute of Biomedical and Life Sciences
University of Glasgow

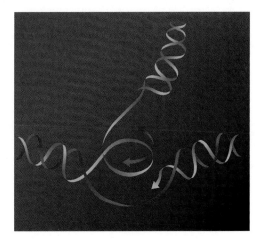

ADDISON-WESLEY
Wokingham, England • Reading, Massachusetts • Menlo Park, California
New York • Don Mills, Ontario • Amsterdam • Bonn • Sydney • Singapore
Tokyo • Madrid • San Juan • Milan • Paris • Mexico City • Seoul • Taipei

Acquisitions Editor: Jane Hogg
Production Manager: Stephen Bishop
Managing Editor: Alan Grove
Development Editor: Jane Bryant
Production Controller: Jim Allman
Marketing: Nicola Lyons
Sales Team Adviser: Samantha Eardley
Permissions Coordinator: Kristin Cooper
Editorial Assistant: Victoria Cook
Cover design: Designers & Partners, Oxford
Text design: Designers & Partners, Oxford;
Scribe Design, Gillingham
Illustrations: Gecko Limited, Bicester; Ian Ramsden,
Glasgow; Scribe Design, Gillingham
Typesetter: Scribe Design, Gillingham
Printed and bound in the United States of America

Text set in 10/12 pt Stempel Garamond and Helvetica

First printed 1995

ISBN 0–201–63187–3

British Library Cataloguing-in-Publication Data
A catalogue record for this book is available from the
British Library.

Library of Congress Cataloging-in-Publication
Data is available

Educational philosophy

Issues of content and style in medical education are currently the subject of intense debate within British education circles. The impetus for this has been the recognition that medical knowledge has grown at a rate which far outstrips the capacity of students to grasp all that is known within a five-year undergraduate curriculum. This has led to the recognition that medical education should focus first and foremost on those issues which are strictly relevant to the understanding of how the human body works and how it responds to disease. Many of the issues, including the integrated function of tissues and organs, the biochemical basis of good nutrition and human-specific mechanisms of growth, inheritance and disease prevention are not given adequate treatment in many current textbooks of biochemistry, which tend to be directed at students within science programmes. It is the purpose of this book to provide for students in medicine and related disciplines a concise but thorough treatment of the biochemical issues relevant to their studies, including a flavour of the most recent relevant scientific findings. We hope that this approach will fire students with an enthusiasm for understanding the human animal as a biological entity, its potential for development, its ability to thrive in the face of threats to its existence, and a curiosity to understand the unfolding biological map of our existence which will continue throughout their career.

It is a legitimate educational aim that students should learn to seek out information from sophisticated sources. However, most undergraduate medical courses are so heavily overloaded with information that students do not have the time this would require. While there are many academic arguments for educating medical students beyond the knowledge they will need as General Practitioners, the basic requirement must be to provide an understanding at the molecular level of how the human body functions: how it uses energy, maintains its structure, recognizes and responds to a wide variety of signals, develops and grows, protects itself against invasion by disease-causing organisms. A sound grasp of these principles should help doctors in all branches of medicine to contribute to improvements in the health of the community, the nutritional status of individuals, disease prevention, and the treatment of disorders afflicting the human body at all ages.

Concurrent with the creation of this textbook, the General Medical Council in Britain has produced its *Recommendations on Undergraduate Medical Education*. These recommendations have the aim of encouraging all medical schools in the UK to devise new approaches to presenting the curriculum and to identify the 'core' of knowledge required of a newly qualified doctor, so that the current 'factual overload' can be reduced. Throughout, the recommendations emphasize the need for knowledge of the 'sciences basic to medicine' and encourage continuing scientific education throughout a student's medical training, rather than confining it to the early years, as is currently common practice. They also encourage additional study in areas of interest to individual students and in-depth exploration of specific subjects.

We believe that this textbook will provide the 'core' knowledge of biochemistry and molecular biology for medical students, and the stimulus for students to seek continuing and expanding knowledge of at least some aspects of these subjects.

Biological knowledge is expanding at a hitherto unimagined rate, and as teachers we will have failed if this new knowledge is not applied to both prevention and treatment in all branches of medicine. Paradoxically, this information explosion is in some ways simplifying the overall picture of how the human body works. Many phenomena traditionally regarded as being within the realms of cell biology, physiology or pharmacology are now explained in molecular terms, bringing clarity of understanding to previously ill-understood processes. We are learning that there are molecular processes common to a diverse range of cellular and tissue-related events. The Human Genome Project is defining normal and abnormal biological structures and functions in molecular terms. Already an improved understanding of disease has come from gene-based identification of proteins present in such small quantities as to be undetectable by conventional approaches. These techniques also underpin exciting developments in understanding the life cycle of disease-causing organisms and how we combat them and are providing new tools for forensic medicine.

Approach and organization

The overall approach of the book is to present the pertinent body of material at a level and in a manner that is appropriate for the target readers. Unnecessarily detailed descriptions of topics have been avoided, as has the use of bacterial examples. The understanding of eukaryotic systems in general and of humans in particular has developed to the point where repeated reference to *Escherichia coli* is no longer necessary.

The general philosophy of the authors has been to progress from the molecular, through the cellular level to consideration of organs and eventually the whole body. In this way, it is hoped that the relevance of understanding the structures and behaviour of molecules in the body will become apparent to the student. We also anticipate that instructors will add their own particular expertise, interests and knowledge to expand on parts of the course as suits their own situation. It is likely that, with the planned development and use of 'problem-based learning' methods, a textbook will increasingly become a 'home base' for the student, rather than the fount of all wisdom.

The organization of the various parts of the book reflects the molecules-to-whole body approach. Part A deals with the molecules that the student will encounter and outlines the salient points of cell biology, with an emphasis on membranes and their roles in cells. The principal features of 'molecular biology' and the transmission of information in biology is the theme of Part B, where the most important aspects of proteins, RNA and DNA are considered in turn. Part C covers the generation of useful energy and the essential aspects of the metabolism of carbohydrates, lipids and nitrogenous molecules in the body. The focus switches more to organs and whole body processes in Part D, where discussion of hormone action leads to consideration of blood, the metabolism of tissues, the structural components of the body and nutrition. Part E deals with biochemical aspects of areas that have traditionally been taught by other disciplines – immunology, genetics and microbiology. The boundaries of biology, however, have become irrelevant to the extent that it is no longer appropriate to ignore these areas in any discussion of biochemistry in medicine.

Boxes have been used throughout the text to discuss areas that are not central to the thread of the chapter, but supplement the text by, for example, giving a flavour of more advanced topics, such as might be pursued in the 'new approaches' to the medical curriculum. Examples of clinical relevance are interspersed throughout the text focusing on topics which can be explained in biochemical detail. Throughout the book extensive use has been made of cross-referencing, using the steroid nucleus as a 'molecular' icon and the rod of Aesculapius as the 'whole body' icon. These cross-references direct students to related topics in other parts of the book. Much of the artwork is illustrative for there are many examples of biological structures which have not yet been described in detail. For some there is sufficient information to allow biochemists to make intelligent guesses about how structures might be formed, but some of these will undoubtedly turn out to be imperfect estimates.

Note to students

Many years of contact with medical and other undergraduates have made us aware that the features that students look for in a textbook are not always those that authors provide. Our motivation in writing this book has been primarily to produce a text that is at an appropriate level and with the correct balance of detailed fact and explanation. Our hope is that students will find that this balance has been struck.

The structure of the book is progressive: broadly, Part A deals with molecules; Part B with cells; Part C with what is often referred to as 'metabolism'; and Part D with tissue and whole body biochemistry. Part E covers topics that are not traditionally seen as 'biochemistry', but these days there are so many biochemical aspects of immunology, genetics and microbiology that these topics cannot be ignored in a text such as this. Clinical teachers will emphasize the aspects of these specialities which are the common experience of doctors in community and hospital practice.

Glossary of frequently-used illustrative features

Cross-references icon – 'whole body' subjects	Heart	Erythrocytes (red blood cells)
Cross-references icon – 'molecular' subjects		Cascade arrow (amplification)
Orthophosphate ion	Liver	Membrane bilayer
Pyrophosphate ion		
Phosphate group attached by a 'low-energy' bond	Brain	Transport in a membrane bilayer
Phosphate group attached by a 'mid-energy' bond		Ion channel in a membrane bilayer
Phosphate group attached by a 'high-energy' bond	Adipose tissue	
Skeletal muscle	Spleen	Acetyl CoA — Activation of an enzyme by acetyl CoA
		ADP — Inhibition of an enzyme by ADP

Molecular biology and cell biology are growing closer together and more aspects of clinical medicine can now be explained in terms of these disciplines. A glossary of molecular biological terms is provided at the end of Part B.

In many medical schools major changes are being made in the way the curriculum is delivered. The Glasgow Medical Faculty, where the authors of this book teach, will introduce a radically new curriculum in which all of the traditionally separate subjects of the medical curriculum are integrated and learning will be through a problem-based approach. A major feature of this curriculum will be the opportunity for students to seek out knowledge for themselves using all available sources of information rather than have all of the information presented to them in lectures. We hope that this textbook, with its new approach to the biochemistry of human life, will be a source of information and understanding not only for our students but for those in other schools, regardless of the approach to curriculum delivery.

Finally, we hope that our colleagues will find this a useful addition to the library of biochemistry and will feel uninhibited about sending us constructive comments and suggestions for future editions.

Jocelyn Dow
Gordon Lindsay
Jim Morrison

October 1995

Division of Biochemistry and Molecular Biology
Institute of Biomedical and Life Sciences
University of Glasgow

We are indebted to colleagues across the world who read all or part of the text of this book and offered us their views. Particularly we thank Professor Gerry Harrington, Dr John Leaver, Dr Mary Cotterell and Dr G. Kenneth Scott from Auckland for their detailed comments and suggestions. Within our own department we thank many colleagues who contributed by answering our questions and discussing aspects of the text. In particular, we thank Dr Bill Cushley for his contribution to Chapter 16.

It has been a pleasure to work with Alan Grove and others at Addison-Wesley, whose high degree of professional skill is evident in this book. Most of all we acknowledge that without the ministrations, organization, perseverance and skill of both Jane Bryant and Jane Hogg this textbook might still be no more than an idea.

Figures

Figure 1.2, *TIBS* March 1993. Figure 2.6c, micrograph by Don Fawcett, M.D. Figures 2.6d, e, 3.5, 3.10, 3.13, 3.23, 4.5, 15.1, 15.2a, 15.11, 15.14, from W.M. Becker, *The World of Cell*, Benjamin/Cummings. Figures 2.17, 15.8, from D. Voet, J.G. Voet, *Biochemistry*, 1990. Reprinted by permission of John Wiley & Sons, Inc. Figures 3.2, 3.5, 6.1, 13.24b, 15.2 b, c, from C.K. Mathews, K.E. van Holde, *Biochemistry*, 1990, Benjamin/Cummings. Figures 2.6a, 6.1, Irving Geis, Figures 3.3, 15.21, 15.24, From L. Stryer, *Biochemistry*, 1988, W.H. Freeman. Figure 3.7, A.L. Lehninger, *Principles of Biochemistry*, 1982, Worth Publishers. Figures 3.8a, 9.28, courtesy of Dr G.B.M. Lindop, University of Glasgow Department of Histopathology, Western Infirmary, Glasgow. Figure 3.8b, Glasgow University Pathology Department, W.I.G., Electron Microscope Unit. Figure 3.9a, courtesy of Barbara Hamkalo. Figure 3.11b, courtesy of Dr Susumu Ito. Figure 3.19b, 3.20, Box Figure 17.3 1, from C.J. Avers, *Molecular Cell Biology*. 1986 Addison-Wesley Publishing Company. Figure 3.21a, from Dr R.A. Milligan, 3.21b, from Dr C.M. Feldherr, 5.17, B. M. Austen, O.M.R. Westwood, *Protein Targetting and Secretion*, 1991, IRL Press. Figure 3.27, J.R. Marx *Science* 238 (1987) 615–616c 1987 by the AAAS provided by R. Lefkowitz. Figure 3.28, reproduced by permission, J. Ramachandran and A Ullrich, *Trends in Pharmacological Science* 8 (1987) 28–31; Elsevier Science Publishing Company Co., Inc., Figure 4.3a, A.G. Amit et al., *Science* 233 (1986) 747–753; © by the AAAS. Box Figure 5.3b, P. Chambon, *Sci. Am.*, 245 (1981) 60–70: © 1981 by Scientific American, Inc. All rights reserved. Figure 5.3, reprinted by permission of Blanchetot, *Nature* 301 (1983) 732–734. Figure 5.26, D.M. Engelman and T.A. Steitz, *Cell* 23 1981, 411. Figure 6.2, reprinted by permission from *Nature* 171 (1953), 740; © 1953 Macmillan Magazines Ltd. Figure 6.5, J. Cairns, *Cold Spring Harbor Symp. Quant. Biol.* 28 (1963):44. Figure 6.8a, reprinted by permission from *Nature* 369 19 May p211 © 1953 Macmillan Magazines Ltd. Figure 7 Box 7.4, Figure 7.26a, A. Tzagoloff, Mitochondria 1982, Plenum. Figure 9.28, 13.6a, 13.16b, c, 14.2, University of Glasgow, Western Infirmary, Glasgow. Figure 12.6, © Lennart Nilsson/Boehringer Ingelheim International, Dr Henry Slayter. Figure 13.1, P. Julien, J.-P. Despres and A. Angel, *J. Lipid Res.*, 30 (1989) 293–299 used by permission. Figure 13.6a, courtesy of Dr

W. M. H. Behan, University of Glasgow Department of Histopathology, Western Infirmary, Glasgow. Figure 13.14, 16.3, adapted with permission from *Molecular Cell Biology* by James Darnell et al., © 1986 Scientific American Books, Inc. Figure 13.16a, courtesy of Dr A K Foulis, University of Glasgow Department of Pathology, Royal Infirmary, Glasgow. Figure 13.24a, R.G. Kessel and R.H. Kardon, *Tissues and Organs: A Text Atlas of Scanning Electron Microscopy*, © 1979 W.H. Freeman and Co. Figure 15.4, courtesy of J. Heuser. Figure 15.5, courtesy of Hugh Huxley, Brandeis University. Figure 15.7, courtesy of T. Pollard. Figure 15.18, J.A. Buckwalter and L. Rosenberg, *Collagen Relat. Res.* 3(1983) 489–504. Figure 16.5, Arthur J. Olsen, PhD. *Computer graphics modelling and photography*, 1986 Research Institute of Scripps Clinic. Figure 16.7, A.G. Amit, R.A. Mariuzza, S.E.V. Phillips and R.J. Poljak, *Science* 233 1986) 749. Figure 16.9, courtesy of Dr Eckhard Podack. Figure 16.18, *Nature* vol. 364 (1993) no 6432, 1st July. Figure 17.2, Courtesy of Brenda Gibson, Figure 17.3b, c, Elizabeth Boyd, Figure 17.3a, Conner M. and Fergusson-Smith, M, *Essential Medical Genetics, 1993, Blackwell Scientific Publications.* Figure 18.4a, *Proceedings of the Natational Academy of Science* 90, 508 (1993) Fig 1d.

Review board

SHORT CONTENTS

■ Preface   v

■ Acknowledgements   ix

■ **PART A**   MOLECULES AND CELLS   1

     Chapter 1   Biochemistry, biology and medicine   3

     Chapter 2   Small molecules   11

     Chapter 3   Cells, organelles and membranes   39

■ **PART B**   INFORMATION IN BIOLOGICAL SYSTEMS: ITS FUNCTION, EXPRESSION AND STORAGE   73

     Chapter 4   Proteins as functional units: sequence, structure and modification   81

     Chapter 5   Genes and their expression: RNA in the generation and translation of messages   109

     Chapter 6   DNA as the store of biological information: the nature and replication of genomes   145

     Glossary of molecular biological terms   173

■ **PART C**   CELLULAR BIOCHEMISTRY   179

     Chapter 7   An introduction to metabolic processes and energy production in mitochondria   181

     Chapter 8   The metabolism of glucose: storage and energy generation   221

     Chapter 9   Lipid metabolism   261

     Chapter 10   Metabolism of nitrogen-containing molecules   293

■ **PART D**   TISSUES OF THE BODY: THEIR STRUCTURE, FUNCTIONS AND INTERACTIONS   329

     Chapter 11   Communication between cells and tissues: hormones, transmitters and growth factors   331

     Chapter 12   Blood: molecular functions of cells and plasma   355

     Chapter 13   The molecular and metabolic activities of multicellular tissues   379

     Chapter 14   From food to energy: molecular aspects of digestion and nutrition   411

     Chapter 15   The structural tissues   455

■ **PART E**   MOLECULAR MEDICINE: IMMUNOLOGY, GENETICS AND MECHANISMS OF INFECTION   481

     Chapter 16   Immunoglobulins and defence mechanisms of the body   483

     Chapter 17   The molecular genetic basis of disease   507

     Chapter 18   Invaders of the body   549

■ Index   585

CONTENTS

Preface v
Acknowledgements ix

PART A MOLECULES AND CELLS 1

Chapter 1 Biochemistry, biology and medicine 3
What is biochemistry? 4
Levels of organization in biology 5
Molecules, disease and medicine 9

Chapter 2 Small molecules 11
The atoms and molecules of life 12
Small molecules as building blocks 13
Box 2.1 The activation of small molecules
for reaction 14
Sugars: mono-, oligo- and polysaccharides 15
Fatty-acid containing lipids and cholesterol 23
Amino acids and polypeptides 28
Box 2.2 The ionization properties of
amino acids and peptides 29
Nucleotides and polynucleotides 32
Box 2.3 Nucleotide terminology 35
Molecules in cells and in extracellular fluids 37

Chapter 3 Cells, organelles and membranes 39
The structure of mammalian cells 40
Organelles and their functions 44
Box 3.1 Endocytosis and exocytosis 49
Cellular membranes 51
Membrane proteins 53
Dynamic behaviour of cell membranes 57
Specialized membrane structures in cells 60
Transport of molecules and ions across
membranes 62
Cell-surface receptors 70

PART B INFORMATION IN BIOLOGICAL SYSTEMS: ITS FUNCTION, EXPRESSION AND STORAGE 73

Function, message and store 76
Information in biological systems and its
transfer 76
Macromolecular sequences and their
significance 78
The Central Dogma and the transfer of
information 78
The medical relevance of the study of
biological information transfer 80

Chapter 4 Proteins as functional units:
sequence, structure and modification 81
Protein function 82
Location of proteins in the body: in solution,
in structures or in cell membranes 82
Box 4.1 Disulphide bonds and
extracellular proteins 83
Globular proteins: the binding of ligands 84
Box 4.2 Association of ligands with
proteins 86
Catalytic proteins: enzymes 88
Box 4.3 Measurement of enzyme
activity and kinetics 92
Box 4.4 Enzyme inhibition 93
Primary structure of proteins – amino acid
sequences 94
Box 4.5 General properties of
polypeptides 96
Box 4.6 Direct determination of the
amino acid sequence of a protein 97
Box 4.7 The identification of protein
sequences from nucleotide sequence data 98
The three-dimensional structure of proteins:
the importance of sequence 98
Box 4.8 The denaturation and
renaturation of proteins 99
Box 4.9 X-ray diffraction in biology 100
Box 4.10 Weak forces involved in
protein structure 102
Box 4.11 α-Helices, β-sheets and turns –
regular features in proteins 104
Functional aspects of protein structure 106

Chapter 5 Genes and their expression: RNA
in the generation and translation of messages 109
The nature and expression of genes 110
Transcription: copying of genetic
information from DNA and RNA 112
Box 5.1 Control of transcription in
bacteria and humans 117
Box 5.2 Study of ribosomal RNA
sequences has changed ideas about
the early evolution of life 120
RNA processing: producing functional RNA
molecules from transcripts 121
Box 5.3 The discovery and significance
of introns 124
mRNA translation: the machinery of protein
synthesis 127

mRNA translation: the genetic code and open reading frames   130

mRNA translation: initiation, elongation and termination of polypeptide chains   134

Targeting of proteins   136

    Box 5.4   Inherited disorders involving targeting defects   137

    Box 5.5   The absence of cholesterol makes the endoplasmic reticulum membrane more permeable than the plasma membrane   140

Gene expression in action: maintaining the level of cholesterol in cells   142

Chapter 6   DNA as the store of biological information: the nature and replication of genomes   145

The nature of genomes   146

The structure of DNA and its implications   147

The organization of DNA in genomes: chromosomes and genes   149

    Box 6.1   Some applications of the information from DNA sequencing   150

Packaging DNA within cells   150

Replication of DNA: initiation, fork movement and the attainment of fidelity   151

Damage to DNA and its repair: maintaining the integrity of the information store   160

    Box 6.2   DNA repair and xeroderma pigmentosum   164

Mutation: changes in DNA and their consequences   166

    Box 6.3   The Ames test – identifying mutagens   169

    Box 6.4   Cigarette tar, metabolic activation and carcinogenesis   169

Recombination of DNA   170

DNA, RNA and the origin of life   170

    Box 6.5   Was RNA the first genetic material in the origin of life?   170

Comparison of sequence information   171

    Box 6.6   Nucleic acid and protein sequences can be used for 'molecular phylogeny'   171

Glossary of molecular biological terms   173

PART C   CELLULAR BIOCHEMISTRY   179

Chapter 7   An introduction to metabolic processes and energy production in mitochondria   181

Energy transfer in metabolism   182

    Box 7.1   Thermodynamics of metabolic processes   189

Metabolism is regulated to increase the efficiency of energy consumption and production   194

Aerobic ATP production: the process of oxidative phosphorylation   199

    Box 7.2   The evolutionary origin of mitochondria   200

    Box 7.3   Iron-sulphur (Fe–S) centres in the electron transport chain   202

    Box 7.4   The chemiosomotic theory highlights the role of the proton gradient in driving energy-linked reactions in mitochondria   204

    Box 7.5   Cytochrome P-450   206

The citric acid cycle: the pathway for complete oxidation of acetyl coenzyme A   213

Chapter 8   The metabolism of glucose: storage and energy generation   221

The metabolic relationships between glucose and glycogen   222

    Box 8.1   Mechanisms for switching off hormone activation of cascades   230

    Box 8.2   Glycogen-storage diseases   230

The glycolysis pathway   231

Control of glycolysis   238

Metabolic fates of pyruvate   241

Gluconeogenesis: the synthesis of glucose from noncarbohydrate precursors   244

Catabolic reactions depend on reoxidation of reduced nicotinamide and flavin nucleotides   254

The pentose phosphate pathway   258

Chapter 9   Lipid metabolism   261

Properties and functions of fatty acids and triacylglycerols   262

Mobilization and oxidation of fatty acids   263

    Box 9.1   Implication of medium chain fatty acid acyl CoA dehydrogenase in 'sudden infant death syndrome'   267

    Box 9.2   Genetic defects in fatty acid oxidation   268

    Box 9.3   The generation and utilization of propionyl CoA   270

    Box 9.4   Ruminants produce and use more ketone bodies than humans   274

Fatty acid biosynthesis   274

Synthesis of triacylglycerols and membrane lipids   279

    Box 9.5   The active component of lung surfactant is a phospholipid   281

    Box 9.6   Snake venoms contain phospholipase A_2 284

Cholesterol: a major structural and functional lipid   284

Contents

Transport of fatty acids, phospholipids and
   triacylglycerols   285
   Box 9.7   Albumin is not essential for
      transport of fatty acids   287
   Box 9.8   Defects in the LDL receptor are
      the cause of hypercholesterolaemia   291

Chapter 10   Metabolism of nitrogen-
containing molecules   293
Catabolism of amino acids - the removal of
   amino groups   294
   Box 10.1   Distribution of specific
      aminotransferases provides diagnostic
      information when tissues are damaged   297
The urea cycle   301
   Box 10.2   Deficiencies in urea cycle
      enzymes lead to serious health problems   304
Biosynthesis of amino acids   306
   Box 10.3   Homocystinuria arises from a
      failure of cysteine metabolism   307
Catabolism of amino acids – the fate of the
   carbon skeletons   309
   Box 10.4   Alkaptonuria is caused by a
      genetic deficiency of homogenistate
      dioxygenase   311
   Box 10.5   Phenylketonuria is an inborn
      error of metabolism which leads to
      mental retardation   312
Amino acids are precursors for a wide range
   of specialized compounds that control
   metabolism or physiological functions   312
Nucleotide metabolism   315
   Box 10.6   A genetic defect in adenosine
      metabolism impairs the immune system   320
Folic acid and 'one carbon' metabolism   323
Porphyrin metabolism   326
   Box 10.7   Clinical jaundice is due to
      hyperbilirubinaemia   326

PART D   TISSUES OF THE BODY:
THEIR STRUCTURE, FUNCTIONS
AND INTERACTIONS   329

Chapter 11   Communication between cells
and tissues: hormones, transmitters and
growth factors   331
Mechanisms used by hormones acting
   through plasma membrane receptors   335
   Box 11.1   Cholera and whooping cough   337
Mechanisms used by hormones acting
   through intracellular receptors   341
   Box 11.2   Nitric oxide is a signal molecule   342
Synthesis, storage, processing, secretion and
   mode of action of selected hormones   342
   Box 11.3   The 'flight or fight' response is
      mediated by adrenaline   343

   Box 11.4   The brain produces insufficient
      dopamine in patients with Parkinson's
      disease   343
   Box 11.5   Graves' disease is an immuno-
      logical disorder affecting the thyroid
      gland   345
   Box 11.6   Goitre is a condition arising
      from dietary deficiency of iodine   345
   Box 11.7   Underactivity of the thyroid   346
   Box 11.8   The benefits of seed oil and fish
      oil fatty acids   349
   Box 11.9   Eicosanoid therapies   350
   Box 11.10   Hypertension and control of
      aldosterone production   353
Growth factors   353

Chapter 12   Blood: molecular functions of
cells and plasma   355
Plasma and its proteins   357
Blood clotting   360
   Box 12.1   von Willebrand factor –
      molecular biology and blood clotting   361
The erythrocyte and oxygen   364
The erythrocyte membrane   366
Haemoglobin – structure and function   369
Metabolism in the erythrocyte   375
Maintenance of the erythrocyte   377

Chapter 13   The molecular and metabolic
activities of multicellular tissues   379
The metabolic functions of adipose tissue   380
The metabolic functions of muscle   384
   Box 13.1   Creatinine production can be
      used to measure the amount of muscle
      tissue   385
   Box 13.2   Lactate dehydrogenase exists
      as several isoenzymes   387
Metabolic functions of the liver   388
   Box 13.3   Glucose-6-phosphate
      deficiency and gout   388
Metabolic functions of the kidney   390
Functions of the pancreas   394
   Box 13.4   C-peptide and insulin
      overdose   398
The molecular basis of electrical activity in
   tissues   401
   Box 13.5   Suxamethonium blocks the
      binding of acetylcholine   404
   Box 13.6   Myasthenia gravis   405

Chapter 14   From food to energy: molecular
aspects of digestion and nutrition   411
Digestion and absorption of nutrients   412
   Box 14.1   Stomach pH is maintained by
      active secretion of HCl   413
Digestion and absorption of protein   415

Box 14.2   Classification of proteolytic
    enzymes   416
Digestion and absorbtion of carbohydrate   417
Digestion and absorbtion of lipids   418
Nutritional aspects of metabolism   422
Fat in the diet   423
 Box 14.3   Lipids and heart disease   424
Carbohydrate in the diet   424
 Box 14.4   Glycogen stores in the neonate   426
Protein in the diet   428
Metabolic cooperation between tissues   431
Distribution of nutrients in the well fed state   432
Distribution of nutrient between meals   434
Distribution of nutrients in starvation   437
Degradation of cellular macromolecules is
 part of a process of repair and maintenance   440
Obesity   442
Thermogenesis   443
The role of vitamins in metabolism   443
Fat-soluble vitamins   445
 Box 14.5   Absorption and transport of
    vitamin E depend on circulating
    lipoproteins   447
Water-soluble vitamins   448
 Box 14.6   Scurvy   450
The role of inorganic minerals   451
 Box 14.7   Renal osteodystrophy   452

Chapter 15   The structural tissues   455
The structure of muscle cells   456
The extracellular matrix – connective tissue   467
 Box 15.1   Hyaluronidase   469
 Box 15.2   Destruction of connective tissue   472
 Box 15.3   Clinical disorders arise from
    errors in collagen synthesis   475
The mineral composition of bone   478
 Box 15.4   Calcification of arterial walls in
    atherosclerosis   478
 Box 15.5   Disorders of bone metabolism   479

PART E   MOLECULAR
MEDICINE: IMMUNOLOGY,
GENETICS AND MECHANISMS
OF INFECTION   481

Chapter 16   Immunoglobulins and defence
mechanisms of the body   483
The nature of the immune response   484
 Box 16.1   Congenital immunodeficiency
    diseases   487
 Box 16.2   Penicillin: hapten and
    immunogen   487
 Box 16.3   Polio virus vaccines   488
The structure of immunoglobulins   490
 Box 16.4   IgG: passive protection and
    damage   492

 Box 16.5   IgE and allergic responses   494
 Box 16.6   The complement cascade   495
The molecular and genetic basis of antibody
diversity   496
The functions of T lymphocytes   499
 Box 16.7   Cytokines   501
 Box 16.8   HLA and disease   503
 Box 16.9   MHC antigens in infection   505

Chapter 17   The molecular genetic basis of
disease   507
Genetics and medicine   509
 Box 17.1   Some genetic terms   509
DNA and chromosomes   513
 Box 17.2   Repeated DNA sequences   514
Chromosomes and cells   515
 Box 17.3   Human cells can be grown and
    studied in culture   515
 Box 17.4   Karyotype analysis of
    chromosomes   516
Chromosomal disorders   518
 Box 17.5   Two types of chromosomal
    translocations   520
Mitochondrial genome disorders   521
Multifactorial diseases   523
 Box 17.6   The link between genetic make
    up and response to environmental
    factors – the P-450 system and
    hydroxylation   524
Single-gene defects   525
Molecular aspects of cancer   528
Oncogenes and proto-oncogenes   529
 Box 17.7   Proto-oncogenes and their
    functions   532
Initiation of tumorigenesis   532
 Box 17.8   Inherited deficiencies in DNA
    repair are associated with increased
    cancer risks   533
Tumour suppressor genes (anti-oncogenes)   535
 Box 17.9   Protein p53, mutations and
    cancer   536
 Box 17.10   Cell proliferation or cell
    death?   537
Control of cancer   537
 Box 17.11   Cancer multidrug resistance
    and the ABC transporter protein
    superfamily   539
'Genetic engineering' and some medical
applications   539
 Box 17.12   Restriction endonucleases   541
 Box 17.13   DNA sequencing methodology   543
 Box 17.14   In vitro expression of human
    genes   544
 Box 17.15   Gene therapy for inherited
    defects   546
 Box 17.16   The human genome project   547

Contents

Chapter 18   Invaders of the body   549

The nature of infectious disease   550

Bacterial infections   551

Box 18.1   Some beneficial aspects of
bacteria   552

The control of bacterial infections   555

Box 18.2   Penicillins interfere with
bacterial cell wall synthesis   557

Box 18.3   The modes of action of some
antibiotics that inhibit bacterial protein
synthesis   558

The nature of viruses   560

Box 18.4   Other noncellular
infectious agents: viroids, plasmids
and prions   561

Box 18.5   Classification and general
features of human viruses   568

Viruses and disease   570

Box 18.6   Viral hepatitis   571

Box 18.7   Influenza A virus: humans,
ducks and pigs   572

Retroviruses, HIV and AIDS   574

The control of viral diseases   578

Box 18.8   Potential targets for antiviral
drugs   580

Box 18.9   Two developments in antiviral
chemotherapy   581

Parasitic diseases   582

Index   585

PART A

MOLECULES AND CELLS

CHAPTER 1
Biochemistry, biology and medicine

CHAPTER 2
Small molecules

CHAPTER 3
Cells, organelles and membranes

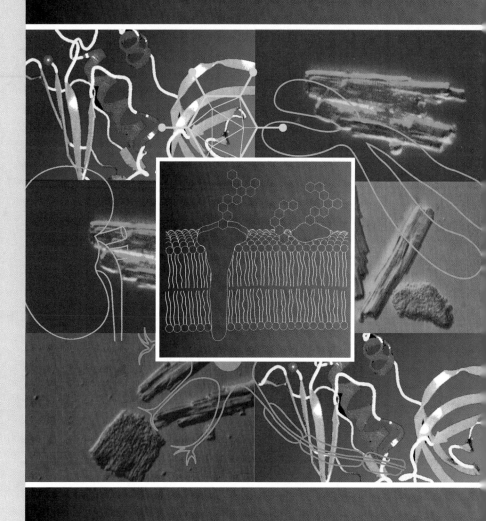

Biochemistry, biology and medicine

In this chapter we introduce the study of biochemistry in relation to biology and medicine. We look briefly at the way in which biochemistry developed and in particular its relationships with medicine and chemistry. Biochemistry plays a central role among the biological sciences, especially in its expanding relationship with genetics in the area of molecular biology. Perhaps the most distinctive feature of biochemistry is the essential contribution it continues to make to all areas of biology and medicine as they become more 'molecular' in their emphasis.

The principal theme of this book is that of considering the progressively increasing degrees of order from small molecules to the whole body. This chapter highlights the fact that organization is the 'key to life' and points out that increasing complexity may be discerned in the progression from small molecules and macromolecules, through organelles, cells and tissues to give a deeper understanding of the complexities of the functioning of the whole body.

What is biochemistry?

Biochemistry is a way of looking at living things

Biochemistry has been defined as 'the study of the molecules of living things'; however, such a definition is misleading because a study of the structures and properties of the molecules themselves gives only a *static* picture – a frozen frame – of what a cell, a plant or an animal is like at the molecular level. A very important aspect of biochemistry deals with dynamic questions such as these.

- How is a particular molecule made in the cell or body?
- What fates may it undergo?
- What is its function?
- How is its function related to its structure?
- How does it interact with other molecules?

Such questions give us clues to the nature of biochemistry. Biological sciences such as botany, zoology and microbiology are disciplines in the sense that they study identifiable subsets of the living world (plants, animals or microorganisms); biochemistry is more of an *approach* to the study of living systems. Thus, the philosophy, concepts and technology of biochemistry may be applied to the study of any aspect of biology – humans, mammals, insects, plants, bacteria, viruses – and biochemists specialize in each of these areas.

Biochemistry has always been closely related to medicine

From its earliest days, biochemistry has had an intimate association with human medicine. Indeed, most of the older-established departments of biochemistry in universities began in medical schools, often as offshoots of physiology departments, and were often called 'Physiological Chemistry' departments! It was realized that many of the 'systems' studied in physiology needed to be studied at the molecular level. Even today, when biochemistry has diversified into almost every area of biology, what might be called 'human biochemistry' remains a major part of the subject. A specialized branch of the subject, usually called 'Clinical Biochemistry', is directly associated with clinical medicine and most of its major specialties. All major hospitals have large and well equipped laboratories for the rapid and precise measurement of a wide range of substances in blood, urine and other samples derived from patients. Determination of the concentration of a body chemical (for example blood glucose) may be valuable in the diagnosis of disease and in monitoring the success of therapeutic measures.

Biochemistry has important links with many other sciences

Biochemistry also has strong connections with virtually every branch of the biological and medical sciences (Figure 1.1). The relationship with chemistry is not as close as might be imagined from the name, although in some areas (for example, study of the physical properties of biological molecules or the design of drugs) the relationship is intimate. Although organic chemists work predominantly with reactions occurring in organic solvents, biochemistry is very much concerned with *water-based* reactions, reflecting the importance of water in the functions of all living material at the molecular, cellular and higher levels of organization. Another distinctive feature of biochemistry is the high degree of interest in the **macromolecules** of living things – these include proteins, nucleic acids and polysaccharides (see below).

Chapter 1. Macromolecules are of central importance in the function of the body and its cells, page 6.

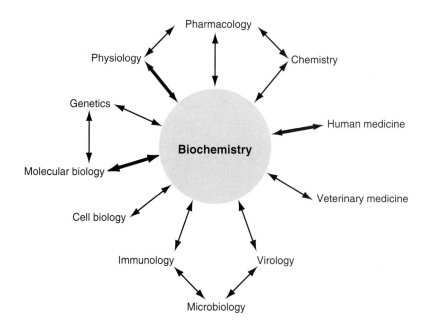

Biochemistry and molecular biology have large areas of overlap

In recent years the term **molecular biology** has been increasingly used in both the scientific and the medical communities. Its importance has been emphasized by developments in gene technology and the growth in 'genetic engineering'.

On the face of it, 'biochemistry' and 'molecular biology' mean essentially the same thing, but in practice 'molecular biology' means those areas of 'biochemistry' concerned with nucleic acids and proteins, but with a strong genetic flavour (see Part B). The very powerful combination of biochemistry and genetics has been used to solve many problems of biology and continues to push back the frontiers of knowledge and understanding of fundamental problems, many of which have great importance for the progress of medicine.

Levels of organization in biology and medicine

The most striking features of the range of living things at the macroscopic level are

- the **diversity** of organisms
- their degree of **organization**
- their **dynamic behaviour**.

As living things are studied in increasingly finer detail the degree of diversity diminishes; many similarities are discernible among very different organisms. Considerable molecular uniformity underlies the apparent diversity. However, the high degree of organization and the dynamic behaviour persist, no matter how minute the scale of observation.

It is possible to distinguish several levels of organization in humans and other animals (Table 1.1).

From the point of view of the study of the biochemistry of the body, the three most important levels (whole animal, cell and molecule) are shown in bold in Table 1.1, although macromolecules and their unique properties occupy a

Table 1.1 Levels of organization in humans and other animals

Population
Whole animal
Organ or tissue
Cell
Organelle
Macromolecule
Molecule
Atom
Subatomic particles

The general approach of this book is to start at the level of the **molecule**, especially the small molecules that form the essential components of the body, and then progressively to examine the ways in which the properties of these molecules give rise to the increasing levels of organization in both structure and function.

special place. The population level is generally beyond the immediate concern of this book, but is relevant to the areas of clinical biochemistry and biochemical genetics.

A wide variety of small molecules and ions exist in the body

The small molecules of living things include a wide range of organic compounds and a smaller number of inorganic ions (Table 1.2 and Figure 1.2). Most of these are found principally within cells but some are also found in extracellular fluids (see Chapter 2).

Many of the functions of cells are concerned with the metabolism of these compounds, breaking them down (**catabolism**) or building them up (**anabolism**). Other functions are involved in linking compounds together (**polymerization**) or moving them between compartments (**transport**) (see also Chapter 3). Nearly all of these functions are performed by proteins (see Chapter 4). Some small molecules also act as **signals**, which initiate, stop or control processes within cells (see Chapter 11).

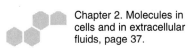

Chapter 2. Molecules in cells and in extracellular fluids, page 37.

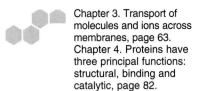

Chapter 3. Transport of molecules and ions across membranes, page 63. Chapter 4. Proteins have three principal functions: structural, binding and catalytic, page 82.

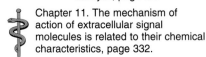

Chapter 11. The mechanism of action of extracellular signal molecules is related to their chemical characteristics, page 332.

An important distinction can be made between small molecules and macromolecules

The types of molecule found in all living things fall mainly into these two size ranges:

- relative molecular mass (M_r) below about 1500
- relative molecular mass (M_r) above about 10 000

and very few occur between these two values.

Molecules in the first group are often referred to as 'small molecules' in distinction to those of the second group (macromolecules), which are all polymers of a subgroup of the small molecules (mainly sugars, amino acids and nucleotides; see Chapter 2). Macromolecules, along with membranes (see below) are responsible for all the activities of cells and many of the properties of organs. Their importance cannot be overstated.

Chapter 2. Macromolecules are usually made from sugars, amino acids or nucleotides but can contain other components, page 13.

Macromolecules are of central importance in the function of the body and its cells

The biological macromolecules – proteins, nucleic acids and polysaccharides – represent the first level of order above that of 'molecule'. The macromolecules are polymers of certain of the small molecules – proteins are polymers of amino acids, nucleotides are polymers of nucleic acids and sugars are polymers of monosaccharides. Occasionally other small molecules are joined to them, as in the case of *glyco*proteins – proteins with covalently attached sugars. The degree of polymerization of proteins, polysaccharides and RNA molecules generally runs from a few tens to a few thousands of units, but DNA molecules may be many millions of units long.

Biological macromolecules fulfil a number of roles – structural (see Chapter 15), catalytic (see Chapter 4) and informational (see Chapters 5 and 6) – and their ability to bind to other molecules is often of great importance. The basic structure of molecules is formed by covalent bonds within them, but macromolecules also exhibit weak interactions, which play key roles in their three-dimensional conformations and in the ways they interact with other molecules. Such weak interactions confer properties on macromolecules that are essential for the dynamic behaviour of living things.

Chapter 4. Catalytic proteins: enzymes, page 88. Part B. Information in biological systems and its transfer, page 75. Chapter 6. The nature of genomes, page 146.

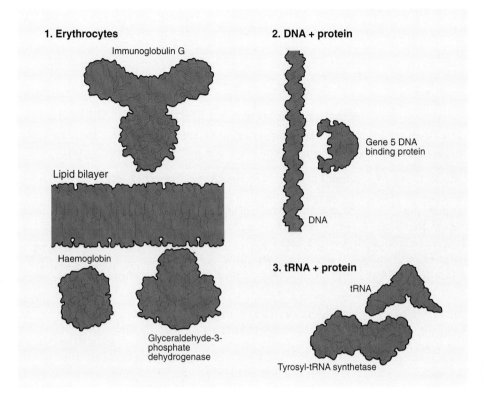

1. Erythrocytes

Immunoglobulin G

Lipid bilayer

Haemoglobin

Glyceraldehyde-3-
phosphate
dehydrogenase

2. DNA + protein

Gene 5 DNA
binding protein

DNA

3. tRNA + protein

tRNA

Tyrosyl-tRNA synthetase

Figure 1.2 Some molecular structures (showing relative sizes).

The formation of specific associations between molecules is vital

The binding properties of many macromolecules have already been mentioned; these include a number of crucial features (see Figure 1.2).

- The binding of small molecules by large ones, for example a substrate binds to an enzyme; a hormone to its receptor.
- The interaction between two or more macromolecules, often with very high specificity – for example, an antibody binds very specifically to its antigen; a restriction enzyme interacts specifically with its cutting site on DNA.
- The formation of more elaborate and complex structures, such as ribosomes, nucleosomes and cytoskeletal elements.

An appreciation of cell biology is of fundamental importance in understanding the function of the body

The cell is the operational unit of all organisms (except viruses: see Chapter 18). It is the highest level of organization in unicellular organisms such as bacteria and protozoans. Every human being starts as a single cell – the fertilized egg – and a complex 'programme' of cell division, growth and differentiation leads eventually to the neonate. This programme continues through childhood and adolescence into adult life. An important aim of cell biology and molecular biology is to understand how this programme is achieved and coordinated. An understanding of cell structure and function (see Chapter 3) is crucial to an appreciation of the normal function of the healthy body and of the pathological states found in disease.

Cell membranes form the boundaries of cells and enclose compartments

One of the principal extended structures found in all cells is the biological membrane, formed of proteins embedded in a **bilayer** of phospholipids and other

Table 1.2 Inorganic components of the body. Major inorganic components expressed as approximate percentage of human body weight

Anions	%	Cations	%
Na^+	0.2	Cl^-	0.2
K^+	0.4	PO_4^{2-}	1.0
Ca^{2+}	1.5	HCO_3^-	0.1
Mg^{2+}	0.1	SO_4^{2-}	0.1
Fe^{2+}	0.01	I^-	trace

Other trace elements: Zn^{2+}, Cu^{2+} Se, Mo, etc.

Chapter 18. The nature of infectious disease, page 550.

Chapter 3. Organelles and their functions, page 44.

lipids. Cell membranes form the boundary between a cell and its external environment and delineate the various cellular compartments, such as the nucleus, mitochondrion and endoplasmic reticulum (see Chapter 3). The components of these membranes are joined by weak noncovalent interactions, which allow a degree of mobility that is essential for membrane function.

Transport and communication are essential for cell and body function and depend on membrane structure and the binding properties of proteins

The existence of a boundary to a compartment means that molecules must be transported across the boundary if the compartment is not to be 'closed' and static. Beyond the cell, transport processes ensure the efficient movement of solutes around the body – again using the specific binding properties of proteins, for example haemoglobin binds molecular oxygen in order to transport it around the body. In multicellular organisms, communication mechanisms are also needed to ensure that the behaviour of the component parts of the organism is coordinated.

Survival of cells depends on their ability to derive energy from food

The fundamental unit of energy in cells is the purine nucleotide **adenosine triphosphate** (**ATP**), sometimes described as the 'energy currency' of the cell or as the 'common intermediate' of metabolism. These terms reflect the fact that energy derived from foodstuffs as they are oxidized is captured during the oxidation reactions and is transferred to a molecule of ATP. The energy of ATP resides in the phosphoanhydride bonds linking the second the third phosphate residues to adenosine monophosphate. This energy is released when these bonds are hydrolysed and may be transferred to other molecules by a variety of reactions or processes:

- *chemical synthesis* creates new molecules required for cell growth and repair
- *cellular processes*, such as the transport of molecules across cell membranes
- *whole body activities*, such as movement or temperature control.

Organs are specialized assemblies of cells; disease states often become apparent at this level

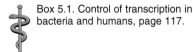
Box 5.1. Control of transcription in bacteria and humans, page 117.

Multicellular organisms may be regarded simply as collections of cells that have differentiated and grouped together into **organs** to perform particular tasks. Selective gene expression (see Chapter 5) plays a vital role in the differentiation of cells during growth and development. Tissue-specific expression of genes means that the organs of the body have distinctive biochemical features that suit them to their tasks. A knowledge of these features is important in understanding the normal functioning of organs and what can go wrong in disease states (see Chapter 13). Studies of epithelial and other specialized cells are beginning to reveal the ways in which cells interact structurally and cooperate metabolically (see Chapter 3).

Chapter 3. Specialized membrane structures in cells, page 60.

Different tissues provide specialist activities, which all contribute to an integrated body function

Many cellular activities are common to all body tissues; for example

- reactions through which cells derive energy from nutrients
- processes of cellular repair and cell division
- maintenance of the distinctive intracellular and extracellular ionic environments.

Each tissue also has specialist functions:

- *muscles* provide mechanical activity
- the *kidneys* control the composition of body fluids
- the *pancreas* synthesizes and controls the distribution of some hormones and digestive enzymes
- *adipose tissue* is an energy store and provides insulation and mechanical protection
- the *liver* has a vast repertoire of biochemical activities, including supply of nutrients to other tissues, synthesis of plasma proteins and disposal of xenobiotics and excess body nitrogen.

These differences are reflected in the structures of the cells in the various tissues, in the organelles they contain, in their molecular compositions and in their responses to hormonal signals.

Integrated function of the body depends on signalling between different cell types

The whole body is immense relative to molecules or cells, yet the enzymes of the metabolic pathways in the liver need to be able to respond rapidly to the energy needs of the brain and of muscle. In general terms, adipose tissue (fat) and the liver provide energy to the other tissues. Adipose tissue is a very efficient store of energy in the form of triacylglycerols. The liver stores relatively little energy but its metabolic processes are largely responsible for production of nutrients (glucose, ketone bodies, triacylglycerols), which it exports to other tissues.

These energy-providing tissues are informed that the other tissues require energy by primary signal molecules (**hormones**) circulating in the bloodstream and signalling the changing demands of one tissue to another. Hormones are produced in many different tissues and have many specialized functions, but all operate by binding to receptor proteins and setting in train processes that change the behaviour of molecules within those cells.

Molecules, disease and medicine

Blood can also be regarded as an organ and is central to body function in health and disease

The blood is not generally considered as an organ of the body because its cells are separate and not joined into a distinctive form but, in biochemical terms, it is just as much an organ as the liver or the brain (see Chapter 12). It occupies a central place in the overall biochemistry of the body – especially in the transport of gases, ions, nutrients and metabolites – and in the body's defence mechanism (see Chapter 16). Measurement of the concentrations of substances in blood is central to the diagnosis and monitoring of disease by the techniques of clinical biochemistry. For instance, the presence of abnormal amounts of tissue-specific molecules such as enzymes in the blood suggests organ-specific tissue damage.

The molecular basis of the body's response to disease is increasingly well understood

Proteins of the immunoglobulin superfamily play a vital part in maintaining the health of the body in the face of external assaults and internal wear and tear (see Chapter 16). Molecules of this class have the unique ability to bind very specifically to other molecules of many types. Soluble forms are important for inactivating circulating bacteria and viruses, while membrane-associated forms are involved

9

in cell-mediated responses to virus infection and to cells that have become cancerous. A complex genetic mechanism is responsible for the great variety of immunoglobulins and their relatives.

Some disorders, including cancer, have a genetic basis

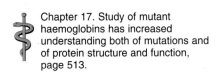

Chapter 17. Study of mutant haemoglobins has increased understanding both of mutations and of protein structure and function, page 513.

The human body is thought to possess no more than 100 000 genes, which function with remarkable fidelity. Errors in these genes can cause genetic disorders, which are fortunately mostly very rare (see Chapter 17). The study of genetic disease has yielded much useful information on molecular function within the normal individual. Most (if not all) cancers are known to be of somatic origin – that is, they arise from a single cell that has mutated, losing control of its growth and acquiring the ability to invade other tissues. The molecular basis of cancer is now almost completely elucidated.

Understanding the molecular nature of agents of infectious disease is important for prevention and cure

Many diseases are caused by infectious agents, especially bacteria and viruses. The molecular nature of these agents, their multiplication and their pathological effects, have been intensively studied and the understanding gained has improved techniques for prevention and cure (see Chapter 18). In particular, a knowledge of the processes of replication, transcription and translation of viruses has revealed the ways in which they multiply. This knowledge is now being applied to the control of viral disease by the development of more effective immunological and chemotherapeutic treatments.

CHAPTER 2

Small molecules

Small molecules have two main functions in the body – they are sources of useful energy and act as 'building blocks' for larger molecules and structures, especially macromolecules and membranes. These small molecules are composed principally of the atoms carbon, hydrogen, oxygen and nitrogen, with lesser amounts of sulphur and phosphorus. Certain groupings of atoms – carboxyl, hydroxyl, carbonyl, amino, sulphydryl and phosphoryl groups – are especially significant for the structure and function of small biomolecules.

The small molecules belong to four major groups: sugars; fatty acids and other lipids (notably sterols); amino acids; nucleotides. Each has its characteristic chemical and biological properties and is used in different ways in the body. In this chapter we outline the most important of these properties in preparation for discussions of metabolism, structure and function in later chapters.

Different forms predominate in the intra- and extracellular fluids of the body, as exemplified by the existence of glucose as its free form in plasma, but largely as phosphate esters within cells. The distribution of ions is also very variable, for example Na^+ is the major extracellular cation, while K^+ predominates within cells.

Table 2.1 Carbon, hydrogen and nitrogen in living things. Major elements expressed as an approximate percentage of human body weight

Oxygen	65
Carbon	18
Hydrogen	9
Nitrogen	4
All others	4

The atoms and molecules of life

The molecules of the body are made from carbon, hydrogen, oxygen and nitrogen, with lesser amounts of sulphur and phosphorus

It has long been known that life is based on carbon compounds; indeed 'organic' chemistry is the study of carbon compounds. *Carbon* is invariably accompanied by *hydrogen* – as in the hydrocarbons – and commonly also by *oxygen* as in the carbo*hydrates*. These three atoms form the sugars and their derivatives, the simple lipid molecules and comprise the bulk of the wet mass of all organisms (Table 2.1). The main mass of most organisms is water; all life is water based.

However, no life form is composed only of carbon, hydrogen and oxygen: *nitrogen* is essential; the growth of plants and microorganisms is often limited by the availability of nitrogen. Nitrogen is an essential component of the amino acids, the nucleotides, the amino sugars and some more complex lipids such as the phospho- and sphingolipids.

Phosphorus, in the form of phosphate (PO_4^{3-}) is also important, both as an inorganic constituent of blood and other fluids and as a component of many molecules, including nucleotides and phospholipids.

Sulphur is required in small amounts, principally in the amino acids cysteine and methionine

Characteristic groupings of atoms are found in many biomolecules

The hydrocarbons referred to above are examples of a simple arrangement of atoms found in many biomolecules. The geological hydrocarbons – oil and natural gas – are believed to have been derived from the material of ancient plants by chemical processes occurring in an anaerobic environment. They are composed of arrays of methylene ($-CH_2-$) and methyl ($-CH_3$) groups, which are predominantly linear, are termed **aliphatic** and are very nonpolar or hydrophobic in nature. Hydrocarbons also occur in closed or **alicyclic** forms. Aliphatic groupings of carbon and hydrogen also form important parts of the molecules of living organisms (Figure 2.1), especially in the fatty acids. Cholesterol is an important alicyclic biomolecule.

Biomolecules obey the laws of chemistry

Oxygen in biomolecules usually occurs as part of distinctive functional groups – hydroxyl, carbonyl, carboxyl, ether or ester (see Figure 2.1) – and the chemical properties of these groups give rise to the important functions and reactivities of biomolecules that will be discussed throughout the book. There is nothing 'magic' about biomolecules – they obey the laws of chemistry and no 'new' laws have been discovered from the study of living things.

Nitrogen occurs mainly in the form of amino and imino groups and is often found in molecules with cyclic structures, such as the purines and pyrimidines of nucleotides. Sulphur occurs in structures similar to those containing oxygen, such as thiols (sulphydryl groups) and thioesters, consistent with its relatedness to oxygen. Phosphorus almost invariably occurs as phosphate derivatives.

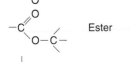

C, H	
$>CH_2$	Methylene
$-CH_3$	Methyl
$>CH$	Methenyl

C, O, H	
$>C=O$	Carbonyl
$>C=O$ (H)	Aldehyde
$-C$ O (H⁺) ⊖ O	Carboxylate
$-C$ O–C	Ester
$-C$ O–C	Ether

C, N, H	
$>C=NH$	Imino
$-C-NH_3^+$	Amino

C, O, N, H	
$-C$ O N–H	Amide

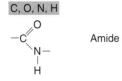

Figure 2.1 Groupings of carbon, hydrogen, oxygen and nitrogen commonly found in biomolecules.

Small molecules as building blocks

Biomolecules are either small ($M_r<10^3$) or large ($M_r>10^4$)

It is striking that, of the many thousands of types of molecules found in living things, there is a sharp distinction between two groups – the small molecules, with relative molecular mass of less than 10^3 (and mostly under 400) and the large molecules, or **macromolecules** (proteins, polysaccharides and nucleic acids), which have masses generally in excess of 10^4, and ranging to many millions. Very few biological molecules fall between these two ranges; those that do are often produced by cleavage from larger molecules by specific enzymic mechanisms.

Many of the small molecules found within cells have dual roles

- as metabolites
- as components of larger molecules.

The larger molecules include both the macromolecules (proteins, polysaccharides and nucleic acids) and the more complex 'small molecules' such as nucleotide and vitamin derivatives, phospholipids and porphyrins.

Macromolecules are usually made from sugars, amino acids or nucleotides but can contain other components

Organic components of macromolecules
The organic components of larger molecules include sugars, amino acids and nucleotides (which are themselves made from three smaller components – bases, sugars and phosphates). They may be modified by the addition of other organic groups such as methyl or acetyl groups, or inorganic components such as phosphate or sulphate (Table 2.2). Some macromolecules are composed of more than one type of building block – for example glycoproteins are proteins to which sugar residues are covalently attached.

Inorganic components of macromolecules
Several inorganic components are essential for the structure and function of biological molecules. Notable among these are phosphate, sulphate and a range of metal ions including iron, zinc, cobalt (see Chapter 14). Phosphate may be an integral part of the structure of the larger molecule, as in the nucleic acids and phospholipids, or it may be added later by **postsynthetic modification**, as in the phosphoproteins.

 Table 14.5. Trace elements and their functions, page 451.

Table 2.2 Components of macromolecules

Macromolecule	Essential 'building blocks'	Other components which may be found
Protein (polypeptide)	Amino acids	Phosphate (phosphoproteins) Sulphate Sugars (glycoproteins) Fatty acids Metal ions (metalloproteins)
Polysaccharide	Sugars	Amino sugars Amino acids (mucopolysaccharides) Sulphate
Nucleic acid (polynucleotide)	Nucleotides	Methyl groups

Activation of molecules for polymerization

When cellular processes lead to the combination of smaller molecules into larger ones during anabolism (biosynthesis), polymerization (or condensation) is typically preceded by the 'activation' of one of the smaller components. An important reason for this is to satisfy the energetic demands of the coupling reaction, as the unactivated components will rarely react efficiently. A second factor is probably to ensure that the components react to yield the desired product. Examples of such activation are shown in Box 2.1.

Box 2.1   The activation of small molecules for reaction

The invariable use of nucleotides as the triphosphate form

When nucleotides are used to form larger structures (nucleotide coenzymes, nucleotide sugars, polynucleotides) these are always in the form of the nucleoside *tri*phosphate. Typically, the high free energy of hydrolysis of the α–β phospho-anhydride (or **pyrophosphate**) bond provides the energy for the reaction and results in the addition of the nucleotide unit to the other components.

Amino acyl tRNAs in protein synthesis

The attachment of amino acids to their cognate transfer RNA before use in protein synthesis has a dual role:

1. It provides part of the energy required for the formation of the peptide bond; and
2. Perhaps more importantly, the presence of the tRNA component ensures that the correct amino acid is added into the growing protein chain (see Chapter 5).

Nucleotide sugars in polysaccharide synthesis

Although it is possible in the test tube to synthesize glycogen from glucose-1-phosphate (G1P) using the enzyme glycogen phosphorylase, this does not occur in the body, because the concentration of G1P required to drive the reaction in this direction is never attained within the cell. The realization of this dilemma led to the discovery of glycogen synthetase and the requirement for the nucleotide sugar uridine diphosphoglucose (see Chapter 8). The bond between the nucleotide moiety and carbon 1 of glucose activates this carbon atom and provides the energy for the formation of the 1→4 α glycosidic link in the polysaccharide chain.

Coenzyme A in activation of carboxyl groups

Carboxyl groups are not particularly reactive in themselves and many of the reactions in the cell that involve carboxyl groups, especially in fatty acid metabolism, require the carboxyl groups to couple to the sulphydryl group of coenzyme A. The resultant thioester bond has a very high free energy, which can be used to drive a range of coupling reactions in the production of acyl carrier protein for fatty acid synthesis and the production of complex lipids (see Chapter 9).

Chapter 5. Transfer RNAs have a distinctive three-dimensional structure, page 129.

Chapter 8. When glucose is plentiful it is stored as glycogen, page 224.

Chapter 9. Malonyl CoA is the key intermediate in fatty acid biosynthesis, page 275.

Chapter 3. Cellular membranes, page 51.
Chapter 9. Properties and function of fatty acids and triacylglcerols, page 262.

Chapter 14. Fat in the diet, page 423.

Lipids are not polymerized into macromolecules, but form part of extensive, noncovalent structures – the cell membranes

In contrast to the proteins, polysaccharides and nucleic acids, the other major structural and functional component of cells – the lipids – are not covalently linked into larger molecules. Fatty acids (in the form of phospholipids and sphingolipids) arrange themselves into lipid bilayers in which the membrane proteins are 'dissolved' to form the extensive sheets of the cell membranes enclosing both cells and organelles (see Chapter 3). Fatty acids are also the principal components of triacylglycerols (or triglycerides), which make up the fat deposits of the body. Triacylglycerols are the major lipid constituent of the diet (see Chapters 9 and 14).

Sugars: mono-, oligo- and polysaccharides

Sugars are the most abundant components of the biosphere

Sugars (or monosaccharides) are aldehyde- or ketone-containing compounds with multiple hydroxyl groups. They have the general formula $(CH_2O)_n$, where most commonly $n = 6$, as in the six-carbon hexoses, typified by glucose. Carbohydrates do occur as free sugars and are polymerized into short chains (oligosaccharides), but are mainly found as long polymers (polysaccharides), which comprise many of the important polymers found in nature – notably, starch, cellulose and glycogen.

Carbohydrates form the bulk of organic matter on earth and are fundamental to life itself because their formation by photosynthetic fixation of atmospheric CO_2 represents the only route by which new organic material can be generated in bulk. All animals, including humans, are ultimately dependent for their continued existence on this reaction, performed by plants and photosynthetic microorganisms. The amount of organic carbon produced from CO_2 by the biosphere is thought to exceed the total industrial output by several orders of magnitude.

Carbohydrates perform a number of important biological roles

Polysaccharides have a diverse range of functions. These are listed below.

Structural components

A range of polysaccharides is found in the connective tissues of the body (see Chapter 15). Cellulose, the principal structural component of plants, is the most abundant organic material on earth; other polysaccharides form major elements of the cell walls of bacteria and the exoskeletons of arthropods (for example, the major component of crab shells is chitin, a linear polymer of N-acetylglucos-amine).

Chapter 15. Mucopolysaccharides make up the carbohydrate content of the ground substance, page 468.

Stores of energy

Glycogen, occurring principally in liver and muscle (see Chapter 8), is the storage form of glucose in animals. Starch, the storage form of glucose in plants, is the major carbohydrate polymer in the typical human diet. Both of these long-chain polymers of glucose may be digested in the gut to yield glucose and thus energy (see Chapter 14). Cellulose is digestible only by herbivores; symbiotic micro-organisms in the gut of these animals degrade cellulose to glucose.

Chapter 8. The storage and mobilization of glycogen is determined by the energy status of the cell and by hormonal responses, page 224.
Chapter 14. Polysaccharides of glucose may take a variety of forms, page 425.

Sources of building blocks

As glucose passes through three major metabolic pathways (glycolysis, the citric acid cycle and the pentose phosphate pathway; see Chapters 7 and 8) metabolic intermediates derived from glucose may be diverted for use in the biosynthesis of a variety of molecules including amino acids, fatty acids and nucleotides. Nucleotides, the major building blocks of RNA and DNA, contain the pentose (five-carbon) sugars ribose and deoxyribose, respectively.

Chapter 7. The citric acid cycle: the pathway for complete oxidation of acetyl coenzyme A, page 213.
Chapter 8. The glycolysis pathway, page 231.
Chapter 8. The pentose phosphate pathway, page 258.

Cell surface functions

Many proteins and lipids on the surface of cells carry short, covalently linked chains of sugars. These glycoproteins and glycolipids are involved in a variety of important 'receptor' activities, associated with cell signalling, cell–cell interactions and recognition. They are also very important in influencing cell behaviour in response to external stimuli (see Chapter 11).

Chapter 11. Mechanisms used by hormones acting through plasma membrane receptors, page 335.

15

D-Glyceraldehyde-3-phosphate
(an aldose)

Dihydroxyacetone phosphate
(a ketose)

(a)

Figure 2.2 Structures and chirality of glyceraldehyde-3-phosphate and dihydroxyacetone phosphate. (a) The phosphorylated derivatives of the simplest monosaccharides that are important intermediates in carbohydrate metabolism. The figure demonstrates the difference between aldoses and ketoses, which are tautomers of each other in this case. Carbon numbering starts at the aldehyde group for aldoses and with the end carbon adjacent to the ketone group for ketoses. (b) Enantiomeric forms of glyceraldehyde. The D and L forms differ in the configuration of groups around the chiral carbon (C-2). Fischer projections of the two chiral forms are shown below.

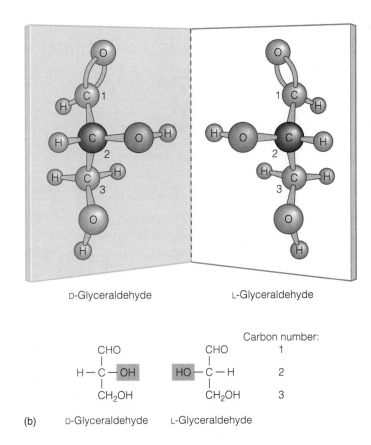

D-Glyceraldehyde   L-Glyceraldehyde

Carbon number:

	CHO	CHO	1
	H—C—OH	HO—C—H	2
	CH$_2$OH	CH$_2$OH	3
(b)	D-Glyceraldehyde	L-Glyceraldehyde	

Monosaccharides have asymmetric carbon atoms and are chiral

Monosaccharides are the basic units from which larger carbohydrate polymers are assembled. In their simplest form (trioses, $n = 3$), the compounds are glyceraldehyde and dihydroxyacetone, which occur in cells as their phosphorylated derivatives (Figure 2.2). Glyceraldehyde is termed an **aldose** sugar because it contains an aldehyde group; dihydroxyacetone, with a ketone group, is a **ketose** sugar.

Carbon atom 2 in glyceraldehyde is asymmetric; it is a chiral centre giving the possibility of two **stereoisomers** – designated D- and L-glyceraldehyde – defined by the absolute configuration of atoms around this carbon. Figure 2.2 shows Fischer projections of these molecules, in which atoms joined to the tetravalent carbon by horizontal bonds project outwards from the page and those by vertical bonds project inwards. D- and L-glyceraldehyde are mirror images, the hydroxyl group being represented on the right in D-glyceraldehyde when the active aldehyde group is at the top. In solution, stereoisomers have the property of rotating plane-polarized light.

Cells contain a variety of monosaccharides

Monosaccharides with four, five, six and seven carbon atoms are common in biological systems and are termed tetroses, pentoses, hexoses and heptoses, respectively. All members of these families of sugars contain two or more asymmetric carbons, so the number of stereoisomers also increases accordingly – a molecule with n asymmetric centres can exist as 2^n stereoisomers. Conventionally, all such sugars with the asymmetric carbon furthest from the aldehyde or ketone group in the same configuration as D-glyceraldehyde belong to the D-series of sugars; members of the L-series have the same configuration as L-glyceraldehyde at the

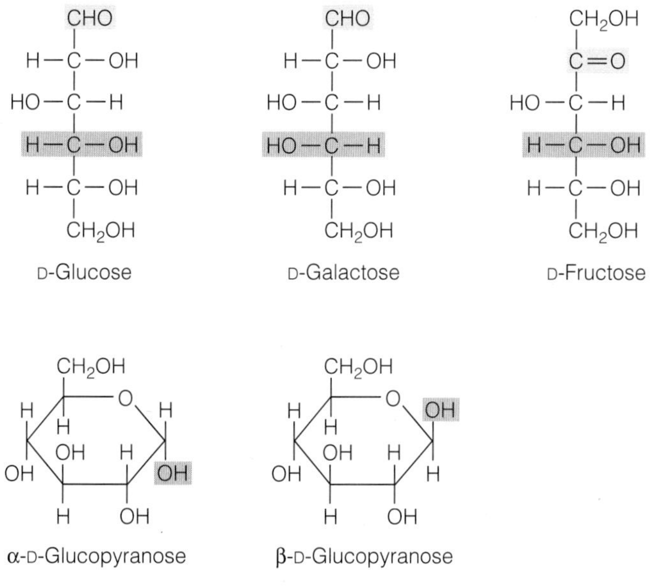

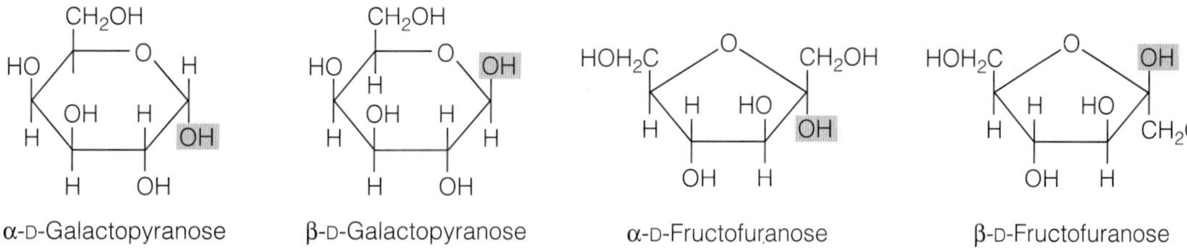

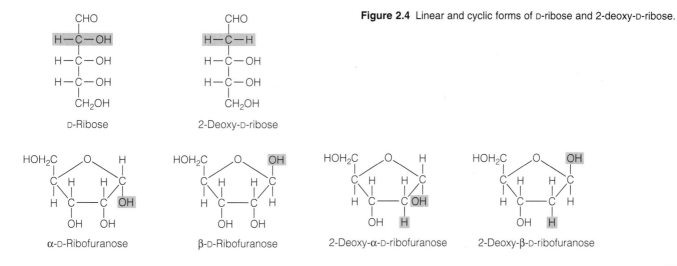

Figure 2.3 Linear and cyclic forms of the common dietary monosaccharides. D-Glucose and D-galactose are epimers, stereoisomers differing at only a single asymmetric centre (C-4). These two sugars are also aldoses arising from the presence of an aldehyde group at C-1, whereas D-fructose, with a ketone group at C-2, is termed a ketose. Intramolecular cyclization of these structures occurs, involving the reactive aldehyde (C-1) and keto (C-2) groups of the aldoses and ketoses and their respective C-5 hydroxyls. The cyclization reaction generates the corresponding pyranose (D-glucose, D-galactose) and furanose (D-fructose) ring forms. In each case, an additional asymmetric centre is created (C-1 in aldoses; C-2 in ketoses), hence the formation of α and β configurational forms.

equivalent carbon atom. The most common **hexoses** in the body (D-glucose and D-galactose) are **epimers**, differing only in the stereochemistry at a single asymmetric centre (Figure 2.3). The major **pentoses** are D-ribose and D-2-deoxyribose; they differ only in the presence or absence of a hydroxyl group at C-2 (Figure 2.4).

Hexoses and pentoses form pyranose and furanose rings

Hexoses (such as glucose, galactose and fructose) and pentoses (such as ribose) usually exist in solution, predominantly as cyclic (rather than linear) molecules.

Figure 2.4 Linear and cyclic forms of D-ribose and 2-deoxy-D-ribose.

The flexibility of the chain in these monosaccharides permits the reactive carbonyl group of the linear form to attack an internal hydroxyl at the other end of the molecule – creating an intramolecular **hemiacetal**.

In glucose the C-5 hydroxyl is the reactive site, generating a six-membered **pyranose** ring by a reaction with the C-1 aldehyde group. Fructose is a ketose, and the ketone group at C-2 reacts with the C-5 hydroxyl to form a five-membered **furanose** ring (Figure 2.3).

For simplicity, the cyclic structures of pentoses and hexoses are normally represented as **Haworth projections,** in which the carbon atoms are omitted; the plane of the ring is nearly perpendicular to the plane of the page (Figure 2.4).

(a)

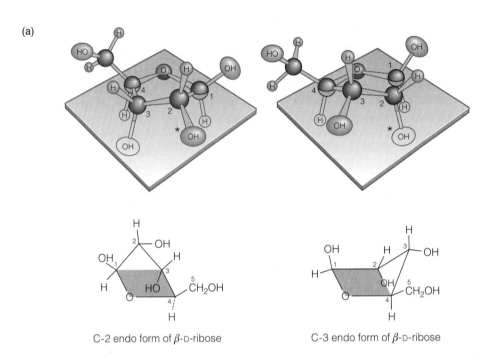

C-2 endo form of β-D-ribose

C-3 endo form of β-D-ribose

(b)

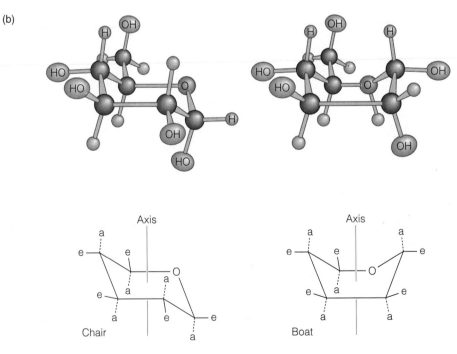

Figure 2.5 (a) Two of the most common ring conformations for ribose and 2-deoxyribose found in nucleic acids. This diagram depicts these two major conformations for β-D-ribofuranose. In these two structures, C-1, O and C-4 lie in a single plane with C-2 above the plane and C-3 in the plane in the C-2 *endo* conformation; C-3 is above this plane and C-2 is in it in the C-3 endo form. A C-3 *exo* conformation would look like the right-hand figure, but the C-3 would lie below the plane. In 2-deoxyribose, found exclusively in DNA, the hydroxyl group at C-2 is replaced by a hydrogen atom. (b) Three-dimensional illustrations of the chair and boat conformations of the pyranose ring structure: a ball and stick model of α-D-glucose, and a skeletal diagram of the bonding. These two molecules are conformational isomers with the same stereochemical configuration but having distinct three-dimensional conformations.

Participation of the C-1 atom of glucose (C-2 of fructose) in the cyclization reaction creates new asymmetric carbons at these centres, producing two possible ring structures, or configurations. In the case of D-glucose, these are called α-D-glucopyranose and β-D-glucopyranose. The hydroxyl group at C-1, the anomeric carbon, is located *below* the plane of the ring in the α form and above the plane of the ring in the β form. Similar considerations apply in the furanose rings of ribose and deoxyribose (Figure 2.4). During nucleotide biosynthesis, the configuration switches from the α form of 5-phosphoribosyl-α-pyrophosphate (PRPP) to the β form of the nucleotides (see Chapter 10).

Chapter 10. Ribose is activated for nucleotide biosynthesis as PRPP, page 315.

Furanose and pyranose rings can exist in different conformations

The five and six-membered rings of these sugars are not planar; this is incompatible with the tetrahedral geometry of the C–C single bonds forming them. As a consequence, the rings adopt *puckered* conformations, the so-called **chair** or **boat** forms in the case of pyranose rings (Figure 2.5). The chair form is the preferred conformation in energetic terms as the proximity of adjacent hydroxyl groups in the boat form leads to a degree of steric hindrance which tends to destabilize this structure.

Substituent groups on the ring carbons may also be of two types: **axial** or **equatorial**. Axial groups are nearly vertical to the average plane of the ring; equatorial groups (depicted horizontally) are almost parallel (Figure 2.5). The pentoses ribose and deoxyribose are present as furanose rings in RNA and DNA, and the rings also exhibit puckering. Furanose rings are very flexible and contribute significantly to the transient alterations in the double helical DNA structure that are vital to various aspects of its function (see Chapter 6).

Chapter 6. The structure of DNA is not completely uniform, but has sequence-dependent local variations, page 147.

Major carbohydrate fuel reserves are polymers of glucose linked by O-glycosidic bonds

Starch and glycogen, the major carbohydrate fuel reserves in plant and animal tissues, are both long-chain polymers of glucose linked by **O-glycosidic bonds**.

Starch is a mixture of two polymers, the linear molecule **amylose** (Figure 2.6a) and the branched chain **amylopectin**, in which a branch point occurs every 24–30 residues. Glycogen is a similarly branched polymer, forming a 'tree-like' structure with branches every 8–12 residues (Figure 2.6b).

In these polysaccharides, the linear arrays of glucose molecules are linked in series by **α,1–4** *O*-glycosidic bonds. At branch points, additional **α,1–6** *O*-glyco-sidic bonds are present, so that a single glucose residue is substituted at both C-4 and C-6 positions. The 'tree-like' structure also provides more compact storage than amylose, although the polar nature of glycogen means that it is highly hydrated and is a much less compact store than fat.

In cellulose, the glucose residues are joined by **β,1–4** linkages (Figure 2.6d). Cellulose is not utilized efficiently by humans as a foodstuff because we have no enzymes in our gut capable of hydrolysing β,1–4 linkages, although some degradation can occur as a result of bacterial action.

Three disaccharides are of nutritional significance

Three disaccharides that are major constituents of the human diet are maltose, lactose and sucrose (Figure 2.7). **Maltose** is the disaccharide of glucose and is derived primarily as an intermediate in the digestion of starch (see Chapter 14). **Sucrose** is usually refined from sugar cane or sugar beet and is employed as a sweetener in a wide variety of food products. It is a disaccharide of glucose and fructose linked by an α,1–2 bond. **Lactose** is the principal disaccharide and carbohydrate source of milk and is composed of a galactose and glucose residue linked by a β,1–4 bond.

Chapter 14. Digestion and absorption of carbohydrate, page 417.

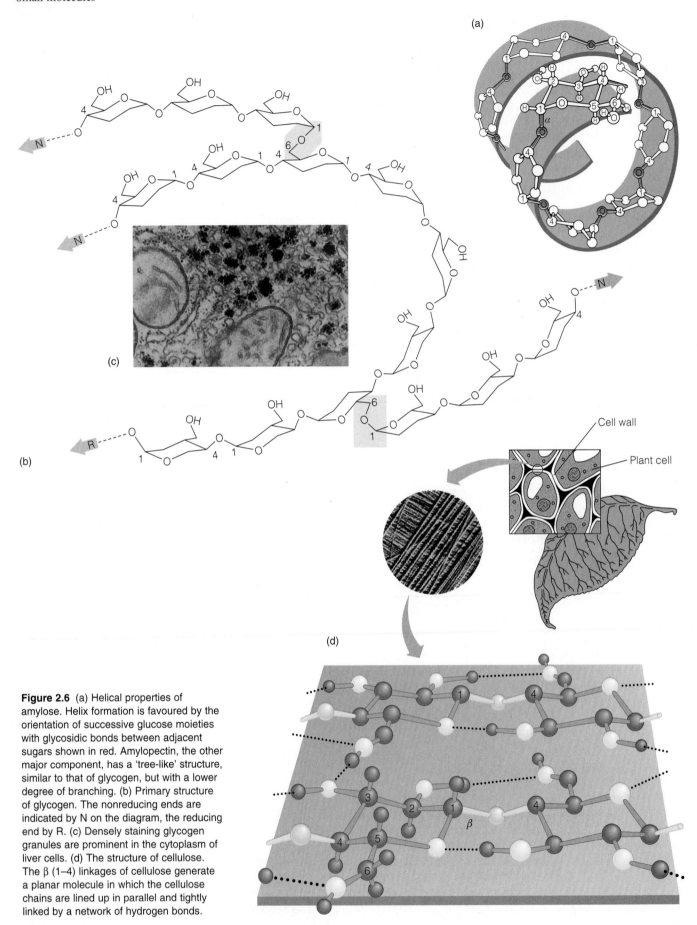

Figure 2.6 (a) Helical properties of amylose. Helix formation is favoured by the orientation of successive glucose moieties with glycosidic bonds between adjacent sugars shown in red. Amylopectin, the other major component, has a 'tree-like' structure, similar to that of glycogen, but with a lower degree of branching. (b) Primary structure of glycogen. The nonreducing ends are indicated by N on the diagram, the reducing end by R. (c) Densely staining glycogen granules are prominent in the cytoplasm of liver cells. (d) The structure of cellulose. The β (1–4) linkages of cellulose generate a planar molecule in which the cellulose chains are lined up in parallel and tightly linked by a network of hydrogen bonds.

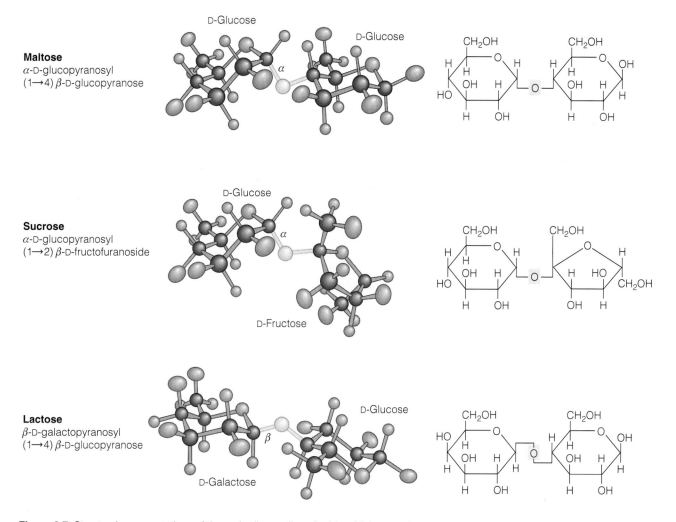

Maltose
α-D-glucopyranosyl
(1→4) β-D-glucopyranose

Sucrose
α-D-glucopyranosyl
(1→2) β-D-fructofuranoside

Lactose
β-D-galactopyranosyl
(1→4) β-D-glucopyranose

Figure 2.7 Structural representations of the main dietary disaccharides. Maltose and sucrose contain two sugars which are α-linked. The anomeric oxygens are highlighted. Lactose contains a β-linkage. Haworth projections of the same molecules are shown.

Oligosaccharide structures are covalently linked to many secretory and membrane proteins and to some lipids

The vast majority of secretory proteins (such as antibodies and other plasma proteins) are glycoproteins (see Chapter 12), as are many of the integral proteins of plasma membranes, where the attached oligosaccharide chains are always asymmetrically distributed, being located on the external surface of the cell (see Chapter 5). Membrane lipids, especially sphingolipids, are also glycosylated (see below). Carbohydrate structures are attached to proteins either by O-glycosidic bonds to the hydroxyl group of serine or threonine residues or by N-glycosidic linkages to asparagine residues (Figure 2.8).

N-glycosylation of proteins always occurs at the sequences asparagine–X–serine or asparagine–X–threonine (see Figure 2.9), where X may be any amino acid, and the first two sugar residues added to the asparagine are always N-acetylglucosamines. These **sequence motifs** are necessary, but not sufficient, for glycosylation because

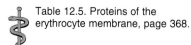

Table 12.5. Proteins of the erythrocyte membrane, page 368.

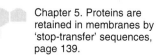

Chapter 5. Proteins are retained in membranes by 'stop-transfer' sequences, page 139.

Figure 2.8 Two types of oligosaccharide linkage to proteins. (a) *O*-Glycosidic linkages are formed usually via the C-1 of *N*-acetylgalactosamine to the hydroxyl group of threonine or serine.
(b) *N*-Glycosidic linkages involve the amide side chain of an asparagine residue, which is covalently attached to an *N*-acetylglucosamine as the first sugar.

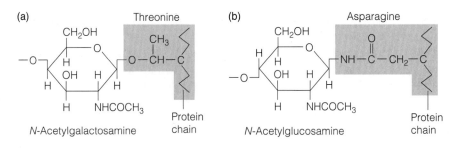

they must be present in a suitable environment of the protein's three-dimensional structure. Three mannose sugars added to the two *N*-acetylglucosamines form a core to which additional monosaccharides are added, generating the great diversity of oligosaccharide structures found on glycoproteins. The major types are the 'high-mannose' type and the 'complex' type (Figure 2.9). In the 'high mannose' type, a number of peripheral mannose units are linked to the core. In the 'complex' type several less common types of sugars are usually present: these are *N*-acetylglucosamine, galactose, fucose and sialic acid.

O-glycosylation involves addition of monosaccharide or disaccharide units, in contrast to the complex, branched oligosaccharides found in *N*-linked structures. The units are usually glucose or galactose and are joined to serine or threonine residues on the protein. Both *N*-linked and *O*-linked glycosylation may be present on the same glycoprotein. The **mucins** (a family of proteins that provide lubrication and protection to epithelial linings of body passages such as the mouth and oesophagus, the bronchi of the lungs and the gastrointestinal tract) are distinctive in being only *O*-glycosylated at multiple sites on the protein.

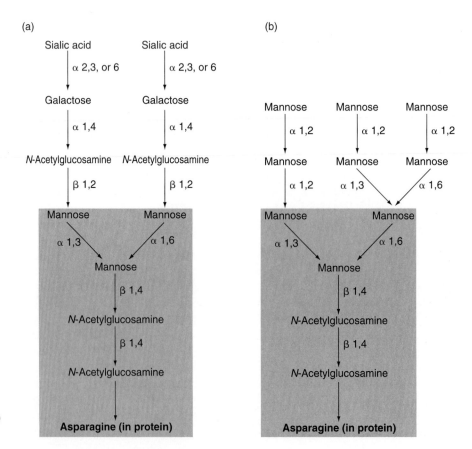

Figure 2.9 The two major classes of oligosaccharide structures located on N-linked glycoproteins. (a) Complex type; (b) high-mannose type.

Protein- and lipid-bound oligosaccharides have the potential to combine in a vast number of structures. Not only are there up to five types of monosaccharides present but also they are joined to adjacent sugars via several kinds of *O*-glycosidic bonds. Different linkages to the same sugar provide the ability to create complex branched structures with an almost unlimited range of conformations. This organizational complexity allows oligosaccharides to be of functional importance in various recognition phenomena, typified by the ABO blood group substances (see Chapter 12).

Chapter 12. The surface of the erythrocyte membrane carries blood group substances, page 369.

Fatty acid-containing lipids and cholesterol

Long-chain fatty acids are amphipathic molecules with both hydrophobic and hydrophilic properties

Fatty acid molecules have two parts:

- an aliphatic chain of methylene groups terminating in a methyl group
- a carboxyl group – ionized and neutralized by a positive ion at physiological pH.

This makes them **amphipathic**, that is, they possess both hydrophilic and hydrophobic properties. It also gives them detergent properties – everyday soaps are the sodium salts of long-chain fatty acids. In mammalian tissues, fatty acids contain even numbers of carbon atoms between 14 and 24, C-16 (palmitate) and C-18 (stearate) fatty acids being the most common (Figure 2.10). In milk, the average length of the fatty acid chains is considerably shorter, down to C-4 (butyrate).

Fatty acids are important as rich sources of energy and as essential components of cell membranes

The two principal roles of fatty acids in the body are:

- as the most energy-rich molecules available for catabolism and ATP production
- as the components of membrane lipids mainly responsible for bilayer formation.

Fatty acids are rich in energy because they are highly reduced compared with carbohydrates and amino acids, which are already partly oxidized. Under aerobic

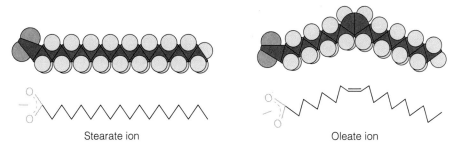

Stearate ion

Oleate ion

Figure 2.10 Molecular representation of the structures of stearate and oleate ions. Both are shown in their maximally extended forms so that the 'kink' induced by the presence of a *cis* double bond in oleate is most evident. Corresponding skeletal structures are included for comparison.

Chapter 9. In the liver accummulating acetyl CoA is converted to ketone bodies, page 271.

conditions, they can be completely oxidized to CO_2 by β-oxidation and the citric acid cycle, with a high energy yield. If fatty acids are not completely oxidized they may be converted to ketone bodies and used as fuels by certain tissues (see Chapter 9). Free (that is, nonesterified) fatty acids are not generally found in substantial quantities; most fatty acids in the body are present as components of triacylglycerols, phospholipids and sphingolipids, where they are described as **fatty acyl chains**. Free fatty acids in the plasma are bound to albumin to protect blood cells from their detergent action.

Fatty acids are components of the triacylglycerols of fat and the phospholipids and sphingolipids of cell membranes

Chapter 14. Triacylglycerols hydrolysed in the gut are reassembled within epithelial cells, page 421.

Chapter 9. Distinct classes of lipoproteins are responsible for the transport of triacylglycerols, phospholipids and cholesterol, page 287.
Chapter 3. Membrane fluidity maintains the appropriate environment for the optimal functioning of membrane proteins and is carefully controlled, page 59.
Chapter 3. Some proteins are attached to membranes by lipid 'anchors', page 55.

Fatty acids in the diet, in circulating lipoproteins and in chylomicrons (see Chapter 14) and in adipose tissue are present principally as triacylglycerols (triglycerides), in which three fatty acids are esterified to a glycerol (Figure 2.11). Mobilization of this fat store requires the action of lipases.

The fatty acyl chains of the phospholipids and sphingolipids of membranes form the hydrophobic interior of the lipid bilayers that make up the basis of biological membranes (see below and Chapter 3). Certain fatty acids, such as myristate (C-14) and palmitate (C-16) are covalently joined to some proteins and act as 'anchors', binding these proteins to the membrane bilayer (see Chapter 3).

Unsaturated fatty acids have *cis* double bonds in specific locations

Saturated acyl chains are quite flexible, with the carbon backbone forming a zigzag arrangement with a bond angle of 120° between adjacent carbons in the chain. Most naturally occurring unsaturated fatty acids have a *cis* Δ double bond, which introduces a rigidity at that bond and causes a major 'kink' in the structure (Figure 2.10), a feature that greatly affects the packing of phospholipids when they are aligned side-by-side in membrane bilayers. High concentrations of unsaturated fatty acids in the triacylglycerols of plants lower their melting temperature and make them oils (as opposed to the solid fats of animals). Some fatty acids have two or more double bonds (polyunsaturates) and cannot be made by animals, which have to obtain these **essential fatty acids** from plant sources.

Trans Δ double bonds, sometimes found in fatty acids of prokaryotic cell membranes and in sphingosine, produce little or no distortion of the acyl chain. *Trans* fatty acids are also produced during the chemical hydrogenation of vegetable oils for the production of margarine and there is concern that these may lead to an increased incidence of heart disease.

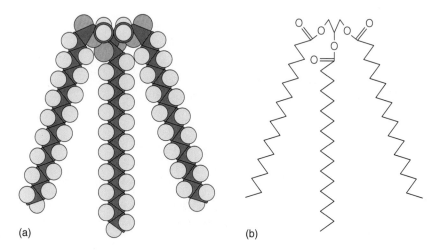

Figure 2.11 Representations of a triacylglycerol. (a) Space-filling model; (b) skeletal structure.

(a)   (b)

The carboxyl group of fatty acids acts as a 'handle' for metabolism and for linking to other molecules

The carboxyl group of fatty acids is more reactive than the aliphatic 'tail' but it is not highly so and needs to be 'activated' (see Box 2.1), usually by coupling the fatty acid to the thiol group of coenzyme A (CoA-SH), producing a **fatty acyl CoA**, which contains a 'high energy' thioester bond. Fatty acyl CoAs are the main metabolic form of the fatty acids in synthesis and degradation, and in the formation of more complex lipids (see Chapter 9).

Chapter 9. Fatty acids must be activated before they enter mitochondria, page 265.

The principal phospholipids of eukaryotic membranes are composed of glycerol, phosphate, fatty acids and several small basic or neutral compounds

The major lipid constituents of biological membranes are the phospholipids. These are composed of a glycerol backbone, with fatty acids attached via ester linkages at C-1 and C-2 and a phosphate head group at C-3. This head group is itself linked to small basic compounds (ethanolamine and choline) or neutral compounds (glycerol or inositol) or the amino acid serine (Figure 2.12) The

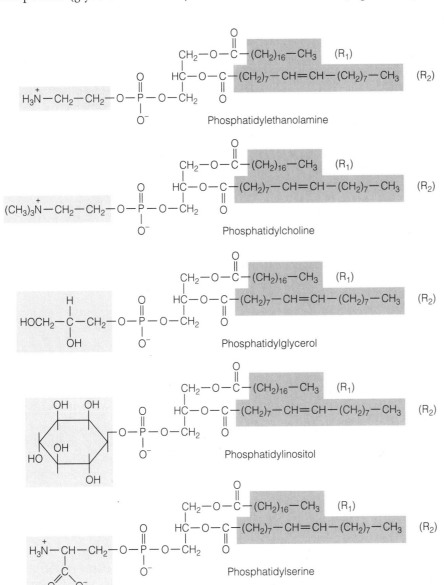

Figure 2.12 Examples of common glycerophospholipids. The hydrophobic R groups, and the very hydrophobic head groups, are highlighted. All may be considered derivatives of phosphatidic acid.

Small molecules

Figure 2.13 Space-filling model of phosphatidylcholine, illustrating the detailed architecture of glycerophospholipid molecules.

Box 7.2. The evolutionary origin of mitochondria, page 200.

general structure of phosphatidylcholine is shown in Figure 2.12, and the molecular (space-filling) model for phosphatidylcholine is depicted in Figure 2.13.

The most important characteristic of phospholipids is their **amphipathic** nature – they possess both hydrophilic and hydrophobic properties (conferred by the phosphate head group region and the long fatty acyl chains, respectively). Phospholipids containing a particular head group make up a family of related molecules because there are variations in the chain lengths and degree of unsaturation of the fatty acyl chains incorporated at C-1 and C-2 in the glycerol backbone. Saturated fatty acyl chains are always located at C-1; unsaturated fatty acids are incorporated preferentially into the C-2 position.

The mitochondrial inner membrane is unusual in possessing large amounts of **cardiolipin**, a phospholipid usually found only in bacterial membranes. This may reflect the prokaryotic origins of mitochondria as propounded in the **endosymbiotic theory** (see Box 7.2). In the mitochondrial inner membrane cardiolipin appears to play a role similar to that of cholesterol in the plasma membrane (see below).

Sphingomyelin, glycolipids and cholesterol are the other major lipids of eukaryotic membranes

Sphingomyelins are also found in mammalian membranes, especially the plasma membrane. Their structures and properties are very similar to those of the phosphatidylcholine family, except that the backbone is derived from the complex base **sphingosine** rather than from glycerol. Sphingosine is an amino alcohol containing a long, unsaturated hydrocarbon chain. It is synthesized *de novo* from palmitoyl CoA and the amino acid serine. In sphingomyelin the primary alcohol group of sphingosine is esterified to phosphorylcholine and its amino group forms an amide bond with a fatty acid (Figure 2.14).

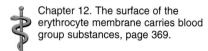

Figure 2.14 Structure (a) and space-filling model (b) of sphingomyelin. Note the resemblance to phosphatidylcholine.

Glycolipids are also based on sphingosine, with one or more sugars replacing the phosphorylcholine group. In the simplest class of glycolipids, the **cerebrosides**, a single glucose or galactose moiety is present; the more complex **gangliosides** contain a complex branched chain of several sugar residues (Figure 2.15). Glycolipids have important functions in a variety of recognition processes at the external surface of the cell. For example, the sugar residues on the lipid components of red blood cells are involved in the determination of A, B and O blood group specificities (see Chapter 12).

Chapter 12. The surface of the erythrocyte membrane carries blood group substances, page 369.

Cholesterol has a characteristic amphipathic, multi-ring structure

Cholesterol has a characteristic structure of four fused alicyclic rings – three six-carbon rings and one five-carbon ring (Figure 2.16). Molecules based on this structure, often called the **steroid ring**, are very important in many functions

(a)

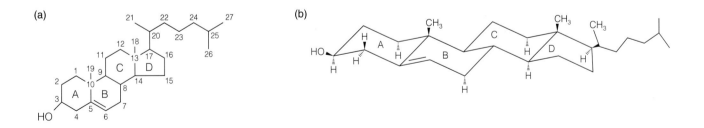

Figure 2.15 Two common types of glycosphingolipids: (a) a cerebroside, which is an important component of neuronal cell membranes in the brain, and (b) a ganglioside, termed G_{m2}, which accumulates in neural tissue of infants with Tay-Sachs disease. This is an inherited defect in which individuals lack the enzyme required to remove the terminal *N*-acetylgalactosamine.

(b)

throughout the body – for example, the bile salts and the many steroid hormones. This importance is reflected in the choice of the steroid ring as the 'molecular icon' in this book.

Cholesterol is synthesized from acetyl CoA (see Chapter 9) and further pathways exist for producing the diverse families of steroid hormones from cholesterol (see Chapter 11). Cholesterol circulates in the plasma as a mixture of cholesterol and cholesterol esters within lipoproteins, which deliver it to the cells of various tissues (see Chapter 9). It is a major component of the plasma membrane and influences its fluidity and function. It is predominantly hydrophobic in nature, but the hydroxyl group attached to one of the six-carbon rings confers amphipathic properties – vital to its role in cell membranes.

Chapter 9. Cholesterol is synthesized in the liver from acetyl CoA, page 284.
Chapter 11. Steroid hormones, page 351.
Chapter 9. Triacylglycerols synthesized within the body are transported in other lipoprotein complexes, page 289.

(a)

(b)

(c)

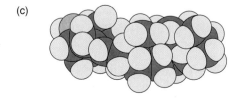

Figure 2.16 Cholesterol. (a) Structural formula. (b) Skeletal model. (c) Space-filling model.

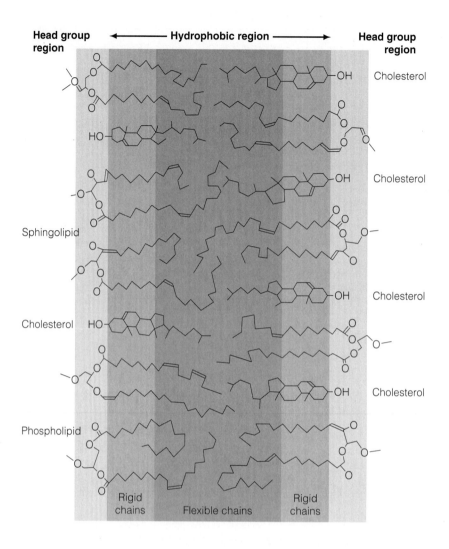

Figure 2.17 Cholesterol in a phospholipid bilayer.

Box 5.5. The absence of cholesterol makes the endoplasmic reticulum membrane more permeable than the plasma membrane, page 140.

Cholesterol has an essential role in the plasma membrane of cells

Cholesterol is capable of intercalating between the fatty acyl chains of phospholipids in the lipid bilayer, with its hydrophilic hydroxyl group facing the external aqueous phase and its hydrophobic part interacting with the ten methylene groups that lie nearest the ester link. Cholesterol is thus able to influence the packing of adjacent phospholipids and to modulate the fluidity of the membrane (Figure 2.17). Of the various membranes of the cell, the plasma membrane contains the highest concentration of cholesterol, which results in it being both thicker and less permeable to small molecules than other membranes of the cell. The low level of cholesterol in the membrane of the rough endoplasmic reticulum is thought to allow the insertion of the signal sequences of nascent polypeptides through the bilayer during protein synthesis (see Chapter 5).

Amino acids and polypeptides

Amino acids are both the structural units of proteins and important metabolites

The amino acids are a group of compounds with both amino and carboxyl groups attached to the same carbon atom, which is referred to as the α-**carbon**, giving rise

Box 2.2 The ionization properties of amino acids and peptides

Ionization is an important property of amino acids. The α-amino groups exhibit pK_a values of about 9.2, while those of the α-carboxyl groups lie around 2.4. Thus, in the pH range 4–8, amino acids exist as **zwitterions** of the form $^+H_3N–CHR–COO^-$; these charges are not present when amino acid residues have been polymerized into polypeptides, because the amino and carboxyl groups form the peptide bond (see below).

Several amino acids have side chains bearing ionizable groups (see Table 2.3) and these are principally responsible for the ionization properties of peptides and proteins. A preponderance of the positively charged amino acids lysine and arginine over the negatively charged aspartate and glutamate gives a **basic** protein, carrying a net positive charge at physiological pH (for example the histones; see Chapter 6). Most proteins of cells are, however, negatively charged or **acidic**.

Chapter 6. A group of basic proteins, the histones, package DNA into nucleosomes, page 151.

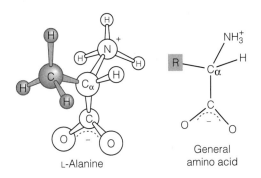

L-Alanine

General amino acid

Figure 2.18 General structure of α-amino acids.

to the term α-**amino acid** (Figure 2.18). Different types of amino acids are distinguished by the nature of the side group attached to the α-carbon atom (Table 2.3).

Amino acids have several important roles in the body.

- They are the structural units of proteins.
- They are major dietary and metabolic components.
- They are precursors of a number of body constituents, including certain hormones (adrenaline) and pigments (melanin).

Most amino acids found in nature are in the L-configuration, but exceptions to this are **glycine**, which does not have a chiral centre, and some amino acids found in certain antibiotics and bacterial cell wall components, which exist in the D-configuration (see Chapter 18). Strictly speaking, **proline**, one of the common 'amino' acids, is an *imino* acid. With the exception of glycine and proline, all the common amino acids have corresponding α-**keto acids** with which they can be metabolically interchanged by the addition or removal of an amino group (see Chapter 10). A number of other amino acids are not present in proteins, but have other roles in the body (see below).

Box 18.2. Penicillins interfere with bacterial cell wall synthesis, page 557. Chapter 10. Aminotransferase reactions facilitate the transfer of amino groups from one compound to another, page 295.

The side chains of amino acids are of crucial importance in protein structure

There are 20 amino acids found in proteins and, during protein synthesis, the selection of each is governed by the rules of the **genetic code** (see Chapter 5). Other amino acids found only in certain proteins, such as hydroxyproline in collagen and iodinated amino acids (thyroxines) in thyroglobulin, are the result of postsynthetic modification of proline and tyrosine respectively.

Chapter 5. The genetic code is the set of rules that governs translation, page 131.

Table 2.3 Properties of the amino acids found in proteins

Amino acid	Three-letter abbreviation	One-letter abbreviation	Side-chain pKa	Side-chain structure
Nonpolar (aliphatic) Isoleucine	ile	I		CH₃–CH₂–CH(CH₃)–
Leucine	leu	L		(CH₃)₂CH–CH₂–
Methionine	met	M		CH₃–S–CH₂–CH₂–
Valine	val	V		(CH₃)₂CH–
Nonpolar (aromatic) Phenylalanine	phe	F		–CH₂–C₆H₅
Tyrosine	tyr	Y	10.1	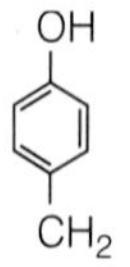 HO–C₆H₄–CH₂–
Tryptophan	trp	W		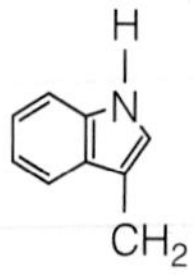 indole–CH₂–
Polar uncharged Alanine	ala	A		CH₃
Asparagine	asn	N		H₂N–CO–CH₂–
Cysteine	cys	C	8.3	HS–CH₂–
Glutamine	gln	Q		H₂N–CO–CH₂–CH₂–

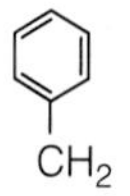

Amino acid	Three-letter abbreviation	One-letter abbreviation	Side-chain pKa	Side-chain structure
Glycine	gly	G		H
Proline	pro	P		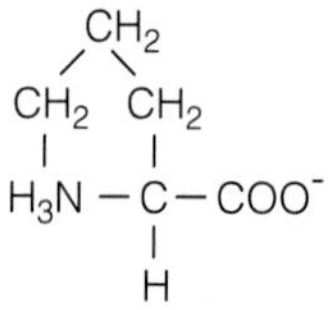 (cyclic structure) H₃N⁺–C(–COO⁻)–CH₂–CH₂–CH₂
Serine	ser	S		HO–CH₂–
Threonine	thr	T		CH₃–CH(OH)–
Charged (positive) Arginine	arg	R	12.5	H₂N–C(=NH₂⁺)–NH–CH₂–CH₂–CH₂–
Histidine	his	H	6.0	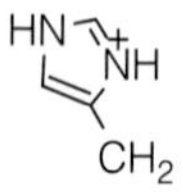 imidazole–CH₂–
Lysine	lys	K	10.0	NH₃⁺–CH₂–CH₂–CH₂–CH₂–
Charged (negative) Aspartate	asp	D	3.9	⁻OOC–CH₂–
Glutamate	glu	E	4.2	⁻OOC–CH₂–CH₂–

The same 20 amino acids are found in the proteins of all organisms; their selection from the range of amino acids possible must therefore have occurred early in the evolution of life. The side chain of each of these amino acids must provide exactly the correct subset of properties for its particular role in the structure and function of all proteins (see Chapter 4).

A useful way of classifying amino acids is by the chemical properties of their side chains – nonpolar, aromatic, polar uncharged, positively charged, negatively charged – as shown in Table 2.3, which also lists the three- and one-letter abbreviations used for each. These properties determine the precise structure and function of each individual type of protein found in the body (see Chapter 4).

There are other important amino acids not found in proteins

As mentioned above, several amino acids in the body are not present in proteins, but function metabolically as individual amino acids (Figure 2.19). These include **ornithine** and **citrulline**, components of the urea cycle (see Chapter 10), ornithine being the next lower homologue of lysine, with only three methylene groups. Arginine, which is found in proteins, is also a component of this cycle. Homocysteine, the next higher homologue of cysteine, is a product of methylation reactions involving methionine (see Chapter 10).

Peptide bonds are formed between two amino acids and have properties central to protein structure

Two amino acids can join, eliminating water between the carboxyl group of one and the amino group of the other to form a **peptide bond**. Peptide bonds play an important role in determining the conformations adopted by polypeptide chains (see Chapter 4). In the cell, peptide bonds are formed on ribosomes and the selection of amino acids is directed by messenger RNA according to the rules of the genetic code (see Chapter 5).

The most important features of the peptide bond are as follows (Figure 2.20):

- Partial double bond character of the C–N bond. Electrons are shared throughout the –CO–NH– system, leading to partial charges on the H and O atoms.

 Chapter 4. Proteins contain 20 different amino acids, page 94.

 Box 4.8. The denaturation and renaturation of proteins, page 99.

 Chapter 10. The urea cycle, page 301.

 Box 10.3. Homocystinuria arises from a failure of cysteine metabolism, page 307.

 Box 4.11. α-Helices, β-sheets and turns – regular features in proteins, page 104.
Chapter 5. The genetic code is the set of rules that governs translation, page 131.

Ornithine

$^+H_3N-CH_2-CH_2-CH_2-\overset{\overset{NH_3^+}{|}}{CH}$
$\underset{\underset{O^-}{|}}{\overset{|}{C=O}}$

Citrulline

$H_2N-\overset{\overset{}{\underset{\underset{O}{||}}{C}}}{}-NH-CH_2-CH_2-CH_2-\overset{\overset{NH_3^+}{|}}{CH}$
$\underset{\underset{O^-}{|}}{\overset{|}{C=O}}$

Homocysteine

$HS-CH_2-CH_2-\overset{\overset{NH_3^+}{|}}{CH}$
$\underset{\underset{O^-}{|}}{\overset{|}{C=O}}$

Figure 2.19 Some important amino acids not found in proteins.

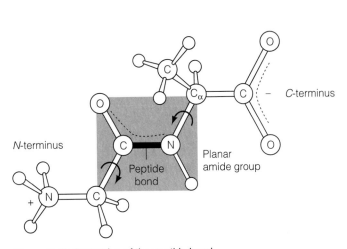

Figure 2.20 Properties of the peptide bond.

- Absence of rotation about the C–N bond because π electron orbitals are shared, as in conjugated double bonds.
- Occurrence of the *trans* configuration in the –CO–NH– system because of the absence of free rotation around the C-N bond.
- Possibility of rotation around the C_α–NH and C_α–CO bonds. This allows the peptide chain to have a degree of flexibility, necessary for the folding of proteins into a compact, globular shape.

The consequences of these features for protein structure are discussed in Chapter 4.

There are oligopeptides with important physiological properties

Box 4.11. α-Helices, β-sheets and turns – regular features in proteins, page 104.

Chapter 5. mRNA translation: the genetic code and open reading frames, page 131.

Most polymers of amino acids found in the body are proteins, or polypeptides with chain lengths of 100–1000 amino acids, which are synthesized on ribosomes according to the instructions on mRNA templates (see Chapter 5). Short oligopeptides are also found, usually of 10 amino acids or fewer, some of which are made by specific cleavage from longer polypeptides and others by nontemplate processes – for example, glutathione (see below).

Physiologically important oligopeptides include:

- Peptide hormones, such as insulin, which are cleaved from longer polypeptide **prohormones** (see Chapter 11).

Chapter 11. Peptide hormones, page 346.

- The tripeptide **glutathione** (γ-glutamyl-cysteinyl-glycine), which is present in millimolar amounts in all cells and is involved in maintaining the oxidation–reduction balance of intracellular components, especially proteins (see Chapter 12). Glutathione is not made by protein synthesis (see Chapter 5); its amino acids are linked by specific enzymes in a process which needs no template direction.

Chapter 12. Detoxification of superoxide needs NADPH and the peptide glutathione, page 378.

Chapter 5. mRNA translation: the genetic code and open reading frames, page 131.

- Certain antibiotics produced by bacteria, such as the polymyxins, which are made by processes not involving templates and often contain D-amino acids (see Chapter 18).

Nucleotides and polynucleotides

Nucleotides are important in a number of respects:

- they are carriers of useful energy (ATP, GTP, etc.)
- they are components of coenzymes and other 'activated' molecules (NAD, coenzyme A, etc.)
- as monomeric units of the polynucleotides, RNA and DNA (ribo- and deoxyribo-nucleotides).

Box 17.13. DNA sequencing methodology, page 543. Chapter 10. Deoxyribonucleotides are synthesized from ribonucleotides, page 319.

Nucleotides are composed of three components – base, sugar and phosphate – linked in that order by *N*-glycosidic and phosphomonoester bonds (Figure 2.21). Molecules lacking the phosphate moiety are called **nucleosides**. Cellular nucleotides normally contain the pentose sugars ribose or 2-deoxyribose (see above) although some nucleotide-like molecules with other sugars, such as arabinose, are found in nature and others have been synthesized in the laboratory (e.g. 2',3'-dideoxyribose, see Box 17.13). These comprise the ribonucleotides and deoxyribonucleotides – those of the *ribo* series are the primary metabolites, while those of the *deoxy* series are present at lower concentrations and are derived from ribonucleotides by special pathways (see Chapter 10).

Block diagram of components
of a nucleotide

Adenosine 5'-monophosphate
(AMP)

Figure 2.21 The general structure of nucleotides.

There are two types of nitrogenous bases – purines and pyrimidines

Two series of nitrogenous bases occur in nucleotides – the **purines** and the **pyrimidines**. The principal purines are **adenine** and **guanine** and the pyrimidines are **cytosine, uracil** and its methylated derivative **thymine** (Figure 2.22). Other bases, including hypoxanthine, 5-methyl cytosine and a variety of other methylated bases in transfer RNA molecules, are found as intermediary metabolites or as minor components of nucleic acids. **Nicotinamide**, a pyridine base, occurs in the nucleotide coenzyme NAD$^+$ (see below).

The important chemical features of the bases are

- the locations of the ring nitrogen atoms
- the position and nature of the substituent atoms
- the conjugated nature of the double bonds.

The important properties of the bases are

- Their flat and nonpolar nature: this means that bases tend to avoid an aqueous environment and are capable of interacting with each other by **base stacking**.

Purines

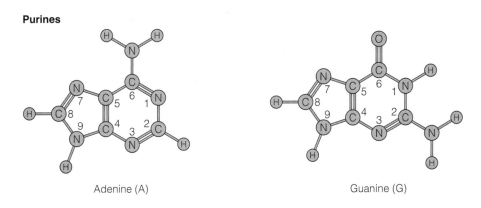

Adenine (A)

Guanine (G)

Pyrimidines

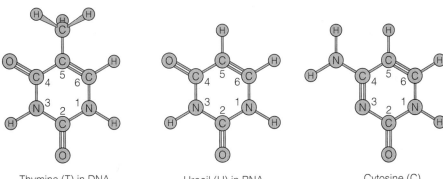

Thymine (T) in DNA

Uracil (U) in RNA

Cytosine (C)

Figure 2.22 Structures of purine and pyrimidine bases.

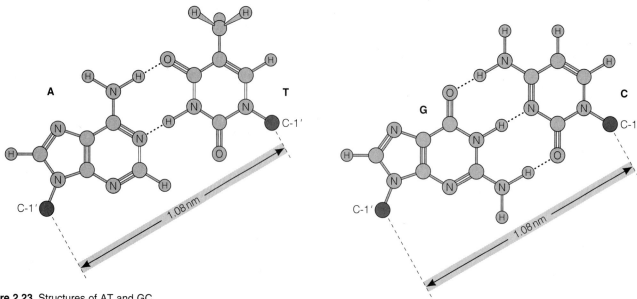

Figure 2.23 Structures of AT and GC base pairs.

Table 5.1. The genome size and gene content of various organisms, page 110.
Table 6.2. Biological implications of the structure of DNA, page 148.
Table 6.6. Damage to DNA, page 161.
Box 6.2. DNA repair and xeroderma pigmentosum, page 164.

Figure 6.1. The double helical structure of DNA, page 147.

- The **hydrogen-bonding potential** of the ring nitrogens and O- and N-substituents: this is perhaps the single most important chemical and biological feature of these biomolecules as it is the basis of hydrogen-bonded **base pairing** (see below and Chapters 5 and 6).
- Their ability to absorb ultraviolet light: this is useful in the detection and measurement of nucleotides and polynucleotides and is also the major cause of the damage done to living things by ultraviolet light (see Chapter 6).

Hydrogen-bonded base pairing

Bases are capable of forming 'edge-on' pairs linked by specific hydrogen bonds according to the rules

A = T(or U) and G = C (Figure 2.23)

Base pairs stack more strongly than bases and are the structural units of **double-stranded** (or **duplex**) nucleic acids, especially in DNA (see Chapter 6). The importance of base pairing cannot be over-emphasized, because it is responsible for the **sequence complementarity** of duplex polynucleotides and lies at the heart of the processes of **replication, transcription** and **translation** (see Part B).

The two pentoses – ribose and deoxyribose – define the two types of nucleic acid RNA and DNA

Ribose is a pentose sugar that exists in the furanose ring configuration, with its 5-carbon protruding from the ring in a hydroxymethyl group. The important status of the 5-carbon in nucleotide structure relates to its being the only *primary* hydroxyl group in the ribose molecule (Figure 2.4). The oxygen atom is missing from the 2-position of 2-deoxyribose; this apparently modest omission is responsible for major differences between *ribo-* and *deoxyribo*polynucleotides. The pentoses are relatively hydrophilic parts of nucleotides and their major importance in nucleotide structure lies in their linkage to the base and in their esterification with phosphate in mono- and poly-nucleotides (see below).

Phosphate and phosphate esters

The phosphate ion is an essential inorganic constituent of all organisms and is derived from the aqueous environment in the *ortho*phosphate form. Its principal features are its ability to form **esters** with organic molecules bearing OH groups

---

Table 2.4 Ribo- and deoxyribo-nucleotides

Ribo series	Deoxyribo series
rAMP rADP rATP	dAMP dADP dATP
rGMP rGDP rGTP	dGMP dGDP dGTP
rCMP rCDP rCTP	dCMP dCDP dCTP
rUMP rUDP rUTP	dUMP dUDP dUTP
* * *	dTMP dTDP dTTP

The r of the ribo-series is commonly omitted, as in AMP, ADP, ATP.

* The ribonucleotides of the T series do not occur in cells

and its ionization properties (see Figure 2.24). When *one* of the OH groups of phosphoric acid is esterified, the molecule is a **phosphomonoester**, bearing approximately *two* negative charges at physiological pH; when *two* OH groups are esterified, the resultant **phosphodiester** has only a single charge. Mononucleotides are phosphomonoesters, polynucleotides are phosphodiesters. The principal position of phosphorylation of mononucleotides in cells is the 5'-carbon. When orthophosphate occurs in the unesterified form, it is termed **inorganic (*ortho*)phosphate (Pi)**.

Box 2.3   Nucleotide terminology

Unless otherwise specified, reference to, for example, an adenine nucleotide implies that it is a 5'-ester (AMP = 5'-AMP = adenosine 5'-monophosphate). The acronym AMP also implies the *ribo* form; the corresponding deoxy nucleotide is termed dAMP (*deoxy-*adenosine 5'-monophosphate); the ribo form can be specified by the abbreviation rAMP. cAMP means 3'–5'*cyclic* AMP, an important intracellular signal (see Chapter 11).

For each of the major bases, a series of six 5'-nucleoside phosphates is possible (Table 2.4). A general type of nucleotide, regardless of base, is specified by the letter N as in rNMP, rNDP, dNMP, dNDP. This is commonly used, for example, when referring to the four rNTPs (rATP, rGTP, rCTP, rUTP) or the four dNTPs (dATP, dGTP, dCTP, dTTP) as substrates for RNA and DNA polymerases (see Part B).

 Chapter 11. Hormonal responses mediated through adenylate cuclase and the production of cAMP, page 338.

 Part B opener Template-directed nucleotide polymerases, page 79.

Pyrophosphate bonds

Two phosphate groups can join with the loss of water to form an anhydride bond, often called a pyrophosphate (or phosphoanhydride) bond. When phosphate is added in this manner to a mononucleotide (or nucleoside monophosphate) it gives rise successively to nucleoside di- and triphosphates, containing one and two pyrophosphate bonds respectively (see below). When pyrophosphate occurs in the unesterified form, it is termed **inorganic pyrophosphate (PPi)** (Figure 2.24).

 Nucleotides are linked in nucleic acids by phosphodiester bonds, page 37.

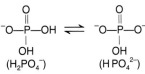

Inorganic ortho-phosphate (Pi)   Phosphomono ester   Phosphodiester   Inorganic pyro-phosphate (PPi)

Figure 2.24 Phosphate esters and pyrophosphates

Nucleoside triphosphates, especially ATP, drive most of the reactions of the cell

The importance of the nucleoside triphosphates, and especially ATP, in energy conservation and transfer and in many enzyme-catalysed reactions, is discussed in Chapter 7. The biosynthesis of nucleotides is described in Chapter 10.

 Chapter 7. ATP is the common intermediate in metabolism, the molecule which conserves energy during catabolism and contributes it to anabolism, page 190.
Chapter 10. Nucleotide metabolism, page 315.

Nucleotide coenzymes participate in many metabolic reactions

The coenzymes are a varied group of important biomolecules, several of which are nucleotide derivatives. The coenzymes NAD, NADP, FAD and Coenzyme A are all derivatives of ADP (Figure 2.25), which has important consequences for the binding of these molecules to enzymes. Other important groups of nucleotide derivatives include

- the nucleotide sugars, 'activated' forms of sugars used in biosynthesis, such as UDP-glucose (see Chapter 8)
- molecules such as CDP-choline, involved in phospholipid biosynthesis (see Chapter 9).

 Chapter 8. When glucose is plentiful it is stored as glycogen, page 224.
Chapter 9. Phospholipids are synthesized by adding 'head groups' to phosphatidic acid, page 281.

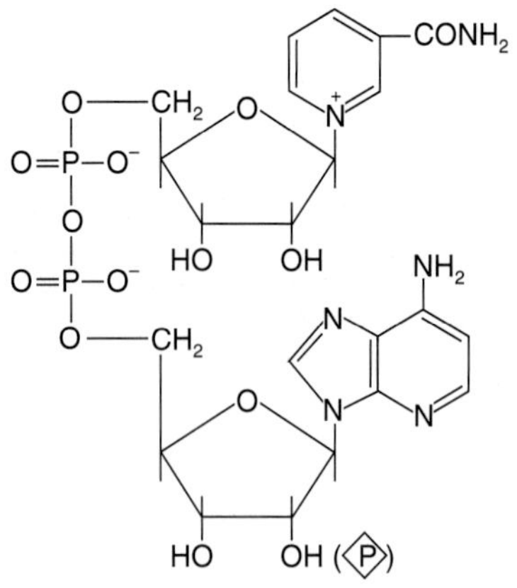

Flavin-adenine dinucleotide (FAD):
oxidation–reduction

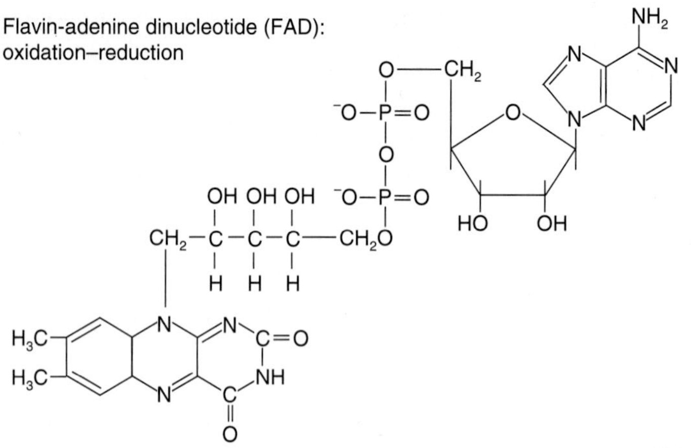

Nicotinamide-adenine
dinucleotide (phosphate) (NAD⁺(P)):
oxidation–reduction

Coenzyme A: acyl transfer

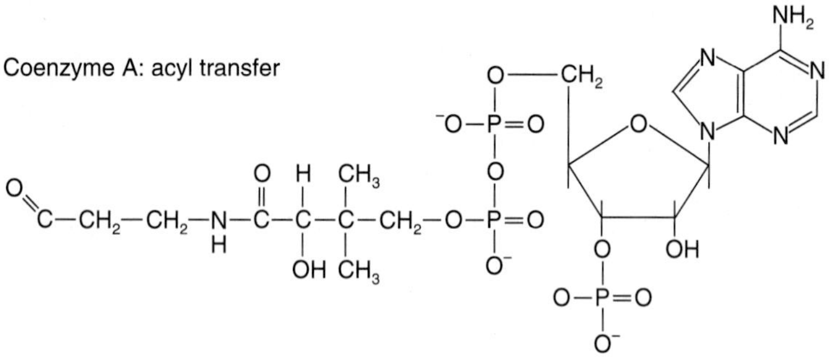

Figure 2.25 Nucleotide coenzymes.

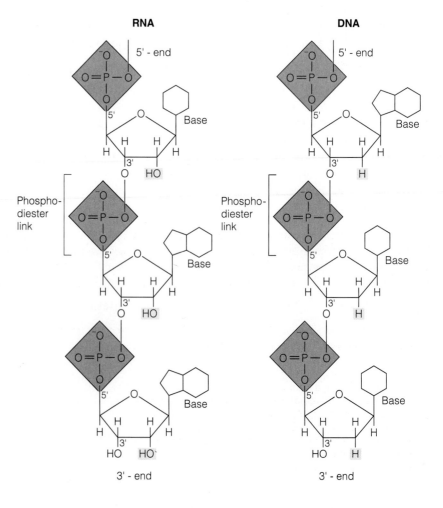

RNA

DNA

5' - end

5' - end

Base

Base

Phospho-
diester
link

Phospho-
diester
link

Base

Base

Base

Base

3' - end

3' - end

Figure 2.26 The phosphodiester bond in
nucleic acids.

Nucleotides are linked in nucleic acids by phosphodiester bonds

The phosphate of most mononucleotides in cells is esterified on the 5'-position (see above); important exceptions are the cyclic nucleotides (cAMP and cGMP), which are important intracellular signals and in which the phosphate is esterified to both 3'- and 5'-positions of the same nucleotide (see Chapter 11). A similar 3',5'-phosphodiester bond occurs between nucleotides in chains of DNA and RNA (Figure 2.26). The principal features of this bond are

Chapter 11. Hormonal responses are mediated through adenylate cyclase and the production of cAMP, page 338.

- The presence of a sugar–phosphate 'backbone', which gives the polynucleotide its continuity and to which the purine and pyrimidine bases are attached by glycosidic bonds.
- The polarity of the chain, with 5'- and 3'-ends or termini, analogous to the amino- and carboxy-termini of polypeptide chains (see above); polynucleotide chains are described as being 'in a 3'–5' (or 5'–3') direction' (see Chapter 6).
- As with all phosphodiesters, the phosphate moiety carries a single negative charge at physiological pH values, so that nucleic acids are polyanions.

Table 6.1. The principal features of the structure of DNA, page 147; Table 6.2. Biological implications of the structure of DNA, page 148.

Molecules in cells and in extracellular fluids

It is characteristic of multicellular organisms that the types of molecules and ions found within cells are generally different from those of extracellular fluids (blood, lymph, cerebrospinal fluid, etc.). The concentrations of these species, both inside and outside the cell, are typically under physiological control – often within close limits. Most molecules are synthesized within cells and so those that appear in extracellular fluids have commonly passed across the plasma membrane of the cell where they were made, usually by specific transport processes (see Chapter 3). Those remaining within cells (for example, most phosphorylated compounds) do so because they are unable to pass across the plasma membrane.

Chapter 3. Transport of molecules and ions across membranes, page 63.

It is difficult to measure solute concentrations within cells, but it is much easier to measure concentrations in fluids such as blood, plasma and urine. The analysis of the levels of such metabolites is the realm of clinical biochemistry and has become an increasingly important aid to clinical medicine in the diagnosis of disease and in assessing the effectiveness of treatment.

The same types of molecules are found in all cells

The different molecules in a typical cell number many hundreds, if not thousands of varieties, although most of these are present at very low concentrations. One of the most striking features of the biochemistry of different organisms – animals, plants, bacteria – is the similarity of the molecules they contain. The principal types of molecules found in cells are listed in Table 2.5.

Table 2.5 Types of molecules found in cells

Sugars and their derivatives (phosphates, nucleotides, etc.)
Amino acids and their derivatives (acyl-tRNAs, α-keto acids, etc.)
Nucleotides and nucleotide coenzymes (NAD, FAD, coenzyme A, etc.)
Fatty acids and their derivatives (acyl-CoAs, phospholipids, etc.)
Other **coenzymes** (mainly vitamin derivatives: PLP, TPP, etc.)
Metabolic **intermediates** (often carboxylic acids, for example those of the tricarboxylic acid cycle)
Fat-soluble **vitamins**
Cholesterol and other steroids

The individual importance of the various metabolites will be found in the appropriate sections of Part C.

There are different molecular compartments within cells

Many metabolites and ions are not uniformly distributed within cells but show local differences, for example between mitochondria and other organelles (see Chapter 7) and the cytosol. These differences have important physiological consequences in processes such as gluconeogenesis and energy generation (see Chapters 7 and 8), and their maintenance depends on the permeability and transport properties of the intracellular membranes.

Molecules are transported across cell membranes

In all metabolic processes, whether within or between cells, the phenomenon of transport is absolutely vital – the properties and behaviour of membrane-bound transport proteins will be discussed in Chapter 3. Molecules can pass between cells via plasma and other tissue fluids (this is especially important in **cell signalling**); within tissues molecules can also pass directly between cells through **gap junctions**.

Molecules and ions in plasma and other fluids differ from those found within cells

As stated above, the types of molecules in plasma differ from those found within cells. For example **glucose**, one of the most important of all metabolites, occurs within cells mainly as its phosphorylated derivatives glucose-1-phosphate and glucose-6-phosphate but free glucose predominates in plasma. The physiological significance of this is discussed in Chapter 8.

Other examples of such differences include the following.

- **Ketone bodies** (acetoacetate and β-hydroxybutyrate) are plentiful in plasma, but are converted to coenzyme A derivatives within cells (see Chapter 9).
- **Long-chain fatty acids** are transported in plasma bound to albumin, but are converted to coenzyme A derivatives after uptake into cells.
- **Sodium** (Na^+) and **potassium** (K^+) ions – in plasma, Na^+ ions predominate; the opposite is true within cells. This difference is maintained by the expenditure of energy by all cells in the operation of the membrane-bound **Na^+/K^+-ATPase** (see Chapter 3).

Chapter 3. Transport of molecules and ions across membranes, page 63.

Chapter 8. Gluconeogenesis: the synthesis of glucose from noncarbohydrate precursors, page 244.
Chapter 9. Ketone bodies are an efficient source of energy, page 274.

Chapter 3. Na⁺/K⁺ ATPase is asymmetrically organized within the membrane, page 64.

Cells, organelles and membranes

Cells are the fundamental units of organisms and a knowledge of their structure, function and organization into tissues and organs is essential for an understanding of the body and its operation. Cells exhibit a high degree of internal organization with numerous subcellular organelles, each equipped for its own specialized functions. Subdivision of cells into nucleus, endoplasmic reticulum, Golgi complex, mitochondrion and the other compartments of eukaryotic cells plays a crucial role in their biochemical operation and their interaction.

The whole cell and its internal compartments are all bounded by membranes. The cell membrane is a lipid bilayer in which a variety of membrane proteins, responsible for most of the functions of the membrane, is embedded. The three-dimensional structures of membrane proteins obey the same general rules as other cell proteins, modified by the effect of the hydrophobic environment of the lipid bilayer. Within the bilayer, proteins behave in a highly dynamic way.

Cell membranes contain a number of specialized structures, especially where there are interactions between cells (for example gap junctions) or cellular compartments (such as the nuclear membrane or envelope). Crucial functions of cell membranes include the transport of ions, small molecules and proteins in and out of cells and cellular compartments, and the binding of signal molecules to receptors on the cell surface.

The structure of mammalian cells

Cells are the fundamental units of all organisms

The fundamental units of all living things are **cells**, small aqueous compartments. Cells were first identified in 1665 by Robert Hooke when he was examining plant tissues under a microscope. He called them 'cellulae'. In 1840, Theodore Schwann proposed that all organisms exist as single cells or aggregates of cells, a theory that is still current. Cells are bounded by **plasma membranes**, which are selectively permeable and regulate the internal cellular composition – which differs markedly from that of the extracellular fluid. The external surface of the plasma membrane is 'the face the cell presents to the outside world'. It contains many receptors that are involved in regulating contacts with neighbouring cells and in monitoring alterations in the environment, so that the individual cell can respond in a manner that is beneficial to the entire organism.

Cells are organized into tissues and organs

The average human is composed of around 1000 million million (10^{15}) cells, organized into multicellular 'cities' – the tissues or organs – each of which performs a specific set of tasks essential to the well-being of the whole body. Integration of functions between tissues is mediated by chemical signals (**hormones**), relatively simple molecules secreted by specialized organs (the **endocrine glands**) in order to elicit the required response from their target tissues. Interactions and communication between cells such as neurones and lymphocytes also make use of local chemical messengers that act as **autocrine** (acting on self) or **paracrine** (acting on near neighbours) factors. By such means the body can respond in a coherent fashion to external and internal stimuli (see Chapter 11).

The plasma membrane marks the boundary of the cell

The boundaries of a cell are delineated by its plasma membrane, a selectively permeable membrane containing transport mechanisms that move nutrients and ions into the cell and remove unwanted products. The specific surface properties of the plasma membrane and the presence or absence of specific protein receptors on individual cell types are of prime importance in regulating cell–cell interactions and the response of individual cell types to hormones, such as insulin, which exert their effects by binding to receptors on the cell surface.

The plasma membrane is a dynamic, fluid structure. Vesicles derived from this membrane in the processes of **phagocytosis** or **endocytosis** convey nutrients that have been engulfed by the cell inwards to the **lysosomes**. Secretory proteins and new plasma membrane proteins travel in the opposite direction in vesicles derived from the **Golgi complex**. These vesicles fuse with the plasma membrane and liberate their contents into the extracellular space in exocytosis (Figure 3.1 and Box 3.1).

The plasma membrane of polarized cells has distinct apical and basolateral surfaces

Most cells of the body are **fibroblastic** (spindle-shaped) or **epithelial** (sheet-like) in nature. Epithelial tissues are formed from sheets of epithelial cells and separate the different compartments of the body. The two surfaces of the epithelial cells are bathed in different fluids and are exposed to different sets of molecules – ions, metabolites, hormones, etc. In order to deal with these different environments such **polarized** cells possess two distinct surfaces on their plasma membrane: one, the **apical** surface, faces externally (for example, into the lumen of a vessel or duct); the other, the **basolateral** surface, faces internally and is bathed by tissue fluids. The two surfaces are separated by the **zonula adherens**, which joins adjacent cells, has a belt of tight junctions (see below and Figure 3.21) on its apical side and contains different populations of membrane proteins (see below), such as cell-surface receptors and transporters.

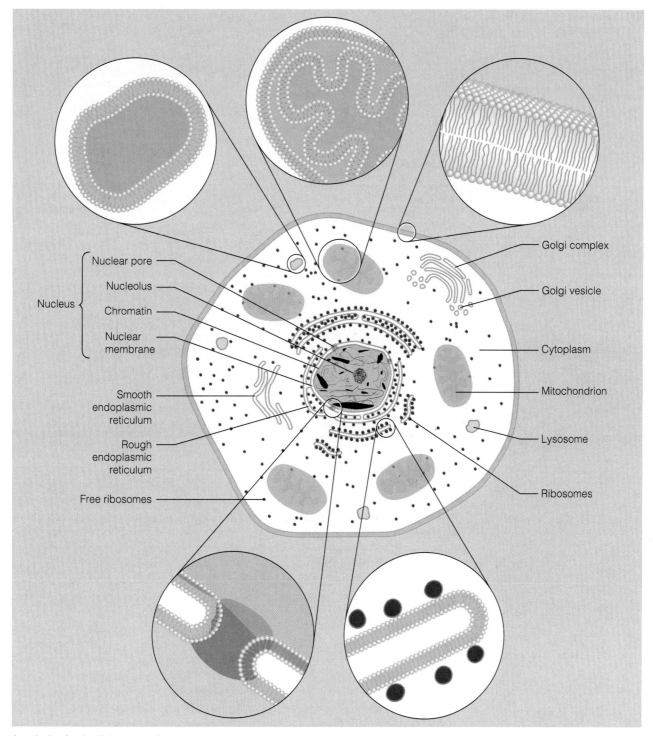

A typical animal cell, its organelles and membranes

Figure 3.1 Involvement of the plasma membrane in phagocytosis and secretion (exocytosis). Sealed vesicles carrying proteins for export (A) bud off from the Golgi complex and fuse with the plasma membrane liberating their contents into the external space. Primary lysosomes, also formed from the Golgi complex, can be used for the digestion of intracellular material, for example in normal turnover of mitochondria (B). They are also involved in digestion of foreign particles, for example bacteria (C).

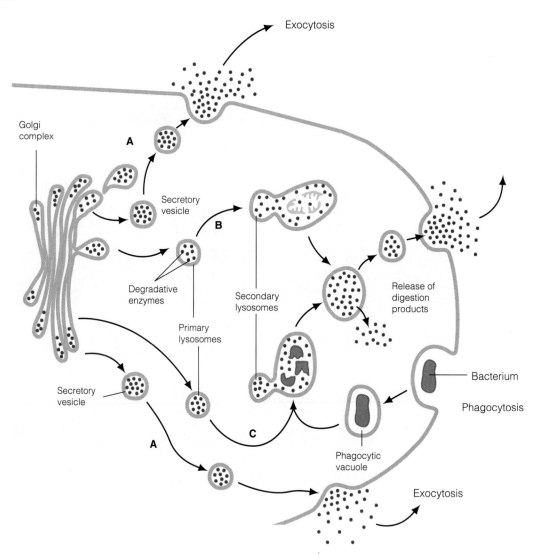

Eukaryotic cells have a distinctive internal organization

Bacteria are small single-cell **prokaryotic** (that is, lacking an organized nucleus) organisms, approximately 1–5 μm in diameter, and do not generally contain organized membranous internal structures. In contrast, human and other **eukaryotic** cells range from about 10 to 80 μm in size and are characterized by a complex internal organization consisting of several types of distinct compartments, or **organelles** (Figure 3.2 and Table 3.1). Each compartment is bounded by a membrane (double in the case of nuclei and mitochondria, but single for all other organelles) and performs a specific set of tasks. More than 90% of the total membrane of a cell is involved in the intracellular structures, and probably at least half of all enzymically catalysed metabolic processes take place on or within membranes.

Eukaryotic intracellular structures include nuclei, mitochondria, endoplasmic reticulum (ER), Golgi complex, lysosomes and peroxisomes. Several morphologically distinct particles or aggregates – such as nucleoli, ribosomes and chromosomes (complexes of DNA or RNA with protein) – are located within compartments. These structures may be regarded as a form of functional compartmentation in that they all perform specific tasks in a defined environment.

Although the cell cytoplasm contains no obvious structural features, many of the enzymes catalysing sequential steps in the major metabolic pathways are

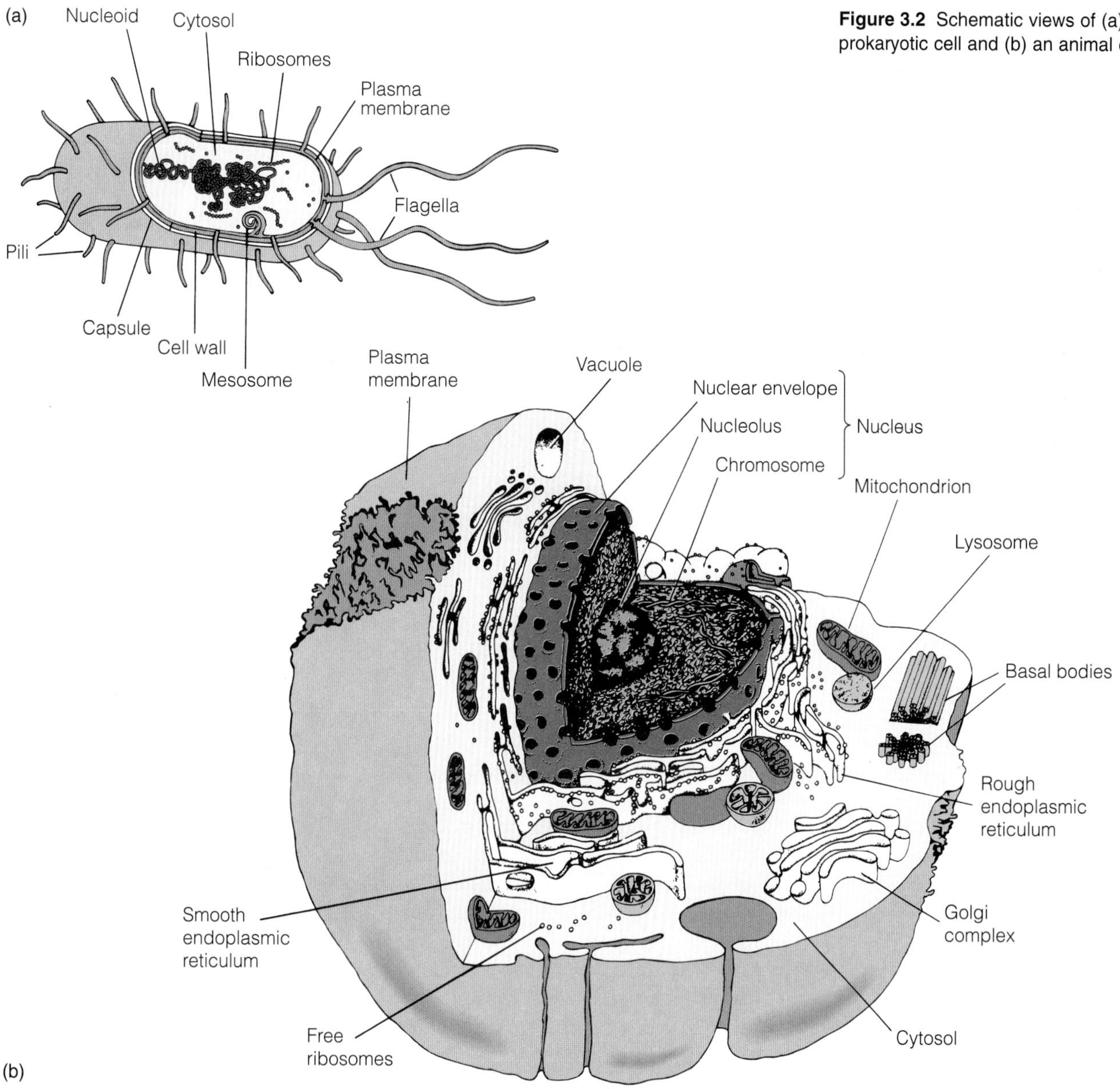

(a)

Nucleoid Cytosol

Ribosomes

Plasma
membrane

Flagella

Pili

Capsule

Cell wall

Mesosome

Figure 3.2 Schematic views of (a) a
prokaryotic cell and (b) an animal cell.

Plasma
membrane

Vacuole

Nuclear envelope

Nucleolus

Nucleus

Chromosome

Mitochondrion

Lysosome

Basal bodies

Rough
endoplasmic
reticulum

Golgi
complex

Cytosol

Smooth
endoplasmic
reticulum

Free
ribosomes

(b)

Table 3.1 Major differences between prokaryotic and eukaryotic cells

	Prokaryotes (Eubacteria*)	Eukaryotes
Nucleus	Absent	Present
Organelles	None	Several†
Genome	One molecule of circular DNA	Several molecules of linear DNA
Histones/nucleosome	Absent	Present (see Chapter 6)
Introns in genes	Absent	Widespread (see Chapter 5)
RNA polymerase	One type	Three types (see Chapter 5)
Ribosomes	70S type	80S type (see Chapter 5)
Translation	Polycistronic	Monocistronic (see Chapter 5)

* This excludes the Archaea (formerly Archaebacteria), which are prokaryotes in the sense of
lacking a nucleus, but are probably closer in evolutionary terms to the eukaryotes (see Box 5.2)
† Some simple unicellular eukaryotes lack mitochondria; others are thought to have lost them
during evolution

Box 5.2. Study of ribosomal
RNA sequences has changed
ideas about the early
evolution of life, page 120.

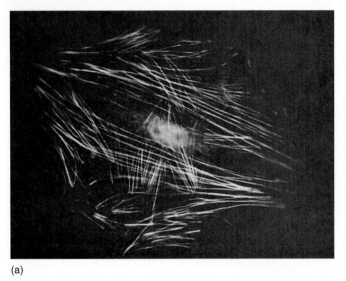

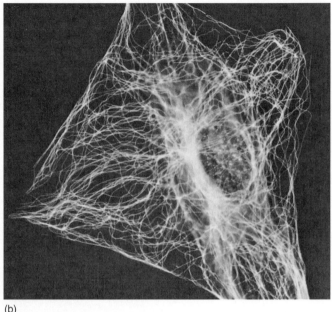

(a)

(b)

Figure 3.3 Microfilament (a) and microtubule (b) networks within cells visualized by fluorescent antibodies.

thought to organize themselves specifically into functional metabolic units, which have been termed **metablons**. This microcompartmentation is usually lost when the cell is disrupted for study in the laboratory. A high degree of cooperation at the level of individual enzymes would have advantages in channelling intermediates, which may be labile or toxic, directly from the active site of one enzyme to that of the next enzyme in the pathway, thereby greatly enhancing overall catalytic efficiency.

Within the cytoplasm of mammalian cells, underlying the plasma membrane and surrounding the nucleus, is a complex network of contractile proteins in the form of **microfilaments, intermediate filaments** and **microtubules**, making up the **cytoskeleton**. This has important functions in controlling the shape, movement and division of cells and in processes such as **endocytosis** and pinocytosis. Such structures are generally visible only in cells that have been stained, for example with fluorescent antibodies against individual protein components such as **actin** or **tubulin** (Figure 3.3). The cytoskeleton contributes to the nature of cells as fluid, dynamic entities in a continuous state of flux (see below and Table 3.2).

Organelles and their functions

The preceding section introduced the general concept of intracellular organelles and their vital role in performing specialized activities within the cellular environment. The sizes of cells and their organelles in relation to the body and to large and small molecules are shown in Figure 3.4.

The nucleus houses the genetic material and is the 'operational centre' of the cell

Under the microscope, the most prominent organelle is the nucleus, which can occupy up to half of the total cell volume. It is present in all cells of the body – with the notable exception of the erythrocyte (red blood cell), which loses its nucleus as it matures (see Chapter 12), although in some species (such as the chicken) the mature erythrocyte remains nucleated. The nucleus is the operational centre for the entire cell and separated from the cytoplasm by the **nuclear**

Chapter 12. The erythrocyte must survive wear and tear, page 365.

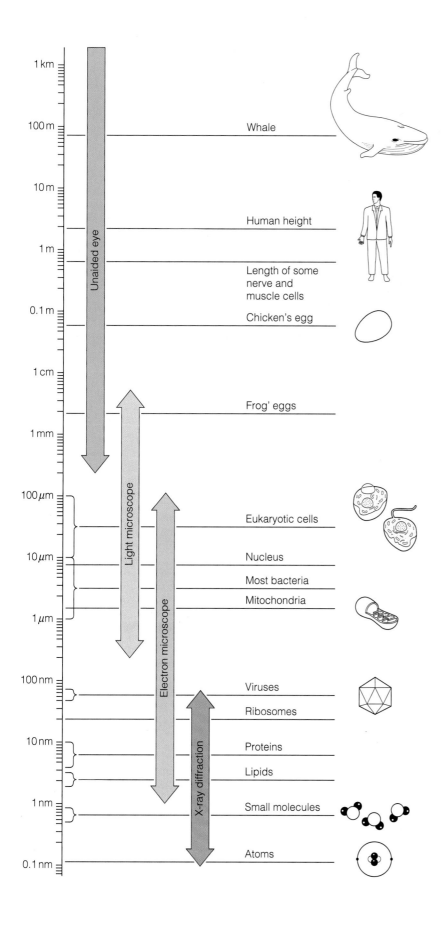

Figure 3.4 Diagram indicating the major cellular structures and their sizes.

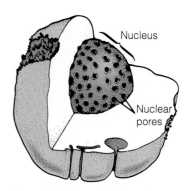

Figure 3.5 Schematic view of nucleus.

Chapter 5. RNAs must be transported to the cytoplasm, page 125. Chapter 5. There are several mechanisms for targeting proteins to the nucleus, page 141.

envelope, two closely spaced membranes containing several hundred **nuclear pore complexes** (see Figures 3.5 and 3.21) that permit the movement of macromolecules into and out of the nucleus (see Chapter 5). Ribosomes and mRNA required for polypeptide synthesis are exported from the nucleus; proteins needed for chromosomal structure (**histones**) and functions such as DNA replication and transcription are imported from their site of synthesis in the cytoplasm. Most cell types contain a single nucleus, but muscle cells and some liver cells are multinucleate.

Nuclei contain chromosomes and nucleoli

Enclosed by the nuclear membrane, the matrix of the nucleus, or nucleoplasm, contains two morphologically distinct particles: **chromosomes** and **nucleoli**. Chromosomes are DNA–protein aggregates which house the complete blueprint for production of new cells and new offspring. Human chromosomes are present as 22 homologous pairs plus the X and Y chromosomes determining sex. The chromosome complement of an organism is referred to as the **species karyotype**; the normal human male karyotype is 46, XY and the female 46, XX (see Chapter 17).

Box 17.4. Karyotype analysis of chromosomes, page 516.

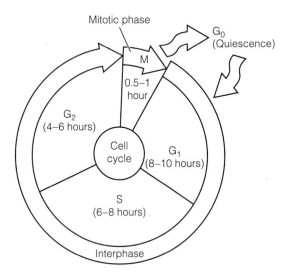

Figure 3.6 The cell cycle. Rapidly growing cells in the body divide approximately once a day and proceed through an orderly cycle in which after cell division there is a prolonged phase (G_1) preceding DNA synthesis (S) which takes approximately 6–8 h. A further period of cell growth (G_2) follows in which the cell enlarges further in preparation for cell division, an event taking only 30 minutes on average. Nondividing cells enter a quiescent phase, termed G_0, and re-enter the cell cycle via G_1 when triggered to do so.

Chromosomes are visible only during the interphase of normal cell division (**mitosis**) or the specialized cell division (**meiosis**) that occurs when the gametes or reproductive cells (sperm and ova) are maximally condensed in preparation for segregation to the daughter cells. For most of the cell cycle (G_1, S and G_2 in Figure 3.6), when the cell is not actively dividing, the chromosomes are unwound and dispersed throughout the nucleus as **chromatin**. Areas in which the DNA is more highly condensed are termed **heterochromatin**; **euchromatin** is less dense and contains most of the active genes.

Several dense regions of nucleoprotein (**nucleoli**) are typically present in each nucleus. The nucleoli are assemblies of RNA and protein involved in the production of new ribosomes for export to the cytoplasm. They are intimately associated with defined regions (the nucleolar organizers) of human chromosomes 13, 14, 15, 21 and 22, which contain multiple copies of the ribosomal RNA genes.

The endoplasmic reticulum and Golgi complex are membrane-bound compartments within the cytoplasm involved in the synthesis of membrane and secretory proteins

The endoplasmic reticulum (ER) comprises an interconnecting network of tubular membranes or flattened sacs (also known as **cisternae**) lying within the cytoplasm,

(a)
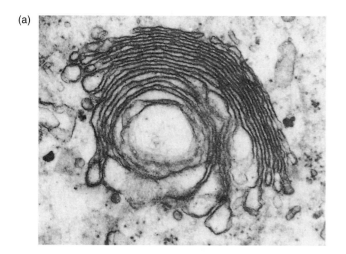

(b)
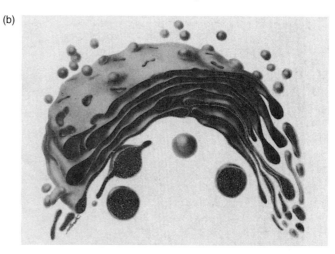

often parallel to the nuclear envelope (see Figure 3.2). Two types of ER are observed: rough and smooth. The rough ER is studded with ribosomes, which are actively engaged in protein synthesis, and generally lies closer to the nucleus. Smooth ER is involved in packaging and delivering proteins to the Golgi complex; it is also the site of biosynthesis of membrane lipids and houses the cytochrome P-450-based enzyme systems involved in the detoxification of drugs and other toxic compounds, especially in liver cells (see Chapter 7).

The rough ER and the Golgi complex are especially prominent in secretory cells (Figure 3.7), an observation that led to the idea that these compartments are involved with the secretory process. Secretory proteins formed on ER-bound ribosomes are transported into the lumen of the ER during synthesis and are subsequently conveyed to the Golgi complex. The Golgi complex consists of a series of flattened stacks of smooth membranes or vesicles in which three regions may be distinguished – the *cis-*, *median-* and *trans*-Golgi. Each component of this complex plays a distinct role in modifying and processing proteins, especially glycoproteins, and ensuring their correct delivery to the plasma membrane or to certain internal organelles such as lysosomes.

Beyond the *trans*-Golgi complex lies the **trans-Golgi network** (TGN), which plays a role in sorting membranous components for their eventual destinations (plasma membrane, lysosome or secretory vesicle) and in recycling material from vesicles. The outward movement of material, from ER to *cis-*, *median-* and *trans*-Golgi complex, then on to the TGN, is balanced by a return inward flow of membrane components, especially lipid.

Figure 3.7 (a) EM of Golgi complex; (b) prominent Golgi complex and rough endoplasmic reticulum in a secretory cell.

 Box 7.5. Cytochrome *P*-450 page 206.

Mitochondria are the 'powerhouses' of cells and are responsible for the aerobic production of ATP

Mitochondria are ellipsoid organelles 2–3 μm by 0.5–1 μm in size. Their characteristic appearance under the electron microscope consists of an inner membrane, which is highly convoluted into **cristae** (Figure 3.8) and encloses the soluble **matrix** surrounded by a smooth, featureless **outer membrane** to create an additional soluble compartment – the **intermembrane space** (Box 7.4). The numbers and location of mitochondria within cells are related to their central role in respiration and ATP synthesis. Cells that need a lot of energy, such as those of cardiac muscle, contain large numbers of mitochondria with densely packed cristae to provide energy to sustain the mechanical contraction of the heart via aerobic ATP production. Neurones, in contrast, contain only a few mitochondria and erythrocytes contain none. Mitochondria in placental tissue have a smaller surface area of inner membrane (Figure 3.8b).

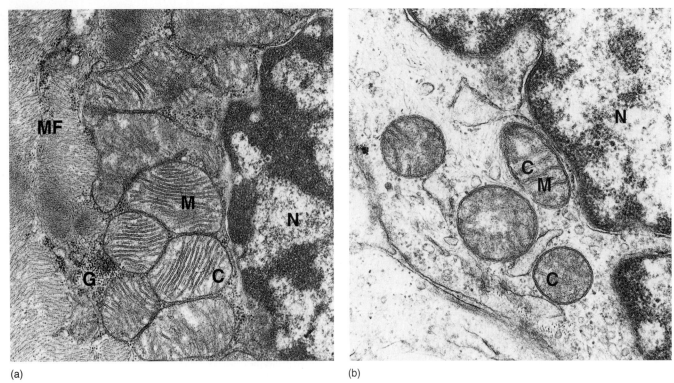

(a)

(b)

Figure 3.8 Mitochondria in a human cardiac myocyte (a) and a stromal cell of human placenta (b), illustrating the difference in density of cristae in mitochondria of tissues with high (cardiac) and relatively low (placental) rates of oxidative metabolism. M (mitochondria), C (cristae), N (nucleus), MF (muscle filaments), G (glycogen granules). Courtesy of Dr G. B. M. Lindop, University of Glasgow, Department of Histopathology, Western Infirmary, Glasgow.

Mitochondria possess a genome with a small number of genes

Chapter 17. The mitochondrial genome encodes 13 polypeptide, page 521.

Chapter 7. The mitochondrial respiratory chain comprises four multisubunit assemblies of proteins, page 199.
Chapter 5. Many mitochondrial proteins are made on cytoplasmic ribosomes, page 139.

Human mitochondria contain a small DNA genome of 16 569 base pairs (bp), which encodes two mitochondrial ribosomal RNA species, 22 species of transfer RNA and 13 proteins (see Chapter 17). These proteins are all components of the integral inner membrane complexes that are involved in the processes of electron transport and oxidative phosphorylation responsible for ATP generation by mitochondria (see Chapter 7). Mitochondria probably evolved from a primitive bacterial ancestor which established a symbiotic relationship with an early eukaryotic cell (see Box 7.2). In the course of evolution most of their original genome has been transferred to the nucleus, so that 95% of mitochondrial proteins are now encoded by nuclear genes and must be imported into mitochondria from their site of synthesis in the cytoplasm (see Chapter 5).

Lysosomes digest ingested macromolecules and participate in the turnover of intracellular components

Lysosomes are small organelles 0.5–1.0 μm in diameter and bounded by a single membrane. They are the scavengers of the cell in that they contain an array of hydrolytic enzymes for degrading carbohydrate, fat and protein that may have entered the cell by phagocytosis. It is clearly advantageous to the cell to maintain such enzymes in a separate compartment so that intracellular structures are not degraded. Lysosomes may be formed *de novo* from vesicles budding off from the Golgi complex; in this case they are referred to as **primary lysosomes**. Initiation of hydrolytic activity occurs only once they fuse with membrane-bounded vesicles, or **endosomes**, containing undigested material; at this stage they are termed **secondary lysosomes**. Digested nutrients are used for synthesizing new

Chapter 5. Ribosomes are directed to the endoplasmic reticulum by signal recognition particles, page 138.

Chapter 14. The hydrophobic character of lipids complicates their digestion and absorption, page 418.

cell materials. Lysosomes also function in the normal turnover of cellular components, for example by degrading proteins that have reached the end of their useful lives.

Several genetic defects give rise to **lysosomal storage diseases**. One example is Sly syndrome (also known as mucopolysaccharidosis VII), in which deficiency of the enzyme β-glucuronidase causes pathological behaviour of a number of tissues because proteoglycans are not being degraded. A wide range of tissues is affected in this condition – liver, spleen, brain, skeleton, kidney and cornea – which has complicated attempts at gene therapy to correct the abnormality.

Peroxisomes are involved in a number of oxidative processes

Peroxisomes are similar in size to lysosomes but contain several enzymes involved in oxidative metabolism, many of which use molecular oxygen and generate hydrogen peroxide (H_2O_2). As H_2O_2 and organic peroxides are all highly toxic, the cell protects itself by isolating the enzymic reactions involved in their generation and rapid destruction within a distinct organelle – the **peroxisome**. Peroxisomes also contain **catalase**, which comprises nearly half the total protein. It is involved in the metabolism of alcohol and other compounds in the liver, and converts excess H_2O_2 to oxygen and water very efficiently. An inherited human disorder (Zellweger syndrome) is associated with deficiency of a 70 kDa peroxisomal protein of as yet unknown function.

The cytoskeleton determines cell shape and movement and regulates the organization of internal structures

The cytoplasm of the cell was once thought of as relatively featureless and containing high concentrations of soluble proteins with little or no obvious organization. The use of immunofluorescence techniques has revealed that the cell contains an underlying cytoskeleton consisting of three classes of filaments: microtubules, microfilaments and intermediate filaments. Each of these is composed of a different type of contractile protein. The structural features, major proteins, intracellular location and functions of these three elements of the cytoskeleton are listed in Table 3.2 and a dramatic picture of the cytoskeletal elements of the cell (as revealed by immunofluorescence) is illustrated in Figure 3.3.

Table 3.2 Cytoskeletal elements

Filamentous structure	Diameter (nm)	Major constituent proteins	Inhibitors of assembly/disassembly	Overall functions
Microfilaments	7	Actin, many regulatory proteins	Cytochalasin B, phalloidin	Involved in cell movements (e.g. membrane 'ruffling', cytoplasmic streaming, nerve axon growth), clot retraction
Intermediate filaments	7–11	Large multigene family of proteins capable of forming two or three α-helical structures, keratins, vimentin, neurofilaments, desmin (muscle), laminins (nucleus)		Contribute to mechanical strength and stability of cells and internal structures (e.g. laminin in formation of fibrous network within the nucleus)
Microtubules	approx. 30	α- and β-tubulins, forming hollow cylinders, accessory proteins (e.g. kinesin, dynein)	Colchicine	Involved in chromosome separation at mitosis, intracellular transport of vesicles and organelles, movement of cilia

(a)

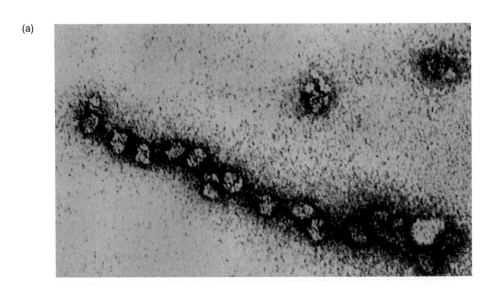

(b)

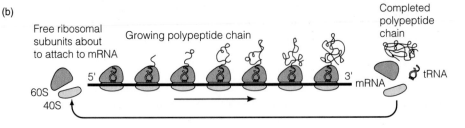

Figure 3.9 A polysome – formed by attachment of several ribosomes, each in the act of translation, proceeding along a single strand of mRNA. (a) Electron micrograph; (b) Diagrammatic representation of above.

The cytoskeleton is responsible for maintaining the cell shape and controlling the movements of cells and of organelles within cells. It is involved in phagocytosis, pinocytosis and receptor-mediated endocytosis and has important functions in the segregation of chromosomes during cell division. Cytoskeletal components, notably tubulin, are able to dissociate and reassociate. This allows the cytoskeleton to change its shape readily, a feature that is particularly important in muscle cells.

Ribosomes provide the machinery for the synthesis of polypeptide chains

Ribosomes are aggregates of RNA and protein and are the sites of protein synthesis in the cell. A description of their role in protein synthesis will be found in Chapter 5. Ribosomes in the cytoplasm often occur as **polysomes** (several ribosomes are attached to a single strand of messenger RNA; Figure 3.9), in which they are actively engaged in protein synthesis. A large number of ribosomes are also found attached to the cytoplasmic surface of the ER, where they produce polypeptides destined for entry into the lumen of the ER (see above). Many of the apparently 'free' cytosolic ribosomes are now known to be attached to the cytoskeleton and may be associated with the 'tracking' of RNA from the nucleus (see Chapter 5).

Chapter 5. mRNA translation: the genetic code and open reading frames, page 130.
Chapter 5. RNAs must be transported to the cytoplasm, page 125.

Cellular membranes

Biological membranes are fluid and vital to cellular integrity and many cellular functions

The membranes of cells are thin, sheet-like structures composed of different species of protein and lipid molecules, many of which have small amounts of carbohydrate covalently linked to them. Membranes are not inert, static barriers; they are fluid, dynamic structures in which both lipids and proteins exhibit considerable lateral mobility. This mobility is evident in aspects of cellular function such as cytoplasmic streaming, receptor-mediated endocytosis, development of pseudopodia and the formation of gap junctions with neighbouring cells. Membrane proteins can be thought of as floating in a sea of lipid and the fluidity of membrane lipids is vital to the proper functioning of the proteins located within them as they undergo conformational changes during their roles in selective transport, energy transduction, cell signalling and generation of electrical and chemical impulses.

Membranes differ in protein, lipid and carbohydrate content

The ratio of protein to lipid varies greatly between different types of membrane, but typical biological membranes contain about 50% protein. The myelin sheath, which acts as an effective insulator lining neuronal cells, has a high lipid (80%) and low protein (20%) content, consistent with its passive role in propagating the transmission of electrical pulses along the axon between adjacent nodes of Ranvier (Figure 3.10). In contrast, metabolically active membranes such as the mitochondrial inner membrane comprise about 75% protein and 25% lipid by weight.

Carbohydrate (1–10% by weight) is present on some, but not all membranes. It is found mainly on the external surface of the plasma membrane (as glycoprotein and glycolipid) and the internal surfaces of the endoplasmic reticulum, Golgi complex and lysosomes. It always assumes an asymmetric distribution, being located on the side of the membrane facing away from the cytoplasmic compartment. The oligosaccharide chains on glycoproteins can be *N*-linked through asparagine or *O*-linked through serine or threonine residues of the protein moiety (see Chapter 2).

The principal lipids of human cell membranes are phospholipids, sphingolipids and cholesterol, with smaller amounts of glycolipids and cardiolipin (see Chapter 2). The various membranes contain different proportions of each, cardiolipin, for example, being confined to the inner membrane of mitochondria. Phospholipids and sphingolipids both form bilayer structures, the behaviour of which is modified by the presence of cholesterol. Glycolipids have important functions in a variety of recognition processes at the external surface of the cell (Figure 3.11).

Chapter 2. Oligosaccharide structures are covalently linked to many secretory and membrane proteins and to some lipids, page 21.
Chapter 2. Sphingomyelin, glycolipids and cholesterol are the other major lipids of eukaryotic membranes, page 26.

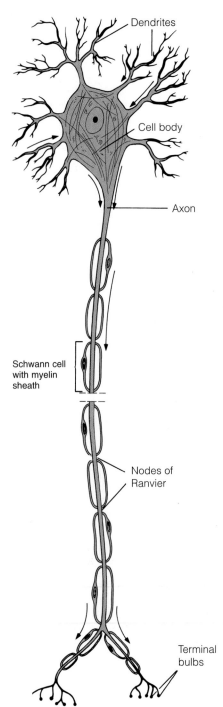

Figure 3.10 Nerve axon at high magnification, illustrating the myelin sheath and nodes of Ranvier.

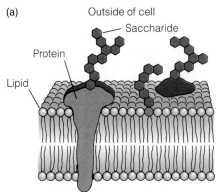

(a)

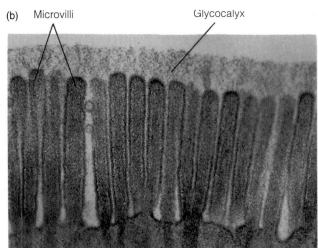

(b)

Figure 3.11 Cell surface recognition factors. (a) Schematic view of a lipid membrane. Oligosaccharides are attached to the outer surface, through either membrane-embedded proteins or special lipid molecules. (b) Electron micrograph of the surface of an intestinal epithelial cell. The projections, called microvilli, are covered on their outer surface by a layer of branched polysaccharide chains attached to proteins in the cell membrane. This carbohydrate layer, called the glycocalyx, is found on many animal cell surfaces.

Phospholipids spontaneously form lipid bilayers in water

The amphipathic nature of phospholipids provides clues as to the most thermo-dynamically favourable arrangements such molecules tend to adopt in aqueous media. In appropriate conditions, phospholipids assemble spontaneously into extensive sheet-like structures composed of a bimolecular leaflet of lipids with their hydrophilic headgroups on the surface and their fatty acyl chains in the internal hydrophobic phase (Figure 3.12). This arrangement is driven by hydrophobic forces which promote the removal of the hydrocarbon tails of the fatty acids from direct contact with water into the nonpolar interior of the bilayer. Bilayers have a strong tendency to be self-sealing and to form closed compartments.

Membranes can be visualized under the electron microscope

Viewed by electron microscopy after treatment with osmium tetroxide, membranes appear as a characteristic trilaminar or 'triple track' structures, with the phospholipid headgroups on the exterior of the membrane forming electron-dense regions and the hydrophobic interior staining only lightly.

The technique of freeze-fracture electron microscopy permits the two halves of the bilayer to be separated and the internal structures that penetrate into or across this region to be visualized. It is possible to observe large protein particles that tend to segregate with the inner half of the bilayer, leaving corresponding pits in the outer leaflet. This suggests that many of these particles may be anchored to internal structures of the cell (Figure 3.13).

(a)

(b)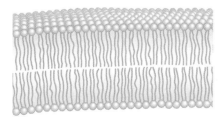

(c)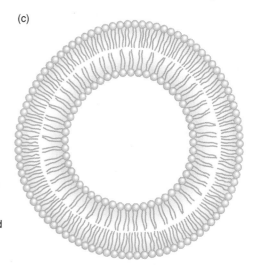

Figure 3.12 Favoured arrangements of fatty acids and phospholipids in solution. Fatty acid and detergents tend to associate into micelles (a), which are limited structures with their aliphatic chains sequestered away from the aqueous environment. (b) Phospholipids tend to assemble spontaneously into extensive bilayer structures as the presence of two fatty acid chains means that, in general, they are too bulky to fit into a micellar structure. Again the major driving force is the requirement to satisfy the appropriate hydrophobic and hydrophilic interactions. (c) Lipid bilayers tend to form closed compartments so that there are no ends with exposed hydrocarbon tails.

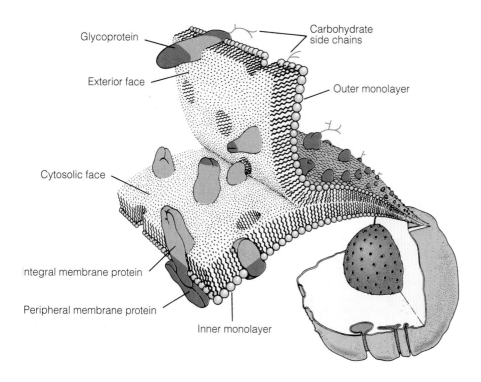

Figure 3.13 Freeze-fracture electron microscopy is useful for examining internal features of a membrane bilayer. Membranes are separated into their two leaflets, revealing differences in the organization of proteins embedded in the bilayer. The figure is a diagrammatic representation of the structures visualized by this technique.

Membrane proteins

Most proteins of the cell are water-soluble and are discussed in Chapter 4, which should be consulted for information on protein structure and function. Membrane proteins obey the same general rules as soluble ones, but their association with membrane lipids gives rise to a number of distinctive features, which will be discussed here.

Membrane-associated proteins are categorized by the nature of their interaction with the lipid bilayer

Two major classes of membrane polypeptide are recognized: peripheral (or extrinsic) proteins and integral (or intrinsic) proteins. Peripheral proteins may be removed from membranes in a water-soluble form by treatment at high ionic strength or alkaline pH without disrupting the lipid bilayer. They are generally similar in nature to water-soluble proteins as they are bound to the surface of the membrane only by electrostatic interactions, often via phospholipid headgroups.

In contrast, integral proteins can be solubilized only by the use of powerful detergents, which rupture the membrane. They tend to contain a high proportion of hydrophobic amino acid residues which interact with the hydrophobic region of the membrane bilayer. For some proteins this interaction is limited to a short stretch of polypeptide chain, often located at their *N*- or *C*-termini, as in cytochrome b_5 of the mitochondrial outer membrane or the aminopeptidases of brush border epithelia in the small intestine. This section of polypeptide acts as a 'membrane anchor' inserted into the lipid bilayer while the catalytic domains of the enzymes are situated outwith the confines of the membrane.

Integral membrane proteins span the membrane once or several times

Glycophorin, a major protein of the red blood cell, is involved in the determination of minor blood group specificities, the M and N antigens. This protein has a single membrane-spanning domain containing a hydrophobic region of about 20 amino acids (forming a transmembrane segment in an otherwise hydrophilic structure). Other membrane proteins have multiple membrane-spanning domains and are markedly more hydrophobic overall (Figure 3.14). Once they have been extracted from their membrane, integral membrane proteins remain in solution only in the presence of detergents, which solubilize membrane proteins by replacing the phospholipids surrounding the hydrophobic parts of the polypeptide.

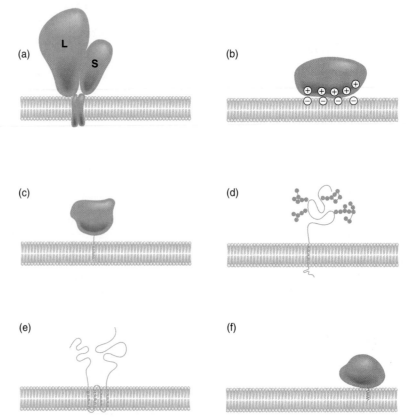

Figure 3.14 Schematic representation of various types of interaction of membrane proteins with the phospholipid layer. (a) Hydrophilic subunits bound directly to integral membrane polypeptides via protein–protein interactions, for example succinate dehydrogenase activity of mitochondrial inner membrane; (b) Peripheral protein bound by electrostatic interactions to phospholipid head groups, for example cytochrome *c*; (c) Soluble protein bound by lipid 'anchor'; (d) Transmembrane protein with single membrane-spanning α-helix, for example glycophorin of erythrocytes; (e) Integral protein with multiple membrane-spanning segments; (f) Protein associated with bilayer by short hydrophilic segment at the *N* or *C*-terminus, for example aminopeptidases of gut epithelia.

The transmembrane organization of an integral protein can be predicted from its primary structure

Analysis of transmembrane proteins such as the family of tissue-specific **glucose transporters** has revealed a high α-helix content, in the form of multiple transmembrane α-helices (see Chapter 4). In contrast, **porin**, which forms aqueous channels in the mitochondrial outer membrane, contains extended stretches of antiparallel β-sheet folded around to generate a β-barrel. Both α-helices and β-pleated sheets are favoured in nonpolar surroundings as they are stabilized by intramolecular hydrogen bonds which are competed for by water.

Although the detailed structure of most membrane proteins is not known, their probable transmembrane organization can often be predicted from their amino acid sequences (see Figure 12.8). To span a phospholipid bilayer, an α-helix needs to be about 20 amino acids long and, because most of these residues must be hydrophobic, such groupings are easily identified in the protein sequence (see Chapter 4). Because it contains more cholesterol, the plasma membrane is thicker than that of the endoplasmic reticulum and so membrane-spanning domains of plasma membrane proteins are about 22 amino acids long, compared with 17 for proteins of the ER (see Box 5.5).

Box 4.11. α-Helices, β-sheets and turns – regular features in proteins, page 104.

Chapter 4. There are many practical applications of amino acid sequence data, page 97.
Box 5.5. The absence of cholesterol makes the endoplasmic reticulum membrane more permeable than the plasma membrane, page 140.

Some proteins are attached to membranes by lipid 'anchors'

The covalent attachment of lipid is responsible for binding a variety of enzymes and plasma membrane receptors to cell membranes. Some proteins undergo covalent addition of the C-14 saturated fatty acid myristate to the *N*-terminal glycine residue (myristylation) or the C-16 saturated fatty acid palmitate to an internal serine, threonine or cysteine (palmitoylation). Addition of a fatty acyl chain causes the acylated protein to associate with membrane by acting as a lipid 'anchor' that inserts itself into the lipid bilayer. A similar effect is achieved in other proteins by the addition of isoprenoid chains (geranyl or farnesyl), or by linking phosphatidylinositol to certain glycoproteins (Table 3.3). Such interactions may be relatively permanent or quite transient. Possible roles for these **post-translational modifications** (see Chapter 5) are in regulation of protein transport, in cell-signalling mechanisms and in anchoring proteins at a particular site to exercise their specific function.

Chapter 5. Proteins are retained in membranes by 'stop-transfer' sequences, page 139.

Table 3.3 Types of covalent protein–lipid membrane 'anchors'

Lipid	Types of linkage	Sequence specificity	Function
Myristic acid (C14:0)	Amide	Gly-X-X-X-Cys N-terminal motif	Membrane localization; protein turnover
Palmitic acid (C16:0)	Thioester to cysteine; oxyester to serine and threonine	Internal residues, various sites in protein	Membrane binding, sometimes in conjunction with neighbouring farnesyl residues; protein turnover
Isoprenyl groups, either farnesyl (C15) or geranyl (C20) (steroid precursors)	Oxyester	Cys-Ala-Ala-X	Membrane anchoring
Glycosyl phosphatidylinositol (GPI anchor)	Amide	C-terminal amino acid	Anchors proteins to outer leaflet of plasma membrane; possible role in protein targeting

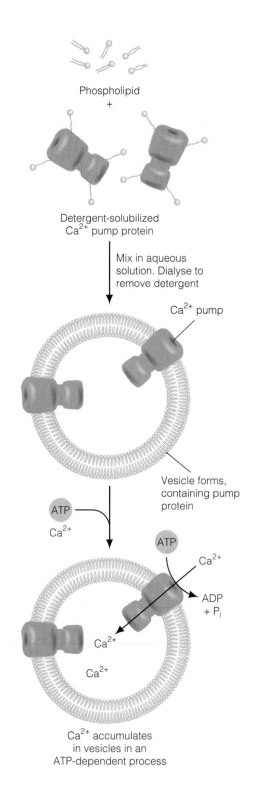

Figure 3.15 Insertion of a membrane protein Ca²⁺-ATPase into artificial phospholipid vesicles.

Many membrane-bound proteins are functional only in a hydrophobic environment

Many important membrane-associated proteins or protein complexes have enzymic activities that can be assayed directly in detergent-containing solutions after extraction from the cell membrane. However, many integral proteins (such as transporters) have vectorial (or directional) functions that can be expressed only by virtue of their location in intact membranes. For example, the activities of membrane channels such as porin in the mitochondrial outer membrane or **connexin** (involved in gap junction formation and cell–cell communication) depend on the presence of sealed membrane compartments. Similarly, the generation of the proton motive force by the mitochondrial **respiratory chain complexes** (see Chapter 7) and the function of specific transport systems such as the **adenine nucleotide translocase** of mitochondria or the **Na+/K+ ATPase** of the plasma membrane are dependent on intact membrane structures.

Purified membrane proteins may be inserted into artificial membranes (liposomes)

The activity of purified integral proteins can be studied in the laboratory by incorporating them into artificial membrane vesicles, or liposomes, thus mimicking their environment in the original biological membrane (Figure 3.15). This is accomplished by adding phospholipids (see Chapter 2) to the purified protein in detergent and then removing the detergent. Sealed phospholipid vesicles form spontaneously, with the protein inserted into the bilayer. Such liposomes behave in a vectorial manner, as in the case of sarcoplasmic reticulum Ca^{2+}-dependent ATPase of muscle cells (see Chapter 15) which, when incorporated into liposomes, will transport Ca^{2+} into the interior at the expense of ATP hydrolysis.

Liposomes have also been used clinically for the delivery of drugs or antibodies to specific targets within the body and have been tested for use in gene therapy as a means of delivering DNA to cells (see Chapter 17).

Chapter 7. The energy for ATP production is derived from the proton motive force, page 208.

Chapter 2. The principal phospholipids of eukaryotic membranes are composed of glycerol, phosphate, fatty acids and several small basic or neutral compounds, page 25.

Chapter 15. The sarcoplasmic reticulum has mechanisms for the uptake, storage and release of calcium, page 466.

Chapter 17. Gene therapy is a potential cure for some genetic disorders, page 545.

Dynamic behaviour of cell membranes

Membranes are fluid, dynamic structures in which both lipids and proteins can rapidly diffuse in the plane of the bilayer

It has already been emphasized that biological membranes are fluid, dynamic structures. Phospholipids are able to exchange rapidly with adjacent molecules in their half of the bimolecular leaflet by lateral diffusion: nearest-neighbour exchange of adjacent phospholipids has been estimated to occur at 10^7 molecules per second, which represents a rate of diffusion of 1–2 μm per second for each phospholipid. Individual phospholipids can therefore diffuse from one end of the cell to the other in a few seconds. This rate has been measured, using fluorescent phospholipids, by bleaching an area of membrane using a laser and then measuring the speed at which the fluorescent phospholipids move back into the bleached area (Figure 3.16).

The lateral mobility of proteins in membranes was first demonstrated using virus-induced fusion of human and mouse cells in tissue culture to produce a hybrid cell. Initially, mouse and human proteins in the newly fused plasma membrane were segregated into opposite halves of the cell, shown by using human antibodies labelled with fluorescein (green) and mouse antibodies labelled with rhodamine (red). When the cells were incubated at 37°C, the membrane proteins mixed completely within an hour, indicating that proteins are capable of diffusing rapidly in the plane of the membrane (Figure 3.17).

The lateral mobility of some proteins is restricted by the fact that they are attached to the cytoskeleton. This was first shown in the erythrocyte (see Chapter 12), but is now known to occur in all cells.

Mobility within the membrane is essential to the proper functioning of many membrane proteins. For example, binding of the hormone glucagon to its receptor induces a conformational change in the receptor, enabling it to interact with a

Chapter 12. Peripheral proteins and the cytoskeleton, page 368.

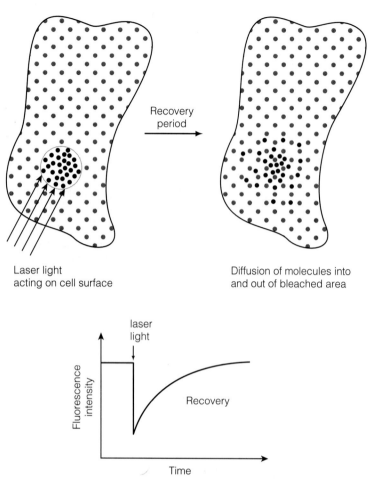

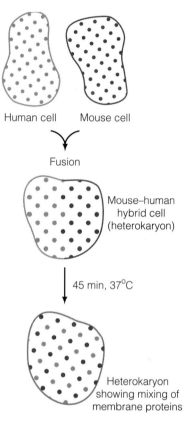

Figure 3.16 Diagram illustrating principles of the fluorescence recovery after photobleaching (FRAP) technique. After introducing fluorescently tagged proteins or lipids into cell membranes, a small area of membrane can be 'bleached' (decolorized) with high-intensity laser light and the rate of diffusion of surrounding fluorescent lipids or proteins backing into the bleached area measured. This gives important information on the mobility of membrane proteins and lipids.

Figure 3.17 Mixing of mouse and human plasma membrane proteins in hybrid cells following virus-induced fusion.

Chapter 11. G-protein-linked receptors mediate a wide variety of processes, page 336.

Chapter 15. The extracellular matrix holds together cells of multicellular tissues, page 467.

trimeric **G-protein** on the internal surface of the membrane. One of the G-protein polypeptides then migrates to interact with the enzyme adenylate cyclase, increasing its activity and raising the intracellular level of cAMP (see Chapter 11). Rapid action of the hormone is also promoted by the lateral mobility of the receptor, which allows a strong transmembrane protein–protein interaction with G-protein.

Not all membrane proteins are laterally mobile

Other integral membrane proteins are involved in activities that require them to remain at a specific location in the membrane: in mitochondria, for example, around 2000 contact points between the inner and outer membranes house the machinery for importing mitochondrial polypeptides from their site of synthesis in the cytoplasm into the matrix of the mitochondrion. An important transmembrane protein of plasma membranes – **fibronectin receptor** – is involved in anchoring cells to the extracellular matrix (see Chapter 15). The external part of this protein binds fibronectin and its inward-facing part binds cytoskeletal actin filaments in the interior of the cell. Fibronectin receptor exhibits extremely low mobility in the membranes of cells.

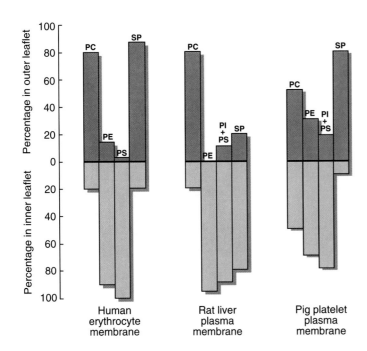

Figure 3.18 Asymmetry of phospholipid organization in plasma membranes. The composition of lipids in the two halves of the bimolecular leaflet are depicted for human erythrocyte membranes, rat liver plasma membranes and pig platelet plasma membranes. For individual membranes, the amounts of each lipid in the outer and inner leaflets are shown in blue, and brown, respectively. PC, phosphatidylcholine; PS, phosphatidylserine; PI, phosphatidylinositol; SP, sphingomyelin.

Proteins and lipids are asymmetrically distributed in cell membranes

Transverse diffusion of phospholipids (or 'flip-flop') through the membrane is extremely rare because transfer of a phospholipid molecule across the lipid bilayer involves the thermodynamically unfavourable transfer of the hydrophilic phosphate head group through the hydrophobic interior of the membrane. It occurs on average only once in several hours, some 10^{11} times slower than the rate of lateral exchange. Such estimates are consistent with the observed asymmetric distribution of different types of phospholipid between the two halves of the bilayer. In red blood cells, for instance, sphingomyelin and phosphatidylcholine are enriched in the outer half of the membrane but phosphatidylserine and phosphatidylethanolamine predominate in the inner half of the bimolecular leaflet (Figure 3.18). Glycolipids show an absolute specificity for the outer half of the bilayer, keeping the carbohydrate portions accessible on the external surface where they are important in the determination of A, B or O blood group specificity (see Chapter 12).

Integral membrane proteins have not been seen to 'flip-flop'. Membrane proteins are asymmetrically distributed because integral membrane polypeptides have extensive polar domains which cannot penetrate the lipid bilayer. An illustration of this is the erythrocyte protein glycophorin (see Chapter 12). Glycosylation sites on plasma membrane proteins are always exposed on the exterior surface of the cell (see above and Chapter 2) (Figure 3.18).

Membrane fluidity maintains the appropriate environment for the optimal functioning of membrane proteins and is carefully controlled

The flexibility of the hydrocarbon tails of membrane lipids is influenced by several parameters:

- length of the fatty acyl chains;
- degree of unsaturation of the fatty acyl chains;
- presence of cholesterol.

Chapter 12. The surface of the erythrocyte membrane carries blood group substances, page 369.

Chapter 12. Integral proteins, page 367.

Chapter 2. Oligosaccharide structures are covalently linked to many secretory and membrane proteins and to some lipids, page 21.

Bacteria are able to regulate the fluidity of their cell membranes in response to changes in ambient temperature, lower temperatures leading to a higher content of the more fluid unsaturated fatty acids. In contrast, mammalian cells, whose temperature is carefully regulated, govern their membrane fluidity primarily by cholesterol content. Cholesterol is a bulky molecule containing a planar steroid ring with a hydroxyl group at one end and a flexible aliphatic chain at the other. It inserts easily into phospholipid bilayers between adjacent fatty acyl chains. Cholesterol is an important moderator of membrane fluidity in mammals, maintaining the bilayer in the appropriate state for optimal operation (see Chapter 2).

Chapter 2. Cholesterol has an essential role in the plasma membrane of cells, page 28.

Specialized membrane structures in cells

Gap junctions between adjacent cells are important in cell–cell communication and other processes

In many tissues, cells interact directly with their neighbours through large aqueous channels in their plasma membranes that connect with similar structures in adjacent cells, thus bridging an intercellular gap of about 4 nm. **Gap junctions**, as these formations are called, are clustered in regions of the membrane in close contact with neighbouring cells. They are composed of hexameric arrays of a 32 000 M_r transmembrane polypeptide (**connexin**) that form stable associations with similar channels in adjacent cells. The channel so formed permits small molecules to transfer from the cytoplasm of one cell to another (Figure 3.19). The hexameric organization of gap

Figure 3.19 Gap junctions. A gap junction is a relatively nonspecific pore that connects two cells. The plasma membranes of adjacent cells contain hexagonal structures made from a protein called connexin. When these are matched up, a connection is made through which small molecules or ions can flow between cells. The gap junction is gated; it can close in response to certain stimuli. In this schematic illustration (a) one is shown closed, another open. (b) An electron micrograph of a gap junction at low resolution.

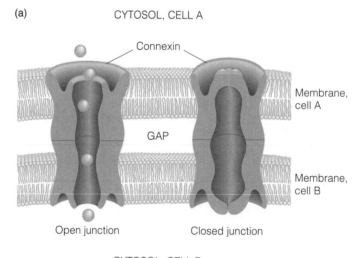

(a)

CYTOSOL, CELL A

Connexin

Membrane, cell A

GAP

Membrane, cell B

Open junction   Closed junction

CYTOSOL, CELL B

(b)

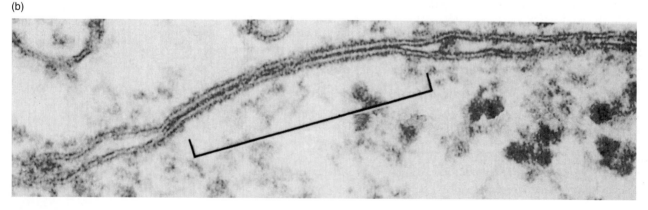

junctions and the presence of a closable 2 nm hydrophilic pore in the centre of the channel has been revealed by high-resolution electron microscopy. Gap junctions will permit the passage of most small polar molecules with M_r of less than 1500: macromolecules cannot gain access to these channels but ions, metabolites, sugars, amino acids and nucleotides can flow easily between cells.

Gap junctions are important in cell communication and in the coordinated regulation of tissue development. Specific examples of the importance of gap junctions are found in cornea and bone, which are not well infiltrated by blood capillaries and rely on the transfer of nutrients from one cell to the next. The rapid transfer of ions through gap junctions also ensures the synchronized contraction of heart muscle. Gap junctions play a key role in embryonic development and tissue differentiation, for example in establishing gradients of mitogens (molecules that bring about concentration-dependent changes in cell growth patterns). Towards the end of pregnancy, the muscular lining of the uterus develops gap junctions. The wall of the uterus is normally involved in protecting the fetus but at this stage it is also required to undergo vigorous and synchronized contractions associated with birth. These contractions are mediated via the newly formed junctions. Gap junctions may remain open for several minutes but they are closed by high concentrations of Ca^{2+} and lowered intracellular pH.

Tight junctions are involved in sealing intercellular spaces in epithelial cells lining various ducts and glands

Epithelial cells lining the surface of glands, ducts and the gastrointestinal tract are bipolar in structure because they are involved in the unidirectional secretion or uptake of enzymes or nutrients from one compartment to another. The apical and basolateral surfaces of these cells have different compositions, contain different complements of integral proteins and are separated by tight junctions (Figure 3.20).

In the intestine sugars, amino acids, electrolytes and vitamins are actively absorbed at the lumenal (apical) surface of epithelial cells (see Chapter 14). The plasma membrane in these regions is specifically adapted for this purpose and contains all the appropriate transporters and enzymes involved in the digestion of foodstuffs. The surface area of the plasma membrane is increased, for maximal absorption of nutrients, by the presence of numerous finger-like microvilli. These processes give rise to the term 'brush border' epithelium, which describes the morphology of the lumenal surface of these cells as observed by electron microscopy (see Figure 14.2).

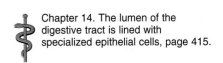

Chapter 14. The lumen of the digestive tract is lined with specialized epithelial cells, page 415.

The interstitial (basolateral) surface of the epithelial cell is in contact with intercellular fluid, blood vessels and lymph and has distinct properties reflecting its role in the active transport of nutrients from the epithelial cell to the portal blood supply and the lymphatic system. The basolateral surface of the membrane also contains integral proteins mediating general cellular functions, such as Na^+/K^+ ATPase and hormonal or neuronal receptors.

Intercellular spaces between epithelial cells are effectively sealed by tight junctions (Figure 3.20), thereby preventing lumenal contents leaking into the interstitial space.

Nuclear pores and mitochondrial contact sites are important in transfer of macromolecules across the double membranes of nuclei and mitochondria

The boundary of the nucleus is defined by the nuclear envelope, which has a characteristic ultrastructure. It consists of two membranes 2 nm apart and containing large pores, about 9 nm in diameter. The pores are important in regulating the

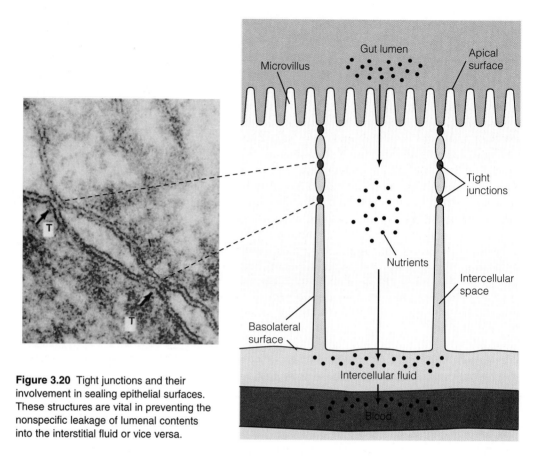

Figure 3.20 Tight junctions and their involvement in sealing epithelial surfaces. These structures are vital in preventing the nonspecific leakage of lumenal contents into the interstitial fluid or vice versa.

Chapter 5. There are several mechanisms for targeting proteins to the nucleus, page 141.
Chapter 5. Many mitochondrial proteins are made on cytoplasmic ribosomes, page 139.

entry and exit of macromolecules, ions and metabolites into the nucleoplasm, the soluble matrix of the nucleus. Nuclear pores have a complex structure (Figure 3.21); they provide the exit site for ribosomes and mRNA to the cytoplasm and the entry for proteins synthesized on cytoplasmic ribosomes and targeted to the nucleus (see Chapter 5).

Mitochondrial contact sites are areas where the inner and outer membranes lie close together – and may even fuse into a single bilayer. They are important in the uptake of nuclear-encoded mitochondrial proteins from their site of synthesis in the cytoplasm. Each mitochondrion has about 2000 such sites, which contain proteins that permit the import of mitochondrial proteins.

Transport of molecules and ions across membranes

Transport processes are crucial for cell function

Cell membranes are selectively permeable barriers, regulating cell volume and internal pH and exercising a crucial role in maintaining the correct intracellular environment of ions and small molecules.

Chapter 11. Hormonal responses are mediated through inositol phosphates and calcium ions, page 339.
Chapter 8. Activation of glycogen phosphorylase and inhibition of glycogen synthase occur in response to a common cascade, page 227.
Chapter 15. Tropomyosin and troponin regulate the interactions between actin and myosin in skeletal and cardiac muscles, page 462.

- Transient alterations in the levels of specific ions such as Ca^{2+} are intimately linked to cell-signalling processes (see Chapter 11), control of enzyme activity (see Chapter 8) and muscle contractility (see Chapter 15).

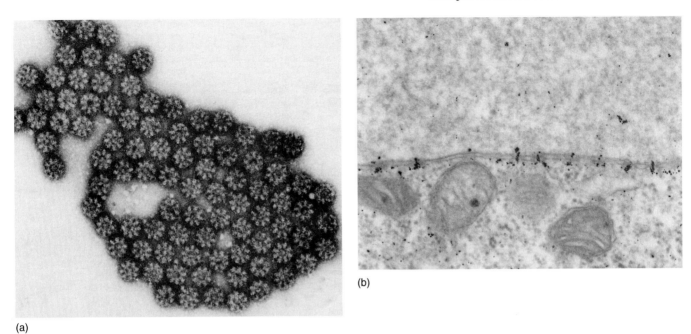

(a)

(b)

Figure 3.21 Structure of nuclear pores. (a) High-resolution electron micrograph showing detailed subunit organization of individual nuclear pores, (b) Low resolution electron micrograph of double nuclear membrane showing gaps corresponding to the presence of nuclear pores in this region.

- Small metabolites such as glucose and amino acids are taken up from the tissue fluids for use as metabolic fuels or as precursors for biosynthetic reactions.
- Toxic products are actively eliminated from cells so that they can be transported to the liver or kidney for detoxification and excretion.
- Ionic gradients are maintained across membranes of all cells – an essential feature of cell function, especially in nerve and muscle (see Chapter 13).

 Chapter 13. Membrane potentials and the generation of action potentials, page 401.

Several mechanisms exist for transporting solutes across membranes

Only water molecules and small uncharged molecules with lipophilic (lipid-soluble) properties are able to travel across lipid bilayers by simple diffusion. Metabolites and macromolecules are transferred across membranes via transporters composed of proteins or via protein complexes that form channels or 'pores' through the bilayer.

The different types of transporters have distinct mechanisms (Table 3.4).

- *Facilitated diffusion* occurs when ions or molecules are transferred across the lipid bilayer down a concentration gradient through a specific transporter or channel. There is no net accumulation of solute within the cell and no requirement for energy.
- *Active transport* is characterized by an input of energy (as ATP or as a membrane potential or pH gradient), leading to the net accumulation of the appropriate solute against a concentration gradient. Such transporters are often referred to as 'pumps'.
- Some channels contain receptors as part of the transporter complex (channel-linked receptors – see Chapter 11).
- Some mechanisms involve transport of pairs of molecules or ions in the same (symport) or opposite (antiport) directions.

 Chapter 11. Channel-linked receptors regulate the permeability of membranes to ions, page 335.

Table 3.4 Categories of membrane transport

Type of transport	Carrier involvement	Saturability	Energy requirement	Rate of transport	Examples of transported molecules
Passive diffusion	No	No	None	Depends on solubility in lipid bilayer	Water, ethanol, organic compounds, for example 2,4-dinitrophenol
Facilitated diffusion	Yes	Yes	None	Depends on V_{max} and K_m of carrier	Glucose
Active transport Ion pumps	Yes	Yes	ATP	Depends on turnover number of carrier	Na^+, K^+, Ca^{2+}
Cotransporters	Yes	Yes	Ion gradients/ membrane potentials	Depends on concentration gradients/properties of carrier	Amino acids/Na^+; Glucose/Na^+; Citrate/isocitrate; ATP/ADP

All transmembrane pumps have three basic features in common

1. They must contain a cavity large enough to accommodate small molecules or ions in a specific binding site.
2. The pump must exist in two distinct conformational states, so that the cavity is accessible to opposite sides of the membrane in the two states.
3. The affinity of the transported metabolites or ions for the pump must differ in the two conformational states.

The Na^+/K^+ ion pump of plasma membranes is driven by hydrolysis of ATP

Mammalian cells have a high internal concentration of potassium ions and a low internal concentration of sodium ions relative to those of the surrounding tissue fluids. The correct ionic balance within cells is vital in regulating cell volume, maintaining nerve and muscle excitability and in establishing membrane potentials. These ionic gradients are maintained at the expense of ATP by the ubiquitous Na^+/K^+ ion pump (Na^+/K^+-ATPase: see below), which is able to exchange three Na^+ ions for two K^+ ions for every molecule of ATP hydrolysed. The ion gradients can be used for promoting active transport of sugars and amino acids into cells. As in all ATP-dependent reactions, Mg^{2+} ions are also essential because the substrate is MgATP rather than free ATP (see Chapter 7).

Hydrolysis of ATP is an intrinsic part of the Na^+/K^+ exchange mechanism and both the ion pump and ATPase activity are susceptible to inhibition by cardiotonic steroids such as digitalis. The ability of plasma membranes of different cell types to transport Na^+ and K^+ ions is closely related to their ATPase activity. Thus nerve cells possess highly active Na^+/ K^+ pumps and high levels of Na^+/K^+-dependent ATPase activity – in contrast to the low activities found, for example, in red blood cells.

Na^+/K^+- ATPase is asymmetrically organized within the membrane

Hydrolysis of ATP is always coupled to the export of Na^+ from the cell in exchange for the import of K^+; similarly, ATP must be present inside the cell. These observations indicate that the Na^+/K^+-ATPase is located asymmetrically in the lipid bilayer. Activation of the ATPase linked to exchange of Na^+ and K^+ requires the presence of Na^+ on the inside of the plasma membrane and K^+ in the external

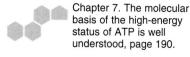
Chapter 7. The molecular basis of the high-energy status of ATP is well understood, page 190.

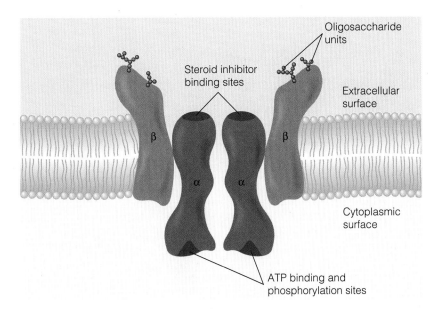

Figure 3.22 Schematic representation of the intramembrane organization of the Na$^+$/K$^+$-ATPase.

medium. Digitalis, digoxin and other cardiotonic steroids, which are inhibitors of the ATP-driven Na$^+$/K$^+$ pump, bind to the external face of the membrane (Figure 3.22).

In the subunit organization of the $\alpha_2\beta_2$ tetrameric Na$^+$/K$^+$ pump, both the α (M$_r$ 112 000) and β (M$_r$ 41 000) subunits are predicted to contain multiple membrane-spanning α-helices, probably seven and four in number, respectively (Figure 3.23). The β-subunits contain several N-linked oligosaccharide chains and are exposed primarily on the external surface of the cell. A specific aspartate residue on the β-subunits becomes transiently phosphorylated during the catalytic cycle, suggesting that it forms part of the active site. In contrast, the α-subunit is predominantly located on the cytoplasmic surface along with the ATP binding site but it also contains the inhibitory site for cardiotonic steroids on the external surface and must be transmembrane in nature.

A simple model for ATP-induced Na$^+$/K$^+$ exchange involves phosphorylation and conformational change

A general mechanism for ion exchange through the Na$^+$/K$^+$ ATPase can be deduced as follows:

- the dephosphorylated and phosphorylated forms of the enzyme are in two separate conformational states;
- the presence of Na$^+$ and Mg^{2+} in the cytoplasm causes phosphorylation of the enzyme by ATP, whereas dephosphorylation is induced solely by the presence of K$^+$ in the external medium;
- although the dephosphorylated form has a high affinity for Na$^+$ and is stabilized by dephosphorylation, the phosphorylated form is stabilized by phosphorylation and exhibits a high affinity for K$^+$.

The importance of phosphorylation reactions is emphasized in another context; as a mechanism for altering the conformation of enzymes and thus regulating their activity (see Chapter 7). Although it would appear from Figure 3.23 that conformational states are dramatically different, only subtle structural alterations may be necessary to produce the desired changes in substrate binding and surface-specific access to the Na$^+$ and K$^+$ binding sites – as is also the case with gap junctions.

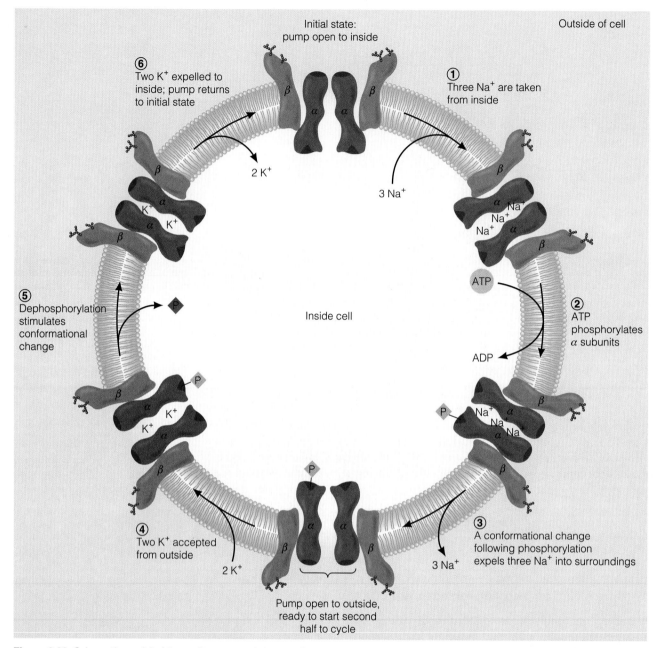

Figure 3.23 Schematic model of the sodium–potassium pump in operation. Steps of the pump's operation are shown in a cycle.

Na+/K+-ATPase has been a clinical drug target for many centuries

The ability of cardiotonic steroids such as digitalis to inhibit the action of the Na+/K+-ATPase is of great clinical relevance. Digitalis may be extracted from foxgloves (*Digitalis purpurea*) and has been used since the time of Ancient Greece to treat congestive heart failure. Many centuries later, it has been established that the increased force of contraction of the heart following administration of digitalis is caused by elevated levels of intracellular Ca^{2+}. This is because the diminished Na^+ gradient across the plasma membrane leads to a lowered rate of Ca^{2+} extrusion via the Na^+/ Ca^{2+} ion exchanger.

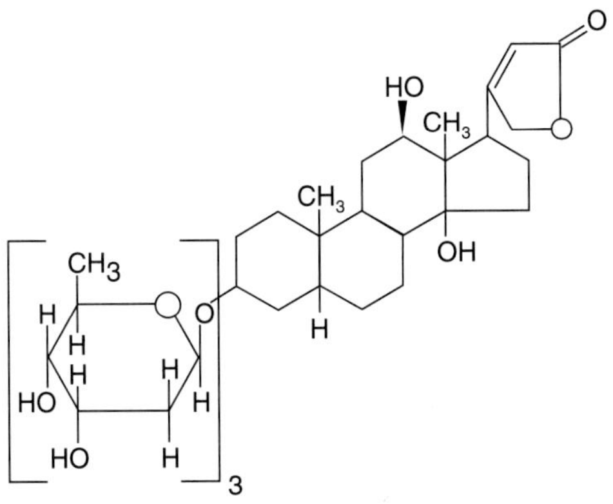

Figure 3.24 The structure of digoxin, one of the major active components in foxglove extracts.

A family of ion-motive pumps probably share an evolutionary origin

A similar mechanism, involving a phosphorylated enzyme species, is present in the H^+/K^+ ion exchanger of the apical membrane of parietal cells lining the stomach wall. This can generate the million-fold gradient of protons across the stomach lining required for maintaining the high acidity of gastric contents essential for digestion of foodstuffs (see Chapter 14). The structural and functional features shared by this family of ion exchangers (Na^+/K^+, Ca^{2+} and H^+/K^+) imply that they have probably been derived from a common ancestral pump.

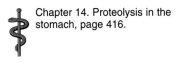

Chapter 14. Proteolysis in the stomach, page 416.

Muscle sarcoplasmic reticulum contains another type of ion-motive ATPase

In skeletal muscle Ca^{2+} ions are rapidly released from intracellular stores within an intricate internal network of membranes (the sarcoplasmic reticulum) in response to contractile stimuli (see Chapter 15). Muscle relaxes as Ca^{2+} is taken up into the sarcoplasmic reticulum through the sarcoplasmic Ca^{2+}-ATPase (two Ca^{2+} ions are transported per ATP hydrolysed). This pump, a single polypeptide (M_r 112 000) resembling the α-subunit of the Na^+/K^+-ATPase, has a mode of action similar to that of the Na^+/K^+-ATPase (see Figure 3.23).

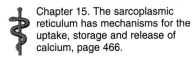

Chapter 15. The sarcoplasmic reticulum has mechanisms for the uptake, storage and release of calcium, page 466.

Intracellular vesicles and mitochondria also have ATPases that pump protons

Eukaryotic cells contain two additional families of ATP-linked ion transporters, classified as V-type and F-type ATPases (Table 3.5).

The V-type enzymes are vital in regulating the acidity of vesicular compartments to which proteins are targeted, such as lysosomes, Golgi complex and intracellular vesicles, and in the operation of **exocytosis** and **receptor-mediated endocytosis**. These are large multisubunit enzymes, in which ATP hydrolysis and H^+-movements are controlled by separate polypeptides.

F-type ATPases are also linked specifically to proton gradients in the process of oxidative phosphorylation in the mitochondrial membrane. These proton-translocating ATPases function in the 'reverse' direction; primarily in ATP synthesis at the expense of dissipating the respiration-driven proton gradient across the mitochondrial inner membrane. They are composed of an integral membrane proton channel, the F_0 unit, attached to a soluble catalytic ATPase with the complete enzyme functioning as a H^+-driven ATP synthase (see Chapter 7).

Chapter 7. Aerobic ATP production: the process of oxidative phosphorylation, page 199.

Table 3.5 Major families of ion-motive ATPases in animal cells

Location	General structure/mechanism	Ion specificity	Inhibitors	Function
P-Type Plasma membranes	All contain a related subunit with multiple membrane-spanning regions, which is phosphorylated on an aspartate residue during the catalytic cycle	Na^+/K^+		Regulation of intracellular ionic balance
Gastric epithelia Sarcoplasmic reticulum		H^+/K^+ Ca^{2+}	Vanadate	Acid secretion Muscle contraction by recapturing Ca^{2+} ions from the cytoplasm
V-Type Vacuoles in yeast fungi, lysosomes, endosomes, secretory vesicles	Large multisubunit enzyme with transmembrane distribution	H^+		Acidification in ligand/receptor dissociation, possible roles in protein targeting and vesicle fusion
F-Type Mitochondrial inner membrane, chloroplast thylakoid membrane	F_0 transmembrane proton channel plus soluble F_1-ATPase	H^+	Oligomycin, rutamycin	Proton-driven ATP production

Active transport mechanisms can be driven by the simultaneous movement of ions down a concentration gradient

In many cases, the active uptake of metabolites is driven directly by the simultaneous movement of Na^+ ions down a concentration gradient. The Na^+ gradients are regenerated by Na^+/K^+-ATPase.

- Uptake of amino acids is coupled to the cotransport of Na^+ in a wide variety of animal cells; for example, brush border cells of the small intestine.
- Although glucose enters cells mainly by facilitated diffusion through glucose transporters (see below), some cell types can also accumulate glucose using a **Na^+/glucose antiporter**. These include brush border cells of the small intestine (see above) and kidney, where the action of the antiporter is involved in the glycosuria associated with diabetes mellitus (see Chapter 13).
- Movement of sodium ions into cells provides the energy for extrusion of Ca^{2+} from the cell. Coupled movement of two ions in the opposite direction is referred to as antiport; symport describes the coupled transport of two ions in the same direction.

Active accumulation of glucose against a concentration gradient is unnecessary in many cell types, as it is trapped on entry by its rapid conversion to glucose-6-phosphate.

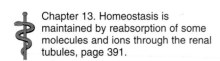

Chapter 13. Homeostasis is maintained by reabsorption of some molecules and ions through the renal tubules, page 391.

Entry of glucose into cells is mediated by several related tissue-specific transporters

The main route of glucose uptake into cells is by facilitated diffusion from the bloodstream through glucose transporters, a family of closely related, tissue-specific proteins whose properties are tailored to the particular requirements of the tissue. Seven distinct glucose transporters, each encoded by a separate gene, have been identified and a summary of their tissue distribution and general properties is shown in Table 3.6.

Based on their amino acid sequences, the transporters are predicted to contain 12 transmembrane segments (Figure 3.25) with an extensive extracellular loop

Table 3.6 Major sites of expression of the various glucose transporters and their functions

Transporter	Tissue-specific expression	Function
GLUT 1	Erythrocytes, placenta, low levels in muscle and adipose tissue	General uptake of glucose from the bloodstream
GLUT 2	Liver, pancreatic β-cells, proximal tubules of kidney, small intestine (basolateral membrane)	High K_m, high turnover transporter involved in glucose sensing, rapid glucose efflux when extracellular levels are low
GLUT 3	Brain, nerve, low levels in placenta, liver, kidney and heart	Supplements action of GLUT 1 in tissues with high energy demand
GLUT 4	Muscle, heart, adipose tissue	Insulin-responsive glucose uptake in these tissues
GLUT 5	Small intestine (apical membrane), adipose tissue, low levels in brain and muscle	Primarily a high-affinity fructose transporter. Has poor ability to transport glucose
GLUT 6	Not-expressed (a pseudogene)	
GLUT 7	Liver (endoplasmic reticulum)	Transport of glucose from endoplasmic reticulum to cytoplasm

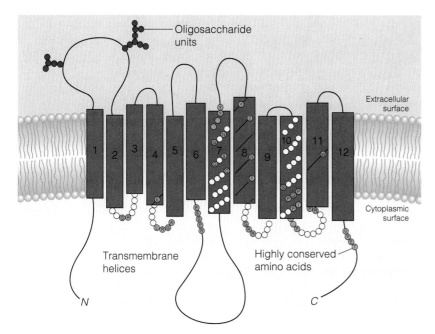

Figure 3.25 Predicted transmembrane organization of glucose transporter family. Each of the family of tissue-specific forms of the glucose transporter has twelve transmembrane α-helices with *N* and *C* termini located in the cytoplasm and a major extracellular glycosylated domain located between membrane-spanning segments 1 and 2. The major areas of sequence which are highly conserved between all members of this family are highlighted, suggesting that these regions of the protein are of general importance in the mechanism of glucose uptake.

containing *N*-linked oligosaccharide between the first and second *N*-terminal helices and the major cytoplasmic domain situated between the sixth and seventh transmembrane segments. Both *N*- and *C*-termini are present within the cytoplasm of the cell.

Changes in the expression and glucose translocating properties of these transporters, especially GLUT 1, GLUT 2 and GLUT 4, may be clinically significant in relation to the occurrence of type I and type II diabetes. GLUT 4, the insulin-responsive glucose transporter of adipose cells, occurs primarily in intracellular vesicles in the absence of insulin but is rapidly translocated to the surface in response to insulin, promoting a 4–7-fold increase in the rate of glucose uptake (Figure 3.26).

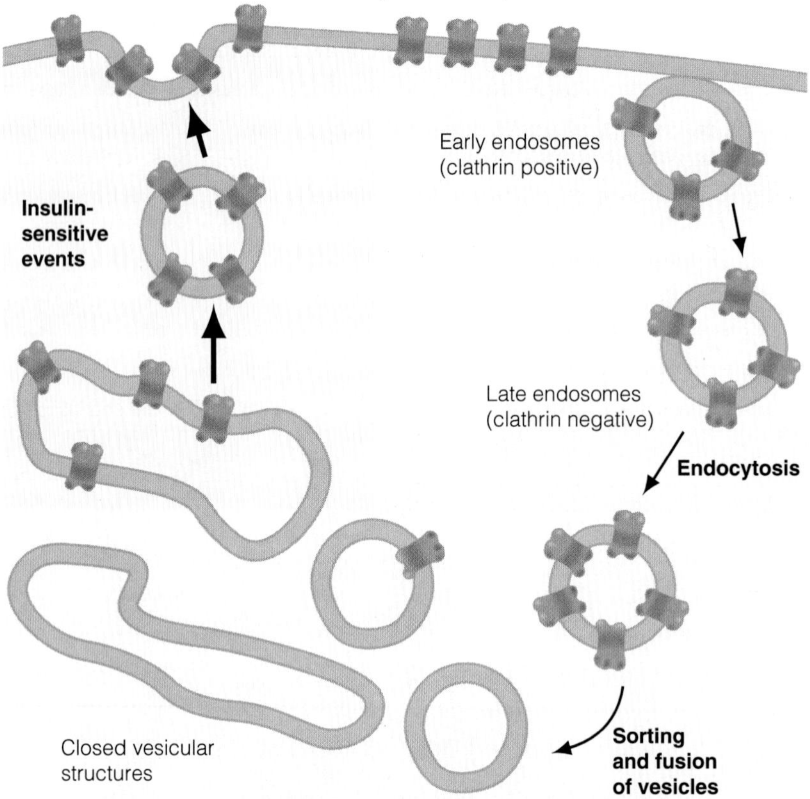

Fully functional
glucose transporters

Early endosomes
(clathrin positive)

**Insulin-
sensitive
events**

Late endosomes
(clathrin negative)

Endocytosis

Closed vesicular
structures

**Sorting
and fusion
of vesicles**

Figure 3.26 The insulin responsive glucose transporter (GLUT 4) in appropriate tissues is cycled continuously between internal tubulo-vesicular structures and the plasma membrane. In the absence of insulin, only a small proportion of glucose transport is expressed at the cell surface. Secretion of insulin promotes the rapid translocation of these internalized transporters to the cell surface leading to a new steady-state in which there is a 20–30 fold increase in the level of glucose transporters in the plasma membrane. Clathrin is a self-assembling protein capable of forming a cage-like structure that is important in producing initial membrane invagination into 'coated pits' and formation of endosomal vesicles.

Cell-surface receptors

Hormone receptors in the plasma membrane are involved in transmission of signals to the interior of the cell

The effects of many hormones are mediated by binding to proteins on the external surface of cells of the target tissue with specific receptor function. Ligand–receptor interaction generates transmembrane signalling events that activate intracellular enzyme systems involved in the regulation of cell metabolism, growth and development. When a hormone binds to a site on the outer cell surface, it leads to conformational changes in proteins on the inner surface, which trigger chemical changes within the cell (Chapter 11).

A family of transmembrane receptors is involved in the G-protein-mediated activation of the adenyl cyclase and phosphoinositide pathways

A number of cell-surface receptors exert their effects via G-proteins, a group of heterotrimeric proteins which exhibit GTPase activity and mediate the interaction of the hormone-receptor complex with **adenylate cyclase** or **phospholipase C** (see Chapter 11). The β_2-adrenergic receptor was the first of this type for which the gene was cloned and sequenced. Analysis of the gene revealed the presence of a large glycosylated N-terminal domain on the external face of the cell, seven transmembrane α-helical segments and an extensive C-terminal 'loop' implicated in interaction with G-proteins (Figure 3.27). A list of the various G-protein linked receptor subtypes is presented in Table 3.7.

Chapter 11. The mechanism of action of extracellular signal molecules is related to their chemical characteristics, page 332.

Chapter 11. G-protein-linked receptors mediate a wide variety of processes, page 336.

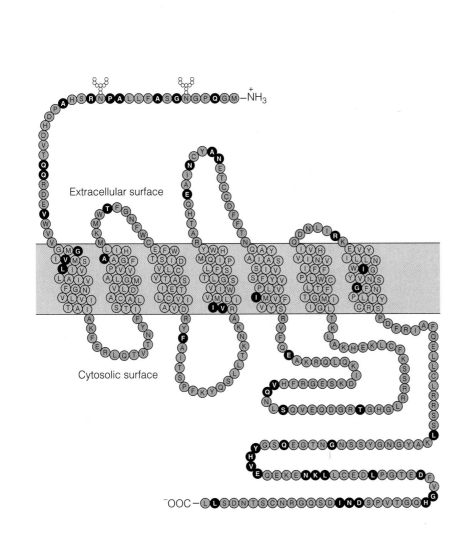

Figure 3.27 Primary sequence and predicted transmembrane organization of the β₂-adrenergic receptor with its seven conserved transmembrane domains (orange) and three extracellular and three cytoplasmic loops (blue). Also represented are the two externally located oligosaccharide units. Receptor interaction with G- proteins is mediated by phosphorylation of specific *C*-terminal serine and threonine residues.

Table 3.7 G-Protein-linked receptors

There are four major families of heterotrimeric G-proteins, each composed of α, β and γ subunits, the α subunit determining the specificity and class of G-protein. These are termed the G_s series, the G_i series, the G_q series and the $G\alpha_{12}$ series. The G_s and G_i classes are inhibited by cholera and pertussis toxins, respectively, which modify the α subunits by ADP ribosylation. Each of these G_α subunits is encoded by a separate gene, but there are also four splice variants of $G\alpha_s$ and two of $G\alpha_o$.

G-Protein	Function	Distribution
$G\alpha_s$	Stimulation of adenylate cyclase	Ubiquitous
$G\alpha_{olf}$		Olfactory sensory neurones
$G\alpha_{i1}$	Undefined	Brain
$G\alpha_{i2}$	Inhibition of adenylate cyclase	Ubiquitous
$G\alpha_{i3}$	Regulation of K⁺ channels	Ubiquitous
$G\alpha_o$	Regulation of Ca²⁺ channels	Neural tissue
$G\alpha_{t1}$ (transducin 1)	Activation of cGMP phosphodiesterase	Retinal rod outer segments
$G\alpha_{t2}$ (transducin 2)		Retinal cone outer segments
$G\alpha_z$	Undefined	Neural tissue
$G\alpha_{15}$	Regulation of phospholipase C	B Lymphocytes
$G\alpha_{16}$		T Lymphocytes
$G\alpha_{14}$		Stromal cells, epithelial cells
$G\alpha_{11}$	Regulation of phospholipase Cγ1	Ubiquitous
$G\alpha_q$		Ubiquitous
$G\alpha_{12}$	Undefined	Undefined
$G\alpha_{13}$	Undefined	Undefined

Growth factors and insulin bind to a family of receptors with intrinsic tyrosine kinase activity

Chapter 11. Catalytic or enzyme-linked receptors have tyrosine kinase activity, page 337.

Chapter 11. Growth factors, page 353.
Chapter 17. Proto-oncogenes are normal cellular genes involved in cellular growth and development, page 530.

A second major family of receptors was discovered with the isolation of the insulin receptor from adipocytes (fat cells) and the subsequent cloning and characterization of the genes encoding its two polypeptide chains (Chapter 11). The receptor is an integral membrane glycoprotein consisting of two α-chains (M_r 135 000) and two β-chains (M_r 95 000) linked by three disulphide bonds. The α-chains are located entirely on the exterior of the cell and so do not interact directly with the membrane but the β-chains possess a single transmembrane-spanning region with an extensive C-terminal domain located within the cytoplasm of the cell (Figure 3.28).

Interaction of insulin with the receptor activates the previously latent tyrosine kinase activity of the receptor. This activity is associated with the internal domain and causes autophosphorylation of the receptor, further increasing the tyrosine kinase activity and its potential for phosphorylating other protein substrates. The receptors for **epidermal growth factor** and **platelet-derived growth factor**, hormones with key roles in the regulation of cell growth and development, are similar to the insulin receptor (Chapter 11). Some receptors of this type are proto-oncogenes (see Chapter 17).

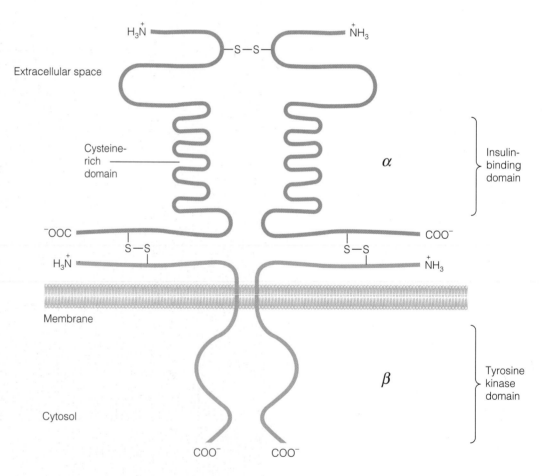

Figure 3.28 Model of the insulin receptor, an α2β2 tetramer. The interior C-terminal portion of the β chains contains the tyrosine kinase activity which is activated by binding of insulin onto the external surface of the cell. Antophosphorylation of the receptor on tyrosine residues in the *C*-terminal region of the β chains further enhances and maintains the intrinsic tyrosine kinase activity in the absence of insulin binding.

PART B

INFORMATION IN BIOLOGICAL SYSTEMS: ITS FUNCTION, EXPRESSION AND STORAGE

CHAPTER 4
Proteins as functional units:
sequence, structure and
modification

CHAPTER 5
Genes and their expression:
RNA in the generation and
translation of messages

CHAPTER 6
DNA as the store of biological
information: the nature and
replication of genomes

GLOSSARY

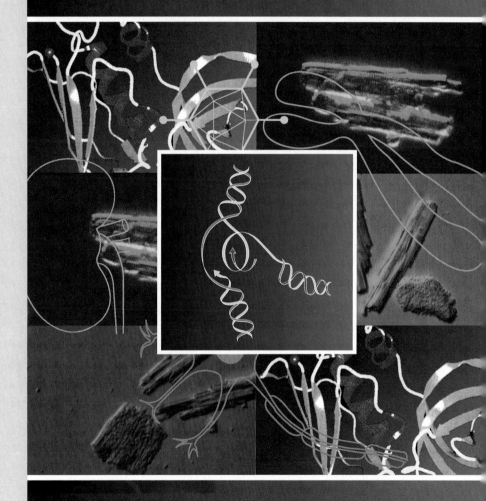

Information in biological systems: its function, expression and storage

This part of the book discusses the structures and 'molecular biology' of **proteins, RNA** and **DNA**. An understanding of the nature of the transactions in which these molecules participate helps to explain how the information present in individual cells and the body as a whole – in the form of the amino acid and nucleotide sequences of proteins and nucleic acids – gives rise to the multiplicity of observed functions. Knowledge of this area is essential to an understanding of the many new developments in 'molecular medicine', 'genetic disease' and the applications of 'genetic engineering' to medicine.

Function, message and store

The theme of this part of the book is summarized by the three words of this heading, and each will be covered in successive chapters.

- **Function** The functions of the human body and its cells are predominantly carried out, directly or indirectly, by proteins. In Chapter 4 the structures and functions of proteins are discussed in the context of the fundamental importance of the amino acid sequence of the distinct molecular species of polypeptides in the body.
- **Message** As a first approximation, the information specifying each protein species resides in a different gene. In Chapter 5 we discuss the mechanisms of gene expression in the context of the message, or messenger RNA, which is copied or transcribed from the DNA of the gene and subsequently translated into the polypeptide that it encodes.
- **Store** If the functions and messages of the cells of the body are to operate correctly, there must be a stable and accessible store of information from which the messages can be retrieved as required. Chapter 6 discusses the structure of DNA and the ways in which the information it contains can be maintained within generations and passed from one generation to the next.

Information in biological systems and its transfer

Among the most striking features of living things are their organization, diversity and dynamic behaviour. Any analysis of the nature of life at the molecular level must address questions such as:

- How is the organization of an organism achieved and maintained?
- What are the mechanisms that bring about the diversity among organisms?
- What are the molecular features that allow dynamic behaviour?

One of the paradoxes of biology is that there is a substantial degree of uniformity at the molecular level that underlies the great diversity at the macroscopic level. This molecular uniformity is one of the strongest arguments in favour of a common evolutionary origin for all life on Earth (see Box 6.5).

An organism is a collection of functions

One starting point for such an analysis is to consider an organism as a collection of **functions**, where each function is the basis for some character or process in that organism. These functions include structural characteristics such as size and number of limbs, in the sense that a structure can be regarded as the product of a function or group of functions, but a particular collection of structures gives rise to the characteristic overall organization of the cell or organism (see Chapters 1 and 3). If organisms are to achieve their dynamic behaviour, they must possess functions that can bring about necessary changes and maintain themselves in the face of external changes.

Proteins are fundamental to life

Recent decades have been marked by enormous advances in understanding of the molecular nature of living things (molecular biology). These have revealed the fundamental importance of proteins in almost every structure and process of the living world. Many of the structural features of organisms contain protein – those which do not, such as those based on polysaccharide, depend on proteins for their formation. As the molecular details of the processes of life have been elucidated it has become apparent that every process depends on the active participation of a specific group of proteins.

Nucleic acids carry information in their sequences

The central importance of proteins has long been recognized and in the first half of this century was undoubtedly over-emphasized, because proteins were thought to be responsible for every aspect of living things, including inheritance. As the importance of the nucleic acids was realized it became apparent that proteins could not direct their own production; that the information allowing the heritable characteristics of organisms to be passed from generation to generation resided in, and was transmitted by, sequences of nucleic acids. The explanation of the mechanisms of classical genetics in molecular terms, now often called molecular genetics, has been an important achievement of molecular biology.

There are direct links between information, structure and function

It is now clear that, as a first approximation, biological function and protein structure are directly linked in that the former is directly dependent on the latter (see Chapter 4). A second direct connection has been demonstrated between protein structure and inherited information in that the former is a direct expression of the latter. For all cellular organisms the chemical nature of the inherited information is DNA (see Chapter 6). DNA is the storage form of this information, which is retrieved as an RNA copy that can either function directly (such as ribosomal RNA) or act as an instruction for the synthesis of protein (such as messenger RNA) (see Chapter 5). An instructive analogy can be drawn between these aspects of biological information transfer and the operation of computers, in which information transfer has become a catch phrase (Table B.1).

Other computer analogies in biology

Just as most computers are operated using only a small fraction of the information in their memory, most cells utilize only a small fraction of the information in the genome and different fractions are used by different cells.

Table B.1 A computer analogy for biological information, its transfer and function

---

- Unless it is able to perform useful functions, a computer is an expensive piece of useless hardware. In order to function, a computer must operate useful and appropriate **programs**.
- To run these programs, the computer requires information in the form of appropriate **software**.
- Unless this information is supplied each time it is required, the computer needs a store of information (**memory**) and the ability to retrieve the information at the appropriate time.

Property	Computer	Cell (or whole body)
Function	Program	Protein
Message	Software	Messenger RNA
Store	Memory	DNA genome
Retrieval	Copying	Replication
	Loading	Transcription
	Running	Translation
	Hardware	Ribosomes, etc.

Some computer programs that replicate themselves have been called 'viruses', but real viruses bring new 'memory' into their host cell and use some of the cell's 'hardware' to operate new functions leading to viral replication and the production of progeny viruses. The behaviour of natural viruses is described in Chapter 18.

Macromolecular sequences and their significance

Molecules of both proteins (polypeptides) and nucleic acids (polynucleotides) in all organisms have characteristic lengths and sequences that are intimately associated with their functions (see Box 4.6). The direct sequencing of proteins is relatively complicated and time consuming, but the development and refinement of the techniques of DNA sequencing have meant that protein sequences can be readily deduced from a DNA sequence using suitable computer programs (see Box 6.1). Computers are also extensively used for comparing sequences, and this has led to the identification and quantitation of relationships between genes and between proteins. Such comparisons have been used to establish **molecular phylogenies**, which have both confirmed and modified relationships between organisms that had been derived by traditional palaeological methods (see Box 6.6).

The Central Dogma and the transfer of information

As the underlying mechanisms of protein and nucleic acid synthesis began to be understood in the late 1950s, it became apparent that the flow or transfer of information was vital and that one type of macromolecule could act as a template for the synthesis of another. These ideas were encapsulated by Crick in 1958 in what came to be known as the **Central Dogma** of molecular biology (Table B.2). This still holds good, but has had to be extended to include the 'reverse transcription' seen in the retroviruses (see Chapter 18). A discussion of the underlying principles and details of the processes of translation, transcription and replication comprise the bulk of this part of the book.

The operation of these mechanisms depends on two crucial processes:

1. the action of **template-directed polymerases** in the synthesis of the polynucleotides, DNA and RNA (see below);
2. the 'decoding' of polynucleotide sequences into polypeptide (protein) sequences by the rules of the **genetic code**.

Essential for both of these processes is the phenomenon of hydrogen-bonded **base pairing** between nucleotides (see Chapter 2).

Table B.2 The Central Dogma and the flow of information

Replication	Transcription	Translation
DNA $\longleftrightarrow$ DNA	$\rightarrow$ RNA	$\rightarrow$ Protein
A sequence of deoxynucleotides	A sequence of ribonucleotides	A sequence of amino acids

Template-directed nucleotide polymerases

At the heart of the copying of nucleic acid sequences is the concept of the template-directed nucleic acid polymerase – an enzyme capable of polymerizing nucleotides into a polynucleotide chain in a sequence-specific manner, in which the sequence of the newly synthesized strand is specified by a complementary strand of nucleic acid acting as a **template**.

The general reaction for these enzymes is:

$$(NMP)_n + NTP = (NMP)_{n+1} + PPi$$

where $(NMP)_n$ represents the nucleic acid chain being synthesized (always in a 5' to 3' direction) and NTP the nucleoside 5'-triphosphate (ribo- or deoxyribo-, always used as the Mg^{2+} salt) being polymerized. Selection of each NTP is directed by the next nucleotide of the template strand, according to the A = T (or U), G ≡ C hydrogen-bonded base-pairing rules, but the template strand is not covalently changed by the reaction.

Because there are two possible types of triphosphate (rNTP or dNTP) and two possible templates (DNA or RNA), there are four possible types of such polymerase (see Table B.3). All four are found in nature, but only DNA polymerase and RNA polymerase appear to occur in cellular organisms, reflecting the DNA nature of their genomes. The other types are found in the retroid and RNA viruses, respectively (see Chapter 18)

Table B.3 Template-directed nucleotide polymerases

NTP	Template	
	DNA	RNA
dNTP	DNA polymerase	Reverse transcriptase
rNTP	RNA polymerase	RNA replicase

The genetic code

As will be explained in detail in Chapter 5, translation of the sequence of nucleotides in a messenger RNA (mRNA) into the sequence of amino acids in the resultant protein requires the action of a group of small RNA molecules called transfer RNAs (tRNAs). Transfer RNAs act as 'adaptors' because each species can bind only one type of amino acid and has a characteristic **anticodon** which is capable of binding to a particular **codon** on an mRNA molecule.

The **genetic code** (Figure 5.12) is the set of rules that relates the 64 possible trinucleotide codons on mRNA to the 20 different amino acids used in the synthesis of proteins and to the signals for starting and stopping the synthesis of a polypeptide chain.

It is important to emphasize that 'the code' is a decoding mechanism and *not* the information itself – a common misconception! It is incorrect to refer to 'the genetic code for globin' (or any other protein) when we mean 'the gene for globin'. The code is often described as universal or applying to all organisms, although there are some minor but significant modifications in the case of some mitochondrial and bacterial proteins.

The medical relevance of the study of biological information transfer

An understanding of the structure and function of proteins and nucleic acids is not something that most medical practitioners will use on a daily basis, unlike many other aspects of biochemistry. It is, however, essential for even a basic understanding of modern biology, which underpins much of medical knowledge. In addition, the recent developments in 'genetic engineering' and their many applications in medicine can be satisfactorily understood only from a firm foundation of molecular biology. An excellent example of this is the discovery of the connection between defects in the human tumour suppressor protein p53 and a wide range of common human malignancies (see Chapter 17). This initial discovery has been followed up by a range of molecular biological analyses to discover the action of p53 at the molecular level and its role in preventing cancer.

CHAPTER 4

Proteins as functional units: sequence, structure and modification

In this chapter we first consider the roles of protein in the body and its cells and then look at the widespread ability of globular proteins to bind other molecules, often with very high specificity. Enzymes are proteins which not only bind other molecules but also catalyse reactions; an understanding of enzymes is crucial to any consideration of molecular medicine.

Underlying the function of proteins is the subtlety and complexity of their structures. The amino acid sequence or primary structure determines the higher levels of protein organization and the various features of the three-dimensional structure of proteins. Finally, certain functional aspects of proteins are discussed in terms of structure, including the effects of mutations in genes encoding proteins.

The structures and functions of proteins are discussed in terms of the relationships between their amino acid sequences, three-dimensional structures and biological functions. The great diversity of possible sequences and structures permits the enormous variety of proteins and their functions. These are discussed in general terms of the 'structural' roles of specialized, fibrous proteins and the 'binding' roles of the globular proteins, both water soluble and membrane bound. Enzymes are treated as a special subgroup of 'binding proteins' which catalyse reactions involving the bound molecules. The importance of interactions between proteins is emphasized.

Protein function

This chapter deals with general aspects of protein structure and function. More specific aspects of the structure and function of individual proteins and groups of related proteins that are of physiological importance are dealt with elsewhere (for details, see Table 4.1).

Table 4.1 Where to find information on specific proteins

Protein	Chapter	Pages
Actin and other muscle proteins	15	461
Collagen and other 'structural' proteins	15	470–77
Haemoglobin and myoglobin	12	369–75
Histones	6	151
Immunoglobulins	16	490–95
Membrane proteins	3	53–57
Nucleic acid binding proteins	5, 6	118, 157
Protein hormones	11	346
Receptor proteins	3, 11	70–72, 335
Ribosomal proteins	5	127
Transport proteins	3	63–69
Viral proteins	18	562–64

Proteins have three principal functions: structural, binding and catalytic

The functions of proteins may be categorized under these three headings.

- **Structural proteins.** The structural roles of fibrous and other proteins in the body are outlined below and are discussed in more detail in Chapter 15.
- **Binding proteins.** The ubiquitous binding properties of globular proteins will be discussed in the next section. They include the many proteins, including hormones and growth factors, which act as 'signals' within the body.
- **Catalytic proteins.** Enzymes are a subset of globular binding proteins that catalyse reactions involving their ligands and are discussed below.

Location of proteins in the body: in solution, in structures or in cell membranes

Proteins in solution are found in both intracellular and extracellular locations

Most of the proteins of both unicellular and multicellular organisms occur as aqueous solutions, either intracellular or extracellular.

Intracellular proteins

Intracellular proteins occur in widely differing concentrations and are involved in interactions with small molecules, proteins and other macromolecules. They are found in the main intracellular volume (the cytosol) or in compartments such as the nucleoplasm or mitochondrial matrix and must be targeted specifically to these locations (see Chapter 5).

Chapter 5. There are several mechanisms for targeting proteins to the nucleus, page 141. Chapter 5. Many mitochondrial proteins are made on cytoplasmic ribosomes, page 139.

Box 4.1 Disulphide bonds and extracellular proteins

The amino acid **cysteine** is one of the less common components of proteins. Its characteristic feature is the **sulphydryl group** of its –CH₂SH side chain (see Figure 4.2). In intracellular proteins, the SH groups of cysteine are typically maintained in the reduced state by the sulphydryl-containing tripeptide **glutathione** (see Chapter 12). Intracellular proteins may be inactivated by oxygen and, when extracted from cells in the laboratory, must be protected and maintained in an active state by the use of gentle reducing agents that also contain sulphydryl groups, for example di*thio*threitol and *mercapto*ethanol.

In contrast, the cysteines in extracellular proteins are often paired by covalent **disulphide bonds** (–S–S–). These bonds, often between distant parts of protein molecules, stabilize the functional, three-dimensional structure of the protein. The cysteine residues involved in disulphide bonds are quite specific and are often highly conserved (see below). Because the disulphide bond keeps cysteine in an oxidized state it is less susceptible to damage by atmospheric oxygen. Disulphide bonds are found in many physiologically important **globular proteins**, including plasma albumin, immunoglobulins and the extracellular domains of many cell surface proteins. Disulphide bonds also contribute to the great stability of **fibrous proteins**, such as keratin of hair and skin. The chemical breaking and reforming of the disulphide bonds of the keratin of hair is the basis for permanent waving of the hair ('perming').

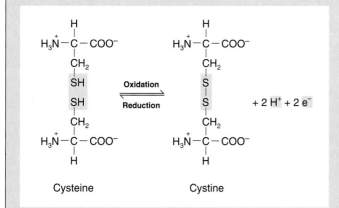

Cysteine   Cystine

$$+ 2\,H^+ + 2\,e^-$$

Box Figure 4.1
A disulphide bond between two cysteine residues.

Chapter 10. Glutathione is a tripeptide with several important biological functions, page 313.

Extracellular proteins

These include proteins secreted by cells into the bloodstream (such as the immunoglobulins and other plasma proteins: Chapter 12) or into extracellular spaces, such as the digestive enzymes in pancreatic exocrine secretions. These proteins are often stabilized against the rigours of the extracellular environment by **disulphide bonds** (Box 4.1). All proteins, including extracellular proteins, are synthesized inside cells.

Chapter 12. There are many types of plasma proteins, page 357.

Proteins are found as components of structures in the body and its cells

Not all proteins are water soluble; the insoluble ones include many with 'structural' functions (Figure 4.1b).

Extracellular proteins

Some structural proteins are extracellular; they are often termed **fibrous** proteins and are found in skin, bone, hair and other structures. They typically have extensive regions of regular secondary structure (see below) and are stabilized by networks of

Chapter 15. Synthesis of procollagen is completed by postsynthetic modifications, page 473.

 Chapter 3. The cytoskeleton determines cell shape and movement and regulates the organization of internal structures, page 49.

 Chapter 5. Ribosomes are made of two subunits, each with its own RNA and protein molecules, page 127.
Chapter 3. Nuclei contain chromosomes and nucleoli, page 46.

 Chapter 12. Thrombin catalyses the conversion of fibrinogen to fibrin, page 363.

 Chapter 3. Membrane proteins, page 53.

 Chapter 3. The transmembrane organization of an integral protein can be predicted from its primary structure, page 55.

 Chapter 11. The mechanism of action of extracellular signal molecules is related to their chemical characteristics, page 332.

 Chapter 5. Targeting of proteins, page 136.

disulphide bonds (Box 4.1). Structural proteins are often synthesized as soluble precursors, which are then modified and converted to the mature structural protein by **postsynthetic modification**, as, for example, collagen (see Chapter 15).

Intracellular proteins

'Structural' proteins are also found inside the cell – these include actin, myosin and the many other cytoskeletal elements (see Chapter 3). The protein components of ribosomes (see Chapter 5) and chromosomes (see Chapter 3) also have important structural properties, as does the fibrin of blood clots (see Chapter 12).

Proteins in cell membranes have distinctive properties

Most of the water-insoluble proteins in cells are present in the various cellular membranes but they generally resemble the water-soluble globular proteins more than the fibrous proteins in terms of structure and function: that is, they are generally compact in shape and function by binding to ligands (Figure 4.1a). The plasma membrane and the various intracellular membranes contain distinct populations of 'integral' proteins which confer a variety of functions (solute transport, signal transduction, cell–cell recognition and communication, endocytosis and exocytosis), upon the particular membrane (see Chapter 3). Part of an integral membrane protein is always hydrophobic and interacts with the nonpolar interior of the lipid bilayer of the membrane.

Most integral membrane proteins have both hydrophilic and hydrophobic **domains** and those that span the membrane commonly have three: one hydrophilic domain on either side of the membrane and a hydrophobic, membrane-spanning domain, usually composed of one or more sections of α-helix (see Figure 3.14). The spanning domains may be recognized from amino acid sequence data because of their content of hydrophobic amino acids (see below and Table 2.3). Other integral membrane proteins lie outside the bilayer but are firmly 'anchored' to it through covalently attached hydrophobic moieties such as fatty acids, diglycerides and isoprenoid derivatives (for example farnesyl and geranyl groups) (see Figure 3.14). Owing to the functional asymmetry of biological membranes, membrane-spanning proteins must be oriented correctly if they are to fulfil their function – receptors for hormones must be on the outer face of the plasma membrane of cells, for instance (see Chapter 3). Special targeting mechanisms ensure that proteins are correctly inserted into the appropriate membrane during its biosynthesis (see Chapter 5).

Globular proteins: the binding of ligands

Most nonstructural proteins are globular in nature

The term 'globular protein' is a useful description of the shape of many proteins, both soluble and membrane bound, that are generally compact in shape because of the precise folding of the polypeptide chain (Figure 4.1a). This folding brings distant parts of the polypeptide into close spatial proximity and forms surface features that act as highly specific 'sites' for the various aspects of the functions of these proteins, especially binding of ligands and interaction with other molecules such as proteins, lipids and nucleic acids.

Most globular proteins are able to bind other molecules

Arguably the most important function of proteins in living things is their ability to interact, usually in a highly specific way, with other molecules, both large and

(a)

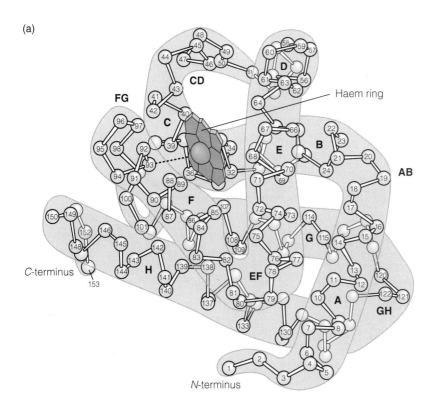

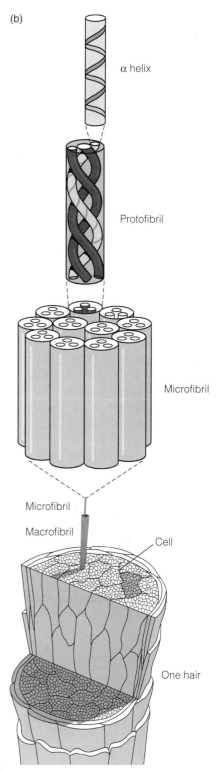

(b)

Figure 4.1 Globular proteins such as myoglobin (a) have binding properties. Myoglobin in muscle binds oxygen at its haem ring. Fibrous proteins such as keratin in hair (b) typically have structural roles.

small. Indeed, it would be difficult, if not impossible, to construct a model of something that we would consider 'alive' without such specific binding properties. Proteins show remarkable versatility in the range of molecules which they are capable of binding (Table 4.2) using only the chemical characteristics of the amino acid side chains (Figure 4.2) and a few other, relatively simple, components such as metal ions and organic cofactors such as NAD, required by many enzymes.

Table 4.2 Binding proteins and their ligands

Protein	Ligand
Plasma proteins	
Plasma albumin	Fatty acids, bilirubin, Cu^{2+}, Zn^{2+}
Ceruloplasmin	Cu^{2+}
Transferrin	Fe^{2+}
Immunoglobulins	Very many, each highly specific
Intracellular proteins	
Haemoglobin	Oxygen
Histones	DNA (not sequence-specific)
Thyroid hormone receptor	Thyroxine, DNA (sequence-specific)
Protein p53	DNA (sequence-specific)
Cell surface receptors	
Insulin receptor	Insulin
LDL receptor	Low-density lipoprotein (LDL)
FGF receptor	Fibroblast growth factor (FGF)
CD4	MHC II, human immunodeficiency virus

Binding of ligands to proteins involves weak forces

The bound molecule (**ligand**) does not usually associate in a vague or amorphous way with the surface of the binding protein as it might, for example, to a chemical catalyst. Instead it associates specifically with a distinct part of the protein – the **binding site** – and is held there by a number of weak interactions such as hydrogen bonds and hydrophobic interactions (Figure 4.3), which are essentially similar in nature to those that maintain the tertiary and quaternary structures of proteins (see Box 4.10).

Proteins differ in the specificity and tightness of ligand binding

All binding of ligands by proteins shows some degree of specificity but this varies considerably. It may be extremely specific, which is a reflection of the versatility of surface detail and associated formation of weak bonds that proteins can achieve. All noncovalent binding by proteins is reversible, but a wide range of affinities between individual proteins and ligands is possible (Box 4.2).

Box 4.2 Association of ligands with proteins

The behaviour of protein–ligand binding conforms to the reversible equilibrium:

[BP] + [Li] = [BP–Li]

where [BP] and [Li] are the concentrations of free binding protein and ligand and [BP–Li] is the concentration of the protein–ligand complex, usually noncovalent in nature. Physiologically important ligands are very varied in nature and include hormones, metabolites, metal ions and many drugs of plant or synthetic origin.

The **dissociation constant** of the process, K_d, is a measure of the affinity of the protein for its ligand:

$$K_d = \frac{[BP]\,[Li]}{[BP-Li]}$$

Ligands with very high affinity for their binding protein may have dissociation constants as low as 10^{-9} or 10^{-10} mol l^{-1}.

The **association constant** (K_a) is the reciprocal of K_d.

The precise nature of the binding is related to the particular function of the protein and may be very tightly controlled, as in the well studied example of haemoglobin (see Chapter 12). Some proteins will bind the same ligand at two distinct sites, which differ in both affinity and specificity towards the ligand; other proteins are capable of binding two different ligands at two discrete sites. For instance, the intracellular hormone receptor proteins for steroid and thyroid hormones (see Chapter 11) bind both a hormone and a DNA sequence with high degrees of specificity, the two sites being located on different **domains** of the protein (Figure 4.7c). Another important example is the tumour suppressor protein p53 (see below and Chapter 17), which has a central specific DNA-binding domain flanked by two other domains that are responsible for the activation of transcription and for binding to other molecules of p53 (oligomer formation).

Binding of ligands can alter protein function

Although each protein has a precisely defined three-dimensional structure (see below), this is dynamic and capable of subtle but significant conformational changes due to the involvement of many weak but highly specific interactions,

Chapter 12. Oxygen binding by haemoglobin is cooperative, page 371.

Chapter 11. Mechanism of action of thyroid and steroid hormones, page 341.

Box 17.9. Protein p53, mutations and cancer, page 536.

Chapter 4. X-ray crystallography can be used to study parts of proteins, page 100.

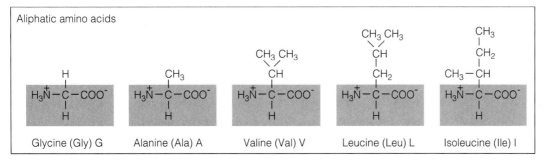

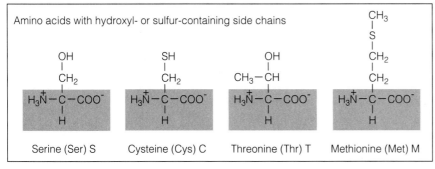

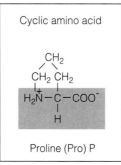

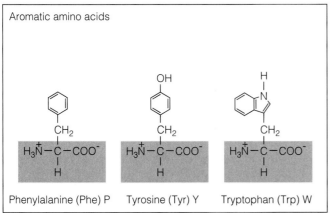

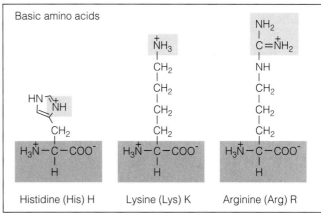

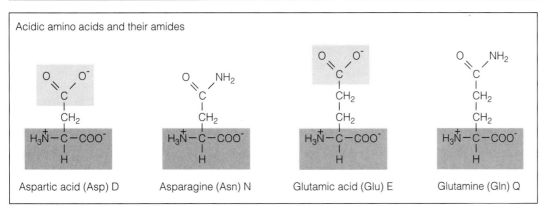

Figure 4.2 The 20 amino acids that are incorporated into proteins.

which can break and reform easily. Binding of a ligand molecule at one site in a protein may provoke a local change in the protein's conformation, which is transmitted to a second binding site (for example the catalytic site of an enzyme) and modifies the properties and behaviour of the second site (Figure 4.3). This is termed **allosteric** behaviour and is associated with the presence of quaternary structure in the protein concerned. The properties of **allosteric enzymes** (see below) are very important in the control of metabolic reactions.

Chapter 4. Many enzymes show allosteric behaviour, page 96.

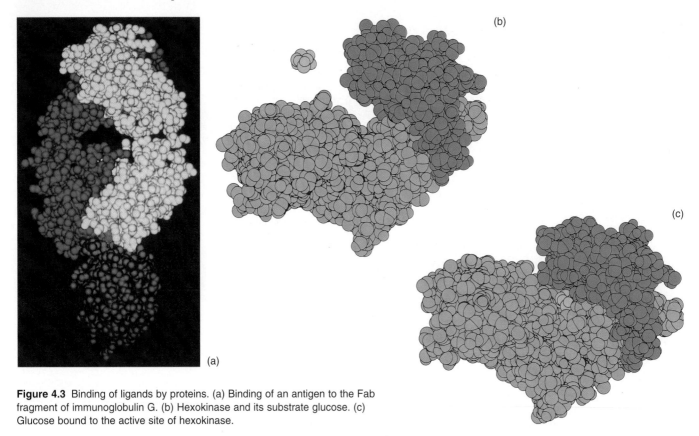

(b)

(c)

(a)

Figure 4.3 Binding of ligands by proteins. (a) Binding of an antigen to the Fab fragment of immunoglobulin G. (b) Hexokinase and its substrate glucose. (c) Glucose bound to the active site of hexokinase.

Binding of a ligand can be followed by catalysis

Almost every globular protein has at least one site capable of binding another molecule. A major subset of these binding proteins, having bound the ligand or ligands, go on to effect the catalysis of a chemical reaction involving these ligands. Proteins with such catalytic properties are termed **enzymes**.

Catalytic proteins: enzymes

Proteins with catalytic functions are found in every organism

Enzymes are proteins that catalyse chemical reactions: almost every reaction in the body is catalysed by an enzyme. Even very simple reactions which can proceed spontaneously, such as the hydration and dehydration of carbon dioxide ($H_2O + CO_2 = H_2CO_3$) use enzyme catalysts (in this case carbonic anhydrase – see Chapter 12), because the spontaneous rate is inadequate for physiological function. The molecules on which enzymes act are called **substrates** – in the example above these are H_2O, CO_2 and H_2CO_3 – and may also be referred to as the **reactants**.

Chapter 12. The anion-exchange protein participates in carbon dioxide metabolism, page 372.

The suffix **-ase** is almost invariably found in the names of enzymes, the few exceptions being enzymes that were identified long ago and where the historical names have been retained – for example, the digestive enzymes trypsin and pepsin. Enzymes are so central to our understanding of the functioning of cells and organisms that it is almost impossible to imagine some 'life system' that does not possess them.

Some RNA molecules have catalytic properties

Until the mid 1980s, all biological catalysis was thought to be effected by proteins, but it was then discovered that certain RNA molecules (**ribozymes**) are capable of catalytic activity. The reactions catalysed by RNA are limited and especially concerned with the cleavage of polynucleotide chains, but their discovery has strongly influenced ideas about the origin of life (see Chapter 6). By using knowledge of these natural ribozymes, synthetic RNA molecules have been manufactured with catalytic activity – for example, with the aim of developing antiviral agents. The vast bulk of catalytic reactions in all living things are, however, catalysed by proteins – the enzymes.

Box 6.5. Was DNA the first genetic material in the origin of life? Page 170.

Enzymes effect catalysis by binding a substrate to an 'active site'

The widespread ability of globular proteins to bind molecular ligands has already been described. Enzymes are also binding proteins in that the first step in enzyme catalysis is the binding of a reactant (or **substrate**, S) by the enzyme (E) at its **active** (or **catalytic**) site to form an enzyme–substrate complex (E–S). Although 'true' binding proteins have only one path available once the ligand binds (dissociation of the complex and release of the ligand), enzymes also have the capability of changing the enzyme–substrate complex to enzyme–product complex; in other words catalysing the conversion of substrate to product (P):

$$E + S \rightleftharpoons E\text{–}S \rightleftharpoons E\text{–}P \rightleftharpoons E + P$$
$$\quad\ A \qquad\quad B \qquad\quad C$$

Where A represents the binding/dissociation of enzyme and substrate, B represents the catalysis brought about by the enzyme – the interconversion of S and P – and C represents the binding/dissociation of enzyme and product.

Most enzymes have several substrates

Most enzymes do not catalyse single-substrate reactions of the type $S \rightleftharpoons P$ shown above, but act on two substrates thus:

$$S_1 + S_2 \rightleftharpoons P_1 + P_2$$

where S_1 and P_1 are chemically related, as are S_2 and P_2. Such enzymes need two closely adjacent binding sites. Many **dehydrogenases** are of this type, where S_2 and P_2 are NAD^+ and NADH, respectively, and the 'cofactor' is a substrate of the enzyme. For example, lactate dehydrogenase, an important enzyme in skeletal muscle, catalyses the reaction

$$\text{lactate} + NAD^+ \rightleftharpoons \text{pyruvate} + NADH + H^+$$

The terms 'substrate' and 'product' can be subjective, as most biological reactions are reversible (at least in theory, and often in practice; see Chapter 7). The physiological substrates for lactate dehydrogenase in the cell are (lactate + NAD^+) or (pyruvate + NADH), depending on the direction of the reaction.

Chapter 7. Reversal of metabolic conversion by a different pathway allows independent controls, page 197.

Enzymes catalyse reactions by lowering the activation energy

Enzyme catalysis is true catalysis in that the catalyst does not alter the nature of the reaction. It differs from chemical catalysis in that most enzymes actively participate in the reaction mechanism but return to their initial state once the reaction is complete – that is, the E–S complex is often a new covalent molecule with a very transient existence. The enzyme does not change the equilibrium of the reaction, but increases the rate of attainment of equilibrium by lowering the activation energy barrier to the reaction, possibly by replacing a single large barrier by several smaller ones.

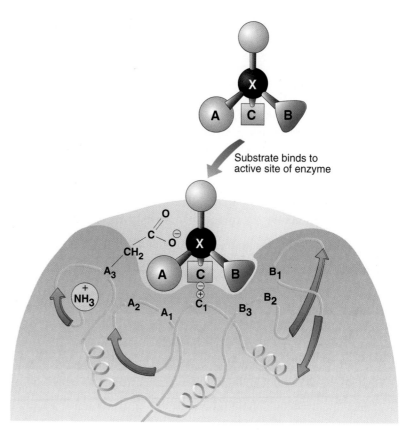

Substrate binds to
active site of enzyme

Figure 4.4 Enzyme catalysis occurs at a
specific site on the surface of an enzyme.

Enzymes bring about catalysis in a very precise and highly specific fashion because their function depends on certain **crucial amino acids** present on the enzyme protein which participate in binding substrate, forming and breaking covalent bonds and other aspects of enzyme activity. The precise reasons for the efficacy of enzymes as catalysts is not known, but certain well studied enzyme-catalysed reactions indicate that it is probably due to a combination of one or more of the following properties at the active site (Figure 4.4):

- achieving proximity of reactants;
- achieving correct orientation of reactants;
- stabilization of a transition state.

Enzyme reactions are stereospecific

A virtually universal feature of enzyme catalysis is its **stereospecificity**. Where a substrate for an enzyme has two possible **stereoisomers** (see Chapter 2) the enzyme will act only on one of them, for example it will act on an L-amino acid, but not a D-amino acid. Some enzymes act on symmetrical compounds in an asymmetric fashion to generate a stereospecific product: this is why natural compounds normally occur as specific stereoisomers (for example, D-glucose) rather than racemic mixtures. Stereospecificity of enzymes is easily understood by reference to a simple active site model with three points of contact with the substrate, as shown in Figure 4.4.

Enzymes have both protein and catalytic properties

Enzymes have the properties shared by all proteins, especially the importance of the native state for biological function (see below) and the ability to form associations with other molecules. Because enzymes are proteins, they are subject to the same kinds of irreversible inactivation as other proteins, including

- *denaturation* by heat, extremes of pH, detergents and organic solvents – the three-dimensional structure of the enzyme protein is destroyed, disrupting the active site;
- *covalent modification* by chemicals – this changes or destroys one or more of the crucial amino acids at the active site.

Chapter 14. Protein turnover is the process by which the structure and function of tissue is maintained, page 440.

These events do not often occur in the body, but their study in the laboratory has contributed to our understanding of enzyme activity. Enzymes, like other proteins, are inactivated within the body, sometimes specifically as part the control of a physiological process, but sometimes in a nonspecific manner as part of the normal turnover of cellular and body components.

Enzymes have a range of distinctive features relating to their catalytic and biological functions (Table 4.3).

Table 4.3 Some catalytic and biological properties of enzymes

Crucial amino acid residues exist within the sequence of the enzyme protein
    These participate in the binding of substrates or regulatory molecules or in catalysis at the active site
    They are brought together in space by the three-dimensional structure of the enzyme protein
    The residues are conserved in the same enzyme from different organisms (see Chapter 6)

Some enzymes are multifunctional and catalyse several reactions
    Two or more active sites can exist on different domains or subunits
    The enzymes may participate in opposing reactions (e.g. DNA polymerase participates in polymerization and hydrolysis during proofreading – see Chapter 6)
    Efficient processing of substrates in pathways (e.g. domains in fatty acid biosynthesis – see Chapter 9; subunits in pyruvate dehydrogenase – see Chapter 8)

Different protein molecules catalyse the same reaction
    In the same cell, but in different compartments (for example, the cytosol and the mitochondrion) (see 'Functional aspects of protein structure')
    Enzyme proteins in different species can have amino acid sequences that are related (divergent evolution from a common ancestor) or unrelated (convergent evolution of function from different ancestors) – see Chapter 6

Chapter 6. Radical substitutions, page 168. Chapter 6. Fidelity of replication is aided by proofreading by DNA polymerase, page 158.

Chapter 9. Fatty acid carbon chains are assembled through a sequence of four reactions, page 276.

Chapter 8. The pyruvate dehydrogenase complex catalyses a sequence of reactions, page 243. Box 6.6. Nucleic acid and protein sequences can be used for 'molecular phylogeny', page 171.

Only a limited range of enzyme-catalysed reaction types occur

Compared with the range of reaction types known in organic chemistry, only a relatively limited number occur in living cells. Reactions have been classified into six groups (see Table 4.4), which have been given systematic names and numbers (Enzyme Commission – EC – numbers, for example EC 2.7.7.7) but this system is often rather cumbersome and the use of trivial rather than systematic names continues. For instance, enzyme EC 2.7.7.7 has the systematic name DNA nucleotidyl transferase but continues to be known, even in the scientific literature, as DNA polymerase (see Chapter 6). The vast majority of enzymes encountered in biochemistry belong to EC groups 1, 2 and 3 (see Table 4.4).

Chapter 6. DNA polymerases are central to DNA replication, page 154.

Many enzymes show allosteric behaviour

Many enzymes conform to Michaelis–Menten kinetics – they show a hyperbolic response to substrate concentration (see Box 4.3) – but some enzymes give a sigmoid (rather than hyperbolic) response to substrate concentration. This is caused by a second or **allosteric** site on the enzyme, distinct from the active site.

Table 4.4 Class of enzyme-catalysed reactions

Enzyme class	Example (common name)
1	Lactate dehydrogenase
2	RNA polymerase
3	Glucose-6-phosphatase
4	Citrate lyase
5	Triose phosphate isomerase
6	DNA ligase

Box 4.3   Measurement of enzyme activity and kinetics

Assay of enzyme activity

Enzyme activity is usually assayed (measured) by measuring the rate at which a substrate disappears or a product appears, typically by monitoring some readily measurable property of substrate or product such as the absorption of light. Other methods involve using radioactive isotopes or antibodies. The conditions within the assay solution (pH, temperature, substrate concentration and cofactors) must be carefully controlled in order to obtain reliable and reproducible results. Enzyme activity is usually expressed as units (moles of substrate converted per unit time) and can be referred to the volume of enzyme solution (**total activity** as units l^{-1}) or to amount of protein (**specific activity** as units mg^{-1} protein). It is important to measure the initial velocity (V_o) of the reaction, because the rate often decreases with time as substrate is depleted or inhibitory products accumulate.

Kinetic parameters

The interaction of enzyme and substrate can be described in many cases by applying the Michaelis–Menten equation, using the parameters velocity (V) and concentration of substrate ([S]) and yielding K_m (the Michaelis constant) and V_m or V_{max} for a particular set of conditions of enzyme and substrate. If a plot of V against [S] gives a hyperbolic curve (a), a plot of $1/V$ against $1/[S]$ gives a linear plot (b), the enzyme obeys Michaelis–Menten kinetics; many enzymes do not. Appropriate forms of the equation are shown above each plot.

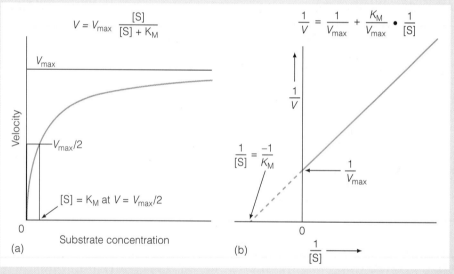

$$V = V_{max} \frac{[S]}{[S] + K_M}$$

$$\frac{1}{V} = \frac{1}{V_{max}} + \frac{K_M}{V_{max}} \cdot \frac{1}{[S]}$$

Box Figure 4.3 Kinetic parameters of enzymes.

Chapter 4. Quaternary (4°) structure: structural and functional features, page 105.

Chapter 12. The quaternary structure of haemoglobin affects oxygen binding, page 370.
Chapter 7. Allosteric regulation allows enzymes to respond to metabolite changes in the intracellular environment, page 194.

Allosteric enzymes are very common and allosteric behaviour is associated with the quaternary structure of enzyme proteins (see 'Protein structure', below). The development of the ideas on the nature and behaviour of allosteric enzymes was profoundly influenced by the cooperative effects and conformational changes observed in the interaction between haemoglobin and oxygen. Both allosteric enzymes and haemoglobin interact cooperatively with their ligands, while nonallosteric enzymes and myoglobin interact noncooperatively. There are many parallels between allosteric/nonallosteric enzymes and haemoglobin/myoglobin (see Chapter 12).

Allosteric enzymes allow rapid modulation of enzyme activity in response to variation in the concentrations of metabolites (see Chapter 7). Molecules that influence the activities of allosteric enzymes are called **allosteric effectors**; they may have positive (stimulatory) or negative (inhibitory) effects. An allosteric effector might be a substrate of the enzyme (a homotropic effector) or another type of molecule (a heterotropic effector); heterotropic effectors are often important metabolites such as ATP, acetyl CoA or citrate.

Measurement of enzyme activity and kinetic parameters yields useful information

The measurement of enzyme activity plays an important role in many areas of biochemistry, biology and medicine. The study of enzyme kinetics can be quite complex, but the principles underlying the measurement or **assay** of enzymes are

---

Box 4.4 Enzyme inhibition

Irreversible inhibition
Many substances inhibit the activities of enzymes irreversibly. They reduce the number of active enzyme molecules in solution, although the surviving molecules behave normally. Many irreversible inhibitors react covalently with specific parts of the enzyme molecule, such as the sulphydryl groups of cysteine residues (these include *p*-chloromercuribenzoate (PCMB) and Hg^{2+} ions), some act specifically on a particular group of enzymes (for example organophosphorus acetyl cholinesterase inhibitors such as di-isopropyl phosphofluoridate react with specific serine residues).

Reversible inhibition
Many other inhibitors, including some that are physiologically significant, act reversibly in the sense that if they are removed from the enzyme the activity of the enzyme is completely restored. Inhibition by such compounds can be analysed using Michaelis–Menten kinetics and two principal types are recognized – **competitive** and **noncompetitive** inhibitors, depending on whether addition of substrate will reverse the inhibition or not (see Figure). It is possible to determine the **inhibition constant** (K_i) for an inhibitor (I) of a particular enzyme–substrate pair. A third group of reversible inhibitors (the allosteric inhibitors) acts in a different fashion, and many of these are important in the control of metabolic pathways.

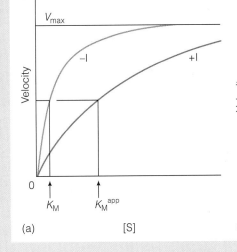
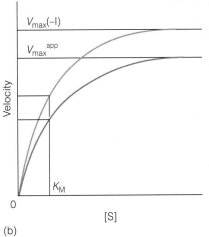

Box Figure 4.4 Competitive (a) and noncompetitive(b) inhibition of enzymes.

---

relatively simple (Box 4.3) and may yield useful information on amounts of enzyme and behaviour towards substrates and inhibitors (Box 4.4).

The study of enzymes has valuable applications in medicine

The study of enzymes has two main applications in medicine, one theoretical and one practical.

At the theoretical level, an understanding of the nature of enzymes is fundamental to a grasp of the way in which cells, organs and the whole body operate at the molecular level. It would not be an exaggeration to say that the concept of 'molecular medicine' would be meaningless without this understanding. More specifically, new drug molecules can be designed only from a detailed knowledge of the protein structure and catalytic function of enzymes. This approach,

Box 10.1 Distribution of specific aminotransferase provides diagnostic information when tissues are damaged. Page 297.

sometimes called 'knowledge-based drug design' brings together studies of enzyme catalysis, protein structure and gene structure and expression (see Chapter 5) with the widespread application of computer technology and graphics.

The main direct practical use in medicine of the study of enzymes is its application in the diagnosis and monitoring of human disease. This depends principally on measuring the levels of various enzymes (such as aminotransferases and alkaline phosphatase) in plasma. Very small amounts of these enzymes are present in the plasma of normal, healthy individuals as a result of the routine turnover of cells within the various organs but the levels are elevated in individuals with a variety of organ-specific diseases (heart, liver, skeletal muscle, etc.) due to damage in these tissues. Measurement of these enzymes in plasma is therefore a valuable diagnostic aid.

Primary structure of proteins – amino acid sequences

There is a direct link between amino acid sequence and biological function

One of the 'secrets of life' is undoubtedly the way in which 20 relatively simple chemicals – the amino acids present in proteins – can be assembled into an almost infinite variety of different linear polypeptides, each of a unique length and containing a unique sequence of amino acids. This sequence defines the three-dimensional conformation that the polypeptide adopts, which, in turn, confers biological function upon the protein.

Proteins contain 20 different amino acids

Chapter 2. Amino acids are both the structural units of proteins and important metabolites, page 28.

Chapter 15. Hydroxylation reactions, page 473.

The 20 amino acids found in proteins are stereochemically in the L-configuration and are *alpha*-amino acids except for glycine (which is symmetrical) and proline (which is an *imino* acid; see Chapter 2). Other amino acids are found in certain specialized proteins, especially structural ones, but these are incorporated during translation as one of the standard set of 20 and modified after translation. In collagen, for example, hydroxyproline and hydroxylysine are derived by hydroxylation of proline and lysine residues in procollagen (see Chapter 15). The amino acids are linked into polypeptide chains (or proteins) by peptide bonds whose properties are important in determining the three-dimensional structure or conformation of polypeptide chains (see Chapter 2).

Many proteins contain nonamino acid components

Chapter 2. Peptide bonds are formed between two amino acids and have properties central to protein structure, page 31.

Chapter 4. Many proteins are post-translationally modified, page 107.

Some proteins are composed solely of amino acids but many also contain nonamino acid components (Table 4.5). Some of these, such as sugars (in *glyco*proteins) and phosphate (in *phospho*proteins), are covalently attached to amino acid residues by post-translational modification (see below). Other components, such as metal ions and organic molecules (cofactors, prosthetic groups) may be bound by covalent or noncovalent interactions.

A polypeptide chain is composed of rigid units linked by 'hinges'

Proteins or polypeptides are polymers of amino acids in which the amino group of one amino acid is linked to the carboxyl of its neighbour to form a **peptide bond**. Chemical features of the peptide bond that are significant in protein structure are

Table 4.5 Components of proteins other than amino acids

Component	Coupled to
Glycoproteins	
Sugars (*O*-glycosylation)	Serine, threonine
Sugars (*N*-glycosylation)	Asparagine
Phosphoproteins	
Phosphate	Serine, threonine, tyrosine
Membrane proteins	
Fatty acids (membrane 'anchor')	Cysteine, glycine
Other proteins	
Acetic acid (histones)	Lysine
Carbon dioxide (haemoglobin)	*N*-terminus
Carbon dioxide (γ-carboxylation)	Glutamate
Porphyrin (cytochrome c)	Cysteine
Metal ions (metalloproteins)*	Various
Porphyrin (haemoglobin)*	Histidine
Retinol (rhodopsin)*	Lysine

*Noncovalently bound; all other components are covalently bound

summarized in Chapter 2 (Figure 2.20). A major consequence of the nature of the peptide bond is that each –CO–NH– unit, in its *trans* configuration, is effectively a rigid unit linked to its neighbours by the α-carbon atoms, which carry the side chains. The repeating unit $-C_\alpha-[CO-NH]-$ is capable of rotation about the $C_\alpha-C$ and the $N-C_\alpha$ bonds and these rotations allow polypeptide chains to adopt their observed structures. A completely unstructured polypeptide chain adopts a 'random coil', but the secret of the structure and function of proteins is their ability to adopt precise three-dimensional conformations, which are determined by their amino acid sequence (see below).

Chapter 2. Peptide bonds are formed between two amino acids and have properties central to protein structure, page 31.
Chapter 4. The three-dimensional structure of proteins: the importance of sequence, page 98; Box 4.8, The denaturation and renaturation of proteins, page 99.

Natural polypeptides have defined lengths and sequences

Each protein of an organism has a distinctive polymer length and sequence (or **primary structure**) that gives it a particular set of properties as well as its biological function. Some general properties of polypeptide chains are shown in Box 4.5. Distinctive features associated with individual proteins or domains within proteins (see below) include

- *hydrophilic/hydrophobic nature*: local concentrations of nonpolar amino acids occur, for example, in integral membrane proteins
- *positive/negative charge*: local concentrations of positively charged amino acids are often found in proteins that bind to nucleic acids, such as the histones
- *sequence regularities*: regular distribution of certain amino acids can give rise to structural features such as the leucine zipper, found in some proteins that bind to specific sequences on DNA (see Chapter 5).

Amino acid sequences can either be determined directly or deduced indirectly from nucleic acid sequences

Before gene manipulation was possible, amino acid sequences could only be determined directly on a pure protein, using amino acid composition, *N*-terminal identification and specific cleavage (Box 4.6). These procedures require relatively large amounts of highly purified protein, something which is often difficult to

Box 4.5 General properties of polypeptides

Amino acid composition

This is determined by analysis of the hydrolysed protein and defines the relative proportions of each of the 20 amino acids in a protein. It gives general information on the nature of the protein – hydrophobicity or hydrophilicity and scarcity of certain amino acids. Some amino acids – for example, tryptophan, cysteine and methionine – are typically present in small amounts; because each (if present) must occur as at least one residue per protein molecule, a **minimum molecular mass** can be calculated from the amino acid composition.

Molecular mass

The relative molecular mass (M_r) of a protein is a multiple of the minimum molecular mass and can be determined by a variety of physical methods, of which SDS–polyacrylamide gel electrophoresis is probably the most widely applied to single polypeptide chains. Proteins found in cells and plasma range from a few amino acid residues up to several thousand, with most containing 100–1000. Small peptides of less than 50 amino acids are usually produced by cleavage from a larger protein, as in the case of the encephalins and the peptide hormones (see Chapter 11) but some, such as glutathione (see Chapter 10), are made directly by special, nontemplate-directed mechanisms.

Amino acid sequence

This is the most important parameter of a polypeptide and consists of the number and linear order of the amino acids that make up the chain. It can be determined experimentally from a pure preparation of a protein (if at least a milligram is available) or can be deduced from the corresponding nucleotide sequence, if known (see below).

Polypeptide chain polarity

Polypeptide chains have polarity, which is defined by the amino (*N*) and carboxy (*C*) termini of the chain.

Amino terminus

The amino acid at one end of the protein carries a free amino group (all other amino groups are part of the peptide bonds of the protein). It can be identified directly and is generally taken as the reference point for the protein – numbering of amino acid residues starts from this point. Proteins, when newly synthesized, have an *N*-terminal methionine residue (see Chapter 5), but the chain is often subjected to specific proteolysis, resulting in the appearance of a different *N*-terminus.

Carboxy terminus

This is at the opposite end of the polypeptide chain to the *N*-terminus. It is the last part of the polypeptide to be synthesized on the ribosome and contains a free carboxylate group at physiological pH.

Net charge

The charge on a protein molecule at a defined pH value is its net charge. It depends on the relative amounts of amino acids with charged side chains – aspartate and glutamate (negative charges); arginine, lysine and histidine (positive charges). The $-NH_3^+$ and $-COO^-$ of the constituent amino acids are involved in peptide bonds and only the two termini carry charges – all other charges are on the side groups. For any protein there is a pH at which the net charge is zero (the **isoionic** or **isoelectric point**) and at which the protein will not migrate in an electric field (when subjected to **electrophoresis**). Most cellular and plasma proteins are negatively charged around neutral pH and are said to be **acidic**. In contrast, **basic** proteins are positively charged, because they contain many lysine and arginine residues. Examples include the histones of the chromosomes (see Chapter 6).

Chapter 11. Peptide hormones, page 346.
Chapter 10. Glutathione is a tripeptide with several important biological functions, page 313.

Chapter 5. Initiation, page 134.

Chapter 6. A group of basic proteins, the histones, package DNA into nucleosomes, page 151.

Box 4.6   Direct determination of the amino acid sequence of a protein

Preliminary steps

In order to determine the amino acid sequence of a protein or peptide, it must be available in pure form and in sufficient amounts, typically of the order of 1 mg or more. Determination of the amino acid composition and identification of the *N*-terminal amino acid (see Box 4.5) is normally done as a preliminary step and some measure of the relative molecular mass may be obtained. Presence of more than one *N*-terminus or of several polypeptides under denaturing conditions may indicate the existence of hetero-oligomeric quaternary structure in the protein (see below), in which case each molecular species of polypeptide present in the protein must be separated and analysed separately.

Automated determination

The **Edman degradation** reaction cleaves polypeptides sequentially from the *N*-terminus, releasing each amino acid in turn. This method forms the basis of operation of automated peptide sequencing machines which can sequence small peptides completely or analyse the first 50–60 amino acids from the *N*-terminus of larger ones.

Sequencing longer polypeptides

The sequencing problem of longer polypeptides cannot be solved by a single analysis: the long polypeptide must be broken down into a series of shorter peptides whose sequences *can* be determined. These sequences are then reassembled to yield the complete sequence.

The strategy depends on the use of specific proteinases to cleave the protein at specific locations, yielding a set of fragments small enough to be sequenced. Proteinases commonly used for this purpose are the pancreatic proteinases trypsin and chymotrypsin, which cleave peptide bonds on the carboxyl side of basic (lysine or arginine) and aromatic (phenylalanine, tyrosine or tryptophan) amino acid residues respectively. The fragments are separated, isolated and their individual sequences determined. This process is repeated with an enzyme with different specificity and the fragments are placed in order by searching for overlapping sequences in the two sets of fragments, thus permitting determination of the complete sequence.

---

obtain. Now that DNA can be handled much more easily and sensitively than protein, many sequences are deduced indirectly from a DNA or cDNA sequence (see Chapter 6 and Box 4.7). It is now possible to determine the sequence of a protein that has not previously been isolated or even recognized. In many circumstances, the direct and indirect approaches can be used in a complementary manner.

Box 6.1. Some applications of the information from DNA sequencing, page 150.

There are many practical applications of amino acid sequence data

Once the primary structure of a protein has been determined it may be put to many practical and experimental uses (see Box 17.6). These include

Box 17.16. The human genome project, page 547.

- Recognition of novel or previously unknown proteins (these may initially be identified as potential open reading frames – see Box 4.7).
- Recognition of similar (or **homologous**) proteins in different organisms (computer programs can compare and align similar sequences).
- Recognition of conserved regions within sequences of homologous protein (conserved amino acids and regions are generally important in protein function).
- Deduction of evolutionary relationships between organisms (often known as molecular phylogeny – see Chapter 6).
- Design of oligopeptides for antibody production.
- Design of oligonucleotide 'probes' for gene detection (see Chapter 17).

Box 6.6. Nucleic acid and protein sequences can be used for 'molecular phylogeny', page 171.
Chapter 4. X-ray crystallography can be used to study parts of proteins, page 100.
Box 17.9. Protein p53, mutations and cancer, page 536.

These applications are well illustrated by studies on the tumour suppressor protein p53 (see this chapter and Chapter 17).

Box 17.13, DNA sequencing methodology, page 543.

Box 4.7   The identification of protein sequences from nucleotide sequence data

- DNA sequencing (see Box 17.13) yields a double-stranded DNA sequence.
- Each strand of DNA contains *three* potential **reading frames**, once copied into a messenger RNA (i.e. there are six in a double-stranded DNA molecule).
- AUG sequences (**codons**) on RNA (ATG on DNA) mean 'start translating'.
- UAA, UAG or UGA codons on RNA (TAA, TAG or TGA on DNA) mean 'stop translating'.
- An AUG sequence, followed by a length of nucleotides and a stop codon *in the same frame as* the start codon constitutes a potential **open reading frame (ORF)**.
- Once determined, DNA sequences are stored in computer files.
- Computer programs are able to find all potential ORFs in any sequence file.
- Other sequence information (such as **promoter** and **polyadenylation** sequences) can support the identification of an ORF.
- Some direct experimental evidence is needed to confirm that the ORF really does encode a protein.

See Chapters 5 and 6 for detailed explanations of the terms used here.

Part B. Glossary, page 173.

The three-dimensional structure of proteins: the importance of sequence

The three-dimensional structure of proteins is important for biological function

Most proteins in their 'native' or biologically active conformation have a precise three-dimensional structure in which each atom occupies a specific spatial position relative to its neighbours. This conformation is essential for the proper functioning of the protein and anything which perturbs it significantly could impair or destroy the function. Perturbation can be caused by external factors, such as temperature or solvent changes, or changes within the protein caused by chemical modification or gene mutation.

The main reason for the importance of the three-dimensional structure in the function of proteins is that it brings distant parts of polypeptide chains into close proximity, thus allowing the chemical properties of these parts to act in concert to form a binding site, an enzyme active site or some other distinct functional feature (Figure 4.4). The folding of a particular type of polypeptide chain into its native conformation is predominantly an intrinsic property of the **primary** structure, or amino acid sequence, of the polypeptide (Box 4.8).

The three-dimensional structure of a protein can be determined by X-ray diffraction or by nuclear magnetic resonance

The three-dimensional structure of a protein may be determined using **X-ray diffraction** (Box 4.9). This is a major undertaking and requires the preparation of pure protein in milligram amounts and its crystallization into a suitable form, either or both of which may be difficult or impossible. The determination can be performed to various levels of precision, but only the highest levels allow location of individual atoms within the molecule. Since the initial resolution of the structures of haemoglobin and myoglobin in the late 1950s, the structures of over 500 proteins have been determined and certain patterns of protein structure have been discerned (see below).

Figure 4.8. Supersecondary structure, page 105.

Box 4.8   The denaturation and renaturation of proteins

Denaturation is the term describing the loss of the proper three-dimensional (or **native**) structure of a protein, leading to the loss of its biological function.

Denaturation of proteins can be caused by a number of agents:

- Heat, where thermal motion disrupts the weak forces maintaining protein structure.
- Extremes of pH, which disrupt hydrogen bonds.
- High concentrations of certain chemicals, such as 8 M urea.
- Organic solvents, which alter the aqueous environment, perturbing hydrophobic interactions.
- Strong detergents, which also perturb hydrophobic interactions.

Only the first two, in burns and stomach acid, are likely to be of clinical importance but the others are important in the study of proteins.

Denatured proteins are often insoluble, as happens in the coagulation of egg white during cooking. This is because denaturation exposes the hydrophobic interior of the protein, which can then interact with other molecules, forming an insoluble, intermolecular network.

Renaturation of proteins can be demonstrated in the laboratory, for example, by slow removal of the denaturing condition (such as dialysis to remove 8 M urea). If done carefully, such procedures can lead to recovery of biological function. Such observations have been very important conceptually because they show that the information for protein folding lies in the primary structure of the polypeptide chain. Recent discoveries of **molecular chaperones** and **chaperonins** (see Chapter 5) have led to some revision of this view. Although the underlying concept remains, it seems likely that the spontaneous folding of proteins in the cell would probably be too slow to be effective and so chaperones are required to accelerate the process and prevent undesirable aggregation.

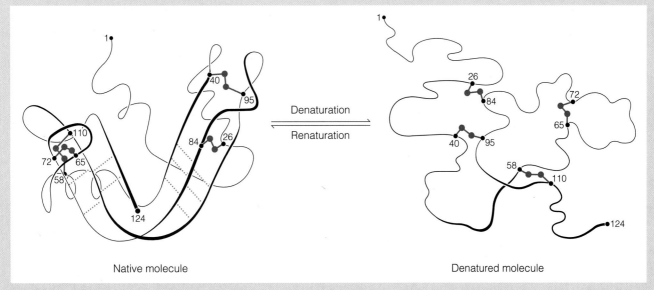

Box Figure 4.8 Renaturation of a protein after heat denaturation, but without breaking disulphide bonds (blue).

The major limitations of X-ray crystallography of proteins are:

- the need for at least several milligrams of pure protein;
- the ability to grow suitable crystals of the protein;
- the substantial amount of work involved.

The need to produce crystals of protein is especially troublesome in the case of membrane proteins and proteins containing markedly 'flexible' portions. These features make proteins difficult to crystallize, even when adequate amounts of material are available.

Chapter 12. The quaternary structure of haemoglobin affects oxygen binding, page 370.

Chapter 6. The structure of DNA has many important biological implications, page 148.

Box 17.9. Protein p53, mutations and cancer, page 536.

Box 4.9   X-ray diffraction in biology

Crystals are regular arrays of atoms, ions or molecules. When a narrow beam of X-rays is passed through a crystal, the X-rays are **diffracted** by this array, much as visible light is diffracted by the lines on a grating. The atoms of a crystal of a particular molecule, such as a protein, diffract the X-ray beam in a characteristic fashion, which can be recorded on a photographic plate in the technique of **X-ray crystallography**.

Mathematical analysis of this pattern can be used to build up a three-dimensional picture of the electron density within the protein and, with further refinements, produce a three-dimensional map of its constituent atoms. This was first performed in 1957 for haemoglobin (see Chapter 12) and has since been carried out for several hundred other proteins and enzymes. The information obtained in this way provides valuable insights into the structure and functions of proteins, both individually and as a group of macromolecules.

X-ray diffraction can also be performed on other non-crystalline arrays of molecules, such as DNA fibres, cell membranes and phospholipid bilayers. This approach, although not crystallography, has also yielded crucial insights into the structures of complex biological molecules, notably in the proposal of the double-helical **Watson–Crick model** for DNA in 1953. True X-ray crystallography of DNA, showing its detailed molecular structure, was not performed until nearly a quarter of a century after the Watson–Crick model was proposed, after the chemical synthesis of short lengths of pure DNA (see Chapter 6).

Nuclear magnetic resonance (NMR) is an alternative to X-ray diffraction, but still requires substantial amounts of material and is really most appropriate for smaller proteins or fragments of larger ones. However, because NMR can be performed in solution, it allows the three-dimensional structures of proteins that cannot be crystallized to be determined.

X-ray crystallography can be used to study parts of proteins

The problem of being unable to crystallize a protein can be circumvented by using part of a protein, especially a discrete domain. This has proved valuable for studying membrane proteins, such as the influenza virus envelope proteins haemagglutinin and neuraminidase (see Chapter 18). Membrane proteins are generally difficult (or even impossible) to crystallize, but it is possible to crystallize their external domains by cleaving off the membrane-spanning domains which comprise most of the proteins.

A similar approach has recently been applied to the study of the tumour suppressor protein p53 (see Chapter 17), a three-domain protein which proved refractory to crystallization. The central, DNA-binding domain of this protein was expressed *in vitro* in a bacterium and the resultant protein crystallized along with a section of synthetic DNA corresponding to its recognition site. This approach has yielded the three-dimensional structure of this crucial part of the p53 protein and its complex with DNA.

There are three higher levels of protein structure

Three higher levels of protein structure have been distinguished – **secondary**, **tertiary** and **quaternary** (Figure 4.5). These are not found in all proteins; some contain very little secondary structure, while many are composed of a single polypeptide chain, and thus possess no quaternary structure. A feature common to all of these higher levels of structure is that they are predominantly

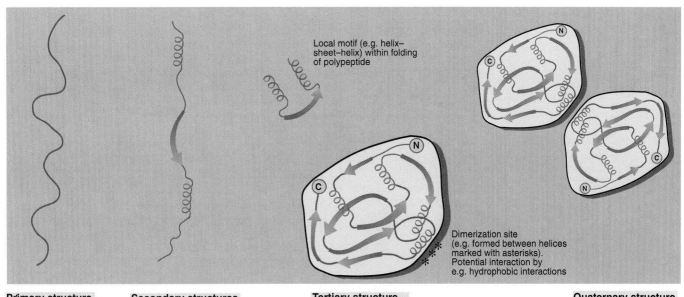

Primary structure
Part of a polypeptide chain in its unstructured (or unfolded) state.

Secondary structures
joined by looped regions. Local organization of chain into helices and sheets.

    ℓℓℓℓ = α-helix

    ➤ = β-sheet

    〜 = Loop

Tertiary structure
Compact folding of helical and sheet regions to give precise, three-dimensional structure of protein.

Quaternary structure
Two or more polypeptides interact in a specific manner to form a dimer (or trimer, tetramer, etc.).

Local motif (e.g. helix–sheet–helix) within folding of polypeptide

Dimerization site (e.g. formed between helices marked with asterisks). Potential interaction by e.g. hydrophobic interactions

Figure 4.5 Higher levels of protein structure. This figure illustrates the four levels of protein structure: primary, secondary, tertiary and quaternary.

maintained by weak forces between specific parts of the protein molecule – mostly hydrogen bonds, ionic and hydrophobic interactions and Van der Waals' forces (Box 4.10).

Secondary (2°) structure: regularities within proteins

Within the overall three-dimensional structure of most globular proteins are sections of regular or 'periodic' structure. The three most common types of periodic structures found in globular proteins are the α-**helix**, the β-**sheet** and the β-**turn** (Box 4.11) The first two of these are found extensively in 'structural' proteins, for example helices in keratin and β-sheets in the crystalline regions of silk fibroin. Regions of proteins with none of these features have precise three-dimensional structure, but do not show any regularity or periodicity. Such regions are described as being in 'coil' or 'random coil' conformation.

The secondary structures are maintained by hydrogen bonds occurring between –C=O and –N–H atoms of the peptide bonds of the polypeptide 'backbone'. The side-chains play no direct role in secondary structure, but the types of amino acid side chain in any region of a protein strongly influence whether that region adopts an α-helix, β-sheet or coil conformation and this can be used for predicting secondary structure (see below). The lengths and locations of regions of secondary structure are typical of each protein, and vary widely from protein to protein.

Tertiary (3°) structure: polypeptide folding and 'domains'

In both proteins with quaternary structure and those comprising a single polypeptide chain the individual atoms of the polypeptide have precise locations in space, which are brought about by the folding of the polypeptide chain. This is the **native** conformation of the protein and is maintained by many weak bonds

Chapter 5. Transfer RNAs have a distinctive three-dimensional structure, page 129.
Chapter 6. The structure of DNA is not completely uniform, but has sequence-dependent local variations, page 147.
Chapter 3. Phospholipids spontaneously form lipid bilayers in water, page 52.
Chapter 2. Hydrogen-bonded base pairing, page 34.

Box 4.10 Weak forces involved in protein structure

The three-dimensional structure of a protein is principally maintained by a large number of weak, that is non-covalent, forces or interactions between different parts of the molecule. These forces are also important in the structures of nucleic acids (see Chapters 5 and 6) and of lipid bilayers and cell membranes (see Chapter 3). There are four principal types.

Hydrogen bonds

Hydrogen (>) bonds occur between an electronegative atom (N or O, bearing a partial negative charge) and a hydrogen atom bearing a partial positive charge because it is attached to another electronegative atom (HO– or HN=). There are thus four types found in biological molecules:

1. =N>HN=, found in base pairs in DNA (see Chapter 2);
2. =N>HO, found in base pairs in DNA;
3. =O>HO, important in the structure of water;
4. =O>HN=, found in α-helices and β-sheets of proteins (see Box 4.11). Besides the 'backbone' hydrogen bonds which maintain secondary structure, they are also found in the tertiary structure of proteins, involving the side-chains of the amino acids, for example between the hydroxyl of serine and the carbonyl of asparagine (see Figure 4.6).

Electrostatic interactions

These also involve interactions between charged groups, but this time complete, not partial. In proteins, they may occur between amino acids with positively charged side chains (lysine, arginine, histidine) and those with negative charges (aspartate, glutamate). Repulsive interactions can also occur between similarly charged groups.

Van der Waals' forces

Such forces operate at very short ranges and serve to ensure that there are few 'holes' in the structure of globular proteins and help to ensure a compact shape.

Hydrophobic interactions

Hydrophobic interactions are extremely important in protein structure and chiefly involve the aliphatic and aromatic amino acids. The interactions are driven by entropic factors because hydrophobic groups in water are surrounded by ordered or 'ice-like' water molecules. Bringing hydrophobic groups together reduces the amount of ice-like water, reducing order and increasing entropy, which is favoured thermodynamically. This explains why water-soluble proteins generally have hydrophobic interiors and also why proteins are susceptible to denaturation by organic solvents and strong detergents.

or interactions, predominantly between the various side-chains of the amino acids of the polypeptide chain (Box 4.10). The nature and location of these residues in the amino acid sequence defines the identity and distinctiveness of an individual molecular species of protein. The folding of a protein is also influenced by its local **environment**, that is whether it is in aqueous solution (hydrophilic) or embedded in a biological membrane (hydrophobic). The hydrophobic amino acids of water-soluble proteins are typically concentrated in the interior of the molecule.

Many proteins contain **domains**, distinct parts of their structure with discrete physical and functional identity. Domains were first recognized in the immunoglobulins, where IgG molecules contain six domains with similar internal folding, but distinct functions (see Chapter 16). Although the term domain has a

Chapter 16. Immunoglobulins have a multidomain organization, page 490.

Type of interaction   Model   Example

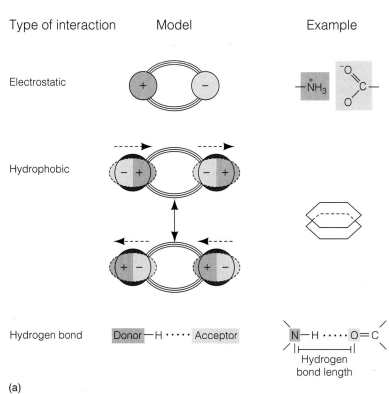

Figure 4.6 (a) Weak forces involved in protein structure. (b) Intramolecular interactions in proteins. The figure shows three kinds of hydrogen bonds within the enzyme lysosome — those between side chain groups (beige), those between side chain groups and backbone amide hydrogens or carbonyl oxygen (dark purple) and those directly between backbone groups (light purple).

(a)

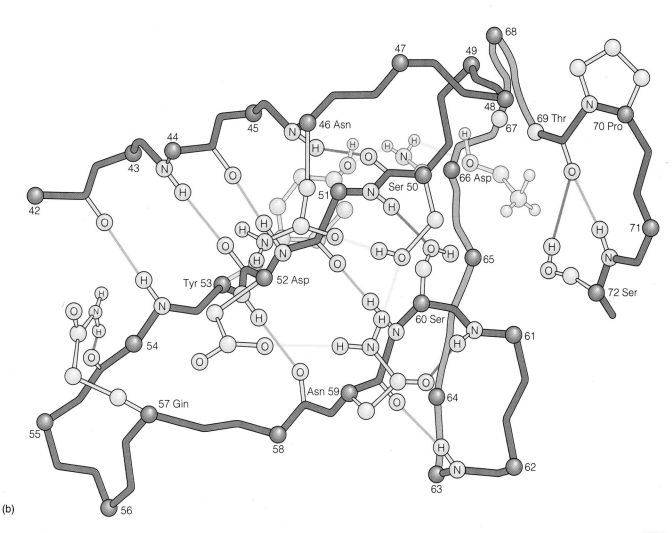

(b)

Box 4.11 α-Helices, β-sheets and turns – regular features in proteins

α-Helices, β-sheets and turns are regular or **periodic** structures found in most proteins and form the **secondary structure**. They are periodic because the spatial relationship of each amino acid to its neighbours is the same, being defined by the angles around the N–C_α and C_α–C bonds (Box Figure). Each protein contains its characteristic proportion of each of the three forms with many proteins having 20–30% of each type. Other proteins are rich in β-sheets (for example, immunoglobulins) or in α-helices (for example, haemoglobin).

The α-helix is the most stable of several possible helical conformations that polypeptide chains may adopt (Box Figure). It is favoured over the others because the hydrogen bonds are least strained, lying nearly parallel to the axis of the helix. The side chains project outwards from the helix.

The β-sheet structures may be parallel or antiparallel and contain the polypeptide chain in a fully extended conformation (as opposed to the coiled form of the helix: Box Figure).

β-Turns are composed of four amino acids which effect a 180° change in the direction of the polypeptide chain. They are often located at the surface of globular proteins.

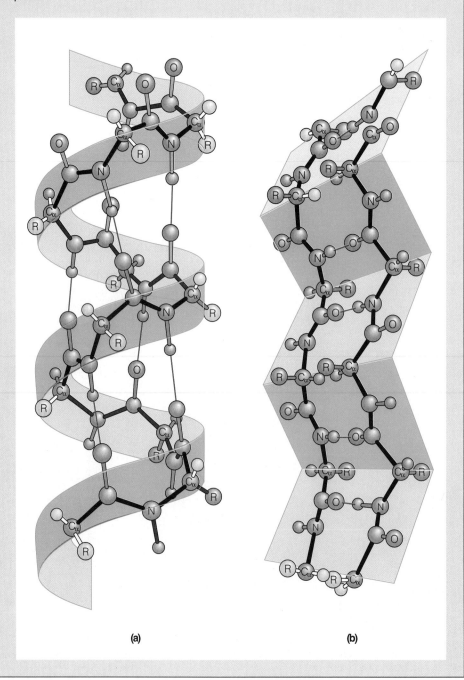

Box Figure 4.11 α-Helices and β-sheets are the major regular secondary structures of proteins. In the α-helix (a) the hydrogen bonds (red lines) occur locally, whereas in the β-sheet (b) hydrogen bonds occur between two chains or distant parts of the same chain. Ⓡ represents the side chains of the amino acids.

(a)   (b)

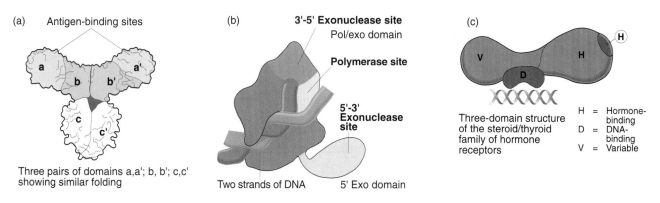

Figure 4.7 Domains in proteins: (a) immunoglobulin; (b) *E.coli* DNA polymerase I; (c) steroid hormone receptors.

fairly well-defined meaning in terms of protein structure, it is often used in a number of looser senses and should be interpreted with some caution (Figure 4.7).

Regions of regularity or secondary structure lie within the tertiary structure of proteins and also within domains. The tertiary structures of different proteins contain different proportions of the three main types (helix, sheet and turn) disposed in a number of characteristic patterns or **supersecondary** structures (Figure 4.8) and associated with particular functional properties, such as the binding of NAD, ATP or other cofactors.

Quaternary (4°) structure: structural and functional features

Many protein molecules are composed of two or more polypeptide species that are held together only by weak interactions of the types which maintain tertiary structure. This level of structure is called quaternary structure.

Probably most globular proteins show this **oligomeric** structure, commonly containing 2–6 polypeptides, forming dimers, trimers, tetramers, pentamers or hexamers. Dimers and tetramers are particularly common among water-soluble proteins; a number of integral membrane proteins are trimeric. Cases where two molecules of one species of polypeptide chain are present are called **homodimers**;

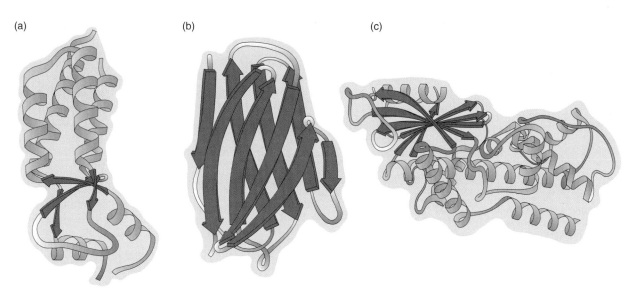

Figure 4.8 Supersecondary structure. The figure shows three of the main types found in globular proteins. (a) Predominantly α-helical (helix bundles): tobacco mosaic virus coat protein. (b) Predominantly β-sheet: immunoglobin domain. (c) Mixed α-helix and β-sheet: hexokinase, domain 2.

Chapter 12. The quaternary structure of haemoglobin affects oxygen binding, page 370.

Box 17.9. Protein p53, mutations and cancer, page 536.

others (**heterodimers**) contain two different polypeptides. The first protein in which quaternary structure was recognized was haemoglobin, an $\alpha_2\beta_2$ heterotetramer (see Chapter 12). Some proteins, such as the tumour suppressor protein p53, have a distinct 'oligomerization domain', which is responsible for interaction between molecules (in this case forming a tetramer).

A characteristic of many proteins showing quaternary structure is their ability to modify their functional behaviour in the presence of small molecule effectors that cause subtle but significant changes in both conformation and function; this is called the allosteric effect and was first described in haemoglobin.

It is theoretically possible to predict the three-dimensional structure of proteins

As many proteins are able to reform their native structures after denaturation (Box 4.8), it ought to be possible to predict the complete three-dimensional structure of a polypeptide from knowledge of its amino acid sequence. Unfortunately this is not yet possible because the rules and mechanisms of protein folding are not adequately understood and because enormous computational effort is required. At the present stage of understanding, it is possible to predict the three-dimensional structure of a protein reliably only if the structure of a similar protein is already known. Reasonably reliable predictions can, however, be made of probable secondary structural features (see above) by a knowledge of frequencies of occurrence of amino acids in proteins of known structure: proline and glycine, for example, are rarely found in α-helical regions. A plot of the occurrence of hydrophobic amino acids (a hydrophobicity or hydropathy plot) can be used to predict transmembrane regions of integral membrane proteins (see Chapter 3 and below).

Chapter 3. The transmembrane organization of an integral protein can be predicted from its primary structure, page 55.

Functional aspects of protein structure

There is no 'unique solution' to the fulfilment of a particular biological function by proteins

Many functions and activities of proteins are found in most organisms. These homologous proteins have both similarities (see above) and differences, which tells us that there is no 'unique solution' to a particular biological function in terms of the protein responsible. Rather there is a range of possible 'solutions' in the form of different polypeptide sequences, some of which will be very similar in function, while others will be slightly, but significantly, different and yet others will be so different as to show no similarity. Thus, there are many positions in the sequence of a typical protein in which a number of different amino acids can be tolerated without alteration of function.

Some proteins contain apparently redundant sequences

Parts of the sequence of a protein appear not to be essential for its function. This surprising finding has emerged because recent technical developments have shown that it is possible to delete quite large parts of some proteins by genetic engineering without apparently affecting their function (see Chapter 17). There may be considerable advantage for a cell or organism in this 'slackness', because a system that demands total precision at *every* amino acid of *every* polypeptide and complete essentiality of *every* amino acid might be too 'perfect' to be practicable.

Chapter 17. Genes can be finely engineered by site-directed mutagenesis, page 545.
Box 5.1. Control of transcription in bacteria and humans, page 117.
Chapter 5. Targeting of proteins, page 136.

Cells contain two or more proteins with similar functions

An important aspect of this phenomenon is that different organisms contain different proteins that perform the same biological function. Even within the same cell, we can find different proteins carrying out the same catalytic role, but tailored to different situations; these are sometimes called **isoenzymes** (or **isozymes**). Generally, each enzyme form is coded by a different gene, which is expressed and targeted in different ways (see Chapter 5).

Three examples of this phenomenon are:

1. **Carbamoyl phosphate synthetase** activity is found both in the mitochondrial matrix and in the cytosol, but is catalysed by completely different proteins. The first acts in the urea cycle, the second is a multienzyme complex carrying out the first steps of pyrimidine biosynthesis (see Chapter 10).
2. **RNA polymerase** is found in the cell nucleus as three molecular species, broadly similar in their complex quaternary structure and in their catalytic behaviour but each adapted to a specific type of RNA synthesis (see Chapter 5).
3. **Lactate dehydrogenase** (see Chapter 8) is found in five different forms in different tissues. The five forms arise because the enzyme is a tetramer with two types of subunits: M (muscle) and H (heart). The five different forms are the two homotetramers – M_4 and H_4 – and three heterotetramers – HM_3, H_2M_2 and H_3M. Different proportions of these isoenzymes are found in different tissues.

Chapter 10. Carbamoyl phosphate provides the first ammonium ion for urea synthesis, page 301.
Chapter 10. Carbamoyl phosphate, page 317.
Chapter 5. RNA polymerases are capable of both initiation and elongation, page 114.
Chapter 8. Pyruvate is a central metabolite which may be oxidized or diverted into anabolic processes, page 241.

The folding of some proteins is assisted by 'molecular chaperones'

Because it has been demonstrated that denatured proteins are capable of refolding into their native, biologically active conformations (see Box 4.8), it has generally been assumed that this process would also occur in cells after polypeptides had been synthesized. However, the situation is not quite as simple as this: there are proteins – called **molecular chaperones** – which assist in the folding process (see Chapter 5). It may be that, although proteins are capable of folding themselves into the correct conformation, the process may be too slow or inefficient to be compatible with physiological requirements and so chaperone mechanisms have evolved. This is rather similar in principle to the evolution of proteins catalysing reactions that proceed spontaneously, such as carbonic anhydrase (see Chapter 12).

Chapter 5. Heat-shock proteins assist protein refolding page 141; Heat-shock proteins are also involved in protein targeting, page 142.

Chapter 12. The anion-exchange protein participates in carbon dioxide metabolism, page 372.

Many proteins are post-translationally modified

Biological macromolecules are made by the enzyme-catalysed polymerization of their constituents, but in many cases the immediate product of this synthesis is not the biologically active form and subsequent covalent modifications must be made by various processes of postsynthetic modification. Proteins are no exception, and undergo a wide range of post-translational modifications (see Table 4.5). The major determinants of the occurrence of a given modification in a particular protein include

- the presence of the appropriate modifying enzymes in the cell involved;
- the presence of an appropriate sequence on the protein which is 'recognized' by the modifying enzyme and to which the modification is covalently attached;
- in some cases, the protein requires to be located in a particular membrane or cellular compartment for modification to occur.

Proteins are targeted to specific intracellular locations

Because cells, especially those of eukaryotes, contain several 'compartments' mechanisms must exist to ensure that proteins get to the correct location in the cell.

Chapter 5. Targeting of proteins, page 136.

A major mechanism appears to be the presence of 'signal sequences' on proteins or their precursors. In some cases, a terminal signal sequence is cleaved off as the protein reaches its target; in contrast, the sequences directing proteins to the nucleus are internal and not cleaved. Such **consensus signal sequences** can be identified within protein sequences and yield clues to the probable intracellular location of a protein. This topic is discussed in greater detail in Chapter 5.

Mutation can have important effects on protein structure and function

Mutation in a particular gene can lead to the production of a protein in which the amino acid sequence has been altered by processes such as substitution, insertion or deletion (see Chapter 6). The nature of the consequent changes in protein structure and function will depend on the nature of the change in the primary structure.

Insertions and deletions are **frameshift** mutations and cause major changes in the primary structure of the protein from the mutation to the C-terminus; generally their effect on structure and function is very damaging, unless they are very near the C-terminus. Surprisingly, some proteins appear to be able to tolerate deletion of large parts of their sequence without apparent effect on function (see above).

Chapter 17. Genes can be finely engineered by site-directed mutagenesis, page 545.

Substitution mutations may be natural or induced and are of considerable interest in the probing of the 'active sites' of proteins and enzymes by the technique of site-directed mutagenesis (see Chapter 17). Substitution by a similar amino acid (such as valine for isoleucine) is termed **conservative; radical** substitutions are caused by a drastic change in the nature of the amino acid side-chain (such as if aspartate is substituted for isoleucine). In general, conservative substitutions have little or no effect on structure or function while the effect of radical substitutions depends on the location. Any change in a crucial amino acid (perhaps at the active site of an enzyme) will lead to near or complete loss of function.

Amino acid substitutions can lead to complete loss of activity, subtle changes in structure (whereby the protein is still functional, but is less stable – for example, temperature-sensitive mutants, which give proteins active only at lower than normal temperatures) or changes in substrate specificity and/or kinetic parameters of an enzyme (for example zidovudine-resistant mutants of HIV reverse transcriptase – see Chapter 18).

Chapter 18. HIV mutates in infected individuals and becomes drug-resistant, page 577.

Naturally occurring mutations can be very informative

Chapter 17. Study of mutant haemoglobins has increased understanding, page 513.
Box 17.9. Protein p53, mutations and cancer, page 536.

The study of naturally occurring mutations has been of great importance in the study of haemoglobin (see Chapter 17), but a recent very exciting finding has been that the tumour suppressor protein p53 (see Chapter 17) is mutated in many common human tumours. These are mainly substitutions of single amino acids, and analysis of the primary structure of this protein shows that most of these lie in its DNA-binding domain. The three-dimensional structure of the DNA-binding domain confirms that the side-chains of the amino acids affected make contact with DNA. This exciting and direct link between molecular biology and common human tumours opens new avenues for potential therapies.

CHAPTER 5

Genes and their expression: RNA in the generation and translation of messages

Genes are the functional units of genomes and each consists of a unique sequence of DNA. For a gene to be expressed, it must be transcribed into an RNA molecule; to produce a protein, the RNA copy must be translated. Information flows in gene expression from DNA to RNA to protein (this is the central dogma of molecular biology).

Transcription of DNA produces the precursors of messenger, ribosomal and transfer RNAs. These primary transcripts must undergo processing events such as capping, tailing, splicing and transport to the cytoplasm in order to produce mature, functional RNA molecules.

Translation of messenger RNAs occurs on ribosomes, which are specific complexes of RNAs and proteins, formed into large and small subunits. Protein synthesis uses not free amino acids but amino acids covalently joined to specific tRNAs – the aminoacyl-tRNAs.

The synthesis of proteins is described, with an emphasis on the central role of messenger RNAs, and the subsequent modifications that convert many proteins to their functional forms and direct them to their appropriate locations. The control of gene expression is discussed in the context of the function of tissues and organs.

The nature and expression of genes

Genomes contain the information for organisms and genes are their functional units

The information for every type of organism is contained within its **genome**. For all kinds of cellular organisms – bacteria, protozoa, fungi, plants and animals – the genome is composed of double-stranded DNA (see Chapter 6); genomes of viruses contain either DNA or RNA and may be either single or double stranded (see Chapter 18).

A **gene** can be defined as a functional unit of a genome. It consists of a sequence of DNA nucleotides containing the information for a **gene product** responsible for the function of the gene. In order to exercise its function, a gene must be **expressed** and to achieve this must be copied or **transcribed** to form an RNA molecule. For most genes, the gene product is a polypeptide (or protein) and the product of transcription – messenger RNA (mRNA) – must be **translated** to yield the polypeptide. Some genes, for example those that encode the various functional RNAs (ribosomal RNA, transfer RNA and the small RNAs involved in splicing and signal recognition), the gene product is RNA itself and the RNA molecules copied from these genes are not translated. Nucleotide structure and the phosphodiester linkage in polynucleotides are described in Chapter 2.

The overall process of **gene expression** may be encapsulated in what is sometimes called the **Central Dogma** of molecular biology:

DNA → RNA → protein

(see the introduction to Part B). In the genomes of humans, other animals and plants, only a small amount of the DNA comprises sequences that are expressed. Functions have been ascribed to a small proportion of the other sequences – for example, in the organization and replication of DNA and chromosomes – but some biologists believe that much of the cell's DNA has no function. This has been termed 'junk' DNA. Perhaps some light will be shed on this dilemma as work proceeds on the Human Genome Project (see Chapter 17). Estimates of the numbers of genes in various organisms are shown in Table 5.1. Note that humans do not appear to contain many more genes than simple vertebrates – the difference in genome size appears to be mainly due to the greater lengths of the introns (see below) in higher vertebrates.

Other sequences are not expressed as RNA or protein, but have a direct role in the processes of gene expression. These sequences include promoters, enhancers and the other regulatory sequences that participate in the transcription of genes and in the control of gene expression. Although these sequences do not encode RNA or protein products, they are essential elements of the genes with which they are associated (Figure 5.1).

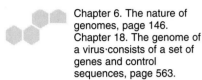

Chapter 6. The nature of genomes, page 146. Chapter 18. The genome of a virus·consists of a set of genes and control sequences, page 563.

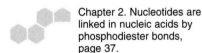

Chapter 2. Nucleotides are linked in nucleic acids by phosphodiester bonds, page 37.

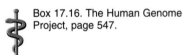

Box 17.16. The Human Genome Project, page 547.

Table 5.1 The genome size and gene content of various organisms

	Genome size	Gene number
Viruses		
Small	3 kbp	3–4
Large	300 kbp	300
Bacteria	3 Mbp	3000*
Invertebrates		
Nematode	100 Mbp	15 000*
Vertebrates		
Fish	400 Mbp	70 000*
Human	3000 Mbp	70–100 000*

kbp = kilobase pairs; Mbp = megabase pairs
*Estimated

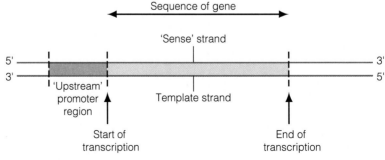

Figure 5.1 A simple picture of a gene.

Gene expression involves transcription, translation and processing

The variety of proteins and their functions were described in Chapter 4. Although proteins have an amazingly versatile repertoire of functions, they are unable to direct their own production: 'protein can't make protein' because even a simple cell requiring, say, a thousand proteins would need another thousand to make the ones it needed, and so on. This led to the idea that there has to be a basic 'assembly plant' for protein synthesis which can make any protein needed as long as it is supplied with an 'instruction tape'. Proteins are themselves integral components of this basic mechanism.

The synthesis of proteins is an essential function within all cells and involves four systems (Figure 5.2).

1. A mechanism for copying RNA molecules from the DNA genome (**transcription**). These RNA species act as messengers (mRNAs) or as components of the

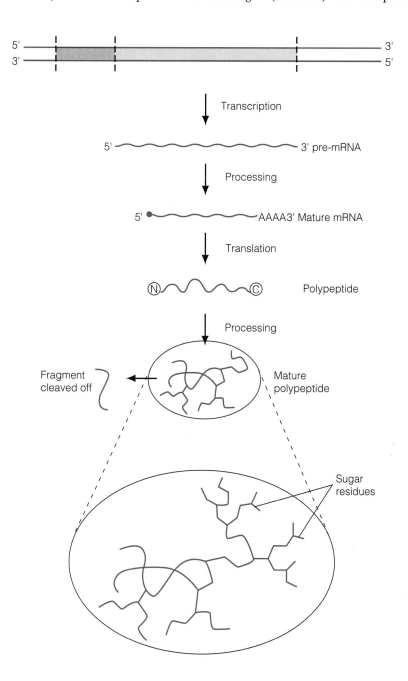

Figure 5.2 The overall process of gene expression.

translation machinery (rRNAs and tRNAs). Transcription must occur from *specific regions* of the genome, at the *proper times*, in *appropriate amounts*. The RNA transcripts are produced by **RNA polymerases**, aided by protein **transcription factors** that ensure proper control of the process.

2. Mechanisms for covalently modifying primary RNA transcripts into the mature species active in translation (**RNA processing**). Post-translational processing of RNA involves a variety of reactions:
   - methylation of bases and ribose in RNA;
   - specific hydrolytic cleavage of RNA chains;
   - addition of nucleotides to and removal from the ends of RNA chains;
   - breaking and rejoining of RNA molecules at specific sites (**splicing**).

   These reactions are carried out by enzymes and small RNA molecules and are followed by the transport of mature RNA molecules to their correct intracellular locations in the cytoplasm and on the rough endoplasmic reticulum.

3. A molecular apparatus (the ribosome – the 'assembly plant') for producing polypeptide chains according to mRNA 'instruction tapes' (**translation**). Other components required for protein synthesis include
   - accessory protein translation factors that interact with the ribosome to achieve proper initiation, elongation and termination of polypeptide chains;
   - a population of tRNA molecules and a set of aminoacyl-tRNA synthetases which 'load' them with their cognate amino acids.

4. Mechanisms for **processing** the completed polypeptide chain into the mature, functional protein and **targeting** it to its correct intracellular location.

   Many proteins undergo post-translational modifications, including specific proteolytic cleavage and glycosylation.

   Mechanisms exist, often involving signal sequences on the protein, which bring about the targeting of the protein to the endoplasmic reticulum, mitochondrion, nucleus or other location.

   Although a protein folds spontaneously into its correct three-dimensional structure as directed by its amino acid sequence, this often appears to be assisted within the cell by **molecular chaperones**.

The remainder of this chapter considers these four systems and their coordination to achieve the appropriate protein synthesis required by cells and the whole body.

Transcription: copying of genetic information from DNA into RNA

The chemistry of nucleotides, phosphodiester bonds and polynucleotides (including RNA) is outlined in Chapter 2, while general aspects of template-directed nucleotide polymerases are discussed in the introduction to Part B.

Transcription of DNA is local and asymmetric

In the context of the cellular DNA genome, transcription is a **local** event that occurs at the level of genes. DNA replication, by its very nature, must be **global** in extent, encompassing the complete genome (see Chapter 6). The local nature of transcription means that an essential part of its operation is the correct identification of signals for starting (or initiation) and stopping (or termination) the synthesis of RNA chains.

The antiparallel and complementary nature of the two strands of duplex DNA (see Chapter 6) means that the information encoded by the two strands is

Chapter 2. Nucleotides and polynucleotides, page 32. Part B. Template-directed nucleotide polymerases, page 79.

Table 6.2. Biological implications of the structure of DNA page 147; Chapter 6 Chromosomal DNA replication is a complex and demanding process, page 151.
Chapter 6. DNA has two polynucleotide chains in a double helix, page 147.

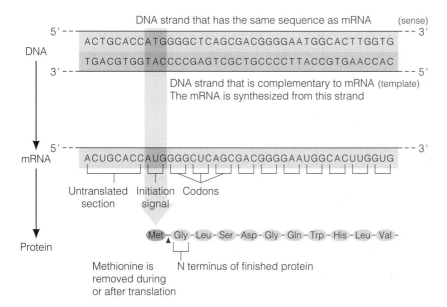

Figure 5.3 Transcription occurs from the template (or antisense) strand of DNA.

entirely different. Thus, transcription is asymmetric and must select and copy the correct (the **template** or **antisense**) strand once the initiation site has been identified (Figure 5.3). Transcription involves recognition of promoter and other sequences usually, but not always, adjacent to the **transcriptional start site**.

There are quantitative differences in the synthesis of rRNA, tRNA and mRNA

Ribosomal (rRNA) and **transfer** (tRNA) RNAs, which form the major components of the translation apparatus of cells, are transcribed as multiple copies made from a relatively small number of genes and which are not subsequently translated.

Messenger RNAs (mRNAs) are copied in relatively small numbers from a large number of different genes. Each mRNA species contains the information for the synthesis of a discrete protein.

There are significant differences in gene expression between bacterial and eukaryotic cells

Eukaryotes
In eukaryotic cells transcription is only part of the production of mRNA: the other RNAs and the **primary transcripts** (pre-rRNA, pre-tRNA and pre-mRNA) undergo extensive covalent modification. Transport out of the nucleus is an essential step in the delivery of functional RNAs to the translational machinery in the cytoplasm. The processes of transcription and translation are thus physically separated.

Prokaryotes
In bacteria both transcription and translation occur in the cytosol and are closely linked; modification of mRNA does not occur and ribosomes can start translating mRNA molecules while they are still being transcribed. Bacterial translation may initiate at internal sites on mRNAs – bacterial mRNAs are therefore often **polycistronic** (encoding several polypeptides) but eukaryotic mRNAs are **monocistronic**. The differences between bacterial and eukaryotic translation and their significance are discussed in Chapter 18.

Chapter 18. Protein synthesis, page 559.

RNA polymerases are capable of both initiation and elongation

DNA-directed RNA polymerase, usually referred to simply as **RNA polymerase**, is an essential enzyme in all organisms, cellular or viral, which have a DNA genome. RNA polymerases have a polymerizing ability similar to that of the other three types of template-directed polymerases – the **elongation** reaction – but are unique in also possessing the ability to initiate RNA chains by polymerizing two rNTPs (see below). The RNA polymerase elongation reaction is:

$$(rNMP)_n + rNTP = (rNMP)_{n+1} + PPi$$

where $(rNMP)_n$ represents the RNA chain being synthesized in a 5' to 3' direction, complementary to the DNA template strand, and rNTP the ribonucleoside 5'-triphosphate (as the Mg^{2+} salt) being added to the growing chain.

The RNA polymerases (RP) of eukaryotic cells are proteins with complex quaternary structures of ten or more subunits (M_r about 1 000 000), where the subunits play different roles in the overall reaction. Bacterial cells have only one RP (five subunits; M_r 450 000); the nuclei of eukaryotic cells contain three different enzymes, RP I, RP II and RP III (Table 5.2), each of which is responsible for the synthesis of only *one* of the major types of RNA (rRNA, mRNA and tRNA) or, more precisely, their precursor transcripts. These assignments were made partly on the basis of the action of the deadly poison α-amanitin, from the mushroom *Amanita phalloides*, a particularly potent inhibitor of mRNA synthesis by RP II. RP I is specifically located in the nucleoli, which are the sites of rRNA synthesis and assembly of ribosomal precursors. RNA polymerases are the central component of the transcriptional apparatus, but require the assistance of other proteins – **transcription factors** (see below).

Initiation of RNA chains creates a 5'-triphosphate terminus

A distinctive features of most RNA polymerases is their ability to initiate the synthesis of polynucleotide chains, in contrast to DNA polymerases, which lack this function (see Chapter 6). The initiation reaction occurs between two rNTPs, usually both purines, because RNA transcripts invariably start with a purine nucleotide:

$$pppN\text{-}3'OH + pppN\text{-}3'OH \rightarrow pppNpN\text{-}3'OH + PPi$$

(pppN-3'OH is a representation of a nucleoside triphosphate (NTP), emphasizing the 5'-triphosphate and 3'-OH groups).

This initiation reaction forms the first phosphodiester bond of a new RNA chain; its product, a **dinucleoside tetraphosphate**, is a dinucleotide with a triphosphate at its 5'-end and a hydroxyl group at its 3'-end, ready to accept the next rNTP in the first elongation reaction. The initiation reaction is unique to RNA polymerases, involves a subset of the subunits of the enzyme and is inhibited by antibiotics of the rifampicin type (see Chapter 18), which do not, however, affect

Chapter 6. DNA polymerases are central to DNA replication, page 154.

Table 18.4. Antibiotics with useful antibacterial properties, page 556.

Table 5.2 RNA polymerases

RNA polymerase	Number of subunits	Type of RNA produced
Bacteria		
RP	5	rRNA, mRNA, tRNA
Mammals		
RP I	10	pre-rRNA
RP II	10	pre-mRNA
RP III	10	pre-tRNA and other small RNAs

the subsequent elongation reaction. One result of this initiation mechanism is that all newly synthesized RNA chains carry a 5'-triphosphate; in the case of mRNA this is important in the addition of the 5'-'cap' (see below). Another important enzyme, DNA primase, which is required for the initiation of DNA chains during DNA replication (see Chapter 6), has a similar ability to initiate chains in this way: it is, indeed, a special type of RNA polymerase.

Chapter 5. 5'-capping, page 123.
Chapter 6. Initiation of DNA chains, page 157.

RNA polymerases must recognize the start sites on genes

If an RP is to transcribe a gene correctly, it must start at the **transcriptional start site** of the gene, because a transcript shorter than the gene would be incomplete and probably nonfunctional. To achieve correct initiation, RP molecules need to be able to bind to DNA so as to recognize the correct sequence (the start of the gene) and direction (to bind 'upstream' of the start site and then proceed in a 'downstream' direction – Figures 5.2, 5.4 and 5.11). Because the DNA double helix is antiparallel, only one of the two strands – the template (or antisense) strand – must be selected for copying. The other DNA chain (the sense strand) has the same sequence as the RNA copy, but contains thymine instead of uracil. The two strands of DNA contain completely different information and any stretch of DNA contains six potential informational sequences (see Figure 5.11).

Initiation of transcription also requires promoter sequences on DNA and protein transcription factors

Sequence recognition by RNA polymerases during transcription is a complex process, which varies between cells and between genes. It is necessarily complex because, in order to function properly, cells must be able to express the correct genes at the appropriate times and to make RNA in suitable amounts, large or small.
·Recognition involves two elements:

1. Specific sequences on DNA; and
2. Proteins that bind to these sequences in a sequence-specific manner (Figure 5.4 and Box 5.1).

Specific sequences on DNA are involved in the initiation of transcription

These sequences (Figures 5.4 and 5.11) include

- **RNA polymerase binding sites**, which lie at the transcriptional start site.
- **The 'TATA box'**, which lies 25 bp from the start site and plays a central role in the initiation and control of transcription of many genes.
- **Promoter sequences**, which typically lie upstream from the TATA box.
- **Enhancers**, controlling sequences which may be quite remote from the start site but also influence the binding of RP to a gene or group of genes.

Proteins bind to specific sequences on DNA and influence the initiation of transcription

These proteins are said to act in *trans* (as in the term *trans*-activation) and bind to specific sites on DNA either directly or in conjunction with other proteins. They can affect transcription positively (activation) or negatively (repression) by influencing formation of the RNA polymerase initiation complex. Bacteria generally 'shut down' unwanted genes by repressor proteins and then induce them in response to some specific stimulus – for example, expression of the *lac* operon of *Escherichia coli* allows the bacterium to grow on lactose.

Animal cells typically 'switch on' many of their genes by a positive mechanism of activation from a low, basal level of transcription (see Box 5.1), probably 'bending' the DNA to bring distant sites near to the transcriptional start site. This type of control is of special importance in multicellular organisms, where RNA transcription must be controlled so as to produce the correct RNAs in different cell types, at different times, in widely differing amounts and in response to different signals (such as those hormones that act via intracellular receptors; see Chapter 11). The study of transcription factors is extremely complex and it will probably be some years before transcription is adequately explained at the molecular level.

Chapter 11. The mechanism of action of extracellular signal molecules is related to their chemical characteristics, page 332.

Elongation of RNA chains requires transient unwinding of the DNA double helix

Once initiation has occurred, the elongation reaction will go on selecting the appropriate rNTPs and forming phosphodiester bonds until the full length of the RNA chain (or transcript) has been made. The template strand of DNA can be copied only if the two strands of DNA are separated; RP is capable of achieving this locally over about two turns of the helix. As elongation continues, this local 'bubble' of strand separation proceeds along the DNA in the direction of RNA synthesis, releasing the nascent RNA chain as it moves (Figure 5.4). Contrast this temporary strand separation with the permanent separation involved in DNA replication.

The error rate of copying into RNA is of the order of one 'wrong' nucleotide incorporated every 10^4–10^5 nucleotides, much higher than that of DNA replication, but adequate for transcription because the RNA copy is 'disposable' and the

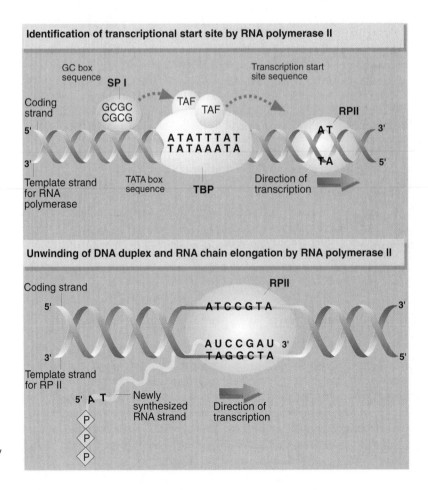

Figure 5.4 The initiation and directionality of transcription (see Box 5.1).

Box 5.1   Control of transcription in bacteria and humans

The control of transcription is one of the outstanding problems in biology. Until the 1980s, it was possible to study this problem only in bacteria, and many of the basic concepts were worked out in the field of repression and induction in bacteria and their viruses. More recently it has proved possible to tackle this problem in eukaryotes and it has emerged that the control mechanisms are quite different from those of bacteria – they are very complex and have by no means been fully elucidated.

Many genes in bacteria and animals are expressed constitutively

In any organism there are genes, sometimes called 'housekeeping genes', whose products are required to be made at all times. There is no need to switch the expression of these genes 'on' and 'off'; they are termed **constitutive** because their RNA transcripts are made continuously. If all genes were expressed in this way it would be very wasteful, as synthesis of RNA and proteins requires both energy and precursors. It thus makes sense for even simple cells to control the expression of their genes.

Control of gene expression is much more subtle than simple 'on/off' mechanisms because it must include parameters such as the level to which a particular gene is expressed and the time of expression. Control of gene expression in bacteria has been studied for over 30 years and is fairly well understood. In animals, detailed work has been carried out for about half that time and the situation is, as might be expected, much more complex. Dissection of the mechanisms of control of gene expression is one of the major problems remaining in molecular biology and has enormous implications for many aspects of human medicine.

Bacteria control expression of many of their genes by repression and induction

Bacteria contain a single type of RNA polymerase, which initiates RNA chains at promoter sites with the aid of an accessory protein (σ factor). The principal mode of control in bacteria is negative, that is, non-constitutive genes are normally 'switched off' or **repressed** and are 'switched on' or **induced** by an appropriate stimulus. The first system to be dissected was the *lac* operon (a group of related genes) of *Escherichia coli*, which enables the bacterium to adapt to growth on lactose by expressing a **lactose permease** and the enzyme β-**galactosidase**, which hydrolyses lactose. Expression of the operon is normally blocked by constitutive synthesis of the lac repressor protein, which stops RNA polymerase from initiating transcription (Box Figure 1). Other bacterial operons are more complex.

Animals control many of their genes in a positive fashion

The *lac* operon was well understood long before we had even the vaguest ideas of how control of gene expression in animal cells was achieved, but it was generally thought to be similar. Control in animals, however, turns out to be quite different. Negative controls *do* occur in animals and are important, but positive control is a widespread phenomenon. Many genes appear to be normally expressed at a low (or basal) level and activated by one or more factors (often called '*trans*-activators') – proteins that respond to various chemical signals and bind to the transcription complex to increase the level of transcription. At least two sets of proteins cooperate with RNA polymerases in animal cells – those that participate in basal transcription and those that activate or 'up-regulate' it.

Humans, in common with all eukaryotes, contain three RNA polymerases (see Table 5.2), each of which having its own set of transcription factors, for example, the basal transcription factors for RNA polymerase II, include TF II D, TF II A and B, TF II F and E (listed in the order in which they join the basal transcription complex in Box Figure 2). Another factor, TF II H, also participates in DNA repair, linking the two processes (see Box 6.2). TF II D is itself a complex of several proteins – the 'TATA-box binding protein' (TBP; a protein that is highly conserved in all eukaryotes) – and several TBP-associated factors (TAFs). Activator proteins up-regulate transcription by binding to the basal complex, either directly or via other factors.

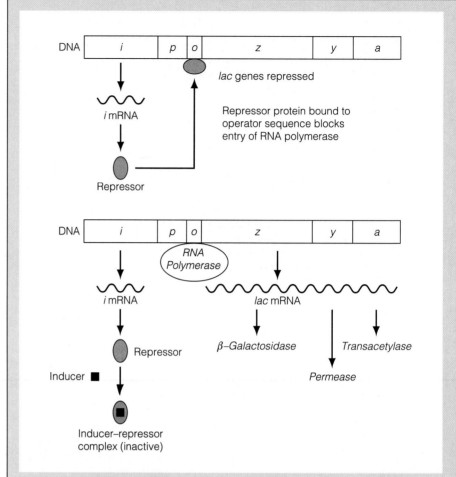

Box Figure 5.1.1 The *lac* operon of *E. coli*.

Control of gene expression involves both proteins and sequences of DNA

Besides the TATA box, which is present in most genes transcribed by RNA polymerase II, there are several other consensus sequences that lie 'upstream' from the transcription start site. Each of these is recognized by a different transcription factor. Many such sequences and factors have been identified – some of the most commonly found examples are shown in Box Table 1 (see also Figure 5.11). The upstream regions of genes contain different combinations of these consensus sequences which presumably gives each gene its appropriate level of transcriptional control. The transcription factors shown in the table are ubiquitous; others are enriched in certain tissues, for example the HNF-I (hepatocyte nuclear factor) and C/EBP (CAAT-box enhancer binding protein) families in the liver.

Box Table 1 DNA-binding transcription factors

Name of control element	Consensus sequence	Transcription factor
TATA box	TATAAAA	TBP (TF II D)
CAAT box	GGCCAATCT	CTF or NF I
GC box	GGGCGG	Sp-I
Octamer	ATTTGCAT	Oct-I
		Oct-2
ATF	GTGACGT	ATF or CREB
Hormone receptor	AGAACA	Steroid hormone receptors
elements	AGGTCA	Thyroid hormone receptors

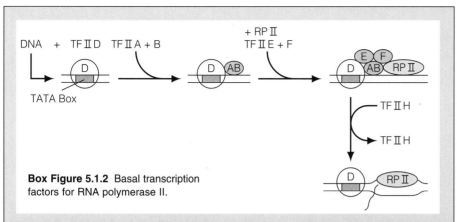

Box Figure 5.1.2 Basal transcription factors for RNA polymerase II.

Transcriptional control is important in the long-term regulation of glycolysis and gluconeogenesis

Glycolysis and gluconeogenesis are central to metabolism and are under both short- and long-term control (see Chapters 7, 8). Analysis of the genes encoding enzymes of these pathways has revealed the presence of a variety of response elements to hormones and other regulatory molecules.

- The gene for the glycolytic enzyme **glyceraldehyde phosphate dehydrogenase** contains two **insulin response elements** located between 250 and 500 nucleotides upstream from the transcription start site.

- The gene for the key gluconeogenic enzyme – **PEP carboxykinase** – has a complex promoter region that stretches at least 450 nucleotides upstream and contains elements responsive to a variety of regulatory molecules, including cyclic AMP, glucocorticoids, insulin and thyroid hormone, as well as consensus sequences for a variety of transcription factors.

Extending analysis of promoter regions to other regulated enzymes should eventually lead to a detailed molecular understanding of hormonal and other types of metabolic control at the level of gene expression (see Chapter 11) .

Gene expression in iron metabolism is also controlled post-transcriptionally

Although it is likely that most control of gene expression in humans occurs at the level of transcription, translational control also occurs and a physiologically significant example is known from iron metabolism. The mRNAs encoding three proteins (see below) important in iron metabolism contain stem-loop structures that act as **iron responsive elements** (IREs). When cells are depleted of iron a regulatory protein (IRE-binding protein – IRE-BP) binds to these mRNAs as follows (Box Figure 3):

- transferrin receptor mRNA has three 3'-IREs and binding of IRE-BP stimulates its translation, in turn stimulating iron uptake, while

- ALA synthetase mRNA and ferritin mRNAs have one 5'-IRE and binding of IRE-BP inhibits translation of these mRNAs, reducing iron use for haem synthesis and iron storage in ferritin.

Repletion with iron causes removal of IRE-BP from the IREs on these mRNAs and reverses the conditions pertaining during iron depletion.

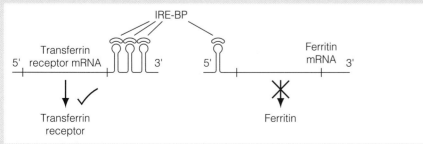

Box Figure 5.1.3 Post-transcriptional control of gene expression.

Chapter 7. Synthesis of new enzyme molecules provides a cell with increased catalytic capacity, in response to increased metabolic demand, page 195.
Chapter 8. Control of glycolysis, page 238.

Chapter 11. Tissues may exhibit both short and long-term responses to hormones, page 333.

Chapter 14. Iron, page 453.

Box 6.6. Nucleic acid and protein sequences can be used for 'molecular phylogeny', page 171.

Box 5.2 Study of ribosomal RNA sequences has changed ideas about the early evolution of life

For many years organisms were divided into the eukaryotes, or those with nucleated cells and the non-nucleated prokaryotes. The latter essentially comprised the bacteria, which were all regarded as rather similar in evolutionary terms. Eukaryotes were divided into kingdoms – plants, animals, fungi and protists – and these were thought to be ancient lineages.

Since the mid-1960s the concept that evolutionary information is recorded in the sequences of modern nucleic acids and proteins has steadily gained ground (see Chapter 6). Evolutionary relationships between organisms may be analysed in this way, but its usefulness in tracing historical evolution depends on rate of change – sequences that change rapidly soon become 'scrambled' and cannot be aligned. To delve back into the ancient history of life requires sequences that are easily obtained and slow to change. The ribosomal RNAs of the small ribosomal subunit of organisms have proved to be such sequences.

Analysis of the sequences of small-subunit rRNAs has yielded some unexpected findings (Box Figures 1 and 2), which can be summarized as follows.

- Eukaryotes and prokaryotes probably diverged about 2000 million years ago.

- A group of 'bacteria' (the Archaebacteria (now Archea)) resemble eukaryotes.

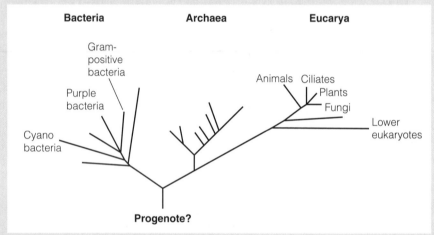

Box Figure 5.2.1 Molecular phylogenetic map of modern organisms based on comparison of ribosomal RNA sequences.

consequent degree of error in the synthesis of proteins is acceptable. The situation is very different in DNA replication, where the copy is permanent; steps are taken to ensure much higher fidelity of copying by DNA polymerase (see Chapter 6).

Chapter 6. Fidelity of replication is aided by proofreading by DNA polymerase, page 158.

Termination: the relationship between gene and transcript

Termination of transcription is associated with secondary structure ('hairpins') in the RNA product. Although transcription termination factors exist, especially in bacterial cells, precise control of termination is probably less important in eukaryotes than in prokaryotes, where transcription often continues well past the 'end' of the gene. Although this is apparently wasteful of energy, energy is probably not a major consideration because much of the RNA made is discarded as the transcript is processed into mature mRNA – for example, the cleavage which occurs near the polyadenylation signal (see below).

Any excess sequence at the 3'-end of mRNA molecules will simply be ignored; translation will terminate at the 'stop' codon. Eukaryotic mRNAs are

- The lineages of true bacteria (Eubacteria) probably diverged about 3000 million years ago.

- There are three major lineages (or 'domains') of organisms – Eubacteria, Archaea and Eucarya.

- The mitochondria and chloroplasts of the Eucarya are descended from eubacterial ancestors, probably purple bacteria and cyanobacteria, respectively.

- The major lineages of the Eucarya (plants, animals and fungi) diverged within the last 1000 million years.

- The unicellular Eucarya (protists) diverged much earlier from the rest of the Eucarya.

- The Archaea are an ancient and diverse lineage (as many as one-third of marine microorganisms may belong to this domain).

- The observations are consistent with the hypothesis that all life is descended from a single cellular ancestor – the so-called 'progenote' (other explanations are, however, possible).

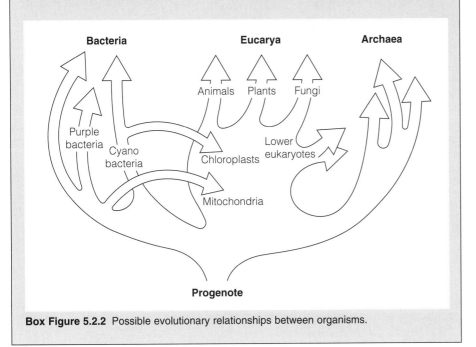

Box Figure 5.2.2 Possible evolutionary relationships between organisms.

monocistronic (or monogenic) – they produce only one polypeptide and translation cannot restart beyond a 'stop' codon (see below). Bacterial mRNAs are often polycistronic; translation can restart several times on a single polycistronic mRNA. This is probably one reason why control of termination of transcription is more important in bacteria than in eukaryotes.

Chapter 5. Initiation, page 134.

RNA processing: producing functional RNA molecules from transcripts

With the exception of the mRNAs of bacteria, all RNAs are initially produced as long unmodified transcripts (the precursors of the mature, functional RNA species) and contain only A, C, G and U nucleotides. The term 'RNA processing' covers the chemical modifications and cleavages performed on RNA by the range

Chapter 5. Pre-mRNAs undergo three processing events: capping, tailing and splicing, page 123.

of specific enzymes described below and leading to the production of mature, functional RNA species.

tRNAs and rRNAs are modified on nucleoside and sugar moieties

Conversion of pre-rRNA and pre-tRNA to their final, functional forms involves three types of covalent changes (Figure 5.5). Messenger RNAs are subject to other types of post-transcriptional modification (see below).

Nucleoside modification

All RNA transcripts contain only the four 'normal' bases – A, G, C and U – but mature, functional tRNA molecules contain a range of 'unusual' nucleoside components such as **inosine, pseudouridine** and **ribo-thymidine**, not found in other RNAs. These are modified post-transcriptionally from the adenosine, uridine and uridine nucleoside residues, respectively, by enzymes that act on pre-tRNA. The modified nucleotides in tRNAs influence various features, including

- the three-dimensional structures of the molecules;
- the function of the **anticodon loop;**
- recognition by aminoacyl-tRNA synthetases.

Ribose methylation

Some of the ribose residues in rRNA and tRNA molecules are also methylated on the 2'-O position; the function of this modification is not known, but as it renders the adjacent phosphodiester bond resistant to alkaline hydrolysis, it may be a way of making these RNAs more stable.

Cleavage of RNA transcripts

In both prokaryotes and eukaryotes, ribosomal and transfer RNAs are transcribed as long precursor RNA molecules (pre-rRNA and pre-tRNA) which are cleaved by specific endonucleases to their mature lengths (Figure 5.5). Production of

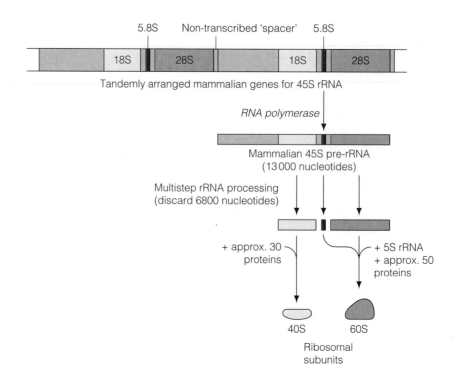

Figure 5.5 Processing of pre-rRNA.

several molecules from a longer precursor is one way of ensuring that the various rRNAs are produced in equimolar amounts before assembly into ribosomes along with the ribosomal proteins (see below).

Ribosomal RNA has been highly conserved during evolution

The sequences of rRNAs have been highly conserved during evolution and analysis of these sequences has allowed the construction of phylogenetic trees which have led to the proposal that there are three domains or major groupings of organisms – the **Bacteria** (often called the Prokaryotes), the **Archaea** (formerly known as the Archaebacteria) and the **Eucarya** (formerly referred to as the Eukaryotes) (see Box 5.2).

The **progenote hypothesis** says that all three domains are descended from the progenote, or first recognizable cellular organisms, the product of the various (unknown) steps in the origin of life (see Chapter 6).

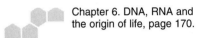

Chapter 6. DNA, RNA and the origin of life, page 170.

Pre-mRNAs undergo three processing events: capping, tailing and splicing

The primary transcripts that are the precursors of mRNA molecules (**pre-mRNAs**) are made by RNA polymerase II. Owing to their nuclear location and heterogeneous size (in contrast to the uniform size of other cellular RNAs), they were termed hnRNAs (*h*eterogeneous, *n*uclear RNAs) long before their role as mRNA precursors was established. Only after the existence of introns and the process of splicing (see Box 5.3) had been recognized was the pathway for conversion of hnRNA to mRNA deduced. True eukaryotic mRNAs had been isolated long before this by the expedient of isolating them from polysomes that were actively synthesizing protein. In 'pre-intron' times researchers were puzzled as to why molecules of mRNA could not be found in the nucleus, but we now understand that the answer lies in the processing steps that occur in the nucleus before the mature mRNA is transported out to the cytoplasm – **capping, tailing** and **splicing**.

5'-capping
Functional eukaryotic mRNAs have unusual nucleotide structures at their 5'-ends which are added to the initial transcript by the stepwise action of a number of transferase enzymes. To the 5'-triphosphate resulting from chain initiation is added a G nucleotide 'in reverse', this guanine is then N-methylated and then none, one or two 2'-O-methyl groups are then added to the first two nucleotides of the RNA (Figure 5.6) to yield the three different types of cap structures. The function of the cap is to ensure efficient initiation of translation (see below) and possibly also to protect the mRNA from premature degradation.

Chapter 5. Ribosomes are made of two subunits, each with its own RNA and protein molecules, page 127.

3'-polyadenylation (tailing)
As stated above, eukaryotic transcription often runs well past the 'end' of the gene. In one of the processing steps a special **RNA endonuclease** recognizes a specific **polyadenylation sequence** (typically AAUAAA) and cleaves the chain some distance beyond it. This new 3'-end is acted on by the enzyme **poly A polymerase**, which adds 200–300 AMP residues to form a 3'-poly A 'tail' (Figure 5.6). In contrast to all the polymerases so far mentioned, this enzyme is *not* template-directed; it can add *only* AMP residues. The 'tail' may have a function in the stability of mRNA but some mRNAs, for example histone mRNAs, do not have them. The 3'-poly A 'tails' of eukaryotic mRNAs act as very useful 'handles' for their isolation from cellular extracts – as in the production of cDNA 'libraries' (see Chapter 17).

Table 17.9. Some important techniques used in genetic engineering, page 540.

Box 12.1. von Willebrand factor –
molecular biology and blood clotting,
page 361.

Box 5.3   The discovery and significance of introns

- When prokaryotic mRNA is annealed to its own DNA, an exact double-stranded RNA–DNA hybrid is formed.
- The same experiment, carried out with eukaryotic mRNA extracted from actively translating ribosomes, showed that an RNA–DNA hybrid was formed, but it was not exact and contained lengths of DNA 'looped out' from the hybrid (see Figure).
- It was eventually deduced that the 'primary' transcript hnRNA (which *does* form an exact hybrid) had been shortened by a 'break-and-join' process called splicing so that mRNA is substantially shorter than the transcript.
- The sequences that were spliced out were called 'intervening sequences' – now usually referred to as introns. The 'coding' sections of genes are termed exons.
- The average length of DNA per gene is much higher in humans and higher vertebrates than in lower eukaryotes (Table 5.1), mainly due to the greater length of introns (1000–20 000 nucleotides) in the higher than the lower (60–1000 nucleotides) vertebrates.
- Many, but not all, eukaryotic genes contain introns – those that do can have up to 50 (see Box 12.1).
- It is not clear whether bacteria originally possessed introns and lost them (the 'introns-early' model) or whether eukaryotes have subsequently acquired them (the 'introns-late' model).

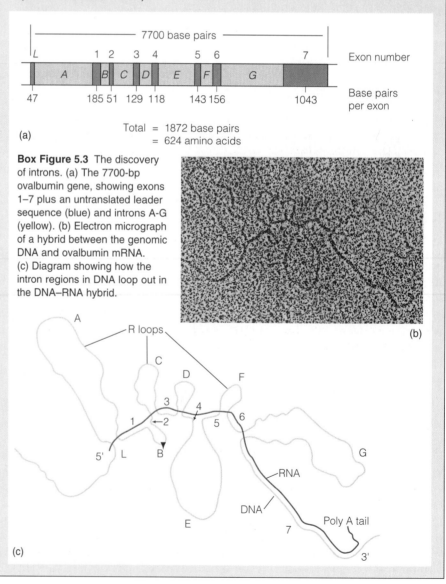

Box Figure 5.3 The discovery of introns. (a) The 7700-bp ovalbumin gene, showing exons 1–7 plus an untranslated leader sequence (blue) and introns A-G (yellow). (b) Electron micrograph of a hybrid between the genomic DNA and ovalbumin mRNA. (c) Diagram showing how the intron regions in DNA loop out in the DNA–RNA hybrid.

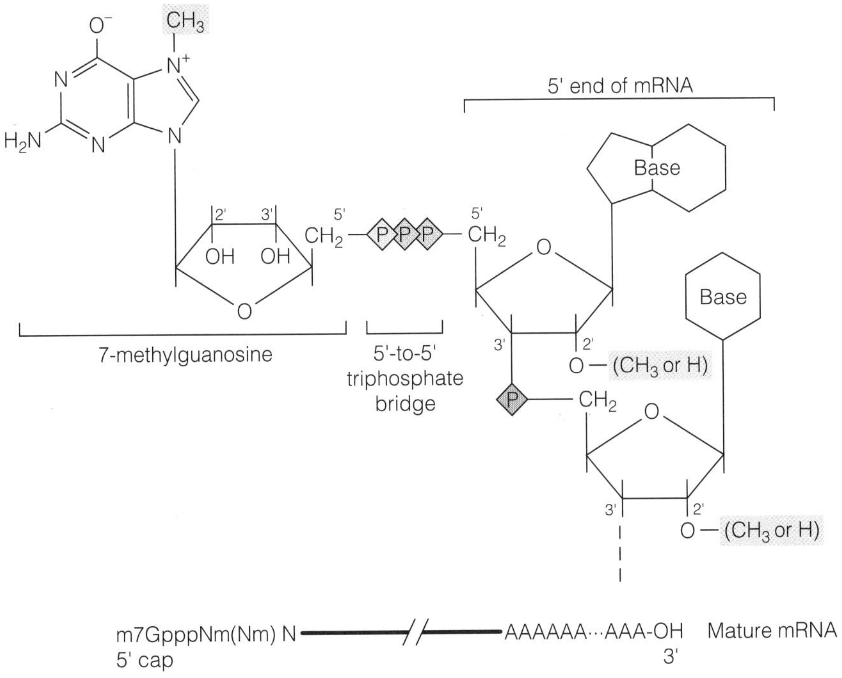

m7GpppNm(Nm) N ——————//——————AAAAAA···AAA-OH   Mature mRNA
5' cap   3'

Figure 5.6 Capping and tailing of pre-mRNAs.

Splicing of eukaryotic messenger RNAs

Eukaryotic mRNA precursors (sometimes called hnRNA) are also shortened, but in a highly distinctive way, involving multiple specific deletions along the length of the molecule. The discovery of this phenomenon was a major conceptual change in biology with many implications (Box 5.3). Splicing occurs at specific sequences or 'splice sites' on modified pre-mRNA and involves a splicing enzyme complex ('spliceosome') with its small nuclear RNAs (snRNAs) (Figure 5.7). The sections of DNA whose RNA copies remain in the mature mRNA are called **exons**; those sections whose RNA copies are 'spliced out' are termed **introns**.

Some RNAs can be spliced in more than one way – alternative splicing

As a first approximation, we can think in terms of simple equivalence of gene/pre-mRNA/mRNA/polypeptide. For many genes and their encoded polypeptides this is probably true. However, a number of situations occur in which more than one polypeptide can be made from one gene – one way is through the splicing process. If appropriate splice sites are present in a pre-mRNA, it may be possible to carry out splicing in more than one way, thus generating two or more mature mRNAs from a single pre-mRNA. These can be translated to yield two different, but related, polypeptides. This occurs in the production of the different G-proteins (see Chapter 3).

Chapter 3. Table 3.7 G-protein-linked receptors, page 71.

RNAs must be transported to the cytoplasm

While all RNAs are synthesized and processed in the nucleus, they function in the cytoplasm, either free in the cytosol or after association with the endoplasmic reticulum. There must, therefore, be mechanisms for the transport of mature mRNA and tRNA molecules across the nuclear envelope; this is achieved by the **nuclear pore** complexes (see Chapter 3). Immature ribosomal subunits, assembled from processed rRNA and ribosomal proteins imported from the cytoplasm, also pass through the nuclear pore complexes and mature in the cytoplasm. Recently,

Chapter 3. The nucleus houses the genetic material and is the 'operational centre' of the cell, page 44.

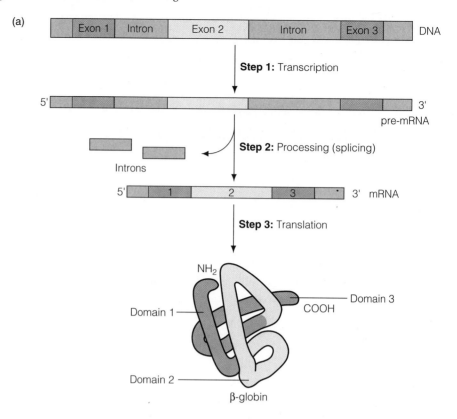

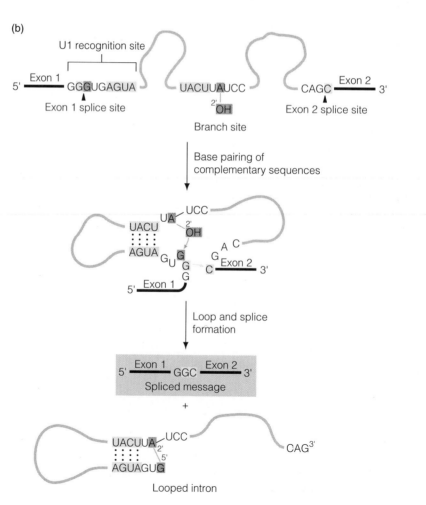

Figure 5.7 The splicing of pre-mRNA
(a) Splicing in the synthesis of β-globin;
(b) Events at the 'spliceosome' during
splicing.

indications have been found of a 'tracking' mechanism, which appears to direct RNAs from the site of transcription to the nuclear pore and beyond. It is probable that the processing of mRNA is physically related to movement along the 'track'.

mRNA translation: the machinery of protein synthesis

The machinery of protein synthesis involves the following principal components:

1. Ribosomes
   - Large ribosomal subunits
   - Small ribosomal subunits
2. Accessory protein factors acting in translation
   - Initiation factors
   - Elongation factors
   - Termination factors
3. Activated amino acids for use in translation
   - A set of tRNA species
   - A set of aminoacyl-tRNA synthetases
   - A set of aminoacyl-tRNAs

(at least one for each of the 20 types of amino acids found in proteins)

Ribosomes are made of two subunits, each with its own RNA and protein molecules

Ribosomes are complex assemblies of RNA and protein molecules. A functional eukaryotic ribosome (80S) is composed of one large (60S) and one small (40S) subunit, each of which comprises a distinctive set of RNA and protein molecules (Figure 5.8). The synthesis in the nucleolus of the various rRNA species by transcription from ribosomal RNA genes by RNA polymerase I, and the subsequent processing of the transcripts, is described above. Ribosomal proteins are translated on cytoplasmic ribosomes and transported to the nucleolus for assembly. At cell division, each daughter cell must receive its share of the pool of ribosomes from the parent, as failure to do so would mean that the deprived daughter would be unable to synthesize proteins and would not be viable.

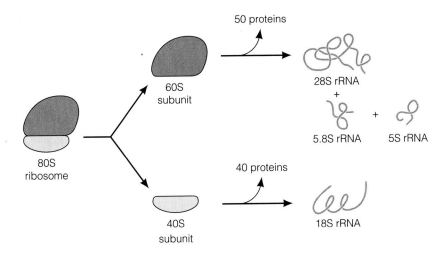

Figure 5.8 The composition of ribosomes.

Chapter 5. Initiation, page 134.

Chapter 5. Ribosomes are directed to the endoplasmic reticulum by signal recognition particles, page 138.

Chapter 18. Protein synthesis, page 559.

Box 7.2. The evolutionary origin of mitochondria, page 200.

Chapter 5. Ribosomes are reused many times within the cell, page 134.

Box 7.2. The evolutionary origin of mitochondria, page 200.

Complete ribosomes are assembled only in the presence of an mRNA molecule during protein synthesis and their component subunits are drawn from pools of large and small subunits in the cytosol (see below). Several ribosomes, typically five or six, may be attached to, and actively translate, a single mRNA molecule. These structures are called polysomes and can be seen under the electron microscope (Figure 3.10). Ribosomes that are actively synthesizing integral membrane proteins or precursors of proteins to be secreted from the cell attach to the endoplasmic reticulum and extrude the newly synthesized protein into the lumen (see below) .

The ribosomes of bacteria are similar in general construction to those of the eukaryotes, but are smaller (70S, with 50S and 30S subunits) and somewhat different in their detailed properties, for example in their initiation of translation and in their sensitivity to certain antibiotics that inhibit protein synthesis (see Chapter 18). Ribosomes similar to the 70S-type are also found in mitochondria and in the chloroplasts of plants – one of a number of features suggesting a prokaryotic origin for these organelles in the evolution of the Eucarya (see Chapter 7).

Specific protein factors are involved in initiation, elongation and termination of translation

Although the ribosome forms the essential framework for the translational mechanism, it is not sufficient in itself for the complete process of protein synthesis. A number of specific protein 'factors' are required for each of the processes of initiation, elongation and termination. These proteins are not permanently associated with ribosomes, but bind to the ribosome at appropriate points in the '**ribosome cycle**' (see below) and dissociate when their task is done (Table 5.3).

Table 5.3 Accessory proteins in translation

Each of the three steps in translation on the ribosome (initiation, elongation and termination) requires its own set of factors or non-ribosomal proteins that associate with the ribosome only for the duration of their particular task. There are similar but distinct sets of factors in bacterial and animal cells, some of which are targets for antibiotics. Mitochondrial protein synthesis resembles that in bacteria (see Box 7.2). Some important eukaryotic factors are listed below; the list is not exhaustive and additional factors are required.

Initiation factors	
met-tRNA$_i$^{met}	Acts as initiator tRNA (met-tRNA$_{met}$ is used for elongation)
eIF-2	Forms ternary complex with met-tRNA$_i$^{met} and GTP
CBP	(Cap binding protein) binds 5'-cap on mRNA
eIF-4A/4B	Remove secondary structure from 5'-end of mRNA, allowing entry of 40S subunit
eIF-3	Binds 40S subunit/ternary complex to 5'-end of mRNA (40S/eIF2/eIF3 complex moves along mRNA from 5'-end to starting AUG codon)
eIF-5	Binds 60S subunit at start codon, releasing eIF-2 and eIF-3
Elongation factors	(both are major cellular proteins)
eEF-1	Brings AA-tRNA to ribosome, with expenditure of GTP (EF-Tu is the equivalent protein in bacteria – sensitive to kirromycin)
eEF-2	GTP-dependent translocation of the ribosome along mRNA; eEF-2 is modified and inactivated by diphtheria toxin (EF-G is the equivalent protein equivalent in bacteria – sensitive to fusidic acid) (only one is required)
Termination factor	
eRF	GTP-dependent release of completed polypeptide

eIF = eukaryotic Initiation Factor; eEF = eukaryotic Elongation Factor; eRF = eukaryotic Release Factor

---

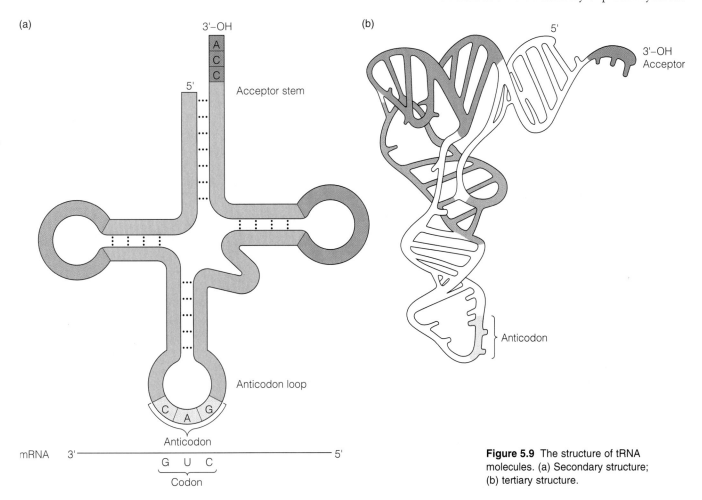

(a)

3'–OH

Acceptor stem

5'

Anticodon loop

C   A   G

Anticodon

mRNA   3'——————————————————5'

G   U   C

Codon

(b)

5'

3'–OH
Acceptor

Anticodon

Figure 5.9 The structure of tRNA molecules. (a) Secondary structure; (b) tertiary structure.

Transfer RNAs have a distinctive three-dimensional structure

Transfer RNA molecules have distinctive sequences which lead to intramolecular base pairing and the formation of 'stems' and 'loops' (secondary structure) and folding of the '**clover leaf**' into a more compact three-dimensional shape (tertiary structure) (Figure 5.9). Perhaps the most crucial loop is the **anticodon** loop, which carries the three nucleotides that base-pair with the complementary triplet codon on mRNA during translation. All tRNAs have the sequence CCA at their 3'-terminus, to which an amino acid is covalently joined by an ester linkage by aminoacyl tRNA synthetases. Enzymes in the cell are able to restore this CCA end by adding the nucleotides in a nontemplate-dependent fashion.

The double-stranded stems of tRNA have some similarities to the structure of DNA (see Chapter 6) but some different types of base pairing are involved, and the overall structure of tRNA is somewhat analogous to the tertiary structure of proteins. Although they differ in detailed structure, all tRNAs have the same general conformation. This is probably because they are all relatively small molecules subject to the constraints of having to interact with a variety of other elements involved in translation – the codons on mRNA, the large ribosomal subunit and their cognate aminoacyl-tRNA synthetase.

Cells contain over 40 species of tRNA

There are 61 codons which specify amino acids during translation (see below). Even allowing for the 'wobble' phenomenon during the codon–anticodon interaction (see below), over 40 species of tRNA molecule are required to carry out

Chapter 6. DNA has two polynucleotide chains in a double helix, page 147.

Chapter 5. The genetic code is the set of rules that governs translation, page 131.
Chapter 6. Silent mutations, page 167.

Chapter 4. Some RNA molecules have catalytic properties, page 89.
Box 6.5. Was RNA the first genetic material in the origin of life?, page 170.

translation by interacting with these codons. As many as four different tRNAs may be specific for a particular amino acid, the more frequently occurring amino acids generally having more tRNAs. Many of the genes for tRNA occur as local clusters of several pre-tRNAs. The precursor transcripts of tRNAs are made by RNA polymerase III, which is also responsible for the synthesis of other small RNAs. Pre-tRNAs and certain other small RNAs have been shown to possess self-cleaving catalytic activity, sometimes called 'ribozymes' (see Chapter 4). This RNA-based catalysis has given rise to the idea of an 'RNA world' which may have preceded DNA and perhaps even protein (see Chapter 6).

tRNAs are 'loaded' with amino acids by specific aminoacyl-tRNA synthetases

In the process of translation, the substrate or 'raw material' for protein synthesis is not a free amino acid, but one which has been covalently joined by an ester linkage through its carboxyl group to the 3'-hydroxyl terminus of a tRNA molecule. For each of the 20 different amino acids used in protein synthesis, one or more species of tRNA molecule is specific (or cognate) for that amino acid (for example $tRNA_{ala}$) and an enzyme, one of the **aminoacyl tRNA synthetases** (such as alanyl tRNA synthetase), which recognizes both the amino acid and its cognate tRNA and catalyses the following reaction (using the examples just given):

$$Alanine + tRNA_{ala} + ATP \rightarrow Alanyl\text{-}tRNA_{ala} + AMP + PPi$$

Each amino acid must be converted into an aminoacyl-tRNA (aa-tRNA) before it can participate in protein synthesis. After it has given up its amino acid for protein synthesis, tRNA can be 'reloaded' and reused many times.

The roles of the aminoacyl-tRNA synthetases are

- To ensure that the correct amino acid is 'loaded' on to its cognate tRNA. This is the important step in ensuring accuracy in protein synthesis; if the 'wrong' amino acid has been coupled with a tRNA, it will be used erroneously in peptide bond formation. Some of these enzymes have a proofreading function and can remove a 'wrong' amino acid and try again.
- To activate the carboxyl group for peptide bond formation. Peptide bond formation is catalysed by the peptidyl transferase activity of the large ribosomal subunit and results from the attack of the amino group of the 'incoming' aminoacyl tRNA on the ester bond between the growing peptide chain and the tRNA to which it is attached (see below).

Chapter 5. Elongation, page 135.

The production of the three RNA components of the translational process: the rRNAs in the ribosomes (the 'assembly plant'), mRNA (the 'instruction tape') and the aminoacyl-tRNAs (the 'raw material') have now been discussed. The next section considers the way in which these act in concert to bring about the synthesis of proteins.

mRNA translation: the genetic code and open reading frames

Translation occurs using the sequence of a specific mRNA molecule, in the environment of the ribosome, to direct the sequential selection of aminoacyl-tRNAs leading to the formation of peptide bonds in the correct polypeptide sequence. The 'rules' for this selection are summarized in 'the genetic code' and are based on the hydrogen-bonded pairing of trinucleotide codons on mRNA with corresponding anticodons on tRNA.

The genetic code is the set of rules that governs translation

The fact that there are 20 different amino acids in proteins but only four types of nucleotide in RNA indicates that at least three nucleotides must 'code' for each amino acid. A 'dinucleotide' code would give only 4^2 (= 16) codons; a trinucleotide gives 4^3 (= 64) codons. As the 'genetic code' has no punctuation between the codons on mRNA and does not overlap, there is a surplus of codons. This surplus results in the code being 'degenerate': there are several codons for amino acids (four for most, and as many as six for leucine, serine and arginine). The amino acids methionine and tryptophan, both of which occur at low frequencies in proteins, usually have only a single codon (Figure 5.10).

What is a codon?

A codon is a trinucleotide sequence in an mRNA, read 5' to 3', which base pairs with nucleotides in the anticodon loop of a corresponding tRNA molecule. In the most commonly used representation of the genetic code (Figure 5.10) the first position of a codon is listed UCAG down the left side, the second position UCAG along the top and the third position UCAG down each of the 16 boxes corresponding to the dinucleotides of the first two positions.

The genetic code is (nearly) universal

Although the genetic code is sometimes described as 'universal' because it applies in all organisms, some minor variations occur in mitochondrial protein synthesis (see Chapter 17) and some other situations: for example, in some organisms UGA acts not as a stop but as a second codon for tryptophan.

Chapter 17. Mitochondria use a modified form of the genetic code, page 521.

One important implication of the near-universality of the genetic code is that all organisms are descended from a common ancestor, because it seems unlikely that such a complex set of rules could have evolved more than once in the same manner.

Second position

First position (5' end)	U	C	A	G	Third position (3' end)
U	Phe	Ser	Tyr	Cys	U
U	Phe	Ser	Tyr	Cys	C
U	Leu	Ser	Stop	Stop	A
U	Leu	Ser	Stop	Trp	G
C	Leu	Pro	His	Arg	U
C	Leu	Pro	His	Arg	C
C	Leu	Pro	Gln	Arg	A
C	Leu	Pro	Gln	Arg	G
A	Ile	Thr	Asn	Ser	U
A	Ile	Thr	Asn	Ser	C
A	Ile	Thr	Lys	Arg	A
A	Met/Start	Thr	Lys	Arg	G
G	Val	Ala	Asp	Gly	U
G	Val	Ala	Asp	Gly	C
G	Val	Ala	Glu	Gly	A
G	Val	Ala	Glu	Gly	G

Figure 5.10 The genetic code. Codons are shown as bases in mRNA (written in a 5'→ 3' direction).

'Start' and 'stop' codons are essential for translation

In addition to the codons determining the selection of amino acids, there is a small but important group of codons that mean start (AUG) or stop (UAA, UAG and UGA). These codons play important roles in the synthesis of polypeptides and they define **open reading frames**, or the polypeptide-defining portions of mRNA sequences.

A functional mRNA must contain an open reading frame

The translation mechanisms of cells can correctly identify and translate the sequence of a protein. With the relative ease of DNA sequencing, the DNA sequence of many thousands of genes is now known, even though the corresponding protein may not

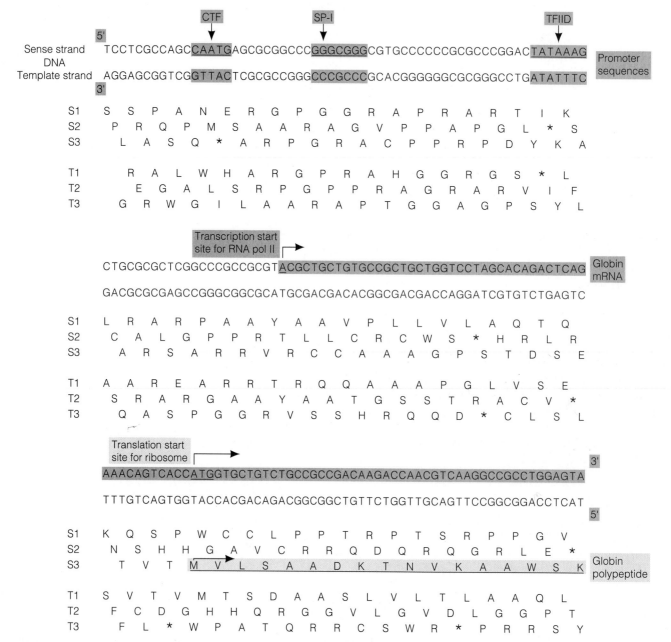

Figure 5.11 Start of a gene (α-globin), showing the selection of one of the six possible reading frames. (TF, SP-1 and TFIID mark the binding sites for these transcription-related protein (see Box 5.1).

have been identified, let alone isolated. It is instructive to consider how a scientist can consciously do what the cell does unthinkingly – namely, identify the sequence of a gene in the midst of many thousands of nucleotides.

Such a search would be very laborious to do manually, but computer programs have been developed to do the searching. What, then, is the computer instructed to look for?

There are three potential reading frames in an RNA sequence

Because of the triplet nature of the code a sequence of RNA has three potential 'reading frames', each of which would give rise to a completely different polypeptide. There are two strands of DNA, so any sequence of DNA has six potential reading frames (Figure 5.11). In most cases only one of these reading frames is used and is selected by an AUG start codon which 'locks' the ribosome on to that frame and causes it to ignore the other two. Any 'frameshift' during translation will completely change the sequence of the remainder of the synthesized protein (see Chapter 6).

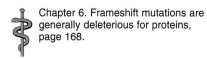

Chapter 6. Frameshift mutations are generally deleterious for proteins, page 168.

A start codon selects the reading frame

Many AUG trinucleotides (ATG in the 'sense' strand of DNA) exist in any substantial length of nucleic acid sequence. These fall into three categories:

1. Most mean 'methionine' in the interior of a protein sequence.
2. Some are followed by an in-frame 'stop' codon within a small number of nucleotides and do not lead to protein synthesis. The function of these is not known.
3. Others, located near the 5'-end of an mRNA molecule, mean 'start a new polypeptide chain with a methionine' and are necessary for chain initiation (see below).

These last are 'start' codons, which are usually surrounded by a characteristic 'consensus sequence' that favours the initiation process. Different sequences within this consensus give different 'strengths' of 'start' signal, for example the sequence ANNAUGG in mRNA acts as a strong start in eukaryotic translation.

The stop codon must be 'in frame' with the start codon

As the ribosome moves along the mRNA, elongating the nascent polypeptide chain according to one of the three possible reading frames, it eventually encounters a 'stop' codon 'in frame' with the initial AUG. It may well have passed several other 'stops' in the two other reading frames, but will not have recognized these. There are no tRNAs with anticodons for 'stop' codons and chain termination occurs followed by dissociation of the two subunits of the ribosome (see below).

The correct combination of the following three elements constitutes an open reading frame (ORF) (Figure 5.12).

Chapter 5. Initiation, page 134.

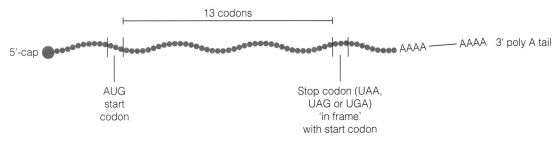

Figure 5.12 An open reading frame in an mRNA molecule.

1. A **start codon** in a suitable consensus lying near the 5'-end of an mRNA.

2. A length of reading frame with a number of codons corresponding to the number of peptide bonds in the encoded protein.

3. A **stop codon** which is in frame with the start codon and lies towards the 3'-end of the mRNA.

Analysis of a sequence of DNA may reveal a large number of potential open reading frames and other information is required to confirm the identity of a hypothetical ORF. This can come from other sequence information relating to transcription and processing, such as the presence of promoter sequences upstream from a start codon and polyadenylation signals downstream from a stop codon (see Figure 5.4 and Box 5.1). Direct evidence for the existence of a functional ORF comes from experiments demonstrating the presence of functional mRNAs and/or polypeptides.

mRNA translation: initiation, elongation and termination of polypeptide chains

Ribosomes are reused many times within the cell

As described above, the ribosome is only fully assembled when functioning in protein synthesis – at other times the subunits are separate entities. Thus, the 'life' of a ribosomal subunit consists of three stages repeated many times (the **ribosome cycle**; Figure 5.13), with each stage consisting of one of the three processes making up the whole of translation and involving its own set of protein accessory factors (see Table 5.4).

1. Assembly into a functional mRNA–ribosome complex (**initiation**).

2. Translation of the ORF of the mRNA (**elongation**).

3. Dissociation into subunits and release of polypeptide (**termination**).

Initiation

The sequence of events leading to successful polypeptide chain initiation on a 'capped' eukaryotic mRNA involves the following sequence of events (Figure 5.13).

1. Binding of a small ribosomal subunit to the 5'-cap to form an initiation complex.

2. ATP-dependent 'scanning' of this complex to the first AUG start codon.

3. Binding of a large ribosomal subunit.

Various initiation factors (IF) participate in these steps and some of these are shown in Table 5.3.

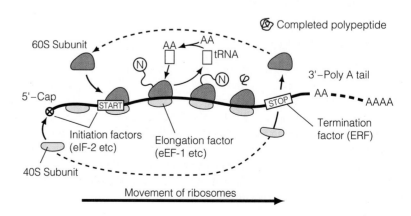

Figure 5.13 The ribosome cycle.

Elongation

This process must occur once for each peptide bond in the completed polypeptide and is shown in Figure 5.14. The peptidyl transferase activity that catalyses the formation of the peptide bond is associated with the large subunit and depends on the participation of both protein and RNA. A major energetic driving force for chain elongation is the hydrolysis of the terminal phosphate of GTP during ribosomal translocation. Various elongation factors (EF) participate in these steps (Table 5.3).

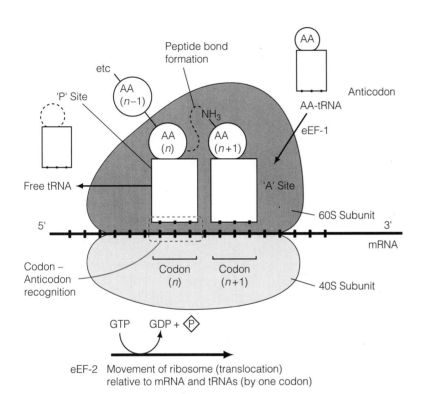

Figure 5.14 Peptide bond synthesis.

Termination

When the stop codon of the ORF of an mRNA enters the acceptor site of the ribosome, there is no tRNA to respond to this situation and this sets in train the events leading to termination of protein synthesis, including release of the completed polypeptide and dissociation of the ribosome–mRNA complex (Figure 5.13) by a termination factor (Table 5.3).

Newly synthesized polypeptides must fold into native proteins

Proteins must fold into their functional, native three-dimensional structure (see Chapter 4). This process probably begins while the nascent polypeptide chain is still being synthesized – the new protein folds up from its *N*-terminus (see Figure 5.13). This process may occur spontaneously (renaturation of denatured proteins can be demonstrated in the laboratory), but it is probably assisted in the cell, for example by the enzyme prolyl isomerase. This enzyme catalyses the isomerization of proline residues in polypeptides from *cis* to *trans* so as to facilitate attainment of the native state, or certain heat shock proteins which act as **molecular chaperones** (see below).

Chapter 4. The three-dimensional structure of proteins: the importance of sequence, page 98.

Chapter 5. Heat-shock proteins are also involved in protein targeting, page 142.

The details of protein synthesis differ in animal and bacterial cells

There are significant differences in protein synthesis between bacterial and animal cells, probably due, at least in part to the smaller bacterial ribosome (70S compared with the animal 80S). Bacterial protein synthesis is also capable of internal initiation on an mRNA molecule but most animal protein synthesis depends on the presence of the 5'-cap. Some of these differences underlie the selective toxicity of several clinically useful antibiotics (see Chapter 18).

Viruses exploit the cell's ability to make proteins

All viruses need to make proteins during their replicative cycle (see Chapter 18) and, because they lack the machinery for protein synthesis, use the host cell's mechanism. Viral mRNAs are translated by the cell's protein-synthesizing machinery to yield the viral proteins necessary for production of new virus particles. Some viruses – for example poliovirus and hepatitis C virus (see Chapter 18) – have been shown to translate RNA in a cap-independent fashion; this also appears to occur in uninfected cells with some cellular mRNAs. Cap-independent initiation depends on stem-loops and other structural features near the 5'-end of the mRNA.

Box 18.3. The modes of action of some antibiotics that inhibit bacterial protein synthesis, page 558.

Chapter 18. Viruses need a number of host cell functions, page 565.

Box 18.5. Classification and general features of human viruses, page 568.

Targeting of proteins

Cells need to target newly synthesized proteins to their correct destinations

The division of eukaryotic cells into separate compartments poses a problem because of the cytoplasmic location of the protein-synthesizing machinery. The several thousand individual protein species made in each cell of the body are synthesized on cytoplasmic ribosomes. Noncytosolic proteins must then be transported to the various organelles of the cell or, in the case of secretory proteins, to extracellular destinations such as the bloodstream. To solve this problem, eukaryotes possess mechanisms for transporting proteins to distant parts of the cell, a journey which involves the transfer of polypeptides across one or more internal membranes. The processes for the delivery of polypeptides to the correct final intra- or extracellular destinations are referred to as **protein targeting**.

Targeting of proteins depends on the presence of signal sequences in polypeptides

At a molecular level, targeting 'signals' are present in newly forming polypeptides, typically as short sequences located near the amino terminus. These molecular 'addresses' have usually been removed by the time the protein reaches its destination. If proteins are 'addressed' incorrectly, misdelivery occurs; a number of inherited genetic disorders are now known to result from targeting of proteins to the wrong location (see Box 5.4).

There are two sites of protein synthesis within cells: the cytosol and the rough endoplasmic reticulum

Protein synthesis occurs on free polysomes in the cytosol and on the membrane-bound ribosomes of the rough endoplasmic reticulum (RER). All ribosomes initiate synthesis of protein in the cytoplasm and either remain there or are directed to

Box 5.4   Inherited disorders involving targeting defects

Lysosomal storage diseases are conditions where harmful degradative enzymes, normally stored in the lysosome, are erroneously secreted into the bloodstream. This results in the accumulation of undegraded products in the lysosomal compartment, for example, glycosaminoglycans or glycolipids in the case of **I-cell disease**, so named because of the appearance of large inclusion bodies in this organelle.

In some patients with **primary hyperoxaluria**, characterized by an inability to metabolize oxalate, there is a targeting defect in which the peroxisomal enzyme L-alanine: glyoxylate aminotransferase aberrantly assumes a mitochondrial location. The molecular basis for this phenomenon appears to be a point mutation in the 5'-region of the gene that contains a cryptic mitochondrial targeting signal. This mutation results in the synthesis of an altered form of the enzyme, in which the *N*-terminal region exhibits a mitochondrial targeting signal.

the RER. From each of these locations, proteins are sent to the various parts of the cell. A scheme illustrating the distribution of specific classes of intracellular proteins is shown in Figure 5.15.

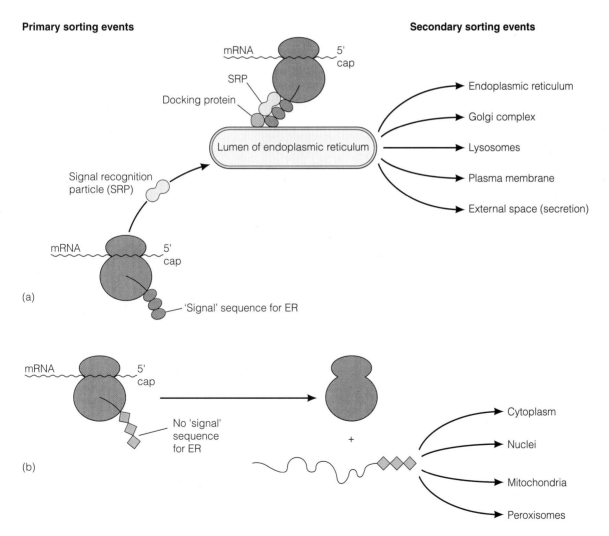

Figure 5.15 General scheme for the intra- and extracellular targeting of proteins in a mammalian cell. (a) Via the RER; (b) from cytosolic ribosomes.

Ribosomes are directed to the endoplasmic reticulum by signal recognition particles

The primary sorting of actively synthesizing ribosomes takes place during translation where newly forming polypeptide chains containing the appropriate *N*-terminal signal sequences are rapidly transferred to the membrane of the ER (Box 5.5). This requires a number of components: a ribonucleoprotein complex, or **signal recognition particle** (SRP; Figure 5.16); a membrane-bound docking protein; and ribosomal anchoring proteins. Proteins synthesized by this route are destined for final locations in the ER, the Golgi complex, the lysosomes, the plasma membrane or for secretion into extracellular space.

As translation proceeds, the protein is either inserted into or translocated across the ER membrane, where secondary sorting signals in the polypeptide chain are responsible for directing it to particular intracellular sites during their passage though the series of membrane-bound compartments (see Figure 5.15) leading from the lumen of the rough ER to the Golgi complex and eventually to the plasma membrane (see Chapter 3). The most basic form of these signals is seen in secreted proteins, which contain only the basic information in the form of a 15–30 amino acid 'signal', usually located at the *N*-terminus. This effects their entry into the secretory pathway by complete translocation into the lumen of the ER. Nonsecreted proteins destined for the various membrane locations *en route* to the plasma membrane are diverted at various stages from the main secretory pathway by the presence of additional structural features on the polypeptides. The secondary structure of the secreted polypeptide is important in determining the functional efficiency of signal sequences in promoting their interaction with SRPs.

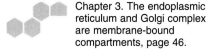

Chapter 3. The endoplasmic reticulum and Golgi complex are membrane-bound compartments, page 46.

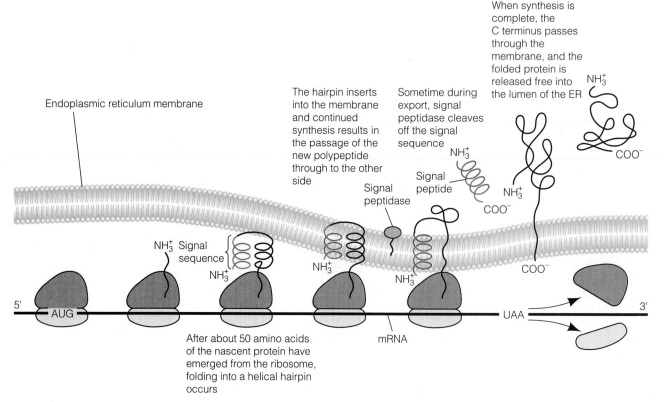

Figure 5.16 Translation accompanied by protein export through membranes. Polyribosomes synthesizing a protein destined for extracellular transport become attached to the cell membrane through *N*-terminal sequences. The completed polypeptide is then passed through the membrane.

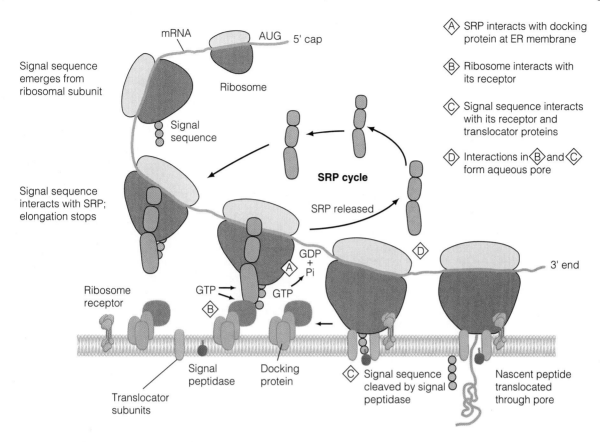

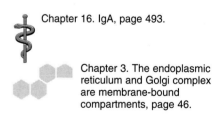

Figure 5.17 Signal recognition particles in the direction of ribosomes to the rough endoplasmic reticulum.

Proteins are retained in membranes by 'stop-transfer' sequences

Ensuring that a nascent polypeptide reaches and remains in its target membrane requires the presence of a hydrophobic 'stop-transfer' sequence of about 20 amino acids within the polypeptide. This arrests its passage as it crosses the ER membrane, leading to the eventual incorporation of such polypeptides into the ER, Golgi, lysosomal or plasma membranes, all of which are derived from the ER, the site of *de novo* membrane synthesis. This process is especially well illustrated in the case of the plasma membrane-bound and secreted forms of immunoglobulin heavy chain which differ only in a short hydrophobic sequence at their *C*-termini (see Chapter 16). Sorting the proteins to lysosomes depends on the presence of a specific mannose-6-phosphate residue on the *N*-linked oligosaccharide chains of these glycoproteins. Addition of the sugar residues that completes glycosylation of the protein occurs in the cisternal compartment of the Golgi complex (see Chapter 3).

Chapter 16. IgA, page 493.

Chapter 3. The endoplasmic reticulum and Golgi complex are membrane-bound compartments, page 46.

Some proteins synthesized on cytoplasmic ribosomes are also targeted

Although many of the proteins produced on cytoplasmic ribosomes remain free in the cytosol, the second major class of targeted polypeptides is produced on these ribosomes before being incorporated into nuclei, mitochondria or peroxisomes (Figure 5.18). The targeting mechanisms are primarily post-translational in nature.

Many mitochondrial proteins are made on cytoplasmic ribosomes

In most cases, proteins are specifically targeted to mitochondrial membranes by an extra sequence of amino acids, often located at the *N*-terminus of the cytoplasmic

Chapter 2. Cholesterol has an essential role in the plasma membrane of cells, page 28.

Box 5.5   The absence of cholesterol makes the endoplasmic reticulum membrane more permeable than the plasma membrane

The plasma membrane contains a high concentration of cholesterol – approximately equimolar with the total phospholipid and sphingolipid (see Chapter 2). The presence of cholesterol makes methylene groups 2–10 of the fatty acyl chains pack more tightly, thus reducing bilayer permeability. It also makes them more perpendicular to the plane of the bilayer, increasing its thickness by about 20%. This makes the plasma membrane thicker than inner membranes of the cell (70 nm, as opposed to 55 nm), such as those of the ER, which contain little or no cholesterol.

The extra thickness and rigidity caused by the cholesterol is thought to have the important function of limiting the permeability of the plasma membrane to small molecules. This ensures that they are retained within the cell, unless actively transported. It is clearly less important for the ER membrane. The ER membrane does, however, have to be capable of accepting the insertion of a polypeptide chain after the docking of the SRP and its nascent polypeptide. It is thought that the absence of cholesterol is crucial for this process, because it would not be possible to insert a polypeptide through a more rigid membrane such as the plasma membrane. In the absence of cholesterol, the fatty acid chains in the hydrophobic interior of the bilayer are sufficiently mobile to allow passage of the polypeptide chain.

Integral membrane proteins generally span the membrane with one or more lengths of α-helix, and the lengths of these helices differ in the proteins of different cell membranes. Membrane-spanning regions of the proteins of the plasma membrane generally contain 22–23 amino acids. In contrast, proteins remaining in the ER have shorter spanning regions of about 17 amino acids, presumably reflecting the thinner nature of the ER membrane and assisting in their retention within the ER.

These considerations are important when considering how the cell is able to regulate its content and distribution of cholesterol (see below)

Chapter 5. Gene expression in action: maintaining the level of cholesterol in cells, page 142.

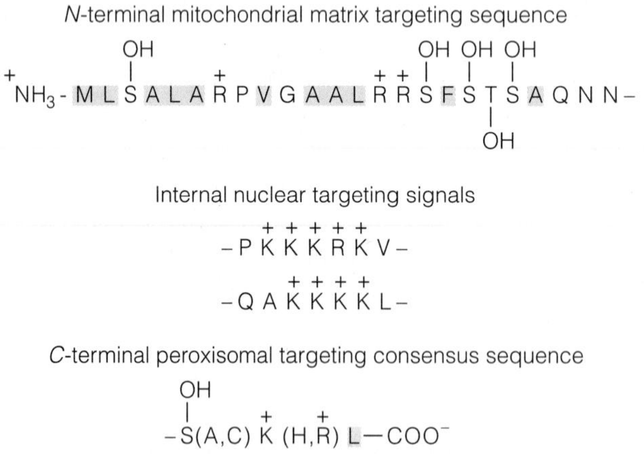

N-terminal mitochondrial matrix targeting sequence

$$^+NH_3\text{-}M\,L\,S\,A\,L\,A\,R\,P\,V\,G\,A\,A\,L\,R\,R\,S\,F\,S\,T\,S\,A\,Q\,N\,N\text{-}$$

Internal nuclear targeting signals

$$-P\,K\,K\,K\,R\,K\,V-$$

$$-Q\,A\,K\,K\,K\,K\,L-$$

C-terminal peroxisomal targeting consensus sequence

$$-S(A,C)\,K\,(H,R)\,L-COO^-$$

Figure 5.18 Targeting of proteins made on cytoplasmic ribosomes. Polypeptides targeting to the mitochondrial matrix have N-terminal cleavable presequences, rich in positively charged and hydroxylated amino acids. These are capable of forming amphiphilic helices. Short segments of positively charged amino acids appear to promote nuclear localization whereas peroxisomal targeting is dependent primarily on a particular type of tripeptide consensus sequence at the C-terminus.

precursor. This is cleaved by specific proteases to yield the mature protein during or immediately after import into the mitochondrion. The signal sequence is usually 15–25 amino acids in length and is rich in basic (mainly arginine) and hydroxylated (serine and threonine) amino acids. The ability of the signal sequence to form an **amphiphilic helix** is important for successful targeting: this is a helix in which positive side chains are localized on one side, hydrophobic side chains on the opposite surface. Additional targeting signals are present in proteins that must be directed to specific locations within the mitochondrion – the inner membrane, outer membrane or the intermembrane space (Figure 5.19).

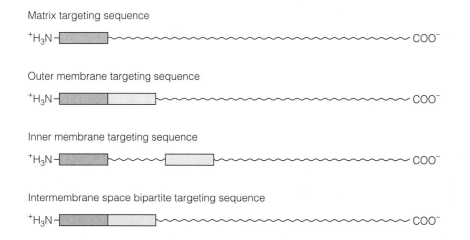

Matrix targeting sequence

Outer membrane targeting sequence

Inner membrane targeting sequence

Intermembrane space bipartite targeting sequence

Figure 5.19 Targeting to specific mitochondrial locations. Basic mitochondrial matrix targeting sequence (grey) is required for initial targeting to the organelle. The disposition of an additional hydrophobic 'stop-transfer' sequence (beige) directs localization to the inner or outer membrane. Intermembrane proteins have a 'bipartite' signal sequence with the second sequence (red) exposed in the mitochondrial matrix rerouting such proteins to the intermembrane space.

There are several mechanisms for targeting proteins to the nucleus

Although all mRNAs are produced in the nucleus, it has no capacity for protein synthesis and so the proteins required must be imported from the cytoplasm. The precise mechanism of targeting is not known, but the route is through the nuclear pore complex (see Chapter 3). At least two types of **nuclear localization signals** have been identified: one with a local concentration of basic amino acids, the other with a bipartite motif. Proteins that have to enter the nucleus include histones for chromosomal assembly and ribosomal proteins (ribosomal assembly occurs within the nucleus).

Chapter 3. The nucleus houses the genetic material and is the 'operational centre' of the cell, page 44.

Heat-shock proteins assist protein refolding

The **heat-shock proteins** (hsps), also called the stress-related proteins, are expressed in large amounts when cells are exposed to heat shock or other forms of stress (Table 5.4). They are capable of promoting the correct refolding of proteins that have been partially denatured, for example by exposure to elevated temperatures. They also exhibit weak ATPase activity and are capable of recognizing hydrophobic regions on aberrantly folded polypeptides and binding to them. It appears that the release of hsps from the complex with the partly denatured protein is an ATP-dependent event, the ATP providing energy to induce the protein back into the correct folding pathway.

Table 5.4 Families of heat-shock proteins in mammalian cells

Subcellular location	Name	Functions
hsp 70 family		
Cytoplasm	hsp 73	Binds nascent polypeptides; stimulates protein transport across organellar membranes
Mitochondria	mt hsp 70	Promotes final stages in translocation of mitochondrial polypeptides; mediates initial stages in assembly and transfer to mt hsp 60
Endoplasmic reticulum	Immunoglobulin binding protein (BiP)	Interacts with unassembled subunits of multimeric endoplasmic reticulum proteins in lumen
hsp 60 family		
Mitochondria	hsp 60	Promotes folding and assembly of imported proteins; prevents aggregation of denatured proteins
TRIC family		
Cytoplasm	TRIC (TCP-1 ring complex)	Promotes folding of actin and tubulin; may be involved in further types of folding reactions

Heat-shock proteins are also involved in protein targeting

Many hsps are also expressed at low levels under normal physiological conditions. It appears that they perform essential functions in the intracellular movement of polypeptides and in their final ordered assembly into oligomeric complexes in the appropriate intracellular compartment. In the cytoplasm, for example, newly synthesized polypeptides *en route* to mitochondria are transported as complexes with **hsp 70**, one of a family of closely related hsps which appear to have overlapping functions.

Members of this family have been termed **molecular chaperones** because they are responsible for maintaining precursors in a loosely folded open conformation believed to be important for insertion into membranes. Molecular chaperones may also prevent the formation of inappropriate aggregates between exposed hydrophobic surfaces and other cytoplasmic proteins. Optimal interaction of targeted proteins with specific receptors at the surface of mitochondria requires that the precursor polypeptides are retained in open, translocation-competent states by their interaction with cytoplasmic molecules of hsp 70. These are subsequently released in an ATP-dependent manner.

The assembly of multisubunit proteins requires chaperonins

Many proteins occur as oligomers or large multisubunit complexes (for example pyruvate dehydrogenase; see Chapter 8). Once the constituent polypeptides of such complexes enter compartments such as mitochondria or the lumen of the ER they must be assembled into the mature, functional complex. This requires the participation of other protein factors, called **chaperonins**. These associate only transiently with the component subunits, directing the ordered assembly of multisubunit enzymes, but are absent in the final complexes. Without chaperonins, nonfunctional (and usually insoluble) aggregates are formed. The function of chaperonins appears to be to prevent the occurrence of inappropriate interactions during the assembly process.

In the ER a chaperonin protein termed BiP associates with free immunoglobulin heavy chains, mediating their proper attachment to light chains before secretion of the complete immunoglobulin. Mitochondria also contain two chaperonin proteins, mt hsp 70 and mt hsp 60: mt hsp 70 appears to promote the final stages in the transfer of precursors across the inner membrane. It then presents the incoming polypeptides to mt hsp 60 in the correct fashion, leading to the ordered formation of active oligomeric complexes.

Intense research is being carried out to further our understanding of the modes of action of the various types of molecular chaperones because the ability to express clinically important proteins from cloned human genes in systems such as yeast or *Escherichia coli* is of considerable medical and commercial significance (see Chapter 17). Owing to the differences between human and microbial cells, it is not always possible to express the desired human proteins in a functional state. Coexpression of proteins and molecular chaperone molecules could, for example, overcome some of the difficulties inherent in some of these genetic engineering procedures.

Chapter 8. The pyruvate dehydrogenase complex catalyses a sequence of reactions, page 243.

Chapter 17. Cloned human genes can be used to make human proteins *in vitro*, page 544.

Gene expression in action: maintaining the level of cholesterol in cells

Cholesterol and closely related sterols are made in the liver, where a small fraction is used locally and goes into hepatocyte membranes. The rest is exported as bile acids and cholesterol esters (ChE) (see Chapter 14), which are transported by low-density lipoprotein (LDL) to tissues where they are stored or used. Tissue cells can obtain their cholesterol in this way by taking ChE up via receptor-mediated endocytosis of LDL. Alternatively, they can synthesize cholesterol *de novo* from acetyl-CoA via mevalonate (see Chapter 9).

Cells need to strike a balance between these two sources in order to conserve mevalonate, which is also required for the synthesis of other isoprenoids – sterols, ubiquinone, farnesyl groups, etc. This is achieved by feedback regulation of several functions, including uptake of ChE from plasma LDL by the LDL receptor and hydroxymethylglutaryl CoA (HMG-CoA) synthase and reductase-dependent synthesis of cholesterol.

This regulation operates through a sterol regulatory element (SRE) – a 10bp sequence upstream of the transcriptional start sites of the genes concerned. This element enhances transcription of the genes for LDL receptor and HMG-CoA synthase in the absence of sterols. The detailed mechanism for this process reveals a subtle combination of molecular events which brings about a very delicate control of cholesterol levels in the ER (Figure 5.20).

SRE-binding protein (SREBP) in nuclei binds to SRE with a high degree of specificity and contains a supersecondary structure feature (see Chapter 4) found in many transcription factors – the **basic-helix-loop-helix** motif, consistent with a

Chapter 14. The hydrophobic character of lipids complicates their digestion and absorption, page 418.

Chapter 9. Cholesterol is syntheized in the liver from acetyl CoA, page 284.

Chapter 4. Tertiary (3°) structure: polypeptide folding and 'domains', page 101.

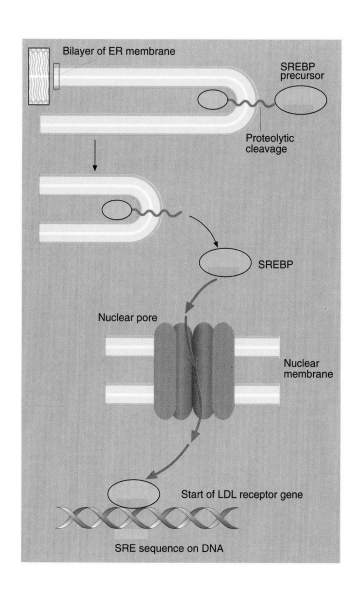

Figure 5.20 Cholesterol regulates cleavage of the membrane-bound SREBP precursor. A decrease in ER membrane cholesterol allows proteolytic cleavage of the membrane-bound SREBP precursor. The 68 ₓDa cleavage product is transported into the nucleus by a nuclear pore complex, binds to SRE, and activates transcription of genes flanked by this element, such as the LDL receptor gene.

Chapter 5. There are several mechanisms for targeting proteins to the nucleus, page 141.

role as a transcriptional regulator. It has a molecular weight of 68 kDa, but sequencing of the cDNA of the mRNA for this protein suggested the existence of a larger protein of 125 kDa as an SREBP precursor.

The situation appears to be that the 125 kDa protein is synthesized and inserted into the ER membrane where it acts as a sterol sensor. The sensing mechanism is sensitive to the levels of sterols in the ER membrane; when these fall below a critical level, the cytosolic domain of the protein becomes susceptible to proteolytic cleavage and is released from its attachment to the ER membrane. The released peptide contains a nuclear localization signal (see above) and travels to the nucleus, where it acts as a transcription factor and 'switches on' the genes increasing the supply of cholesterol by the LDL and *de novo* pathways.

DNA as the store of biological information: the nature and replication of genomes

The genomes of cellular organisms consist of lengths of double-stranded (duplex) DNA made of distinctive sequences. DNA is described in terms of the structural features which suit it for its role in the storage of information and its organization into genes and genomes for the purposes of heredity and gene expression.

DNA replication is the process whereby a parental duplex is converted into two progeny (or daughter) duplexes before cell division. It occurs in a semiconservative and highly faithful fashion. DNA replication is discussed in terms of the underlying principles, rather than by very detailed consideration of the mechanism.

DNA, due to its inherent chemical instability, is subject to spontaneous damage and damage induced by external agents, both physical and chemical. Unrepaired damage can lead to cell death, mutation or carcinogenesis. Description of damage to DNA and its repair by cellular mechanisms leads to a discussion of mutation, carcinogenesis and the importance of mutagens and carcinogens in the environment.

The nature of genomes

The **genome** of all cellular organisms is in the form of a sequence of double-stranded DNA and is arranged into one or more molecules, or **chromosomes**. A chromosome comprises a linear series of functional units (**genes**), each of which contains the information for a single function or discrete part of a more complex function. Most, but not all, genes are expressed, via mRNA, as proteins which carry out most of these functions.

Much of the DNA of human and other higher animal cells does not appear to encode proteins (see Table 5.1) and some is present as introns (see Chapter 5). It is likely that only 2–3% of the human genome consists of protein-coding sequences (see Chapter 17). Mammalian genomes contain many simple repeating sequences, the functions of which are largely unknown, but some are components of the **centromeres** of chromosomes (see Chapter 17). The highly repeated *Alu* sequences, which are important in the technique of **DNA fingerprinting** (see Chapter 17) may comprise as much as a quarter of the human genome, but again their function is unknown.

The replication of genomes

The genome must be doubled or duplicated, in the process of DNA replication, before cell division. This replication must be both **global**, that is the whole genome must be copied, and highly **faithful** in its copying. In this way, each daughter or progeny cell obtains a complete and accurate copy of the parental genome.

Genomic maintenance and information exchange

Other processes performed on cellular genomes include

- **DNA repair**, a group of activities found in cells, which remove and replace damaged components from DNA, thus maintaining the integrity of the genome.
- **DNA recombination**, a process which allows exchanges of material to occur between genomes or parts of genomes.

These will be discussed later in this chapter.

'Noncellular' genomes exist in nature

A variety of subcellular or noncellular entities possess genomes or have genome-like properties. These include the subcellular organelles mitochondria and chloroplasts as well as viruses, plasmids and transposons. With the exception of some types of virus, all resemble cellular organisms in having genomes of double-stranded DNA that encode a specific set of proteins. Viral genomes take a variety of forms and not all viral genomes are composed of double-stranded DNA: some are of single-stranded DNA; others are made of RNA; yet others alternate between DNA and RNA forms (see Chapter 18). Many 'noncellular entities' do not possess genomes: **prions**, agents of neurodegenerative disease (see Box 18.4), appear to consist only of protein and **viroids** (infectious RNA agents of plants) do not encode proteins.

Two of the major subcellular organelles of eukaryotes – mitochondria (see Chapter 3) and the chloroplasts of plants – both contain DNA genomes that encode rRNAs, tRNAs and a small number of the proteins they require. The remainder of the organellar proteins are encoded by genes in the main nuclear genome of the cell. Mitochondrial (see Chapter 17) and chloroplast genomes are thought to be descended from bacterial ancestors (see Chapter 7) and it is likely that genes have transferred from the organellar to the nuclear genome in the distant past.

Box 5.3. The discovery and significance of introns, page 124.

Chapter 17. The human genome probably does not contain many more genes than that of the simplest vertebrate, page 513.

Chapter 17. Satellite sequences form the basis for DNA fingerprinting, page 514.

Chapter 18. There are three main types of virus, page 564.

Box 18.4. Other noncellular infectious agents: viroids, plasmids and prions, page 561.

Chapter 17. The mitochondrial genome encodes 13 polypeptides, page 521. Box 7.2. The evolutionary origin of mitochondria, page 200.

The structure of DNA and its implications

DNA has two polynucleotide chains in a double helix

The structure of DNA is shown in Figure 6.1 and its principal features listed in Table 6.1. Nucleotide structure, the properties of bases and the phosphodiester linkage in polynucleotides are discussed in Chapter 2. The **double-helical model** for DNA was proposed by Watson and Crick in 1953 on the basis of X-ray diffraction analysis of stretched DNA fibres performed by Franklin, Kendrew and Wilkins (Figure 6.2). Although their model has been refined in detail, the basic features of the structure were correct and it has proved to be perhaps the most influential model of 20th-century biology.

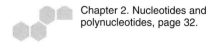

Chapter 2. Nucleotides and polynucleotides, page 32.

The structure of DNA is not completely uniform, but has sequence-dependent local variations

The X-ray analysis of DNA fibres is not true crystallography and only gives an *averaged* structure for the DNA duplex, in which all base pairs are equivalent. The true molecular structure of DNA, including the detailed structures of AT and GC

Table 6.1 The principal features of the structure of DNA

- DNA consists of two helical polydeoxyribonucleotide strands (a **duplex**)
- The sugar–phosphate backbones of the strands are **antiparallel**, that is, they run 3'→5' and 5'→3' respectively
- The strands are joined noncovalently by **hydrogen bonding** between bases
- Bases always pair according to the rules **A=T, G≡C** (Figure 6.3).
- Base pairs show **stacking interactions** because they are flat and nonpolar
- Stacking confers **stiffness** on the DNA duplex
- The sequences of the two strands are **complementary**
- DNA normally occurs in the B-form, with the following dimensions:
    10 base pairs per turn of the helix
    3.4 nm per turn of the helix (0.34 nm per base pair)
    2.0 nm diameter
- The two types of base pair have similar, but significantly different dimensions
- Each nucleotide carries a **negative charge** on the phosphate

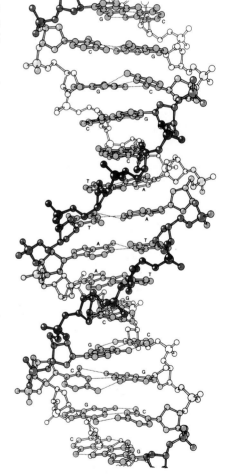

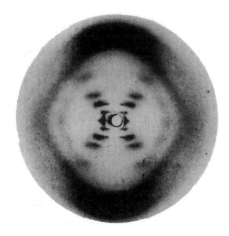

Figure 6.2 The X-ray diffraction pattern produced by wet DNA fibres. This photograph, obtained by Rosalind Franklin, played a key role in the elucidation of the DNA structure. A helical structure is indicated by the 'cross' pattern. The strong spots at top and bottom correspond to the helical rise of 0.34 nm. The fact that the layer line spacing is one-tenth of the distance from the centre to either of these spots indicates that there are 10 base pairs per repeat.

Figure 6.1 The double-helical structure of DNA. A space-filling model of B-DNA is shown here, with each atom given its van der Waals radius.

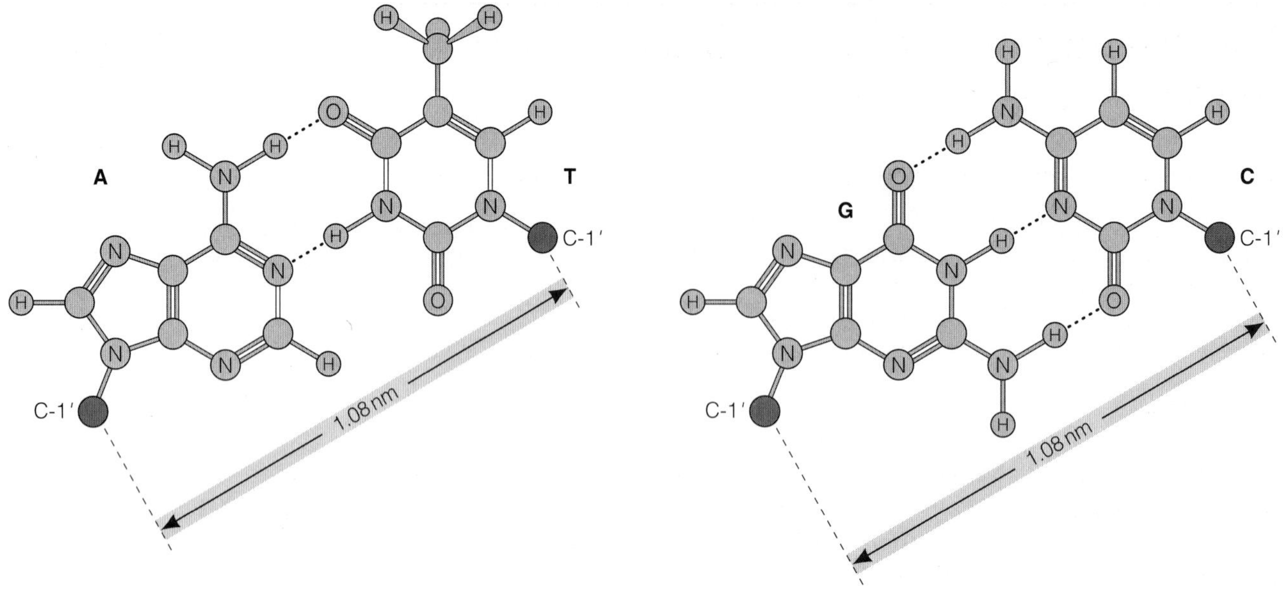

Figure 6.3 The structures of AT and GC base pairs.

base pairs was not obtained for another quarter century. The oligonucleotide AATTGCGCAATT (a dodecanucleotide) is self-complementary and forms a 12-base pair duplex in solution. It proved possible to crystallize this compound and determine its complete three-dimensional structure (Figure 6.3). This showed that base pairs are not completely flat, but show a 'propeller twist' and also that the precise atomic structure of each base pair is influenced by its neighbours. These differences in detail are of great biological significance, as they allow DNA-binding proteins (including origin-binding proteins, transcriptional factors and restriction endonucleases) to recognize specific sequences in DNA and bind to them.

The structure of DNA has many important biological implications

The proposal of the double-helical model for the structure of DNA was one of the most remarkable landmarks of biology of the past century. One of the most significant aspects of the model is the way it has stimulated a plethora of ideas and experimentation in biology and medicine over the past 40 years. Some of the principal biological implications arising from the model are listed in Table 6.2.

Table 6.2 Biological implications of the structure of DNA

- The base pairing ability of a DNA strand gives it a template function
- DNA can act as a template for synthesis of a DNA or an RNA strand
- The copied strand is antiparallel and complementary to the template
- Copying of DNA requires strand separation to expose the template
- Copying of DNA gives replication, of RNA gives transcription
- In genomic terms, replication is a global event, transcription a local one
- Replication needs permanent strand separation
- Only temporary separation of strands occurs during transcription
- DNA replication must be semiconservative (i.e. the single strand, rather than the duplex, is the conserved structure)

The organization of DNA in genomes: chromosomes and genes

Genomes of bacteria and eukaryotes are organized differently

Bacterial genomes are composed of a single molecule of DNA, that is their genomes and chromosomes are identical. In eukaryotes, however, the genome is divided into a number of component parts, or **chromosomes**, each of which contains a single molecule of DNA. (Note that 'chromosome' is applied to both the nucleoprotein structure and to the DNA molecule it contains.) The number of chromosomes is a characteristic of the haploid organism, while diploid cells contain them in pairs – the human cell contains 23 pairs (see Chapter 17). One probable reason for this division of the genome of eukaryotes into chromosomes is that their large size would make an unwieldy or unstable entity if the genome were a single molecule. The total size of the human genome is 3000 million base pairs; that of the bacterium *Escherichia coli*, for example, is a thousand times smaller (see Tables 5.1 and 6.3).

Table 5.1. The genome size and gene content of various organisms, page 110.

Genes are the functional units of genomes

In organizational terms, the next level below genome and chromosome is that of the **gene**, which probably constitutes the basic functional unit of all genomes (see Chapter 5). Each gene corresponds to a single function of an organism or, in the case of a complex function such as the determination of height, to a single identifiable part of that characteristic, for example a gene involved in the growth of the long bones.

Most, if not all, genes are believed to exert their influence by giving rise to a product which carries out the relevant function in the cell or whole body. In the processes of **gene expression** (see Chapter 5), the genes of all types of organisms undergo **transcription** (the copying of an RNA molecule from a stretch of DNA) and, with the exception of the 'functional RNAs' (rRNA, tRNA and other small RNAs of the cell), subsequent **translation** to yield the proteins which are the products of most genes.

Chapter 5. Genomes contain the information for organisms and genes are their functional units, page 110.

Chapter 5. Gene expression involves transcription, translation and processing, page 111.

Genomic organization differs between bacteria and eukaryotes

There are major organizational differences between the genomes of the Eucarya and the Bacteria (see Chapter 18) in that many bacterial genes exist in groups (or **operons**), which may be transcribed as a single, large **polycistronic** (or polygenic) mRNA encoding several polypeptides. Eukaryotic genes are generally **monocistronic** because the translation mechanism is unable to initiate internally (see Chapter 5). No gross physical manifestations mark the boundaries of genes, but they can be distinguished within long sequences of DNA by the presence of specific sequence features such as promoters, start and stop codons and polyadenylation sites (see Chapter 5). Cells have been able to recognize these features for billions of years, but it is only in the last decade that scientists have been able to recognize and interpret these signals within the sequences of DNA molecules. This has become possible with the advent of **DNA sequencing** (see Box 17.13).

Chapter 5. Initiation, page 134.

Chapter 5. RNA polymerases must recognize the start sites on genes, page 115.

Box 17.13. DNA sequencing methodology, page 543.

The sequences of DNA molecules can be determined

Once biologists developed the ability to manipulate and clone specific regions of DNA, it became possible to determine the sequences present in these regions (see Chapter 17) and to identify the genes they contained. DNA may be sequenced

Box 6.1 Some applications of the information from DNA sequencing

The huge amounts of data are generated by DNA sequencing need to be handled and stored in databases using computer methods. Important uses of this technology and information include the following.

The indirect sequencing of proteins

Once a stretch of DNA has been sequenced, it is easy to apply a computer search for open reading frames (ORF: a stop codon in frame with a start codon – see Chapter 5) and obtain *deduced* protein sequences. The ORFs may then be compared with known sequences in databases to see whether they correspond to or resemble a known gene or protein. Most protein sequences are now determined in this way, as it is so much easier than direct protein sequencing and does not require purification of the protein, which is often a difficult task.

The comparison of different genomes and their constituent genes

A wide range of computer programs is available for handling nucleotide and amino acid sequence data and many are used for comparing two or more sequences. Such comparisons are used to identify functionally important parts of proteins which tend to be conserved between species. They can also be used to measure evolutionary distances between proteins (see Box 6.6).

Applications of sequence information

The information from sequence comparisons can be used for many purposes, including site-directed mutagenesis, the rational design of drugs and understanding physiological aspects of protein behaviour, such as targeting to specific intracellular locations (see Chapter 17).

Chapter 5. The stop codon must be 'in frame' with the start codon, page 133.

Box 17.16. The human genome project, page 547.

using chemical or enzymic methods and was first achieved in small DNA-containing viruses of a few thousand nucleotides. What was once an enormous task is now routine and many large DNA viruses (herpesviruses and poxviruses) have been sequenced as well as over half the *E. coli* genome and substantial parts of the genomes of yeast, the mouse and the nematode worm *Caenorhabditis elegans*, a very simple multicellular eukaryote. An international project to sequence the human genome is now under way (The Human Genome Project; Box 17.16). Box 6.1 shows some practical applications of the information obtained from DNA sequencing; no doubt many more will be discovered.

Packaging DNA within cells

The length and stiffness of DNA presents a problem to the cell

The purine and pyrimidine bases of DNA are relatively nonpolar molecules and in base pairs the polar parts of the bases are mainly tied up in hydrogen bonding (Figure 6.3). This makes the base pairs even more nonpolar in nature and with a greater tendency to form **stacking interactions** (see Chapter 2), so that the interior of the double helix resembles a stack of coins. As a result, duplex DNA molecules are long, thin and stiff and are very susceptible to shearing when in unprotected solution. All DNA-containing structures – bacterial cells, the nuclei of eukaryotic cells and DNA virions – contain DNA molecules very much longer than the greatest dimension of the structure (Table 6.3). Because of this, a mechanism is required to package the long stiff DNA molecule into a compact, condensed form and to protect it from shear and other damage.

Table 6.3 Lengths of DNA molecules

- 10 base pairs = 3.4 nm
  5 kilobase pairs – a small DNA virus genome (SV40) – nearly 2 µm in a particle of about 5 nm diameter
- 10 kilobase pairs = 3.4 mm
  3000 kilobase pairs – a bacterial genome (*E. coli*) – about 1 mm in a cell about 10 mm long
- 10 000 kilobase pairs = 3.4 mm
  3 000 000 kilobase pairs – the human genome – about 1 m in a nucleus about 10 mm in diameter
- 10 000 000 kilobase pairs = 3.4 m

Table 6.4 Properties of histones

Histone	M_r	% arginine	% lysine	% positively charged residues
H1	21 000	30	1.5	31.5
H2a	14 000	11	9	20
H2b	14 000	16	6	22
H3	15 000	9	13	22
H4	11 000	11	14	25

A group of basic proteins, the histones, package DNA into nucleosomes

Chromosomes of human and other eukaryotic cells can be isolated from nuclei and purified to yield **chromatin**, which is composed of DNA and protein with some RNA. Among the proteins of chromatin, the **histones** have long been recognized by their distinctive basic character caused by their high arginine and lysine content (Table 6.4). The positive charges on these amino acids neutralize the negative charges on DNA phosphates.

Partial digestion of isolated chromatin with DNases produces discrete particles with characteristic contents of histones and constant lengths of DNA. Refinement of these studies led to the discovery of **nucleosomes**, octameric complexes of four of the five histones (H2a, H2b, H3 and H4) around which a length of DNA duplex (140bp) wraps itself twice. The nucleosome is the first step in the process of compacting DNA for packaging into chromosomes – subsequent steps involve the fifth histone (H1) and the 'inter-nucleosomal' lengths of DNA, which vary in length from 40 to 100bp. A remarkable feature of histones is that their sequences have been highly conserved during evolution – they are among the most highly conserved proteins in eukaryotes, plant histones being able to substitute for human ones in some experimental systems.

The nucleosome is the first level of condensation of DNA in the overall structure of the chromosome, giving the 10-nm fibre, which coils into the 30-nm fibre – the basic packing unit of the chromosome (see Chapter 17). This structure must be unravelled to allow transcription and replication, and new histones must be synthesized following DNA replication to allow nucleosomes to reform on the progeny duplexes.

Replication of DNA: initiation, fork movement and the attainment of fidelity

Genomic duplication is a prerequisite of cell division so that each daughter cell receives a complete and faithful copy of the parental genome (see Box 3.1). Present views on the origin of life (see Box 6.5) favour the idea that the most primitive precellular genomes may have consisted of RNA. In this model, the change from RNA to DNA genomes is seen as an early and crucial event that allowed high-fidelity genomic replication and thus the appearance of stable cellular species.

Chromosomal DNA replication is a complex and demanding process

Consider the overall process of replicating the single, large double-stranded DNA molecule that comprises an individual chromosome, either human or bacterial. To

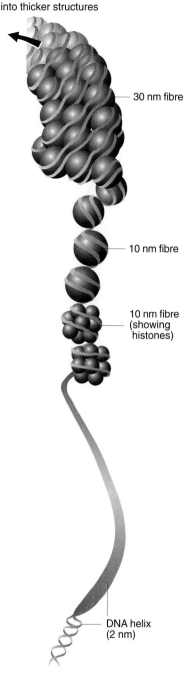

30 nm fibre coils into thicker structures

30 nm fibre

10 nm fibre

10 nm fibre (showing histones)

DNA helix (2 nm)

Figure 6.4 The compaction of DNA by histones. The figure shows the DNA helix at the bottom, wrapping itself round the eight histones of the nucleosome and folding into the 30 nm fibre.

produce two complete versions for distribution to the daughter cells, it is necessary to accomplish the following:

- **unwinding** of the parental DNA duplex and separation of the two strands;
- **copying** of both strands of the parental DNA duplex to the very last nucleotide;
- provision of an adequate supply of all four dNTPs for synthesis of the new strands (see Chapter 10);
- provision of a supply of new histone molecules to reform the replicated DNA into nucleosomes for packaging into chromosomes.

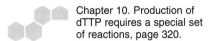 Chapter 6. Helix unwinding, untangling and protection, page 156.

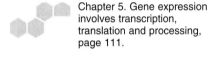 Chapter 10. Production of dTTP requires a special set of reactions, page 320.

The Watson–Crick model of DNA suggested the existence of fork structures in replicating DNA

The Watson–Crick model implied a semiconservative mode of DNA replication and suggested a general mechanism for the achievement of the first of the tasks listed above – the movement of a **replication fork** into the parental duplex, generating two daughter duplexes in the process. The fork was suggested long before the molecular details of DNA replication had been worked out and indeed were observed under the electron microscope (Figure 6.5). The concept of **fork movement** is central to any detailed model of DNA replication and the molecular events occurring at the fork are discussed below. Replication forks are established by special mechanisms which we are only beginning to understand, but which appear to occur only during initiation of replication (see below).

Chapter 6. Molecular events at the replication fork are centred on DNA polymerase, page 156.

Specific mechanisms initiate DNA replication

In common with the mechanisms for transcription and translation (see Chapter 5), the bulk of DNA replication involves an **elongation** reaction, discussed below. The process of initiation of chromosomal replication is less well understood. Our knowledge of the initiation of DNA replication has largely come from studies in bacteria and viruses, but the principles learned from these systems probably also apply to human cells, at least in general terms.

Chapter 5. Gene expression involves transcription, translation and processing, page 111.

Replication of bacterial genomes initiates at a single origin

Bacterial cells generally have a single DNA chromosome, the circularity of which has been demonstrated both directly and by genetic mapping. Genetic studies of *E. coli* have revealed that not only is the genetic map circular but also that replication occurs from a unique **origin of replication** locus (*ori*). Biochemical studies showed that initiation of replication at this origin results in the production of **two** replicating forks which move bidirectionally from the origin (Figure 6.5). There is also a **termination** locus, at which the two forks eventually meet, leading to separation of the two daughter duplexes.

Eukaryotic chromosomes initiate replication at multiple sites

Although multiplying bacteria replicate their DNA continuously throughout growth of the population, DNA replication in dividing eukaryotic cells is confined to about one-third of the cell cycle – the **S phase** (see Figure 3.6) – typically lasting 6–8 hours. During that period, each chromosome of the genome must be replicated – and this is a much longer stretch of DNA than that in the bacterial genome. As there must be a physical limit to the rate of fork movement, whatever the type of cell, eukaryotes have had to find a solution to this 'time-and-motion' problem: the existence of **multiple origins of replication** within the single DNA molecule that forms each eukaryotic chromosome.

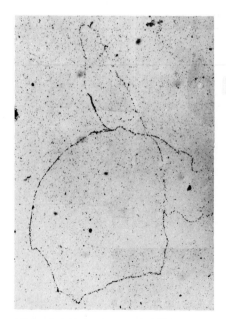

Figure 6.5 Fork structures in DNA replication.

Bacterial DNA replication

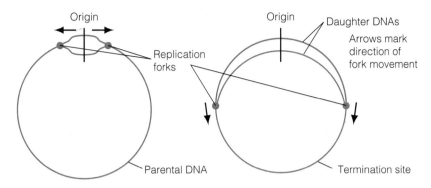

Eukaryotic DNA replication

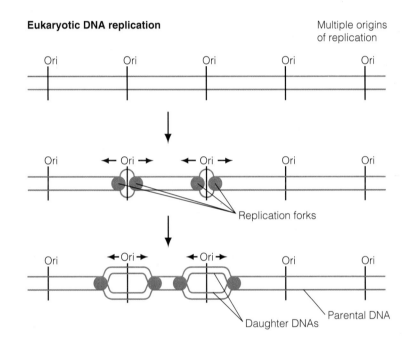

Figure 6.6 The replication of bacterial and eukaryotic chromosomes.

The fundamentals of the replicative procedure in eukaryotes are similar to those in bacteria (bidirectional movement of two forks from an origin), but each fork eventually meets a fork that has set out from the adjacent origin, not the other fork from the same origin (Figure 6.6). Each of the lengths of DNA between origins is called a **replicon**. Replicons are approximately 'gene-sized'.

There are unique sequences at origins of replication

Sequencing of the DNA of bacteria and DNA viruses has revealed that origins of replication have distinctive sequence features, often involving repeats of 10–20 nucleotides, and the ability to form unusual tertiary structures such as cruciforms (cross-like forms). Such features are probably recognized by proteins involved in the initiation of chromosomal replication. Work on such proteins has so far concentrated on bacteria and viruses, but common properties include

- sequence-specific DNA binding (origin-binding proteins) and
- the separation of DNA strands (DNA helicase – see below).

The initiation mechanisms of several human and animal DNA viruses are relatively well understood and they differ significantly from each other. The polyomavirus SV40 mechanism probably works in a manner very similar to that of the human cell, but the functions of one or more of the cellular proteins involved in DNA replication are replaced by a multifunctional viral protein, the **large T antigen** (see Chapter 18).

Control of DNA replication is important for normal cell behaviour

As most cells of the adult body are not actively dividing the control of DNA replication is clearly important. Little is yet known of the control of initiation of human DNA replication, but defects in the genes responsible can lead to uncontrolled cell growth and even cancer. The products of other cellular genes (tumour suppressor genes) are important in preventing cells from becoming cancerous. Important among these is p53, a protein that seems to be important in stopping improper DNA replication. It has been called 'the guardian of the genome' (see Chapter 17). Normal growth is probably due to a balance between opposing factors, perturbation of any one of which may be deleterious.

DNA polymerases are central to DNA replication

Once DNA replication has been initiated at an origin of replication, fork movement involves many repetitions of the same sequence of nucleotide polymerizing events until the whole replicon has been duplicated. Template-directed nucleotide polymerization (see Part B Introduction) is catalysed by **DNA polymerase**, a reaction necessary for, but not sufficient to achieve, fork movement. In this section the properties and limitations of DNA polymerases will be discussed, the next section will describe the roles of other proteins and enzymes necessary for overall DNA replication which act in concert with DNA polymerase.

The general DNA polymerase reaction is

$$(dNMP)_n + dNTP = (dNMP)_{n+1} + PPi$$

where $(dNMP)_n$ represents the **growing strand** of DNA and dNTP the deoxyribonucleoside 5'-triphosphate (in the form of the Mg^{2+} complex) being polymerized. As in the case of other template-directed polymerases, selection of each dNTP is directed by the next nucleotide on the **template** strand of DNA, according to the A=T, G≡C hydrogen-bonded base-pairing rules, but the template strand is not itself covalently changed by the reaction. The 3'-terminal nucleotide of the growing chain must bear a free 3'-hydroxyl group and must be correctly hydrogen-bonded to its complementary nucleotide on the template strand (see below).

DNA polymerases are found in all types of cells and in the larger DNA viruses. The genes for many of these have been sequenced, revealing certain similarities which suggest that DNA polymerase genes are all descended from a very ancient ancestor. Although the detailed aspects of behaviour of the individual DNA polymerases differ, they are remarkably similar in their general properties:

- DNA polymerases copy DNA chains with a **high degree of fidelity** – indeed, they are the only known enzymes with this property, which is their forte and is closely tied in with the role of DNA as the stable repository of the genome.

- DNA polymerases can **only synthesize chains in a 5' to 3' direction**. This is a serious limitation because of the problem of replicating antiparallel DNA strands during fork movement. It is solved by the participation of other enzymes and proteins (see below).

Box 17.9. Protein p53, mutations and cancer, page 536.

Part B. Template-directed nucleotide polymerases, page 79.

Chapter 6. Fidelity of replication is aided by proofreading by DNA polymerase, page 158. Chapter 6. Molecular events at the replication fork are centred on DNA polymerase, page 156.

- DNA polymerases are **unable to initiate DNA chains**. This contrasts with RNA polymerases (see Chapter 5) and chain initiation is achieved by the participation of primase (see below).

Chapter 5. RNA polymerases are capable of both initiation and elongation, page 114.

The situation as it exists in present-day organisms is thought to be the result of evolution of replicative DNA polymerases as specialists in the very accurate copying of polynucleotide sequences (see Box 6.5). The price of this is that they lack the ability to initiate. In contrast, RNA polymerases can initiate replication, but have a higher error rate (see Chapter 5).

Chapter 5. RNA polymerases are capable of both initiation and elongation, page 114.

Bacterial and eukaryotic cells contain several different DNA polymerases

Table 6.5 shows some important properties of the DNA polymerases of bacteria, animals and a DNA virus.

Bacterial DNA polymerases

E. coli contains three distinct types of DNA polymerase, and similar sets of enzymes appear to be present in most other bacteria. Most of detailed understanding of DNA polymerases comes from study of these enzymes.

pol I is the enzyme which appears to carry out some of the 'finishing' activities during DNA replication (see below). *E. coli* pol I is the best understood of all DNA polymerases, mainly through the work of Arthur Kornberg and his collaborators. This knowledge has contributed enormously to the study of DNA polymerases from all sources, including humans and their viruses. The protein comprises two **domains**.

1. A larger domain, which can be cleaved from the whole protein by controlled proteolysis and isolated as the 'Klenow fragment'. It contains both polymerase and 3'→5' exonuclease activities.
2. The smaller domain contains a 5'→3' exonuclease activity involved in the removal of RNA primers from replicating DNA (see below).

Chapter 6. Fidelity of replication is aided by proofreading by DNA polymerase, page 158.
Chapter 6. Removal of RNA and ligation, page 157.

The three-dimensional structure of the Klenow fragment has been determined, both on its own and complexed with DNA. It shows a 'thumb' structure wrapping round the DNA duplex (Figure 6.7).

Table 6.5 DNA polymerases of bacterial and animal cells

DNA polymerase	M$_r$	Exonuclease		Function
		3'→5'	5'→3'	
Bacterial cells				
pol I	100 000	+	+	Removal of primer RNA; completion of DNA chains
pol II	90 000	+	–	DNA repair
pol III	900 000	+	–	DNA replication
Animal cells				
pol α	300 000	–	–	Initiation of DNA chains; has associated primase
pol γ	200 000	–	–	Mitochondrial DNA replication
pol δ	300 000	+	–	Nuclear DNA replication. Needs accessory protein PCNA
Virus				
HSV	140 000	+	–	Herpes simplex virus (HSV) DNA replication

Chapter 6. Removal of RNA and ligation, page 157.

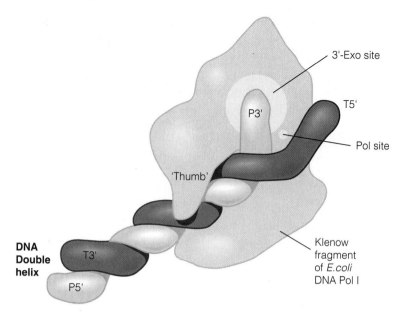

Figure 6.7 The detailed structure of a DNA polymerase. The figure shows the three-dimensional structure of the Klenow fragment of DNA polymerase I of *E. coli*, based on X-ray crystallography. Pink, enzyme; dark blue, template DNA strand; pale blue, primer DNA strand.

pol II is less well understood than pol I and III, but may be involved in DNA repair

pol III is the enzyme responsible for DNA chain elongation during replication of *E. coli* DNA, acting in concert with a variety of other enzymes and proteins (see below). pol III is a multisubunit enzyme containing about ten different types of polypeptide and functions as a dimer. It shows a very high degree of **processivity** (it polymerizes many thousands of nucleotides in a single encounter with DNA). This enzyme also has an associated **3'–5' exonuclease** activity in one of its polypeptides.

Eukaryotic DNA polymerases

Of the several types of DNA polymerase found in eukaryotic cells **DNA pol α** and **DNA pol δ** appear to be the most important for DNA replication. Each has associated enzyme activities: **primase** (see below) and **3'–5' exonuclease**, respectively. These two DNA polymerases appear to act together as part of the eukaryotic DNA replication complex (see Figure 6.8). Their importance in replication was recognized partly by the use of the inhibitor aphidicolin, which specifically inhibits these enzymes, but not DNA pol β, DNA pol γ or the bacterial DNA polymerases I and III. Highly specific inhibitors such as aphidicolin (and α-amanitin – see RNA polymerase, Chapter 5) have proved invaluable in the molecular dissection of many complex biological processes.

Molecular events at the replication fork are centred on DNA polymerase

The complex series of molecular events at the replication fork is centred on the action of DNA polymerase, but also involve a number of other processes (Figures 6.8 and 6.9).

Helix unwinding, untangling and protection

Purified DNA can be unwound and the chains separated *in vitro*, but it requires energy in the form of heat (temperatures approaching 100°C). The cell achieves the same effect at physiological temperatures by using energy in the form of ATP and the enzyme **DNA helicase**, which unwinds the helix ahead of the replicating fork. If this unwinding proceeded in isolation, the DNA would become hopelessly tangled. This tangling is relieved by the action of **DNA topoisomerases**, which act

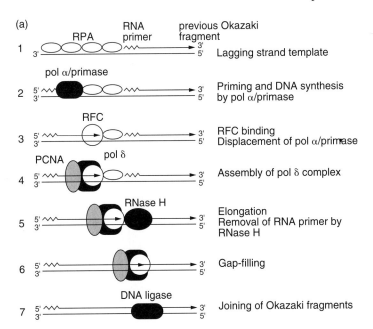

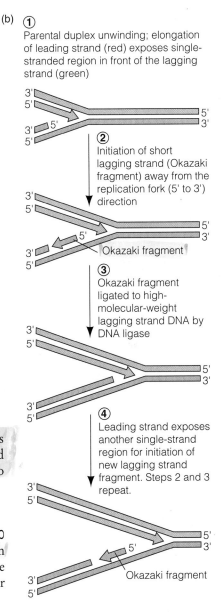

Figure 6.8 Molecular events at the DNA replication fork. (a) Mechanism of lagging strand synthesis at a eukaryotic cell DNA replication fork. RPA, RFC and PCNA are proteins that participate in DNA replication. (b) The Okazaki model of discontinuous chain growth in DNA replication.

by a breaking and rejoining mechanism. In the vicinity of the fork, the DNA chains are transiently single-stranded, and thus vulnerable to cleavage. They are protected by the presence of a **single-stranded DNA binding protein** (ssb), which also serves to facilitate strand separation (protein RPA, Figure 6.8a).

Initiation of DNA chains

A special type of RNA polymerase, **primase**, has the ability to make short (10–20 nucleotide) chains of RNA using the lagging strand as template. Primases are often found in association with helicases or DNA polymerases. The oligoribonucleotide that they synthesize is subsequently used by DNA polymerase as a primer for DNA elongation during discontinuous synthesis.

Leading and lagging strand synthesis

As the fork proceeds, the parental strand lying with its 3'–5' polarity colinear with the direction of fork movement can be copied directly by DNA polymerase, extending the daughter chain in the 5'–3' direction (this is the **leading** strand). The other parental strand, however (the **lagging** strand), has 5'–3' polarity (Figure 6.8b). It must be copied in the opposite direction discontinuously because DNA polymerase cannot extend the chain in the 3'–5' direction. A current model for DNA replication has the two strands being made simultaneously by a dimeric form of DNA polymerase acting on a looped structure in the fork region (Figure 6.9).

Removal of RNA and ligation

DNA polymerase continues along the lagging strand template until it encounters the initiating RNA oligonucleotide of the preceding DNA 'unit'. RNA is not normally found in DNA, because these nucleotides are removed by a 5'–3' **RNase H** activity, which may be associated with DNA polymerase, as is the case in *E. coli* pol I. DNA polymerase fills in the resultant gap until the two DNA chains are exactly adjacent, whereupon they are joined by the enzyme **DNA ligase**, which uses ATP (or NAD⁺ in some bacteria) to bring about the reaction shown in Figure 6.10.

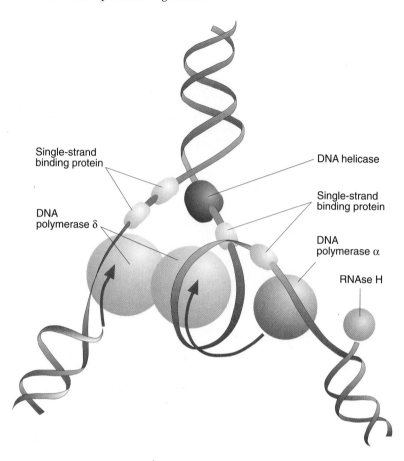

Figure 6.9 A model for the simultaneous synthesis of both strands of DNA.

This action of DNA ligase completes each section of discontinuous DNA synthesis on the lagging strand which is continued until the replicating fork encounters another going in the opposite direction (see Figure 6.9).

Fidelity of replication is aided by proofreading by DNA polymerase

The fidelity of copying by any template-directed polymerase is ultimately directed by the hydrogen-bonded pairing between bases. This is not in itself a very faithful

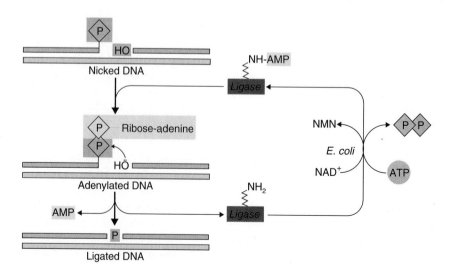

Figure 6.10 The reaction catalysed by DNA ligase.

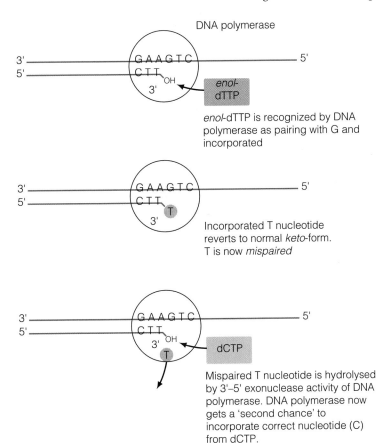

DNA polymerase

enol-dTTP is recognized by DNA polymerase as pairing with G and incorporated

Incorporated T nucleotide reverts to normal *keto*-form. T is now *mispaired*

Mispaired T nucleotide is hydrolysed by 3'–5' exonuclease activity of DNA polymerase. DNA polymerase now gets a 'second chance' to incorporate correct nucleotide (C) from dCTP.

Figure 6.11 Proofreading during the action of DNA polymerase.

process, because each base spends about 0.01% of its time in the *enol* or *imino* rather than the predominant *keto* or *amino* tautomeric forms. Because of this, there is about a 1% chance of any base pair being wrongly matched. Even relatively inaccurate polymerases, such as reverse transcriptases, perform about a hundred times better than this and so must be contributing to accurate base pair selection, perhaps by rejecting the minor forms.

DNA polymerases are a further hundred-fold more accurate in dNTP selection and this level of fidelity of copying is further enhanced by the **proofreading** action of a **3'–5' exonuclease** activity associated with many DNA polymerases (see above). DNA polymerase may incorporate a 'wrong' nucleotide because it is in the minor tautomer at the instant of selection (for example, *enol*-dTTP might base pair with a G on the template strand). A fraction of a second later, the nucleotide reverts to the major form and so becomes mispaired (in this example *keto*-T/G). The 3'–5' exonuclease activity of DNA polymerase enzyme hydrolyses the mispaired nucleotide at the 3'-end and its polymerase activity then inserts the correct one (Figure 6.11).

Proofreading is not the whole explanation for high fidelity, because overall DNA replication is about a hundred times more accurate even than DNA polymerases. Other factors, probably involving further ATP expenditure, must be involved. The consequences of unfaithful copying are discussed below.

Nucleosomes must reform on newly replicated DNA before cell division

The DNA of the daughter cells must reform into nucleosomes and chromosomes. Figure 6.4 shows that 'naked' newly replicated DNA is assembled into nucleosomal structures. This requires the synthesis of new histone molecules and is one reason why DNA replication in eukaryotes requires continuous protein synthesis.

DNA replication requires a continuous supply of deoxyribonucleotides

As DNA polymerase must copy all the nucleotides in parental DNA during genomic replication and deficiency in any one of the four dNTPs will cause the enzyme to stop at the position of the 'missing' nucleotide, thus blocking DNA replication. Cells in S phase, and thus replicating DNA, actively synthesize deoxyribonucleotides, both *de novo* and by thymidine kinase 'salvage'.

Completion of DNA replication is a prerequisite for cell division; interference with the supply of dNTPs will block cell division. This forms the basis of action for several important anticancer drugs, such as methotrexate and 5-fluorouracil (see Chapter 10). Note that this type of anticancer therapy is not specific for cancer cells; DNA replication in all dividing cells, including those of the skin, hair, intestinal lining and blood cells – is blocked, giving rise to the well known and unpleasant side-effects associated with cancer chemotherapy.

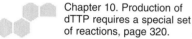

Chapter 10. Nucleotide components can be salvaged, page 322.

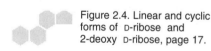

Chapter 10. Production of dTTP requires a special set of reactions, page 320.

Damage to DNA and its repair: maintaining the integrity of the information store

DNA can sustain both spontaneous and induced damage

Figure 2.4. Linear and cyclic forms of D-ribose and 2-deoxy D-ribose, page 17.

DNA is relatively stable compared with many biological molecules – markedly more so than RNA because of the lack of the 2-oxygen in the pentose (see Chapter 2) – but is still susceptible to a variety of covalent alterations, both spontaneous and induced by external agents. Much of the damage, if uncorrected, would corrupt the biological information encoded in DNA, leading to the consequences listed below.

Spontaneous damage
DNA can sustain several types of chemical changes under physiological conditions, even without exposure to any exogenous factors. Three of the most important (Figure 6.12) are listed below.

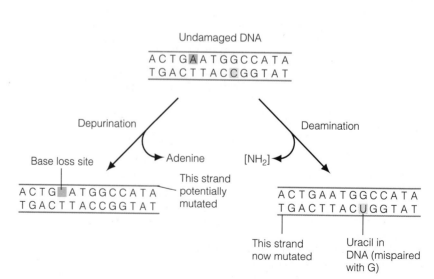

Figure 6.12 Spontaneous changes to DNA within cells.

Loss of purine bases
The bond between deoxyribose and the purine ring is the most labile in the DNA molecule; it has been estimated that breakage of these bonds occurs 10^4 times in 24 h in each human cell. Loss of a purine base results in the production of **apurinic DNA**. Such 'base loss' sites (apurinic or apyrimidinic) do not cause the phosphodiester backbone of the DNA to break, but the absence of the purine base impairs the template function of DNA during transcription and, more importantly, replication.

Loss of pyrimidine bases
The bond between deoxyribose and the pyrimidine ring is less labile than that of the DNA molecule but measurable breakage does occur, with similar consequences.

Deamination
Three of the four bases in DNA (A, C and G) have amino groups on their rings, loss of which changes the hydrogen-bonded pairing properties of the bases. In particular, spontaneous deamination of cytosine in DNA to uracil has been estimated to occur about 400 times per 24 h per human cell. It is potentially mutagenic if not corrected because the resultant uracil will behave like a thymine the next time the DNA strand is replicated.

 Chapter 2. Hydrogen-bonded base pairing, page 34.

Damage induced by external agents
DNA may also be damaged by a variety of external physical and chemical agents. Some of the more important of these agents and their effects on DNA are shown in Table 6.6. Although ionizing radiation is a physical agent, many of its effects on DNA are chemical in nature, because ionizing radiation generates highly reactive chemical species such as hydroxyl radicals from water. Many environmental and industrial pollutants damage DNA and many of them are **mutagens** and/or **carcinogens** (see below) and can pose a considerable potential threat to public health. Some are also **teratogens** – substances that cause fetal abnormalities.

 Chapter 6. Most carcinogens are mutagens, page 168.

Table 6.6 Damage to DNA

Spontaneous damage (occurs to DNA molecules, even in the absence of external agents)

Type of damage	Event	Product	Consequence	Frequency[†]
Depurination	Loss of purine base	AP DNA*	Mutation	10 000
Deamination	Loss of amino group from base	e.g. C →U	Mutation	400

Induced damage

Agent	Effect on DNA	Biological consequence
Physical agents		
Ultraviolet light	e.g. pyrimidine dimer formation	Inhibition of DNA replication
X-rays	e.g. backbone breakage	Inhibition of DNA replication
		Loss of integrity of genome
Chemical agents		
Nitrous acid	Deamination (e.g. C→U)	Mutation
N-alkylation	*N*-alkylated base (e.g. N^7-methyl G)	Depuration (low mutation rate)
O-alkylation	*O*-alkylated base (e.g. O^6-methyl G)	High mutation rate

*AP DNA = DNA that has lost a base (apurinic or apyrimidinic)
[†]Estimated no. of events per day per human genome

Damage to DNA can have a range of consequences

There are three possible consequences for a cell subjected to DNA-damaging agents:

1. The agent may have no effect;
2. The cells may die;
3. Mutation may occur.

No effect

Cells are capable of sustaining damage to their DNA without any apparent harmful effect. This can be due to one or more of the following reasons.

- The damage causes a change in a 'nonessential' part of the cell's DNA.
- The damage causes a change in an essential part of the cell's DNA, but does not alter the cell's information (see below).
- The damage causes a change which is repaired by the cell's mechanisms before it can exert any harmful effect (see below).

Cell death

Cell death is catastrophic for a unicellular organism, but the death of cells in multicellular organisms is normal and, unless extensive or in a vital organ, is unlikely to be of great consequence. Cell death can arise in two main ways: **necrosis**, or simple degeneration of the cell caused by damage; and **apoptosis**, sometimes called 'programmed cell death', which appears to be a natural part of the biology of multicellular organisms, involved, for example, in growth and development (see Chapter 17). Both types may be induced by agents that damage DNA, but gross cellular damage is likely to be the result of necrosis.

Box 17.10. Cell proliferation or cell death?, page 537.

Mutation

In multicellular organisms, and especially long-lived ones such as humans, this is a major concern in at least two important areas.

Germ cells

In individuals with the potential to produce offspring, mutation in germ cells (ova or spermatocytes) has the potential to damage future progeny by causing chromosomal damage or mutations. This is particularly so for ova, because of their long life, and so women of child-bearing age should be particularly careful to avoid exposure to DNA-damaging agents. It is likely that the higher incidence of various birth defects in mothers over 35 is partly due to accumulation of damage to ova.

Cancer

The somatic theory of cancer is now widely accepted and mutations to individual cells of the body, whether spontaneous or induced by chemicals in the environment, radiation or viruses, are probably the major (and perhaps the only) direct cause of the initiation of malignant disease (see Chapter 17). Ionizing radiation, ultraviolet radiation and certain chemicals have long been known to be carcinogenic in humans (see below).

Chapter 17. Initiation of tumorigenesis, page 532.
Chapter 6. Most carcinogens are mutagens, page 168.

Cells can repair their DNA after sustaining damage

The cells of the body are far from defenceless in the fight to protect their genome from damage and contain a battery of **DNA repair enzymes** and mechanisms. Some forms of DNA damage can be rectified directly: for example, pyrimidine dimers formed by ultraviolet radiation can be cleaved by the process of **photoreactivation**. Most DNA repair, however, involves a combination of **incision** (cutting into the damaged DNA) and **excision** (cutting out of the damaged part). This results in **gapped DNA**; completion of repair is achieved by filling the gap by DNA polymerase and joining of the resultant nick by DNA ligase (Box 6.2).

Chapter 6. Mutations arise from replication errors or damage to DNA and can have a wide range of effects, page 166.

There are separate repair pathways for small and bulky lesions in DNA

Cells contain a number of DNA repair mechanisms, but the two best-understood pathways (Box 6.2) are described below.

Small lesions that do not affect the overall shape of the DNA duplex

Examples of lesions of this type are the abnormal bases – uracil and hypoxanthine – which can arise by the deamination of cytosine and adenine, respectively. This pathway uses one of a range of **DNA glycosylases** specific for the particular lesion followed by the action of an apurinic endonuclease (specific for sites that have lost a base).

Bulky lesions that distort the shape of the DNA duplex

Examples of such lesions are pyrimidine dimers and polycyclic hydrocarbon adducts (see Box 6.4). This pathway uses complex **incision endonucleases** to initiate the repair process by making a 'nick' in DNA near the lesion.

Defects in incision into DNA near bulky lesions are found in the genetic disorder xeroderma pigmentosum, a condition that causes abnormal sensitivity to sunlight and a high incidence of skin cancers. This is not a single gene defect (see Chapter 17) but a family of related defects, each member affecting one of a number of different genes on the human ultraviolet repair pathway (Box 6.2) yet producing essentially similar pathological effects. Some of these genes encode transcription factors (see Chapter 5), supporting the idea of a link between repair and transcription (see below).

Chapter 17. Single-gene defects, page 525.
Box 6.2. DNA repair and xeroderma pigmentosum, page 164.

Box 5.1. Control of transcription in bacteria and humans, page 117.

Damage to DNA can be repaired by several other processes

Error-prone repair

This an emergency mechanism in bacteria and introduces many errors into the repaired DNA. Similar mechanisms may exist in humans.

Mismatch repair

These mechanisms are capable of correcting mismatched 'base pairs', such as G–T, which may have arisen during DNA replication and escaped the proofreading mechanism of DNA polymerase (see above).

DNA recombination

This mechanism is particularly important in the repair of damage that has destroyed the linear integrity of DNA, such as double strand breaks (which can be caused by ionizing radiation) and lesions that have occurred in regions of DNA while they were temporarily single-stranded (as, for example, during DNA replication).

Chapter 6. DNA can sustain both spontaneous and induced damage, page 160.

Box 17.10. Cell proliferation or cell death?, page 537.

Chapter 6. Cells can repair their DNA after sustaining damage, page 162.

Box 6.2   DNA repair and xeroderma pigmentosum

Enzymic repair of damaged cellular DNA

Damage that occurs to cellular DNA, whether spontaneous or induced by external agents (see Table 6.6) can have a variety of harmful effects These include

- mutation – resulting in changes in any daughter cells arising from the damaged cell;
- inhibition of DNA replication – resulting in failure of the damaged cell to divide or even in cell death from apoptosis or necrosis (see Chapter 17).

Fortunately cells are not helpless in the face of these insults but possess a variety of pathways for the repair of the damaged DNA and the restoration of the genome to its pristine condition. One of the most important of these is nucleotide excision repair, in which damaged sections of DNA are cut out and replaced by new, undamaged nucleotides (see below). There are other mechanisms, including mismatch repair which repairs erroneous base 'pairs', such as G–T or A–C, arising from mistakes in DNA replication.

The excision repair pathways can deal with two broad groups of lesions.

1. Small lesions, such as base changes caused by deamination, that do not perturb the overall shape of the DNA duplex.
2. Bulky lesions, such as pyrimidine dimers caused by ultraviolet light, that do perturb the DNA duplex

Excision repair is understood in detail only in bacteria, where only a handful of gene products are required. In eukaryotes, it appears that at least a dozen (and perhaps many more) proteins are needed (see below).

Altered bases are typically removed by specific **DNA glycosylase**, such as uracil DNA glycosylase which recognizes the product of cytosine deamination leaving a base loss (or AP site). Such sites are recognized and cleaved by an AP endonuclease, causing an incision into the DNA backbone.

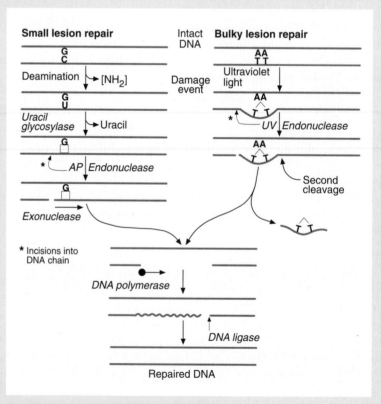

Box Figure 6.2 Cellular DNA repair processes.

Box 6.2   (continued)

The distortion caused by bulky lesions is recognized by complex repair endonucleases, which make an incision near the distortion.

The incised DNAs produced by either of these mechanisms is then acted on by exonuclease action, yielding a gap in the DNA duplex which is then filled in by a repair DNA polymerase and finally joined by DNA ligase (Box Figure).

Xeroderma pigmentosum – DNA repair genes are similar in humans, hamsters and yeast

In the rare genetic disorder xeroderma pigmentosum (XP), deficiency in DNA repair is associated with a variety of symptoms, notably acute sensitivity to sunlight and a high incidence of skin cancers. Detailed analysis of this condition has revealed that it is not a single gene defect, but is caused by a defect in one of a variety genes involved in excision repair of DNA – it is thus a family of closely related defects. Other rare genetic defects (Cockagne syndrome, Brittle hair syndrome) involve other genes for excision repair processes.

The functions of these genes are very difficult to study directly in humans, but cloning and sequencing of the human genes involved in XP has allowed the use of other approaches.

- Cultured hamster cells that are UV-sensitive because of defects in DNA repair genes can be made normal in their response to UV light by treatment with cloned human genes. Some of these genes are homologous to XP genes.
- Examination of the proteins encoded by human XP genes revealed that they resemble yeast proteins known to be involved in DNA repair (see Box Table 1).

This illustrates the usefulness of studying simple eukaryotes such as yeast. The biochemistry and genetics of yeast are simple enough and sufficiently well understood to allow a more detailed analysis of complex processes such as DNA repair than is possible in humans.

Box Table 1 DNA repair genes in humans and yeast

XP gene	Yeast gene	Function
XP-A	RAD14	Damage recognition
XP-E		Damage recognition
XP-D	RAD3	Helicase
XP-B	RAD25	Helicase (TFIIH)
XP-C	RAD4	ATPase
XP-F	RADI	DNA endonuclease
	RAD10	DNA endonuclease
XP-G	RAD2	ssDNA endonuclease

Interestingly, some of these proteins, such as XP-B (a component of TFIIH – transcription factor H of RNA polymerase II), are also transcription factors (see Box 5.1). This reveals a previously unsuspected link between the processes of transcription and DNA repair. In retrospect, this is a sensible arrangement for the cell, because it means that while DNA is undergoing transcription it is also being screened for harmful lesions. It also seems likely that in the complex genomes of hamsters and humans, DNA repair mechanisms are more selective than in bacteria, concentrating on areas of greatest need – for example, those areas of the genome that are being actively transcribed. This may partly explain the greater number of genes involved in eukaryote DNA repair.

 Box 5.1. Control of transcription in bacteria and humans, Box Table 1, page 117.

DNA repair is probably an ancient process

This wide range of repair mechanisms possessed by cells indicates the importance of the process. It is almost certain that, in evolutionary terms, the ability to repair DNA is a very ancient function. Damage to DNA is as old as life itself and even the earliest organisms (see Box 6.5) must have had some repair capability in order to prevent their genomes from mutating too fast and to exist as recognizable species. DNA repair enzymes are found in all modern cells.

There are links between DNA repair and transcription

It has recently been discovered that some proteins involved in DNA repair are also transcription factors and that there appear to be functional connections between repair and transcription (Box 6.2). This seems a 'sensible' arrangement, because it means that, while the cell is performing one essential transaction on its DNA – transcription – it can check out and make good any defects, rather like a workman doing some preventive maintenance on a potential defect spotted while doing some other task.

Mutation: changes in DNA and their consequences

Mutations arise from replication errors or damage to DNA and can have a wide range of effects

Mutations are changes in the DNA sequence of the genome of an organism and arise from two main sources:

- uncorrected errors arising during DNA replication, repair or recombination;
- unrepaired damage sustained by DNA, either spontaneous or induced.

Chapter 5. Gene expression involves transcription, translation and processing, page 111.
Chapter 4. Mutations can have important effects on protein structure and function, page 108.

An understanding of the nature of mutations and their effects on the mutated organism requires consideration of a number of areas of molecular biology, including DNA structure and replication (this chapter), gene expression (Chapter 5) and protein structure and function (Chapter 4). The effects of mutations range from none at all to lethal.

Many mutations have an effect because they occur within the protein-encoding part of a gene. Other mutations occur in DNA sequences that do not code for proteins (which act, for example, as promoters, origins of DNA replication and other types of regulatory or control elements) and mutations may exert their effects by changing the functions of these elements. An example of this is a mutation in a promoter sequence of a gene altering the ability of the promoter to bind a transcription factor. The gene product itself would not be altered, but might be produced in larger or smaller amounts due to a change in level of expression of the gene caused by the mutation.

Three main types of mutations affect proteins

Three principal types of mutations affect proteins, their structure and their biological functions (see Chapter 5).

Substitutions

In substitution mutations a nucleotide is substituted for the correct (or wild type) nucleotide in a DNA sequence, leading to an altered codon on the resultant

Wild type sequence (* is the transcribed strand)

DNA   3' -TACCGGAAAACC-5' *
         5' -ATGGCCTTTTGG-3'
mRNA   5' -AUG\GCC\UUU\UGG-3'
Protein N -met-ala-phe-trp-

Nucleotide substitution giving amino acid substitution

DNA   3' -TAC**T**GGAAAACC-5' *
         5' -ATG**A**CCTTTTGG-3'
mRNA   5' -AUG**ACC**\UUU\UGG-3'
Protein N -met-**thr**-phe-trp-

Nucleotide substitution giving no amino acid change
(silent mutation)

DNA   3' -TACCGGAA**G**ACC-5' *
         5' -ATGGCCTT**C**TGG-3'
mRNA   5' -AUG\GCC\UU**C**\UGG-3'
Protein N -met-ala-**phe**-trp-

Nucleotide substitution leading to premature chain
termination (nonsense mutation)

DNA   3' -TACCGGAAAA**T**C-5' *
         5' -ATGGCCTTTT**A**G-3'
mRNA   5' -AUG\GCC\UUU\U**A**G-3'
Protein N -met-ala-phe-(**stop**)

Nucleotide insertion leading to frameshift mutation

DNA   3' -TACG**C**GGAAAACC-5' *
         5' -ATGC**G**CCTTTTGG-3'
mRNA   5' -AUG**C**GC\CUU\UUG\G-3'
Protein N -met-**arg-leu-leu**-

(NB: nucleotide deletion has an analogous
frameshifting effect)

Figure 6.13 Mutation in DNA and its effect
on proteins.

mRNA and possibly to a change of one amino acid in the translated polypeptide
(Figure 6.13).

Insertions
One or more nucleotides is inserted into a DNA sequence, leading to a change in
the reading frame of the resultant mRNA from that point onwards.

Deletions
One or more nucleotides is deleted from a DNA sequence, leading to a change in
the reading frame of the resultant mRNA from that point onwards.

Insertions or deletions may be very large, comprising whole sections of a
chromosome. Certain types of flat nonpolar molecules, such as acridines, which
intercalate between base pairs of DNA, have a tendency to cause such
mutations.

Substitution mutations have a range of effects on proteins

Three main types of DNA substitution mutations can affect the structure and/or
function of proteins (see Chapters 4 and 5).

Silent mutations
Because of the relative unimportance of the third or 'wobble' position of the
codon–anticodon interaction, mutations in this position often have no effect, as the
same amino acid is incorporated and the protein is identical to the wild type (for
example, UUU → UUC still encodes phenylalanine).

Conservative substitutions
When a mutation leads to the substitution of a similar amino acid (perhaps
isoleucine for valine, aspartate for glutamate – see Chapter 4), the resultant change
in the function of the protein may be very small or undetectable. The degenerate
nature of the genetic code itself tends to ensure that many substitutions are of this

Chapter 4. Mutations can
have important effects on
protein structure and
function, page 108.
Chapter 5. The genetic code
is the set of rules that
governs translation,
page 131.

type. The organization of the code means that mutations in the first position of codons often behave in this way (for example, AUU → GUU leads to substitution of valine for isoleucine, both branched-chain aliphatic amino acids).

Radical substitutions

If a mutation leads to the substitution of an amino acid of a very different type, the effect on the function of the protein is more likely to be significant (for example, GUU → GAU leads to substitution of aspartate, a negatively charged amino acid, for valine).

The effects of substitutions are highly dependent on the nature of the protein and the location of the mutation within the protein. Substantial portions of many proteins seem to be relatively unimportant and have been deleted experimentally without apparent effect on function. Clearly substitutions in such regions, even when radical in nature, are unlikely to have any significant effect on the function of the protein. However, individual amino acids that are crucial for the function of a particular protein (for example at the active site of an enzyme) are usually highly sensitive to substitution and any change may completely destroy biological function.

Chain-terminating mutations

When a nucleotide substitution changes a codon into a stop codon (UAA, UAG or UGA) the effect is to cause premature termination of the polypeptide chain during protein synthesis. Such a truncated protein is very likely to be nonfunctional, unless the truncation occurs near the carboxy terminus and this region of the protein is not critical for function. Truncated proteins are usually degraded very rapidly within cells, probably by the **ubiquitin pathway** (see Chapter 14). They do not generally occur naturally, but only in mutations, either experimental or as a result of genetic disease (see Chapter 17). The opposite effect (a substitution leading to a change from a stop codon to one for an amino acid) leads to the production of a longer than normal polypeptide, as occurs in some haemoglobinopathies (see Chapter 17).

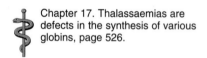

Chapter 17. Thalassaemias are defects in the synthesis of various globins, page 526.

Frameshift mutations are generally deleterious for proteins

The effects of substitutions are often innocuous but insertion and deletion mutations are generally very deleterious to the function of a protein because they cause **frameshift**. This means that every codon in the mRNA on the 3' side of the mutation – and thus every amino acid in the resultant protein towards the C-terminus – is altered, leading to a complete change in the primary and three-dimensional structures of the mutant protein compared with the wild type. In most cases, this will drastically impair the function of the protein.

Most carcinogens are mutagens

Mutagens are substances that can cause mutations and include many environmental contaminants. Most of the agents that induce damage in DNA (see above) have varying mutagenic effects and there is a correlation between the ability to cause damage that affects the hydrogen-bonding properties of bases and the degree of mutagenicity of a particular mutagen. For example, dimethylsulphate methylates guanine in the N-7 position, well away from the hydrogen-bonding atoms, and is a very weak mutagen but methylmethane sulphonate – a very powerful mutagen – modifies guanine nucleotides in DNA to O^6-methylguanine, which, unlike N^7-methylguanine, can no longer base pair with cytosine nucleotides (Table 6.6). The **Ames test** (Box 6.3) is widely used as a test of the mutagenicity of chemicals.

Box 6.3 The Ames test – identifying mutagens

There is much concern about toxic chemicals in the environment, especially if they are associated with causation of cancer (carcinogenesis). Proving that a substance is carcinogenic can be a difficult, expensive and time-consuming process but most known carcinogens are also mutagens and it is thus wise to minimize exposure of human populations to mutagens. Identifying agents that are mutagenic to humans is also difficult (and direct testing is unethical), so a more practical method is needed. This has been provided by the **Ames test** in its various forms.

In the Ames test, a mutant of the bacterium *Salmonella typhimurium* is treated with various concentrations of the potential mutagen to be tested and the rate of reversion of the mutant back to the wild type form is measured. This shows whether an agent is mutagenic for this bacterium. Many agents, however, are not directly mutagenic in humans, but become so after metabolic activation in the liver (see Box 6.4). In order to test for this, the potential mutagen is incubated with a liver extract before being added to the bacterium (the **indirect Ames test**). By using these tests, an indication of the likely mutagenic (and thus potential carcinogenic) effects of compounds can be quickly and cheaply evaluated in order to decide whether more expensive and complicated investigations are worthwhile.

Carcinogens are substances that cause cancer and one of the main reasons for interest in environmental mutagens is that most carcinogens are also mutagens, and there is a strong link between mutagenicity and causation of malignancies (see Chapter 17). The **polycyclic hydrocarbons** are a well studied group of compounds, some of which are potent carcinogens and medically important because of their connection with malignancies, especially lung cancer. These very harmful components of cigarette tars illustrate the principles of **metabolic activation** in the liver and mutagenic action involving intercalation into DNA (Box 6.4).

Chapter 17. Chemicals can participate in both initiation and promotion of tumours, page 532.

Box 6.4 Cigarette tar, metabolic activation and carcinogenesis

There is much public concern over the connection between cigarette smoking and the causation of a variety of diseases, most notably lung cancer (see Chapter 17). The tars produced by smoking cigarettes are known to be hazardous, and among their constituents are polycyclic hydrocarbons (PHCs). Study of these chemicals, their metabolism and interaction with DNA illustrates a range of important points.

1. Not all PHCs are carcinogenic.
2. PHCs do not themselves damage DNA.
3. PHCS can be metabolically activated by the liver.
4. Activation is performed by the cytochrome *P*-450 system (see Chapters 7 and 17).
5. PHCs with a 'bay region' in their structure are most harmful.
6. The diol-epoxide formed by activation reacts with guanines in DNA.
7. Intercalation of activated PHCs between base pairs potentiates their reactivity.

The ability of an individual to carry out activation of PHCs in the liver depends on the cytochrome *P*-450 proteins present and has an important influence on their risk of developing lung cancer.

Recombination of DNA

DNA recombination breaks and rejoins pairs of DNA duplexes

Chapter 5. Transcription: copying of genetic information from DNA into RNA, page 112.

The third major cellular transaction involving covalent change in DNA chains is **DNA recombination** (transcription does not change DNA, which acts only as a template – see Chapter 5). Recombination of DNA leads to exchanges between DNA duplexes and can occur between different DNA molecules or between different parts of the same DNA molecule. Recombination was originally recognized from genetic analysis, where frequency of genetic recombination is a method for measuring the distances between genes and the mechanism was suggested to operate by 'crossing over' occurring between two genomes or parts of genomes. It is now known that this occurs at the level of the DNA duplex, by processes of **breaking and rejoining** and **strand exchange**. The details are not completely understood, but involve unique structures in DNA called **Holliday junctions** and various enzymes and proteins, including **resolvases** and **strand-exchange proteins** (Figure 6.14).

Chapter 16. The molecular and genetic basis of antibody diversity, page 496.
Chapter 17. 'Genetic engineering' and some medical applications, page 539.

DNA recombination plays a very important role in the generation of diversity of antibodies (Chapter 16) and can also bring about the repair of damaged DNA. The term 'recombinant DNA' is widely used in genetic engineering (see Chapter 17) and is used to describe DNA that has been artificially recombined in the laboratory.

DNA, RNA and the origin of life

Studies on DNA genomes indicate that cellular life in its modern form probably originated more than 3500 million years ago. There is considerable evidence to favour the view that so-called **prebiotic** life forms may have had RNA genomes and that DNA was developed as a more stable form of genome that would allow the existence of relatively stable 'species' of organisms. Some of the lines of evidence suggesting that RNA was the primitive genetic material are outlined in Box 6.5.

Chapter 10. Deoxyribo-nucleotides are synthesized from ribonucleotides, page 319.

Chapter 18. Retroviruses mutate rapidly, page 576.

Chapter 4. Some RNA molecules have catalytic properties, page 89.

Box 6.5   Was RNA the first genetic material in the origin of life?

- Ribonucleotides and their synthesis have metabolic primacy over deoxyribonucleotides, even in modern organisms (see Chapter 10). It seems likely that the first available nucleotides were ribonucleotides and that deoxyribonucleotides made a later appearance.
- Modern 'organisms' with RNA genomes – the RNA viruses – are much more 'plastic' than cells or DNA viruses (see Chapter 18). Stable genomes, required for well defined species, may have appeared only after the putative RNA-to-DNA 'switch'.
- Template-directed copying involving RNA is inherently less faithful than DNA-to-DNA copying.
- The removal of RNA primers from newly synthesized DNA may be important for retaining genetic integrity. The use of RNA primers in the replication of modern DNA genomes may be an echo of primitive RNA genomes.
- Some modern RNA molecules have catalytic activity, leading to the idea that RNA may be more 'primitive' than proteins and that the earliest prebiotic forms may have constituted an 'RNA world' in which RNA acted both as a repository of information and as catalytic agents for primitive metabolism
- If these ideas are correct, the first step from the RNA world would probably have been the appearance of proteins and enzymes, made under the direction of RNA templates. Later, once enzymes capable of making the dNTPs and acting as DNA polymerases had made their appearance, the switch to DNA genomes would have become possible.

Comparison of sequence information

Sequences of nucleotides on DNA or RNA or of amino acids in proteins can be compared by making alignments of the sequences under consideration and then analysing the alignments mathematically. This provides a quantitative assessment of how close the sequences are or how many changes are required to get from one sequence to another. This type of comparison has provided useful information on many facets of biology:

- The evolution of the various types of globins (see Chapter 12)
- Sequences of mitochondrial DNA (which is maternally inherited) have supported the hypothesis of an African origin for *Homo sapiens*;
- The recent evolution of human immunodeficiency viruses (see Chapter 18) and their simian relatives
- 'Molecular phylogeny' (see Box 6.6) has become an important area in the study of evolution.

Box 6.6   Nucleic acid and protein sequences can be used for 'molecular phylogeny'

Classical views on the evolution of organisms are based on study of the comparative morphology of fossils and of modern organisms on a timescale based on the geological record. Such comparisons can lead to the construction of **phylogenetic trees**, which illustrate the probable relationships between groups of organisms based on divergence from common ancestors.

In the last 30 years, sequence information from proteins and nucleic acids has provided another way of looking at relationships between organisms This provides a quantitative way of looking at evolutionary distances between modern organisms. It was realized, however, that present-day sequences also contain a record of the past and that 'molecular phylogenetic trees' can be constructed.

Such trees are based on the appropriate alignment of homologous sequences from different organisms followed by computer-based evaluation of the number of mutational changes required to produce the observed differences (that is, an estimate of the evolutionary distance). These molecular methods are more easily quantitated than the classical methods of morphological comparisons, but involve assumptions about the rates of evolutionary change – sometimes called the 'molecular clock'.

In many cases, molecular phylogenies have supported the classical ones, but some radical changes have been proposed, such as the division of life into three 'domains' – the Eubacteria, Archaea and Eucarya (see Chapter 5) or the proposal that fungi are closer to animals than to plants. One of the most exciting developments is the determination of DNA sequences from fossilized remains of up to 100 million years of age – leading to the popular science fiction fantasy of *Jurassic Park*!

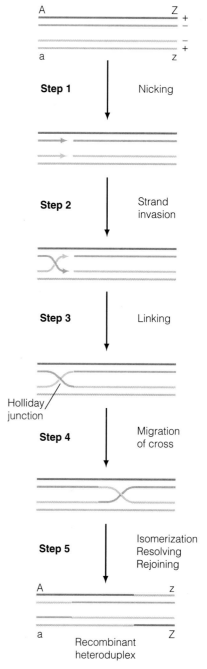

Figure 6.14 Recombination events in DNA.

Glossary of molecular biological terms

Aminoacyl tRNA (aa-tRNA) A *transfer RNA* molecule carrying a specific (or cognate) amino acid at its 3'-end. The amino acid is added by highly specific aa-tRNA synthetase and aa-tRNAs are the substrates for *translation* or protein synthesis.

Anticodon A *sequence* of three *nucleotides* on a specific stem-loop of a *transfer RNA* molecule which interacts with a *codon* on *messenger RNA* at the *ribosome* during translation.

Base pair (bp) A specific interaction between the bases of two *nucleotides*, one pyrimidine and one purine, normally between adenine and thymine or uracil (an AT base pair) or between guanine and cytosine (a GC base pair). Forms the basis of the copying that occurs during the processes of *DNA replication*, *transcription* and *translation*. Strictly speaking, it should be termed a 'nucleotide pair', but 'base pair' is the common usage.

Cap (or 5'-cap) A structure found at the 5'-end of most *messenger RNAs* of eukaryotes and added during *processing*. There are three types, containing a N^7-methyl G nucleotide and 0, 1 or 2 methyl groups on the ribose moieties of the first two nucleotides of the RNA (for example, Type II: $^7mGpppA_mA_m$).

Central Dogma A description of the flow of information in biological systems, represented as DNA ($\rightarrow$ DNA) $\rightarrow$ RNA $\rightarrow$ protein, and involving *template-directed polymerization* in the processes of *DNA replication*, *transcription* and *translation*.

Chromatin A term describing the association of DNA and protein in eukaryotic nuclei and in *nucleoprotein* extracted from them. See *Chromosome*; *Nucleosome*.

Chromosome (1) A *nucleoprotein* structure, found in the nuclei of eukaryotes, which contains one of the DNA molecules that comprise the *genome* of the organism. The proteins are the basic *histones* and other, acidic, proteins. (2) The DNA contained within a particular chromosome. (3) In prokaryotes, which generally contain a single DNA molecule, the term is roughly synonymous with *genome*. Chromosomal DNA can be used to distinguish the bacterial genome from any *plasmid* DNA.

Codon A *sequence* of three *nucleotides* on a *messenger RNA* molecule that interacts with an *anticodon* on an *aminoacyl tRNA* molecule in the *ribosome* during *translation*. There are 64 possible codons which between them specify 'start', 'stop' and the 20 amino acids used in translation. See also *Genetic code*.

Deoxyribonucleoside triphosphate (dNTP) The 5'-triphosphates of the deoxyribonucleotides containing adenine, guanine, cytosine and thymine (dATP, dGTP, dCTP, dTTP). They are substrates for *DNA polymerases* and *reverse transcriptases*.

DNA binding protein A wide variety of proteins function by binding to DNA, but do so in distinctive ways. Some, such as many *transcription factors*, show specificity towards particular sequences, while others bind in a nonspecific fashion. Others bind to single- rather than double-stranded DNA – these single-strand binding (ssb) proteins participate in *DNA replication*.

DNA polymerase (DP) An enzyme that catalyses the *template-directed polymerization* of DNA chains using a DNA template. DNA polymerases bring about *DNA replication* in concert with a variety of other enzymes and proteins. Eukaryotes have several different DPs, including DP α, which has an associated *primase* activity, and DP δ, with an associated 3'–5' *exonuclease*.

DNA repair Cellular mechanisms that reverse or remove and replace regions of DNA that have been damaged by covalent alteration by physical (for example,. ionizing radiation, UV light) or chemical agents (for example, alkylating or deaminating agents).

DNA replication The process by which a *duplex* DNA molecule is copied to produce two daughter duplexes from one parental duplex. The process is semi-conservative, with each daughter duplex acquiring one strand from the parental duplex, the other being newly synthesized. It proceeds by a replication fork invading the parental duplex. Replication is achieved by *DNA polymerases* acting in concert with a number of other enzymes and proteins to achieve a very faithful (one error in 10^9) copying of the genome. See also *DNA binding protein*; *Helicase*; *Ligase*; *Primase*; *RNase H*; *Topoisomerase*.

Duplex A term indicating the double-stranded nature of a polynucleotide, e.g. duplex DNA.

Elongation The process of increasing the length of a biological polymer (*polypeptide*, *polynucleotide* or polysaccharide) by one unit. The elongation step has to be repeated (*n*-1) times for a polymer of length *n*, in contrast to *initiation* and *termination*, which each occur only once per chain.

Endonuclease An enzyme that catalyses the hydrolysis of *nucleic acids* at internal locations on the *polynucleotide* chain. Endonucleases can act in a sequence-specific or nonspecific manner and can be DNA-specific, RNA-specific or nonspecific. See also *Exonuclease*; *Restriction endonuclease*.

Exon A *sequence* of *nucleotides* in the DNA of a eukaryotic *gene* whose RNA copy persists into the mature *messenger RNA*, in contrast to the *introns* which are removed by *splicing*.

Exonuclease An enzyme that catalyses the hydrolysis of *nucleic acid* in a stepwise fashion from one end of the *polynucleotide* chain. They can act in a 3'–5' or 5'–3' manner and can be DNA-specific, RNA-specific or nonspecific. See also *Endonuclease*.

Frame shift A *mutation* in the DNA of a *gene*, involving insertion or deletion of *nucleotides,* which results in a change of *reading frame* and thus of the sequence of the C-terminal part of the encoded protein.

Gene A length of DNA *sequence* that comprises a functional unit of a *genome*.

Gene expression The process by which the product(s) of a *gene* are made within a cell. All genes undergo *transcription* and the transcripts are *processed* into functional RNAs. Some gene products (e.g. *ribosomal* and *transfer RNAs*) function as RNAs, but expression of most genes leads to the production of *messenger RNA*, which undergoes translation to yield a *polypeptide* gene product. For most genes, therefore, gene expression is the outcome of both transcription and translation.

Genetic code The rules relating the 64 possible trinucleotide *codons* of *messenger RNAs* to the 20 amino acids used in *translation* and start and stop signals. The codons are written 5'–3' and lead to the selection of the correct *aminoacyl tRNA* by interacting with its *anticodon* in the *ribosome*.

Genome The total genetic information for an organism. In all cellular organisms, this is in the form of a *sequence* of DNA and is organized into a number of *genes*.

Helicase An enzyme activity, which uses the energy from the hydrolysis of ATP to ADP and phosphate to *unwind* a *polynucleotide* in the *duplex* state. DNA helicases are part of the mechanism of *DNA replication*.

Histone A type of basic protein that associates with DNA in *chromatin*. There are five types of histone, four of which (H2a, H2b, H3 and H4) are components of *nucleosomes* and a fifth (H1) which associates with internucleosomal regions. All contain a high proportion of the positively charged amino acids arginine and lysine, which help to neutralize the negative charges on DNA.

Initiation The process that starts the synthesis of biological polymers such as *polypeptides* and *polynucleotides*, generally occurring once per complete chain. See also *Elongation*; *Termination*.

Intron A *sequence* of *nucleotides* in the DNA of a eukaryotic *gene* whose RNA copy is removed by *splicing* during the *processing* of mature *messenger RNA*, leaving the *exons*.

Kilobase, kilobase pair (kb, kbp) A measure of the length of *genomes*, *genes* or *polynucleotide* molecules in thousands of bases, or *base pairs* for *duplex* polynucleotides. (Megabase is used for millions of bases.) Strictly speaking, they should be termed 'nucleotide' or 'nucleotide pair', but 'base' and 'base pair' are the common usage.

Ligase A type of enzyme that joins two molecules, with the expenditure of the energy in ATP or equivalent. DNA ligase joins two suitably aligned chains of DNA and catalyses the last step in the processes of *DNA replication* and *DNA repair*.

Messenger RNA (mRNA) An RNA molecule capable of acting as the *template* for *translation*. Eukaryotic mRNAs are made as long pre-mRNA *transcripts* by *RNA polymerase* II and subsequently *processed* into mature mRNA. Each species of mRNA directs the synthesis of a specific *polypeptide*. See also *Cap*; *Codon*; *Exon*; *Intron*; *Open reading frame*; *Polyadenylation*; *Splicing*; *Start codon*; *Stop codon*.

Mutation A heritable change in the *genome* of an organism. Mutations are the results of specific alterations in the DNA of the genome leading to altered *gene expression*, for example failure to produce a particular protein, or production of a faulty protein.

Nuclear pore complex A structure found on the membrane or envelope of the nucleus of eukaryotes which bridges the double bilayer of the nuclear membrane. There are several hundreds of these per nucleus and they have a characteristic and complex ring-like structure though which macromolecules, especially proteins made in the cytoplasm (see *Translation*) and RNAs made in the nucleus (see *Processing*; *Transcription*), pass between nucleus and cytoplasm.

Nucleic acid See *Polynucleotide*.

Nucleoprotein A complex, usually noncovalent, of *nucleic acid* (DNA or RNA) and protein. See also *Chromatin*; *Nucleosome*; *Ribosome*.

Nucleosome The fundamental packing unit of DNA in the *chromosomes* of eukaryotes, comprising an octameric complex of *histones* (two molecules each of histones 2a, 2b, 3 and 4) with 160 bp of *duplex* DNA wrapped around it and forming the 10 nm fibre of *chromatin*. Internucleosomal DNA associates with histone H1, leading to the formation of 30 nm fibres, the folding unit of the chromosomes of higher eukaryotes. Nucleosomes influence *transcription* and have to be disassembled and reassembled during *DNA replication*.

Nucleotide A molecule made up of three components: (1) a nitrogenous base, either a pyrimidine (adenine [A] or guanine [G]) or a purine (cytosine [C], uracil [U] or thymine [T]); (2) a pentose sugar, either D-ribose or 2-deoxyribose; (3) *ortho*-phosphate.

Open reading frame (ORF) A *sequence* of *nucleotides* on a functional *messenger RNA* molecule encoding a *polypeptide* and comprising (from 5' to 3') a *start codon*, a number of *codons* equal to the number of peptide bonds in the polypeptide and a *stop codon* in frame with the start.

Origin of replication A *sequence* of *nucleotides* in DNA within which *DNA* (or genomic) *replication* is initiated.

Plasmid A small DNA *genome* present in cells in addition to the genome of the cell. In prokaryotes, plasmids are small, circular *duplex* DNAs that replicate independently of *chromosomal* DNAs and carry a number of *genes* which confer properties on the bacterium such as resistance to antibiotics or to heavy metal ions. Plasmids of this type have been widely used in the laboratory for gene manipulation.

Polarity (of polymer chains) A term used to indicate that the two ends, or termini, of linear biological macromolecules are different. *Polynucleotides* have

3'- and 5'- ends, *polypeptides* have amino- and carboxy- termini (*N*- and *C*-). These molecules can be viewed (or attacked by an enzyme), for example, in a 5'-3' or in a 3'-5' direction. The two chains of *duplex* polynucleotides are antiparallel, or in opposite polarities.

Polyadenylation (poly A) The nontemplate-directed addition of 50–300 AMP residues to the 3'- end of RNA molecules during their *processing* into mature *messenger RNAs*. Occurs near a polyadenylation sequence (typically AAUAAA) and involves cleavage of the RNA chain and action of poly A polymerase.

Polynucleotide A polymer of *nucleotides*. Often used synonymously with nucleic acid. Occurs as DNA or as RNA.

Polypeptide A polymer of amino acids. This term is often used synonymously with protein.

Primase An activity that catalyses *template-directed polymerization* of short chains of RNA before the *initiation* of synthesis of DNA chains. See also *DNA polymerase*.

Processing A term describing covalent changes that are made to macromolecules, especially RNA and proteins, subsequent to their initial *template-directed polymerization*. See also *Cap*; *Messenger RNA*; *Polyadenylation*; *Signal sequence*; *Splicing*.

Promoter A *sequence* of nucleotides on DNA that influences the *transcription* of a specific *gene*. Often lies *upstream* from the transcription start site. See also *RNA polymerase*; *TATA box*; *Transcription factor*.

Reading frame Because the *genetic code* is triplet-based, any *sequence* of RNA can be *translated* in three different ways or possesses three reading frames, each encoding a different *polypeptide*. Normally a *messenger RNA* is translated in only one of these (but see *frame shift*) and the *ribosome* is directed to this by the *start codon*. A length of duplex DNA has six possible reading frames, three on each strand. See also *Open reading frame*.

Restriction endonuclease A member of a group of enzymes isolated from many different species of bacteria which have the property of hydrolysing duplex DNA at specific sequences. These sequences are 4–6 *base pairs* in length and are characteristic of the individual enzyme. Their function in the bacterium is thought to be protection against invasion of the bacterial cell by foreign DNA, but they are widely used in the laboratory for gene manipulation, which would not be possible without them.

Reverse transcriptase (RT) An enzyme that catalyses the *template-directed polymerization* of DNA chains using a RNA template. RTs are present in the particles of retroviruses and play a crucial role in the distinctive replication of these viruses. RTs are widely used in the laboratory for making cDNAs (DNA molecules complementary to RNAs).

Ribonucleoside triphosphate (rNTP) The 5'-triphosphates of the ribonucleotides containing adenine, guanine, cytosine and uracil (rATP, rGTP, rCTP, rUTP). They are substrates for *RNA polymerases* and *RNA replicases*.

Ribosomal RNA (rRNA) The RNA present in *ribosomes*. The small ribosomal subunit contains a single molecule of RNA (16S in prokaryotes and 18S in eukaryotes), the large subunit contains one large (23S in prokaryotes and 28S in eukaryotes) and one or two small molecules of RNA (5S in prokaryotes and both 5S and 5.8S in eukaryotes).

Ribosome A particle, made of RNA and protein and formed of one large and one small subunit. Prokaryotic ribosomes (70S) are made of 30S and 50S subunits; the larger eukaryotic ribosomal (80S) subunits are 40S and 60S. The subunits come together only during initiation of translation and dissociate at termination. Many of the ribosomes of eukaryotic cells are bound to the rough endoplasmic reticulum. See *Targeting*.

RNA polymerase (RP) An enzyme that catalyses the *template-directed polymerization* of RNA chains using a DNA template. RNA polymerases bring about *transcription* in concert with a variety of *transcription factors*. Eukaryotes have three multisubunit RPs, RP I, RP II and RP III, which catalyse the synthesis of pre-rRNA, pre-mRNA and pre-tRNA *transcripts* respectively, each with its own set of *transcription factors*.

RNA replicase (RNA transcriptase) An enzyme that catalyses the *template-directed polymerization* of RNA chains using a RNA template. Enzymes of this type are probably absent from cellular organisms but are necessary for the replication of 'true' RNA viruses, all of which have one or more *genes* encoding RNA replicase components. Some RNA viruses contain RNA transcriptase activity in their virus particles.

RNase H An enzyme activity that hydrolyses only the RNA strand of an RNA–DNA hybrid *duplex*. RNase H activities are involved in the removal of the RNA made by *primase* during *DNA replication* and during the conversion of the RNA form of retroviral *genomes* into duplex DNA form by *reverse transcriptase*.

Sequence The order and number of amino acids in a *polypeptide*, or of nucleotides in a *polynucleotide*. Also known as primary structure, it represents the biological information content of organisms, which is stored in the sequence of the DNA of the *genome*. The direct link between sequence, three-dimensional structure and biological function lies at the heart of the understanding of molecular biology. The enormous growth of DNA sequence information has been the greatest development in biology in recent decades.

Signal sequence (or localization sequence) A *sequence* of amino acids, usually short, within a *polypeptide* necessary to direct the molecule to its proper location within the cell. See also *Targeting*.

Splicing The process by which sections of RNA sequence (*introns*) are cut out of pre-mRNA molecules as part of the *processing* into mature *messenger RNA*. To preserve the message, this must occur with precision at exact consensus sequences, and is catalysed by spliceosomes (complexes of RNA and proteins). Alternative splicing can lead to the production of more than one type of mRNA (and thus *polypeptide*) from a single *transcript*.

Start codon A *codon*, normally AUG, in a *messenger RNA* molecule, recognized by the *ribosome* and associated *initiation* factors and leading to the selection of the first amino acid (methionine) of the new *polypeptide* chain. In eukaryotic mRNAs, it normally lies near the 5' *cap* and within an appropriate consensus sequence, although cap-independent initiation can also occur. Bacterial protein synthesis can initiate at internal start codons on an mRNA molecule

Stop codon A *codon* in *messenger RNA* that is recognized by the *ribosome* and associated *termination* factors and leads to release of the completed *polypeptide* from the ribosome. There are three possible stop codons – UAA, UAG and UGA – and they must be in the same *reading frame* as the *start codon*. See also *Open reading frame*.

Tail A *sequence* of 50–300 AMP residues covalently attached to the 3' end of most mature eukaryotic *messenger RNAs* during *processing*. See also *Polyadenylation*.

Targeting The processes that bring about the direction of newly synthesized macromolecules, especially proteins, to their correct intracellular location. See also *Signal sequence*.

TATA box A consensus sequence of *nucleotides* in DNA 25 bp *upstream* from the transcriptional start site of eukaryotic *genes* copied by *RNA polymerase II*. It is recognized by a protein – the TATA box binding protein – which binds to both the sequence and a number of other *transcription factors* in order to form a pre-initiation complex with RNA polymerase

Template A molecule with a defined *sequence* that can direct the synthesis of another molecule. All known biological templates are *polynucleotides*, either DNA or RNA, and act in the processes of *DNA replication*, *transcription* and *translation*.

Template-directed polymerization The process by which polynucleotides of specific are synthesized according to the instruction of a complementary polynucleotide template, using nucleoside triphosphates (rNTPs or dNTPs) as precursors. See also *DNA polymerase*; *RNA polymerase*; *RNA replicase*; *Reverse transcriptase*.

Termination The process that completes the synthesis of biological polymers such as *polypeptides* and *polynucleotides*, occurring once per complete chain and leading to the release of the completed molecule. See also *Elongation*; *Initiation*.

Topoisomerase An enzyme that can relax supercoiling of closed circular *duplex* DNA molecules by breaking and rejoining *polynucleotide* chains. Topoisomerases relieve tangling generated by the unwinding of DNA during *DNA replication* and *transcription*.

Transcript The immediate product of the action of *RNA polymerase* in the *transcription* of one strand of a *duplex* DNA *template*. Such transcripts (pre-rRNA, pre-mRNA or pre-tRNA) are subsequently processed into the mature, functional RNA species (rRNA, mRNA or tRNA). See also *Processing*; *Splicing*.

Transcription The process by which a *sequence* of *nucleotides* in a *duplex DNA* molecule acts in the *template-directed polymerization* of *rNTPs* into an RNA molecule. Transcription in eukaryotes is catalysed by one of three *RNA polymerases* acting in concert with protein *transcription factors*.

Transcription factor (TF) A member of a group of proteins that act in concert with *RNA polymerase* to carry out *transcription*. They appear to act by binding, directly or indirectly via other proteins to specific *sequences* on DNA and so regulating the *initiation* of transcription positively (activation or up-regulation) or negatively (repression or down-regulation). TFs that interact at the *TATA box* effect basal transcription; those interacting with more distant sequences are involved in control of transcription.

Transfer RNA (tRNA) A member of a group of small (75–90 *nucleotides*) RNA molecules to which amino acids can be added in a specific fashion by *aminoacyl-tRNA* synthetases for use in *translation*. Molecules of tRNA all have similar three-dimensional structures, with three stem–loop structures, one of which carries the *anticodon*. Cells contain over 40 species of tRNA, at least one for each of the 20 amino acids used in translation.

Translation The process by which the sequence of *codons* in a *messenger RNA* molecule acts in the *template-directed polymerization* of amino acids into a specific *polypeptide* molecule. It occurs on a *ribosome* and requires *aminoacyl-tRNAs*, energy in the form of GTP and a range of protein factors for the individual processes of *initiation*, *elongation* and *termination*.

tRNA See *Transfer RNA*.

Unwinding The energy-dependent separation of the two *polynucleotide* strands of *duplex* DNA, required temporarily during *transcription* and permanently during *DNA replication*. See also *Helicase*.

Upstream sequence A *sequence* on DNA on the 5'-side of the *transcription* start site of a *gene*, to which *transcription factors* can bind in a specific manner, thus influencing *gene expression*.

PART C

CELLULAR BIOCHEMISTRY

CHAPTER 7
An introduction to metabolic processes and energy production in mitochondria

CHAPTER 8
The metabolism of glucose: storage and energy generation

CHAPTER 9
Lipid metabolism

CHAPTER 10
Metabolism of nitrogen-containing compounds

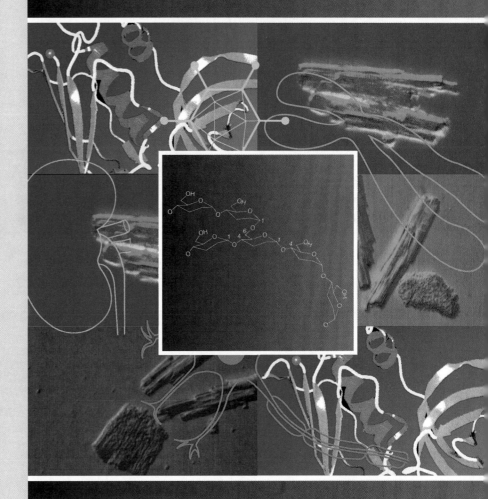

CHAPTER 7

An introduction to metabolic processes and energy production in mitochondria

Nutrient molecules are a source of energy, which is extracted through reactions catalysed by enzymes. A series of enzymes that cooperate to degrade or to synthesize specific molecules forms a metabolic pathway. Energy in the form of reducing equivalents, released from nutrients as they are oxidized, is harvested as the reduced nucleotides NADH and $FADH_2$.

Within the mitochondrial **electron transport chain** the reducing equivalents are transferred to a series of carrier molecules. The flow of electrons generated in this process is linked to the formation of a proton gradient across the inner mitochondrial membrane. The energy of this gradient can be captured as protons flow back through the membrane, by coupling it to the phosphorylation of ADP to form ATP (the process of **oxidative phosphorylation**).

The catabolism of carbohydrates, fatty acids and some amino acids produces acetyl coenzyme A, an activated form of acetate which is completely oxidized to CO_2 and H_2O through the **citric acid cycle** (Krebs cycle or tricarboxylic acid cycle) within the mitochondrial matrix.

Cells have many mechanisms through which they regulate energy production and utilization, and some of these will be introduced in this chapter.

The survival of biological species depends on their ability to derive energy from nutrients

Animals of all species devote a high proportion of their time to providing food for themselves and their dependants. This food is the source of energy for movement, thought, reproduction, growth, repelling invading organisms, repair of body tissues and every other activity needed or desired by animals. In the modern world the human animal has new ways of collecting and providing food. Most of us have replaced direct hunting with 'going to the office, factory, or institute of learning' to indulge in activities that bring in an income we can use to buy food, so that we spend a smaller fraction of our time in the supermarket exchanging the income for food. However, this is only a sophistication of the time-honoured process, and it is as important for us as it was for primitive humans that the food we consume should be converted efficiently to other useful forms of energy, so that the time we spend 'collecting' food is no greater than is compatible with the other requirements of survival.

Energy transfer in metabolism

The complex molecules of foodstuffs contain energy, which can be released by oxidation

Nutrient molecules (carbohydrates, fats and proteins) have potential energy arising from their chemical properties. As these energy-rich foodstuffs are oxidized and degraded to smaller molecules energy is transferred to **nucleotide coenzymes**. Many of the smaller molecules produced can be reused as substrates for the synthesis of new structural or nutrient storage macromolecules designed to meet the specific requirements of the cell. Synthetic processes consume energy. Efficient metabolic utilization of foodstuffs therefore requires that as much as possible of the energy released during degradation is harnessed for use in biosynthesis. However, many biological processes are less than 50% efficient and energy is lost, largely as heat.

Energy can not be passed directly from energy-generating to energy-utilizing reactions

In biological systems energy must be passed from one process to another in the form of specialized 'energy-rich' molecules, notably purine nucleotide triphosphates (ATP and GTP), flavin adenine dinucleotide ($FAD(H_2)$) and nicotinamide adenine dinucleotides (NAD(H) and NADP(H)).

The drive for energy efficiency is further complicated by the fact that the degradation of nutrients does not always occur at the same time, the same rate, or even in the same location at which energy is required. This means that energy released during degradation of nutrients must be stored in a form that allows it to be transferred to the energy-utilizing processes, possibly in a different tissue.

Throughout metabolism varying proportions of the chemical energy are released as heat. This heat is important in enabling us to control our body temperature within very narrow limits, even when we are exposed to relatively wide variations in environmental temperature.

Adenosine triphosphate (ATP) is the immediate form of energy used by most processes

Adenosine triphosphate (Figure 7.1; see also Chapter 2) is described as having a 'high phosphate transfer potential'. This means that a large amount of free energy

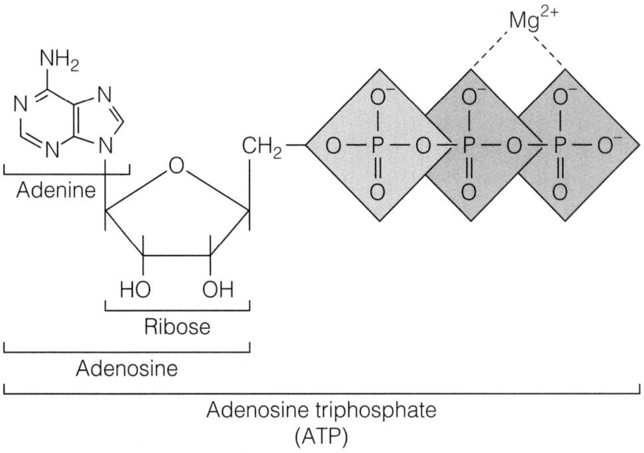

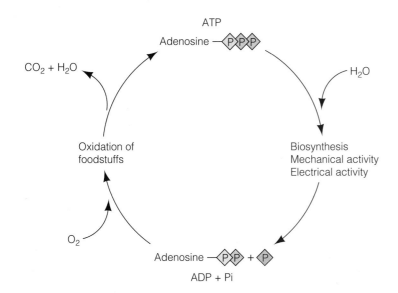

Figure 7.1 The ATP–Mg^{2+} complex.

becomes available if the terminal phosphate group of ATP is hydrolysed or transferred. The product of this reaction is the corresponding diphosphate, **adenosine diphosphate (ADP)**, and the standard free energy change ($\Delta G^{o'}$) associated with this hydrolysis is 30.5 kJ mol^{-1} (7.3 kcal mol^{-1}).

$$ATP + H_2O \rightarrow ADP + Pi + H^+$$

The energy released in this way may be coupled to a variety of energy-requiring processes including biosynthesis, generation of mechanical energy for muscle contraction or generation of electrical energy, which permits transmission of impulses in brain, nerve and muscle cells.

Cells have only small amounts of ATP

Even muscle cells, which have potentially the greatest demand for ATP, have only about 5 µmol g^{-1}: other tissues have a great deal less. The energy inherent in each ATP molecule is consumed almost as soon as it is available so that, in order to sustain the function of cells, ATP must be regenerated continuously by adding a phosphate group to ADP. The energy required for this phosphorylation, the generation of an energy-rich bond, is derived from the oxidation of foodstuffs. There is therefore a continuous cyclic process by which the phosphoanhydride bonds of ATP molecules are hydrolysed to make energy available and reformed using energy extracted from the oxidation of nutrients (Figure 7.2).

Figure 7.2 Hydrolysis of the γ-phosphate group of ATP releases energy, which is coupled to energy-using processes. ADP is rephosphorylated through metabolic process that derive energy from the oxidation of foodstuffs.

An adult at rest consumes an amount of energy equivalent to 40 kg ATP per day. Of course, we do not have 40 kg of ATP in our bodies and it would be more accurate to say that in the course of a day we consume an amount of energy equivalent to that made available when the terminal phosphate group is hydrolysed from about 80 mol ATP (the molecular weight of ATP is about 500 Da). Most of the ADP produced will be rephosphorylated immediately, restoring ATP so that the same molecule is continually cycled between the diphosphate and the triphosphate forms. At times of increased physical activity there is a huge increase in the energy required by muscle cells. Energy consumption may rise to the equivalent of 1 mol (approximately 0.5 kg) of ATP per minute, and mechanical activity will continue only while ATP is regenerated at the same rate.

Endogenous nutrient stores sustain our energy requirements between meals

We absorb more nutrient during a meal than is required to sustain our immediate energy needs. The excess energy is used to form intracellular (endogenous) nutrient stores which take the form of complex carbohydrates (in mammals this is a polymer of glucose called glycogen) or complex lipids (mainly triacylglycerols or triglycerides).

Storage of nutrients is important because the frequency of food intake varies between individuals as well as between species. The molecules we store as endogenous nutrients provide an energy reserve, a source of nutrient from which the body can draw energy between meals. Between meals we maintain ATP concentrations in cells by oxidizing the fats and carbohydrates that have been stored in the immediate aftermath of a meal. Even at rest, cells are constantly replacing structural and functional molecules, and maintaining ion gradients across cell membranes, both processes that require ATP.

Chapter 2. Major carbohydrate fuel reserves are polymers of glucose linked by *O*-glycosidic bonds, page 19.

The molecules of intracellular nutrient stores frequently differ in chemical form from those of dietary nutrients because stored molecules must be in a state from which energy can be efficiently and rapidly recovered. For example, most of the carbohydrate in our diet is in the form of starch but the carbohydrate stored in tissues is in the form of glycogen. Both starch and glycogen are polymers of glucose and differ only in the way the glucose units are linked (see Chapter 2).

Metabolic pathways consist of sequences of enzyme-catalysed chemical reactions

Chapter 4. Only a limited range of enzyme-catalysed reaction types occur, page 91.

Chapter 4. Catalytic proteins: enzymes, page 88.

The chemical changes of metabolism are catalysed by a special class of proteins called **enzymes** (see Chapter 4). Each enzyme is capable of catalysing only a single type of chemical change. A single enzyme-catalysed step may involve no more than the removal or addition of an atom. Some functional groups such as phosphate groups or amino groups may be treated as a chemical unit in enzyme-catalysed reactions (Chapter 4). In order to achieve the substantial chemical changes needed to completely oxidize glucose or a fatty acid many enzymes must cooperate to produce, by small sequential steps, a large overall chemical change. Close cooperation between the enzymes of a pathway ensures that the production of energy when nutrients are degraded is maximized, and that the expenditure of energy during synthesis of new molecules is kept to a minimum.

Metabolic pathways may be linear, branched or circular. In a **linear** pathway the substrate is converted through a sequence of reactions to one or more products that do not re-enter the pathway. Several intermediate products may be derived from a linear pathway in addition to the obvious end product.

$$a \rightarrow b \rightarrow c \rightarrow d$$

Glycolysis (Figure 7.3) is an example of a linear pathway. Glucose, a six-carbon sugar, enters the pathway and after ten separate enzyme-catalysed reactions the main product is two molecules of the three-carbon compound, pyruvate.

A **branched** pathway has an initial linear sequence of reactions common to two or more separate linear pathways that diverge from a single product arising at the branch point.

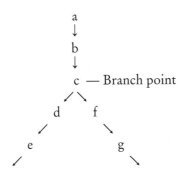

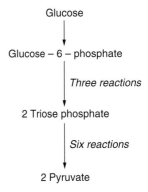

Figure 7.3 Outline of glycolysis.

In a **circular** pathway the intermediates form a circular array. A substrate enters the pathway by condensation with one of the intermediates. As the reactions proceed, product(s), not necessarily derived directly from the substrate entering the cycle, is (are) released. The Krebs or citric acid cycle (Figure 7.4) is representative of this type of pathway: a two-carbon unit, the acetyl group of acetyl CoA, enters the cycle by condensation with oxaloacetate and two molecules of CO_2 are released with each turn of the cycle. The carbon released as CO_2 is derived from the carbon skeleton of the original molecule of oxaloacetate rather than from the molecule of acetyl CoA entering the cycle. The cycle is completed with the regeneration of a molecule of oxaloacetate, ready for the condensation reaction that will incorporate another acetyl unit from acetyl CoA.

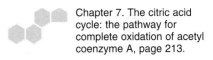

Chapter 7. The citric acid cycle: the pathway for complete oxidation of acetyl coenzyme A, page 213.

Metabolism has two components: anabolism and catabolism

In **catabolic** processes larger molecules (macromolecules) are degraded to smaller molecules. Catabolism is associated with the liberation of energy, and generally involves oxidative processes. In metabolic processes molecules are oxidized

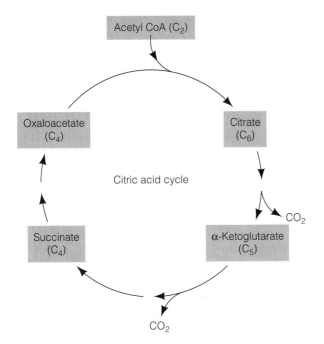

Figure 7.4 Outline of the citric acid cycle. The citric acid cycle is a cyclic pathway in which a C_2 molecule (acetyl CoA) condenses with a C_4 molecule (oxaloacetate) to produce the C_6 intermediate citrate, from which the cycle takes its name. In a series of oxidation and decarboxylation reactions two molecules of carbon dioxide are evolved and a new molecule of oxaloacetate is generated.

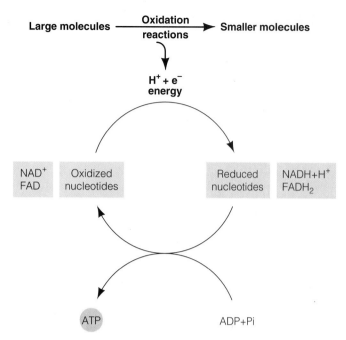

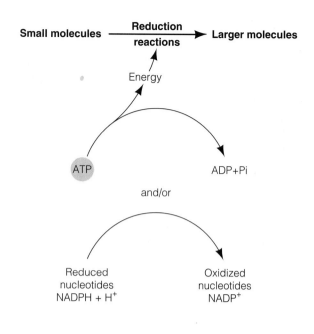

Figure 7.5 In some catabolic reactions ADP is phosphorylated to produce ATP directly but, in most, nicotinamide (NAD$^+$) or flavin (FAD) coenzymes are reduced using the energy generated by the catabolism of large molecules. The reduction of these coenzymes is coupled to the phosphorylation of ADP to ATP.

Figure 7.6 The energy for anabolic reactions may be derived from hydrolysis of ATP and reducing power from the oxidation of the reduced nucleotide NADPH.

through the loss of reducing equivalents and in many cellular reactions these are transferred to the coenzymes **nicotinamide adenine dinucleotide (NAD$^+$)** or **flavin adenine nucleotide (FAD),** which are reduced to NADH or FADH$_2$ respectively (Figure 7.5).

The biosynthesis of more complex molecules from smaller precursors is called **anabolism**. This consumes energy from high-energy molecules such as ATP. Biosynthesis usually involves reduction reactions, the addition of reducing equivalents being contributed by the reduced nucleotides. Anabolism is important for the synthesis of structural components of cells and molecules to be stored as endogenous energy reserves (Figure 7.6).

In biological systems electrons may be transferred in several different ways

Electrons are lost from a substrate during its oxidation and electrons are gained by the substrate being reduced. By definition, oxidation of a substrate must be linked to reduction and reduction of a substrate must be linked to oxidation: neither occurs in isolation.

Direct combination of electrons with molecular oxygen is rare in biological systems but does occur in a small number of reactions catalysed by **oxidases**, including **xanthine oxidase**, which converts hypoxanthine through xanthine to uric acid (see Chapter 10):

Chapter 10. The catabolism of purine nucleotides can have important clinical consequences, page 322.

$$\text{Hypoxanthine} \xrightarrow[\textit{Xanthine oxidase}]{O_2 \quad H_2O_2} \text{Xanthine} \xrightarrow[\textit{Xanthine oxidase}]{O_2 \quad H_2O_2} \text{Uric acid}$$

Transfer of electrons alone occurs in reactions linked to changes in the oxidation state of metal ions such as copper or iron.

$$Cu^{2+} + e^- \rightleftharpoons Cu^+$$
$$Fe^{3+} + e^- \rightleftharpoons Fe^{2+}$$

Examples of this type of oxidation–reduction coupling are found in the function of the cytochromes of the electron transport chain, and of those enzymes that contain iron-sulphur centres.

Transfer of a hydrogen atom, equivalent to a hydrogen ion plus an electron, occurs in dehydrogenase-catalysed reactions linked to FAD. In these reactions FAD accepts two hydrogen atoms from the substrate, and is reduced to FADH$_2$.

The general type of reaction is illustrated below:

$$R.CH_2.CH_2.R' + FAD \rightleftharpoons R.CH = CH.R' + FADH_2$$

Transfer of a hydride ion, which is a negatively charged species carrying two electrons, occurs in dehydrogenase-catalysed reactions linked to NAD$^+$, in which the coenzyme either accepts or donates the hydride ion. The enzyme-catalysed reaction removes two hydrogen ions and two electrons (effectively two hydrogen atoms) from the substrate. NAD$^+$ accepts two electrons and one hydrogen ion (effectively a hydride ion, H$^-$) into the nicotinamide ring, forming NADH. The second hydrogen is released into the cytosol or mitochondrial matrix as a proton. In these reactions the reducing equivalent is the electron carried by the hydrogen atom and transferred to the coenzyme, not the hydrogen ion released into the cell. The coenzymes involved in oxidation–reduction reactions are specialized for carrying electrons.

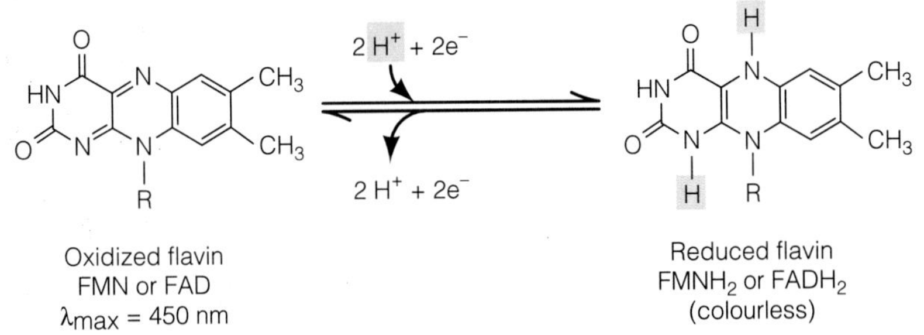

Box 7.3 Iron-sulphur centres in the electron transport chain, page 202.

Oxidized flavin
FMN or FAD
λ_{max} = 450 nm

Reduced flavin
FMNH$_2$ or FADH$_2$
(colourless)

Figure 7.7 The reduction of flavin nucleotides FAD and FMN to FADH$_2$ and FMNH$_2$ in simplified form. Two reducing equivalents are taken up in reducing the isoalloxazine ring system. (See Figure 7.10 for the full structure of FAD.)

Reduced and oxidized forms of the nicotinamide ring of the coenzyme nicotinamide adenine dinucleotide are shown in Figure 7.8.

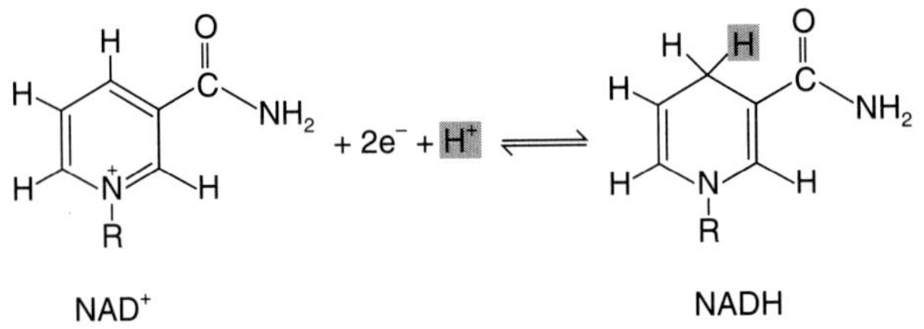

NAD$^+$

NADH

Figure 7.8 During NAD$^+$-dependent oxidation reactions the nicotinamide ring accepts a proton, forming the reduced coenzyme NADH. R represents the remainder of the NAD$^+$ molecule (see Figure 7.10 for the full structure).

The typical form of these NAD(H)-dependent reactions is illustrated by the reaction catalysed by lactate dehydrogenase, although the electrons which are transferred are not apparent in this conventional presentation of the reaction (Figure 7.9).

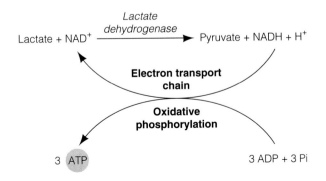

Figure 7.9 Reoxidation of NADH is linked to energy production in mitochondria.

Nicotinamide adenine dinucleotide (NAD$^+$)

Flavin adenine dinucleotide (FAD)

Figure 7.10 Structures of NAD$^+$ and FAD.

Chapter 7. Aerobic ATP production: the process of oxidative phosphorylation, page 199.

The reduced coenzymes serve as electron donors

The process of oxidation of nutrients is completed when these reduced coenzymes transfer their electrons to the mitochondrial electron transport chain (which is described in detail later in this chapter) and are thereby reoxidized. Reoxidation of one NADH to NAD$^+$ through the electron transport chain supplies sufficient free energy to generate three phosphoanhydride bonds, phosphorylating three molecules of ADP to ATP. Reoxidation of one FADH$_2$ to FAD produces sufficient energy to generate two phosphoanhydride bonds.

As the coenzymes cycle between reduced and oxidized forms the overall effect is that energy liberated by the oxidation of nutrients is transferred to the synthesis of a β–γ phosphoanhydride bond as ADP combines with Pi, producing ATP.

Within eukaryotic cells, oxidation and reduction processes are often segregated by intracellular membranes

Mitochondria are subcellular structures specialized for those aspects of metabolism requiring oxidation (catabolism) of nutrient molecules. Oxidation of nutrients involves enzyme-catalysed transfer of electrons from the nutrient molecules to the NAD$^+$ and FAD coenzymes: most of the NAD$^+$ and FAD within the cell is found in the mitochondrial compartment. By contrast, reductive reactions involve the enzyme-catalysed transfer of electrons from the reduced coenzyme **nicotinamide adenine dinucleotide phosphate** (**NADPH**) to the substrate molecules. Reactions involving reduction of substrates occur predominantely in the cytoplasm, and it is here that we find the highest concentration of NADPH. This coenzyme contributes electrons to reductive reactions involved in anabolic reactions including the synthesis of fatty acids and of steroid molecules

The structure of NADP$^+$ differs from that of NAD$^+$ only in that an additional phosphate group is esterified to the 2'-OH of the adenine-linked ribosyl ring (Figure 7.11). NADPH carries electrons in the same way as NADH, but is able to donate these electrons to reductive reactions such as are involved in the synthesis of fatty acids; NADH passes its electrons predominantly to the synthesis of ATP. A typical example of the type of reaction in which NADPH participates is the reduction of a carbon–carbon double bond.

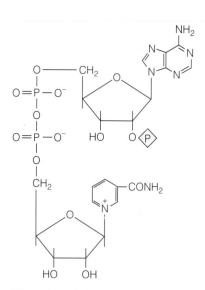

Figure 7.11 Structure of NADP$^+$.

$$R.CH = CH.CO.R' + NADPH + H^+ \rightleftharpoons R.CH_2.CH_2.CO.R' + NADP^+$$

Box 7.1 Thermodynamics of metabolic processes

The relationship between free energy change and the equilibrium constant of reactions

Before we consider how cells control their energy balance we must look briefly at the relationships between free energy change and the direction of chemical reactions. **Free energy** is the energy in a system available for useful work and is measured in calories mol^{-1} or joules mol^{-1} (1 calorie = 4.185 J). In molecular terms free energy may be seen as the energy stored within the structure of the molecule, and is a measure of the capacity of the molecule to release energy in the course of chemical reaction. The caloric value of foods is a measure of the amount of free energy that may be derived from the complete oxidation of the food to CO_2 and H_2O.

The direction in which a reaction proceeds is indicated by whether the **free energy change** (ΔG) for the reaction is negative or positive. The magnitude of ΔG indicates how far the process is from equilibrium. A reaction exhibiting a negative free energy change releases energy (exergonic reaction) and will proceed to equilibrium. A reaction exhibiting a positive free energy change absorbs energy (endergonic reaction) and will not proceed unless it is provided with energy. In considering biological processes it is important to remember that the change in free energy associated with conversion of one substance to another is the same, irrespective of the route through which that change is produced.

The free energy change associated with a reaction is related to the equilibrium constant for that reaction

The equilibrium constant is the ratio of the molar concentrations of products and reactants at equilibrium.

$$A + B \rightleftharpoons C + D$$

For the reaction above this ratio would be expressed as:

$$K_{eq} = \frac{[C][D]}{[A][B]}$$

where [A], [B], [C] and [D] represent the molar concentrations of the substances A, B, C, and D when the reaction has reached equilibrium.

The **equilibrium constant** is independent of the rate of reaction and would be the same whether the reaction were enzyme-catalysed or spontaneous. However, the concentrations of the individual components of the reaction at equilibrium will depend on the initial concentrations of all the participating species.

The free energy change (ΔG) for a reaction is related to the equilibrium constant by the relationship

$$\Delta G = \Delta G^\circ + RT \ln K_{eq}$$

where ΔG° is the standard free energy change measured in kJ mol^{-1} (or kcal mol^{-1}). ΔG° is a constant for each reaction and is calculated for conditions where temperature is 25°C, pressure 1 atmosphere, and the initial concentration of each of the reactants and products is 1M. The ΔG° therefore refers to conditions that do not prevail within cells but it allows us to compare the free energy change that would be associated with different reactions if they were occurring under equivalent conditions. $\Delta G^{\circ'}$ refers to an environment where the pH is 7.0 and is by convention used for biological systems.

At equilibrium, ΔG is by definition 0, so the equation simplifies to

$$\Delta G^{\circ'} = -RT \ln K_{eq}$$

$\Delta G^{\circ'}$ is the standard free energy change associated with the reaction occurring at pH 7.0. It can perhaps best be understood as the difference between the energy content of the reactants and the energy content of the products under standard conditions and at pH 7.0.

In essence, the standard free energy change of a chemical reaction is simply a mathematically different way of expressing its equilibrium constant. If the concentrations of reactants and products at equilibrium can be measured, the standard free energy change can be calculated using the above equation. If the equilibrium constant is greater than 1, the standard free energy change will be negative; conversely, if the equilibrium constant is less than 1, the standard free energy change will be positive.

The ratio of oxidized to reduced nucleotide determines the direction of oxidation–reduction reactions

The total concentration of NAD^+ plus NADH is about 10 μM in most mammalian cells, and the ratio of NAD^+ to NADH is maintained at about 500:1. This ratio favours the acceptance of hydrogen ions from metabolites, making NAD^+ an ideal coenzyme for oxidation processes. In contrast, the total concentration of $NADP^+$ plus NADPH is about 1 μM, and the reduced form predominates. This favours hydride transfer from NADPH to metabolites, making $NADP^+$ the ideal coenzyme for reductive reactions.

Reactions with unfavourable free energy changes can be driven by those with favourable free energy changes

For any sequence of reactions, the total free energy change is equal to the sum of the individual free energy changes of the component reactions. This explains how coupling of a thermodynamically favourable reaction to an unfavourable reaction allows the latter to proceed: indeed, one reaction with a large negative free energy change could in theory drive an entire pathway. In metabolism, thermodynamically unfavourable reactions are often coupled to, and driven by, energy released by the hydrolysis of phosphoanhydride bonds of ATP.

It is important when looking at tables of standard free energy changes to remember that the 'standard' thermodynamic conditions do not prevail within biological systems. The chemical components of biological reactions are not present at 1 M concentrations, and the pH varies. The standard free energy changes provide useful comparisons, but should not be used for a quantitative analysis of the actual energy changes occurring in metabolism.

In this section we have concentrated on the concept of energy inherent in molecular structures, energy which can be released in the course of chemical reaction. In addition, energy may be inherent in concentration gradients, in which a substance is maintained at different concentrations on two sides of a biological membrane. If the concentrations on the two sides of the membrane are equalized there is a concentration equilibrium across the membrane and the energy is dissipated.

ATP is the common intermediate in metabolism, the molecule which conserves energy during catabolism and contributes it to anabolism

ATP serves as the principal immediate donor of free energy in biological systems. It is often called the **common intermediate** or **energy currency** in metabolism because it stores energy in an amount intermediate between that of the highest energy molecules and that of low-energy molecules in biological systems. ATP is therefore ideally positioned in the biological energy spectrum to transfer energy between reactions.

The molecular basis of the high-energy status of ATP is well understood

ATP is called a high-energy compound because it has two energy-rich phosphoanhydride bonds. The energy derives from interactions between atoms forming the phosphoanhydride bonds and is released when the phosphate groups are removed by hydrolysis or transferred to other molecules.

ATP has three phosphate groups; the α, β, and γ phosphates. Thermodynamic calculations indicate that when the terminal, γ, phosphate is hydrolysed by the reaction:

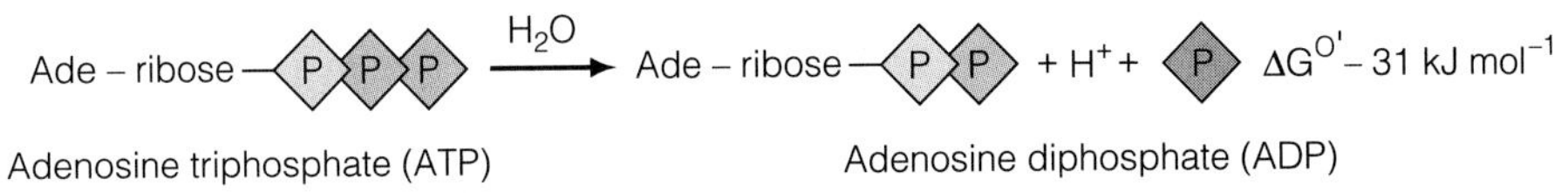

Ade – ribose $\Diamond$P$\Diamond$P$\Diamond$P $\xrightarrow{\text{H}_2\text{O}}$ Ade – ribose $\Diamond$P$\Diamond$P + H$^+$ + $\Diamond$P $\Delta G^{O'}$ – 31 kJ mol^{-1}

Adenosine triphosphate (ATP)   Adenosine diphosphate (ADP)

31 kJ (7.3 kcal) energy is released per mol ATP. This theoretical value is calculated for ATP hydrolysis under 'standard' thermodynamic conditions. The actual amount of energy released when this reaction occurs in a cell may be as great as 50 kJ (12 kcal) per mol, depending on the pH and the buffer capacity of the environment, and the concentrations of nucleotides, inorganic phosphate and Mg^{2+} and Ca^{2+} ions. These ions are important because they interact electrostatically with the charged oxygens of the phosphate groups. As magnesium ions bind much more tightly between the β and γ phosphates of ATP than between the α and β phosphates, the fraction of cellular Mg^{2+} remaining free will increase as ATP is hydrolysed to ADP. This may have consequences for other Mg^{2+}-dependent processes. The pH contributes to the energy change because hydrogen ions participate in the reaction.

Hydrolysis of the β phosphate group of ADP

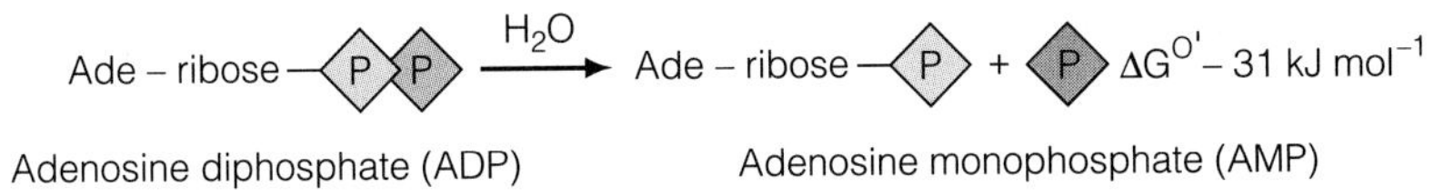

Ade – ribose $\Diamond$P$\Diamond$P $\xrightarrow{\text{H}_2\text{O}}$ Ade – ribose $\Diamond$P + $\Diamond$P $\Delta G^{O'}$ – 31 kJ mol^{-1}

Adenosine diphosphate (ADP)   Adenosine monophosphate (AMP)

releases an almost identical amount of energy, 31 kJ/mol.

Hydrolysis of the α phosphate group releases considerably less energy, only about 14 kJ/mol. This reflects the fact that the α phosphate group is linked to the ribose residue by a phosphoester bond rather than a phosphoanhydride bond. The α phosphate is therefore attached through a relatively 'low energy' bond.

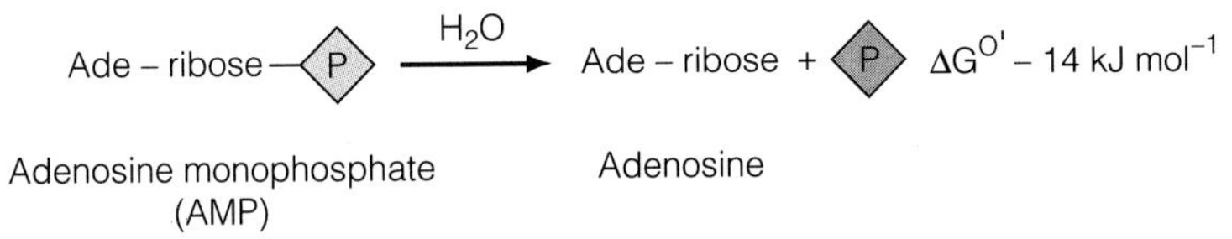

Ade – ribose $\Diamond$P $\xrightarrow{\text{H}_2\text{O}}$ Ade – ribose + $\Diamond$P $\Delta G^{O'}$ – 14 kJ mol^{-1}

Adenosine monophosphate   Adenosine
(AMP)

This is summarized in Figure 7.12.

Some metabolic reactions derive energy from coupling to the hydrolysis of the pyrophosphate group (removal of the β–γ diphosphate) of ATP:

This reaction has a standard free energy change of 31 kJ mol^{-1}, but the pyrophosphate is immediately subjected to pyrophosphorolysis by a pyrophosphatase enzyme producing two orthophosphate ions and releasing additional energy amounting to approximately 30 kJ mol^{-1}. The total energy yield is therefore equivalent to the hydrolysis of two high-energy phosphate bonds.

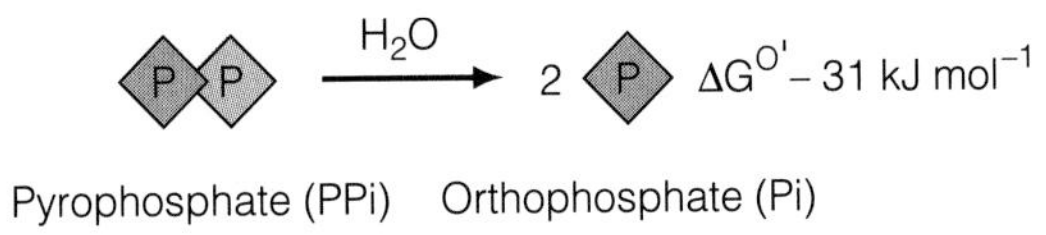

$\Diamond$P$\Diamond$P $\xrightarrow{\text{H}_2\text{O}}$ 2 $\Diamond$P $\Delta G^{O'}$ – 31 kJ mol^{-1}

Pyrophosphate (PPi)   Orthophosphate (Pi)

Adenylate kinase transfers phosphate between adenine nucleotides

As the free energy of hydrolysis of the β and γ phosphates of ATP is so similar, the reaction catalysed by **adenylate kinase** operates freely in either direction and provides a mechanism for temporarily sustaining the ATP concentration by transferring a phosphate group from one molecule of ADP to another and generating AMP in the process. This is particularly important in muscle cells for, although AMP is a relatively low-energy compound, it is an extremely important metabolic

Figure 7.12 A summary of the standard free energy changes ($\Delta G^{O'}$ in kJ mol^{-1}) associated with the hydrolysis of each of the γ and β phosphoanhydride bonds and of the α phosphoester bond of ATP. Throughout this book $\langle P \rangle$ is used to represent a phosphate group where the phosphate group of a molecule is not central to the reaction being illustrated.

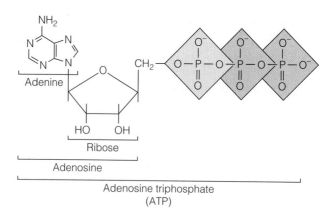

Adenosine triphosphate
(ATP)

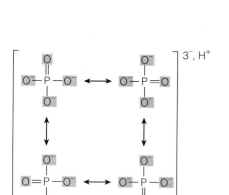

(a) Structures of phosphate ion contributing to resonance stabilization

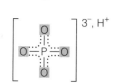

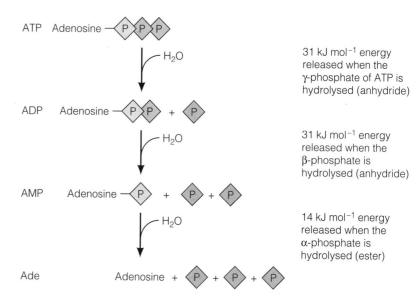

31 kJ mol^{-1} energy released when the γ-phosphate of ATP is hydrolysed (anhydride)

31 kJ mol^{-1} energy released when the β-phosphate is hydrolysed (anhydride)

14 kJ mol^{-1} energy released when the α-phosphate is hydrolysed (ester)

(b) Resonance hybrid with delocalized electrons

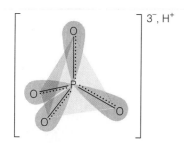

(c) Molecular orbitals of tetrahedral phosphate ion

Figure 7.13 Delocalization of electrons in the orthophosphate group ($HPO_3{}^{2-}$) results in four possible forms (a) without the H$^+$ being assigned to any of the oxygens. The hybrid of these four possibilities is presented in (b). (c) shows the molecular orbitals of the tetrahedral phosphate ion.

signal molecule. Increases in AMP concentration indicate that the cellular concentrations of ATP and ADP are low. For this reason AMP, which is an activator of some enzymes that provide energy through metabolism, is an effective signal molecule.

$$AMP + ATP \rightleftharpoons 2ADP$$

There is a structural basis for the energy status of the phosphoanhydride bonds of ATP

The free energy of hydrolysis of phosphoanhydride bonds is determined by the structural differences between the substrate and the product. Several factors contribute to the amount of energy made available when the bonds between either the α–β or the β–γ phosphate groups of ATP are hydrolysed.

- At pH 7.0 ATP carries four negative charges on the oxygen atoms and ADP carries three negative charges. The electrostatic repulsion between these negative charges is reduced when a phosphate group is removed by hydrolysis.

$$ATP^{4-} + H_2O \rightarrow ADP^{3-} + HPO_4{}^{2-} + H^+$$

- A free inorganic phosphate residue has a number of resonance forms in which the delocalization of electrons is a stabilizing force. Within ATP the phosphate group has fewer resonance forms, and therefore less stability. This increases the tendency for the phosphate group to be released (Figure 7.13).

- Hydrolysis of the phosphoanhydride bond liberates a proton. The buffer capacity of a cell is substantial because of the presence of proteins, so the liberated proton will be absorbed by one of these buffer molecules. Removal of the proton drives the hydrolysis reaction, increasing the negative free energy change associated with hydrolysis of the phosphoanhydride bond.

Standard free energy changes associated with hydrolysis of phosphate groups from some metabolic intermediates are shown in Figure 7.14.

Figure 7.14 The standard free energy changes ($\Delta G^{O'}$ in kJ mol^{-1}) associated with the hydrolysis of phosphate groups of central metabolites. Note that the phosphate bonds of some of these compounds have greater free energy than the β and γ bonds of ATP; others have less.

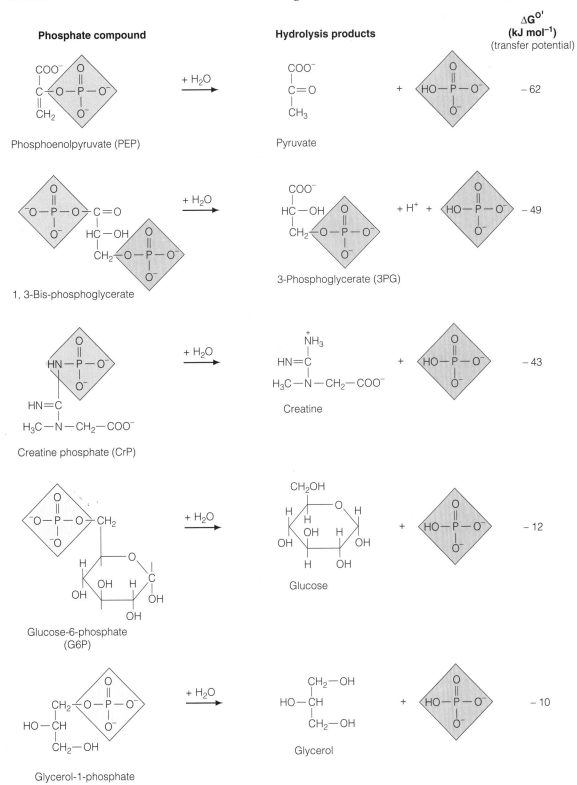

| Phosphate compound | | Hydrolysis products | | $\Delta G^{O'}$ (kJ mol^{-1}) (transfer potential) |

Metabolism is regulated to increase the efficiency of energy consumption and production

Metabolic processes are never completely efficient, but many regulatory mechanisms ensure that the maximum amount of energy is recovered from the oxidation of nutrients and that the minimum energy is used for biosynthesis of new molecules or in unnecessary heat generation. In addition, all cells have complex control mechanisms that allow them to recognize changing patterns of metabolism, and to channel energy to meet developing needs.

Animals also have to ensure that energy is made available preferentially to those tissues that either have the greatest energy requirement or which are at most risk of functional failure if their energy requirements are not met. To cope with the complex balancing act demanded by these requirements, multitissue organisms have developed a battery of regulatory molecules to act as indicators of energy status, some within cells and others in the whole body.

Subdivision of cells into compartments both separates and concentrates cellular processes

Division of the cells of eukaryotes into specialized subcellular compartments (discussed in Chapter 3) offers several advantages.

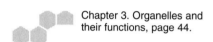
Chapter 3. Organelles and their functions, page 44.

- All the enzymes of a single pathway are generally restricted within a common compartment so that the distances that metabolites have to diffuse from one reaction to the next are short.
- Processes with competing requirements, for example oxidation and reduction reactions, are separated into different areas of the cell.
- Compartmentation of proteolytic enzymes within lysosomes ensures that the activity of these enzymes is controlled so that the normal repair processes proceed without risk of indiscriminately destroying useful molecules.
- Molecules that are synthesized and stored in one tissue but act on a different tissue are stored in specialized subcellular vesicles until they are required.

Rates of reaction are modified through several distinctive enzyme regulatory mechanisms

The rates of most metabolic processes may be altered by changing the catalytic activity of typically only a single rate-limiting enzyme in the pathway. Across the metabolic spectrum there are a relatively small number of processes through which the rates of enzyme-catalysed reactions are altered. These include

- allosteric regulation of enzyme activity;
- covalent modification of enzymes;
- control of synthesis of new enzyme molecules.

Allosteric regulation allows enzymes to respond to metabolite changes in the intracellular environment

Chapter 4. Many enzymes show allosteric behaviour, page 91.

Allosteric regulation of enzymes is one of the most important and most flexible mechanisms available for regulating metabolic pathways (see Chapter 4). Allosteric control produces rapid and readily reversible changes in the activity of enzymes by the binding of allosteric effectors to specific regulatory sites on enzymes.

The allosteric effector is often a product of the pathway, and the enzyme subject to allosteric control is located at, or near, the beginning or branch point of a pathway. Allosteric mechanisms provide highly specific controls, and ensure that energy is not wasted by partial operation of a metabolic pathway. Commonly an allosteric enzyme is inhibited by negative feedback by an end-product metabolite of the pathway in which the committed reaction is located. Many nucleotides can act as positive (activating) or negative (inhibitory) effectors. For example an energy-producing pathway might be inhibited by ATP and activated by ADP or AMP. In this example the allosteric enzyme is 'sensing' the energy status of the cell. The mechanism ensures that ATP is generated only in the amounts required.

Some allosteric enzymes are subject to control by the substrates or products of other pathways, which provides a mechanism for coordinating the activities of pathways. Pathways may be branched, the intermediates generated before the branch point being common to two or more subsidiary pathways. In these pathways allosteric control enzymes are often found immediately after a branch point so that each of the metabolic branches can be regulated independently. Regulating the activity of the initial steps common to more than one pathway inhibits or accelerates the rates of all branch pathways (see Figure 7.16).

Covalent modification of existing enzyme molecules can change their activity

Many enzymes are synthesized in inactive forms, or in forms subject to activation or inactivation by subsequent covalent modification. The most common forms of covalent modification involved in metabolic regulation are addition or removal of phosphate groups.

The conformation of a protein, and its functional activity, may be altered by covalently attaching a phosphate group (Figure 7.15). Protein phosphorylation reactions are catalysed by **protein kinase** enzymes, the phosphate group being derived from the terminal (γ) phosphate of ATP. Phosphate groups may be added to serine, threonine or tyrosine residues in proteins. Phosphorylation may result in activation or inhibition of catalytic activity or may alter the sensitivity of the enzyme to other effectors. The effect is reversed by removing the phosphate group, which is catalysed by **protein phosphatases**. Many of the phosphorylation and dephosphorylation reactions occur in response to hormonal stimulation and may be part of the system of **cascade amplification** (see below). Protein phosphorylation is used as a control mechanism, in which sustained but reversible changes in metabolic activity are required.

 Chapter 7. Cascades are multistage activations which amplify a response, page 198.

Synthesis of new enzyme molecules provides a cell with increased catalytic capacity, in response to increased metabolic demand

Within a particular cell or tissue, the rate at which substrate is consumed may be altered by increasing the number of enzyme molecules in the cell. The amount of

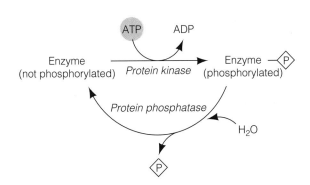

Figure 7.15 Regulation of enzyme activity by phosphorylation. Phosphorylation often results in activation of the enzyme. Many enzymes are inhibited by phosphorylation, the inhibition being relieved by removal of the phosphate group.

195

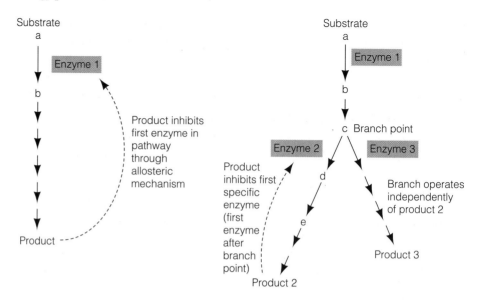

Figure 7.16 Regulation of the rates at which pathways operate by allosteric enzymes.

any protein in a cell reflects the balance between its rate of synthesis and degradation. Under normal conditions the rate at which each type of protein is degraded is fixed, and increases *only* if the activity of lysosomal enzymes ceases to be controlled or if the cell is irreparably damaged. Thus, most enzymes of the central metabolic pathways are present in approximately constant amounts within a particular cell, irrespective of changes in metabolic demand. Exceptions do occur, where cells respond to fluctuating metabolic demand by increasing (and subsequently decreasing) the number of molecules of specific enzymes in the cell (Figure 7.16).

Enzyme induction is a process through which transcription and translation of a specific gene product is accelerated (Chapter 5). The signal molecules that activate the gene are often substrates for the reaction or pathway in which the enzyme participates, but some hormones also induce the synthesis of specific proteins. The liver is the principal tissue in which enzyme induction occurs.

Some examples of enzyme induction reflect responses to nutritional changes in our diet:

Box 5.1. Control of transcription in bacteria and humans, page 117.

- If the diet is low in protein, ingested amino acids are directed towards protein synthesis, but on a high-protein diet amino acids also provide a useful energy source. The liver responds to an increase in protein intake by synthesizing more of the enzymes that convert amino acids to intermediates of carbohydrate metabolism, so that they can be used for energy generation.

- High amounts of protein in the diet also increase the amount of ammonia to be removed from the body. Enzymes of the cycle that converts ammonia to the safe, disposable molecule urea are located primarily in the liver (Chapter 10). These enzymes are also subject to enzyme induction, apparently by ammonium ions, thus increasing the capacity of the liver to dispose of ammonia.

Chapter 10. The urea cycle, page 301.

- Synthesis of the enzyme **galactokinase** can be induced in the liver. In infants the main source of carbohydrate is the milk disaccharide lactose. Lactose in the diet is hydrolysed in the intestine, releasing the monosaccharides glucose and galactose which are absorbed into the bloodstream. The enzyme in the liver which allows galactose to enter the carbohydrate-metabolizing pathways is galactokinase. At birth galactokinase is present in the liver at very low amounts, but the amount increases by induction very rapidly in response to the ingestion of milk.

Chapter 14. Lactose provides an important source of carbohydrate, page 426.

The synthesis of new molecules of enzyme consumes energy. Mechanisms that regulate the number of enzyme molecules in the cell therefore enable the cell to maximize its metabolic activity by increasing the number of enzyme molecules only when increases in catalytic rate are needed, preventing the wasteful use of amino acids when the specific metabolic function met by that enzyme is in less demand.

Cooperative binding of substrate ensures that as substrate concentration increases reaction rate is accelerated

Many proteins are multimers, made up of more than one polypeptide subunit. In some cases the subunits are identical; in others the subunits differ in both structure and function. The subunits of multimeric proteins often act in a cooperative manner, augmenting the response over that which would be achieved by individual subunits. This **cooperativity** also increases the sensitivity of the process to changes in the concentration of a ligand (perhaps a substrate) bound by the protein because as ligand binds to one subunit a conformational change is induced in adjacent subunits, the effect of which is to increase the affinity with which other subunits bind the ligand. A plot of the relationship between ligand concentration and ligand binding generates a sigmoid curve characteristic of these processes. Examples of cooperative binding are seen in the binding of oxygen to haemoglobin (Chapter 12), the binding of cyclic GMP to sodium channels as part of the light detection mechanism in the eye (Chapter 13) and in the binding of the substrate phosphoenol pyruvate to pyruvate kinase in glycolysis (Chapter 8). Allosteric enzymes usually demonstrate cooperative binding.

Chapter 12. Oxygen binding by haemoglobin is cooperative, page 371.
Chapter 13. Signal transduction in rod cells is mediated by cGMP through a cascade process, page 407.

Chapter 8. Pyruvate kinase responds to energy status through allosteric regulation, page 240.

Reversal of metabolic conversion by a different pathway allows independent controls

Many metabolic pathways are reversible, the direction being determined by nutritional demands. For example, when energy is required glycogen is degraded to glucose by glycogenolysis; conversely, under nutrient-replete conditions glucose is converted to glycogen by glycogenesis (Chapter 8). Metabolic efficiency requires that the pathways involved in the degradation and the synthesis of glycogen should have some independently controlled reactions so that the cell is not operating a futile energy-depleting cycle in which glycogen is degraded as quickly as it is synthesized. This is achieved by ensuring that one or more reactions in the two opposing pathways are catalysed by different enzymes. In many cases these unique enzymes are both regulated, so that a common regulatory mechanism may operate on enzymes of two pathways to slow the process in one direction while accelerating it in the other.

Chapter 8. Activation of glycogen phosphorylase and inhibition of glycogen synthase occur in response to a common cascade, page 227.

Another example of reversibility of reactions is seen in glycolysis, in which ten enzymes are involved in converting glucose to pyruvate. Seven of these enzymes catalyse freely reversible reactions and also function as part of the gluconeogenesis pathway by which pyruvate is converted to glucose. The other three enzymes operate essentially unidirectionally in glycolysis, and an enormous input of energy would be required to catalyse the reaction in the reverse direction. For gluconeogenesis, reversal of these chemical changes is achieved by four different enzymes catalysing different types of reaction (see Chapter 8). In both directions it is these pathway-specific enzymes which provide the regulatory sites determining the overall rate of the pathways.

Chapter 8. Gluconeogenesis: the synthesis of glucose from noncarbohydrate precursors, page 244.

Hormones are secreted in response to changes in the state of the body and mediate developmental, nutritional, electrical and mechanical responses

Hormones are small molecules, mostly peptides, amino acid, fatty acid or steroid derivatives, produced by endocrine tissues and released into the circulation in very

Chapter 3, Cell surface receptors, page 70.

low concentrations. The endocrine tissues monitor changes in nutrients or other molecules or ions circulating in the bloodstream, and release hormones in response to the changes they detect. They also control differentiation and developmental changes in tissue structure and function. Target tissues have specific binding sites, receptors, to which hormones must bind in order to exert their effects. Some hormones act by binding to receptors on the plasma membrane and do not enter the cell, others cross the plasma membrane to bind their receptors in the cytoplasm or nucleus of the cell. Hormones that act through plasma membrane receptors commonly trigger short-term responses which regulate metabolism and ionic balance, although some are also implicated in longer term changes. Those that bind to intracellular receptors commonly mediate longer-term responses involving protein synthesis, cell differentiation and tissue development.

Hormones also regulate nutrient distribution throughout the body and modulate metabolism to meet exceptional energy demands. The distinctive feature of hormonal control is that it allows for integration of function between different tissues of the body. Individual hormones and their modes of action are considered in Chapter 11.

Cascades are multistage activations which amplify a response

Cascade processes generally involve a series of enzymes, one activating the next progressively in a sequence so that an initial small response of perhaps a few molecules is amplified into a very large change at the end of the cascade. Cascades are often initiated by hormones and provide mechanisms for achieving massive changes in catalytic activity. The effectiveness of a cascade depends on the fact that, at each step in the cascade, one component activates a large number of molecules of the next component. It is not uncommon for each step to produce a several hundred-fold amplification of the signal received (Figure 7.17).

Cascade activation mechanisms can work in one of two ways:

Chapter 12, Specific and sequential proteolytic clearage gives a 'cascade' effect, page 362.

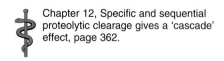

1. Limited proteolysis, partial digestion of a protein, may be the activating factor. This type of modification is irreversible and the system remains permanently activated. Proteolysis is the cascade activation mechanism responsible for the blood clotting mechanism.

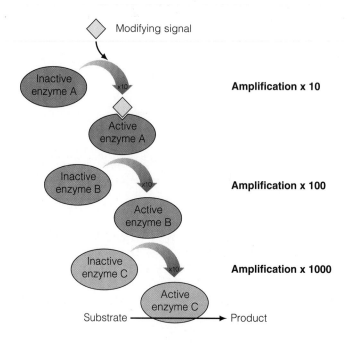

Figure 7.17 Representation of a cascade process. The cascade is triggered by a signal molecule, often a hormone. Each component of the cascade then sequentially activates the next, creating an amplification of the initiating signal. This amplification may be several million-fold overall.

2. Phosphorylation of protein (enzyme) components of a cascade, commonly by covalent binding of a phosphate group. This type of activation can be simply reversed by phosphatase-catalysed removal of the phosphate group.

Aerobic ATP production: the process of oxidative phosphorylation

The mitochondrion is the site of aerobic ATP production in eukaryotic cells

Mitochondria are subcellular organelles present in the cytoplasm of the vast majority of eukaryotic cells (Chapter 3). They contain the enzymes required for several different metabolic processes including the oxidation of acetyl CoA, the reoxidation of the reduced forms of NADH and FADH$_2$ and the catabolism of fatty acids (Chapter 9).

Chapter 3. Mitochondria are the 'powerhouses' of cells and are responsible for the aerobic production of ATP, page 47.
Chapter 9. β-oxidation progressively decreases the length of fatty acid carbon chains by removing two carbon atoms, page 266.

The mitochondrial respiratory chain comprises four multisubunit assemblies of proteins

One of the major functions of the mitochondrion is the transfer of reducing equivalents (electrons) from the coenzymes NADH and FADH$_2$ to molecular oxygen, regenerating the oxidized forms of these coenzymes. This multistage process is catalysed by four physically separate, integral membrane complexes located in the mitochondrial inner membrane. The entire assembly is called the **electron transport** (or **transfer**) chain. Each complex is responsible for the transfer of electrons along a specific segment of the electron transport chain.

The four complexes are

- NADH-ubiquinone reductase (complex I);
- Succinate-ubiquinone reductase (complex II);
- Ubiquinone-cytochrome *c* reductase (complex III);
- Cytochrome *c* oxidase (complex IV).

Within these complexes electrons are transferred through a series of protein-bound cofactors, namely flavins, iron–sulphur proteins, quinones and a series of haemoproteins or cytochromes. These cofactors are able to undergo oxidation–reduction reactions through which they generate an electrochemical potential gradient across the mitochondrial inner membrane.

ATP synthase harnesses a proton gradient generated by the respiratory chain

A fifth integral inner membrane complex, ATP synthase, is responsible for harnessing the proton gradient generated by the electron transfer reactions and using the energy of the gradient to produce ATP by adding a third phosphate group to ADP through the formation of a high-energy phosphoanhydride bond.

A simple representation of the mechanism of mitochondrial ATP synthesis (Figure 7.18) shows the passage of electrons down the thermodynamic gradient from NADH or FADH$_2$ to molecular oxygen leading to initial energy conservation in the formation of a **proton gradient** (more accurately termed an **electrochemical potential gradient**), which can be employed to drive ADP phosphorylation.

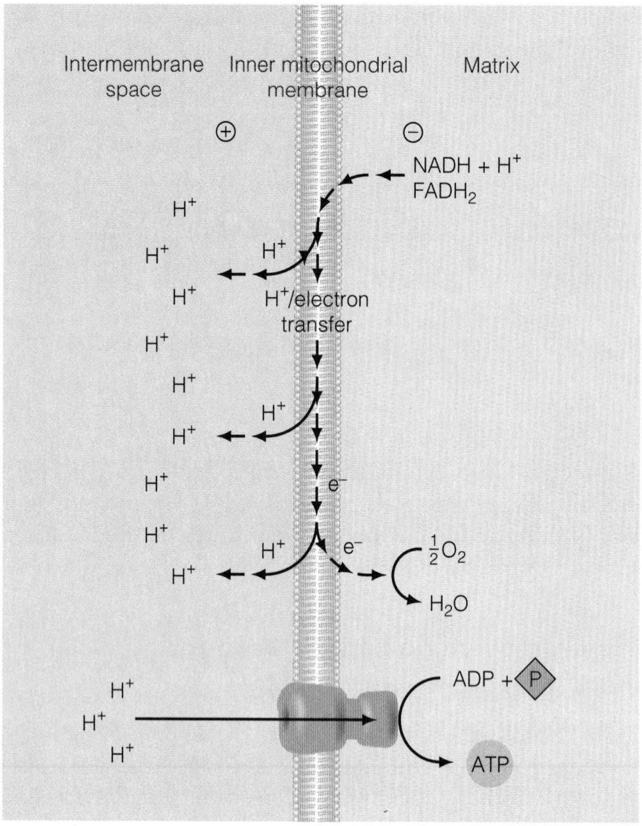

Figure 7.18 Schematic representation of the link between respiration and ATP production. The reoxidation of the reduced cofactors NADH and FADH$_2$, via an integral membrane electron transport pathway with molecular oxygen as the ultimate electron acceptor, leads to the pumping of protons from the mitochondrial matrix into the intermembrane space. The combined pH gradient and membrane potential thus generated constitutes an electrochemical potential gradient (the proton-motive force, or PMF), which is used in driving ATP formation via a vectorial proton-translocating ATP synthase.

Chapter 8. The glycolysis pathway, page 231.

Chapter 17. The mitochondrial genome encodes 13 polypeptides, page 521.

Chapter 3. Membrane fluidity maintains the appropriate environment for the optimal functioning of membrane proteins and is carefully controlled, page 59.

Box 7.2 The evolutionary origin of mitochondria

Glycolysis (see Chapter 8) is a mechanism for extracting a limited amount of metabolically usable energy as ATP from molecules and probably appeared at a very early stage in the evolution of life on earth. Primitive prokaryotes derived their energy exclusively through glycolysis or similar fermentations that represented adaptations to environmental conditions with an abundance of organic nutrients, including sugars, and a reducing atmosphere. As photosynthetic prokaryotes became plentiful the level of oxygen in the atmosphere rose and some organisms evolved to harness molecular oxygen directly in the more efficient oxidation of fuel molecules, resulting finally in development of a complete catabolic pathway for oxidizing sugars to CO_2. Direct generation of ATP during the individual reactions of glycolysis is termed **substrate-level phosphorylation** to distinguish it from **oxidative phosphorylation**, the generation of ATP coupled to the transfer of electrons along the respiratory chain.

Many features of present-day mitochondria suggest that they have a prokaryotic origin, and this idea is called the **endosymbiotic theory**.

- The mitochondrion has its own genome – a small, circular double-stranded DNA (see Chapter 17).

- The mitochondrial translation machinery has many prokaryotic features including sensitivity to chloramphenicol and resistance to cycloheximide (see Chapter 18).

- The mitochondrial inner membrane is unique among the membranes of mammalian cells in containing **cardiolipin**, a normal component of bacterial membranes. It also lacks cholesterol, a ubiquitous membrane component in eukaryotes (see Chapter 3).

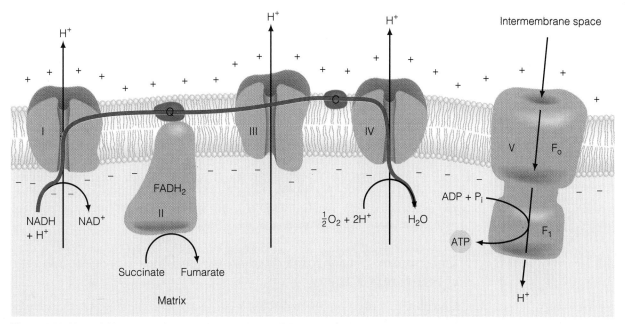

Figure 7.19 Vectorial transport of protons by complexes of the mitochondrial respiratory chain. Protons are pumped by complexes I, III and IV, which are all multisubunit transmembranous assemblies. Complex II, which is involved in the FAD-linked oxidation of succinate, does not translocate proteins. CoQ_{10} and cytochrome c act as 'mobile carriers', facilitating electron transfer between the various complexes. Re-entry of protons into the matrix through the F_0 channel in the ATP synthase complex (sometimes called complex V) provides the energy to drive ATP synthesis.

The complexes of the electron transport chain diffuse freely in the lipid bilayer

The complexes of the electron transport chain are present in molar ratios of approximately 1:2:3:7, but do not form stable associations with one another. Instead, they diffuse freely in the plane of the bilayer, interacting by random collisions to permit the flow of electrons. In addition, the protein **cytochrome c** and the lipid **ubiquinone** (Figure 7.19) are present in large amounts on the inner mitochondrial membrane (approximately 50-fold and ten-fold molar excesses respectively). They act as 'mobile carriers' promoting the rapid and reversible physical and functional interactions between complexes which occur during the course of electron transfer.

NADH-ubiquinone reductase (complex I) catalyses a series of electron transfer reactions involving FMN and iron–sulphur proteins

This proton-translocating complex, which catalyses the transfer of electrons in the early stages of the respiratory pathway, contains over 40 different types of polypeptide chains as well as several cofactors and prosthetic groups that are active in oxidation–reduction reactions – flavin mononucleotide (FMN) and 6–7 separate iron–sulphur (Fe–S) centres. The final electron acceptor is the lipid-soluble ubiquinone (coenzyme Q) (Figure 7.20).

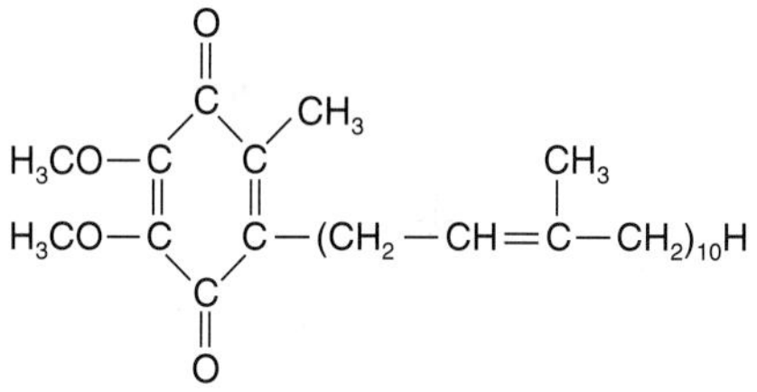

Oxidized coenzyme Q_{10} (CoQ)

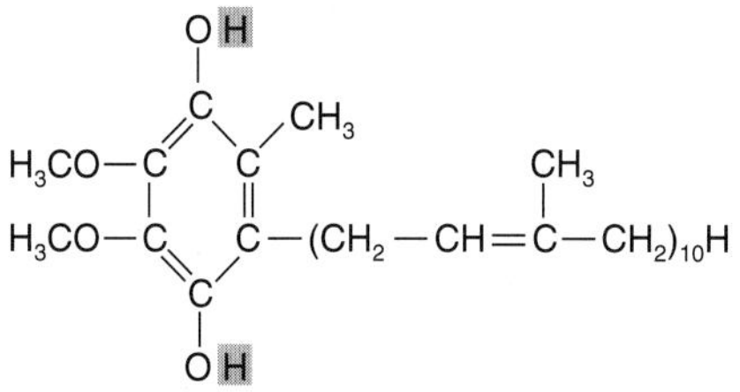

Reduced coenzyme Q_{10} (CoQH₂)

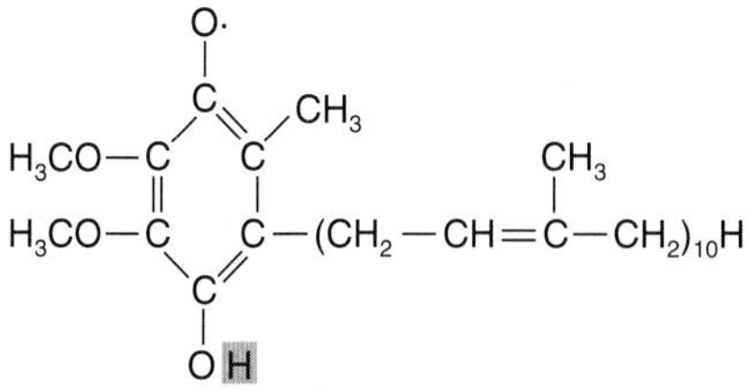

Semiquinone form of coenzyme Q_{10}

Figure 7.20 The structure and oxidation–reduction states of coenzyme Q_{10}. This cofactor is extremely hydrophobic because it possesses a long 'tail' of 10 isoprenoid units, enabling it to diffuse freely in the plane of the lipid bilayer. CoQ_{10} can participate in both one and two-electron transfer reactions. The semiquinone form of the cofactor is an important intermediate in the oxidation–reduction reactions of the mitochondrial electron transport chain.

Succinate-ubiquinone reductase (complex II) and other FAD-linked enzymes transfer electrons directly into the respiratory chain via ubiquinone

Chapter 8. The malate–aspartate shuttle is a mitochondrial mechanism for sustaining the oxidizing power of the cytoplasm, page 255.

Several FAD-requiring enzymes from different metabolic pathways transfer electrons into the respiratory chain via ubiquinone. These include **succinate dehydrogenase** of the citric acid cycle, **electron transferring flavoprotein** (involved in β-oxidation of fatty acids) and *sn*-**glycerophosphate dehydrogenase**, a component of a mitochondrial 'shuttle' mechanism for transferring reducing equivalents from cytoplasmic NADH into the mitochondrial electron transport chain (Chapter 8). Only small free energy changes are associated with any of these reactions, insufficient for proton translocation. Succinate-ubiquinone reductase contains a covalently attached FAD cofactor and three Fe–S centres, all of which participate in the electron transfer process (Box 7.3).

Ubiquinone-cytochrome c reductase (complex III) contains cytochromes b and c_1

Complex III has a complicated subunit organization containing 12–13 polypeptides. Many of these are of unknown function and some do not appear to be involved directly in the oxidation–reduction reactions. The three best characterized components of the complex are cytochrome b, an Fe–S protein and cytochrome c_1. Cytochrome b is a hydrophobic, integral membrane protein which is predicted on the basis of its primary structure to contain eight transmembrane α-helices. It contains two haem groups.

Chapter 3. Membrane-associated proteins are categorized by the nature of their interaction with the lipid bilayer, page 53.

Within this complex, electrons are eventually transferred to the only soluble cytochrome, cytochrome c. This peripheral protein (see Chapter 3) is located on the outer surface of the inner mitochondrial membrane. Its structure has been studied in great detail; the haem group is located within a deep hydrophobic crevice in the molecule, with the iron atom covalently bound to the sulphur atom of a methionine residue and the nitrogen atom of a histidine residue within the protein and complexed with the four nitrogen atoms of the porphyrin ring. In

---

Box 7.3   Iron–sulphur (Fe–S) centres in the electron transport chain

Iron–sulphur (Fe–S) centres are present in complexes I, II and III and are usually of the 2Fe–2S or 4Fe–4S type in mammalian mitochondria. These centres contain two or four atoms of inorganic sulphur linked to iron atoms which are also bound to the polypeptide chain, through cysteine or histidine residues. More than one type of Fe–S centre may be present on a single polypeptide chain. Despite the varying degrees of complexity of these centres, each cluster can accommodate only one electron; however, it has been shown experimentally that all these centres are active participants in oxidation–reduction reactions of the electron transport chain.

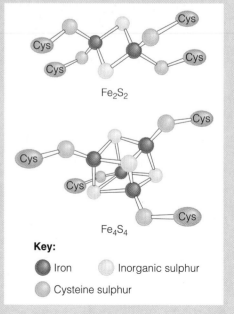

Key:
● Iron ○ Inorganic sulphur
○ Cysteine sulphur

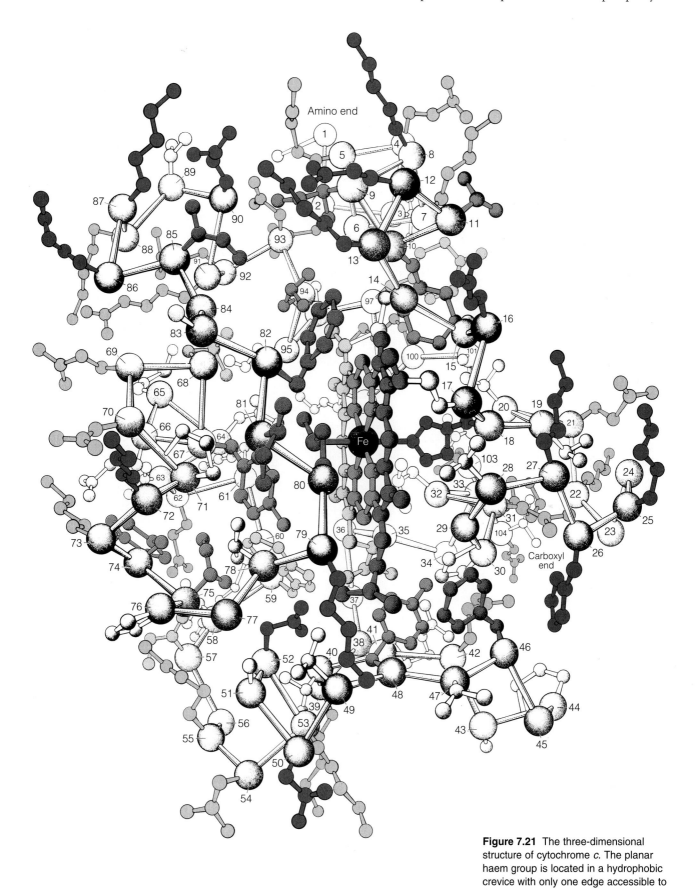

Figure 7.21 The three-dimensional structure of cytochrome *c*. The planar haem group is located in a hydrophobic crevice with only one edge accessible to the aqueous phase. Hydrophobic residues are highlighted.

Chapter 12. An alternative interaction between haemoglobin and oxygen is harmful, page 377.

Box 6.6. Nucleic acid and protein sequences can be used for 'molecular phylogeny', page 171.

contrast to haemoglobin, which functions as a carrier of molecular oxygen only in the ferrous state (see Chapter 12), the iron atoms of the haem groups of cytochrome *c* cycle between the ferrous and ferric states during its catalytic cycle.

The sequences and three-dimensional structures of cytochrome *c* (Figure 7.21) from a wide variety of aerobic organisms have marked similarities, showing that it has been highly conserved throughout evolution. The study of cytochrome *c* has been very important in formulating ideas about the evolution of genes and proteins (see Chapter 6).

Box 7.4   The chemiosmotic theory highlights the role of the proton gradient in driving energy-linked reactions in mitochondria

Recognition of the central role of the proton gradient in energy-linked uptake of solutes and ATP production is formalized in the four basic tenets of the **chemiosmotic theory**:

- The mitochondrial inner membrane (termed the coupling membrane) must remain essentially impermeable to protons.

- Energy derived from the oxidation of the reduced cofactors NADH and FADH$_2$ in the mitochondrial electron transport chain is directly linked to the formation of a proton gradient across the energy-transducing membrane.

- ATP synthesis is catalysed by the action of a vectorial proton-translocating ATP synthase complex located in the inner membrane.

- Movement of charged molecules including citric acid cycle intermediates such as citrate, succinate and α-ketoglutarate, ATP, ADP, Pi and small ions in and out of mitochondria must be carefully regulated so as not to disturb the electrochemical potential gradient across the membrane which provides the driving force for ATP synthesis.

In simple terms, therefore, the ATP synthesizing machinery can be viewed as two asymmetrical proton pumps working in conjunction across the inner membrane, with generation of the proton gradient by the respiratory chain coupled to its dissipation via the ATP synthase complex as an integral part of its catalytic mechanism (Box Figure).

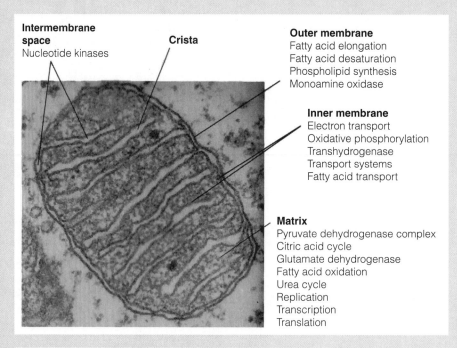

Intermembrane space
Nucleotide kinases

Crista

Outer membrane
Fatty acid elongation
Fatty acid desaturation
Phospholipid synthesis
Monoamine oxidase

Inner membrane
Electron transport
Oxidative phosphorylation
Transhydrogenase
Transport systems
Fatty acid transport

Matrix
Pyruvate dehydrogenase complex
Citric acid cycle
Glutamate dehydrogenase
Fatty acid oxidation
Urea cycle
Replication
Transcription
Translation

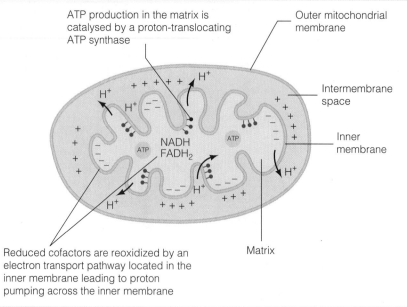

ATP production in the matrix is catalysed by a proton-translocating ATP synthase

Outer mitochondrial membrane

Intermembrane space

Inner membrane

Matrix

Reduced cofactors are reoxidized by an electron transport pathway located in the inner membrane leading to proton pumping across the inner membrane

Cytochrome *c* oxidase (complex IV) reacts with molecular oxygen

Cytochrome *c* oxidase is also a multisubunit protein complex of about 13 polypeptides. It contains two chemically identical, but functionally distinct, type 'a' haems, termed cytochromes *a* and *a₃*, which are linked functionally with two copper atoms. In cytochrome *a₃* the sixth coordination position of the iron atom within the haem group is vacant, which explains its reactivity to ligands such as oxygen and provides a site of reaction for the poisons carbon monoxide and cyanide, which are potent inhibitors of respiration.

The function of complex IV is to transfer electrons from the pool of cytochrome *c* molecules (a one-electron carrier) to molecular oxygen, O_2, which requires four electrons for complete reduction to water. Electrons are transferred via the first copper atom and haem *a* to the second copper and haem *a₃*, which are thought to form a **binuclear centre** (this means that both these components cooperate in oxygen binding and both must be reduced before interacting with oxygen). Thus, O_2 can be reduced initially in a two-electron step to form a tightly bound peroxide intermediate, preventing the formation of damaging oxygen free radicals (see Chapter 12).

The composition and general properties of the mitochondrial respiratory complexes are shown in Figure 7.22.

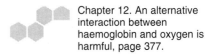

Chapter 12. An alternative interaction between haemoglobin and oxygen is harmful, page 377.

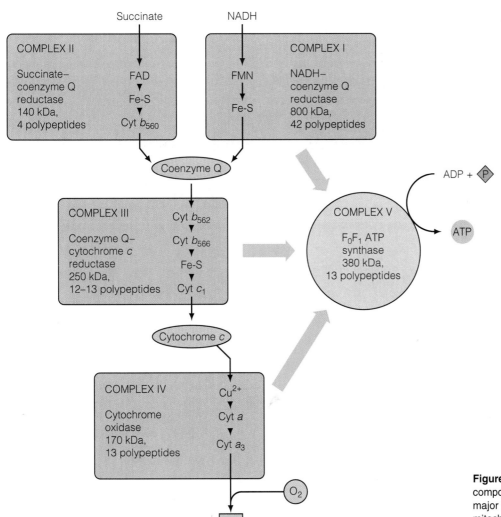

Figure 7.22 Approximate subunit composition and general properties of the major multimeric complexes involved in the mitochondrial respiratory chain and ATP production.

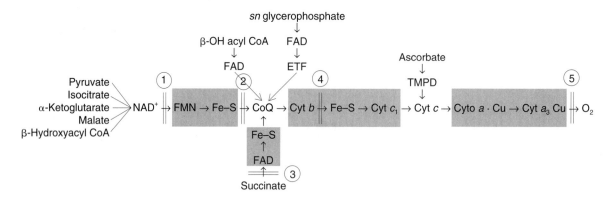

Figure 7.23 Route of electron transfer through the various cofactors and prosthetic groups of the mitochondrial respiratory chain complexes. The oxidation–reduction centres located within specific respiratory assemblies (outlined) are highlighted, as are the sites of action of the inhibitors amobarbital (1), rotenone (2), malonate (3), antimycin A (4) and cyanide or carbon monoxide (5). In terms of iron–sulphur centres (Fe–S) the scheme is oversimplified, as there are at least six distinct centres within complex I, three in complex II, but only one in complex III.

Use of inhibitors helped to establish the order of the electron transport chain complexes

Our understanding of the organization and function of the electron transport chain and of the coupling of oxidation and phosphorylation has developed from investigations with a variety of inhibitors, which may be classified broadly into two classes (Figure 7.23):

1. Inhibitors of respiration, which arrest electron flow down the respiratory chain at a specific point between NADH and O_2;
2. Inhibitors of ADP phosphorylation, which affect respiration rates indirectly because of the close coupling between the two processes.

Transfer of reducing equivalents down the electron transport pathway generates a proton gradient across the inner membrane

The appearance of a proton gradient linked directly to respiration was first demonstrated conclusively when anaerobic suspensions of rat liver mitochondria were pulsed with small amounts of oxygen in the presence of NAD^+- or FAD-linked mitochondrial substrates such as β-hydroxybutyrate or succinate, while the pH of the medium was monitored with a sensitive, rapidly responding pH meter.

Box 7.5   Cytochrome *P*-450

Cytochrome *P*-450 is a specialized cytochrome that acts as the terminal acceptor in a short electron transport chain found in liver microsomes or the outer membrane of adrenal mitochondria which is devoted to activation of oxygen for hydroxylation purposes rather than in ATP generation. In mitochrondria, NADPH transfers its high potential electrons via a membrane-bound flavoprotein to adrenodoxin, a small iron–sulphur protein which reduces the haem group of cytochrome *P*-450 permitting its activation for participation in hydroxylation reactions (type I system). In microsomes, only two components are involved with NADPH, providing electrons to an FAD and FMN-containing reductase which inactivates the cytochrome *P*-450 haemoprotein before hydroxylation of substrate (type II system).

The hydroxylation reactions of the cytochrome *P*-450 system also have an important role in normal cellular metabolism in the biosynthesis and degradation of fatty acids and steroids. All of these hydroxylations require NADPH and O_2, one atom of oxygen being incorporated into the hydroxyl group of the modified substrate and the other reduced to water. Thus the enzymes catalysing these steps are referred to as **mono-oxygenases** or **mixed function oxidases**.

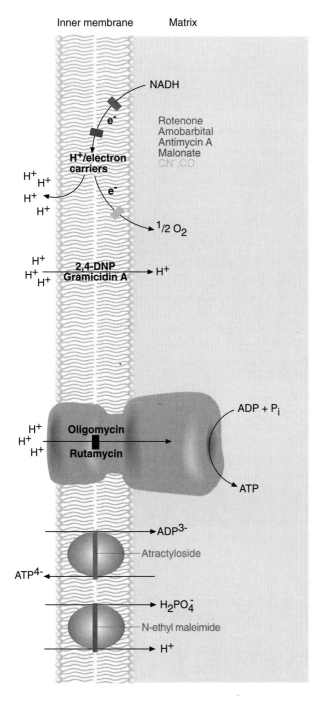

Figure 7.24 General scheme indicating modes of action of various inhibitors of mitochondrial respiration and ATP production. Sites of inhibition are shown by coloured blocks. Energy transduction inhibitors also affect respiration indirectly as the result of the close linkage between the two processes (see Figure 7.29).

- Initiation of respiration was accompanied by rapid ejection of protons into the external medium.
- After consumption of oxygen and cessation of respiration, there was a slow decay of protons back across the inner membrane via the proton channel of the ATP synthase complex.
- If specific inhibitors of the electron transport chain (rotenone for NAD⁺-linked substrates and antimycin A for FAD-linked substrates) were included in the medium before pulsing with oxygen no proton gradient formed, confirming the involvement of electron transport in this process.
- If uncouplers of oxidative phosphorylation such as 2,4-dinitrophenol were added during the slow back-decay of protons following oxygen depletion, the rate of dissipation of the proton gradient increased significantly.

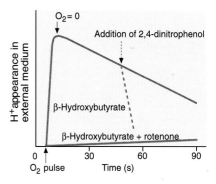

Figure 7.25 Formation of the proton gradient during respiration in a solution of rat liver mitochondria, using β-hydroxybutyrate (NAD⁺ linked) as substrate. Note the slow decay of the proton gradient, via the F_0 channel of the ATP synthase, on anaerobiosis. This decay is stimulated greatly by the addition of uncouplers.

The energy for ATP production is derived from the proton motive force

Respiration is linked to the formation of a proton gradient (or more accurately an electrochemical potential gradient) termed the **proton motive force**. In essence, the proton motive force is composed of two components: (1) as a proton is a charged species, development of a proton gradient generates a membrane potential which is positive on the outside and negative on the inside of the mitochondrial inner membrane; (2) additionally, the unequal distribution of H^+ ions across the bilayer creates a pH difference across the membrane. Both of these components represent conserved energy, which is utilized in the phosphorylation of ADP to ATP.

Mitochondrial inner membranes contain a vectorial proton-translocating ATP synthase which catalyses the final step in oxidative phosphorylation

The enzyme complex that catalyses the terminal step in mitochondrial ATP production is seen in electron micrographs as regular arrays of 'knobs' and 'stalks' lining the matrix surface of the inner membrane or the outer surface of mitochondrial inner membrane fragments. ATP synthase is a multisubunit enzyme composed of several hydrophobic integral proteins (F_0) and an 11-polypeptide ($\alpha_3,\beta_3,\gamma_3,\delta,\epsilon$) structure for the catalytic F_1-ATPase (Figure 7.26).

It is still not clear how protons entering the active site of the ATP synthase from the proton channel are coupled to ATP production. One possibility is that proton movements through the membrane channel in F_0 into the active site of the ATP synthase induce conformational changes in the enzyme which result in tightly bound ADP and Pi being sequestered into a hydrophobic pocket within the complex. In the absence of water molecules, ADP and Pi can link to form ATP without further energy input. This scheme (Figure 7.27) neatly completes the proton circuit in that conformational changes in the respiratory chain oxidation–reduction pumps appear largely responsible for generation of the proton gradient, which can be converted into ATP via H^+-driven conformational effects within the ATP synthase complex.

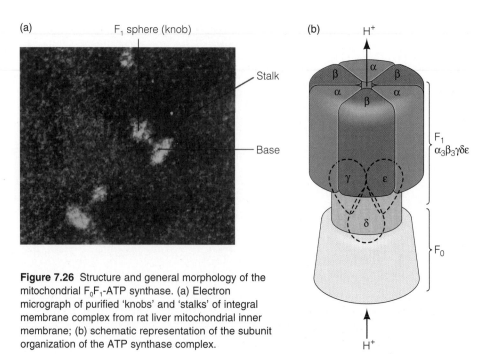

Figure 7.26 Structure and general morphology of the mitochondrial F_0F_1-ATP synthase. (a) Electron micrograph of purified 'knobs' and 'stalks' of integral membrane complex from rat liver mitochondrial inner membrane; (b) schematic representation of the subunit organization of the ATP synthase complex.

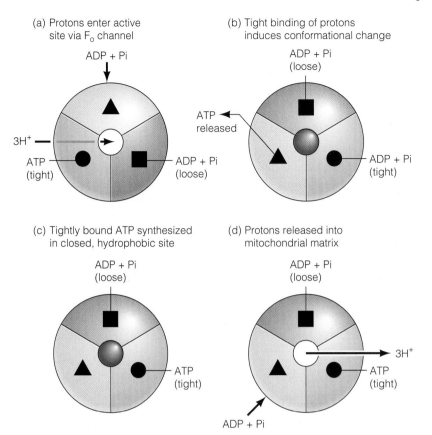

(a) Protons enter active site via F$_0$ channel

ADP + Pi

3H$^+$

ATP (tight)

ADP + Pi (loose)

(b) Tight binding of protons induces conformational change

ADP + Pi (loose)

ATP released

ADP + Pi (tight)

(c) Tightly bound ATP synthesized in closed, hydrophobic site

ADP + Pi (loose)

ATP (tight)

(d) Protons released into mitochondrial matrix

ADP + Pi (loose)

3H$^+$

ATP (tight)

ADP + Pi

Figure 7.27 Possible mechanism for linking proton movements into the active site of ATP synthase to ATP production. Transfer of protons to high-affinity sites in the active centre of the enzyme is thought to induce conformational changes in the catalytic subunits (probably the three β subunits, which contain ATP, ADP and Pi binding sites) as shown. Three different conformational sites are proposed: the open, loose and tight states. In the tight state water is excluded from the active site of the enzyme so that ATP synthesis may proceed with no further energy requirement.

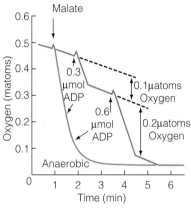

Figure 7.28 Effects of 'uncoupling' oxidative phosphorylation on the rate of respiration in carefully isolated mitochondria. In actively phosphorylating mitochondria (red line) the rate of respiration is controlled by the levels of adenine nucleotides, so oxygen uptake is stimulated only slightly on addition of substrate alone. Addition of ADP (in the presence of Pi) leads to a rapid burst of respiration until all the ADP is converted to ATP. This phenomenon is termed 'respiratory control' and is the mechanism whereby mitochondria respond to the energy requirements of the cell. In 'uncoupled' mitochondria (blue line) the rate of respiration is rapid and uncontrolled as the proton gradient is being dissipated continuously.

Electron flow and phosphorylation of ADP can be 'uncoupled'

Mitochondria become 'uncoupled' when their inner membrane becomes nonspecifically permeable to protons, through either physical damage or the use of chemical uncouplers: 2,4 dinitrophenol is commonly used as a chemical uncoupler in experimental studies (Figure 7.28). In 'uncoupled' mitochondria, oxidation continues but the energy of the proton gradient is dissipated as heat rather than being 'coupled' to the phosphorylation of ADP. As the rate of oxygen consumption is normally restrained by coupling to the phosphorylation of ADP, higher than normal rates of respiration are seen in 'uncoupled' mitochondria.

Uncouplers stimulate respiration by making the inner membrane permeable to protons so that the proton gradient is dissipated as heat. In contrast, compounds that prevent the ADP-induced increase in respiration are preventing the dissipation of the proton gradient; hence they are preventing the coupling of H$^+$ movements to ATP formation. Oligomycin and rutamycin inhibit ATP synthase directly, by blocking the passage of H$^+$ ions from the transmembrane channel into the catalytic site of the enzyme (Figure 7.29).

Uncoupling is of physiological importance in the brown adipose tissue of infants and hibernating animals, where regulation of body temperature is vital. In these tissues the mitochondrial membrane contains an ADP-regulated proton channel, which permits energy dissipation as heat in a controlled fashion (see Chapter 13).

Chapter 13. Thermogenin dissipates mitochondrial proton gradients, page 383.

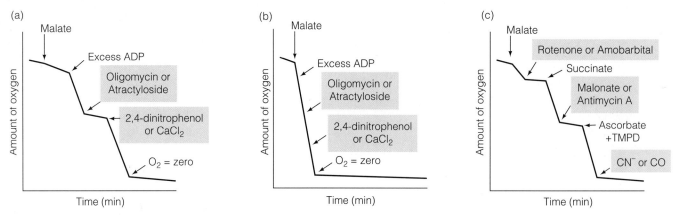

Figure 7.29 Modes of action of major inhibitors of mitochondrial respiration and energy transduction. Oligomycin, an inhibitor of the proton translocating ATP synthase, and atractyloside, an inhibitor of ATP export from and ADP import into mitochondria, both inhibit mitochondrial respiration indirectly only in 'coupled' mitochondria (a). They do this by inhibiting ATP synthesis, causing a build up of the proton-motive force. In contrast, 2,4 DNP and CaCl₂ stimulate respiration in 'coupled' mitochondria as they dissipate the proton motive force by making the membrane nonspecifically permeable to protons (2,4 DNP) and using the membrane potential to drive its own active transport (CaCl₂). Uncoupled mitochondria (b) do not generate a proton motive force so respiration is not affected by energy transfer inhibitors. Direct inhibitors of respiration (see Figure 7.25) are equally effective in 'coupled' and 'uncoupled' mitochondria (c).

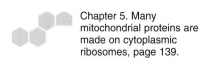

Chapter 5. Many mitochondrial proteins are made on cytoplasmic ribosomes, page 139.

ATP is exported from the mitochondrial matrix in exchange for ADP taken up into the organelle

The mitochondrial inner membrane has an **adenine nucleotide translocase**, the function of which is to transport the ATP produced in the matrix to the cytoplasm in exchange for ADP. Movement of the two nucleotides is tightly linked in a 1:1 ratio. Mitochondria also possess a related **phosphate transporter**, which replenishes the supply of inorganic phosphate within the mitochondrial matrix. The adenine nucleotide translocator is inhibited by **atractyloside**, a toxic compound found in Mediterranean thistles. This compound prevents the import of ADP into mitochondria and the concomitant export of ATP. Inhibition of adenine nucleotide transport stops the supply of ADP to ATP synthase, and therefore blocks not only ATP export but also, indirectly, further ATP synthesis.

The proton gradient is employed in ATP formation through the proton-translocating ATP synthase complex, but its energetic potential is also required for the ATP/ADP translocation and phosphate transport (Figure 7.30). A significant amount of the proton motive force is expended in these processes. Other energy-requiring reactions which utilize the mitochondrial membrane potential include Ca²⁺ uptake and the translocation of nuclear-encoded mitochondrial proteins into the organelle from their site of synthesis in the cytoplasm (Chapter 5).

Rates of respiration in mitochondria are controlled by the concentrations of ATP, ADP and orthophosphate in the cell

Many important features of mitochondrial respiration linked to ATP production may be studied in the laboratory by measuring rates of oxygen uptake by isolated mitochondria. Intact isolated mitochondria exhibit very low rates of oxygen uptake in the presence of citric acid cycle intermediates unless ADP and inorganic (ortho)phosphate (Pi) are present. ADP and Pi together stimulate respiration but respiration rates return to basal levels as the ADP is converted to ATP and the phosphate potential (the ratio of [ATP] to [ADP]×[Pi]) is restored to a high value (see Figure 7.29). This illustrates the fact that mitochondrial respiration rates *in*

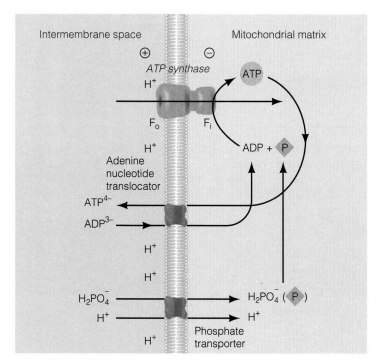

Figure 7.30 Energy requiring steps in mitochondrial ATP production. Generation of ATP requires movement of protons through the specific F_0 proton channel of the ATP synthase into the active site of the enzyme in the matrix. The product, ATP^{4-}, must then be exported to the rest of the cell and fresh substrate, ADP^{3-}, imported by the action of the adenine nucleotide translocase. This dissipates the membrane potential by one charge unit. Correspondingly, fresh Pi (as $H_2PO_4^-$, the other substrate required) must be imported in conjunction with a proton, leading to a decline in the pH gradient.

vivo respond to the energy demands on the tissue, a phenomenon with which we are all familiar from the breathlessness brought on by vigorous exercise.

Mitochondria exhibit **respiratory control**, whereby rates of respiration are closely geared to the energy requirements of the tissue in response to the concentrations of adenine nucleotides. Physiologically, this is an extremely important regulatory mechanism as there is a fine balance between our energy production and utilization processes. The average ATP molecule is 'turning over' about 2000 times a minute. In vigorous exercise, this rate increases substantially and respiration must increase accordingly.

The integrity of the mitochondrial inner membrane is essential to aerobic ATP production

Biophysical studies have revealed that the mitochondrial inner membrane has one of the lowest conductivities of all biological membranes, reflecting its impermeability to small ions in general and protons in particular. The inner membrane acts as an insulator preventing the nonspecific leakage of protons across the lipid bilayer. Thus, the proton movements leading to ATP production can be viewed as a 'proton circuit' with protons appearing in the intermembrane space from specific locations on the outer surfaces of the respiratory chain complexes which are functioning as 'proton pumps' and flowing back across the membrane via the proton channels of the ATP synthase. In this directed transmembrane flow of protons, the inner membrane acts as an insulator because its integrity is essential to prevent the dissipation of the proton gradient as heat: hence its apt title of the **coupling membrane**, which highlights its central role in linking oxidation reactions to ATP formation (Figure 7.31).

(a) Intermembrane space
Matrix
H+
NO2
NO2
O−
OH
+ H+
2,4-Dinitrophenol (DNP)
(b) Intermembrane space
Matrix
Gramicidin A
Gramicidin A

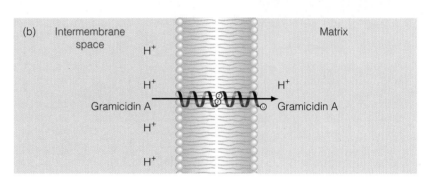

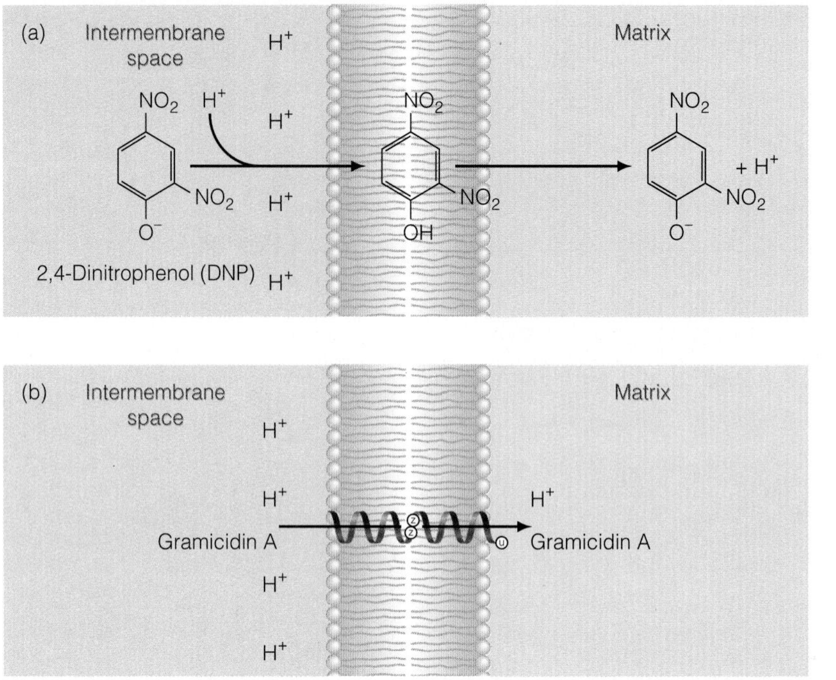

Figure 7.31 Modes of action of two types of uncouplers of oxidative phosphorylation. Uncoupling agents short circuit the proton gradient by making the inner membrane nonspecifically permeable to protons. (a) The uncoupler 2,4 DNP is a weak lipophilic anion which can diffuse across the lipid bilayer in its neutral, unprotonated, state, promoting equilibration of the proton gradient. Two molecules of the α-helical peptide gramicidin A (b) can associate transiently in the lipid bilayer as shown to form a transmembrane channel down the centre of the helix, through which protons can readily diffuse.

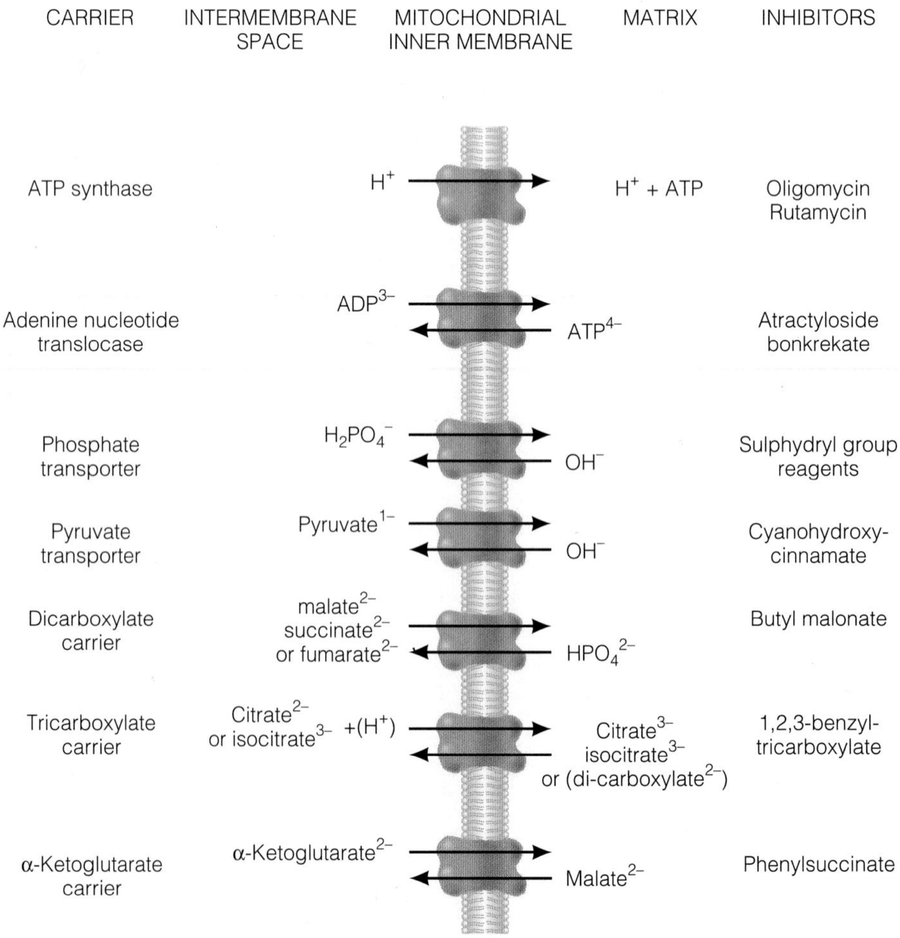

CARRIER
INTERMEMBRANE SPACE
MITOCHONDRIAL INNER MEMBRANE
MATRIX
INHIBITORS
ATP synthase
H+
H+ + ATP
Oligomycin
Rutamycin
Adenine nucleotide translocase
ADP3−
ATP4−
Atractyloside
bonkrekate
Phosphate transporter
H2PO4−
OH−
Sulphydryl group reagents
Pyruvate transporter
Pyruvate1−
OH−
Cyanohydroxy-cinnamate
Dicarboxylate carrier
malate2−
succinate2−
or fumarate2−
HPO42−
Butyl malonate
Tricarboxylate carrier
Citrate2−
or isocitrate3− +(H+)
Citrate3−
isocitrate3−
or (di-carboxylate2−)
1,2,3-benzyl-tricarboxylate
α-Ketoglutarate carrier
α-Ketoglutarate2−
Malate2−
Phenylsuccinate

Figure 7.32 The major metabolite (and ion) transporters of the mitochondrial inner membrane. The proton-motive force (PMF) is used directly in production of ATP. One element of the PMF, the membrane potential, is involved in ATP/ADP exchange and the other element, the pH gradient, is involved since they are linked directly to the PMF. Many other mitochondrial substrates such as citric acid cycle intermediates are transferred across the inner membrane in a 1:1 electroneutral exchange to minimize any disruption of the PMF. These carriers are often classified as secondary or tertiary carriers because their operation is ultimately dependent on the action of the primary carriers.

Exchange diffusion carriers regulate traffic of charged ions and metabolites across the inner membrane

The mitochondrial inner membrane is also involved in mediating the influx and efflux of a wide range of charged metabolites and ions, the transfer of which must be regulated carefully to prevent disruption of the electrochemical potential gradient. This is achieved by a series of electroneutral exchange-diffusion carriers (Figure 7.32). As their name implies, the catalytic mechanism of these transporters promotes the coupled 1:1 exchange of two metabolites or ions of equal charge, thereby preserving the electrical balance across the mitochondrial membrane.

The **chemiosmotic hypothesis** (Box 7.4) highlights the central role of proton gradients and membrane integrity in energy transduction processes leading to ATP generation. However, ion gradients are ubiquitous in nature and are implicated generally in a host of cellular responses including solute uptake, protein targeting, ligand–receptor interactions, hormone response mechanisms, muscle excitability and nerve impulse transmission.

The citric acid cycle: the pathway for complete oxidation of acetyl coenzyme A

Acetyl coenzyme A is a key intermediate in both catabolic and anabolic processes

Acetyl coenzyme A (acetyl CoA) is a high-energy thioester produced from the catabolism of carbohydrate, fatty acids, ethanol and some amino acids. It is also the key biosynthetic substrate for many molecules including fatty acids, various complex lipids including cholesterol, ketone bodies and porphyrins (Figure 7.33).

An important limitation on the usefulness of acetyl CoA in animals is that it can not be converted into carbohydrate because the reaction catalysed by the pyruvate dehydrogenase complex is strictly unidirectional, and animals lack the alternative glyoxylate pathway that is available to plants.

$$\text{Pyruvate} + NAD^+ + \text{CoASH} \xrightarrow{\textit{Pyruvate dehydrogenase complex}} \text{Acetyl CoA} + NADH + CO_2$$

As befits its central role in metabolism acetyl CoA is also an important regulator. High concentrations of acetyl CoA in a cell indicate that the cell is in a state

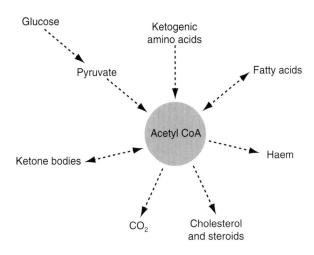

Figure 7.33 The central role of acetyl CoA in metabolism.

of energy sufficiency and that the rate of production of energy through the catabolic pathways may be slowed to conserve complex nutrients without threatening the functions of the cell. Acetyl CoA inhibits the enzymes pyruvate kinase and pyruvate dehydrogenase, both of which catalyse reactions that catabolize carbohydrate, and the enzyme thiolase in the β-oxidation pathway through which fatty acids are oxidized. Acetyl CoA also activates the enzyme pyruvate carboxylase which directs pyruvate towards the synthesis of glucose or glycogen (see Chapter 8).

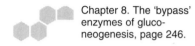

Chapter 8. The 'bypass' enzymes of gluco-neogenesis, page 246.

The citric acid cycle generates energy from the complete oxidation of acetyl CoA

The citric acid cycle is also called the tricarboxylic acid cycle, or Krebs cycle. Its major purpose is to extract energy from acetyl CoA in a series of oxidation and decarboxylation reactions in which the free energy is captured initially in the form of the reduced cofactors NADH and $FADH_2$ and the high-energy phosphoanhydride bond of GTP. Each molecule of acetyl CoA oxidized through the citric acid cycle yields three NADH, one $FADH_2$ and one molecule of GTP. As the cycle proceeds reducing equivalents are passed from cycle intermediates to the oxidized forms (NAD^+ and FAD) of the coenzymes in a series of dehydrogenase reactions. The reduced forms of these nucleotides (NADH and $FADH_2$) have high electron-transfer potential and the energy can be conserved as ATP through the mitochondrial electron transport or respiratory chain and oxidative phosphorylation. Energy for the phosphorylation of GDP to GTP is derived from the hydrolysis of the high-energy thioester bond of the cycle intermediate succinyl CoA.

For every molecule of acetyl CoA oxidized two molecules of CO_2 are produced, not from the carbon atoms of the acetate group but from the oxaloacetate with which it condenses to enter the cycle. Carbon atoms from the acetate group are incorporated into a new molecule of oxaloacetate generated through the cycle.

Acetate enters the citric acid cycle by condensation with oxaloacetate

Acetyl CoA enters the cycle by condensation with oxaloacetate. The 'high-energy' thioester bond of acetyl CoA provides the energy for a condensation reaction with oxaloacetate. The reaction is catalysed by **citrate synthase** and the product is the six-carbon compound citric acid.

$$CH_3-\overset{\overset{O}{\|}}{C}-S-CoA \ + \ \overset{\overset{O}{\|}}{\underset{\underset{CH_2-COO^-}{|}}{C}}-COO^- \ + \ H_2O \ \xrightarrow[synthase]{Citrate} \ HO-\overset{\overset{CH_2-COO^-}{|}}{\underset{\underset{CH_2-COO^-}{|}}{C}}-COO^- \ + \ CoA-SH + H^+$$

Acetyl-S-CoA   Oxaloacetate   Citrate

This reaction is accompanied by a large negative free energy change ($\Delta G^{\circ\prime} = -32$ kJ mol^{-1}), ensuring that the reaction is driven strongly in the forward direction. This is important because the concentration of oxaloacetate in the cell is usually low.

In the next reaction of the cycle **aconitase** catalyses the conversion of citrate to isocitrate. The reaction proceeds through the intermediate formation of *cis*-aconitate requiring first a dehydration and then a rehydration step. Aconitase is an Fe–S (nonhaem iron) protein, which contains four Fe and four S atoms in a cubical cluster that is covalently linked to the enzyme through four cysteine residues linked to the Fe atoms (see Box 7.3). This cluster participates in the dehydration and rehydration reactions:

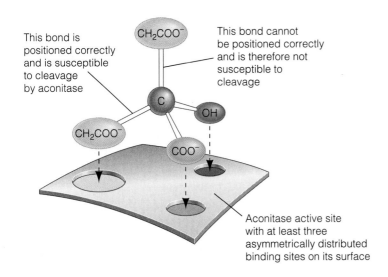

Citrate   [*cis*-Aconitate]   D-isocitrate

The purpose of this isomerization is to transfer the hydroxyl group on the central carbon of citrate to a carbon located next to a peripheral carboxyl group, rendering these regions susceptible to oxidation and decarboxylation. The equilibrium is in the direction of citrate, with only a small fraction of isocitrate, but the reaction is pulled strongly towards isocitrate formation by the next reaction in the cycle. Another important aspect of this isomerization is that it occurs in an asymmetric fashion. Aconitase can distinguish between the two equivalent –CH$_2$COOH groups in the citrate molecule and transfers the hydroxyl group onto one particular CH$_2$ group only (Figure 7.34).

This bond is positioned correctly and is susceptible to cleavage by aconitase

This bond cannot be positioned correctly and is therefore not susceptible to cleavage

Aconitase active site with at least three asymmetrically distributed binding sites on its surface

Figure 7.34 Citrate, which appears to be a symmetrical molecule, binds to aconitase in an asymmetric manner. The binding of citrate to three sites on the enzyme ensures that only one of the groups undergoes the reaction.

The first oxidation reaction is catalysed by isocitrate dehydrogenase

Isocitrate is converted to α-ketoglutarate in a reaction linked to the reduction of NAD$^+$.

Isocitrate   α-Ketoglutarate

There are both NAD$^+$- and NADP$^+$-dependent forms of this enzyme, the latter being important in providing NADPH for reductive biosynthetic reactions.

215

Oxidation of α-ketoglutarate is linked to the generation of a thioester bond

α-Ketoglutarate is subsequently oxidized by the **α-ketoglutarate dehydrogenase complex**, thus:

α-Ketoglutarate   Succinyl-CoA

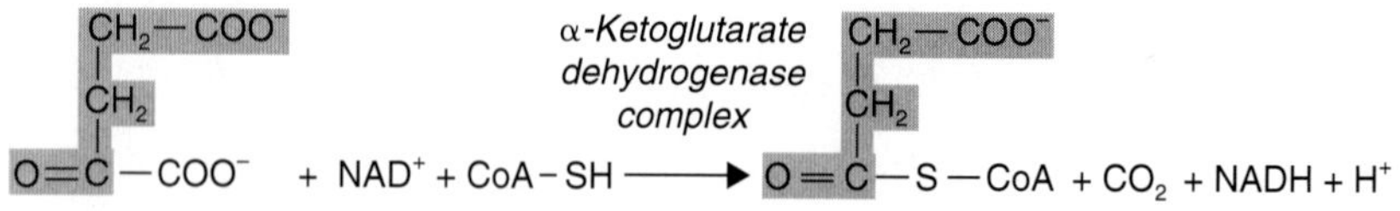

Chapter 8. The pyruvate dehydrogenase complex catalyses a similar sequence of reactions, page 243.

The reaction mechanism is almost identical to that of the pyruvate dehydrogenase complex (Chapter 8) and requires the same cofactors. These two complexes have multisubunit structures and one of the enzymes, the FAD-requiring dihydrolipoamide dehydrogenase, is identical in both.

The energy of the thioester is used to generate a high-energy phosphate bond

Succinyl CoA is, like acetyl CoA, an energy-rich thioester, with a large negative free energy of hydrolysis. In the next reaction of the cycle this energy is coupled to the phosphorylation of GDP, creating a high-energy phosphoanhydride bond with the formation of GTP. The reaction is catalysed by **succinyl CoA synthetase:**

Succinyl-CoA   Succinate

The γ-phosphate of GTP can be transferred to ADP by the enzyme **nucleoside diphosphokinase** so that in energetic terms production of GTP is equivalent to ATP production. In practice, GTP has a number of important specific cellular functions as a phosphoryl group donor, for example, in cellular signalling pathways (Chapter 11) and in protein synthesis (Chapter 5).

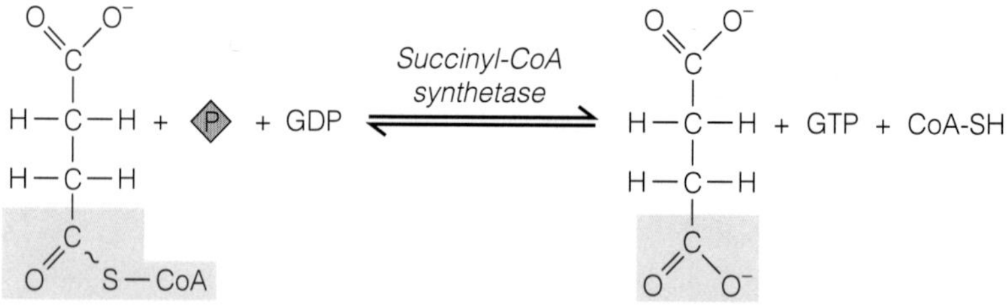

Chapter 11. G-protein α subunits catalyse GTP hydrolysis, page 336.

Succinate dehydrogenase catalyses an iron–sulphur dependent oxidation

Succinate dehydrogenase oxidizes succinate to fumarate through the introduction of a carbon–carbon double bond:

Succinate   Fumarate

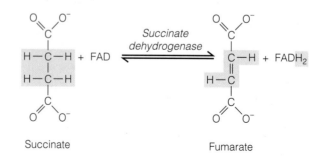

The enzyme possesses a large FAD-containing subunit and a small subunit with three distinct Fe–S centres. It is the only enzyme of the citric acid cycle which is tightly bound to the mitochondrial inner membrane. Fumarate is in the *trans* configuration, another example of an enzyme handling a symmetrical substrate in a specific fashion.

The double bond of fumarate is hydrated in preparation for the final oxidation reaction

The *trans* carbon–carbon double bond of fumarate is hydrated by the enzyme **fumarase** to form malate. Fumarase is stereospecific, being unable to hydrate the double bond of the *cis* isomer malate.

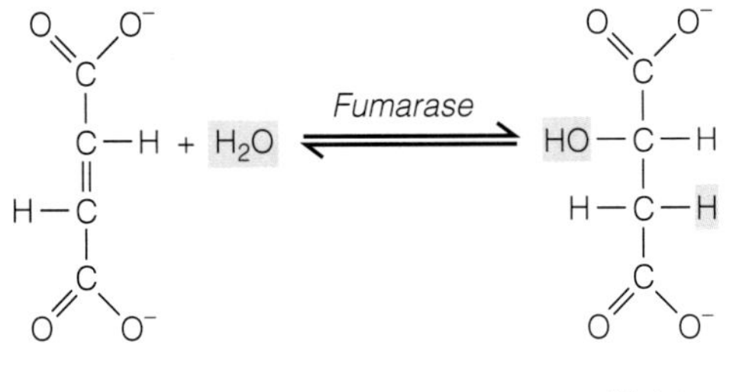

Fumarate   L-Malate

Malate dehydrogenase completes the cycle, producing oxaloacetate through an NAD^+-dependent reaction.

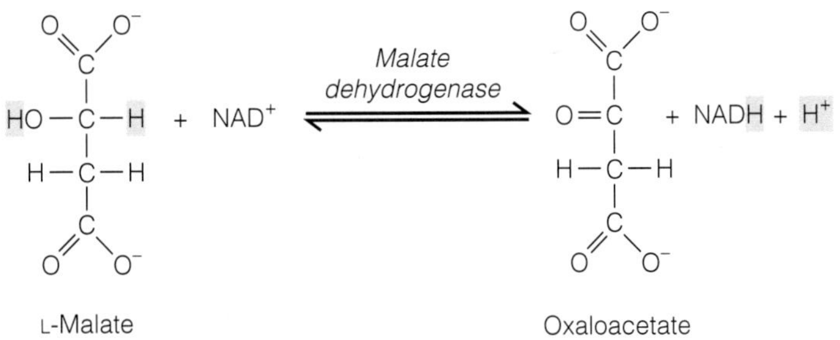

L-Malate   Oxaloacetate

Although this final reaction is not favoured thermodynamically, net conversion of malate to oxaloacetate is maintained by the removal of the product through its condensation with acetyl CoA entering the cycle.

The overall reaction for one turn of the cycle is shown in Figure 7.35.

Reduced nucleotides produced in the citric acid cycle are oxidized through the electron transport chain

The oxidation reactions of the citric acid cycle produce reduced coenzymes, NADH and $FADH_2$, compounds of high electron transfer potential which are

Table 7.1 Energy production from the complete oxidation of pyruvate

Reaction catalysed by	High-energy molecules	Equivalent ATP production
Pyruvate dehydrogenase	NADH	3 ATP
Isocitrate dehydrogenase	NADH	3 ATP
α-Ketoglutarate dehydrogenase	NADH	3 ATP
Succinyl CoA synthetase	GTP	1 ATP
Succinate dehydrogenase	$FADH_2$	2 ATP
Malate dehydrogenase	NADH	3 ATP

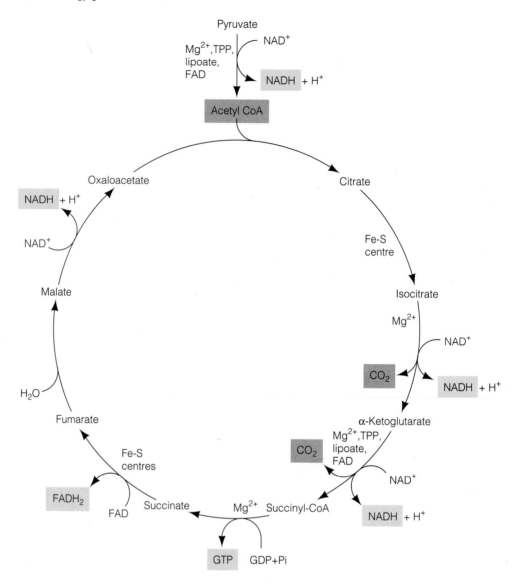

Figure 7.35 The intermediates and cofactors of the citric acid cycle.

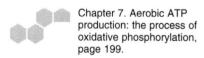

Chapter 7. Aerobic ATP production: the process of oxidative phosphorylation, page 199.

reoxidized in the mitochondrial respiratory chain in a series of oxidation–reduction steps with oxygen as the final electron acceptor (see above). Transfer of reducing equivalents along the respiratory chain yields three molecules of ATP for each NADH reoxidized to NAD^+ and two molecules of ATP for each $FADH_2$ reoxidized to FAD. The total yield of energy generated during the complete oxidation of one molecule of pyruvate is equivalent to 15 molecules of ATP. In the absence of oxygen the operation of the citric acid cycle is rapidly arrested as NADH and $FADH_2$ can not be reoxidized.

Many of the citric acid cycle enzymes catalyse reversible reactions

As presented above, reactions of the citric acid cycle proceed in the direction in which the cycle operates. Although the cycle is not fully reversible, some individual reactions can operate in either direction. The importance of this is reflected in the fact that intermediates of the cycle are withdrawn for other metabolic purposes, particularly in conditions of nutrient repletion (see above). Free energy changes associated with individual reactions of the cycle are listed in Table 7.2.

Table 7.2 Reactions of the citric acid cycle

Reaction	Enzyme	$\Delta G^{\circ\prime}$ (kJ mol^{-1})
Acetyl CoA + oxaloacetate + H_2O → Citrate + CoASH + H^+	Citrate synthase	−32.2
Citrate ⇌ cis-Aconitate + H_2O	Aconitase	
cis-Aconitate + H_2O ⇌ Isocitrate	Aconitase	+6.3
Isocitrate + NAD^+ ⇌ α-Ketoglutarate + CO_2 + NADH + H^+	Isocitrate dehydrogenase	−20.9
α-Ketoglutarate + NAD^+ + CoASH ⇌ Succinyl-CoA + CO_2 + NADH + H^+	α-Ketoglutarate dehydrogenase complex	−33.5
Succinyl-CoA + Pi + GDP ⇌ Succinate + GTP + CoASH	Succinyl-CoA synthetase	−2.9
Succinate + FAD (enzyme bound) ⇌ Fumarate + $FADH_2$ (enzyme bound)	Succinate dehydrogenase	0
Fumarate + H_2O ⇌ L-Malate	Fumarase	−3.8
L-Malate + NAD^+ ⇌ Oxaloacetate + NADH + H^+	Malate dehydrogenase	+29.7
	Net	**−57.3**

The citric acid cycle has an amphibolic role, serving in both catabolism and anabolism

The citric acid cycle operates not only as a mechanism for extracting energy from nutrients (the catabolic role) but also as a source of precursors for the synthesis of other molecules (the anabolic role). This is the **amphibolic** character of the citric acid cycle: the cycle provides both energy and carbon skeletons for biosynthetic purposes including the synthesis of amino acids, nucleotides and porphyrins (Figure 7.36).

Removal of intermediates from the citric acid cycle may reflect the need to synthesize other essential molecules, but it also occurs in nutrient-replete cells when the number of molecules of cycle intermediates is greater than that required to meet the energy demand on the cell. In this situation the excess nutrient will be used for the synthesis of fatty acids (Chapter 9) or glucose (Chapter 8), which can be stored within the cell and mobilized when the energy requirement increases. In some tissues, notably the liver, nutrient excess leads to the export of glucose or fatty acids to other tissues which may have a greater energy demand or, as in the case of adipose tissue, a greater capacity to store nutrient.

Chapter 9. Fatty acid biosynthesis, page 274.
Chapter 8. In an energy-replete cell nutrient can be stored by withdrawing metabolites from the citric acid cycle, page 252.

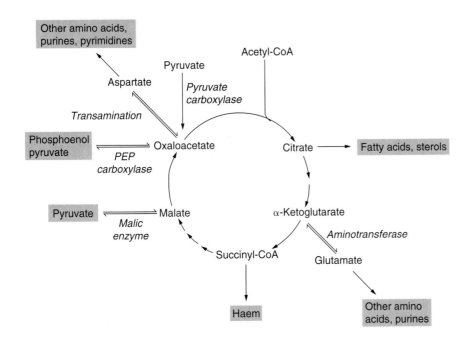

Figure 7.36 Many of the intermediates of the citric acid cycle are also substrates for other reactions. Some of these reactions provide mechanisms for withdrawing nutrients from the cycle for storage; others lead to the synthesis of proteins, nucleic acids and other essential molecules.

Anaplerotic mechanisms permit the energy production from the citric acid cycle to be increased

The **anaplerotic** reactions (from the Greek *ana pleros* meaning 'to fill up' or 'to replenish') provide mechanisms for replenishing intermediates of the cycle, such as oxaloacetate and α-ketoglutarate, which have been removed from the cycle as substrates for aminotransferase reactions (Chapter 10). In humans, the major anaplerotic reaction involves the introduction of more oxaloacetate into the cycle, but the amount of α-ketoglutarate in the cycle is also commonly increased from glutamate.

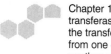
Chapter 10. Amino-transferase reactions facilitate the transfer of amino groups from one compound to another, page 295.

$$\text{Pyruvate} + \text{Biotin} - CO_2 + \text{ATP} \xrightarrow[\substack{\textit{Pyruvate} \\ \textit{carboxylase}}]{} \text{Oxaloacetate} + \text{ADP} + \text{Pi}$$

$$\begin{aligned}\text{Glutamate} + H_2O \\ + NAD^+\end{aligned} \xrightarrow[\substack{\textit{Glutamate} \\ \textit{dehydrogenase}}]{} \begin{aligned}\alpha\text{-Ketoglutarate} + NH_4^+ \\ + NADH + H^+\end{aligned}$$

Chapter 8. The 'bypass' enzymes of gluco-neogenesis, page 246.

Important aspects of the pyruvate carboxylase reaction for gluconeogenesis (Chapter 8) are that it requires biotin as a cofactor for the carboxylation reaction, and that it has an absolute requirement for acetyl CoA as an allosteric activator. The latter requirement means that this anaplerotic response is switched on when acetyl CoA accumulates, a sign that the citric acid cycle flux is slowing down, reducing the rate of energy production in the cell.

The citric acid cycle is regulated by three enzymes

Three of the enzymes of the citric acid cycle, citrate synthase, isocitrate dehydrogenase and α-ketoglutarate dehydrogenase, are subject to regulatory effects that determine the rate at which substrate is oxidized through the cycle. In broad terms the cycle responds to the need for energy in the forms of NADH, as a substrate for the electron transport chain, and ATP.

Under normal conditions most of the acetyl CoA entering the citric acid cycle is derived from fatty acid oxidation, with some derived from glucose through pyruvate and some from the metabolism of amino acids. Within the citric acid cycle the rate of oxidation of acetyl CoA (the overall rate of the cycle) is regulated by **citrate synthase, isocitrate dehydrogenase** and **α-ketoglutarate dehydrogenase**. Citrate synthase is sensitive to the availability of its substrates acetyl CoA and oxaloacetate. The concentration of oxaloacetate is usually low and at times there may be a high demand on oxaloacetate as substrate for gluconeogenesis or for the synthesis of aspartate. Citrate synthase is also inhibited by ATP, an effect which is relieved by ADP acting as an allosteric activator. Both isocitrate dehydrogenase and α-ketoglutarate dehydrogenase are regulated by the NADH:NAD$^+$ ratio.

Citrate provides a link between the citric acid cycle and the glycolytic pathway

The rates of the citric acid cycle and glycolysis are normally closely linked to ensure that glucose is oxidized to form pyruvate, and subsequently acetyl CoA, only at the rate at which the acetyl CoA is required by the citric acid cycle. This link is mediated through citrate, which acts as an inhibitor of the glycolytic enzyme phosphofructokinase. This means that if acetyl CoA is available from other substrates, such as fatty acids, the glycolysis pathway will be inhibited, conserving the supplies of glucose (and glycogen). An important aspect of the interaction between these two pathways is that the citric acid cycle is located in the mitochondrial matrix but the glycolysis pathway is in the cytosol. However, the mitochondrial inner membrane has a specific transporter for citrate, allowing efflux of citrate to the cytosol where it acts as a regulator of glycolysis and a substrate for citrate lyase. These processes are considered in more detail in Chapter 8.

Chapter 8. Phospho-fructokinase is the key regulatory enzyme of glycolysis, page 239.

CHAPTER 8

The metabolism of glucose: storage and energy generation

In this chapter we examine the processes by which glucose entering the body through the digestive tract is stored as a nutrient reserve or metabolized to produce energy. The glucose absorbed after a meal is usually more than sufficient to meet immediate energy requirements and so most is converted to glycogen through glycogenesis in liver and muscle, providing a reserve which can be mobilized through glycogenolysis at a later time. Catabolism of glucose is initiated through the glycolytic pathway, which enables energy to be extracted in the form of reduced nucleotide (NADH) and the high-energy compound ATP. Glucose is also catabolized through the pentose phosphate pathway, the purpose of which is to generate ribose-5-phosphate for nucleotide biosynthesis and the reduced nucleotide NADPH for reductive biosynthesis. Gluconeogenesis enables a variety of metabolites to be converted to carbohydrate. In the liver gluconeogenesis has the special role of producing free glucose which is then released to the bloodstream, providing nutrient for glucose-dependent tissues. The interrelationships between these processes are illustrated in Figure 8.1. The mechanisms by which these pathways are regulated to ensure efficient use of nutrient are described.

The metabolic relationships between glucose and glycogen

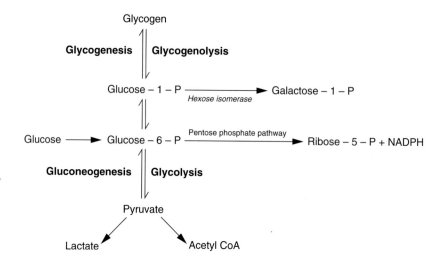

Figure 8.1 Glucose can be stored as glycogen for later reconversion to glucose, metabolized to lactate or acetyl CoA as part of the energy-producing pathways or converted to other sugars (including glucose and ribose) to meet other metabolic requirements of the cell.

The major source of carbohydrate in the human diet is in the form of polysaccharides of D-glucose, mostly starch. Polysaccharide is degraded to monosaccharides and disaccharides in the digestive system (see Chapter 14). Humans store carbohydrate in their bodies as glycogen. Additional dietary sources of carbohydrate are the disaccharides lactose and sucrose. The metabolism of these sugars is described in Chapter 14.

Glycogen is a highly branched polymer of D-glucose, which forms linear arrays joined by α,1–4-glycosidic bonds with branches introduced through α,1–6-glycosidic bonds (Figure 8.2). This is the form in which carbohydrate is stored in animal tissues. In the immediate aftermath of a meal, most tissues take up glucose from the bloodstream and some convert it to glycogen. Between meals, when there is no glucose entering the bloodstream from the digestive tract, the glycogen is degraded and the glucose metabolized to provide energy. The total amount of glycogen stored in the body is about 190 g, more than one mole of polymerized glucose. Of this, most is found in muscle cells, with a smaller fraction in the liver.

Glycogen stored in muscles provides a local energy supply which is used within the muscle cells. Muscle glycogen can not be used to maintain the blood glucose concentration because muscle lacks the enzyme **glucose-6-phosphatase**. Glycogen stored in the liver can be metabolized as an energy source for liver cells, but most is exported from the liver to other tissues as glucose because the liver contains large amounts of glucose-6-phosphatase which allows the release of free glucose from hepatocytes (Figure 8.3). Although the amount of glycogen stored in the liver is only a small proportion of the total stored in the body it is sufficient to meet the requirements of other tissues that do not store glucose, for several hours.

Chapter 14. Digestion and absorption of carbohydrate, page 417.

Nonreducing ends

Figure 8.2 Glycogen is an α,1–6 branched polymer of α,1–4 linked D glucose. The structure of polysaccharides is described in detail in Chapter 2.

α,1–6 branch points

Reducing end

Chapter 3. Entry of glucose into cells is mediated by several related tissue-specific transporters, page 68.

Phosphorylation of glucose serves to retain it within the cell

Glucose is absorbed from the digestive tract, entering the bloodstream whence it is absorbed into cells through a glucose transporter in the plasma membrane (Chapter 3). The first step in the metabolism of glucose is an enzyme-catalysed phosphorylation (Figure 8.4) forming glucose-6-phosphate (G6P). With the addition of a phosphate group, the molecule acquires a negative charge, and is then unable to diffuse out of the cell across the plasma membrane. This reaction

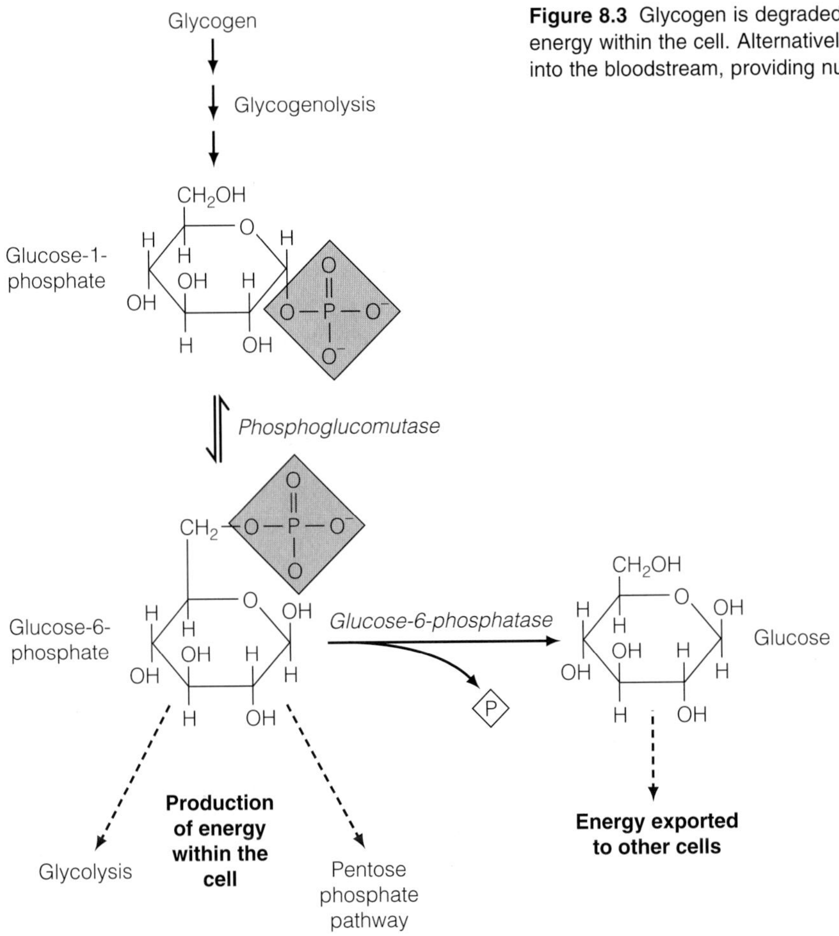

Figure 8.3 Glycogen is degraded through G1P to G6P, which is oxidized to provide energy within the cell. Alternatively, but only in the liver, free glucose may be released into the bloodstream, providing nutrient to other tissues.

therefore has the effect of trapping glucose within the cell, where it is available for catabolism or storage.

Phosphorylation of glucose is catalysed by **glucokinase** in hepatic and pancreatic islet β cells, and by **hexokinase** in other tissues. The differences in kinetic parameters of these two enzymes are important in determining how glucose is distributed between tissues of the body (see Table 8.1).

Glucokinase in the liver helps to regulate the distribution of glucose between different tissues

Glucose enters liver cells through an insulin-independent low-affinity transporter (GLUT 2). Most other tissues have specific insulin-activated glucose transporter systems.

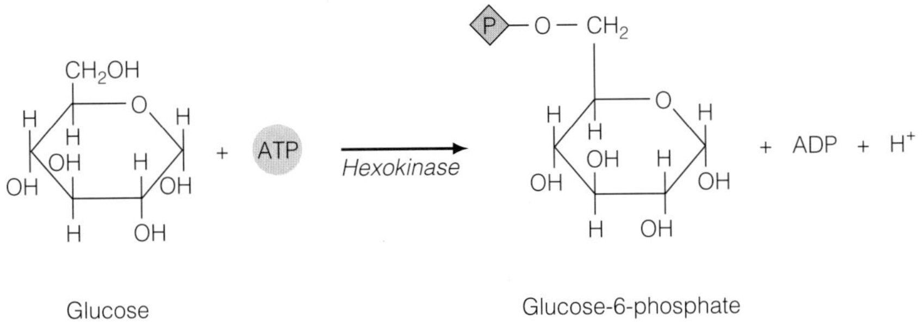

Table 8.1 Properties of glucokinase and hexokinase

Hexokinase	Glucokinase
Broad specificity for hexoses	Specific for glucose
Inhibited by G6P	Not inhibited by G6P
K_m for glucose < 0.1 mM	K_m for glucose 5 mM
Not inducible	Inducible in the liver

Figure 8.4 Conversion of glucose to G6P.

The tissue-specific distribution of glucokinase and hexokinase ensures that

- In tissues other than the liver glucose will not be phosphorylated unless it is being further metabolized, because an accumulation of G6P inhibits phosphorylation of more glucose molecules.
- At low blood glucose concentrations the liver is not competing with other tissues for the supply of glucose because the high affinity (low K_m) of hexokinase for glucose in glucose-dependent tissues such as the brain and red blood cells ensures that they have a prior claim on the circulating glucose. The relatively low affinity (high K_m) of glucokinase for glucose means that the liver is prevented from utilizing glucose until the nutrient requirements of other tissues are satisfied.
- At blood glucose concentrations high enough to meet the needs of other tissues, the free permeability of liver cells to glucose results in increased uptake of glucose. Glucose in liver cells reaches concentrations sufficient to drive the glucokinase reaction, and glycogen is synthesized and stored.

Glucokinase has another interesting control feature: it is subject to inhibition by a protein that binds to the enzyme in a manner regulated by the relative concentrations of fructose-6-phosphate (F6P) and fructose-1-phosphate (F1P) in the cell. F6P promotes binding of the inhibitory protein and inhibits glucokinase activity; F1P inhibits the interaction of the binding protein with glucokinase. This control mechanism effectively enables glucokinase to monitor the utilization of glucose through the later stages of glycolysis. Insulin also has a long-term effect on liver glucokinase, by promoting transcription of the glucokinase gene.

Within cells glucose may be used as an energy source or may be converted to glycogen and stored as an endogenous energy reserve for mobilization to provide energy between meals. Alternatively, it may be converted to fat and stored as triacylglycerol (Chapter 9).

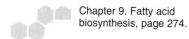

Chapter 9. Fatty acid biosynthesis, page 274.

The storage and mobilization of glycogen is determined by the energy status of the cell and by hormonal responses

Synthesis and degradation of glycogen is regulated through two types of processes:

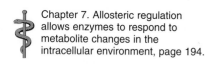

Chapter 7. Allosteric regulation allows enzymes to respond to metabolite changes in the intracellular environment, page 194.

1. Changes in the cellular concentration of specific metabolites activate or inhibit allosteric enzymes that determine the rates of glycogen synthesis and degradation. The concentrations of these metabolites reflect the energy status of the cell (Chapter 7).
2. The same enzymes are regulated by a cellular response to the binding of hormones to receptors on the outer surface of the plasma membrane. Examples are the binding of adrenaline to muscle cell receptors and the binding of glucagon to liver cell receptors (Chapter 11).

Chapter 11. Glucagon, page 348.

When glucose is plentiful it is stored as glycogen

The process by which monosaccharides are added to an existing glycogen polymer is called **glycogenesis**. Extending a glycogen polymer in this way requires the sequential action of three enzymes. The rate-limiting enzyme in this sequence, **glycogen synthase**, is an allosteric enzyme regulated by metabolites; its activity is also regulated by hormone-mediated phosphorylation.

The glucose units stored in a glycogen polymer are increased by adding glucose units one at a time through the formation of α,1–4-glycosidic bonds to the nonreducing end of a glycogen branch, which must be at least four glucose units in length. Addition of glucose units to the polymer requires energy, so each monosaccharide is firstly activated as a **uridine diphosphate** (UDP) derivative.

UDP-Glucose pyrophosphorylase

This catalyses the activation of glucose units:

$$G1P + UTP \rightarrow UDP\text{--glucose} + PP_i$$

UDP-glucose is a high-energy compound, and the energy is used to catalyse the formation of a glycosidic bond. Subsequent hydrolysis of the pyrophosphate by a pyrophosphatase draws the reaction towards synthesis of UDP-glucose.

Glycogen synthase

Glycogen synthase adds the activated glucose unit to the glycogen, (glucose)$_n$, branch, releasing UDP.

$$(\text{glucose})_n + UDP\text{--glucose} \rightarrow (\text{glucose})_{n+1} + UDP$$

The $\Delta G^{o'}$ for this reaction is -13 kJ mol^{-1}, allowing the reaction to proceed in the direction of elongating the glycogen polymer.

UTP is regenerated from UDP, by transfer of a phosphate group from ATP

$$UDP + ATP \rightarrow UTP + ADP$$

so that the overall addition of one glucose unit derived from G1P requires the expenditure of one ATP molecule for every monosaccharide added to the glycogen polymer. As the synthesis of glycogen commonly begins with glucose entering the cell, two molecules of ATP are consumed in adding each new monosaccharide unit to the polysaccharide.

Branching enzyme

Branches are introduced into the structure of glycogen by the enzyme **amylo-1,4 →1,6-transglycosylase**, more commonly called **branching enzyme**. This enzyme breaks α,1–4 glycosidic linkages and reforms α,1–6 glycosidic linkages. Branching enzyme does not move single glucose units, but a chain of about seven glycosyl units with each reaction (Figure 8.5).

The formation of new glycogen polymers is initiated by a primer protein, **glycogenin**, which acts as an attachment point for the first glucose residue and also catalyses the formation of a seven-residue polysaccharide, which is then extended in length by glycogen synthase.

When the energy status of a cell is low monosaccharides will be released from glycogen stores

Three enzymes operate sequentially to provide G6P from glycogen through a process called **glycogenolysis**. The rate-limiting enzyme in this sequence is **glycogen**

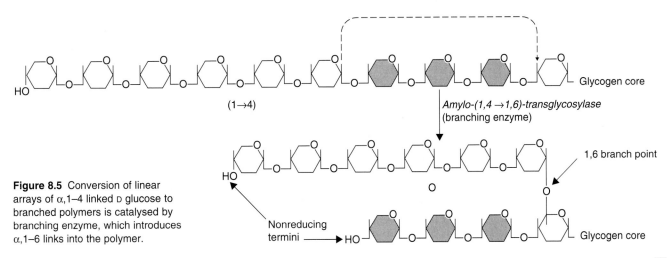

Figure 8.5 Conversion of linear arrays of α,1–4 linked D glucose to branched polymers is catalysed by branching enzyme, which introduces α,1–6 links into the polymer.

Chapter 15. ATP is hydrolysed to provide energy to power the mechanical cycle, page 464.

Table 8.2 Allosteric regulators of glycogen phosphorylase and glycogen synthase

Allosteric regulator	Phosphorylase	Synthase
ATP	–	Activator
AMP	Activator	–
G6P	Inhibitor	Activator

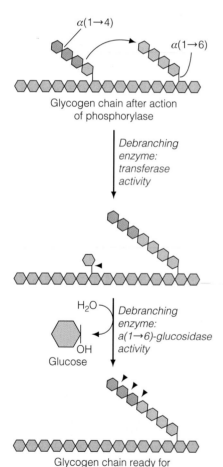

Glycogen chain after action of phosphorylase

Debranching enzyme: transferase activity

Debranching enzyme: a(1→6)-glucosidase activity

Glycogen chain ready for continued action of phosphorylase

Figure 8.6 Once phosphorylase has reduced the length of glycogen branches to four glucose residues, debranching enzyme transfers the remaining residues to another branch so that the action of phosphorylase can proceed.

phosphorylase. As is the case with the rate-limiting enzyme in glycogen synthesis, this enzyme is subject both to allosteric regulation and to regulation through a hormone-stimulated phosphorylation mechanism. The allosteric regulators are listed in Table 8.2. AMP is produced in muscle cells when a high rate of hydrolysis of ATP to support muscle contraction (Chapter 15) combines with the activity of adenylate kinase to convert the resultant ADP to ATP and AMP. When energy utilization is great enough for AMP to accumulate, energy must be produced by the catabolism of glycogen.

Glycogen phosphorylase

This enzyme splits monosaccharide units from glycogen, one at a time from each of the nonreducing ends of the branched polymer, breaking the α,1–4 glycosidic bonds by substitution of a phosphate group in a phosphorolysis reaction.

$$(glucose)_n + P_i \rightarrow (glucose)_{n-1} + G1P$$

Each sugar unit is released as G1P, ready to enter the catabolic pathways. Glycogen phosphorylase will release only those glucose units that are at least five units from a branch point and requires the coenzyme pyridoxal phosphate – a derivative of vitamin B_6.

Debranching enzyme

Debranching enzyme removes residual short branches in the glycogen structure through a transferase activity, forming α,1–4 glycosidic links, lengthening other short branches so that glycogen phosphorylase can complete its task. Together phosphorylase and debranching enzyme ensure that 90% of the glycogen molecule is converted to G1P. The remaining glucose units, those at branch points, are released as glucose (see Figure 8.6).

Phosphoglucomutase

Phosphoglucomutase converts G1P to G6P, providing substrate for two of the energy-producing pathways within the cell: **glycolysis** and the **pentose phosphate pathway**. For glucose to leave the liver cell and enter the bloodstream, thus providing nutrient to other tissues, the phosphate group must be removed by the enzyme **glucose-6-phosphatase**. This activity is found principally in the liver, with a small amount in the kidney. Other tissues lack this enzyme and therefore can not release glucose into the bloodstream.

Glycogen phosphorylase and glycogen synthase are both subject to allosteric regulation

Glycogen phosphorylase and glycogen synthase, respectively the rate-limiting enzymes for glycogenolysis and glycogenesis, are both regulated by metabolites and nucleotides that are indicators of the energy status of the cell. The control of these enzymes ensures that carbohydrate is available as an energy source only when the cell needs more energy.

AMP is displaced from the allosteric site of glycogen phosphorylase by ATP, which is usually present in cells at much higher concentration than AMP. Activation by AMP is therefore seen only in energy-deficient cells. G6P also inhibits glycogen phosphorylase by binding to the AMP site. These controls ensure that glycogen is not wasted if the cell has sufficient energy and is deriving G6P from other sources.

Allosteric activation of glycogen synthase occurs through a site that binds G6P. This means that the enzyme will respond to an increase in the concentration of G6P by storing more glycogen. This enzyme is also activated by ATP, the prime indicator that cells have sufficient energy to meet their needs.

Glycogen metabolism is also regulated by hormones

Glycogen must be degraded in muscle cells at times of increased exercise or in the liver when the glucose concentration in the bloodstream falls, for example between meals and in starvation.

- Increased muscular activity is accompanied by an increase in the secretion of adrenaline. This binds to receptors on the plasma membrane of muscle cells, signalling to the interior of the cell the need to provide energy by degrading glycogen and metabolizing the glucose released. The ATP produced through metabolism provides the energy to support the contractile activity of muscle fibres.

- A decrease in the glucose concentration in the bloodstream is detected by the pancreas, which responds by secreting more glucagon (Chapter 13). This binds to receptors on the plasma membrane of liver cells, signalling the need to degrade glycogen in order to release glucose into the bloodstream.

 Chapter 13. Synthesis and secretion of glucagon, page 397.

The intracellular response to glucagon or adrenaline is mediated through changes in cyclic AMP concentration

The intracellular response to the binding of glucagon or adrenaline to their receptors is the activation of a transmembrane protein, with adenylate cyclase acting on its cytoplasmic domain. This enzyme catalyses the formation of **3',5' cyclic adenosine monophosphate (cyclic AMP or cAMP)** from ATP (Chapter 11).

$$\text{ATP} \xrightarrow[\textit{Adenylate cyclase}]{} \text{cAMP} + \text{PPi}$$

 Chapter 11. Hormonal responses are mediated through adenylate cyclase and the production of cAMP, page 338.

Cyclic AMP is described as a **second messenger**, because it is produced within the cell in response to activation by the primary messenger, the hormone. Activation of adenylate cyclase is mediated by G-proteins through a mechanism which is dealt with in detail in Chapter 11. The increase in cAMP concentration within the cell stimulates the cascade that controls both degradation and synthesis of glycogen. Once the need to activate this cascade is relieved, the concentration of cAMP must be reduced. The hydrolysis of cAMP to AMP is catalysed by **cAMP phosphodiesterase**.

$$\text{cAMP} + \text{H}_2\text{O} \xrightarrow[\textit{cAMP phosphodiesterase}]{} \text{AMP}$$

Increased cAMP concentration activates a 'cascade' of events that amplify the hormonal response

Cyclic AMP activates a multistep cascade (Chapter 7), in which each step amplifies the effect of the previous step so that a small number of hormone molecules binding to the outer surface of the cell produce a very large change in the activity of the regulated enzymes at the end of the cascade. In the case of glycogen metabolism two enzymes are regulated simultaneously: the rate-limiting enzyme for glycogen degradation is accelerated and the rate-limiting enzyme for glycogen synthesis is inhibited. Most of the steps in the cascade involve enzymes changing from their inactive to active forms through changes in quaternary protein structure, the spatial relationships between subunits of the protein (see Chapter 4). The change in quaternary structure may be triggered by phosphorylation (Chapter 7) of the protein or may be a response to the binding of calcium ions to the protein.

 Chapter 7. Cascades are multistage activations which amplify a response, page 198.

Chapter 4. Quaternary (4°) structure: structural and functional features, page 105. Chapter 7. Covalent modification of existing enzyme molecules can change their activity, page 195.

Activation of glycogen phosphorylase and inhibition of glycogen synthase occur in response to a common cascade

The fact that glycogen phosphorylase and glycogen synthase respond to the same cascade, one being activated under conditions that inhibit the other, ensures that the synthesis and degradation of glycogen do not occur simultaneously to any significant extent.

The sequence of events in this cascade is as follows:

1. Adrenaline (or glucagon) binds to the plasma membrane receptor and activates, through G-proteins located in the membrane, adenylate cyclase which catalyses the production of cAMP from ATP

2. Cyclic AMP binds to the regulatory subunits of **cAMP-dependent protein kinase**, activating the enzyme by dissociating the two catalytic units from the two regulatory subunits (Figure 8.7).

3. Activated cAMP-dependent protein kinase catalyses two changes: one leading to inhibition of glycogen synthase and the activation of glycogen phosphorylase. It phosphorylates the synthase, switching it off, and preventing further synthesis of glycogen. At the same time it converts **phosphorylase kinase** from its inactive form to its active form through an ATP-dependent phosphorylation reaction.

   Phosphorylase kinase is only fully active when it also interacts with a small protein called **calmodulin**. Calmodulin responds to small changes in the cytoplasmic Ca^{2+} concentration by assuming a more compact form which enables it to interact with a variety of target enzymes, including phosphorylase kinase. Only after the phosphorylation and Ca^{2+} dependent changes is phosphorylase kinase fully active.

4. Finally, the active form of phosphorylase kinase activates glycogen phosphorylase by an ATP-dependent phosphorylation.

These events are summarized in Figure 8.8.

The hormone-mediated chemical modification of glycogen phosphorylase involves the binding of two phosphate groups, contributed by ATP, to serine residues in the enzyme. Phosphorylase exists in two forms: a 'b' form which is not phosphorylated and is inactive, and an 'a' form which is phosphorylated and active. This activation of glycogen phosphorylase is reversed by the action of a protein phosphatase (Figure 8.9).

Phosphorylation of inhibitor-1 maintains the active state of phosphorylase

Maximum glycogen phosphorylase activity will occur when phosphorylase kinase is activated and phosphoprotein phosphatase is inactive. The activity of the phosphatase is controlled by another protein, **phosphoprotein phosphatase inhibitor-1** (referred to simply as **inhibitor-1**). Inhibitor-1 is phosphorylated by cAMP protein kinase, and in this phosphorylated form it blocks the activity of the phosphatase. It is, in turn, dephosphorylated by the same phosphatase and is then no longer effective as an inhibitor. Inhibitor-1 thus provides the final link in a control system that ensures maximum rates of glycogen degradation after hormonal stimulation. As cAMP concentration increases cAMP-dependent protein kinase is activated, and simultaneously glycogen phosphorylase is activated and inhibitor-1 inhibits phosphoprotein phosphatase (which would otherwise tend to reduce the activity of glycogen phosphorylase). These two responses operate together to maintain maximal activity of glycogen phosphorylase.

Glycogen synthase also exists in two forms

Glycogen synthase exists in two forms: a phosphorylated 'b' form which has low activity, and an active, nonphosphorylated 'a' form. The phosphate groups are removed by the same protein phosphatase that acts on glycogen phosphorylase.

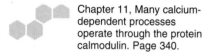

Chapter 11, Many calcium-dependent processes operate through the protein calmodulin. Page 340.

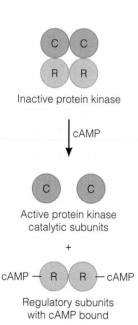

Inactive protein kinase

cAMP

Active protein kinase
catalytic subunits

+

cAMP — R   R — cAMP

Regulatory subunits
with cAMP bound

Figure 8.7 Protein kinase has four subunits, two regulatory (R) and two catalytic (C). Cyclic AMP binds to the R subunits, dissociating them from the C subunits, which are thereby activated and phosphorylate the next component of the cascade.

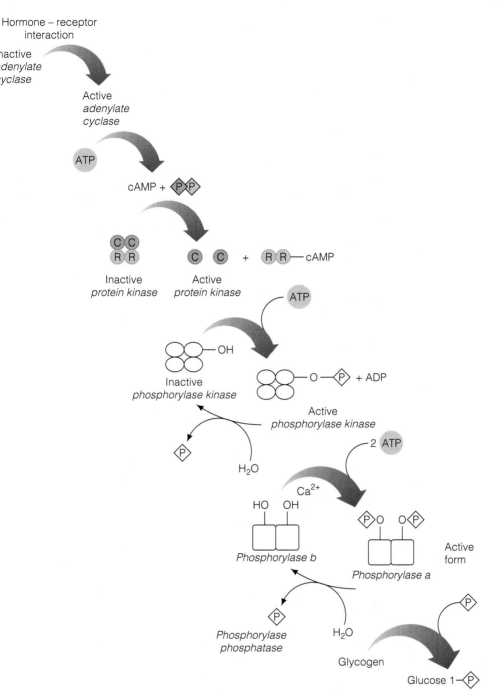

Hormone – receptor
interaction

Inactive *adenylate cyclase*

Active *adenylate cyclase*

ATP

cAMP + P P

Inactive *protein kinase*

Active *protein kinase*

+ R R — cAMP

ATP

Inactive *phosphorylase kinase* — OH

Active *phosphorylase kinase* — O — P + ADP

H_2O

P

HO OH

Phosphorylase b

Ca^{2+}

2 ATP

P O O P

Active form

Phosphorylase a

Phosphorylase phosphatase

P

H_2O

Glycogen

P

Glucose 1 — P

Figure 8.8 The phosphorylase cascade is triggered by binding of a hormone to a specific receptor linked through G-proteins to adenylate cyclase. At each step in the cascade the response is amplified by the activation of a component of the cascade. For details see text.

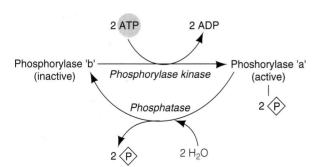

2 ATP   2 ADP

Phosphorylase 'b'   Phoshorylase 'a'
(inactive) ————————————————— (active)
            Phosphorylase kinase

            Phosphatase   2 P

2 P   2 H_2O

Figure 8.9 Glycogen phosphorylase exists in two forms. The active form is phosphorylated and can be inactivated by phosphatase-catalysed hydrolysis of the phosphate group. Two phosphate groups, derived from two ATP molecules, are required to activate this enzyme. Activation is catalysed by glycogen phosphorylase kinase.

229

Box 8.1 Mechanisms for switching off hormone activation of cascades

Because of the many steps in a cascade, there are many mechanisms for switching off the response.

- Inactivation of the hormone molecule, usually through chemical change.
- Releasing the hormone molecule from the receptor.
- Removal of receptor molecules from the cell surface by endocytosis.
- Inactivation of the second messenger, for example hydrolysis of cAMP.
- Direct reversal of the cascade activation steps, for example by a phosphatase-catalysed removal of phosphate groups from cascade intermediates.

The importance of the common elements of the phosphorylation mechanism are that the synthase is inactivated by phosphorylation, the phosphorylase is activated by phosphorylation and vice versa. This reciprocal relationship, in which reaction in one direction is activated by the same factors that inhibit the reverse process, is a common regulatory device in metabolism.

The balance between the synthesis and degradation of glycogen reflects the local needs of the tissues storing glycogen and in the liver also reflects the needs of tissues such as brain and red blood cells, which depend upon a supply of glucose from the liver to sustain their energy needs between meals.

Box 8.2 Glycogen-storage diseases

McArdle's disease
The fact that glycogen synthesis and degradation are catalysed by different enzymes was first established with the recognition of the cause of a disorder known as McArdle's disease. This is an inherited glycogen-storage disease, in which exercise leads to painful muscle cramps. Although the muscles of these patients contain glycogen, showing that muscle cells must have the enzymes for synthesis of glycogen, they lack **glycogen phosphorylase** and are therefore unable to mobilize the glycogen to provide the energy needed to support exercise.

von Gierke's disease
This disease is characterized by a genetic deficiency of **glucose-6-phosphatase** or of **G6P transporter**. As the liver can not convert G6P to glucose for release into the bloodstream, its concentration increases in liver. Glycogen phosphorylase is inhibited and glycogen synthase activity increases. Consequently, glycogen accumulates in the liver and kidneys. The hypoglycaemia seen in these patients results from the failure of glycogen-metabolizing enzymes to respond to glucagon or adrenaline. The condition may be treated with drugs that inhibit glucose uptake by liver, although in many cases liver transplantation is necessary.

Cori's disease
Cori's disease is characterized by deficiency of **glycogen debranching** enzyme. As a result, liver and muscle accumulate glycogen with very short chains and glycogen can not be fully degraded. These patients tend to be hypoglycaemic, although the extent of the hypoglycaemia is not as severe as in von Gierke's disease. Cori's disease is treated by controlling the diet and frequency of meals. The aim is to minimize the amount of glycogen being stored.

Genetic deficiencies of most of the enzymes of glycogen metabolism have been recognized, but some of them occur only rarely.

Chapter 8. Glucose-6-phosphatase, page 249.

The glycolysis pathway

Glycolysis is a sequence of enzyme-catalysed reactions that enables some of the energy inherent in glucose to be recovered as ATP and NADH. At the end of the pathway each molecule of glucose has been converted to two molecules of pyruvate. The net yield of ATP is two molecules from each glucose converted to pyruvate. The enzymes of glycolysis are found in the cytoplasm of all mammalian cells.

Glycolysis has two phases: 'energy investment' followed by 'energy generation'

Retrieving energy from glucose requires initial investment of energy (Figure 8.10). ATP is used in two reactions to phosphorylate the hexose, converting it to fructose-1,6-bisphosphate which can be split into two interconvertible triose-phosphate fragments, glyceraldehyde-3-phosphate and dihydroxyacetone phosphate.

In the energy-generating phase the triose phosphates are converted through two sequences of reactions, each of which first generates a compound with a high phosphoryl transfer potential (Figure 8.11). These compounds readily transfer their phosphate groups to ADP, generating a high-energy phosphoanhydride bond (ATP) in a process referred to as **substrate-level phosphorylation**.

Two molecules of ATP are invested per molecule of glucose in the first phase of glycolysis, and four molecules of ATP, two for each of the triose phosphate molecules, are generated in the second phase; a net gain of two ATP per glucose.

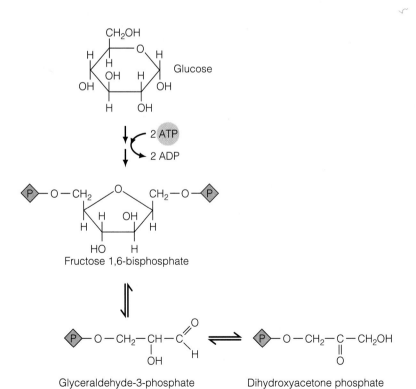

Figure 8.10 The energy-investment phase of glycolysis uses two ATP molecules and produces two triose phosphates as substrates for the energy-generating phase.

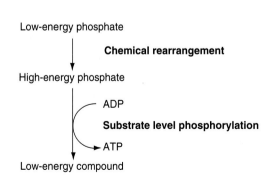

Figure 8.11 A 'low-energy' phosphate bond can be converted to a 'high-energy' phosphate bond by chemical rearrangement. This type of reaction occurs twice in the glycolytic pathway, in both cases generating compounds with sufficient energy to form the β–γ phosphate bond of ATP by transferring their phosphate groups to ADP.

231

The energy-investment phase of glycolysis involves three reactions

Hexokinase

Glucose is phosphorylated in a reaction catalysed by hexokinase or glucokinase, investing energy from transfer of the terminal phosphoryl group of ATP.

Glucose   Glucose-6-phosphate

The hexose phosphate formed in this reaction is relatively low energy. The reaction has the advantage of introducing a phosphate group which will, by subsequent chemical rearrangements, acquire a high phosphoryl transfer potential.

Phosphoglucoisomerase

The aldose sugar is converted to a ketose, fructose-6-phosphate (F6P).

Glucose-6-phosphate   Fructose-6-phosphate

This transfer of the ring oxygen bond from C-1 to C-2 of the sugar exposes the OH group on C-1 to phosphorylation.

Phosphofructokinase

The second energy-investment reaction follows with the phosphorylation of this 1-OH group of fructose. ATP provides the phosphoryl group and the energy to drive this reaction.

Fructose-6-phosphate   Fructose-1,6-bisphosphate

Phosphofructokinase is the main control point of glycolysis. The details of the control mechanisms will be described later in this chapter.

Aldolase

Fructose-1,6-bisphosphate is cleaved to produce two molecules of triose phosphate. The name of the enzyme that catalyses this reaction reflects the fact that it is the reverse of a type of chemical reaction called an aldol condensation.

Fructose-1,6-bisphosphate → Dihydroxyacetone phosphate + Glyceraldehyde-3-phosphate (Aldolase)

Under standard conditions the free energy change for this reaction would be positive, requiring an input of energy. Under the conditions in a cell, the concentrations of products and reactants are such that the reaction proceeds with a small negative free energy change.

Triose phosphate isomerase

The two three-carbon phosphates, one an aldotriose and the other a ketotriose, form an equilibrium mixture catalysed by this isomerase. At equilibrium the ratio of these triose phosphates is 96 (ketose): 4 (aldose).

Dihydroxyacetone phosphate ⇌ Glyceraldehyde-3-phosphate (Triose phosphate isomerase)

Glyceraldehyde-3-phosphate is the substrate for the energy-generating part of glycolysis. Utilization of glyceraldehyde-3-phosphate constantly draws the isomerase reaction in the direction converting dihydroxyacetone phosphate to glyceraldehyde-3-phosphate.

The energy-generating phase of glycolysis involves the formation of phosphate groups with high phosphoryl transfer potential

Glyceraldehyde-3-phosphate dehydrogenase

This catalyses the formation of 1,3-bisphosphoglycerate, a molecule with high phosphoryl transfer potential. This is a complex two-stage reaction in which the aldehyde group of glyceraldehyde-3-P is oxidized to a carboxyl group; oxidizing power being provided by reduction of NAD^+ to NADH. The energy generated by this oxidation is then coupled to the addition of a phosphate group, derived from inorganic phosphate, to the carboxyl group (forming a mixed anhydride).

Glyceraldehyde-3-phosphate + NAD^+ + Inorganic phosphate ⇌ 1,3-Bisphosphoglycerate + NADH + H^+ (Glyceraldehyde-3-phosphate dehydrogenase)

Phosphoglycerate kinase

The 1-phosphate of 1,3-bisphosphoglycerate is an acyl phosphate, a compound with a standard free energy of hydrolysis (-49.4 kJ mol^{-1}) substantially greater than that of the terminal phosphoanhydride bond of ATP (-30 kJ mol^{-1}).

The phosphate group and the energy inherent in it are therefore readily transferred to ADP, producing ATP.

1,3-Bisphosphoglycerate   3-Phosphoglycerate

This is the first of the two substrate-level phosphorylation reactions in the pathway.

Phosphoglycerate mutase

In preparation for the production of the next high-energy compound, 3-phosphoglycerate is converted to 2-phosphoglycerate by this Mg^{2+}-dependent enzyme. The reaction is a simple rearrangement: mutases catalyse the shift of an atom or group, in this case a phosphate group, within the molecule.

3-Phosphoglycerate   2-Phosphoglycerate

Enolase

Enolase catalyses the formation of the second compound with a high phosphoryl transfer potential. This enzyme, which is also Mg^{2+} dependent, catalyses the dehydration of 2-phosphoglycerate. Loss of water converts the low-energy carbon–phosphate bond of the substrate into a high-energy **enol phosphate** bond in the product, phosphoenol pyruvate.

2-Phosphoglycerate   Phosphoenolpyruvate

Pyruvate kinase

The phosphate bond in phosphoenolpyruvate has a much higher energy potential (-61.9 kJ mol^{-1}) than the phosphoanhydride bond of ATP (-31 kJ mol^{-1}) The transfer of the phosphate group to generate ATP from ADP therefore occurs readily.

Phosphoenolpyruvate   Pyruvate

This is the second of the substrate-level phosphorylations.

An overview of glycolysis is given in Figure 8.12.

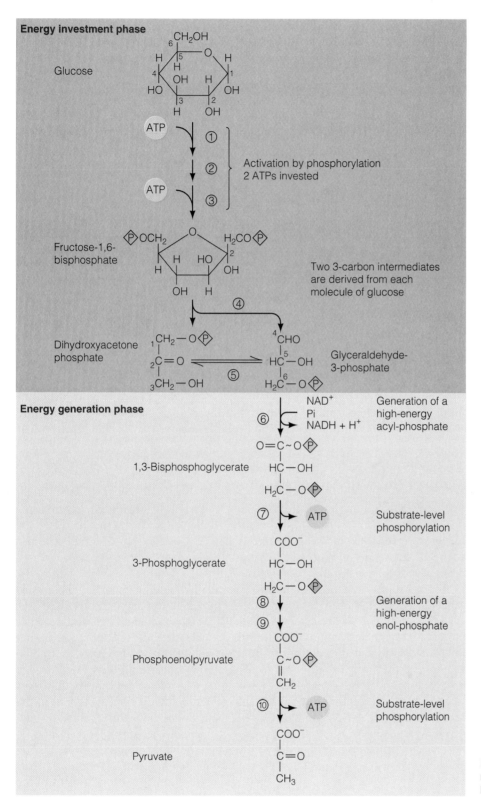

Figure 8.12 Overview of glycolysis. The numbers ① to ⑩ refer to the enzymes listed in Table 8.3.

The energy yield from glycolysis is in the form of ATP and NADH

Glucose enters the glycolytic pathway either directly from the bloodstream or as G1P derived from glycogen stored within the cell. In the former case the phosphorylation of glucose consumes ATP. The net ATP yield from the pathway is then

2 ATP per glucose. If glucose enters the pathway as G1P the initial energy investment reaction is circumvented and the overall energy yield is greater, 3 ATP per G1P. Glycolysis also generates two reducing equivalents, in the form of NADH from each glucose oxidized.

Glycolysis is inhibited if the ratio of reduced to oxidized nicotinamide nucleotide increases

It is essential for the continued function of glyceraldehyde-3-phosphate dehydrogenase that NADH is reoxidized either aerobically through the mitochondrial process or, if oxygen is limiting, through the reaction catalysed by **lactate dehydrogenase**. The function of lactate dehydrogenase is to facilitate energy production from glycolysis by reoxidizing NADH under anaerobic conditions; however, there is a cost to the cell associated with the production of lactate which is a carboxylic acid with a pK_a of 3.8. If large amounts of lactate accumulate within cells, the buffering capacity of the cell may be exceeded and the increased acidity will disturb the optimal function of many molecules. This problem is commonly experienced in severe exercise when proton accumulation disturbs calcium handling in muscle cells (Chapter 15) and may lead to severe muscular cramp.

Chapter 15. The sacroplasmic reticulum has mechanisms for the uptake, storage and release of calcium, page 466.

$$\text{Pyruvate} + \text{NADH} + \text{H}^+ \xrightarrow{\textit{Lactate dehydrogenase}} \text{Lactate} + \text{NAD}^+$$

In an adequately oxygenated cell all of the NADH is reoxidized through the mitochondrial system and the pyruvate produced from glycolysis is converted to acetyl CoA and fed into the citric acid cycle (see below).

Chapter 8. The pyruvate dehydrogenase complex converts pyruvate irreversibly to acetyl CoA, page 242.

Many of the reactions of glycolysis are reversible

In the preceding description the reactions of glycolysis are shown as unidirectional, operating in the forward direction of the pathway. Many of them are, however, reversible: indeed, all except hexokinase, phosphofructokinase and pyruvate kinase are reversible and operate in the reverse direction as part of the process of **gluconeogenesis** (see below).

Chapter 8. Gluconeogenesis: the synthesis of glucose from noncarbohydrate precursors, page 242.
Box 7.1. Thermodynamics of metabolic processes, page 189.

Reversibility of reactions is indicated by the free energy changes. The significance of free energy changes in reactions is discussed in Chapter 7; those for the reactions of glycolysis are listed in Table 8.3. Standard free energy changes have limited relevance to reactions within cells, because metabolites are present within cells at relatively low concentrations. In addition, if the reaction involves a hydrogen ion, it may drive a reaction from a thermodynamically favourable to an unfavourable position, or vice versa. A striking example of the caution that must be applied to interpreting free energy changes in biological systems is seen in the reaction catalysed by phosphoglycerate kinase. Reversal of this reaction, which involves expenditure of ATP to produce a much higher energy compound, is possible because tight coupling between the kinase and the immediately preceding dehydrogenase reaction ensures that the energy requirement of one reaction is provided by coupling to the other. Thus, although the energy of hydrolysis of the high-energy bond of 1,3-bisphosphoglycerate is greater than that of the phosphoanhydride bond of ATP, the difference is met by the energy difference between the substrate and product of the dehydrogenase reaction. A significant factor in this is the relative concentrations of reduced and oxidized forms of the nicotinamide nucleotide. The ratio of [NAD+] to [NADH] within cells is usually maintained at about 500:1.

Table 8.3 Free energy changes associated with the reactions of glycolysis

Reaction	Enzyme	ATP yield	$\Delta G^{\circ\prime}$ (kJ mol^{-1})
1. Glucose, ATP → ADP + H$^+$; Glucose-6-phosphate	Hexokinase	−1	−16.7
2. → Fructose-6-phosphate	Phosphoglucoisomerase		+1.7
3. ATP → ADP + H$^+$; Fructose-1,6-bisphosphate	Phosphofructokinase	−1	−14.2
4. → Glyceraldehyde-3-phosphate + dihydroxyacetone phosphate	Aldolase		+23.8
5. → 2 Glyceraldehyde-3-phosphate	Triose-phosphate isomerase		+7.5
6. NAD$^+$ + Pi → NADH + H$^+$; 2 1,3-Bisphosphoglycerate	Glyceraldehyde-3-phosphate dehydrogenase		+6.3
7. ADP + H$^+$ → ATP; 2 3-Phosphoglycerate	Phosphoglycerate kinase	+2	−18.8
8. → 2 2-Phosphoglycerate	Phosphoglyceromutase		+4.6
9. → H$_2$O; 2 Phosphoenolypyruvate	Enolase		+1.7
10. ADP + H$^+$ → ATP; 2 Pyruvate	Pyruvate kinase	+2	−31.4
Net		+2	−35.5

Glycolysis is a source of molecules for the synthesis of other substances

Many of the intermediates of glycolysis are substrates or intermediates for other processes or pathways. These relationships are illustrated in Figure 8.13.

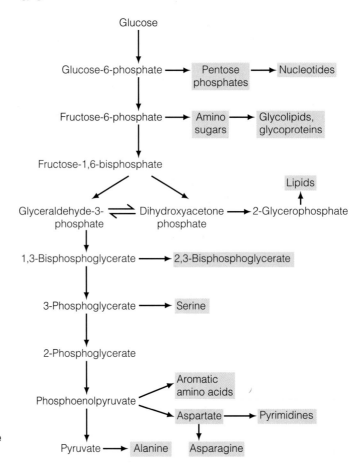

Figure 8.13 Intermediates of glycolysis are precursors for many other processes.

Control of glycolysis

The rate of glycolysis is coordinated with other energy-producing and energy-storing pathways

A complex set of controls ensures operation of the glycolytic pathway at a rate that reflects the energy status of the cell and responds to changes in the cellular concentrations of other nutrients, metabolites from other pathways and high-energy compounds. In this way, the rate of glycolysis is closely coordinated with the rates of gluconeogenesis, the citric acid cycle, the pentose phosphate pathway and glycogen synthesis and degradation. This coordination ensures that in an energy-deficient cell nutrients will be metabolized to produce energy, but in cells with sufficient energy nutrients will be stored. Some details of the regulation of glycolysis differ between tissues, reflecting the different contributions of tissues to metabolic balance throughout the body.

The regulatory enzymes of glycolysis respond to the energy status of the cell

The flux through glycolysis is inhibited if the energy status of the cell is high, as indicated by high concentrations of ATP and low concentrations of ADP and AMP. Glycolysis is also inhibited by high concentrations of other metabolites such as pyruvate, fatty acids, ketone bodies or citrate. The enzymes through which these regulatory signals operate are **phosphofructokinase, hexokinase** and **pyruvate kinase**. The reactions catalysed by these enzymes are essentially irreversible. As in all metabolic processes, the enzymes catalysing irreversible reactions provide the best sites at which to regulate flux through the pathway.

Phosphofructokinase is the key regulatory enzyme of glycolysis

Phosphofructokinase is an allosteric enzyme made up of four identical protein subunits which dissociate into dimers as part of the mechanism of allosteric control. This is the primary regulatory enzyme of the glycolytic pathway.

Pyruvate, fatty acids and ketone bodies, following conversion to acetyl CoA, are further metabolized through the citric acid cycle, and high concentrations of these precursors increase the amount of citrate in the mitochondria. In a cell with more than enough energy, citrate passes from the mitochondria through a specific transporter into the cytoplasm, where it inhibits phosphofructokinase. Citrate thus prevents cells from metabolizing glucose while supplies of other metabolites are adequate.

The activity of phosphofructokinase is inhibited allosterically by high concentrations of ATP which has the effect of lowering the affinity (increasing the K_m) of the enzyme for the substrate, F6P. Citrate acts as an inhibitor because it enhances the binding of ATP to phosphofructokinase. By contrast, if the energy status of the cell is low the AMP concentration is high, and ATP will be displaced from the allosteric binding site by AMP. In this situation the affinity of the enzyme for its substrate is increased, and the flux of glucose through glycolysis is accelerated.

The activity of phosphofructokinase is reduced in an acidic environment. This protects the cell against the accumulation of pyruvate and lactate, which create an undesirable acidosis.

Table 8.4 Activators and inhibitors of phosphofructokinase

Activators	Inhibitors
AMP	ATP
F-2,6-bisP	Citrate
	H^+

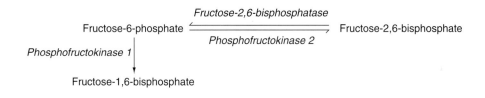

A recently recognized regulator of phosphofructokinase is **fructose-2,6-bisphosphate** (F-2,6-bisP), an allosteric activator. F-2,6-bisP increases the affinity of phosphofructokinase for its substrate (F6P) and diminishes the inhibitory effect of ATP. Fructose-2,6-bisphosphate is synthesized from F6P by **phosphofructokinase-2 (PFK-2)**, when the F6P concentration is high. The action of PFK-2 is reversed by **fructose-2,6-bisphosphatase**. This is a major regulator of carbon flux through glycolysis and gluconeogenesis (see below and Figure 8.14).

Figure 8.14 Fructose-2,6 bisphosphate (F2,6 bisP), an allosteric activator of PFK, is produced by the enzyme PFK(2) when the concentration of F6P is high. When the concentration of F6P drops as it is consumed by PFK, a phosphatase removes the 2-phosphate from F2,6 bisP, and PFK activity returns to basal levels.

Hexokinase, a second regulatory enzyme in glycolysis, is inhibited by its product

Hexokinase responds indirectly to the inhibition of phosphofructokinase. As G6P isomerase catalyses a readily reversible reaction, F6P is in equilibrium with G6P. If phosphofructokinase is inhibited F6P accumulates and there will be an increase in the concentration of G6P, inhibiting hexokinase (Figure 8.15). This control mechanism does not operate in the liver because the corresponding liver enzyme, glucokinase, is not inhibited by G6P (see Table 8.1).

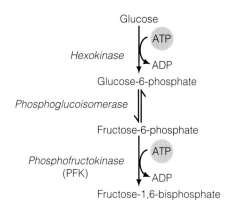

Figure 8.15 The reaction catalysed by phosphoglucoisomerase is readily reversible, so that if F6P accumulates the G6P concentration increases and hexokinase is inhibited.

239

Pyruvate kinase exists as three isoenzymes

Pyruvate kinase is regulated both by allosteric effectors and by covalent modification through a phosphorylation reaction. This is a four-subunit enzyme which exists in animal tissues as three different and tissue-specific isoenzymes (Chapter 4). The L form of pyruvate kinase is found in the liver, the M form in muscle, and the A form in other tissues. Variations in the structure of these isoenzymes gives rise to varying susceptibility to regulation through the glucagon-stimulated cAMP-dependent phosphorylation reaction (Chapter 11).

Pyruvate kinase responds to energy status through allosteric regulation

Pyruvate kinase binds its substrate phosphoenolpyruvate (PEP) to the four subunits in a cooperative manner. As the concentration of PEP increases, the molecules bind with greater affinity and the enzyme exhibits increased catalytic activity. The mechanism is analogous to that exhibited in the binding of oxygen to haemoglobin (Chapter 12). All isoenzymes of pyruvate kinase are inhibited by acetyl CoA. Elevated fructose-1,6-bisphosphate (F-1,6bisP) levels activate the enzyme, signalling the availability of precursors earlier in the pathway. Both ATP and alanine are allosteric inhibitors.

Liver pyruvate kinase is regulated through hormone-stimulated phosphorylation

Phosphorylation of the L form of pyruvate kinase occurs in response to hormonal changes reflecting reduced blood glucose concentration. The pancreas increases its secretion of glucagon in response to reduced blood glucose and glucagon, acting on liver receptors, increases cAMP concentrations (see Chapter 11). The resulting phosphorylation of pyruvate kinase inhibits its activity. This diverts PEP into gluconeogenesis, the process by which glucose can be provided and exported into the bloodstream by the liver (see Figure 8.16).

Chapter 4. Cells contain two or more proteins with similar functions, page 107.
Chapter 11. Hormonal responses are mediated through adenylate cuclase and the production of cAMP, page 338.

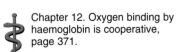

Chapter 12. Oxygen binding by haemoglobin is cooperative, page 371.

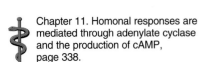

Chapter 11. Homonal responses are mediated through adenylate cyclase and the production of cAMP, page 338.

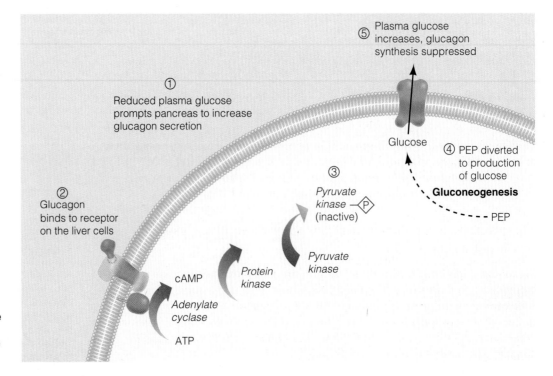

Figure 8.16 The liver (L isoform) of pyruvate kinase is regulated by glucagon-stimulated cAMP-dependent phosphorylation catalysed by protein kinase. When the blood glucose concentration is low, L pyruvate kinase is inactivated by phosphorylation, diverting PEP into gluconeogenesis to produce glucose, which is exported to restore the blood glucose concentration to normal.

① Reduced plasma glucose prompts pancreas to increase glucagon secretion

② Glucagon binds to receptor on the liver cells

③ Pyruvate kinase (inactive)

④ PEP diverted to production of glucose

Gluconeogenesis

⑤ Plasma glucose increases, glucagon synthesis suppressed

Glucose

PEP

Pyruvate kinase

Protein kinase

cAMP

Adenylate cyclase

ATP

The muscle isoenzyme is not subject to phosphorylation. Muscle can not generate glucose through gluconeogenesis and all of the PEP formed in muscle cells is converted to pyruvate to provide energy.

The A form of pyruvate kinase is subject to an intermediate level of control by phosphorylation.

Glucokinase and liver pyruvate kinase contribute to the control of blood glucose

The mechanism controlling pyruvate kinase activity in the liver and the kinetic characteristics of glucokinase both contribute to preventing the liver from metabolizing glucose while other tissues, particularly muscle, brain and red blood cells, require glucose for their energy needs. Glucokinase does not phosphorylate glucose in the liver unless the blood glucose concentration is high; low blood glucose indirectly stimulates liver pyruvate kinase to accelerate gluconeogenesis and increase the output of glucose from the liver to the bloodstream.

The sites at which glycolysis is regulated are determined by factors other than the need to produce ATP or pyruvate

Enzymes through which metabolic pathways are regulated are usually located at the start of the pathway. This ensures that energy is not wasted in partial metabolism of substrates. Why, then, is glycolysis regulated primarily by the third enzyme in the pathway, phosphofructokinase?

Glucose is the starting molecule for several processes (such as synthesis of glycogen and production of ribose-5-phosphate (R5P) and NADPH through the pentose phosphate pathway), and the demands for these processes may conflict. To enter any of the possible routes, with the exception of efflux from the liver, glucose must first be phosphorylated. It is therefore important that the phosphorylation of glucose should proceed independently of any possible route through which it might be metabolized. Thus, hexokinase would not be a suitable enzyme on which to concentrate all of the regulatory control of glycolysis.

Metabolic fates of pyruvate

Pyruvate is a central metabolite which may be oxidized or diverted into anabolic processes

Depending on the energy balance of the cell there are several possible metabolic fates for pyruvate (Figure 8.17).

In well-oxygenated cells most of the pyruvate generated from glucose will be converted to acetyl CoA which will then be oxidized through the citric acid cycle.

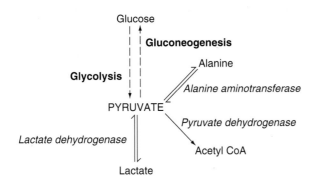

Figure 8.17 Pyruvate is one of the central metabolites through which carbon sources can be directed to immediate energy production (the citric acid cycle) or to energy storage (gluconeogenesis or lipogenesis via the formation of acetyl CoA). Alternatively, it may provide the substrate for lactate dehydrogenase in a reaction that maintains NAD$^+$ concentrations in the cytoplasm.

The reaction, catalysed by the pyruvate dehydrogenase complex, is irreversible: it is not possible to convert acetyl CoA to pyruvate.

$$CH_3-\overset{\overset{\displaystyle O}{\|}}{C}-COO^- + NAD^+ + CoA-SH \xrightarrow[\text{\textit{complex}}]{\text{\textit{Pyruvate dehydrogenase}}} CH_3-\overset{\overset{\displaystyle O}{\|}}{C}\sim SCoA + NADH + CO_2$$

If the energy demand of the cell exceeds the capacity of available oxygen for oxidative metabolism, pyruvate will be converted to lactate in order to maintain the supply of NAD^+, on which energy production through glycolysis depends.

$$\text{Pyruvate} + NADH + H^+ \xrightarrow[\text{\textit{Lactate dehydrogenase}}]{} \text{Lactate} + NAD^+$$

Chapter 12. Metabolism in the erythrocyte, page 375.

Chapter 13. Sustaining energy production in white muscle fibres depends on coupled dehydrogenase reactions, page 387.

Chapter 14. In a well-fed state the liver replenishes its own energy reserves, page 433.

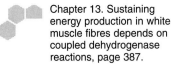

Chapter 9. Fatty acid biosynthesis, page 274.
Chapter 9. In the liver accumulating acetyl CoA is converted to ketone bodies, page 271.
Chapter 10. Aminotransferase enzymes are specific and catalyse readily reversible reactions, page 295.

Chapter 10. The urea cycle, page 301.

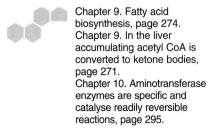

Chapter 9. Complete oxidation of fatty acids depends on the operation of the citric acid cycle, page 271.

Some tissues lack mitochondria, notably erythrocytes (Chapter 12), and others have so few mitochondria that their metabolism is effectively anaerobic. The amount of lactate produced continuously by these 'anaerobic' cells, together with the variable rate of lactate production from muscle cells (Chapter 13) is sufficient to maintain a normal concentration of lactate in the bloodstream of about 1 mM. The reaction catalysed by lactate dehydrogenase is readily reversible and so circulating lactate can be taken up by other tissues and converted to pyruvate for further metabolism. Most of the circulating lactate enters the liver cells (Chapter 13) where it is converted to glucose or glycogen through gluconeogenesis and glycogenolysis. Heart muscle and some skeletal muscle cells are able to oxidize lactate through the citric acid cycle after converting it to pyruvate via the lactate dehydrogenase reaction.

If cells have sufficient other metabolites, pyruvate may be converted to G6P (glucose in the liver) through gluconeogenesis (see below). Alternatively, the acetyl CoA produced from pyruvate may be used for fatty acid synthesis (Chapter 9) or converted in the liver to ketone bodies (Chapter 9) as nutrients for other tissues.

Pyruvate can act as the substrate for the aminotransferase reaction (Chapter 10), which generates alanine either for use in protein synthesis or (in skeletal muscle cells) as part of the mechanism for transporting pyruvate and ammonia groups to the liver where they become substrates for gluconeogenesis (see below) and the urea cycle (Chapter 10) respectively.

The pyruvate dehydrogenase complex converts pyruvate irreversibly to acetyl CoA

Pyruvate produced in the cytoplasm is transferred into the mitochondrial matrix through a specific transporter. Catabolism of pyruvate begins in the mitochondria with an irreversible reaction, an oxidative decarboxylation catalysed by the **pyruvate dehydrogenase complex**. This enzyme complex produces acetyl CoA, which subsequently enters the citric acid cycle for complete oxidation.

The two-carbon fragments derived from catabolism of fatty acids also enter the citric acid cycle as acetyl CoA (see Chapter 9). An important consequence of the irreversibility of pyruvate dehydrogenase is that the products of fatty acid catabolism can not be converted to pyruvate, and therefore carbon derived from fatty acids can not be used for the synthesis of glucose. In contrast, pyruvate converted to acetyl CoA can, under suitable conditions, be used for the synthesis of fatty acids. Because of its strict irreversibility, the reaction catalysed by the pyruvate dehydrogenase complex is sometimes described as a *committed* step in metabolism.

The pyruvate dehydrogenase complex catalyses a sequence of reactions

Pyruvate dehydrogenase is a multienzyme complex, a large multisubunit protein array containing many copies of each of three distinct catalytic subunits, **pyruvate dehydrogenase (E1)**, **dihydrolipoamide acetyltransferase (E2)** and **dihydrolipoamide dehydrogenase (E3)**. It also contains two regulatory enzymes. Five coenzymes, all derived from water-soluble vitamins, participate in the reactions catalysed by this enzyme complex (Table 8.5). The complex is located on the inner surface of the inner mitochondrial membrane. Its activity is regulated both by allosteric modulators which reflect the energy status of the cell and by phosphorylation reactions initiated by hormonal stimuli reflecting nutritional and metabolic changes in the body as a whole.

The overall reaction catalysed by this complex is

$$CH_3-\overset{\overset{\displaystyle O}{\|}}{C}-COO^- + NAD^+ + CoA-SH \xrightarrow[\text{complex}]{\text{Pyruvate dehydrogenase}} CH_3-\overset{\overset{\displaystyle O}{\|}}{C}\sim SCoA + NADH + CO_2$$

Five consecutive reactions are required to achieve the decarboxylation and dehydrogenation described by this equation. Carbon dioxide is derived from the C-1 of pyruvate and the C-2 of pyruvate is oxidized from an aldehyde to a carboxylate group. Throughout the process the intermediates remain bound to the enzyme surface. Free energy liberated by the oxidative decarboxylation is harnessed in two ways: the formation of a high-energy thioester bond, acetyl CoA, and the reduction of NAD^+. Pyruvate dehydrogenase activity is strongly inhibited allosterically by products of the reaction, acetyl CoA, NADH and by long chain fatty acids. Allosteric activators include CoA, NAD^+ and AMP.

Table 8.5 Coenzymes and vitamin precursors required by the pyruvate dehydrogenase complex

Subunit activity	Coenzymes	Vitamin precursor
Pyruvate decarboxylase	Thiamine pyrophosphate	Thiamine
Dihydrolipoamide transacetylase	Lipoate	
	Coenzyme A	Pantothenate
Dihydrolipoamide dehydrogenase	FAD	Riboflavin
	NAD	Niacin

Vitamin deficiencies, particularly thiamine deficiency, can limit the oxidation of pyruvate. This is particularly important in the brain which must complete the oxidation of glucose to provide sufficient energy to sustain its function. Apparent thiamine deficiency is seen in alcoholics, but this is commonly due to a failure of the pyrophosphorylation reaction which converts the dietary vitamin (thiamine) to the active coenzyme (thiamine pyrophosphate) in the liver.

Pyruvate dehydrogenase is inactivated by phosphorylation catalysed by a (tightly bound) kinase that phosphorylates three specific serine residues in the E1 enzyme. This inactivation mechanism is opposed by a specific, loosely bound Ca^{2+}-dependent phosphatase. Each of these enzymes is subject to allosteric regulation by a variety of metabolites.

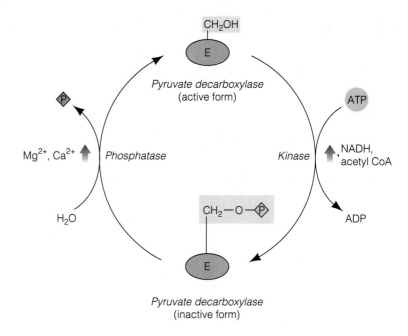

Pyruvate decarboxylase
(active form)

Pyruvate decarboxylase
(inactive form)

The structure and catalytic mechanism of the pyruvate dehydrogenase complex are very similar to those of the α-ketoglutarate dehydrogenase complex of the citric acid cycle.

Complete oxidation of glucose generates energy from glycolysis and the citric acid cycle

The metabolism of pyruvate to acetyl CoA converts the product of glycolysis to a substrate for the citric acid cycle.

In order to calculate the energy yield from the complete oxidation of glucose to CO_2 and H_2O it is important to remember that every molecule of glucose generates two molecules of pyruvate through glycolysis. In the process two ADP are phosphorylated and two NAD^+ are reduced to NADH. Conversion of two molecules of pyruvate to two molecules of acetyl CoA also generates two NADH. Two acetyl CoA completely oxidized through the citric acid cycle generate six NADH, two $FADH_2$ and two GTP. The net yield of ATP from each molecule of glucose oxidized to CO_2 with all of the reduced nucleotide reoxidized through the electron transport chain is 38 ATP. If the glucose is derived from glycogen, one fewer ATP molecule is required in the energy-investment phase of glycolysis, so that the ATP yield from G1P increases to 39. In tissues where reducing equivalents produced in the cytoplasm are conveyed into mitochondria mainly through the FAD-linked α-glycerophosphate shuttle (see below) the net yield of ATP from glucose oxidation is 36.

Gluconeogenesis: the synthesis of glucose from noncarbohydrate precursors

The purpose of gluconeogenesis is to conserve nutrients in cells that are generating sufficient energy for their needs and, in the specific case of the liver, to synthesize glucose for export to other glucose-dependent tissues.

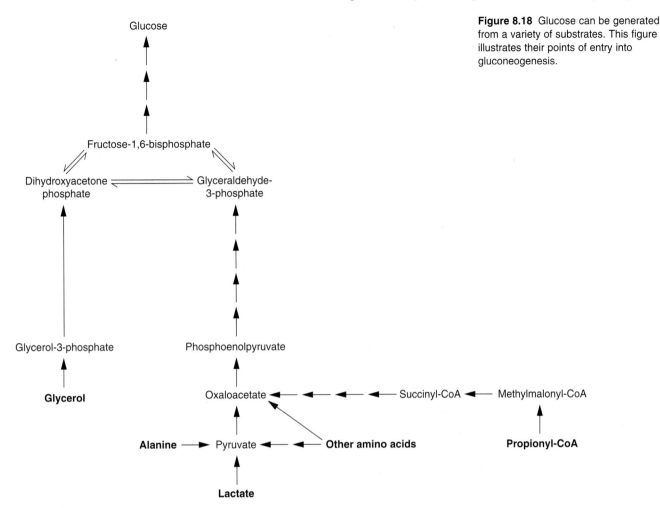

Figure 8.18 Glucose can be generated from a variety of substrates. This figure illustrates their points of entry into gluconeogenesis.

Gluconeogenesis operates as a pathway for the synthesis of glucose-phosphate from a variety of noncarbohydrate precursors including pyruvate, oxaloacetate, lactate, glycerol, and glucogenic amino acids (Chapter 10). Acetyl CoA and metabolites that are metabolized to acetyl CoA can not be converted to oxaloacetate or pyruvate, and therefore can not be substrates for gluconeogenesis. Similarly the two amino acids, lysine and leucine, which are purely ketogenic (their carbon skeletons are converted to acetyl CoA or acetoacetyl CoA) can not be used as substrates for gluconeogenesis. Most of the enzymes of gluconeogenesis are located in the cell cytoplasm, although some are found within the mitochondrial matrix. The G6P may be used through the pentose phosphate pathway (see page 258) or converted to glycogen through G1P. Only in the liver, and to a limited extent in the renal cortex, is G6P converted to glucose (see Figure 8.19).

Chapter 10. Amino acids are converted to one of seven metabolites of the central metabolic pathways, page 390.

Chapter 8 The pentose phosphate pathway, page 257.

Gluconeogenesis occurs by modified reversal of glycolysis

Most of the reactions of gluconeogenesis are simple reversals of glycolytic reactions and are catalysed by the same enzymes. However, three of the reactions, those catalysed by hexokinase, phosphofructokinase and pyruvate kinase, are exergonic and therefore operate in one direction only. These steps are circumvented by other enzymes, specific to gluconeogenesis, which reverse the reactions through different mechanisms.

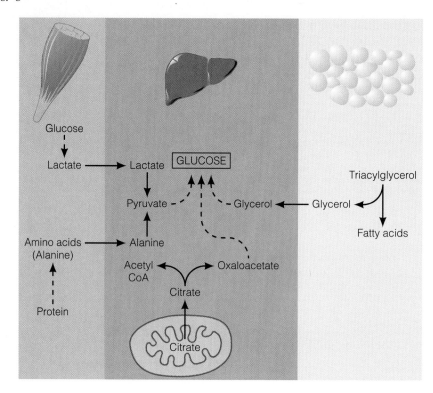

Figure 8.19 Many metabolites can serve as precursors for gluconeogenesis in the liver. Most commonly, the substrate is returned to the liver from other tissues. Substrates include glycerol from adipose tissue, alanine from muscle cells and lactate from muscle and erythrocytes. In some circumstances, oxaloacetate generated in the liver from citrate may be channelled into gluconeogenesis.

Catalysing the opposite directions of a reaction by different enzymes has the advantage of introducing sites at which glycolysis can be accelerated without accelerating gluconeogenesis. An overview of the relationship between glycolysis and gluconeogenesis is presented in Figure 8.20.

The 'bypass' enzymes of gluconeogenesis

The reactions of glycolysis catalysed by hexokinase, phosphofructokinase and pyruvate kinase are reversed respectively by **glucose-6-phosphatase, fructose-1,6-bisphosphatase** and the two enzymes **pyruvate carboxylase** and **phosphoenolpyruvate carboxykinase**, which together catalyse the conversion of pyruvate to phosphoenol pyruvate.

Pyruvate carboxylase

Pyruvate carboxylase is an allosteric enzyme which adds a biotin-activated carboxyl group to convert pyruvate to oxaloacetate. This is the first regulatory enzyme in gluconeogenesis.

$$\text{Pyruvate} + CO_2 + \text{ATP} \xrightarrow[\textit{Pyruvate carboxylase}]{} \text{Oxaloacetate} + \text{ADP} + \text{Pi}$$

The carboxyl group of biotin is linked to the ε-amino group of a lysine residue in the carboxylase to form an amide bond. In many respects the structure of biotin and its mode of attachment to the enzyme via a long flexible chain is similar to the action of lipoic acid which transfers acyl groups as part of the action of pyruvate dehydrogenase and α-ketoglutarate dehydrogenase complexes.

Pyruvate carboxylase is found only in the mitochondrial matrix and provides an important control point, balancing the utilization and storage of energy. The oxaloacetate produced in this reaction may be fed into the gluconeogenesis pathway if the cell has sufficient energy or the citric acid cycle if the cell requires

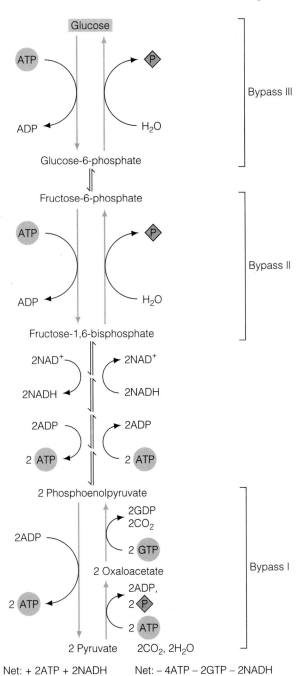

Figure 8.20 Glycolysis and gluconeogenesis have several enzymes in common. Those enzymes of glycolysis that catalyse irreversible reactions (shown in grey) are circumvented by enzymes specific to gluconeogenesis (pink).

energy. The latter route is important because an increase in the number of molecules of oxaloacetate in the cycle increases the rate at which acetyl CoA is incorporated into citrate. It is relevant to both of these metabolic processes that pyruvate carboxylase is activated by acetyl CoA, being essentially inactive in the absence of this metabolite.

The need to synthesize oxaloacetate to facilitate the metabolism of acetyl CoA through the citric acid cycle is signalled when acetyl CoA is abundant but the energy status of the cell is low. This is one of the **anaplerotic** reactions (Chapter 7) that help to maintain the concentration of citric acid cycle intermediates within the mitochondria and therefore to maintain the rate of energy production.

Chapter 7. The citric acid cycle: the pathway for complete oxidation of acetyl coenzyme A, page 213.

247

In the energy-replete cell the presence of abundant acetyl CoA signals pyruvate carboxylase to store, through gluconeogenesis, any pyruvate produced from lactate, amino acids or other substrates.

As pyruvate carboxylase occurs only within mitochondria and the remaining enzymes of gluconeogenesis are found in the cytoplasm, the oxaloacetate produced is transferred indirectly to the cytoplasm by the mitochondrial and cytoplasmic forms of malate dehydrogenase. Malate leaves the mitochondria through the malate–α-ketoglutarate transporter (see below), but oxaloacetate is unable to cross the inner mitochondrial membrane.

Conversion of mitochondrial-derived pyruvate to phosphoenolpyruvate in the cytoplasm is shown in Figure 8.21.

Pyruvate carboxylase uses ATP at a time when the cell needs to produce more energy, but the entry of additional molecules of both acetyl CoA and oxaloacetate into the citric acid cycle ensures that a far greater number of ATP molecules will be produced.

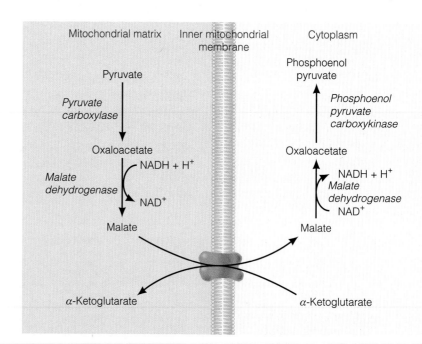

Figure 8.21 Because pyruvate carboxylase is located in the mitochondrial matrix and the other enzymes of gluconeogenesis are in the cytoplasm, the malate:α-ketoglutarate antiporter of the inner mitochondrial membrane is used if the substrate for gluconeogenesis is pyruvate derived from PEP.

Phosphoenolpyruvate carboxykinase

This enzyme is found in both the cytoplasm and in the mitochondrial matrix of liver cells. Oxaloacetate in either compartment of the cell may serve as substrate for the reaction

$$\text{Oxaloacetate} + \text{GTP} \xrightarrow[\substack{\textit{Phosphoenolpyruvate}\\\textit{carboxykinase}}]{} \text{Phosphoenolpyruvate} + CO_2 + \text{GDP} + \text{Pi}$$

In this reaction the carboxyl group added when pyruvate was converted to oxaloacetate is removed. Energy is provided by hydrolysis of the terminal phosphoanhydride bond of GTP.

Fructose-1,6-bisphosphatase

The reaction catalysed by phosphofructokinase is reversed by fructose-1,6-bisphosphatase:

$$\text{Fructose-1,6-bisphosphate} + H_2O \xrightarrow{\textit{Fructose-1,6-bisphosphatase}} \text{Fructose-6-phosphate} + Pi$$

This allosteric enzyme is inhibited by AMP and by fructose-2,6-bisphosphate, and is activated by 3-phosphoglycerate and citrate. As already discussed, AMP, fructose-2,6-bisphosphate and citrate have opposite influences on phosphofructokinase. The consequence is that accelerated flux through glycolysis occurs simultaneously with inhibition of gluconeogenesis (and vice versa). This reciprocal control mechanism ensures that nutrients are not consumed in a futile cycle.

Glucose-6-phosphatase

In mammals this enzyme is found in substantial amounts only in the liver, with small amounts in the cortex of the kidney. For this reason glucose can be produced from gluconeogenesis *only* in liver and kidney.

Glucose-6-phosphate   Glucose

The production of glucose from G6P involves a number of proteins because the catalytic site of glucose-6-phosphatase is located on the inner face of the endoplasmic reticulum, facing the lumen. G6P is produced in the cytoplasm, and is transferred into the endoplasmic reticulum by a specific transport protein to gain access to the phosphatase. The glucose produced requires a second transport protein to transfer it back into the cytoplasm (Figure 8.22). Glucose-6-phosphatase activity may appear to be deficient in individuals who have normal amounts of the enzyme but lack either the G6P transporter or the glucose transporter in the endoplasmic reticulum membrane. In the former case the substrate would not gain access to the enzyme, and in the latter case product inhibition would occur as glucose accumulates within the

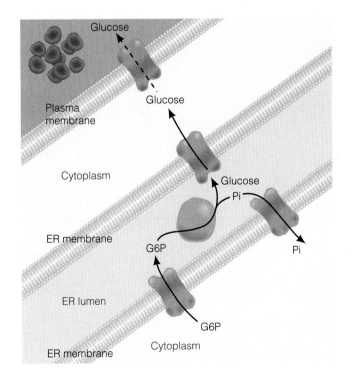

Figure 8.22 Glucose-6-phosphatase (blue) is located on the inner surface of the endoplasmic reticulum and its substrate (G6P) and both products (glucose and Pi) must be transported across the endoplasmic reticulum through specific transporters.

lumen of the endoplasmic reticulum. There is a third transporter, which returns phosphate to the cytoplasm. Lack of one or other of these transport proteins, rather than lack of glucose-6-phosphatase, appears to be the basis of some forms of glycogen-storage disease (see Box 8.2).

Enzymes metabolizing pyruvate balance energy utilization and storage

Both pyruvate kinase and pyruvate carboxylase act as monitors of the energy status of the cell, the activity of both being allosterically regulated by energy metabolites. Pyruvate kinase is activated by fructose-1,6-bisphosphate and inhibited by ATP. Pyruvate carboxylase and PEP carboxykinase are both inhibited by ADP. The consequence is that high concentrations of ADP, which necessarily reflect low ATP concentrations, prevent pyruvate being converted to glucose by inhibiting both of these reactions.

The other modulator of the balance between gluconeogenesis and pyruvate catabolism is acetyl CoA. Pyruvate kinase and pyruvate dehydrogenase are inhibited by acetyl CoA, but pyruvate carboxylase is activated by acetyl CoA. These responses ensure that when acetyl CoA is available from other sources, such as oxidation of fatty acids, more is not provided by glucose metabolism. If the cell is producing sufficient acetyl CoA from the oxidation of fatty acids, inhibition of pyruvate dehydrogenase diverts pyruvate through pyruvate carboxylase, which is activated by acetyl CoA, to produce oxaloacetate. The oxaloacetate can be used as the substrate for gluconeogenesis so that nutrient is conserved, or it can be used to facilitate the incorporation of acetyl CoA into the citric acid cycle (Figure 8.23).

Gluconeogenesis uses more energy than is produced in glycolysis

The energy consumed in gluconeogenesis is more than that produced by glycolysis, largely because the phosphate groups of G6P and fructose-1,6-bisphosphate are hydrolysed rather than recovered as high-energy molecules. It would therefore

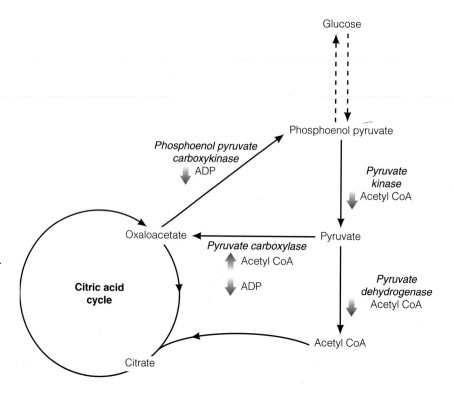

Figure 8.23 ADP and acetyl CoA regulate the metabolic fate of pyruvate through their effects on the enzymes pyruvate kinase, pyruvate dehydrogenase, pyruvate carboxylase and PEP carboxykinase. ADP is a general indicator of low energy status in a cell and a high concentration of acetyl CoA indicates the cell should conserve carbohydrate while there is sufficient fatty acid available for energy production.

Table 8.6 Energy-consuming reactions of gluconeogenesis

Enzyme	High-energy compound used
Pyruvate carboxylase	$ATP \rightarrow ADP$
PEP carboxykinase	$GTP \rightarrow GDP$
Phosphoglycerate kinase	$ATP \rightarrow ADP$
Glyceraldehyde-3-P dehydrogenase	$NADH \rightarrow NAD^+$

be energetically wasteful to convert pyruvate to glucose if the glucose remained within the cell to be metabolized through glycolysis at a later time.

The total energy consumed in converting two molecules of pyruvate to one of glucose amounts to the equivalent of 12 ATP. Subsequent glycolysis of this glucose molecule would yield only the equivalent of 8 ATP. A benefit of gluconeogenesis in a cell replete with metabolites is that the carbohydrate can be stored as glycogen. However, one purpose of gluconeogenesis in the liver is to export glucose to other tissues requiring nutrient, and simple analysis of ATP yields does not reflect the benefits of nutrient redistribution between tissues.

Gluconeogenesis enables the liver to scavenge certain metabolites released from other tissues and to return them as glucose to the bloodstream. We now come to a consideration of how the liver cooperates with other tissues to maintain nutritional supplies to them through its capacity for gluconeogenesis.

Liver and muscle cooperate to conserve energy through the Cori cycle and the alanine cycle

Muscle cells have enough lactate dehydrogenase activity to ensure that pyruvate produced through glycolysis under anaerobic conditions is readily reduced to lactate. This reaction enables muscle cells to continue drawing energy from glycolysis. Lactate diffuses from muscle cells, is taken up by the liver (where oxygen is not in deficit) and oxidized to pyruvate before entering gluconeogenesis. The glucose produced is then returned to muscle or other peripheral tissues. This process is called the **Cori cycle** (Figure 8.24). In converting lactate to pyruvate the

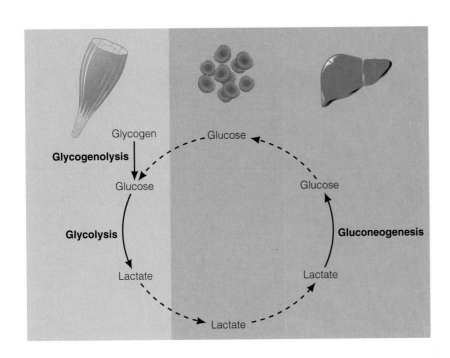

Figure 8.24 The Cori cycle depicts the process by which the liver conserves the carbon source of lactate through gluconeogenesis. The glucose produced is exported from the liver to glucose-dependent tissues.

Chapter 14. In starvation muscle protein is degraded in a controlled manner to support the nutritional needs of other tissues, page 441.

Chapter 9. Fatty acids are released from triacylglycerol stores by the action of a hormone-sensitive lipase, page 263.

reaction catalysed by lactate dehydrogenase consumes a proton, thus contributing to the control of acidity.

The **alanine cycle** ensures that alanine, which is released from muscle at times of nutrient depletion (Chapter 14) or heavy exercise, provides a substrate for liver gluconeogenesis. In a manner analogous to the Cori cycle, some of the alanine-derived glucose is returned to muscle cells.

Adipose tissue and liver cooperate to use glycerol

Degradation of triacylglycerol in adipose tissue releases fatty acids and glycerol (Chapter 9). The glycerol liberated is transported to the liver, where **glycerol kinase** produces glycerol-3-phosphate, which can then enter glycolysis or gluconeogenesis through dihydroxyacetone phosphate (Figure 8.25). Intertissue cooperation between adipose and hepatic tissue is important because adipose tissue lacks glycerol kinase and is therefore unable to reuse the glycerol liberated from triacylglycerols.

In an energy-replete cell nutrient can be stored by withdrawing metabolites from the citric acid cycle

A specific transporter in the inner membranes of mitochondria allows the efflux of citrate from the mitochondrial matrix to the cytoplasm. This occurs when energy production through the central metabolic pathways exceeds the cell's requirement. In the cytoplasm the enzyme **citrate lyase** cleaves citrate, producing one molecule each of oxaloacetate and acetyl CoA:

$$\text{Citrate} + \text{ATP} + \text{CoASH} \xrightarrow[\textit{Citrate lyase}]{} \text{Oxaloacetate} + \text{acetyl CoA} + \text{ADP} + \text{Pi}$$

Chapter 9. The synthesis of fatty acids begins with the attachment of acetyl CoA and malonyl CoA to the acyl carrier region, page 277.

The acetyl CoA produced may be used for fatty acid biosynthesis and stored ultimately as triacylglycerol through lipogenesis (Chapter 9). Depending on the metabolic balance and needs of the cell oxaloacetate may be used as

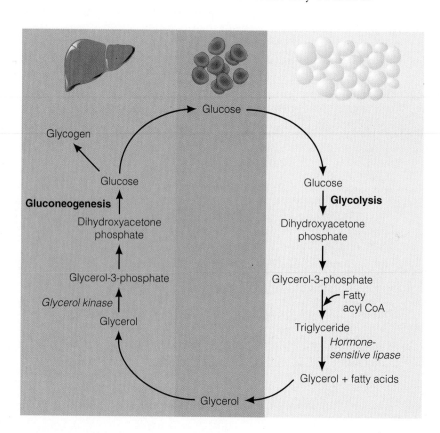

Figure 8.25 Adipose tissue and liver cooperate to conserve the nutrient value of the glycerol released during lipolysis. Adipose tissue is unable to reuse glycerol because it lacks glycerol kinase. This enzyme is abundant in the liver, which either oxidizes the glycerol (through glycolysis) or uses it to produce glucose (through gluconeogenesis).

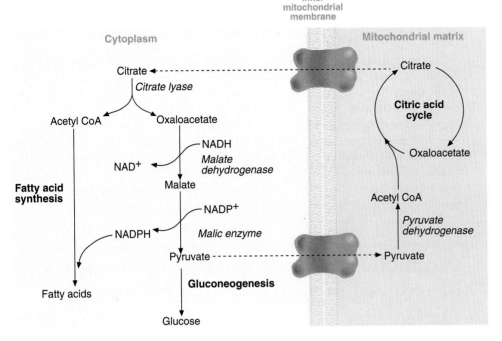

Figure 8.26 A citrate transporter in the inner mitochondrial membrane facilitates the efflux of citrate to the cytoplasm. The products of citrate lyase, oxaloacetate and acetyl CoA, can be converted to carbohydrate or fatty acids, respectively, although part of the carbon source represented by citrate may be returned to the matrix (as pyruvate or malate) to maintain the carbon content of the citric acid cycle.

- a precursor for gluconeogenesis;
- a substrate for malate dehydrogenase to transfer reducing equivalents into mitochondria (see below);
- a substrate for aminotransferase, forming aspartate for protein synthesis or as a mechanism to incorporate amino groups into urea synthesis (Chapter 10);
- a precursor for pyruvate and malate through a series of reactions that also generate NADPH.

Chapter 10. The urea cycle converts ammonia to urea, page 301.

See Figure 8.26.

The reaction catalysed by malic enzyme generates NADPH, which provides the reducing power for fatty acid biosynthesis (Chapter 9). This ensures that the acetyl CoA produced from the citrate released from the mitochondria can be used to synthesize fatty acids.

Chapter 9. Fatty acid carbon chains are assembled through a sequence of four reactions, page 276.

Ethanol metabolized in the liver produces acetate and depletes NAD$^+$

Ethanol consumption is high enough in some individuals that it makes a significant contribution to their energy intake. Commonly individuals who consume excessive amounts of alcohol have an inadequate diet and are in a poor state of nutrition. They develop liver function problems and many of the biochemical processes that normally occur predominantly in the liver fail to meet the needs of the other body tissues.

Ethanol enters the liver, where alcohol dehydrogenase oxidizes it to acetaldehyde, coupled to the reduction of NAD$^+$ to NADH.

$$\text{Ethanol} + \text{NAD}^+ \xrightarrow[\textit{Alcohol dehydrogenase}]{} \text{Acetaldehyde} + \text{NADH} + \text{H}^+$$

A further NAD$^+$-dependent dehydrogenase (acetaldehyde dehydrogenase) converts the aldehyde to acetate, which may subsequently be coupled to coenzyme A.

$$\text{Acetaldehyde} + \text{CoASH} + \text{NAD}^+ \xrightarrow[\substack{\textit{Acetaldehyde} \\ \textit{dehydrogenase}}]{} \text{Acetyl} + \text{CoA} + \text{NADH} + \text{H}^+$$

Acetyl CoA may be used for the synthesis of fatty acids. In a nutrient-replete individual persistent consumption of large amounts of alcohol is therefore likely to lead to obesity or to fatty liver, which can develop without obvious general obesity.

If a large amount of alcohol is metabolized in the liver, the ratio of NADH to NAD^+ increases. This pushes the equilibrium of the lactate dehydrogenase reaction in the direction of lactate production, making pyruvate unavailable for gluconeogenesis. For the same reason, the reaction catalysed by malate dehydrogenase will be pushed in the direction of malate production, making oxaloacetate also unavailable for gluconeogenesis. If this coincides with a period when blood glucose concentration is falling, hypoglycaemia will result because gluconeogenesis will be unable to sustain the blood glucose concentration. The parts of the brain most immediately vulnerable to hypoglycaemia include that controlling body temperature. Body temperature will therefore drop, resulting in hypothermia.

Catabolic reactions depend on reoxidation of reduced nicotinamide and flavin nucleotides

Catabolic processes are categorized as aerobic (dependent on the presence of molecular oxygen to reoxidize coenzymes) or anaerobic (the coenzymes are reoxidized in the absence of oxygen). It is important to remember that molecular oxygen is not involved directly in the reactions by which nutrients are oxidized; **reducing equivalents** are passed from the substrate to the coenzyme(s), and it is the reoxidation of these coenzymes through the mitochondrial electron transport chain that requires molecular oxygen (Chapter 7).

Chapter 7. The mitochondrial respiratory chain comprises four multisubunit assemblies of proteins, page 199.

Cells have cytoplasmic mechanisms for regenerating oxidized nucleotides

In the cytoplasm of cells lactate dehydrogenase and glycerol-3-phosphate dehydrogenase catalyse the reoxidation of NADH, coupled to the reduction of a metabolite (Figure 8.27).

$$\text{Pyruvate} + \text{NADH} + \text{H}^+ \longrightarrow \text{Lactate} + \text{NAD}^+$$

Figure 8.27 Reoxidation of NADH.

$$\text{Dihydroxyacetone-phosphate} + \text{NADH} + \text{H}^+ \longrightarrow \text{Glycerol-3-phosphate} + \text{NAD}^+$$

Large quantities of lactate dehydrogenase are found in mammalian cells; glycerol-3-phosphate dehydrogenase is particularly important in the kidney, in the adrenal glands and in the flight muscles of birds and insects. The presence of these enzymes provides the facility for regenerating oxidized nucleotides in the cytoplasm so that the dehydrogenase reactions of the catabolic pathways can continue to function. However, there is an energy cost to a cell which needs to exploit this mechanism, for NADH reoxidized in this way bypasses the mitochondrial oxidation processes which would otherwise generate three molecules of ATP from each NADH reoxidized.

The energy yield from catabolism is greater if nucleotides are reoxidized aerobically

NADH and $FADH_2$ may be generated through several catabolic pathways, including glycolysis (through which glucose is partially oxidized to the three-carbon

Table 8.7 ATP yields from glycolysis, β-oxidation and the citric acid cycle

	Glycolysis	Citric acid cycle	β-Oxidation
Substrate	Glucose (C-6)	Acetyl CoA (C-2)	Palmitate (C-16)
Product	2 Pyruvate (C-2)	2 CO_2	8 Acetyl CoA (C-2)
High-energy phosphate			
ATP	+2	0	−1
GTP	0	+1	0
Reduced nucleotides			
NADH	+2	+3	+7
$FADH_2$	0	+1	+7
Total potential ATP per substrate carbon atom			
Aerobic	8/6 = 1.33	12/2 = 6	35/16 = 2.19
Anaerobic	2/6 = 0.33		

ATP equivalent yield is based on the fact that each molecule of $FADH_2$ generates 2 ATP and each molecule of NADH generates 3 ATP when reoxidized through the electron transport chain. GTP is a high-energy nucleotide similar to ATP. The difference in energy yield from glycolysis in aerobic and anaerobic conditions reflects the difference between cytoplasmic and mitochondrial reoxidation of NADH. The reduced nucleotides generated from both β-oxidation and the citric acid cycle must be reoxidized aerobically: these pathways fail in an anaerobic environment.
Note that this calculation does not give a complete picture of the energy yield from the complete catabolism of either fatty acids or glucose, simply a comparison of the energy yield from each of the listed metabolic pathways considered separately.

compound pyruvate), the β-oxidation cycle (through which fatty acids are oxidized to two-carbon acetate units), and the citric acid cycle (which completes the oxidation of the acetyl CoA derived from fatty acids, carbohydrates and some amino acids). The β-oxidation cycle is described in Chapter 9. Glycolysis and the citric acid cycle generate high-energy phosphate compounds as well as reduced nicotinamide nucleotides. Table 8.7 lists the total energy yields of all three processes under aerobic and anaerobic conditions.

 Chapter 9. β-oxidation progressively decreases the length of fatty acid carbon chains by removing two carbon atoms, page 266.

It is important to remember that the energy yield per molecule of substrate is only one factor determining the rate of energy production in a cell. Other factors include the flux rate through the pathway, determined by the rate-limiting enzyme, and the number of molecules of each enzyme present in the cell. For example, during prolonged exercise white muscle cells may depend solely upon ATP produced by glycolysis, which yields a relatively small amount of ATP from each molecule of glucose. However, ATP can be produced in large amounts over a short time because these cells store large quantities of glycogen, have very high concentrations of phosphorylase to degrade the glycogen and high concentrations of glycolysis enzymes. Likewise, it has been calculated that in rat heart the maximum rate of ATP production from glycolysis is about 75% of the maximum rate of ATP production from aerobic metabolism. Although the amount of glycogen stored in the heart is sufficient to provide energy for only a few seconds, the activity of hexokinase in the heart is sufficient to sustain the full rate of glycolysis provided that the glucose supply to the myocytes is maintained.

The malate–aspartate shuttle is a mitochondrial mechanism for sustaining the oxidizing power of the cytoplasm

Most of the NAD$^+$ and NADH in a cell is in the mitochondrial compartment, and the inner mitochondrial membrane is essentially impermeable to this coenzyme. Despite this, in an adequately oxygenated cell, NADH produced in

Figure 8.28 The malate–aspartate shuttle provides a mechanism for regenerating NAD⁺ from NADH by transferring reducing equivalents into the mitochondrial matrix. It depends on the existence of both mitochondrial and cytoplasmic forms of the enzymes malate dehydrogenase and aspartate amino-transferase and on the presence of both a malate:α-ketoglutarate exchange transporter and an aspartate: glutamate exchange transporter in the inner mitochondrial membrane.

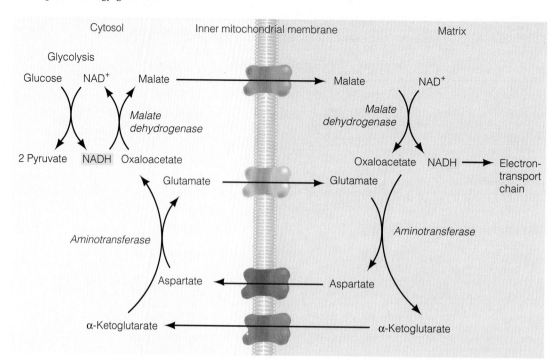

the cytoplasm or in the mitochondrial matrix is reoxidized through the electron transport chain. The explanation of this apparent paradox is that reducing equivalents, rather than reduced nucleotides, are transported into mitochondria in the form of reduced metabolites for which the mitochondrial membrane has specific transporters. Through the malate–aspartate shuttle the reducing equivalents from cytoplasmic NADH are carried into mitochondria as malate (Figure 8.28). The carbon source is returned to the cytoplasm as aspartate. Operation of the malate–aspartate shuttle depends on the existence of different mitochondrial and cytoplasmic forms of both malate dehydrogenase and aspartate aminotransferase.

The shuttle operates as follows:

- NADH produced from glycolysis in the cytoplasm is oxidized to NAD⁺ by the cytoplasmic form of malate dehydrogenase, which reduces oxaloacetate to malate.

- Malate enters the mitochondrial matrix through a transporter in exchange for α-ketoglutarate.

- Within the matrix the mitochondrial form of malate dehydrogenase oxidizes malate to oxaloacetate, generating NADH which can be reoxidized through the electron transport chain within the mitochondria. The mitochondrial membrane is essentially impermeable to oxaloacetate.

- Aspartate, which is produced from oxaloacetate through an aminotransferase reaction, readily crosses to the cytoplasm in exchange for glutamate derived from transamination of α-ketoglutarate through a transporter.

- The cytoplasmic form of the aminotransferase regenerates oxaloacetate from aspartate.

The importance of aminotransferase catalysed reactions is described in Chapter 10. From this shuttle the net result is

- one NADH reoxidized to NAD⁺ in the cytoplasm;

- one NADH made available for energy generation through oxidative phosphorylation in the mitochondria.

Chapter 10 Aminotransferase reactions facilitate the transfer of amino groups from one compound to another, page 295.

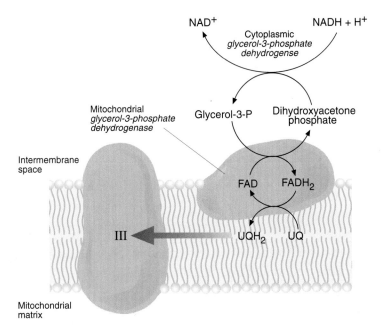

Figure 8.29 The glycerol phosphate shuttle differs from the malate–aspartate shuttle in that it does not involve membrane exchange transporters. It depends on the presence of both cytoplasmic and mitochondrial forms of the enzyme glycerol-3-phosphate dehydrogenase. The mitochondrial form transfers reducing equivalents from FADH directly to ubiquinone in the electron transport chain.

As the malate transporter is linked to the transport of α-ketoglutarate and the aspartate transporter is linked to the transport of glutamate, the net change in the concentrations of either glutamate or α-ketoglutarate in the cytoplasm or mitochondrial matrix is zero.

The α-glycerophosphate shuttle performs the same function as the malate–aspartate shuttle in some tissues

In the adrenal glands and kidney, reducing equivalents are conveyed into mitochondria mainly via the glycerol-3-phosphate shuttle (Figure 8.29). Mitochondrial glycerol-3-phosphate dehydrogenase is an FAD-linked enzyme, so that the net gain of ATP from each cytoplasmic NADH in these tissues is only two molecules.

The pentose phosphate pathway

The main purpose of the pentose phosphate pathway is the production of R5P for incorporation into nucleotides, and NADPH to provide reducing power for reductive biosynthesis of fatty acids, steroids and some amino acids and for other reductive processes within the cell. The enzymes of the pentose phosphate pathway are located in the cytoplasm. The pentose phosphate pathway and glycolysis are closely integrated, having common intermediates and the capacity to exchange metabolites in either direction.

An ingenious feature of the association between glycolysis and the pentose phosphate pathway is that it allows parts of the latter to be exploited to produce specific metabolites without deploying the entire pathway. Few tissues have equal

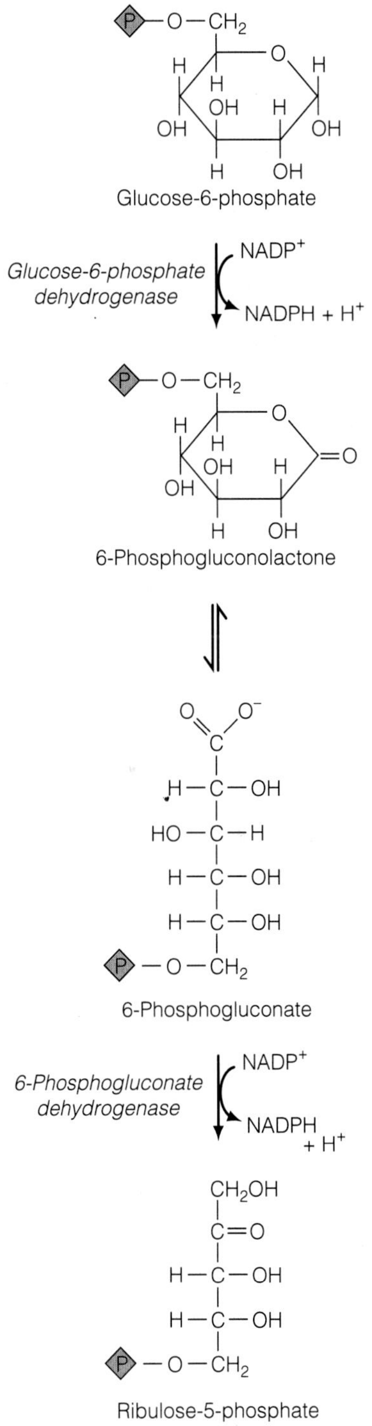

Figure 8.30 Two NADP$^+$-dependent dehydrogenases oxidize G6P to ribulose-5-phosphate. This phase of the pathway effectively achieves its objectives of generating NADPH and pentose phosphates.

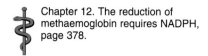
Chapter 12. The reduction of methaemoglobin requires NADPH, page 378.

requirements for R5P and NADPH: cells undergoing cell division or rapid synthesis of RNA for protein synthesis will require large amounts of ribose but may have a much smaller requirement for NADPH; red blood cells have no requirement for pentose phosphate because they do not participate in nucleic acid or protein synthesis, but have a large requirement for NADPH for maintaining the reduced state of ferrous ions in haemoglobin (Chapter 12) and for reducing peroxides.

The oxidative phase of the pentose phosphate pathway generates NADPH

The substrate for this pathway is G6P. The first two reactions are catalysed by NADP⁺-dependent dehydrogenases which generate successively 6-phosphogluconate and ribulose-5-phosphate (Figure 8.30).

Other five-carbon sugars are generated by epimerase- and isomerase-catalysed reactions in the second phase

Ribulose-5-phosphate is converted to xylulose-5-phosphate by an epimerase-catalysed reaction, or to R5P by an isomerase-catalysed reaction. These reactions effectively achieve the purpose of the pathway. However, cells that do not require R5P must make effective use of the nutrient represented by these sugars.

Transketolase and transaldolase reactions convert pentoses to intermediates of glycolysis

The subsequent reactions of the pentose phosphate pathway are a series of transketolase and transaldolase reactions which transfer C-3 and C-2 units from one five-carbon sugar to another, generating a series of C_3, C_4, C_6 and C_7 sugars including F6P and glyceraldehyde-3-phosphate which can enter glycolysis.

These reactions are all readily reversible, so it is also possible for cells with a high demand for pentose phosphates (required for RNA or DNA synthesis) to

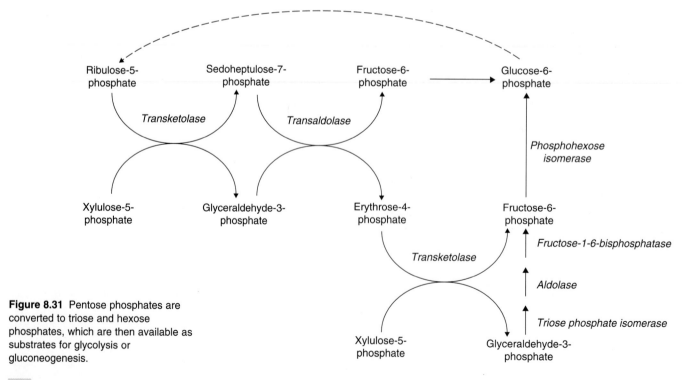

Figure 8.31 Pentose phosphates are converted to triose and hexose phosphates, which are then available as substrates for glycolysis or gluconeogenesis.

extract intermediates from glycolysis to produce R5P, utilizing only those enzymes of the pentose phosphate pathway that convert F6P and glyceraldehyde-3-phosphate to ribose phosphate. This avoids generation of NADPH in cells that may have little demand for it (Figure 8.31).

The pentose pathway, like the citric acid cycle, has an **amphibolic** role in that it generates important intermediates for cell growth and development. It is highly active in rapidly growing cells and in organs such as liver and adrenals with a high biosynthetic capacity.

Metabolic pathways exchange metabolite and share regulatory mechanisms

This chapter has discussed the operation of the two major pathways for the catabolism of glucose in mammals: glycolysis and the pentose phosphate pathway. These pathways are interlinked through common intermediates, and ultimately serve a common catabolic purpose: the recovery of metabolically utilizable energy, either directly as ATP from substrate-level phosphorylation or initially as the reduced cofactors (NADH and $FADH_2$), which yield ATP via oxidative phosphorylation, or as NADPH which provides energy for biosynthesis. When energy levels in a cell are high these pathways also interact with the biosynthetic processes gluconeogenesis and lipogenesis, through which energy is stored until such time as additional energy is required. The other major catabolic pathway, the β-oxidation pathway through which fatty acids are oxidized, is described in Chapter 9.

Chapter 9. β-oxidation progressively decreases the length of fatty acid carbon chains by removing two carbon atoms, page 266.

Lipid metabolism

Triacylglycerols, fatty acids, cholesterol and cholesterol esters are major lipid components of the body. These molecules are derived from the diet but many cells also have the capacity to synthesize them and membrane phospholipids. Fatty acids are synthesized through condensation of acetyl CoA units and are stored as triacylglycerols. Phospholipids, major components of membranes containing two fatty acyl chains, are generated by reactions similar to those used to assemble triacylglycerols.

Fatty acids, the most energy-rich nutrients, are degraded through the mitochondrial β-oxidation pathway by successive removal of two-carbon units to produce acetyl CoA. Condensation of two molecules of acetyl CoA produces the ketone bodies β-hydroxybutyrate and acetoacetate. Ketone bodies, produced in the liver, are a good source of nutrient for other tissues, but excess production of ketone bodies can cause clinical problems.

Cholesterol, an important lipid in biological membranes of higher eukaryotes where it is involved in the regulation of bilayer fluidity, is also synthesized from acetyl CoA. All steroid-based molecules, including hormones, bile salts and vitamin D, are derived from cholesterol.

The body possesses elaborate mechanisms, utilizing specialized types of lipoprotein particles, for transferring triacylglycerols, fatty acids, phospholipids and cholesterol from their sites of absorption, and to and from sites of storage and metabolism.

Lipids are a diverse group of molecules that share the property of being poorly soluble in water. Included in this group are fatty acids, mono-, di- and triacylglycerols, phospholipids, cholesterol and cholesterol esters. In animals, fatty acids are the nutrient molecules which provide the greatest amount of energy for cellular functions. Fatty acids in excess of those immediately required as an energy source are stored as triacylglycerols, mostly in adipocytes but also in small quantities in several other tissues. Phospholipids and cholesterol are the major components of cell membranes. All of these molecules have other important functions which will be discussed in this chapter. The functions of those lipids that act as intracellular signal molecules is described in Chapter 11.

Chapter 11. Hormones derived from fatty acids, page 348.

Properties and functions of fatty acids and triacylglycerols

Fatty acids are versatile molecules that are important participants in a diverse range of cellular activities

Fatty acids are a major source of energy in most mammalian tissues. Complete oxidation of fatty acids is a two-stage process; β-oxidation degrades fatty acids to two-carbon acetyl CoA units, and the citric acid cycle completes the oxidation of acetyl CoA to CO_2 (see Chapter 7). Both pathways occur exclusively within the mitochondrial matrix and yield the reduced nucleotides NADH and $FADH_2$, which can be reoxidized through the respiratory chain in the mitochondrial inner membrane. Cells lacking mitochondria (such as erythrocytes), and cells lacking the enzymes of the β-oxidation cycle (brain cells) can not oxidize fatty acids.

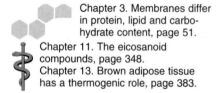

Chapter 7, The citric acid cycle: the pathway for complete oxidation of acetyl coenzyme A, page 213.

Fatty acids have several important biological functions in addition to their role as an energy source.

- They are constituents of the phospholipids and glycolipids that form the basic bilayer of biological membranes (Chapter 3).
- They are precursors of the eicosanoid class of hormones and intracellular messengers implicated in cell-signalling mechanisms (Chapter 11).
- They are constituents of triacylglycerols, which act as heat insulators and which also provide physical protection for organs of the body (Chapter 13).

Chapter 3. Membranes differ in protein, lipid and carbohydrate content, page 51.
Chapter 11. The eicosanoid compounds, page 348.
Chapter 13. Brown adipose tissue has a thermogenic role, page 383.

The properties of fatty acids, including their involvement in membrane structures, are considered in Chapters 2 and 3.

Triacylglycerols have the properties required of a long-term metabolic fuel storage molecule

Long-term storage of excess nutrients in mammals takes the form of large lipid deposits in the cytoplasm of **adipocytes**, the principal cells of adipose tissue (Chapter 13), found in a number of specific subcutaneous and juxtaorgan sites throughout the body. Adipocytes are specialized cells, adapted for the synthesis and degradation of **triacylglycerols**, which are uncharged fatty acid esters of glycerol and are also termed **neutral** fats or **triglycerides**. Triacylglycerols represent highly concentrated stores of metabolic energy that can be mobilized rapidly to release fatty acids for transport to appropriate tissues via the bloodstream.

Triacylglycerols are efficient fuel-storage products, with greater energy per gram than either glycogen or protein. In addition, glycogen is bulky, hydrophilic and usually heavily hydrated, adding to the water content of cells; triacylglycerols are insoluble and therefore osmotically inert. Triacylglycerols are highly reduced compared with carbohydrate or protein so that oxidation of fatty acids yields

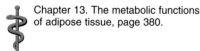

Chapter 13. The metabolic functions of adipose tissue, page 380.

much more energy than can be obtained from the oxidation of a corresponding mass of glycogen. The energy yield from the oxidation of fats is about 39 kJ g^{-1} compared with 18 and 17.5 kJ g^{-1} for carbohydrate and protein respectively.

Triacylglycerol stores constitute about 20% of body weight for the average well nourished individual. These molecules also provide mechanical protection and have important insulating properties that contribute to protection against cold and in maintenance of body temperature (Chapter 13).

Chapter 13. Brown adipose tissue has a thermogenic role, page 383.

Mobilization and oxidation of fatty acids

Fatty acids are released from triacylglycerol stores by the action of a hormone-sensitive lipase

Degradation of triacylglycerol in adipocytes occurs in response to activation of **hormone-sensitive lipase**, which can be triggered by any of three different hormones under different circumstances (Table 9.1). The lipase catalyses hydrolysis of triacylglycerols to produce one molecule of glycerol and three of free fatty acid.

Hormone-sensitive lipase is activated through a cAMP-dependent cascade amplification process (Figure 9.1) similar to that for the activation of glycogenolysis

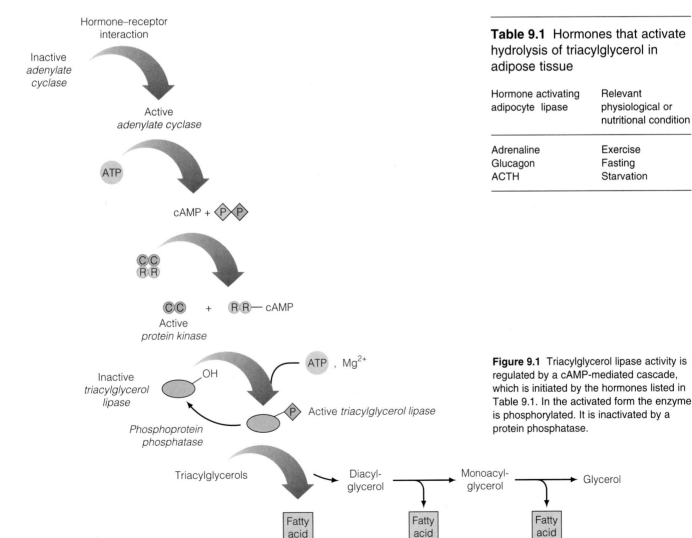

Table 9.1 Hormones that activate hydrolysis of triacylglycerol in adipose tissue

Hormone activating adipocyte lipase	Relevant physiological or nutritional condition
Adrenaline	Exercise
Glucagon	Fasting
ACTH	Starvation

Figure 9.1 Triacylglycerol lipase activity is regulated by a cAMP-mediated cascade, which is initiated by the hormones listed in Table 9.1. In the activated form the enzyme is phosphorylated. It is inactivated by a protein phosphatase.

263

Lipid metabolism

Chapter 8. Increased cAMP
concentration activates a 'cascade' of
events that amplify the hormonal
response, page 227.
Chapter 11. Hormonal responses are
mediated through adenylate cyclase
and the production of cAMP,
page 338.

Chapter 14. Triacylglycerols
hydrolysed in the gut are
reassembled within epithelial cells,
page 421.

(Chapter 8). The cascade is initiated when the hormone binds to a receptor on the plasma membrane of the adipocyte linked, through G-proteins (Chapter 11), to the activation of adenylate cyclase. The increased production of cAMP within the adipocyte activates a **cAMP-dependent protein kinase**, which in turn activates hormone-sensitive lipase through phosphorylation.

The glycerol released when triacylglycerols are degraded can not be reused in adipocytes, and is normally returned to the liver, where it enters either glycolysis or gluconeogenesis as glyceraldehyde-3-phosphate (Figure 9.2).

The fatty acids mobilized from triacylglycerols enter the bloodstream (see Chapter 14), where they bind to **albumin**, the principal protein of plasma, for transportation to the tissues that can oxidize them to provide energy.

The major tissues in which fatty acids are oxidized in the human body are liver and skeletal muscle. However, the purposes for which fatty acids are oxidized in these tissues differ. In muscle cells fatty acids can be oxidized completely to CO_2 in order to produce the maximum amount of ATP, to support the contractile and electrical activities of the muscle cells. In the liver, however, most of the fatty acid is not oxidized beyond acetyl CoA; the energy released from this partial oxidation of fatty acids provides ATP and reduced nucleotides for use in synthetic processes including gluconeogenesis, protein synthesis and steroid synthesis. Most of the acetyl CoA serves as substrate for ketogenesis, providing energy to other tissues.

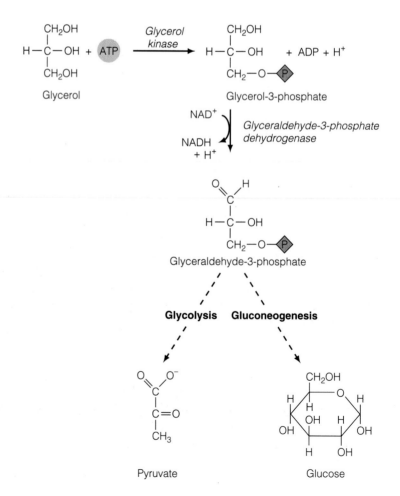

Figure 9.2 Glycerol kinase in the liver converts glycerol to glycerol-3-phosphate, which is then reduced to glyceraldehyde-3-phosphate, an intermediate of both glycolysis and gluconeogenesis.

Fatty acids must be activated before they enter mitochondria

Fatty acids are composed of long hydrocarbon chains, usually 14–24 carbons in length, with a terminal carboxyl group. Most have even numbers of carbon atoms, dictated by the fatty acid synthetic reactions that elongate the carbon chain by successively adding two-carbon units. There are also fatty acids containing uneven numbers of carbon atoms and unsaturated fatty acids containing one or more carbon–carbon double bonds.

Free fatty acids require no special transport systems to enter cells; they simply diffuse across plasma membranes and the rate of uptake into tissues reflects their concentration in the circulation. In contrast, entry of fatty acids into the mitochondrial matrix (where they are oxidized), requires a specific mechanism. Fatty acids are activated by an **acyl CoA synthase** located on the outer surface of the outer membrane of the mitochondrion. Activation is a two-step process, requiring both ATP and CoA. Initially the fatty acid reacts with ATP, forming a high-energy acyl adenylate intermediate and releasing pyrophosphate. This intermediate then reacts with CoA, which displaces the AMP to produce a high-energy thioester (Figure 9.3). The reaction is essentially irreversible due to the expenditure of two high-energy phosphate bonds to produce the activated fatty acyl CoA.

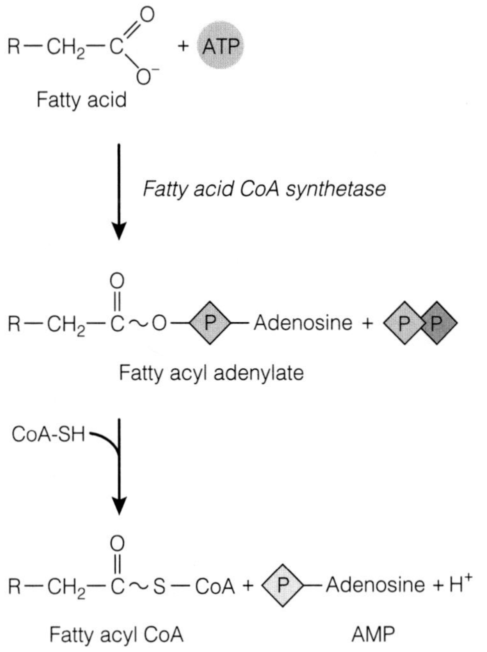

Figure 9.3 The formation of a fatty acyl CoA derivative occurs through a fatty acyl adenylate in which the carboxyl group of the fatty acid is activated.

Fatty acyl CoA derivatives enter the intermembrane space freely but are not transported across the inner mitochondrial membrane: to do so, CoA derivatives must be converted to carnityl derivatives in a reaction catalysed by **acyl carnitine transferase I**. Acyl carnitine transferase I appears to form a pore across the outer mitochondrial membrane.

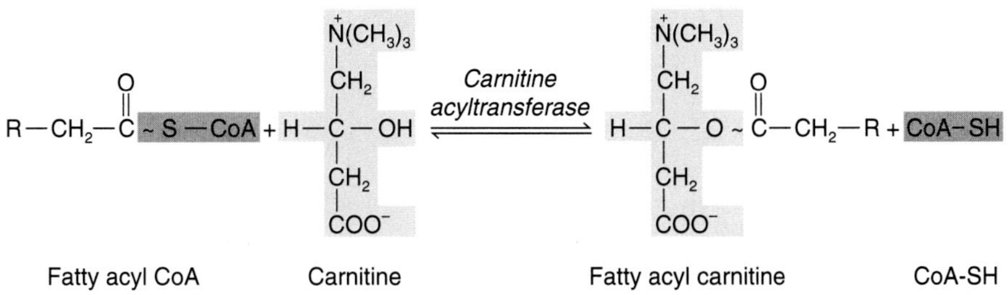

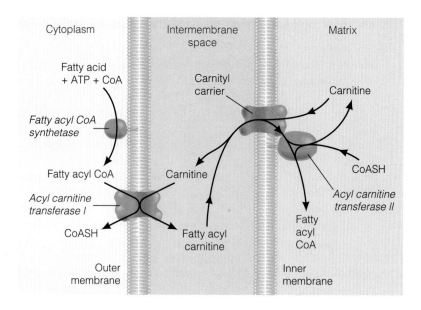

Figure 9.4 Fatty acyl CoAs are unable to cross the mitochondrial membrane, so fatty acids must be transferred from CoA to carnityl groups before they enter the mitochondrial matrix through a specific transporter. Within the matrix, fatty acyl groups are transferred back to CoA for oxidation and the liberated carnitine exchanges across the membrane with incoming fatty acyl carnitine.

Carnitine is a cofactor derived from the amino acids lysine and methionine. Its unique function is to facilitate the transfer of fatty acids into mitochondria. The inner membrane of the mitochondrion possesses a carrier which transfers free carnitine out of the matrix in exchange for fatty acyl carnitine derivatives. At the matrix surface fatty acyl carnitine is reconverted to the acyl CoA derivative by **acyl carnitine transferase II** and free carnitine is liberated (Figure 9.4).

The apparently complicated **carnitine exchange carrier** system by which fatty acids enter the mitochondrial space provides the cell with a mechanism for regulating the concentrations of CoA in the cytoplasm and in the mitochondrial matrix. The separate cytoplasmic and mitochondrial pools of CoA have different metabolic functions, and a reduction in CoA concentration in either compartment would restrict the rate of the processes for which it is required. In the cytoplasm CoA derivatives are used for reductive biosynthesis; within mitochondria they are required for oxidative reactions.

The carnitine exchange carrier in the liver is inhibited by malonyl CoA, an early intermediate in the synthesis of fatty acids from acetyl CoA. This arrangement enables the cell to recognize, from the presence of elevated concentrations of malonyl CoA, that fatty acid synthesis is occurring and prevents a wasteful cycle in which fatty acyl CoA might be transported into the mitochondria and oxidized.

β-oxidation progressively decreases the length of fatty acid carbon chains by removing two carbon atoms

Successive removal of two-carbon units from the carboxyl end of fatty acids proceeds by a sequential cycle of four reactions, in a process called β-oxidation. The substrate is a fatty acyl CoA derivative. Each cycle generates acetyl CoA and an acyl CoA that is two carbon units shorter than the initial substrate. For every acetyl CoA molecule released energy is produced in the form of one molecule of FAD reduced to $FADH_2$ and one molecule of NAD^+ reduced to NADH. The overall process consumes energy because one molecule of ATP is converted to AMP as CoA is used to activate each fatty acid molecule. The acetyl CoA molecules released may be further oxidized to CO_2 through the citric acid cycle or may be used to synthesize other molecules, including ketone bodies (produced in the mitochondrion) and steroid-based molecules (produced in the cytosol).

The first reaction shown in Figure 9.5 is catalysed by the flavoprotein **acyl CoA dehydrogenase**, in which a *trans* double bond is introduced between carbon atoms

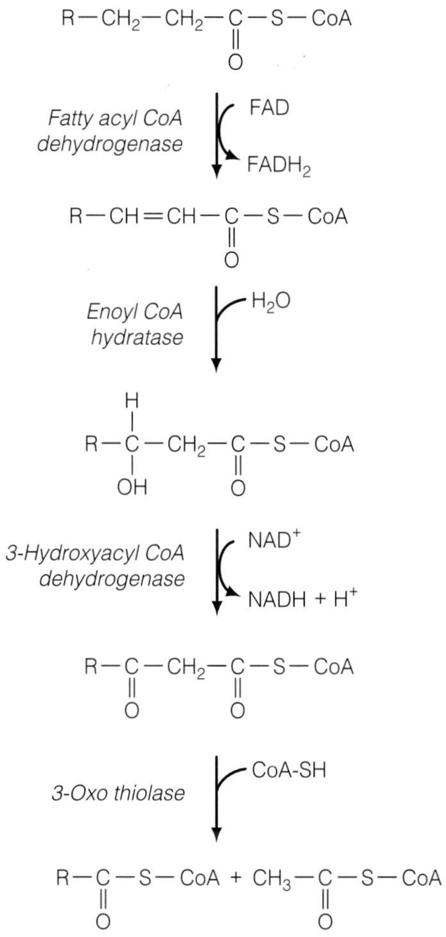

$$R-CH_2-CH_2-\underset{\underset{O}{\|}}{C}-S-CoA$$

Fatty acyl CoA dehydrogenase $\quad\left\{\begin{array}{l}FAD\\ \searrow FADH_2\end{array}\right.$

$$R-CH=CH-\underset{\underset{O}{\|}}{C}-S-CoA$$

Enoyl CoA hydratase $\quad\left\{ H_2O \right.$

$$R-\underset{\underset{OH}{|}}{\overset{\overset{H}{|}}{C}}-CH_2-\underset{\underset{O}{\|}}{C}-S-CoA$$

3-Hydroxyacyl CoA dehydrogenase $\quad\left\{\begin{array}{l}NAD^+\\ \searrow NADH + H^+\end{array}\right.$

$$R-\underset{\underset{O}{\|}}{C}-CH_2-\underset{\underset{O}{\|}}{C}-S-CoA$$

3-Oxo thiolase $\quad\left\{ CoA\text{-}SH \right.$

$$R-\underset{\underset{O}{\|}}{C}-S-CoA + CH_3-\underset{\underset{O}{\|}}{C}-S-CoA$$

Figure 9.5 The β-oxidation spiral. The figure shows one round of the spiral releasing one two-carbon acetate group from the fatty acid substrate.

2 and 3, forming *trans* Δ² enoyl CoA. Three different acyl CoA dehydrogenases are specific for short, medium and long-chain fatty acyl CoAs. All three must be active to allow complete oxidation of long-chain fatty acids.

The next reaction involves stereospecific hydration of the enoyl CoA derivative and is catalysed by **enoyl CoA hydratase**. Introduction of a hydroxyl group permits the second oxidation step, in which the hydroxyl is oxidized to a 3-*oxo* group, catalysed by **3-hydroxyacyl CoA dehydrogenase**. The conversion of a methylene group at C-3 to a 3-*oxo* group (hence the name β-oxidation) renders the bond between C-2 and C-3 susceptible to cleavage by **3-*oxo* (β-*oxo*) thiolase**, a reaction in which CoA is added to produce the shortened acyl CoA. Acetyl CoA is released in readiness for the next round of β-oxidation.

The individual reactions of β-oxidation for short- and medium-chain fatty acids are catalysed by separate enzymes but these must be in close proximity, because intermediates of the cycle are almost undetectable. This means that the product of one reaction is passed to the next enzyme with considerable efficiency. In rat liver, the oxidation of long-chain fatty acids is carried out by two membrane-bound proteins that catalyse all four reactions, shortening acyl CoA chains from C_{18} to C_8. It remains to be established whether the same multifunctional enzymes exist in other animals.

Special mechanisms oxidize unsaturated or odd-numbered fatty acids

Fatty acids with odd-numbered carbon chains and with a variety of different unsaturated carbon–carbon bonds are present in the human diet. Fatty acids with one or more double bonds require modification to convert them into intermediates of the

Box 9.1 Implication of medium chain fatty acyl CoA dehydrogenase in 'sudden infant death syndrome'

It has been suggested that one of the possible causes of the distressing syndrome resulting in the sudden death of infants is the lack of the fatty acyl CoA dehydrogenase acting on medium-chain fatty acids. If this enzyme is absent, or present in inadequate amounts, only a limited amount of energy is derived from the oxidation of fatty acids because the β-oxidation cycle does not proceed to degrade fatty acids beyond medium chain length. In this situation glucose is the major nutrient, and as glycogen reserves are depleted between meals, blood glucose falls to abnormally low concentrations inducing coma and death.

Box 9.2   Genetic defects in fatty acid oxidation

For starving individuals, for those on a high fat diet, or for those engaging in prolonged vigorous exercise, fatty acids are the main nutrient source. Symptoms related to defects in fatty acid oxidation tend to manifest themselves in these conditions, usually as muscle cramps owing to the abnormal accumulation of intermediates of the β-oxidation pathway. The defective enzyme may be identified by analysis of these intermediates and the appropriate enzymes. The importance of the presence of all three **fatty acyl CoA dehydrogenases** (short-, medium- and long-chain specific enzymes) is highlighted in patients with specific genetic defects in only one of the dehydrogenases, in whom intermediates of characteristic chain length corresponding to the rate-limiting step accumulate.

A rare genetic disease has been detected, which involves a specific defect in the **carnitine exchange carrier**, preventing access of long-chain fatty acids to the enzymes of β-oxidation. This unusual defect highlights the importance of good communications between the various intracellular compartments in normally functioning cells.

Carnitine is synthesized in the liver, and a deficiency in production of this cofactor has been recognized in some individuals.

β-oxidation pathway. Oleate, a C_{18} fatty acid with a *cis* double bond between C-9 and C-10 (*cis* Δ^9) is oxidized normally until, after three reaction cycles, the resultant acyl CoA intermediate contains the *cis* double bond between C-3 and C-4. This prevents the normal dehydrogenation reaction, which involves formation of a *trans* double bond between C-2 and C-3. A specific **enoyl CoA isomerase** converts the *cis* Δ^3 bond to a *trans* Δ^2 bond. As acyl CoA intermediates containing a *trans* Δ^2 double bond are normal components of the β-oxidation cycle this compound is processed normally through β-oxidation (Figure 9.6).

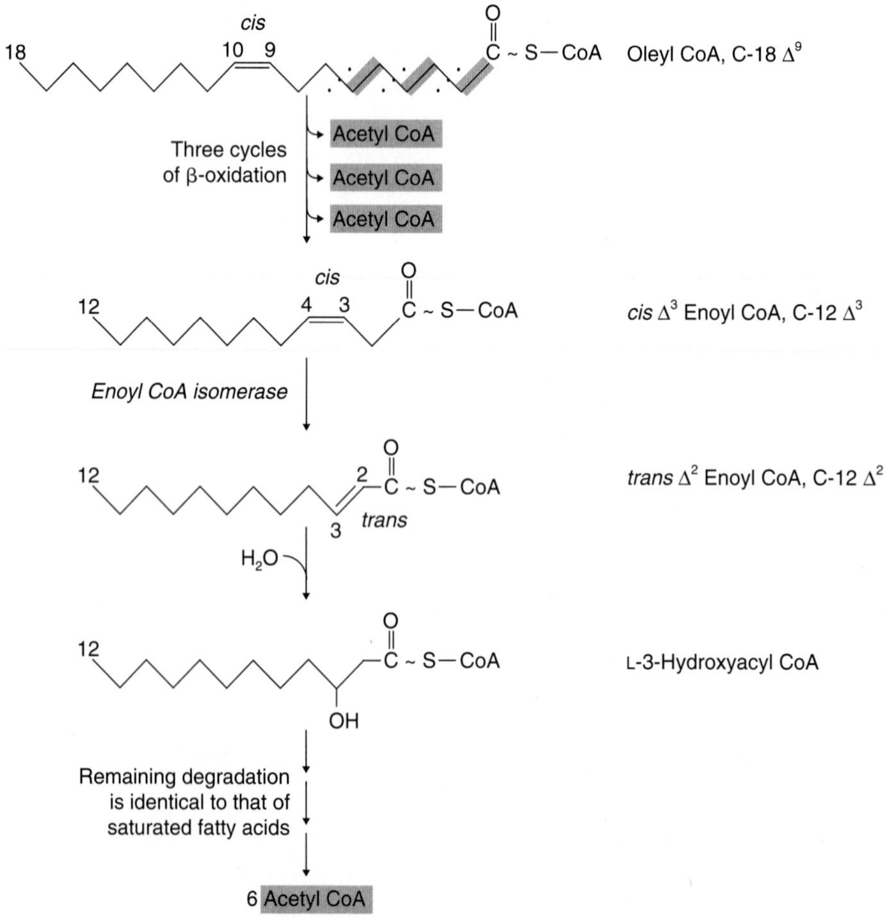

Figure 9.6 Oxidation of monounsaturated fatty acyl CoA derivatives depends on enoyl CoA isomerase-catalysed conversion of a *cis* double bond to a *trans* double bond.

The situation is somewhat more complex for fatty acid intermediates generated during oxidation of polyunsaturated fatty acids such as linoleate acid, which has cis Δ^9 and cis Δ^{12} double bonds. After three reaction cycles the first cis double bond is now at C-3, with the second cis double bond at C-6. The action of enoyl CoA isomerase generates a trans Δ^2 cis Δ^6 intermediate and a *trans* Δ^2 *cis* Δ^4 enoyl CoA

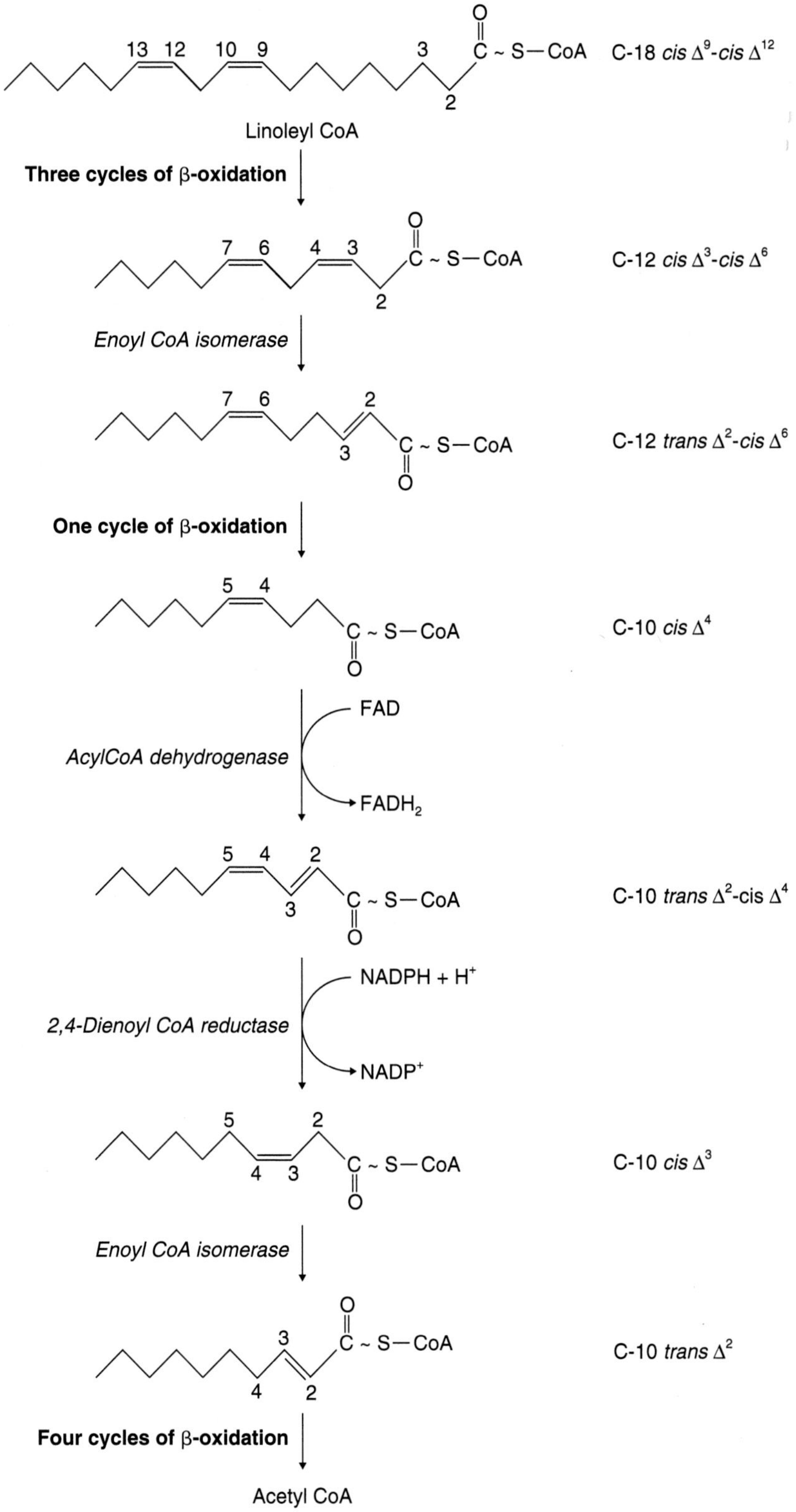

Figure 9.7 The oxidation of polyunsaturated fatty acyl CoA derivatives depends on both the enoyl CoA isomerase and an NADPH-dependent dienoyl CoA reductase.

Box 9.3 The generation and utilization of propionyl CoA

Saturated fatty acids of odd chain length are present in the diet in small amounts, mostly derived from vegetable matter. When these fatty acids are processed through β-oxidation the final cycle yields propionyl CoA as well as acetyl CoA. Propionyl CoA is also produced by the catabolism of the amino acids methionine and isoleucine. Carboxylation of propionyl CoA by the biotin-containing enzyme **propionyl CoA carboxylase** produces methylmalonyl CoA. This reaction is of particular significance in ruminants because they absorb very little of their dietary carbohydrate directly, and instead produce glucose from propionate generated by fermentation reactions in the rumen (see Box 9.4).

Methylmalonyl CoA is finally rearranged to succinate, an intermediate of the citric acid cycle, in an isomerization catalysed by **methylmalonyl CoA mutase**, which uses vitamin B_{12} as cofactor.

Propionyl CoA carboxylase has a mode of action similar to that of **acetyl CoA carboxylase** involved in fatty acid biosynthesis (see later in this chapter) and **pyruvate carboxylase**, an anaplerotic enzyme which also promotes the first step in gluconeogenesis from pyruvate to oxaloacetate (Chapter 8).

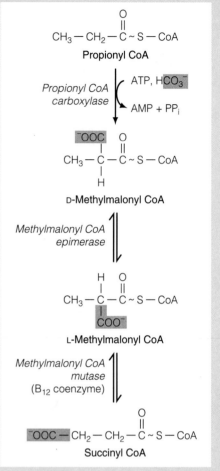

Box Figure 9.3 Catabolism of odd-numbered fatty acids.

Chapter 8. Pyruvate carboxylase, page 246.

is formed in the next step in the process. This is reduced by **2,4 dienoyl CoA reductase** (using NADPH as cofactor) to the *trans* Δ^3 derivative. This intermediate is now a substrate for enoyl CoA isomerase, and is converted to the *trans* Δ^2 intermediate of the standard β-oxidation pathway.

β-oxidation produces relatively large amounts of energy

Degradation of stearic acid (an 18-carbon molecule) requires eight successive cycles of β-oxidation to produce nine molecules of acetyl CoA. Each cycle also produces one NADH and one $FADH_2$. The acetyl CoA produced then enters the citric acid cycle to be oxidized to CO_2 and H_2O with the production of additional reduced nucleotides. Two high-energy bonds are used to activate the fatty acid as ATP is converted to AMP + PPi.

The net energy yield from the complete oxidation of one molecule of the C_{18} fatty acid stearic acid is equivalent to 146 molecules of ATP (Table 9.2), eight for each carbon of the fatty acid. By comparison the complete oxidation of glucose yields 38 ATP from a six-carbon sugar, six per carbon atom derived from glucose.

Table 9.2 Energy yield from complete oxidation of stearic acid

Process or reaction	Substrate	Reduced nucleotides	ATP equivalents
Fatty acid activation	Stearic acid (C_{18})		−2
β-oxidation	Stearyl CoA (C_{18})	8 NADH	+24
		8 $FADH_2$	+16
Citric acid cycle	9 Acetyl CoA (C_2)	9 × 3 NADH	+81
		9 × 1 $FADH_2$	+18
		9 × 1 GTP	+9
Total			**146**

Unusual fatty acids may be oxidized by α-oxidation or ω-oxidation

α-**Oxidation** occurs in the endoplasmic reticulum and appears to be important in dealing with methylated fatty acids ingested from the diet. This pathway was first observed during attempts to explain the metabolic basis of a congenital disease called Refsum's disease. Sufferers accumulate a molecule called **phytanic acid**, which is derived from phytol, a constituent of chlorophyll. α-Oxidation releases CO_2 from the carboxy terminus of methylated carboxylic acids, providing a substrate that can be metabolized through β-oxidation (Figure 9.8). Individuals lacking the enzyme catalysing α-oxidation accumulate phytanic acid, with deleterious neurological effects. These patients should avoid consuming chlorophyll by not eating green vegetables and meat from herbivores fed on grass.

ω-**Oxidation** involves the conversion of the terminal methyl group of fatty acids to a carboxylic acid, forming a dicarboxylic acid. This reaction also occurs in the endoplasmic reticulum. The purpose of ω-oxidation, which is active in the liver, is to solubilize substituted or other unusual fatty acids that can not be degraded by β-oxidation, so that they can be excreted. This is one route through which the prostaglandins are converted to excretable forms.

Complete oxidation of fatty acids depends on the operation of the citric acid cycle

In the normal physiological state, a balance is maintained between the pathways catabolizing carbohydrate and fat. In particular, entry of acetyl CoA into the citric acid cycle requires oxaloacetate. In starvation, the concentration of oxaloacetate is reduced because it is diverted into gluconeogenesis in order to maintain blood glucose. Oxaloacetate concentration will be low in diabetes because the lack of glucose within cells blocks the anaplerotic reaction augmenting the supply of oxaloacetate from pyruvate. When oxaloacetate concentration is reduced, limiting the amount of acetyl CoA that can be oxidized through the citric acid cycle, the reduced nucleotides produced from the oxidation of fatty acids through the β-oxidation cycle become the major source from which ATP can be produced by oxidative phosphorylation.

In the liver accumulating acetyl CoA is converted to ketone bodies

In the liver acetyl CoA in excess of the amount consumed by the citric acid cycle is converted to the ketone bodies **acetoacetate** and D-3-hydroxy (β-hydroxy) **butyrate** (Figure 9.9). Liver is the only tissue with a significant capacity to synthesize ketone bodies: other tissues lack the necessary enzymes. Two of the enzymes that catalyse the production of the ketone bodies (thiolase and the dehydrogenase that converts acetoacetate to β-hydroxybutyrate) are the same enzymes responsible for reactions of the β-oxidation cycle. Acetoacetyl CoA is formed by condensation of two molecules of acetyl CoA, catalysed by thiolase, the terminal enzyme of β-oxidation. A third acetyl CoA is added in a reaction catalysed by **3-hydroxy, 3-methylglutaryl CoA synthase** to produce 3-hydroxy, 3-methylglutaryl CoA, which is not only an intermediate in ketone body synthesis but also a precursor of steroid compounds. Acetyl CoA and acetoacetate are formed by cleavage of this intermediate. Acetoacetate is reduced to 3-hydroxybutyrate by **3-hydroxybutyrate dehydrogenase**. The synthesis of ketone bodies uses NADH, and high ratios of NADH to NAD$^+$ inhibit the citric acid cycle and divert acetyl CoA into synthesis of ketone bodies.

Synthesis of ketone bodies occurs within mitochondria. Acetoacetate and β-hydroxybutyrate subsequently leave the mitochondria through an inner membrane transporter in exchange for pyruvate. Acetoacetate is a relatively unstable compound

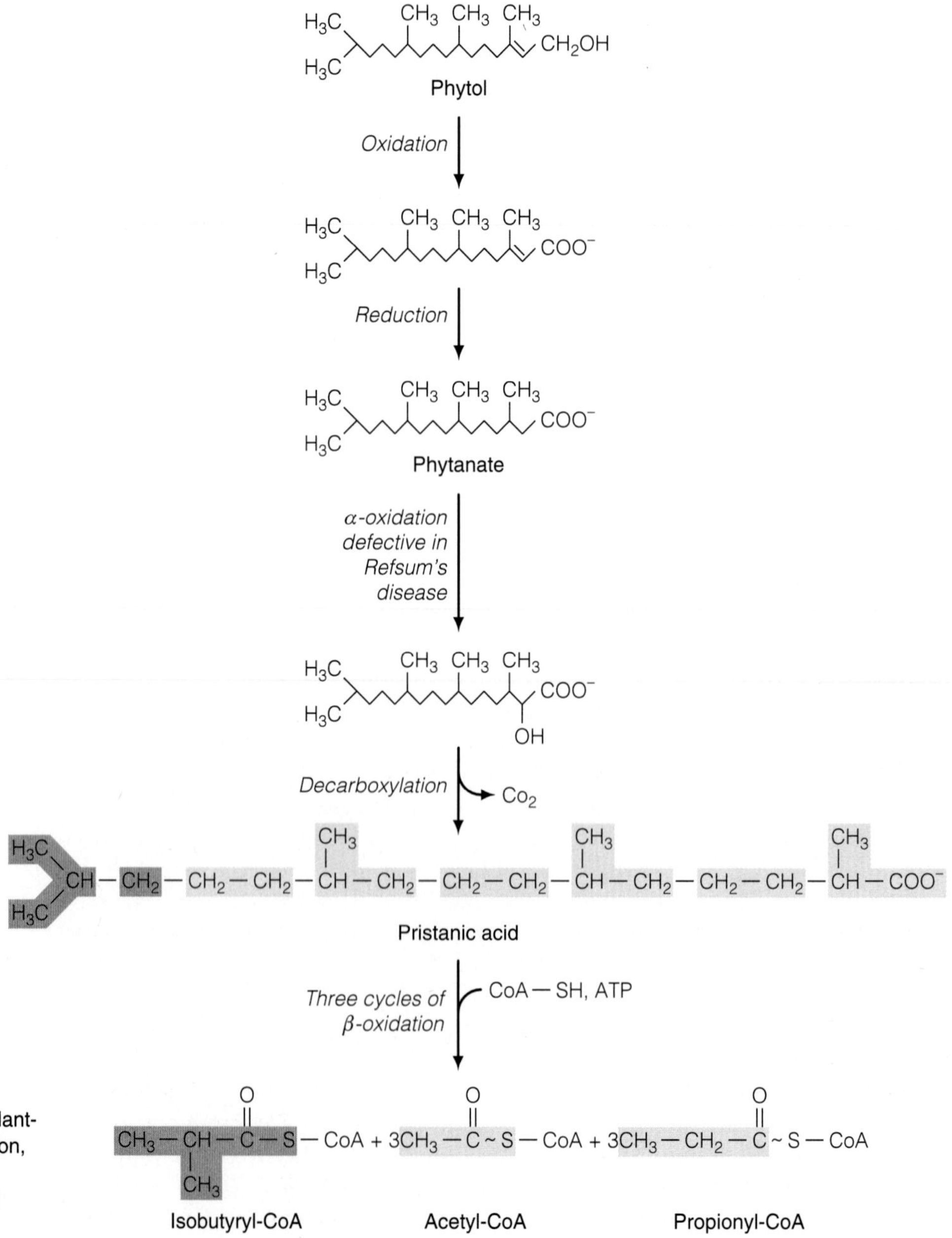

Figure 9.8 Phytol is an example of a plant-derived methylated fatty acid. α-Oxidation, a decarboxylation reaction, provides a product that can be further metabolized through β-oxidation.

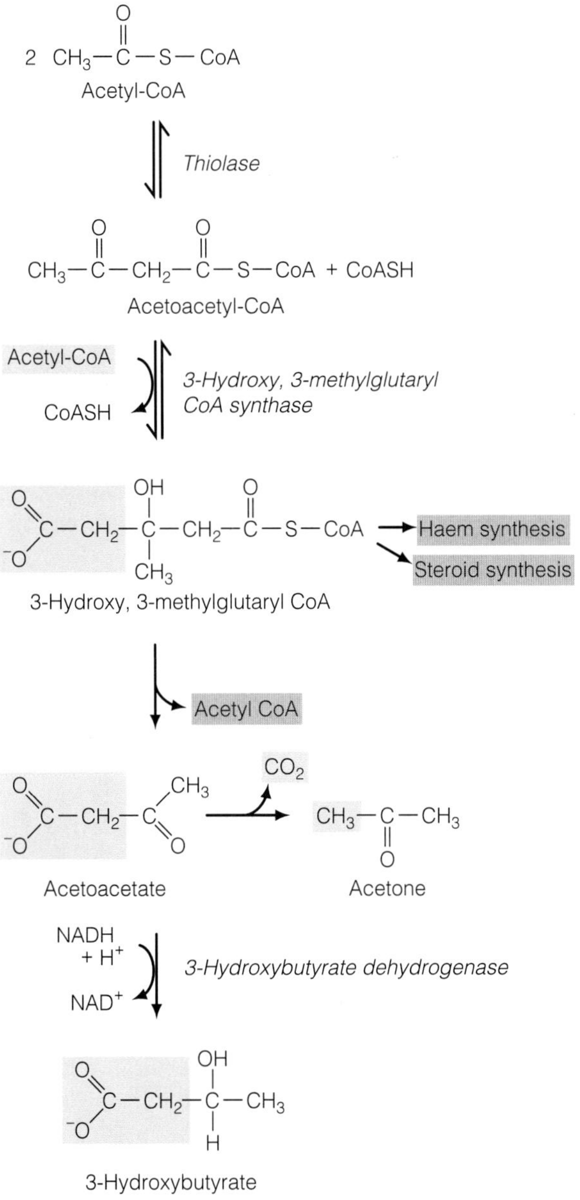

Figure 9.9 Formation of ketone bodies.

and some may be decarboxylated spontaneously in the circulation to produce **acetone**, which is metabolically inert and sufficiently volatile to be exhaled. Acetone may be detected in the breath of patients with severe and untreated diabetes.

Acetoacetate and β-hydroxybutyrate are important nutrients

Ketone bodies are normal metabolites that are produced continuously in the liver and exported to other tissues. Acetoacetate represents a convenient water-soluble compound for transferring acetyl units from the liver to other tissues. The normal concentration of ketone bodies in the bloodstream is low (0.1 mM) because they are efficiently taken up and oxidized by other tissues, particularly the heart and adrenal glands.

In response to nutritional or metabolic imbalance, ketone bodies are produced in greater quantities and circulate in the bloodstream at higher concentrations (as high as 20 mM in extreme conditions). Abnormal rates of ketogenesis occur when the balance between metabolism of glucose and fatty acids is disturbed by a relatively high rate of fatty acid oxidation. During starvation the circulating concentrations of

---

Box 9.4   Ruminants produce and use more ketone bodies than humans

Ketone body concentrations in the bloodstream are normally greater in ruminants than in other mammals because ketone bodies are produced more efficiently in these animals. Ruminants absorb significantly less glucose from their diet than other mammals because glucose is fermented by microorganisms in the rumen. The resulting volatile acids, acetate, propionate and butyrate, are absorbed from the digestive system. These processes contribute to a different balance of ketone bodies and glucose in the bloodstream of ruminants; indeed, the normal ruminant blood nutrient balance would be regarded as more characteristic of starvation if it occurred in humans.

Chapter 14. The brain is the major user of ketone bodies, page 438.

Figure 9.10 Ketone bodies are used as nutrients by tissues that can convert acetoacetate to acetoacetyl CoA. The CoA for this reaction is derived from succinyl CoA, an intermediate of the citric acid cycle.

ketone bodies rise dramatically and the brain can adapt so as to derive up to two-thirds of its energy requirements from ketone bodies (Chapter 14).

Liver can not utilize ketone bodies because it lacks β-ketoacyl CoA transferase. The absence of this enzyme in the liver is one of several mechanisms contributing to the important role of the liver in distributing nutrients to other tissues of the body.

Ketone bodies are converted to acetyl CoA through the reactions shown in Figure 9.10.

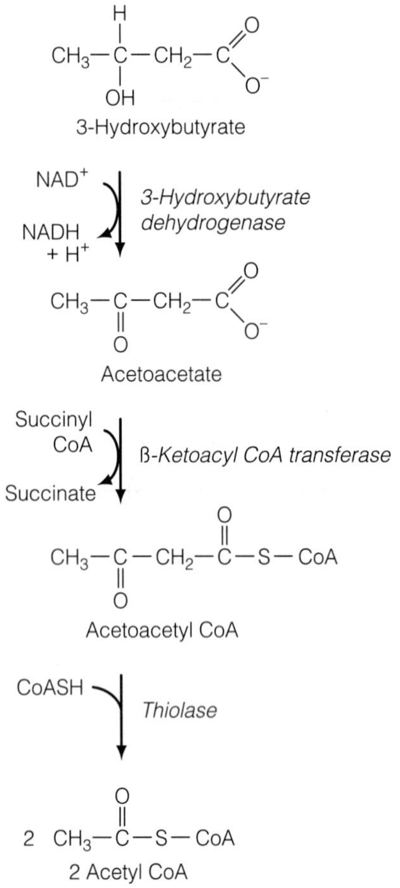

Ketone bodies are an efficient source of energy

Conversion of β-hydroxybutyrate to acetyl CoA produces one NADH. Oxidation of the two acetyl CoA molecules through the citric acid cycle yields a further six NADH, two FADH$_2$ and one GTP. At first sight it might appear that two GTP are produced, but account must be taken of the utilization of succinyl CoA, providing the activated CoA unit for conversion of ketone bodies to acetyl CoA. The total energy yield from one molecule of β-hydroxybutyrate is thus equivalent to 26 ATP: compare this with the direct oxidation of two molecules of pyruvate, which yields the equivalent of 30 molecules of ATP. The true energy yield from ketone bodies should, of course, include the energy required for their synthesis in the liver. However, such simple calculations do not take account of the survival advantage for other tissues when the liver exports nutrients to them. In addition, no energy investment is required to initiate catabolism of ketone bodies, which makes them the preferred substrate in some tissues even if glucose is available. Metabolism of ketone bodies by the brain is particularly important in starvation because this reduces the need to convert body protein to glucose.

Fatty acid biosynthesis

Synthesis of fatty acids is catalysed by a pathway located in the cytoplasm

Fatty acid synthesis is a reductive process, dependent on the availability of NADPH, which is found in highest concentration in the cytoplasm. This is also the location of the enzymes required for the synthesis of fatty acids and their modification to form unsaturated fatty acids.

Fatty acids are synthesized if

- nutrient is available in excess of requirement and needs to be stored as triacylglycerol;

- fatty acids are required to maintain phospholipid components of cell membranes;
- fatty acids are required for the synthesis of specialized molecules such as prostaglandins (Chapter 11).

Although most of the chemical reactions involved in fatty acid synthesis are effectively the reverse of those in β-oxidation, the enzymes are different and present in different locations: fatty acid synthesis occurs in the cytosol, β-oxidation in the mitochondrial matrix. The complete separation of β-oxidation and fatty acid synthesis contrasts with the partial separation of glycolysis and gluconeogenesis, pathways that operate in opposite directions but share some enzymes, only some being specific to one process (Chapter 8).

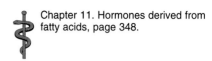

Chapter 11. Hormones derived from fatty acids, page 348.

There are four distinct phases in the synthesis of fatty acids

These are

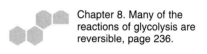

Chapter 8. Many of the reactions of glycolysis are reversible, page 236.

- synthesis of malonyl CoA;
- synthesis of fatty acids by the sequential addition of two-carbon units derived from malonyl CoA, creating carbon chains up to C_{16};
- elongation of fatty acyl chains from 16 to 22 carbon atoms;
- introduction of carbon–carbon double (unsaturated) bonds.

Malonyl CoA is the key intermediate in fatty acid biosynthesis

Fatty acids are synthesized by the progressive addition of two-carbon units to acetyl CoA. With the exception of the first unit, all the carbon atoms are derived via malonyl CoA – *not* directly from acetyl CoA. The three-carbon malonyl units are synthesized by carboxylation of acetyl CoA in a reaction catalysed by a biotin-dependent enzyme, **acetyl CoA carboxylase**, which also requires ATP (Figure 9.11). The reaction is essentially irreversible.

In eukaryotes, acetyl CoA carboxylase is a dimer of two identical subunits. Citrate is an allosteric activator. Activation occurs when the cytoplasmic concentration of citrate is high, and able to provide acetyl CoA as a carbon source from which fatty acids can be synthesized. Palmityl CoA maintains the enzyme in the low-activity form, providing feedback inhibition of fatty acid synthesis.

Acetyl CoA carboxylase is also controlled by adrenaline- or glucagon-mediated phosphorylation, which inactivates it, and insulin, which activates it. These hormonal effects are relatively small and probably occur in times of energy depletion.

The rate-limiting reaction in the synthesis of fatty acids is synthesis of malonyl CoA. The fact that this enzyme is regulated by citrate enables it to detect the efflux of citrate from mitochondria that occurs when cells need to store metabolites as fatty acids (Figure 9.12).

Efflux of citrate from mitochondria is mediated by a specific transporter in the inner membrane, the **citrate–malate antiporter** (see Chapter 8), so named because

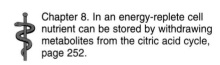

Chapter 8. In an energy-replete cell nutrient can be stored by withdrawing metabolites from the citric acid cycle, page 252.

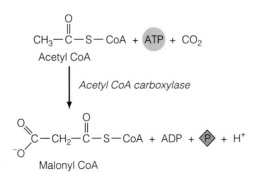

Figure 9.11 Synthesis of malonyl CoA

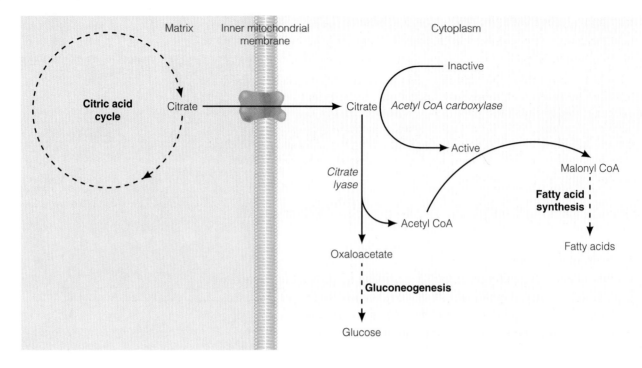

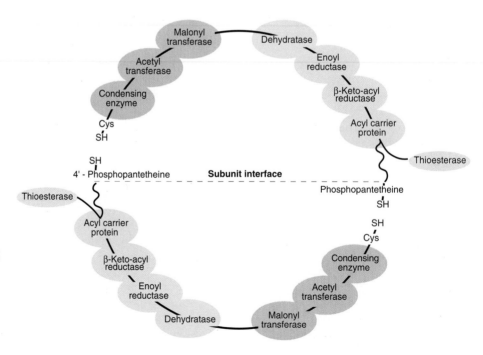

Figure 9.12 When the metabolite concentration in the citric acid cycle is high, citrate is transferred from mitochondria to the cytoplasm where it both activates acetyl CoA carboxylase and is the substrate for citrate lyase, which forms both acetyl CoA and oxaloacetate. The acetyl CoA formed becomes the substrate for the activated carboxylase, which converts the excess nutrient to malonyl CoA and fatty acids.

it exchanges citrate for malate. Malate enters the citric acid cycle at a later stage, thereby maintaining the flux of intermediates around the cycle.

Fatty acid carbon chains are assembled through a sequence of four reactions

The reactions of fatty acid synthesis are catalysed by the **fatty acid synthase** complex, a large multifunctional protein with seven distinct enzymic activities. In animal cells all of these activities are found within a large protein formed from a single polypeptide chain (Figure 9.13).

Figure 9.13 The mammalian fatty acid synthase complex is a dimer (two identical subunits), each subunit having three distinct structural domains and seven separate catalytic activities. The long flexible arm of the phosphopantetheine moiety is essential to the activity of the complex.

As each fatty acid molecule grows it is attached to an **acyl carrier region** of the protein through a phosphopantotheine residue attached to the protein. After each sequence of condensation, reduction, dehydration and reduction the length of the carbon chain is extended by two, up to a maximum of 16 carbon atoms. The reducing power is derived from NADPH generated in the pentose phosphate pathway or from the oxidation of malate to pyruvate by malic enzyme (Chapter 8).

The functional advantage of such a highly organized array of enzymes is the enhancement of overall catalytic efficiency, effected by rapid transfer of the product of one reaction to its substrate-binding site on an adjacent domain. Such a mechanism also helps to protect labile, short-lived intermediates and prevents competing reactions with other enzymes that may utilize similar metabolites.

Chapter 8. The oxidative phase of the pentose phosphate pathway generates NADPH, page 258.

The synthesis of fatty acids begins with the attachment of acetyl CoA and malonyl CoA to the acyl carrier region

One molecule each of acetate and malonate is transferred from its CoA derivative to the acyl carrier (acyl carrier protein, ACP) region of the enzyme through the phosphopantetheine group in reactions catalysed by transacylase enzymes (Figure 9.14). The energy of the thioester bonds of acyl CoA is almost identical to that of acyl-ACP.

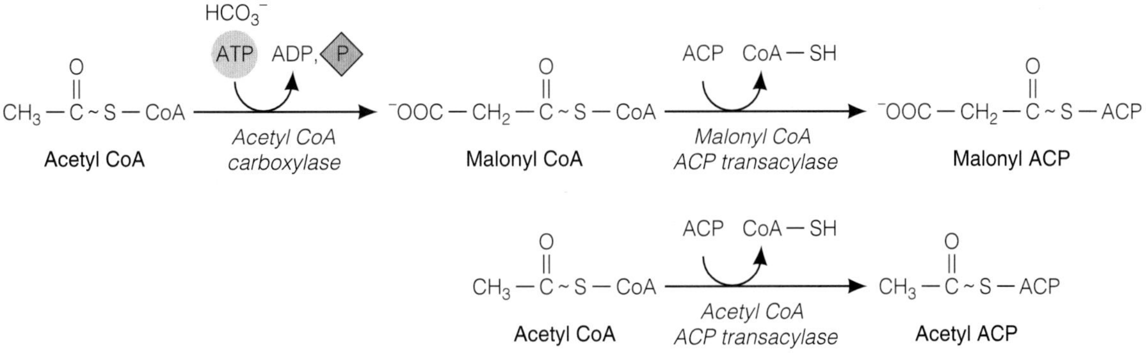

Figure 9.14 There are two transacylase enzymes, which transfer acetyl CoA or malonyl CoA to appropriate sites on the acyl carrier protein.

Carbon chain elongation begins with the condensation of these acetyl and malonyl groups and the elimination of CO_2 from the malonyl group. The product of the first reaction cycle is **butyryl ACP** (C_4), bound at the site originally occupied by the acetyl group and now available to enter a second cycle initiated by condensation with another molecule of malonyl-ACP. The sequence of four reactions is outlined in Figure 9.15. Both reduction reactions require NADPH.

This sequential addition of two-carbon units continues until palmitoyl (C_{16})-ACP is produced. Fatty acid synthase is unable to accommodate longer acyl chains, so no further condensation occurs. Hydrolysis yields palmitate (Figure 9.16) and frees the phosphopantothenic acid prosthetic group linked to the acyl carrier region in preparation for synthesis of the next fatty acid molecule.

The chemical similarities between the reactions of fatty acid synthesis and the β-oxidation cycles are shown in Figure 9.16.

The overall reaction for the synthesis of palmityl CoA from acetyl CoA can be summarized by the equation

$$8 \text{ Acetyl CoA} + 14\text{NADPH} + 14\text{H}^+ + 7\text{ATP} \rightarrow \text{Palmityl CoA} + 14\text{NADP}^+ + 7\text{ADP} + 7\text{Pi} + 7\text{CoA} + 6\text{H}_2\text{O}$$

Lipid metabolism

Figure 9.15 The four reactions of the first cycle of fatty acid synthesis, leading to the formation of the four-carbon intermediate butyryl ACP.

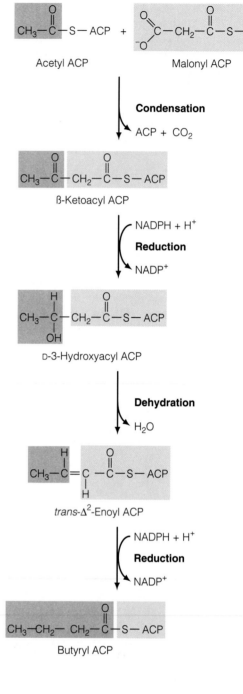

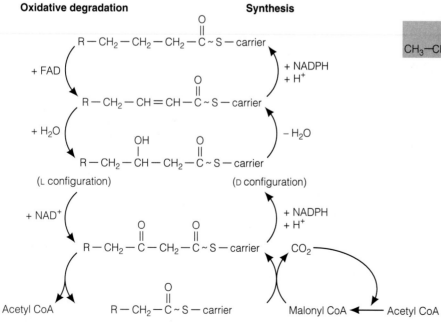

Figure 9.16 The four reactions of a cycle of fatty acid synthesis are very similar to a reversal of those involved in fatty acid oxidation. Note, however, that different coenzymes are used for the two directions of the dehydrogenase reactions.

Longer chain and unsaturated fatty acids are synthesized from palmityl CoA

In humans 60% of fatty acids have chain lengths of 18 carbon atoms or more, C_{20}, C_{22} and C_{24} fatty acids being common. The elongation process by which these larger molecules are produced occurs mainly in the endoplasmic reticulum. The process is remarkably similar to that producing shorter chain fatty acids in that the two-carbon units are donated by malonyl CoA and the same four-reaction sequence occurs. The major difference is that palmitate participates as a CoA rather than an ACP derivative.

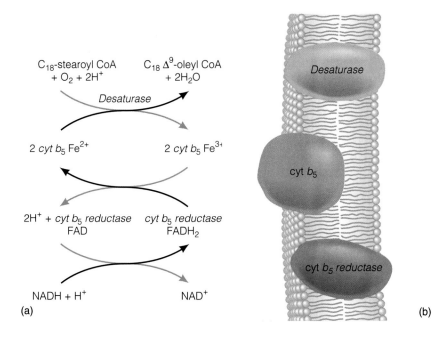

Figure 9.17 (a) The role of the cytochrome b_5 in desaturation of fatty acids. Desaturation requires both NADH and FADH. (b) A model indicating the probable arrangement of the three proteins in the membrane.

Introduction of double bonds into fatty acid carbon chains also occurs in the endoplasmic reticulum catalysed by a complex of three membrane-bound enzymes (Figure 9.17). In animal tissues the common unsaturated fatty acids are **oleic acid** (18:1 *cis*-Δ^9) and **palmitoleic acid** (16:1 *cis*-Δ^9). The **desaturase** enzymes that catalyse the production of these fatty acids are activated by insulin, although the precise mechanism by which this occurs is not clear. Mammals do not possess enzymes that would enable them to introduce double bonds further from the carboxyl group than Δ^9. This means that **linoleic** (18:2 *cis*-Δ^9,Δ^{12}) and **linolenic** acids (18:3 *cis*-Δ^9,Δ^{12},Δ^{15}), important precursors for the synthesis of the **eicosanoid hormones** (Chapter 11), are classified as **essential fatty acids** (Chapter 14) and must be provided in the diet. Sources of essential fatty acids and the synthesis of eicosanoids are dealt with in Chapter 11.

Chapter 11. Hormones derived from fatty acids, page 348.
Chapter 14. Some lipids are necessary in the diet to sustain health, page 423.

Synthesis of triacylglycerols and membrane lipids

Fatty acids can be metabolized as a source of energy, stored as triacylglycerols for future metabolism or incorporated into membrane phospholipids. The amount of triacylglycerol stored in an adult may be 15 kg or more, sufficient to support energy needs for several weeks. The balance between the uses of fatty acids depends on the competing needs for energy or for tissue growth or repair. Structural phospholipids may be synthesized from carbohydrate or protein precursors. The pathways for the synthesis of triacylglycerols and phospholipids have many features in common.

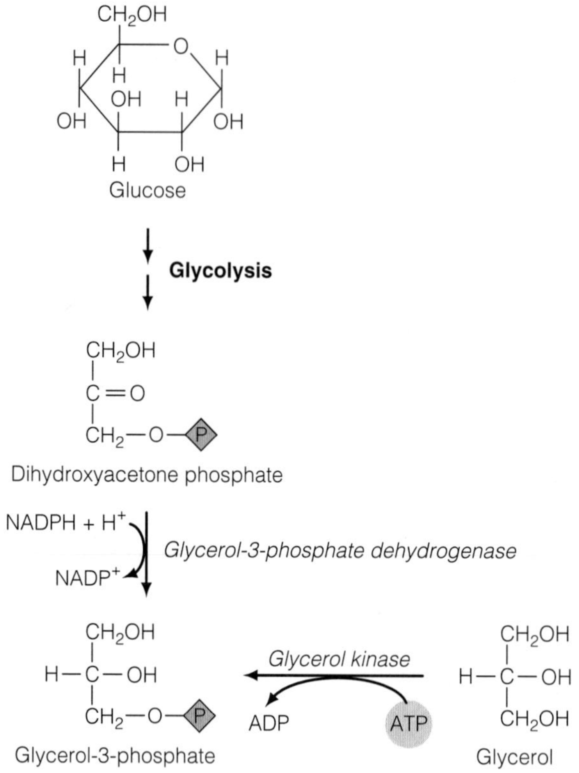

Figure 9.18 Glycerol-3-phosphate, the backbone on which triacylglycerols and phospholipids are assembled, can be synthesized from glucose, using part of the glycolytic pathway, or from glycerol in the liver.

Phosphatidic acid is a common intermediate in the production of phospholipids and triacylglycerols

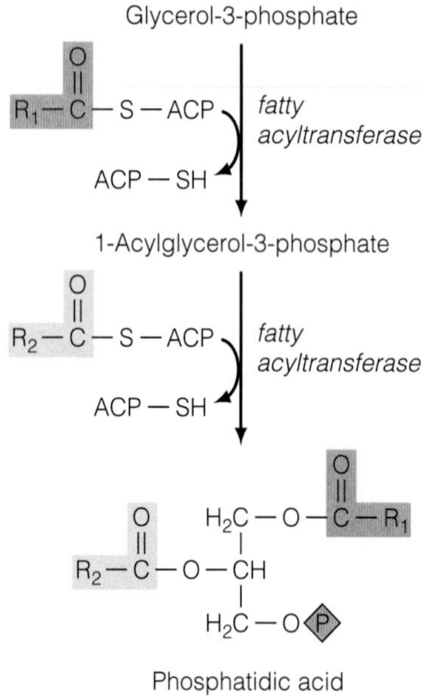

Glycerol forms the backbone of triacylglycerols and some phospholipids. Their biosynthesis requires glycerol-3-phosphate as the substrate, which can be synthesized directly from glycerol (but only in the liver as other tissues lack measurable **glycerol kinase** activity) or through glycolysis from glucose (dihydroxyacetone phosphate is reduced to glycerol-3-phosphate in a reaction catalysed by **glycerol-3-phosphate dehydrogenase**) (Figure 9.18).

Acyl groups are added to the free hydroxyl groups of glycerol through ester linkages formed by **fatty acyl transferases**, first at C-1 of the glycerol backbone to form a **lysophosphatidate** and then at C-2, forming a **phosphatidic acid (phosphatidate)** (Figure 9.19). Before this can happen the fatty acids must be converted to activated CoA derivatives in reactions catalysed by acyl CoA synthetase, thus:

$$\text{R.COO}^- + \text{ATP} + \text{CoASH} \xrightarrow[\textit{Acyl CoA synthetase}]{} \text{R.CO.SCoA} + \text{AMP} + \text{PPi}$$

There is a marked preference for the insertion of an unsaturated fatty acid at the C-2 position.

A specific phosphatase is responsible for removing the phosphate from phosphatidic acid so that a third fatty acid may be linked at the C-3 hydroxyl of the diacylglycerol, completing synthesis of the triacylglycerol.

Figure 9.19 Synthesis of phosphatidic acid.

Box 9.5   The active component of lung surfactant is a phospholipid

Dipalmitoyl phosphatidylcholine is produced in large amounts in the epithelial cells of lung alveoli and is secreted to the exterior surface of the alveoli. This phospholipid, which is unusual in having a saturated fatty acid at C-2 of the glycerol backbone, combines with sphingomyelins and proteins to form **pulmonary surfactant**. This substance, by reducing the surface tension, facilitates the opening of the alveoli and aids the exchange of gases. Synthesis of the dipalmitoyl phospholipid begins before birth, ready for the time when the neonate must begin breathing.

In the epithelial cells of the lungs of preterm infants the synthesis of this compound is not complete, and they may suffer **respiratory distress syndrome** as the lungs fail to inflate. This is one reason for maintaining preterm infants in an atmosphere of high oxygen tension.

Phospholipids are synthesized by adding 'head groups' to phosphatidic acid

Phospholipids are synthesized on the surface of the smooth endoplasmic reticulum through reactions that add a 'head group' to the 3-position of diacylglycerol through a phosphodiester bond. The head group is most commonly serine, ethanolamine, choline or inositol. There are two main routes for the synthesis of these compounds, both involving **CDP** (**cytidine diphosphate**) activation of one of the substrates. The subsequent displacement of CMP provides the energy for the head group to condense to phosphatidic acid. Uridine nucleotides play a role in glycogen synthesis (see Chapter 8) similar to that of CTP in phospholipid synthesis. In each case an activated intermediate is formed (UDP-glucose or CDP-diacylglycerol) from a phosphorylated intermediate. The activation process provides the energy necessary to drive glucose addition to the 4-hydroxy termini of glycogen or phosphatidic acid to the hydroxyl groups of serine or inositol.

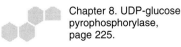

Chapter 8. UDP-glucose pyrophosphorylase, page 225.

The route of phospholipid assembly is specific to the head group being added

In one route diacylglycerol is activated by the addition of CDP to the C-3 position, and the head group is then added, displacing CMP. This is the route used by mammalian cells for the synthesis of the acidic phospholipids **phosphatidylglycerol**, **phosphatidylinositol** and **cardiolipin** (Figures 9.20 and 9.21).

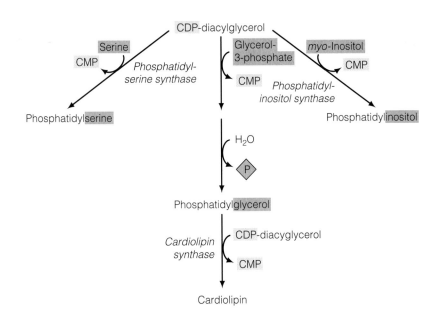

Figure 9.20 Outline biosynthesis of acidic phospholipids

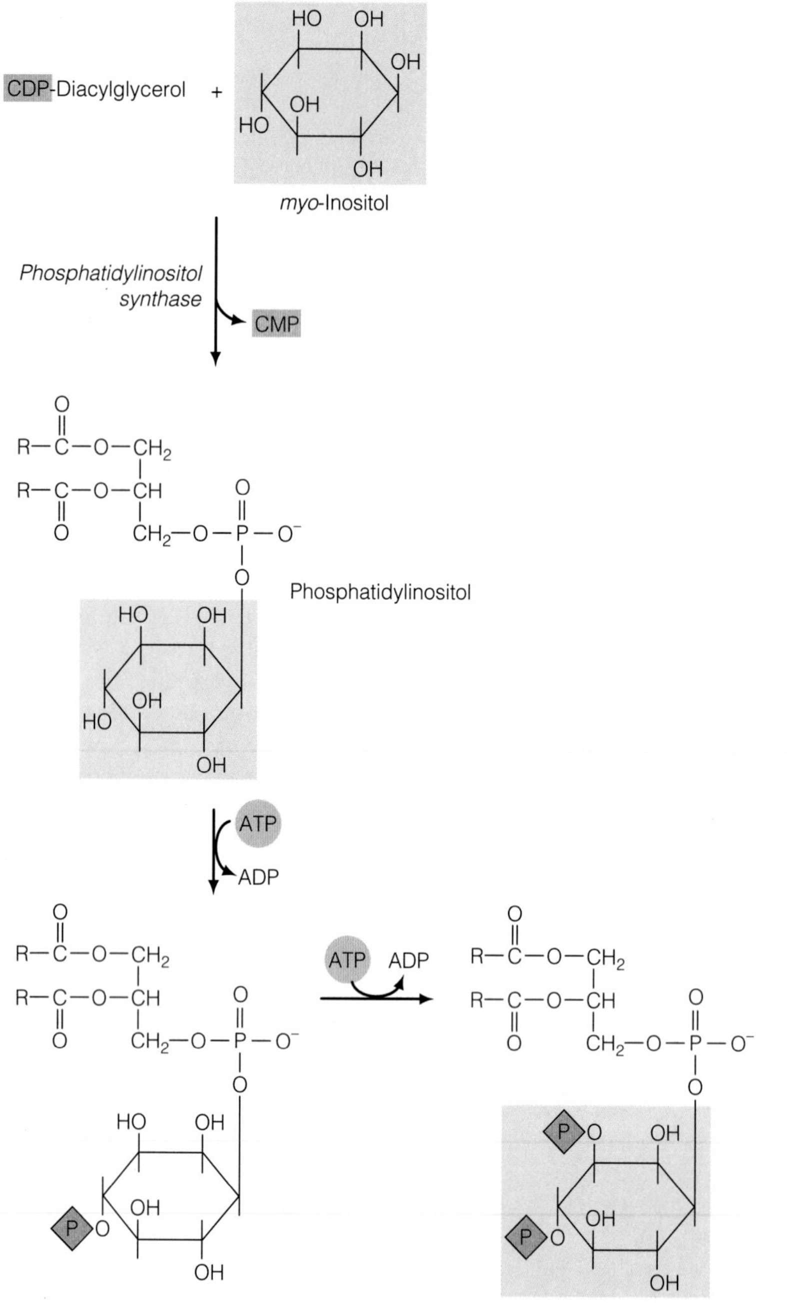

Figure 9.21 Synthesis of phosphatidylinositols.

The second route involves activation of the head group to form the CDP-derivative before condensation with diacylglycerol. In mammals all nitrogen-containing phospholipids, including **phosphatidylserine, phosphatidylethanolamine** and **phosphatidylcholine (lecithin)** are synthesized through this route (Figure 9.22).

Some phospholipids are produced by **head group exchange** reactions, a typical example of which is the exchange of ethanolamine by serine in a reaction catalysed by **phosphatidylethanolamine serinetransferase**.

Phosphatidylinositol is an important constituent of biological membranes and its derivative phosphatidylinositol-4,5-bisphosphate, formed by the action of specific kinases, occupies a key role in the transmission of hormone-mediated stimuli across the plasma membrane (Chapter 11).

Chapter 11. Hormonal responses mediated through inositol phosphates and calcium ions, page 339.

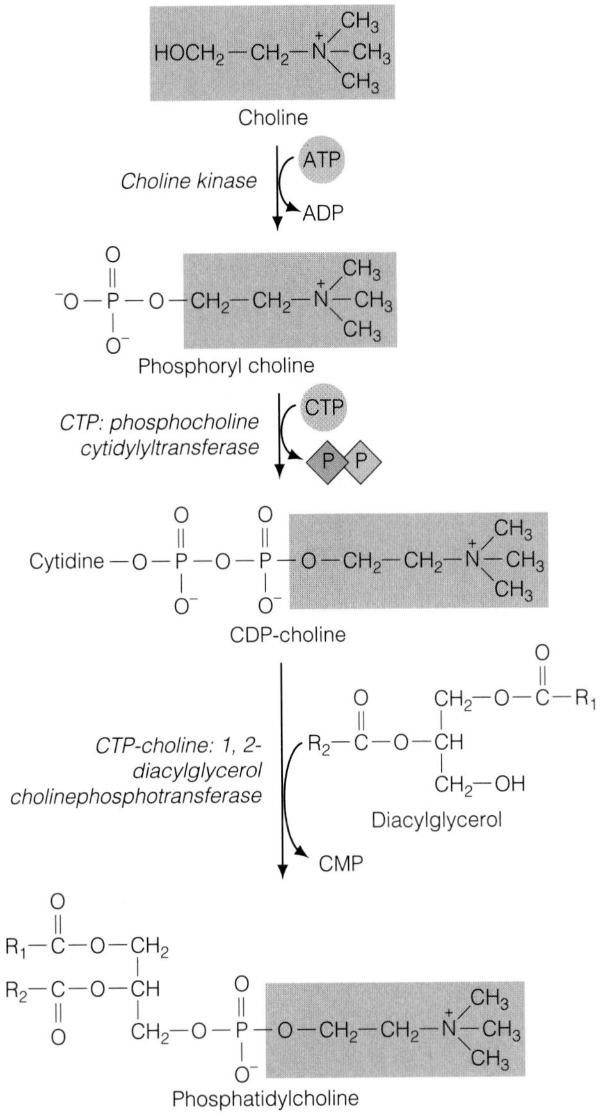

$$\text{HOCH}_2-\text{CH}_2-\overset{+}{\text{N}}\overset{\text{CH}_3}{\underset{\text{CH}_3}{\big|}}\text{CH}_3$$

Choline

Choline kinase — ATP → ADP

Phosphoryl choline

CTP: phosphocholine cytidylyltransferase — CTP → P~P

CDP-choline

CTP-choline: 1, 2-diacylglycerol cholinephosphotransferase — Diacylglycerol → CMP

Phosphatidylcholine

Figure 9.22 Synthesis of phosphatidylcholine from choline. A corresponding pathway converts ethanolamine to phosphatidyloethanolamine.

Plasmalogens are ether-linked fatty alcohol derivatives

Plasmalogens make up 50% of the phospholipid content of heart muscle plasma membranes, and are present at high concentrations in the membranes of other electrically active tissues. Plasmalogens are synthesized by displacement of a fatty acyl group, generally from the C-1 of the glycerol backbone, by a long-chain fatty alcohol through an ether linkage. The characteristic double bond of the plasmalogen is introduced through reactions similar to those involved in the desaturation of fatty acids. In the heart, the C-2 of the glycerol backbone of plasmalogens is usually substituted by arachidonic acid, providing a huge reservoir of this fatty acid that may be mobilized by phospholipase to form the substrate for prostaglandin synthesis (Chapter 11). Platelet-activating factor is also a plasmalogen (Chapter 12).

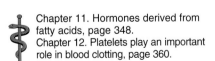

Chapter 11. Hormones derived from fatty acids, page 348.
Chapter 12. Platelets play an important role in blood clotting, page 360.

Ceramide is the basic structural unit of sphingomyelin

The structure of sphingomyelin is based on **sphingosine**, which is derived from palmitoyl CoA and serine in a series of reactions involving a decarboxylation step with pyridoxal phosphate as cofactor and two reduction stages with NADPH and FADH$_2$ as cofactors. In all sphingolipids the amino group of sphingosine is substituted with a long-chain fatty acyl chain, yielding **ceramide**. The terminal hydroxyl

> ### Box 9.6   Snake venoms contain phospholipase A_2
>
> One of the enzymes injected in some snake venoms is a phospholipase A_2 that releases the fatty acid from the C-2 of phosphatidylcholine (lecithin), producing lysolecithin, which has strong detergent properties. It is responsible for lysis of cell membranes, particularly of erythrocytes.

is also modified (with phosphoryl choline in sphingomyelin). In the **glycolipids**, **cerebrosides** and **gangliosides**, carbohydrate residues are linked to the same terminal hydroxyl group. The structures and probable membrane functions of these molecules are discussed in Chapters 2 and 3.

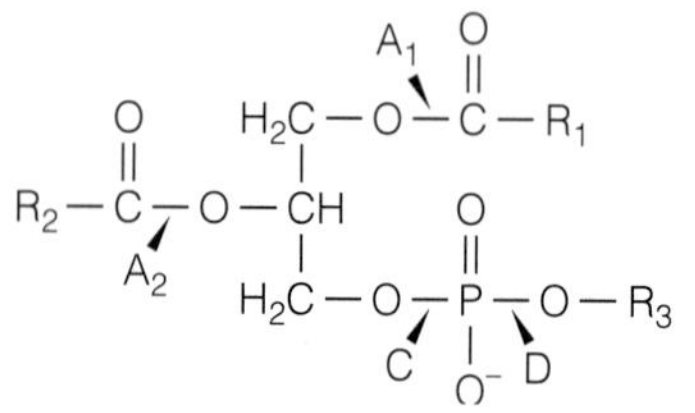

Hydrolysis of phospholipids is catalysed by phospholipases with differing specificities for particular bonds

The turnover of membrane phospholipids is an active process catalysed by intracellular phospholipases which are classified according to their preferences for hydrolysing specific bonds in phospholipid molecules. Phospholipases are also intimately associated with the production of the second messengers **diacylglycerol** and **inositol 1,4,5 trisphosphate** from phosphatidyl inositol, and with the release of arachidonic acid for eicosanoid hormone production (Chapter 11) (Figure 9.23).

Intracellular phospholipases should not be confused with the array of phospholipases A_1, A_2, C and D present in digestive juices, which catalyse the degradation of dietary phospholipids to fatty acids, glycerol, phosphate and the appropriate head group before they are absorbed from the gut (see Chapter 14).

Figure 9.23 The specific bonds hydrolysed by phospholipases A_1, A_2, C and D are indicated by arrows.

Chapter 11. Hormones derived from fatty acids, page 348.
Chapter 14. The hydrophobic character of lipids complicates their digestion and absorption, page 418.

Cholesterol: a major structural and functional lipid

Cholesterol is an important lipid in mammals. It participates in a diverse range of processes

- as a component of the structure of plasma membranes;
- as the precursor of steroid hormones;
- as the precursor of bile salts, which have a key role in the digestion and absorption of fats;
- as the precursor of vitamin D, which regulates calcium and phosphate metabolism; and
- as a component of plasma lipoproteins.

Cholesterol is synthesized in the liver from acetyl CoA

Cholesterol is a four-ring, 27-carbon compound, synthesized from acetyl CoA. Most cholesterol synthesis in animals occurs in the liver. Although the liver is capable of making all the cholesterol needed by the body most diets also contain cholesterol. Only a small fraction of the cholesterol synthesized in the liver is required for liver membrane structure; most leaves the liver as **bile acids** or as **cholesterol esters**. Cholesterol esters are synthesized when a fatty acid is transferred from acyl CoA or from lecithin to the C-3 hydroxy group of cholesterol and are stored in the liver for release into the circulation. Bile acids are secreted from the liver into the digestive system as **bile salts**, conjugated with glycine or taurine (Chapter 14).

Chapter 14. Bile salts are constantly recycled through the enterohepatic circulation, page 419.

The five-carbon compound isopentenyl pyrophosphate is a key intermediate in the production of cholesterol

In the initial phase of cholesterol formation, **isopentenyl pyrophosphate** is synthesized from three molecules of acetyl CoA. The first step is the condensation of acetyl CoA and acetoacetate to form **3-hydroxy, 3-methylglutaryl CoA** (also an intermediate in the formation of ketone bodies). In the cytoplasm this intermediate is reduced to mevalonic acid in a reaction which is the committed step in cholesterol biosynthesis and is catalysed by the regulatory enzyme **3-hydroxy, 3-methylglutaryl CoA reductase**. Isopentenyl pyrophosphate (the activated isoprene unit) is formed by three successive ATP-dependent phosphorylations, a decarboxylation and a dehydration reaction. The activity of the reductase is decreased by glucagon and increased by insulin. This pathway is presented in outline in Figure 9.24.

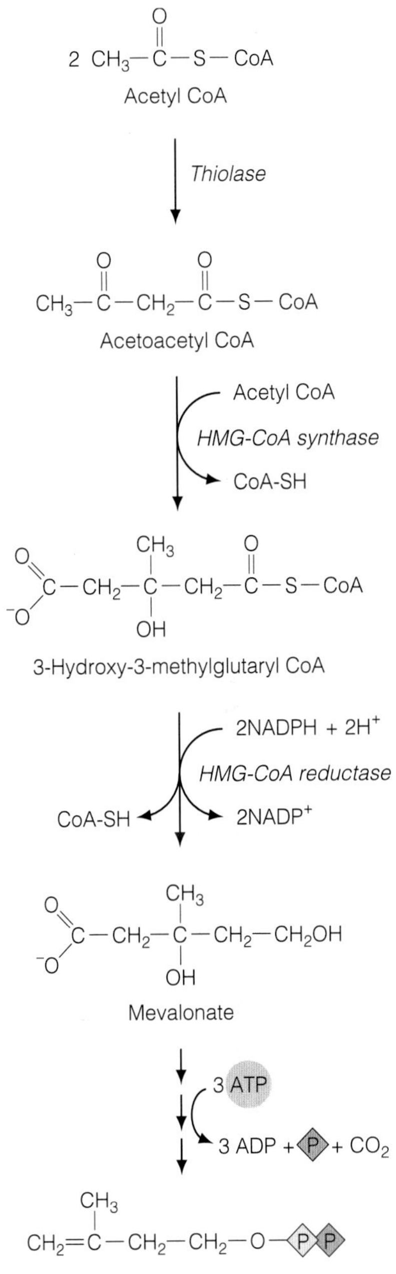

Isopentenyl pyrophosphate units are condensed to form squalene

Isopentenyl pyrophosphate and its isomer **dimethylallyl pyrophosphate** undergo head-to-tail, NADPH-dependent reductive condensation reactions to generate first a C_{10} intermediate (**geranyl pyrophosphate**) and then the C_{15} intermediate **farnesyl pyrophosphate**. Pyrophosphate groups are eliminated in these reactions. Two molecules of farnesyl pyrophosphate are condensed in a head-to-head reaction to form the linear C_{30} intermediate **squalene** (Figure 9.25).

Oxygen is essential for the formation of the ring structure of cholesterol

Conversion of squalene to cholesterol takes place through a sequence of reactions requiring molecular oxygen and NADPH and involving the intermediate **squalene epoxide**. The epoxide is converted to **lanosterol**, an intermediate which has the familiar four rings of cholesterol. This cyclization process involves the concerted movement of electrons through four double bonds and the migration of two methyl groups.

Finally, lanosterol is modified by the removal of three methyl groups, the NADPH-linked reduction of a double bond and the migration of another to yield the final product, cholesterol (Figure 9.26).

Transport of fatty acids, phospholipids and triacylglycerols

Fatty acids in the circulation are derived from triacylglycerol stores of adipose tissues; triacylglycerols are absorbed from cells lining the digestive tract. Fats are not soluble in the aqueous environment of the bloodstream, and must be complexed to proteins to facilitate their transport. Fatty acids, as the sodium salts, are transported bound to albumin and triacylglycerols are transported as lipoproteins, which are aggregates of lipid and protein.

Fatty acids are transported bound to albumin

Albumin is the most abundant protein in plasma, representing 50% of total plasma protein; (Chapter 12), and can bind up to ten molecules of fatty acid, although

Figure 9.24 Synthesis of isopentenyl pyrophosphate from acetyl CoA.

Chapter 12. Albumin is important in osmoregulation and in binding small molecules, page 359.

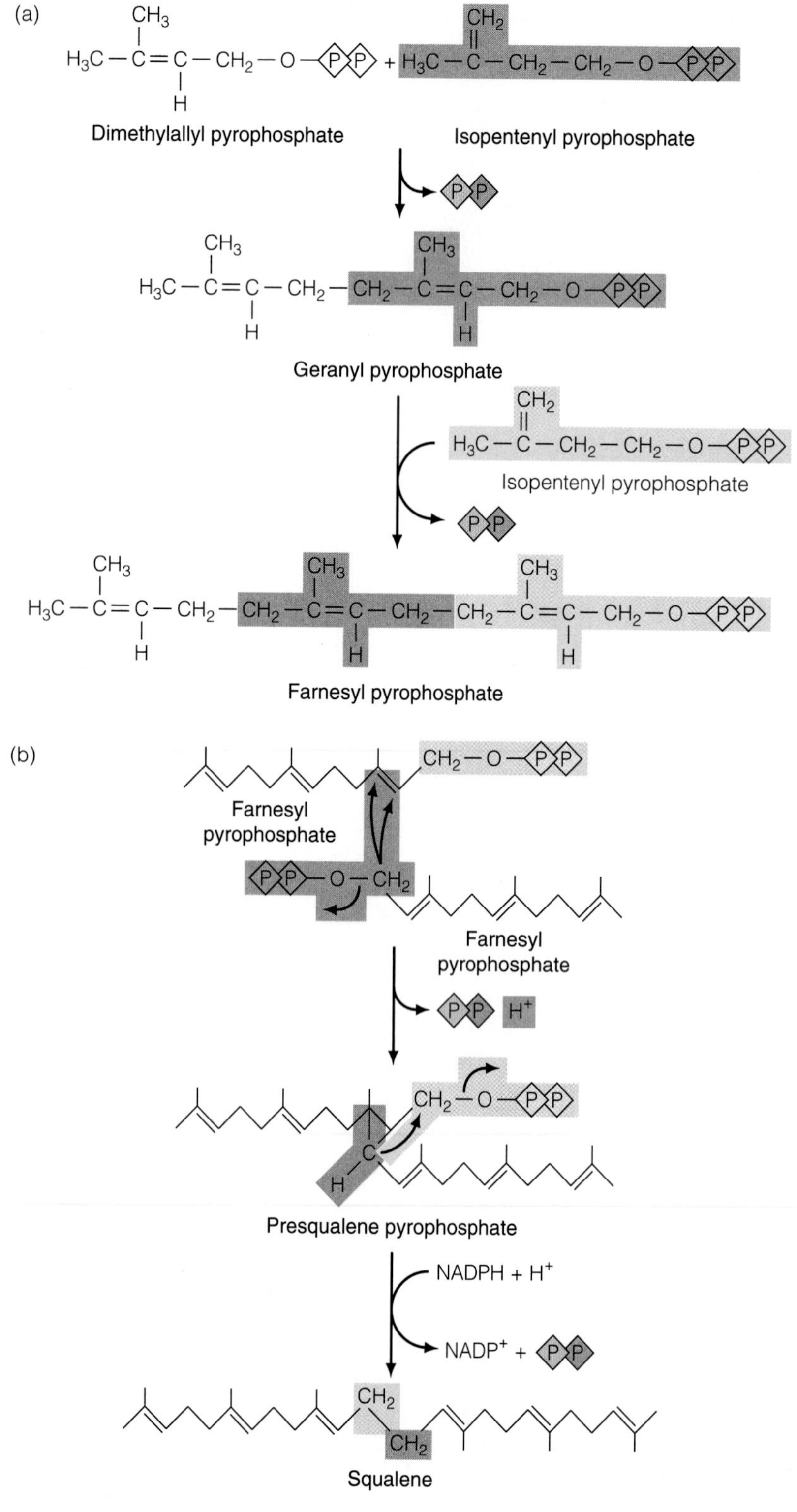

(a)

Dimethylallyl pyrophosphate   Isopentenyl pyrophosphate

Geranyl pyrophosphate

Isopentenyl pyrophosphate

Farnesyl pyrophosphate

(b)

Farnesyl
pyrophosphate

Farnesyl
pyrophosphate

PP H^+

Presqualene pyrophosphate

NADPH + H^+

NADP$^+$ + PP

Squalene

Figure 9.25 Synthesis of squalene, a C-30 precursor of cholesterol formed by condensation of compounds derived from acetyl CoA.

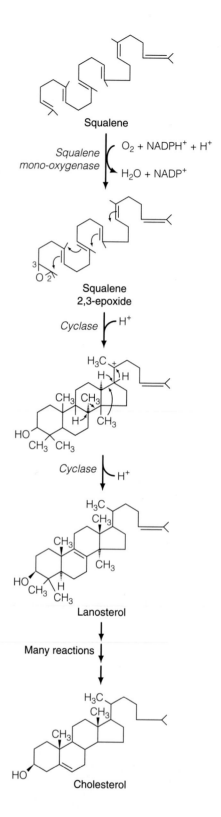

Squalene

Squalene
mono-oxygenase

O_2 + NADPH$^+$ + H^+

H_2O + NADP$^+$

Squalene
2,3-epoxide

Cyclase H^+

Cyclase H^+

Lanosterol

Many reactions

Cholesterol

Figure 9.26 Conversion of squalene to form cholesterol. Closure of the four rings involves a series of electron shifts. The first steroid intermediate formed is lanosterol, and after a further series of reactions cholesterol is produced.

Box 9.7 Albumin is not essential for transport of fatty acids

Surprisingly, individuals with **analbuminaemia** (who have albumin levels only 0.1–1.0% of normal) suffer no ill effects, indicating that albumin is not essential for transport of fatty acids. Such individuals do have elevated triacylglycerol levels, because usage of lipoproteins to carry esterified forms of fatty acid as triacylglycerols or phospholipids is increased. As with many biological activities, fatty acid transport appears to have 'back-up' systems that are able to undertake similar functions, an important safeguard in minimizing the effects of deleterious mutations (see Chapter 12).

Chapter 12. Binding of small molecules, page 359.

normally only one or two are bound. Saturation of all albumin-binding sites occurs when the total circulating concentration of fatty acids reaches its normal upper limit of about 2 mM. Albumin appears to have two or three primary binding sites that are selective for longer chain fatty acids, and seems to have a preference for monounsaturated fatty acids. Thus oleate binds with higher affinity than palmitate. By binding to albumin the fatty acid molecules are prevented from exerting detergent effects that would disrupt the structure of circulating proteins and the lipid bilayers of cell membranes.

Fatty acids enter cells by simple diffusion down a concentration gradient, so that the rate of uptake into cells depends on the relative concentrations of free fatty acids across the cell membrane. Within cells they bind to proteins with high affinity for fatty acids. The concentrations of free fatty acids within cells are thus kept very low and diffusion of fatty acids into the cell is sustained.

Distinct classes of lipoproteins are responsible for the transport of triacylglycerols, phospholipids and cholesterol

Phospholipids, triacylglycerols and cholesterol are transported around the body as plasma lipoproteins, which are classified according to their differing densities. Lipoproteins are aggregates of specific carrier proteins (**apolipoproteins**) with triacylglycerols, phospholipids, cholesterol and cholesterol esters. The proportion of lipid to protein in these complexes determines their density and they are presented here in increasing order of density:

1. Chylomicrons and chylomicron remnants;
2. Very-low density lipoproteins (VLDL);
3. Intermediate-density lipoproteins (IDL);

Table 9.3 General composition of major human lipoprotein classes

Lipoprotein article	Apoproteins present	Composition (% by weight)			
		Protein	Free and esterified cholesterol	Phospholipids	Triacylglycerols
Chylomicrons	B48, Apo C2, C3 and E	1–2	3–7	3–8	90–95
VLDL	B100, Apo C1, C2, C3 and E	10.4	19.7	15.2	53.4
IDL	B100, Apo E	17.8	11.0	21.7	31.4
LDL	B100	25.0	50.5	20.9	3.5
HDL	Apo A1, A2, C1, C2, C3, D, E	48.8	22.1	27.6	1.8

The bulk of cholesterol in all classes of lipoprotein particle is present as cholesterol esters with free cholesterol contents of 3.9–8.6% of total weight. VLDL, very low density lipoproteins; IDL, intermediate density lipoproteins; LDL, low-density lipoproteins; HDL, high-density lipoproteins. Some classifications include IDL and LDL as a single category, with a density range of 1.006–1.063.
Data are adapted from Vance, D. E. and Vance, J. E. (ed), *Biochemistry of Lipids and Membranes*. Menlo Park, CA: Benjamin Cummings.

4. Low-density lipoprotein notable for its high cholesterol content (LDL);
5. High-density lipoproteins (HDL).

Each of these is a discrete class of complex, characterized by different protein (or apolipoprotein) and lipid compositions (see Table 9.3).

Lipoproteins have a dual function:

1. To maintain the solubility of lipids in plasma;
2. To control the movement of particular lipids into and out of appropriate target tissues.

There are several major types of apolipoprotein: A-1, A-2, A-4, B-48, B-100, C and E. Specific signals incorporated into the sequences of individual apolipoproteins are responsible for regulating the import and export of lipids by recognizing specific receptors on the external surfaces of target cells. Overall, the lipoprotein particle is composed of a hydrophobic core of lipids surrounded by a layer of polar lipids with the protein component on the outside (Figure 9.27).

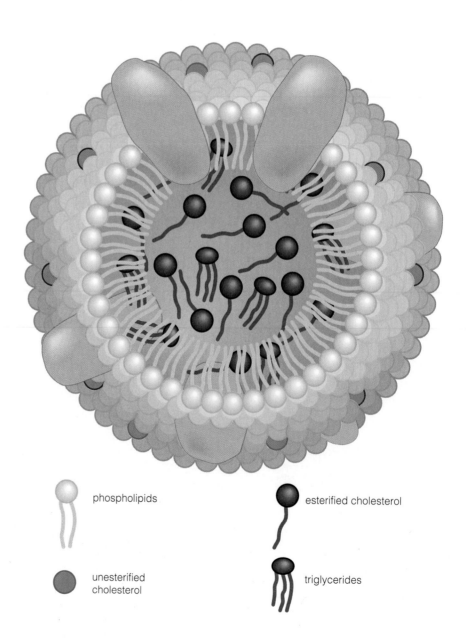

phospholipids

esterified cholesterol

unesterified cholesterol

triglycerides

Figure 9.27 Generalized structure of a lipoprotein particle. The major classes of proteins present in various types of lipoprotein particles and their overall compositions are summarized in Table 9.3. Protein is shown in brown.

Chylomicrons transport triacylglycerols that have been absorbed from the gut

Chylomicrons are large triacylglycerol-rich particles (100–500 nm in diameter) containing less than 2% protein by weight. They are responsible for transporting dietary lipids (including small amounts of phospholipid, cholesterol and cholesterol esters) from the intestine to adipose tissue and the liver (Chapter 14). Triacylglycerols in chylomicrons are rapidly hydrolysed by the lipoprotein lipase on the surface of capillary endothelial cells of peripheral tissues, especially adipose tissue (Chapter 13), producing cholesterol-rich particles (**chylomicron remnants**), which are subsequently internalized in liver cells by receptor-mediated endocytosis (Chapter 3); see also Figure 9.29.

Chapter 14. Triacylglycerols hydrolysed in the gut are reassembled within epithelial cells, page 421.
Chapter 13. Triacylglycerol from the diet is directed towards adipose tissue, page 381.

Chapter 3. The endoplasmic reticulum and Golgi complex are membrane-bound compartments, page 46.

Triacylglycerols synthesized within the body are transported in other lipoprotein complexes

Triacylglycerols synthesized *in situ*, principally by the liver, are incorporated into **VLDL** for export to other tissues. Lipases in the endothelial cells of peripheral blood vessels promote release of triacylglycerols from VLDL, converting them to

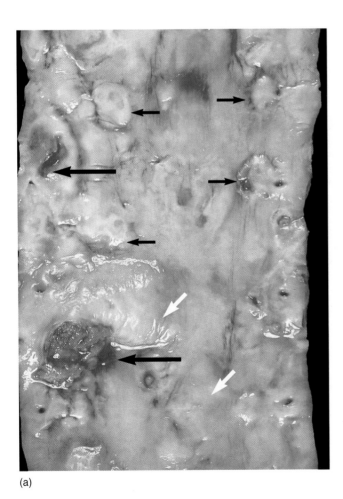

(a)

(b)

Figure 9.28 (a) Atheromatous human aorta, opened to show the endothelial lining. Normal tissue is smooth and pinkish-grey. The lipid in the atheromatous plaques appears yellow (small arrows). Some plaques are ulcerated (large arrows) and appear rough because of thrombus on the surface. Other plaques have thick fibrous caps, which appear white (white arrow). (b) A histological section through one of the fibrous atheromatous plaques seen in (a). Normal structure is replaced by lipid (stained red). The pink tissue is densely collagenous fibrous tissue, which surrounds the lipid deposits and obstructs the lumen (L) of the vessel. Photographs courtesy of Dr G.B.M. Lindop, University of Glasgow Department of Histopathology, Western Infirmary, Glasgow.

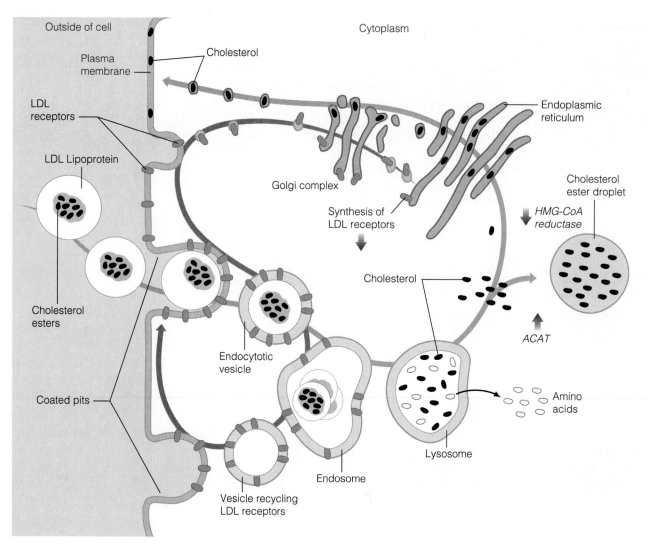

Figure 9.29 Detailed overview of LDL receptor participation in cholesterol uptake and metabolism. On interaction with LDL particles at the cell surface the receptor is internalized by receptor-mediated endocytosis via clathrin-coated vesicles. These are converted to endosomal vesicles where low pH promotes receptor–LDL dissociation. The LDL receptor is recycled to the plasma membrane; LDL apoproteins are degraded in lysosomes.

IDL in the process. Particles of IDL (which are rich in cholesterol ester) have two separate fates: they can be taken up and metabolized by liver cells or modified further in the plasma to form **LDL**, the major carrier of cholesterol in the bloodstream. These lipoprotein particles (roughly 20 nm in diameter) contain a core of over 1000 cholesterol ester molecules, linoleate being the most common fatty acyl chain. Outwith the core lies a shell of phospholipid and unesterified cholesterol in addition to a single molecule of apoprotein B-100, an extremely large protein with a M_r of 514 000. Low-density lipoproteins are responsible for cholesterol transport to peripheral sites; they also have an important role in the regulation of cholesterol synthesis in peripheral tissues (as described in the next section).

High-density lipoproteins have a distinctive role in scavenging cholesterol released from dying cells or from membrane turnover. Free cholesterol is esterified by **lecithin: cholesterol acyltransferase (LCAT)**, an enzyme secreted into the bloodstream by the liver.

Cholesterol biosynthesis must be regulated to accommodate its many uses

Cholesterol is an essential precursor of many important molecules and is an integral component of plasma membranes. In the liver, the principal site of synthesis, dietary cholesterol reduces the activity and also the synthesis of 3-hydroxy,

Box 9.8 Defects in the LDL receptor are the cause of hypercholesterolaemia

The physiological importance of the LDL receptor is evident from analysis of families who suffer from **familial hypercholesterolaemia**, a single-gene defect marked by high serum concentrations of cholesterol. In most cases, the source of the defect has been traced to mutations in the LDL receptor which is encoded by a gene located on chromosome 2. In normal individuals, cholesterol concentration in the circulation is 5.5 mM or less but homozygous individuals (who possess two copies of the defective gene) have levels up to three to four times above normal. Cholesterol deposits appear in the skin, tendons and other body tissues – although the harmful accumulation of cholesterol in arterial plaques is the primary cause of heart disease, strokes and peripheral vascular disease. These effects are very severe in homozygotes, who generally die of coronary artery disease in childhood. Heterozygotes have milder symptoms and a more variable prognosis.

Severely affected patients have few, if any, receptors for LDL and entry of LDL into liver and peripheral tissues is greatly impaired, leading to increased circulating levels of LDL and IDL. Several classes of LDL receptor mutations have been identified, reflecting the multidomain nature of the protein (Figure 9.30):

- patients may lack the ability to synthesize receptor;

- receptors may be synthesized, but not delivered to the plasma membrane if they do not fold correctly or lack the appropriate signals for intracellular transport;

- mutations in the LDL-binding domain result in failure of receptor to recognize LDL;

- defects in the C-terminal domain prevent uptake of LDL into cells, due to inability of the receptor to participate in receptor-mediated endocytosis.

The only effective treatment for homozygous individuals with familial hypercholesterolaemia is liver transplantation. In heterozygotes, it is more appropriate to use approaches designed to lower the elevated cholesterol levels (which are about twice the normal maximum). The expression of LDL receptors on the cell surface reflects the need for cholesterol: thus, inhibition of cholesterol biosynthesis by mevinolin, a potent inhibitor of the committed step in this pathway, increases the numbers of receptors at the cell surface. In addition, oral administration of certain positively charged polymers inhibits reabsorption of bile salts required for cholesterol uptake from the diet. The reduced levels of cholesterol entering the body again promote an increase in receptor numbers.

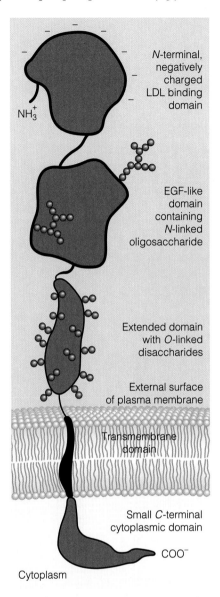

Figure 9.30 Schematic representation of the transmembrane distribution and domain organization of the LDL receptor.

3-methylglutaryl CoA reductase, the enzyme catalysing the committed step in cholesterol production.

Elevated concentrations of cholesterol in the bloodstream are implicated in the formation of atherosclerotic plaques on the walls of the arteries leading to constriction of blood vessels (**atherosclerosis**) and the restriction of blood flow (Figure 9.28). This is a major cause of premature death in the developed world.

Most tissues rely on plasma cholesterol being delivered to them via LDL. Specific receptors on the surface of target cells recognize the B-100 protein of LDL. Binding of this protein initiates **receptor-mediated endocytosis** (Figure 9.30), in which the receptor–LDL complex, now localized in special areas of the plasma membrane (**coated pits**) is internalized by invagination of the plasma membrane. The **clathrin-coated vesicles** thus formed are converted into **endosomes**, specific vesicles no longer containing clathrin. Endocytotic vesicles fuse with lysosomes, the degradative enzymes of which digest the B-100 protein and hydrolyse the cholesterol esters. The LDL receptors are usually recycled to the plasma membrane within 10 minutes, so the same receptor can mediate the uptake of many LDL particles.

Released cholesterol may be used immediately for membrane biosynthesis or may be reesterified by **acyl CoA: cholesterol acyltransferase**, which is activated by the presence of free cholesterol. Stores of cholesterol esters in the liver contain mainly oleate and palmitoleate, in contrast to those in LDL, in which linoleate is the major component.

The LDL receptor is a multifunctional transmembrane polypeptide

Much information regarding the properties of this transmembrane protein has been gleaned from analysis of its primary sequence following cloning and sequencing of the gene for human LDL receptor. This large (M_r 115 000) polypeptide has five distinct domains (Figure 9.30). The mechanism of regulation of gene expression of LDL receptor is described in Box 5.5.

- Its *N*-terminal region is composed of seven tandem repeats of a cysteine-rich sequence. This operates as an LDL-binding domain in which a cluster of negatively charged residues interact specifically with a positive site on the B-100 lipoprotein.
- The adjacent domain in the LDL receptor contains two *N*-linked oligosaccharide chains.
- The third domain, rich in serine and threonine, contains a series of *O*-linked oligosaccharides. These oligosaccharides are thought to enable the LDL receptor to maintain an extended conformation away from the membrane so that its amino terminal region is not hindered from interacting with the large LDL particles.
- A small hydrophobic transmembrane domain of 22 amino acids keeps the fifth, *C*-terminal, domain within the cytoplasmic compartment, where it regulates the formation of endocytotic vesicles.

CHAPTER 10

Metabolism of nitrogen-containing molecules

Most of the nitrogen in cells is present in proteins, but other important nitrogenous compounds occur in smaller amounts. These include nucleic acids, nucleotide coenzymes, hormones, neurotransmitters and other small regulatory molecules and the porphyrin structure in proteins such as the cytochromes and haemoglobin. The major emphasis of nitrogen metabolism is on regulating the amount of nitrogen in the body and ensuring that it is in an appropriate form to be used for synthesis of nitrogen-containing molecules.

The body absorbs nitrogen mostly in the form of amino acids from dietary protein. Nitrogen in excess of requirements is released as ammonium ions. To prevent accumulation of ammonia, which is toxic at high concentrations, the urea cycle converts ammonium ions to urea, a soluble, nontoxic, readily excreted compound.

This chapter briefly considers the metabolic fate of the carbon skeletons released when amino groups are removed from amino acids, and discusses aspects of the biosynthesis and functional roles of amino acids and other important nitrogenous compounds. Nucleotide biosynthesis, through both *de novo* and salvage pathways, is described, and its importance for nucleic acid synthesis and cell division discussed. Purine metabolism leads to the production of uric acid, a metabolite of clinical importance.

Finally the synthesis and degradation of haem are described and some consequences of abnormal haem metabolism are considered.

Nitrogen is an essential constituent of a diverse range of biologically important molecules

Most of the nitrogen in mammalian cells is found in the **amino acids**, the constituents of proteins, or in the **nucleotide bases** of nucleic acids and nucleotide coenzymes. We also need small amounts of other vital nitrogen-containing molecules such as **neurotransmitters** and the **haem nucleus** of proteins such as haemoglobin. Nucleotides and **porphyrins**, which are found in the cells of all organisms, can not be taken up intact from the bloodstream. Therefore, all cells of the body must be able to synthesize them and must possess the appropriate enzymes.

Because we tend to consume more protein than we need, a major emphasis of amino acid metabolism is on the disposal of excess absorbed nitrogen. In well fed individuals most of the amino acids required for protein synthesis are provided in the diet. In those whose protein consumption is less than adequate, the emphasis of protein metabolism will be on conserving enough of the dietary amino acids and amino acids released when tissue proteins are degraded to preserve and restore tissue structure through protein synthesis. In this situation the ability to transfer amino groups from one carbon skeleton to another enables cells to alter the proportions of different amino acids, to meet the demands of protein synthesis.

Catabolism of amino acids – the removal of amino groups

Amino groups must be removed from amino acids if either of the following occurs.

1. Amino acids are ingested in excess of amounts required for synthesis of proteins or other specialized nitrogenous compounds. In this situation the carbon skeletons can be used as an energy source as many are, or are readily converted to, intermediates of the energy-producing pathways.

2. The diet contains insufficient nutrient so that body proteins are degraded, releasing the carbon skeletons of amino acids as an energy source.

Amino groups that are not required for the synthesis of amino acids or other nitrogenous molecules must be excreted because ammonia is toxic above certain concentrations in the bloodstream.

Amino groups are transferred to molecules that can be directly deaminated

Few amino acids are deaminated directly in mammalian cells. The first steps in the catabolic utilization of amino acids therefore involves transferring amino groups to one of the few molecules from which they can be liberated. Only glutamate, glutamine and the amino acids with -OH side chains (threonine and serine) are able to liberate amino groups as ammonium ions. For most amino acids a specific enzyme, an **aminotransferase**, catalyses the transfer of the α amino group to α-ketoglutarate and generates glutamate, which can be deaminated.

Some amino groups are derived from glutamine. Excretion of ammonium ions is an important mechanism for controlling body pH by reducing the hydrogen ion concentration as ammonia is protonated and the ammonium ions excreted in the urine (Chapter 13):

$$NH_3 + H^+ \rightleftharpoons NH_4^+$$

Chapter 13. The kidney regulates acid–base balance by altering the amount of acid or bicarbonate it excretes, page 391.

Most amino groups liberated in tissues are converted to **urea** in the liver. The synthesis of urea provides a mechanism for neutralizing the charge on amino groups and converting toxic ammonium ions to a water-soluble nontoxic substance. This facilitates the rapid and safe excretion of excess nitrogen in urine.

In patients with liver disease such as cirrhosis, which can be caused by alcoholism, hepatitis or biliary obstruction, the capacity of the liver to take up and detoxify ammonia is reduced. **Hyperammonaemia**, an abnormally high concentration of circulating ammonia, results from the release of ammonia from tissues into the bloodstream or the failure of liver cells to synthesize urea at an appropriate rate.

Aminotransferase reactions facilitate the transfer of amino groups from one compound to another

If some amino acids are present in excess of the amount required for protein synthesis but others are in relatively short supply, aminotransferase reactions allow amino groups to be transferred from one carbon skeleton to another, thus generating different amino acids. These reactions also provide a mechanism for pooling excess amino groups in forms, mostly glutamate, in which they may be removed by deamination.

In reactions catalysed by aminotransferases the donor amino acid, by losing its amino group, is converted to a keto acid; the acceptor keto acid acquires the amino group, being converted to an amino acid as shown in Figure 10.1.

All amino acids participate in aminotransferase reactions.

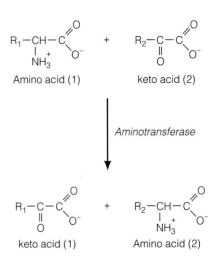

Figure 10.1 Generalized form of the aminotransferase reaction

Aminotransferase enzymes are specific and catalyse readily reversible reactions

Aminotransferases are specific, and named for, the amino acid donating the amino group in one direction of the reaction. For example, **alanine aminotransferase** uses alanine as the amino group donor, converting alanine to the corresponding keto acid, pyruvate.

The **equilibrium constant** for aminotransferase reactions is close to unity. This means that the reactions will proceed with equal ease in either direction, the actual direction being determined by the relative concentrations of substrates and products.

Pyridoxal phosphate is the essential coenzyme for aminotransferases

Aminotransferase reactions depend on the coenzyme **pyridoxal phosphate**, a derivative of vitamin B_6. The mechanism of action of aminotransferases is characterized by the covalent binding of this coenzyme through a Schiff base to the enzyme or the substrate.

- In the absence of substrate, pyridoxal phosphate is bound, through its aldehyde group, to the ε-amino side chain of a specific lysine group of the aminotransferase.
- When a substrate amino acid approaches the catalytic site of the enzyme it displaces the lysine and forms a Schiff base linkage between the amino group of the substrate and the aldehyde group of the coenzyme.

Because pyridoxal phosphate can exist as an aldehyde (**pyridoxal phosphate**) or an amine (**pyridoxamine phosphate**) it has the capacity to act as a carrier of amino groups (Figure 10.2). During catalysis the aldehyde form of the coenzyme accepts an amino group from the substrate amino acid. The product, a keto acid, is released and the coenzyme, now in its amine form, attaches to the second substrate, the acceptor keto acid, transferring the amino group to this second substrate. The keto acid is thereby converted to an amino acid, and the coenzyme reverts to its aldehyde form. Another substrate amino acid may bind to the coenzyme, or in the absence of substrate, the aldehyde form of the coenzyme reforms the Schiff base linkage with the lysine side chain of the enzyme.

Glutamate acts as the major common pool for amino groups

Amino acids can donate their amino group, through aminotransferase reactions, to a variety of keto acids generated in metabolism. Most commonly α-ketoglutarate serves as the acceptor:

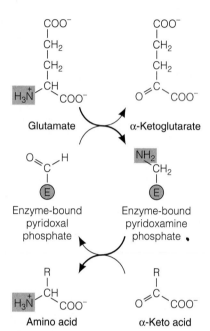

Figure 10.2 Aminotransferase reactions can be seen as having two steps. In the first, an amino acid (glutamate in this example) donates its amino group to the pyridoxal form of the coenzyme, producing α-ketoglutarate. In the second, a keto acid accepts this amino group from the pyridoxamine form of the coenzyme, forming an amino acid and returning the coenzyme to the aldehyde form. In this Figure E represents an aminotransferase enzyme with bound coenzyme in either the aldehyde or amine form. The full structure of the coenzyme is illustrated in Chapter 14.

This use of α-ketoglutarate establishes a common pool in which amino groups collect as a single molecular species, glutamate. This provides an economic way of utilizing and/or disposing of nitrogen.

Only one reaction, catalysed by **glutamate dehydrogenase**, is required to release all the amino groups accumulated as glutamate:

Box 10.1   Distribution of specific aminotransferases provides diagnostic information when tissues are damaged

Aminotransferases for most of the amino acids are found in the liver but other tissues, such as muscle and heart, contain only a few types of aminotransferase. The distribution of enzymes is exploited in clinical tests that assist the diagnosis of cell damage in particular tissues.

The information derived from these assays depends on the following facts.

- Aminotransferases are normally found *within* cells.

- If the plasma membrane of a cell is damaged, enzymes within the cell will leak into the plasma. Elevated levels of tissue enzymes in the circulation are indicative of tissue damage.

- Some aminotransferases are found only in certain tissues, rather than throughout the body.

By measuring the activity of enzymes in serum or plasma it is possible to deduce which, if any, tissues are damaged. For example, it may be possible to determine whether a patient is suffering from liver damage, has had a myocardial infarction or is suffering from pancreatic disease simply by taking blood samples and measuring enzyme activities, which should be found only within those tissues, in the serum. It is often necessary to identify two or more different enzymes, or to identify tissue-specific isoenzymes, to obtain a reliable diagnosis of cell damage.

Glutamate can provide amino groups for synthesis of all other nonessential amino acids if they are in short supply from the diet. Although the aminotransferase reactions utilize α-ketoglutarate extracted from the citric acid cycle, the glutamate dehydrogenase reaction regenerates α-ketoglutarate so that there is no overall depletion of this citric acid cycle metabolite.

It is important to remember that aminotransferase reactions do not always involve α-ketoglutarate: amino groups may be transferred directly to other keto acids. Synthesis of some of the nonessential amino acids occurs by direct transfer of amino groups to other keto acids, including pyruvate and oxaloacetate.

There are deaminases capable of removing the amino groups from threonine and serine

The two amino acids with hydroxyl groups on their side chains can be deaminated directly. The enzymes catalysing these reactions (serine deaminase and threonine deaminase) also use pyridoxal phosphate as a coenzyme. The deaminations are two-step reactions: in the first step, a dehydration, an unstable intermediate is created which reacts with water to liberate ammonia and a keto acid (Figure 10.3).

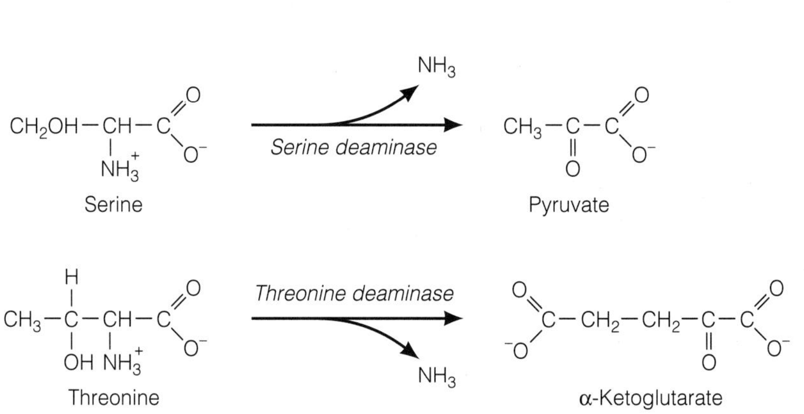

Figure 10.3 Serine and threonine have specific deaminases through which NH_4^+ is released directly. These reactions also require pyridoxal phosphate.

Mechanisms are needed for safe disposal of ammonia because of its toxicity

A high-protein diet generates a substantial amount of ammonia in the body. The removal of amino groups from amino acids can occur in any tissue, so the production of ammonia is widespread throughout the body. Most ammonia is produced from reactions involving glutamate dehydrogenase and the reactions by which neurotransmitters are inactivated.

Ammonia liberated in the liver can be incorporated into urea molecules and eliminated from the body in a soluble and safe form. Free ammonium ions can be excreted in the urine. Because ammonia is toxic it is important to minimize the amount of ammonia liberated within cells that are not capable of converting it to urea.

Safe mechanisms for transporting ammonia from peripheral tissues to the liver involve incorporating it into amino acids

In good health blood ammonium ion concentrations should not exceed 25–40 μM. At concentrations greater than this ammonia enters the brain rather than being removed from it. To ensure that ammonia produced in peripheral tissues is transported to the liver in a form which cannot diffuse into the brain, it is first incorporated into amino acids, principally glutamate, glutamine or alanine.

Free ammonia is toxic because it depletes the citric acid cycle metabolite pool in brain cells and reduces the rate of energy production

Ammonia in solution exists in the equilibrium:

$$NH_3 + H^+ \rightleftharpoons NH_4^+$$

The equilibrium constant for this reaction is such that, at the pH of blood (7.4), most of the circulating ammonia exists in its protonated form, NH_4^+. In common with other charged species NH_4^+ does not readily cross cell membranes. However, those molecules that exist as the uncharged species, NH_3, *are* free to cross cell membranes. This means that ammonia may diffuse from the cells in which it is produced into the circulation, and may enter other cells, including brain cells. Within a cell the equilibrium reforms and a cell contains both charged and uncharged species.

Brain cells have a high and constant demand for energy in the form of ATP to sustain electrical activity (Na^+/K^+ gradients) across cell membranes, for which they depend on oxidative metabolism. The major nutrient for brain cells is glucose, but glucose metabolism provides sufficient ATP for the brain only if the citric acid cycle and oxidative phosphorylation are operating efficiently.

Ammonia in the brain reacts with α-ketoglutarate, in the reaction catalysed by glutamate dehydrogenase, to produce glutamate. The α-ketoglutarate participating in this reaction is derived from the citric acid cycle and its removal from the cycle reduces the rate at which ATP is produced in the brain.

$$\text{α-Ketoglutarate} + NH_4^+ + NADPH + H^+ \xrightarrow[\substack{\textit{Glutamate} \\ \textit{dehydrogenase}}]{} \text{Glutamate} + NADP + H_2O$$

Most of the ammonia in the brain is derived from inactivation of neurotransmitters

Some ammonia enters the brain from the arterial blood supply, and some is generated from the catabolism of protein within brain cells, but most of the ammonium

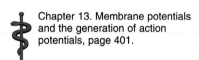

Chapter 13. Membrane potentials and the generation of action potentials, page 401.

ion in the brain is generated when the neurotransmitter γ-**aminobutyric acid** (**GABA**) is inactivated by deamination (see Chapter 11), thus:

Chapter 11. γ-Amino butyric acid (GABA), page 344.

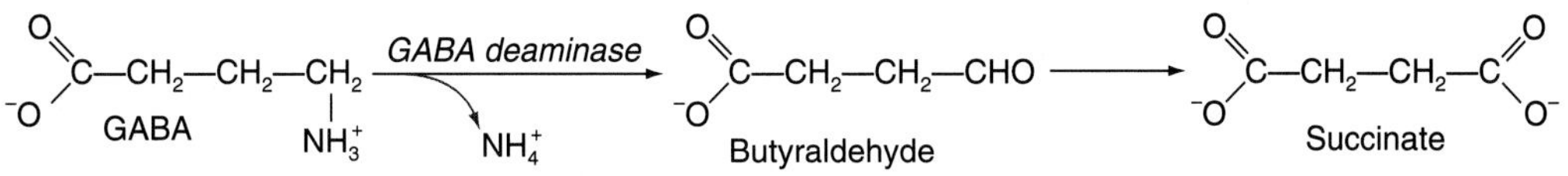

Ammonia is transported from the brain as glutamine

The brain, which is rich in **glutamine synthetase**, uses this enzyme to take up free ammonium ions.

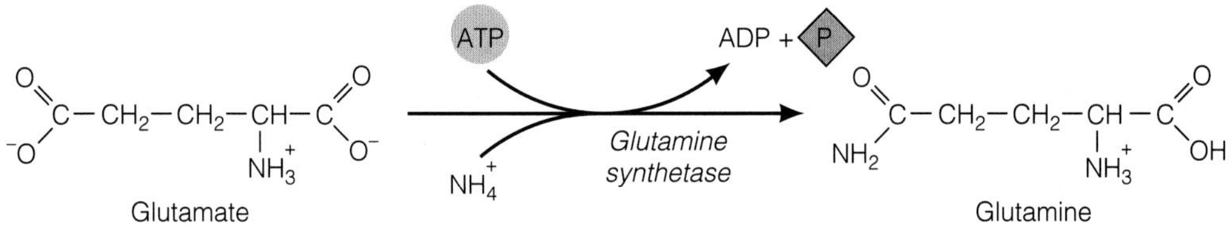

The efficiency of this reaction can be seen if the concentrations of glutamine are measured in arterial and venous blood samples entering and leaving the brain. Normally the concentration of glutamine is substantially higher in venous than in arterial blood.

These reactions remove α-ketoglutarate from the brain, but for every molecule of α-ketoglutarate removed, two ammonium ions are transported safely to the liver as glutamine (Figure 10.4). The metabolite deficit in the brain is constantly replenished from glucose, which is exported from the liver to the brain.

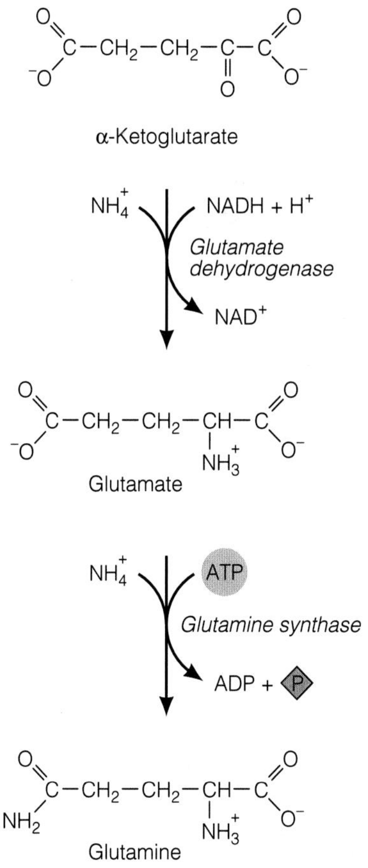

Ammonium ions are liberated from glutamine in the liver

Glutamine exported from the brain is taken up by the liver, where the two amino groups are successively removed in reactions catalysed by the enzymes glutaminase and glutamate dehydrogenase.

These ammonium ions are used to synthesize urea in the liver.

Ammonia is transported from skeletal muscle to the liver as alanine

Ammonia is produced in muscle from two main sources.

1. Amino acids are liberated during the normal turnover of muscle proteins. In starvation substantial quantities of muscle protein may be degraded. In both cases, amino acids that are not incorporated into new protein undergo aminotransferase reactions and their carbon skeletons are either catabolised in muscle or are transported to the liver as substrates for gluconeogenesis. Some glutamate is deaminated in muscle cells.

2. Muscular activity is sustained by using ATP. In severe exercise ATP may be expended more rapidly than the ADP can be rephosphorylated by oxidative or substrate-level phosphorylation. In this situation ADP will accumulate and

Figure 10.4 Synthesis of glutamine by enzymic addition of two ammonium ions to produce α-ketoglutarate.

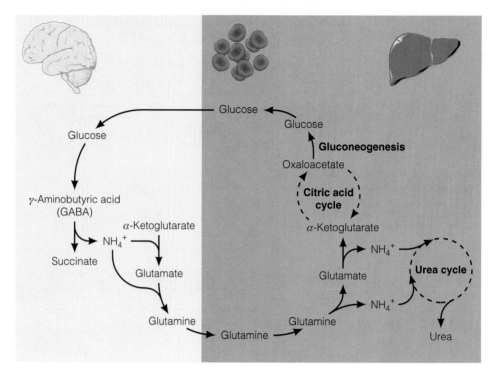

Figure 10.5 Ammonia generated in the brain is removed as glutamine and released within the liver where it is incorporated into urea for safe disposal. The key to this process is the large amount of glutamine synthetase in the brain. The α-ketoglutarate produced in the liver enters the citric acid cycle, adding to the carbon pool from which ultimately more glucose can be distributed to the brain.

adenylate kinase converts two molecules of ADP to one each of ATP and AMP:

$$2 \text{ Adenosine} \langle P \rangle \langle P \rangle \xrightarrow{\text{Adenylate Kinase}} \text{Adenosine} \langle P \rangle \langle P \rangle \langle P \rangle + \text{Adenosine} \langle P \rangle$$

As AMP accumulates it is deaminated to inosine monophosphate (IMP), liberating NH_4^+.

$$\text{AMP} \xrightarrow{\text{AMP deaminase}} \text{IMP} + NH_4^+$$

In severe exercise, muscles produce pyruvate more rapidly than it can be metabolized through the citric acid cycle and so accumulating pyruvate is converted to lactate. The disadvantage of this is that both lactate and pyruvate increase the acidity of muscle cell. Reactions catalysed by glutamate dehydrogenase and alanine aminotransferase facilitate the removal of free ammonium ions. This provides a safe mechanism for transporting ammonia to the liver while preventing the build-up of hydrogen ions within muscle cells.

Ammonium ions are combined with α-ketoglutarate to produce glutamate:

$$\alpha\text{-Ketoglutarate} + NADH + H^+ + NH_4^+ \xrightarrow[\text{dehydrogenase}]{\text{Glutamate}} \text{Glutamate} + NAD^+$$

The amino group is then transferred from glutamate to pyruvate:

$$\text{Pyruvate} + \text{Glutamate} \xrightarrow[\substack{\text{Alanine} \\ \text{aminotransferase}}]{} \text{Alanine} + \alpha\text{-Ketoglutarate}$$

Alanine is released into the bloodstream and transported to the liver, where it is used as a substrate for gluconeogenesis. A useful aspect of this process is that the α-ketoglutarate taken from the citric acid cycle pool by glutamate dehydrogenase is regenerated by the aminotransferase reaction. There is therefore no overall depletion of citric acid cycle intermediates in the muscle cell.

The urea cycle

Healthy adults are in **nitrogen equilibrium** (or **nitrogen balance**), which means that an amount of nitrogen equal to the amount ingested is excreted each day (see Chapter 14). A substantial proportion of this waste nitrogen is released from amino acids as ammonium ions in the liver. Some of the nitrogen is derived from catabolism of purines, pyrimidines and other nitrogenous compounds. About 80% of nitrogen is excreted by humans as urea, some is excreted as ammonium ions and creatinine makes up a significant fraction of excreted nitrogen.

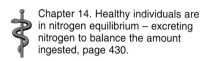

Chapter 14. Healthy individuals are in nitrogen equilibrium – excreting nitrogen to balance the amount ingested, page 430.

Most urea is synthesized in the liver, although the kidneys also produce some. The rate at which urea is produced is carefully regulated to ensure that ammonia does not accumulate in the body.

The urea cycle converts ammonia to urea

A key feature of the urea cycle is that, for each turn of the cycle, a molecule of **ornithine** serves as a base on which the urea molecule is built, and at the end of the cycle is regenerated, containing the same atoms as the original molecule of ornithine. For every molecule of urea synthesized and excreted, the equivalent of two amino groups are eliminated from the body: one taken up as an ammonium ion, the other from the α-amino group of aspartate. Of the three amino acid intermediates of the urea cycle (ornithine, **citrulline** and **arginine**) only arginine is found in proteins (see Chapter 4).

Carbamoyl phosphate provides the first ammonium ion for urea synthesis

The conversion of ornithine to citrulline requires the addition of amino and carbonyl groups. The carbon atom is contributed directly as a molecule of carbon dioxide or a bicarbonate ion, HCO_3^-, generated by oxidative metabolism in the liver. The amino and carbonyl groups are added to ornithine through a single condensation reaction. This requires the preliminary synthesis of a high-energy intermediate, **carbamoyl phosphate**, from CO_2 and NH_4^+, through a reaction catalysed by **carbamoyl phosphate synthetase**.

$$CO_2 + NH_4^+ + 2\,ATP \xrightarrow{\textit{Carbamoyl phosphate synthase}} NH_2-\overset{\overset{\displaystyle O}{\|}}{C}-O-\boxed{P} + 2\,ADP + \boxed{P}$$

Carbamoyl phosphate

The reaction, which is essentially irreversible, consumes energy from two molecules of ATP. One of the phosphate groups is incorporated into the product, the other liberated as inorganic phosphate. Carbamoyl phosphate is a very high-energy compound, with a standard free energy ($\Delta G^{\circ\prime}$) for hydrolysis of the phosphate bond of -49 kJ mol^{-1}. This represents almost twice the energy inherent in the terminal phosphoanhydride bond of ATP, indicating that most of the energy of two ATP phosphoanhydride bonds is incorporated into the anhydride bond of carbamoyl phosphate.

The incorporation of CO_2 into carbamoyl phosphate contributes to the control of body pH by reducing the acidity of the environment from which it is derived.

Carbamoyl phosphate synthetase is an allosteric enzyme, regulating the rate at which nitrogen enters the urea cycle

Carbamoyl phosphate synthase occurs in two forms: one form, located in the mitochondrial matrix, produces carbamoyl phosphate for the urea cycle; the other

Chapter 4. Cells contain two or more proteins with similar functions, page 107.

form, located in the cytoplasm, contributes to the synthesis of pyrimidines (see below). The mitochondrial form, **carbamoyl phosphate synthase I**, is an allosteric enzyme, relatively inactive in the absence of its allosteric activator *N*-acetyl glutamate. Ultimately the activity of this enzyme is controlled by the concentration of arginine in the liver cell. Thus, the concentration of arginine, an intermediate of the cycle, is one of the factors determining the rate at which ammonia is eliminated from the body as urea.

The sequence of reactions by which arginine exerts its effects is illustrated in Figure 10.6.

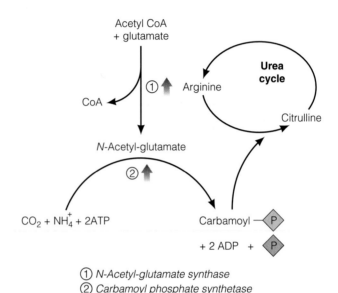

Figure 10.6 Activation of carbamoyl phosphate synthase by arginine.

① *N-Acetyl-glutamate synthase*
② *Carbamoyl phosphate synthetase*

Arginine activates **N-acetyl glutamate synthetase**, which catalyses the synthesis of acetyl glutamate from acetyl CoA and glutamate. Energy is derived from the 'high-energy' thioester bond of acetyl coenzyme A.

Acetyl CoA   +   Glutamate

N-acetylglutamate synthase

N-acetylglutamate + CoA

N-acetyl glutamate is the allosteric activator of carbamoyl phosphate synthetase I. The activator is removed by a specific **deacylase**, which is activated when the amount of ammonia available in the liver is reduced.

Carbamoyl phosphate enters the urea cycle by condensation with ornithine

The energy of the anhydride group of carbamoyl phosphate is used to drive the reaction by which carbamoyl phosphate reacts with ornithine, transferring the carbamoyl group to the δ-amino group of ornithine to form citrulline, and releasing the phosphate group. The reaction is catalysed by **ornithine transcarbamoylase**, an enzyme located in the mitochondrial matrix.

The other enzymes of the cycle are located in the cytoplasm. To accommodate the fact that some enzymes of the cycle are in the cytosol and others are in the mitochondrial matrix, transporters in the mitochondrial inner membrane enable ornithine to enter the mitochondrial matrix and citrulline to leave.

The second amino group is derived from aspartate

The second amino group is introduced into the cycle when citrulline reacts with aspartate. This is a two-stage process, involving a condensation reaction between the α-amino group of aspartate and the carbamoyl group of citrulline, forming argininosuccinate. This reaction is catalysed by **argininosuccinate synthetase**. Energy is provided by hydrolysis of both of the high-energy bonds of ATP (Chapter 7).

Chapter 7. The molecular basis of the high-energy status of ATP is well understood, page 190.

Argininosuccinase then catalyses the cleavage of argininosuccinate, releasing fumarate and arginine. The carbon atoms of this molecule of fumarate enter the cycle as aspartate, and the α-amino group of aspartate is incorporated into arginine.

Argininosuccinate

Argininosuccinase

Arginine   Fumarate

Ornithine is regenerated when urea is released from arginine

The cycle is completed as **arginase** hydrolyses the urea group from arginine, regenerating the original molecule of ornithine. Arginase is a Mn^{2+}-dependent enzyme. Urea, an uncharged molecule, diffuses freely into the bloodstream and is excreted through the kidneys.

The sum of change catalysed by one turn of the urea cycle can be written as:

$$NH_4^+ + CO_2 + 3ATP + Aspartate + 2H_2O \longrightarrow Urea + Fumarate + 2ADP + AMP + 2Pi + PPi$$

Box 10.2   Deficiencies in urea cycle enzymes can lead to serious health problems

Deficiencies of all of the urea cycle enzymes have been recognized clinically, and can cause serious health problems, depending on the extent of the deficiency. Children born with a complete failure to synthesize any of these enzymes are unlikely to survive. Some alleviation of the problems arising from these deficiencies is possible through reduction of the amount of protein in the diet; however, because of the constant normal turnover of protein and the need for cell growth and repair, some protein must be provided in the diet.

Deficiency of *N*-acetyl glutamate synthetase
In this condition, the enzyme carbamoyl phosphate synthase does not respond to increases in the arginine concentration by stimulating the rate of synthesis of carbamoyl phosphate. This problem can not be overcome by administration of acetyl glutamate because this compound does not enter liver cells. The problem may be partially alleviated by the administration of carbamoyl glutamate, which is absorbed into the liver and activates carbamoyl phosphate synthetase.

Deficiency of carbamoyl phosphate synthetase
If the activity of carbamoyl phosphate synthase is inadequate, ammonia is not incorporated into carbamoyl phosphate at a rate sufficient to prevent it diffusing into the bloodstream. The result is an increase in the circulating concentration of ammonia (hyperammonaemia), ammonia entering brain cells and, eventually, mental retardation.

Deficiency of ornithine transcarbamoylase
This is the most common of these problems. It prevents carbamoyl phosphate condensing with ornithine. The gene for this enzyme is located on the X-chromosome, so males are more seriously affected than females.

The cycle is essentially irreversible because of the amount of energy, in the form of ATP, that is necessary to drive it. Reactions in which aspartate contributes an amino group to the synthesis of other molecules, and releases fumarate, also occur in the synthesis of nucleotide bases.

An overview of the urea cycle is given in Figure 10.7.

The urea cycle interacts with metabolites that also occur in the citric acid cycle

A key feature of the urea cycle is the extent of its interaction with intermediates common to the citric acid cycle. However, it appears that these intermediates

Figure 10.7 The urea cycle operates partly within the mitochondrial matrix and partly in the cytoplasm of liver cells. Shading indicates the origin of atoms making up the urea molecule. The rate of the cycle is regulated to ensure that all of the ammonia released in the liver is converted to urea for excretion.

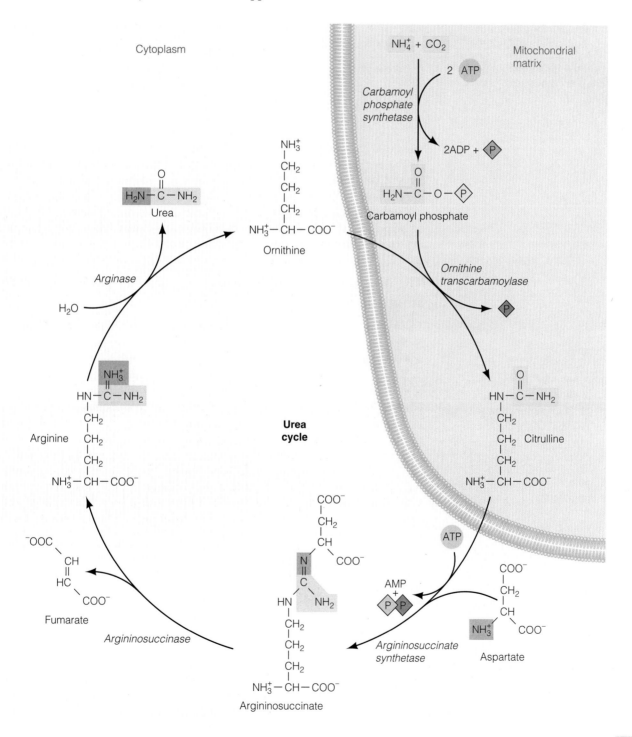

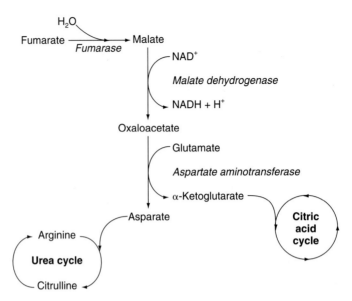

Figure 10.8 The urea cycle and its interactions with intermediates of the citric acid cycle.

remain in the cytoplasm and are kept separate from those of the citric acid cycle. As the enzymes **fumarase**, malate dehydrogenase and aspartate aminotransferase are all found in the cytoplasm, the fumarate released during the urea cycle is probably converted to aspartate in the cytoplasm, the α-amino group being derived from glutamate entering the cytoplasm of liver cells from other tissues. The α-ketoglutarate produced from the aminotransferase reaction may enter mitochondria, to be incorporated into the citric acid cycle (Figure 10.8).

The rate of production of urea is regulated by the amount of ammonia in the liver

The rate of the urea cycle is controlled by the amount of ammonia formed in the body, and specifically by the concentration in liver cells of two substances, arginine and ammonia. The role of arginine has been described above.

The concentration of ammonia in liver cells regulates the number of molecules of the enzymes that catalyse reactions of the cycle. These enzymes are subject to **enzyme induction** (see Chapter 5), accelerated gene expression in response to increased concentration of ammonia. Changes in the amount of protein ingested in the diet, and therefore the amount of nitrogen to be excreted, can result in 10–20-fold changes in the number of molecules of urea cycle enzymes in the liver. Likewise in starvation, when muscle protein is broken down to produce nutrients, the amount of ammonia to be disposed of increases. This situation also results in an increase in the number of molecules of urea cycle enzymes in the liver, although prolonged or severe starvation will limit the capacity of liver cells to induce the synthesis of these enzymes.

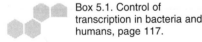

Box 5.1. Control of transcription in bacteria and humans, page 117.

Biosynthesis of amino acids

A balanced diet provides most of the amino acids required for protein synthesis and for the production of other low-molecular weight nitrogenous compounds. However, our cells do have the enzymes to synthesize about half of the amino acids, the **nonessential amino acids**. The **essential amino acids** must, by definition, be provided in the diet, for we do not have the enzymes required for their synthesis (Chapter 14). It is not the intention of this text to explore the pathways

Chapter 14. Essential amino acids are those that cannot be synthesized in the body, page 429.

for the synthesis of all of the nonessential amino acids; we will discuss only the general principles and present a few examples, particularly where abnormalities of the pathways have clinical implications.

Some general principles govern the synthesis of many amino acids

Most amino acids are synthesized from the corresponding keto acids by aminotransferase reactions, and generally glutamate supplies the amino group. The liver contains the enzymes for the synthesis of the carbon chains of the nonessential amino acids. Two mechanisms control rates of amino acid synthesis.

- Allosteric inhibition is the most common mechanism by which the synthesis of amino acids is controlled (Chapters 4 and 7). Generally, the first reaction in the biosynthetic sequence is inhibited by the product of the pathway, so that it operates only if there is a deficiency of the product.
- Enzyme induction (gene activation (see Chapter 5)) allows the liver to increase the amount of specific enzymes if the demand for specific reactions increases. Amino acid biosynthesis is thus regulated by changes in enzyme availability.

Chapter 4. Many enzymes show allosteric behaviour, page 91.
Chapter 7. Allosteric regulation allows enzymes to respond to metabolite changes in the intracellular environment, page 194.
Box 5.1. Control of transcription in bacteria and humans, page 117.

Glutamate, glutamine and proline are synthesized from α-ketoglutarate

The reactions catalysed by glutamate dehydrogenase and glutamine synthetase have already been described. To summarize:

$$\alpha\text{-Ketoglutarate} + NADH + H^+ + NH_4^+ \xrightarrow[\substack{\textit{Glutamate} \\ \textit{dehydrogenase}}]{} Glutamate + NAD^+$$

$$Glutamate + NH_4^+ + ATP \xrightarrow[\textit{Glutamine synthetase}]{} Glutamine + ADP + Pi$$

The conversion of glutamate to proline involves a cyclization process, which is catalysed by three enzymes. Proline is an allosteric inhibitor of the first of these enzymes. The synthesis of proline requires NADPH.

Aspartate and asparagine are synthesized from oxaloacetate

Aspartate is synthesized from oxaloacetate through an aminotransferase reaction. The production of asparagine is through a synthetase reaction analogous to that required for the synthesis of glutamine (Figure 10.9).

Serine is synthesized from 3-phosphoglycerate, an intermediate of the glycolytic pathway

This reaction (Figure 10.10) emphasizes the importance of intermediates of the central metabolic pathways as substrates for many other synthetic processes.

Glycine is synthesized from serine

The human body has a substantial requirement for glycine (see Table 10.2) for processes in addition to protein synthesis, and may require 10–50 times more glycine than is normally provided in the diet. The synthesis of glycine (Figure 10.11) requires tetrahydrofolate.

Cysteine is produced from methionine and serine

Cysteine, a nonessential amino acid, is synthesized from methionine, an essential amino acid, and serine, a nonessential amino acid (Figure 10.12). Serine provides

**Box 10.3
Homocystinuria arises
from a failure of cysteine
metabolism**

In humans, a deficiency of the enzyme catalysing the condensation of serine with homocysteine produces a condition called **homocystinuria**, characterized by excretion of homocysteine in the urine. The clinical symptoms of this disorder include severe mental retardation, dislocation of the lens of the eye and osteoporosis. Both the condensation of homocysteine with serine and the subsequent cleavage of this condensation product to release cysteine depend on pyridoxal phosphate. In some patients administration of very large doses of pyridoxal phosphate alleviate the symptoms of homocystinuria.

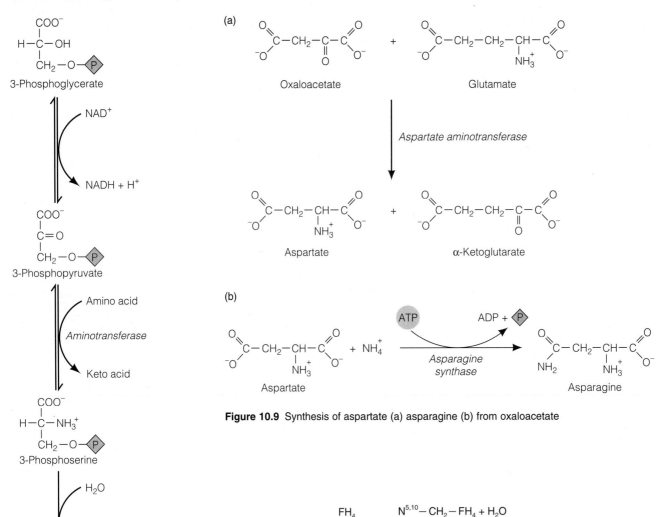

Figure 10.9 Synthesis of aspartate (a) asparagine (b) from oxaloacetate

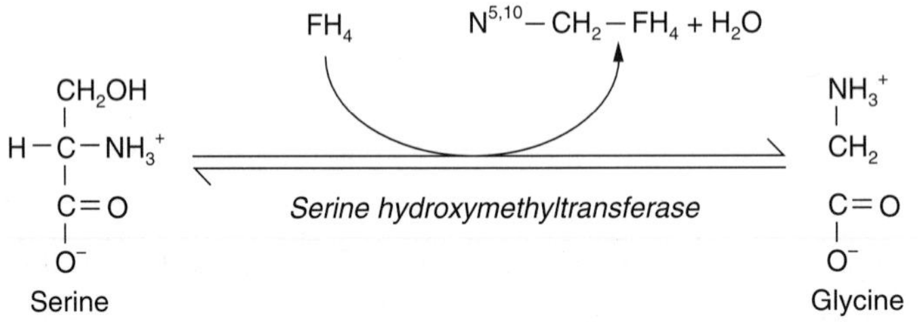

Figure 10.10 Synthesis of serine.

Figure 10.11 Synthesis of glycine.

Figure 10.12 The synthesis of cysteine requires both methionine, which contributes the sulphydryl group, and serine, which contributes the carbon skeleton and α-amino group. In order to participate in this synthesis methionine must be activated as the S-adenosyl derivative (SAM) and this has the added advantage that an activated methyl group (—CH$_3$) is generated, which can contribute to other synthetic reactions.

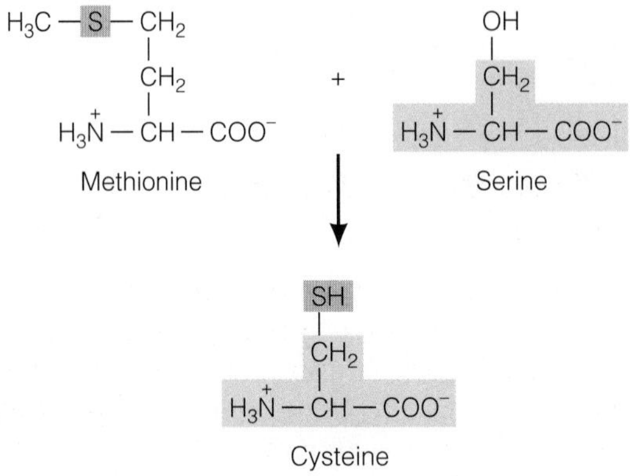

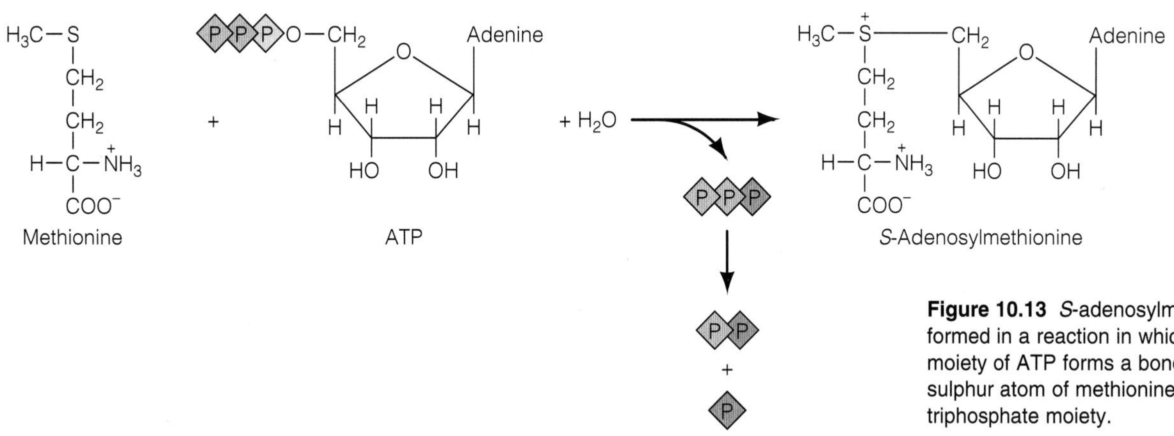

Figure 10.13 *S*-adenosylmethionine is formed in a reaction in which the adenosyl moiety of ATP forms a bond with the sulphur atom of methionine, displacing the triphosphate moiety.

the carbon skeleton, methionine the –SH group. Cysteine is therefore one of the amino acids which is strictly nonessential (Chapter 14) only under conditions when provision of methionine is adequate.

An additional value of this pathway is that it generates a reactive methyl group in the form of *S*-adenosyl methionine (SAM) (Figure 10.13), which can be used in other processes such as the methylation of nucleic acids (Chapter 5).

Chapter 14. Essential amino acids are those that cannot be synthesized in the body, page 429.
Chapter 5. Ribose methylation, page 122.

S-Adenosylmethionine contributes reactive methyl groups

Activated methyl groups, derived from SAM (Figure 10.13), are required for the synthesis of the many important compounds such as

- phosphatidylcholine;
- adrenaline;
- creatine;
- cysteine;
- acetylcholine.

Catabolism of amino acids – the fate of the carbon skeletons

A substantial proportion of the amino acids derived from a high-protein diet are used as nutrients rather than for protein synthesis. Aminotransferase reactions remove the amino groups, leaving carbon skeletons to be converted to intermediates of the central metabolic pathways. It is not the purpose of this chapter to examine all of the pathways by which amino acids are catabolised, only to describe the general principles governing these processes and to present some examples.

If the diet has a high-protein content, enzymes metabolizing amino acids are induced in the liver

Amino acids usually make a relatively small contribution to energy provision in the body. There is therefore little need for the liver to maintain large quantities of the enzymes that would enable it to metabolize amino acids. Exceptions occur in individuals on a high-protein diet or during starvation, when structural proteins are degraded and are a significant portion of the available nutrient.

Each amino acid follows a distinct pathway during its conversion to intermediates of the central metabolic pathways, and for each of these pathways several enzymes may be required. It therefore represents an economy that synthesis of these specialized enzymes should be regulated to provide precisely the amounts required *only* when they are required. The liver responds to a high-protein diet by inducing the synthesis of more of the enzymes required to convert amino acid carbon skeletons to metabolic intermediates. This control appears to operate at the level of gene transcription (Chapter 5).

The liver can extract energy more efficiently from glucose than from amino acids, so high concentrations of circulating glucose inhibit the induction of liver enzymes that degrade the carbon skeletons of amino acid.

Amino acids are converted to one of seven metabolites of the central metabolic pathways

Amino acids are divided into two classes depending on the products of their catabolism (Table 10.1).

1. The **ketogenic amino acids** are converted to one of two potential precursors of ketone bodies.
2. The **glucogenic amino acids** are converted to one of five intermediates of the carbohydrate metabolism pathways and are potential precursors of glucose through gluconeogenesis.

Some amino acids, for example phenylalanine and tyrosine, are degraded to metabolites of both classes, and therefore are both ketogenic and glucogenic (Figure 10.14).

Box 5.1. Control of transcription in bacteria and humans, page 117.

Table 10.1 The metabolic derivatives of amino acid carbon skeletons

Glucogenic products	Ketogenic products
Pyruvate	Acetyl CoA
Oxaloacetate	Acetoacetyl CoA
Fumarate	
Succinyl CoA	
α-Ketoglutarate	

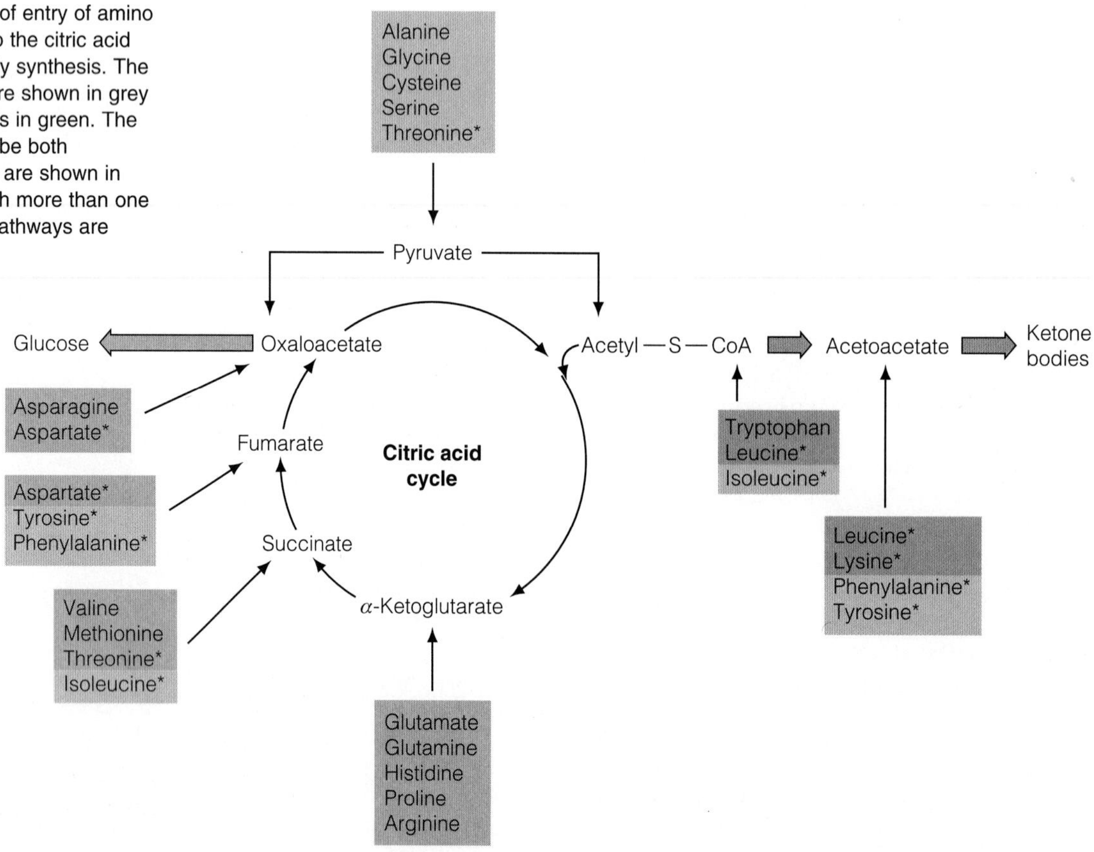

Figure 10.14 The points of entry of amino acid carbon skeletons into the citric acid cycle and into ketone body synthesis. The glucogenic amino acids are shown in grey and ketogenic amino acids in green. The few amino acids that can be both glucogenic and ketogenic are shown in blue. The amino acids with more than one point of entry into these pathways are marked with an asterisk.

Many enzymes catabolising amino acids require vitamin derivatives as coenzymes

Deficiencies of vitamins can arise, despite adequate dietary intake

- in disorders of the gut where absorption process are inefficient;
- in alcoholics;
- in patients with liver failure, in whom vitamins may be converted to the active coenzyme form.

Patients with these types of disorder excrete partially metabolized amino acids, but are likely to have other metabolic problems as well. The vitamins (Chapter 14) prominent in the amino acid metabolizing pathways include

- tetrahydrofolate (folic acid);
- cyanocobalamin (vitamin B_{12});
- biotin;
- pyridoxal phosphate (vitamin B_6);
- nicotinamide nucleotides.

The pathway for catabolism of phenylalanine and tyrosine illustrates some important aspects of amino acid metabolism

Some inherited defects in the metabolism of phenylalanine and tyrosine have been well characterized. For this reason both the normal and the abnormal pathways of phenylalanine and tyrosine metabolism will be examined in some detail.

Phenylalanine, an essential amino acid, is converted to tyrosine, a nonessential amino acid. These amino acids are usually degraded to two metabolic intermediates: one molecule each of acetoacetate and fumarate, and are therefore categorized as both ketogenic and glucogenic. The pathway is illustrated in Figure 10.15.

The conversion of phenylalanine to tyrosine is catalysed by phenylalanine hydroxylase, which requires tetrahydrobiopterin, an analogue of folic acid, as its coenzyme, providing the reducing power for the reaction. Tetrahydrobiopterin is regenerated by a specific reductase which utilizes the reducing power of NADH. Thus the overall reaction is:

$$\text{Phenylanine} + O_2 + \text{NADH} + H^+ \xrightarrow[\text{\textit{Phenylalanine hydroxylase}}]{} \text{Tyrosine} + \text{NAD}^+ + H_2O$$

An aminotransferase reaction then converts tyrosine to **p-OH-phenyl-pyruvate**. This enzyme is inducible, produced in response to increased tyrosine concentration in the liver or to glucocorticoid hormones. The phenyl ring is cleaved by a series of reactions requiring ferrous (Fe^{2+}) ions, ascorbic acid (vitamin C) and molecular oxygen. The ring opening reaction is catalysed by **homogentisate dioxygenase**.

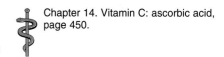

Chapter 14. Vitamin C: ascorbic acid, page 450.

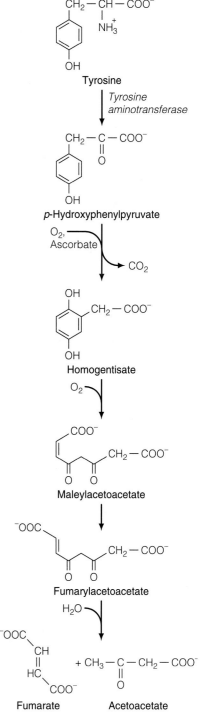

Figure 10.15 The normal metabolism of phenylalanine and tyrosine produces fumarate and acetoacetate.

Box 10.4 Alkaptonuria is caused by a genetic deficiency of homogentisate dioxygenase

Alkaptonuria is an inherited condition. In affected individuals homogentisate accumulates and is excreted in the urine. On exposure to air this compound spontaneously polymerizes, forming a black pigment. It causes no serious impairment of health and has no serious effects for the individual concerned, except perhaps fright at the sight of the black pigment!

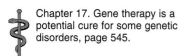
Chapter 17. Gene therapy is a potential cure for some genetic disorders, page 545.

Box 10.5 Phenylketonuria is an inborn error of metabolism which leads to mental retardation

Phenylketonuria is an inborn error, a genetic deficiency of phenylalanine hydroxylase, characterized by the excretion of abnormal products of phenylalanine metabolism. As phenylalanine accumulates in these people, aminotransferase action forms **phenylpyruvate**, which is then converted to either **phenyllactate** or **phenylacetate**. All three compounds are excreted in the urine.

The symptoms of phenylketonuria also occur in patients deficient in **tetrahydrobiopterin** (the coenzyme for the hydroxylase), or of **tetrahydrobiopterin reductase**, the enzyme that restores the reduced state of the coenzyme.

The symptoms of phenylketonuria are severe neurological disorders, and mental retardation sets in within a few days of birth. Brain weight is abnormally low and nerves are inadequately myelinated. This problem is believed to arise because phenylpyruvate competes with tyrosine and tryptophan for entry into brain cells. This leads to a shortage of substrate for the synthesis of the neurotransmitters dopamine and noradrenaline. There is also evidence that some specific nerve bundles fail to develop properly.

Phenylketonuria occurs in about 1 in 10 000 births in the UK and, because brain damage is irreversible, it is important that this metabolic defect is detected within a few hours of birth. Screening of neonates is routine in the UK. Current treatment consists of maintaining these individuals on diets low in phenylalanine. Successful treatment requires a delicate balance because phenylalanine is the precursor for tyrosine, so diets low in phenylalanine may need to be supplemented with tyrosine. Phenylalanine is required for protein synthesis and so some must be provided in the diet. The gene for phenylalanine hydroxylase has recently been identified and cloned, opening the prospect for gene therapy (see Chapter 17).

It is important for people suffering from this disorder to be aware that one of the constituents of some of the modern sweeteners (aspartame, Nutrasweet) introduced to replace sugar in food and in 'diet' drinks is a phenylalanine derivative.

Amino acids are precursors for a wide range of specialized compounds that control metabolism or physiological functions

Detailed consideration will be given here to the synthesis and mechanism of action of a small number of physiologically important compounds that are synthesized from amino acids.

Melanin, the pigment of skin and hair and eye, is derived from tyrosine

The synthesis of melanin, a polymer of metabolites of tyrosine, occurs in melanocytes of the skin. The only enzyme recognized as participating in melanin synthesis is **tyrosinase**, which converts tyrosine to **3,4-dihydroxyphenylalanine**. This is superficially the same reaction as that catalysed by tyrosine hydroxylase but the two enzymes are different structurally and tyrosinase requires Cu^{2+} for its activity while the hydroxylase depends on Fe^{2+}. Tyrosinase is found only in melanin-producing granules. Tetrahydrobiopterin, ascorbate (vitamin C), molecular oxygen and **S-adenosylmethionine (SAM)** are also required for

Table 10.2 Amino acids as precursors of biologically important compounds

Precursor amino acid	Product	Biological role
Tryptophan	Serotonin	Vasoconstrictor
	Nicotinamide compounds (NAD and NADP)	Coenzymes for dehydrogenases
Tyrosine	Dopamine	Neurotransmitter
	Adrenaline and noradrenaline	Hormones and transmitters
	Melanin	Skin, hair and eye pigment
	Thyroxine	Hormone
Histidine	Histamine	Local inflammatory agent. Regulates gastric secretions
Serine	Choline	Component of the transmitter acetylcholine and phospholipids
Glycine	Porphyrins	Iron-binding structures in proteins
	Creatine	Phosphocreatine acts as energy reserve in muscle and brain
	Glutathione	Intracellular reducing agent
	Bile salts	Digestion of lipids
	Purines	Nucleotide coenzymes, RNA, DNA
Glutamate	GABA	Neurotransmitter

melanin synthesis. If tyrosinase is deficient or absent, inadequate amounts of melanin are produced and the skin and hair have little or no colour. These people are poorly protected against even normal exposure to ultraviolet light, suffer skin burns readily, and skin cancer is common among them. They also lack pigment in the eyes, leading to photophobia, but there is generally no impairment of eyesight.

Glutathione is a tripeptide with several important biological functions

Glutathione (GSH), γ-glutamyl-cysteinyl-glycine, is abundant in most cells. It functions as:

- an intracellular reducing agent (this function is described in Chapter 12);
- a scavenger of free radicals;
- a component of one of the mechanisms by which amino acids are transported across cell membranes.

Chapter 12. Detoxification of superoxide needs NADPH and the peptide glutathione, page 378.

Glutathione scavenges radicals and peroxides

Glutathione has the capacity to 'scavenge' hydrogen peroxide or organic peroxides and free radicals that would otherwise cause peroxidation of membrane lipids, thereby increasing the permeability of membranes and eventually leading to lysis of cell membranes. This role of glutathione is particularly important in erythrocytes (Chapter 12) where, without this protection, lysis of erythrocyte membranes would shorten the life of the cells and reduce the amount of haemoglobin available to carry oxygen.

Chapter 12. Detoxification of superoxide needs NADPH and the peptide glutathione, page 378.

$$2GSH + H_2O_2 \rightarrow GSSG + 2H_2O$$

$$2GSH + R{-}OOH \rightarrow GSSG + H_2O + ROH$$

Glutathione drives amino acid transport through the γ-glutamyl cycle

Glutathione is also the key molecule in the γ-glutamyl cycle, the process by which some amino acids are transported across cell membranes. This cycle is particularly active in the kidney. An incoming amino acid is converted to a γ-glutamyl amino acid dipeptide when its α-amino group forms a peptide bond with the γ-carboxyl residue of the glutamate component of glutathione:

$$\text{GSH} + \text{Amino acid} \xrightarrow[\substack{\gamma\text{-Glutamyl} \\ transferase}]{} \gamma\text{-Glutamyl amino acid} + \text{Cysteinyl glycine}$$

This reaction is catalysed by the membrane-bound enzyme **γ-glutamyl transferase**. As the γ-glutamyl amino acid complex crosses the membrane, the cysteinyl glycine dipeptide is also released into the interior of the cell. Further enzymes are required to release the amino acid from the dipeptide, and to resynthesize glutathione, maintaining its concentration within the cell. Overall, three molecules of ATP are required to transport one amino acid across the membrane by this mechanism and to resynthesize glutathione (Figure 10.16).

Recognized abnormalities arise from deficiencies of enzymes responsible for the synthesis of glutathione. Individuals with deficiencies of the transferase excrete large amounts of glutathione in the urine.

Carnitine transports fatty acids across mitochondrial membranes

Carnitine is synthesized from protein-bound lysine residues, through reactions requiring SAM (see Figure 10.13). The role of carnitine in the transport of long-chain fatty acids across the inner mitochondrial membrane for oxidation is described in Chapter 9.

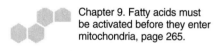
Chapter 9. Fatty acids must be activated before they enter mitochondria, page 265.

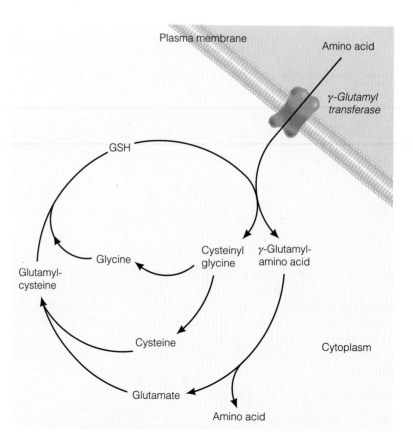

Figure 10.16 The γ-glutamyl cycle.

Nucleotide metabolism

Nucleotide biosynthesis must occur in every cell of the body

The cells of the body contain four sets of nucleotides: the **purine** and **pyrimidine ribo-** and **deoxy-** ribonucleotides. Each of these can exist as a mono-, di- or triphosphate (see Chapter 2). There are therefore 27 principal species of nucleotide, in addition to those that occur as biosynthetic intermediates, but the eight **nucleoside triphosphates** (NTPs) – rATP, rGTP, rCTP, rUTP and dATP, dGTP, dCTP, dTTP – are most directly involved in metabolism, especially in the synthesis of RNA (see Chapter 5) and DNA (see Chapter 6).

Nucleotides of the ribo- series are present at much higher concentrations than those of the deoxy- series, which are derived from ribonucleotides and are used only for DNA replication (see Chapter 6). The ability to synthesize ribonucleotides is a property of all known cells and requires two metabolic pathways, for the production of purine and pyrimidine nucleotides respectively. Cells can make nucleotides *de novo* from simple components or by 'salvaging' preformed purines or pyrimidines (see below). An outline of nucleotide metabolism is shown in Figure 10.17.

Chapter 2. Pyrophosphate bonds, page 35.
Chapter 5. RNA polymerases are capable of both initiation and elongation, page 114.
Chapter 6. DNA polymerases are central to DNA replication, page 154.

Chapter 6. DNA replication requires a continuous supply of deoxyribonucleotides, page 160.

Chapter 10. Nucleotide components can be salvaged, page 322.

Ribose is activated for nucleotide biosynthesis as PRPP

An essential building block for nucleotide synthesis is ribose, which is mainly obtained as a by-product of the pentose phosphate pathway (see Chapter 8) in the form of **ribose 5-phosphate** (R5P). To be used as a substrate for nucleotide biosynthesis, this molecule has to be converted to a form in which the C-1 is 'activated' by addition of pyrophosphate to permit the synthesis of the *N*-glycosidic bond of nucleotides. This activated molecule is **5-phosphoribosyl-1-pyrophosphate**

Chapter 8. The pentose phosphate pathway, page 258.

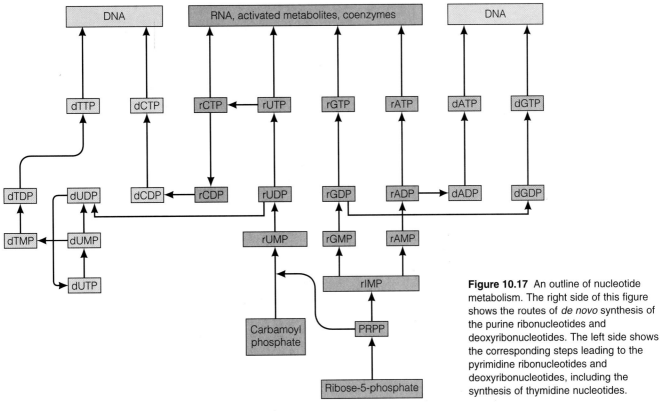

Figure 10.17 An outline of nucleotide metabolism. The right side of this figure shows the routes of *de novo* synthesis of the purine ribonucleotides and deoxyribonucleotides. The left side shows the corresponding steps leading to the pyrimidine ribonucleotides and deoxyribonucleotides, including the synthesis of thymidine nucleotides.

(PRPP) and is made from R5P by **PRPP synthetase.** PRPP is used in the biosynthesis of purine and pyrimidine nucleotides and also in the 'salvage' of purine bases (Figure 10.18).

An important illustration of the stereospecificity of enzyme-catalysed reactions (see Chapter 4) is seen with PRPP and its metabolism. Although the ribose–base glycosidic bond in nucleotides is in the β-form, the pyrophosphate of PRPP is linked to ribose in the α-form. Thus all the reactions shown in Figure 10.17 involve an α-to-β configurational change.

Chapter 4. Enzyme reactions are stereospecific, page 90.

Cells need to phosphorylate nucleotides to the triphosphate level

The key roles of the nucleoside triphosphates have been highlighted above and the ability of cells to generate these molecules from nucleosides, nucleoside monophosphates (NMPs) and nucleoside diphosphates (NDPs) is vital. This is achieved by the transfer of the terminal phosphate from ATP by a number of **kinases** in the general reaction:

$$X + ATP \rightarrow X - P + ADP$$

where X is a nucleoside, NMP or NDP.

Nucleoside kinases

These enzymes catalyse the reaction

$$Nucleoside + ATP \rightarrow NMP + ADP$$

and are usually base-specific. Perhaps the most significant of these is thymidine kinase (TK), which catalyses

$$Thymidine + ATP \rightarrow dTMP + ADP$$

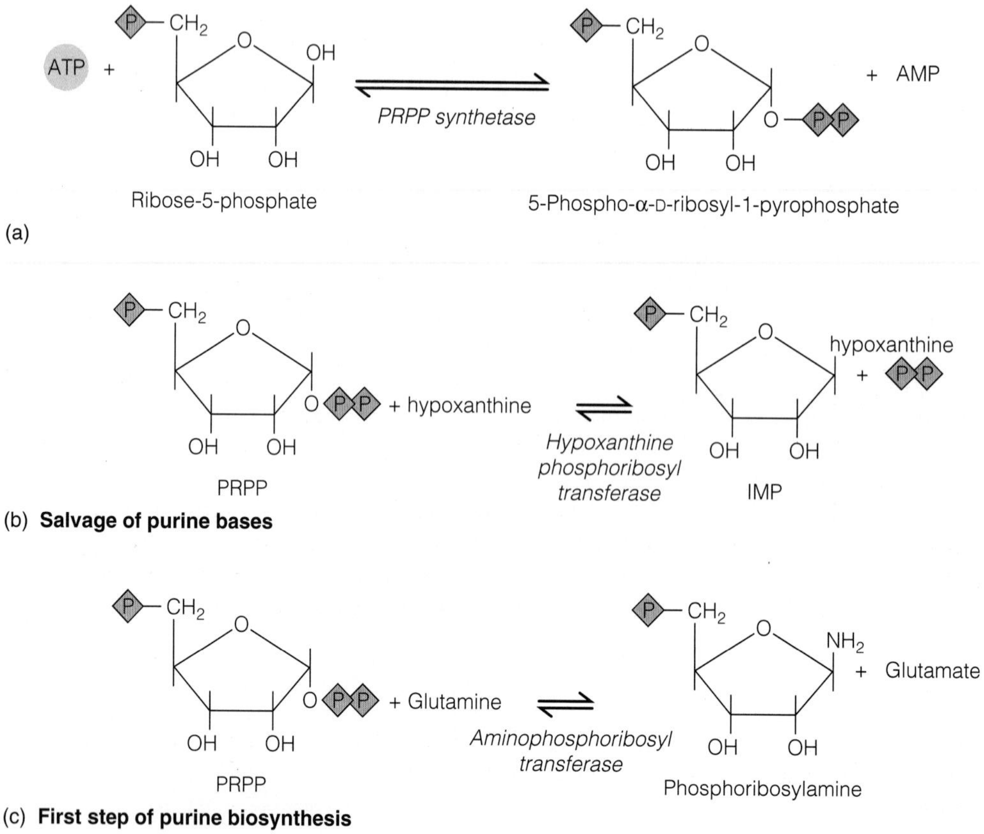

Ribose-5-phosphate   5-Phospho-α-D-ribosyl-1-pyrophosphate

(a)

(b) **Salvage of purine bases**

Figure 10.18 PRPP and its reactions. The transfer of a pyrophosphoryl group to ribose-5-phosphate activates the 1-position so that it can be used in the formation of the glycosidic links in the first step of *de novo* purine biosynthesis and in purine salvage.

(c) **First step of purine biosynthesis**

and appears to be important in DNA replication as it is always present in cells that are actively replicating DNA (see Chapter 6). This is probably because of the need for an adequate supply of thymidine nucleotides (see below).

Chapter 6. DNA replication requires a continuous supply of deoxyribonucleotides, page 160.
Chapter 10. Production of dTTP requires a special set of reactions, page 320.

NMP kinases

These enzymes catalyse the reaction

$$NMP + ATP \rightarrow NDP + ADP$$

and are also usually base-specific. Their importance lies in ensuring an adequate supply of NDPs for conversion to NTPs. An important enzyme of this type is adenylate kinase, which catalyses

$$AMP + ATP \rightleftharpoons ADP + ADP$$

and can thus generate one ATP from two ADPs or prevent accumulation of excessive levels of AMP, which is a metabolic 'backwater' (see Chapter 7).

Chapter 7. Adenylate kinase transfers phosphate between adenine nucleotides, page 191.

NDP kinases

These enzymes catalyse the reaction

$$NDP + ATP \rightarrow NTP + ADP$$

and are usually nonspecific for the base and pentose of the NDP being phosphorylated. They act to ensure adequate supplies of the various nucleoside triphosphates.

Phosphatases

Some **phosphatases**, often nonspecific, are capable of removing inorganic phosphate and thus lowering the level of phosphorylation by one. An important enzyme, **dUTPase**, removes *pyro*phosphate from dUTP, yielding dUMP which can be converted to dTMP. This prevents incorporation of uracil nucleotides into DNA (see below and Chapter 6).

Chapter 10. dUTPase, page 322.
Chapter 6. DNA replication requires a continuous supply of deoxyribonucleotides, page 160.

The pyrimidine biosynthetic pathway produces UMP

The primary end product of the pyrimidine biosynthetic pathway (Figure 10.19) is UMP. This is phosphorylated to UDP and **UTP**; CTP is, in turn, derived by the amination of UTP from glutamine.

Notable features of pyrimidine biosynthesis are described below.

Carbamoyl phosphate

Carbamoyl phosphate is one of the starting components for pyrimidine synthesis but is also an important metabolite of the **urea cycle**. Whereas the **carbamoyl phosphate synthetase** (CPS) for urea production resides in the mitochondrial matrix, the pyrimidine-producing enzyme (which is a different protein) is found in the cytosol.

Chapter 10. Carbamoyl phosphate synthetase is an allosteric enzyme, regulating the rate at which nitrogen enters the urea cycle, page 301.

Multifunctional proteins

In animal cells, but not in bacteria, the cytosolic CPS is part of a multifunctional enzyme complex catalysing all three steps from carbamoyl phosphate to dihydroorotic acid. A second enzyme complex, **UMP synthetase**, catalyses the last two reactions to UMP. Inherited defects in this enzyme can give rise to the clinical condition of **orotic aciduria**.

PRPP

In contrast to its role in purine synthesis, PRPP enters the pyrimidine pathway at the penultimate step, being coupled to orotic acid to yield orotidine monophosphate.

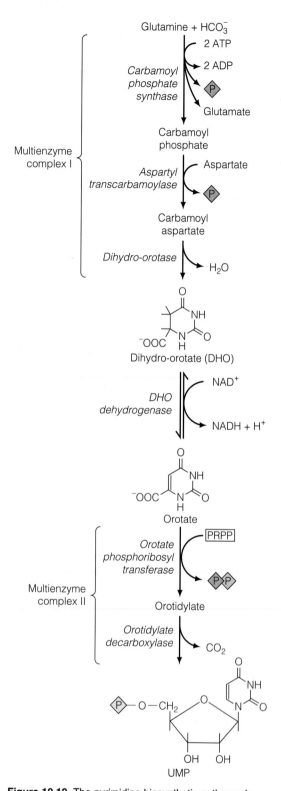

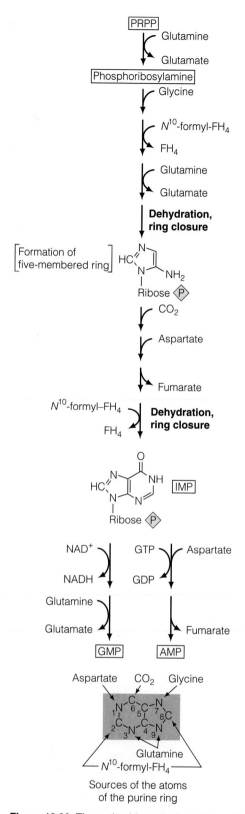

Figure 10.19 The pyrimidine biosynthetic pathway. In human cells, the reactions from glutamine and bicarbonate to dihydro-orotate are catalysed by one multienzyme complex. A second complex synthesizes UMP from orotate; dihydro-orotate dehydrogenase is a separate enzyme.

Figure 10.20 The purine biosynthetic pathway. The top part of the figure shows, in outline, the steps leading from PRPP and glutamine to IMP, highlighting the two steps that involve tetrahydrofolate (FH₄) derivatives. The lower part shows the reactions that generate GMP and AMP from IMP. The numbered purine ring shows the origins of the atoms of the ring.

The purine biosynthetic pathway produces IMP

The primary end product of the purine biosynthetic pathway (Figure 10.20) is **inosine monophosphate** (IMP), the ribonucleotide of hypoxanthine.

The purine biosynthetic pathway is more complicated and energy-intensive than that of the pyrimidines and has several important features, which are described below.

PRPP

In contrast to the situation in pyrimidine synthesis, PRPP enters the purine pathway at the outset, being aminated from glutamine to introduce ring nitrogen atom N-9 on to the 1-position of ribose (producing phosphoribosylamine; see Figure 10.17). From this, the purine ring is built up in a stepwise manner, starting by linking a glycine residue to the amino group through an amide bond.

Sources of nitrogen atoms

The second ring nitrogen atom (N-3) is also derived from glutamine, N-1 comes from an ATP-requiring aspartate-to-fumarate reaction and N-7 comes from glycine (along with C-4 and C-5).

Requirement for folic acid

Two of the ring carbon atoms (C-2 and C-8) are added from the folic acid derivative N^{10}-**formyl-tetrahydrofolate**, one of the reasons for the importance of this vitamin for nucleotide metabolism, the other being thymidylate synthesis (see below); C-6 comes from CO_2.

Chapter 10. Production of dTTP requires a special set of reactions, page 320.

Energy needs

Four ATPs are consumed in the reactions converting PRPP to IMP.

Production of AMP

The 6-amino group of adenine is added by amination of IMP from an aspartate-to-fumarate reaction and the expenditure of one GTP in a reaction similar to that which adds ring nitrogen N-1.

Production of GMP

The C-2 of the ring of IMP is first oxidized using NAD and forming **xanthosine monophosphate** (**XMP**) and then aminated from glutamine in a reaction similar to that which adds ring nitrogen N-3, using another ATP.

Generation of ATP and GTP

AMP and GMP are phosphorylated in two stages, via ADP and GDP respectively, to yield ATP and GTP.

Deoxyribonucleotides are synthesized from ribonucleotides

The ribonucleotides are the major nucleotides of all cells. Cells make and utilize these according to their needs, the pool sizes of the various nucleotide species being under refined control. Most cells of the body are not dividing and so not replicating their DNA; their requirements for deoxyribonucleoside triphosphates (dNTPs) are therefore very small. Cells that are dividing by mitosis need an adequate supply of *all four* dNTPs; failure in the supply of *any one* of these will cause a failure to pass through S phase (see Chapters 3 and 6). As cell division requires doubling of the amount of DNA, each dividing cell needs to produce a total amount of dNTP at least equivalent to its DNA content.

Figure 3.6. The cell cycle, page 46.
Chapter 6. DNA replication requires a continuous supply of deoxyribonucleotides, page 160.

The dNTPs are derived from cellular ribonucleotides, but the *de novo* supply can be supplemented by preformed components taken up from the plasma (see below: 'Nucleotide components can be salvaged'):

- **nucleosides,** especially thymidine, which is phosphorylated after uptake by thymidine kinase;
- **purine bases,** adenine, guanine and hypoxanthine, which are coupled to PRPP by phosphoribosyl transferases (see Figure 10.18).

The pathways of dNTP production from rNDPs are shown in Figure 10.21. Production of dATP, dGTP and dCTP is achieved directly from the corresponding ribonucleotides, the key enzyme involved being **ribonucleotide reductase.**

Ribonucleotide reductase

This ubiquitous and crucial enzyme uses the small protein **thioredoxin** as reductant in the conversion of all rNDPs to the corresponding dNDPs. The oxidized thioredoxin is reduced by thioredoxin reductase and NADPH. Ribonucleotide reductase is a large, multisubunit enzyme under complex allosteric control influenced by the concentrations of a number of nucleotides (notably rATP, dATP, dGTP and dTTP) with both positive and negative effects being observed (see Box 10.6).

Production of dTTP requires a special set of reactions

As deoxythymidine nucleotides are unique to DNA and essential for its synthesis, an adequate supply of dTTP is absolutely required for the continued action of DNA polymerase in DNA replication (see Chapter 6) and several enzymes are crucially involved in the production of dTTP. Interference with the activities of these enzymes impairs DNA replication and thus blocks cell division; they therefore present an attractive 'target' for the design of cytotoxic drugs such as some of those used in anticancer chemotherapy (see Chapter 17).

Key enzymes involved in this pathway are thymidylate synthetase, dihydrofolate reductase and dUTPase (Figure 10.21).

Thymidylate synthetase

This catalyses the reaction

$$dUMP + N^5,N^{10} - CH_2 - FH_4 \rightarrow dTMP + FH_2$$

dUDP, produced from rUDP by ribonucleotide reductase, is dephosphorylated to dUMP, the substrate for thymidylate synthetase, an enzyme characterized by a

Box 10.6. A genetic defect in adenosine metabolism impairs the immune system, page 320.

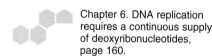
Chapter 6. DNA replication requires a continuous supply of deoxyribonucleotides, page 160.

Chapter 17 Anticancer drug therapy is useful, but has limitations, page 538.

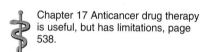
Chapter 10. Ribonucleotide reductase, page 320.

Box 16.1. Congenital immunodeficiency diseases, page 487. Chapter 17. Gene therapy is a potential cure for some genetic disorders, page 545.

> ### Box 10.6   A genetic defect in adenosine metabolism impairs the immune system
>
> Adenosine deaminase (ADA) catalyses the deamination of adenosine to inosine and of 2'-deoxyadenosine to 2'-deoxyinosine. An inherited defect in the gene for ADA causes a condition called **severe combined immunodeficiency** (SCID). Deficiency of this enzyme appears to make itself felt in lymphocytes by causing high intracellular concentrations of certain nucleotides, especially dATP. This in turn shuts down synthesis of the other deoxyribonucleotides by inhibiting ribonucleotide reductase (see above). The net result is inhibition of DNA replication in B lymphocytes and hence a failure to mount effective immune responses (see Chapter 16). Treatment of this condition with gene therapy (see Chapter 17) has had some success.

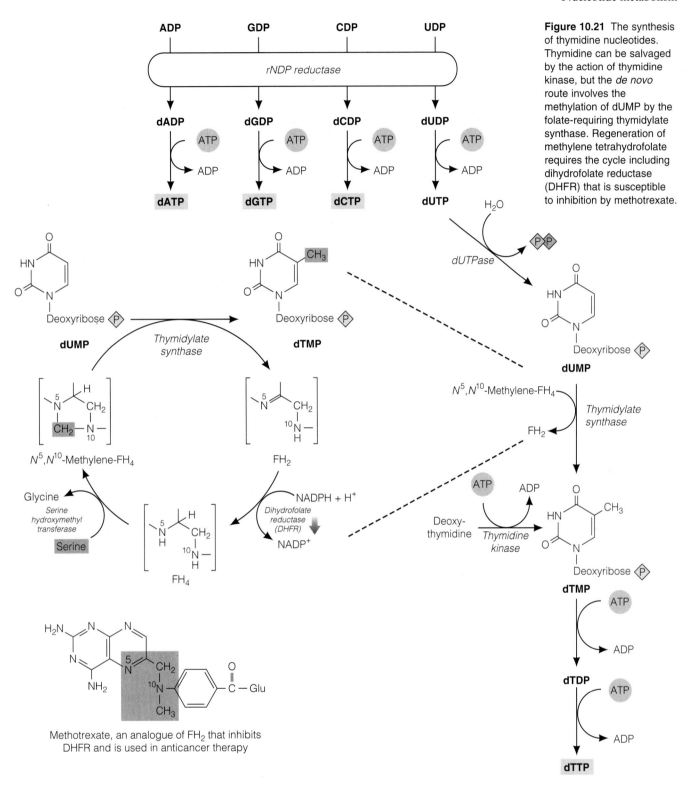

Figure 10.21 The synthesis of thymidine nucleotides. Thymidine can be salvaged by the action of thymidine kinase, but the *de novo* route involves the methylation of dUMP by the folate-requiring thymidylate synthase. Regeneration of methylene tetrahydrofolate requires the cycle including dihydrofolate reductase (DHFR) that is susceptible to inhibition by methotrexate.

Methotrexate, an analogue of FH_2 that inhibits DHFR and is used in anticancer therapy

very high affinity (K_m about 10^{-7} M) for its substrate and a requirement for the folic acid derivative N^5,N^{10}-methylene-tetrahydrofolate (CH_2–FH_4). dTMP is then phosphorylated to dTDP and dTTP in the usual manner but, unless CH_2–FH_4 is continuously regenerated, dTTP production will quickly stop. Thymidylate synthetase is very sensitive to inhibition by FdUMP, a metabolite of the anticancer drug 5-fluorouracil.

Dihydrofolate reductase (DHFR)

The other product of thymidylate synthetase action is dihydrofolate (FH_2), which must be reduced to FH_4 by DHFR, using NADPH as reductant, before the CH_2 group can be added from the amino acid serine to produce CH_2-FH_4.

$$FH_2 + NADPH + H^+ \rightarrow FH_4 + NADP^+$$

Interference with FH_4 regeneration, for example with the anticancer drug methotrexate (a folate analogue; Figure 10.21), will also stop dTTP synthesis.

dUTPase

The dUDP produced by ribonucleotide reductase is phosphorylated to dUTP by the nonspecific NDP kinase:

$$dUTP \rightarrow dUMP + PPi$$

It may be incorporated into DNA, because DNA polymerases appear to be unable to discriminate between dUTP and dTTP. In order to prevent too much uracil being incorporated into DNA (a small amount – about 1 U per 500 Ts – does go in), the pool of dUTP is usually kept very low by the action of dUTPase. The resultant dUMP can then be utilized by thymidylate synthetase.

NADPH supply

Note that the actions of ribonucleotide reductase (through its partner thioredoxin reductase) and thymidylate synthetase (through its partner dihydrofolate reductase) require a constant supply of NADPH as a reductant, so that cells replicating their DNA have a high requirement for a source of reducing power.

Nucleotide components can be salvaged

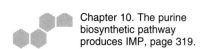

Chapter 10. The purine biosynthetic pathway produces IMP, page 319.

As the *de novo* synthesis of nucleotides, especially purines, is energy-expensive (see above), there is presumably an advantage for cells in possessing mechanisms for the reutilization of nucleotide components. The role of thymidine kinase has already been mentioned, but cells also contain the enzymes capable of joining purine bases to PRPP – the **phosphoribosyl transferases** (**PRTases**), specific for adenine (**APRTase**) and hypoxanthine/guanine (**HPRTase**) respectively, catalyse the reactions of the type (Figure 10.17):

Base + PRPP = Nucleotide + PPi

Of these two enzymes, HPRTase seems to be the more important and appears to be particularly important in brain metabolism, as individuals with a hereditary deficiency in HPRTase are severely mentally retarded and die young – the **Lesch-Nyhan syndrome**. The role of this enzyme is closely tied in with the degradation of purine nucleotides.

The catabolism of purine nucleotides can have important clinical consequences

Degradation of pyrimidine nucleotides proceeds by dephosphorylation to the nucleosides and eventually yields β-alanine and malonyl CoA; few medical problems appear to arise from defects in this pathway. Catabolism of purine nucleotides, on the other hand, may present the body with a series of difficulties which can have unfortunate clinical consequences.

Although it proceeds further in most animals, humans and other primates degrade purines only as far as **uric acid** (Figure 10.22). The key enzyme in this process is **xanthine oxidase**, a molybdenum metalloenzyme that oxidizes both

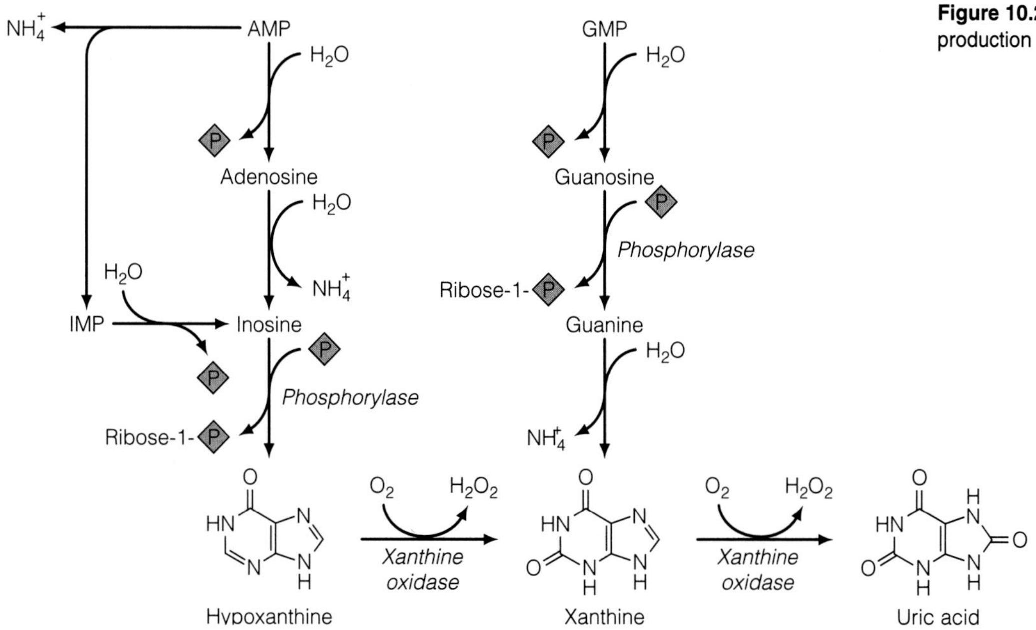

Figure 10.22 Degradation of purines and production of uric acid.

xanthine and hypoxanthine and uses molecular oxygen as its oxidizing agent, releasing H_2O_2. Problems of purine catabolism arise mainly from the relative insolubility of uric acid and the fact that normal levels of plasma urate are near saturation. Thus any impairment of the system is likely to cause **hyperuricaemia** and precipitation of urate crystals, notably in the joints, where gouty arthritis may result. The urate analogue allopurinol is an inhibitor of xanthine oxidase and is used as a drug to control this condition.

Hyperuricaemia is often related to PRPP metabolism due to one or more of the following events.

- Decreased levels of HPRTase give reduced purine salvage.
- Increased levels of PRPP synthetase give increased purine breakdown.
- Loss of feedback inhibition of the first step of purine biosynthesis leads to uncontrolled synthesis of purine nucleotides and results in increased purine breakdown.
- Excess production of PRPP.

Such high normal levels of plasma urate might be considered bad organization, but plasma urate shows antioxidant properties – an important factor in a long-lived species such as *Homo sapiens*. Urate levels are generally lower in shorter-lived species.

Folic acid and 'one-carbon' metabolism

Folic acid derivatives are involved in reactions that add single carbon atoms

Folic acid (anionic form **folate**) is an important vitamin (see Chapter 14) for the human body and, as is the case for most of the water-soluble vitamins, the dietary vitamin is converted to an active form after absorption. Several different, interconvertible folate derivatives are active in what is often termed 'one-carbon' metabolism – a group of anabolic reactions that involve the transfer of a group containing a single carbon atom.

 Chapter 14. Folic acid, page 449.

Figure 10.23 Outline of folate metabolism. (a) The ingested form of the vitamin (folate) is shown as a three-component molecule. (b) The highlighted structures show the reduction of folate to di- and tetrahydrofolate. The remainder of the diagram shows the interconversions of the forms of folate mentioned in the text.

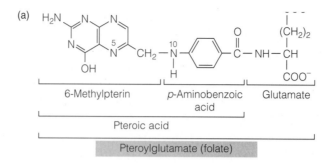

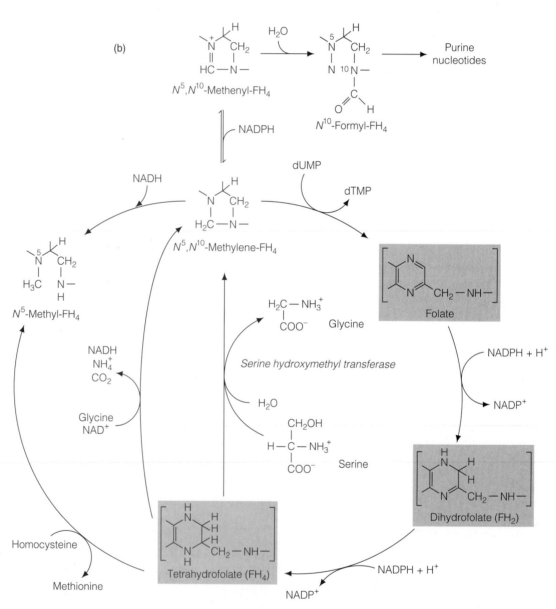

Folate comprises a **pterin** ring coupled through p-amino benzoate to glutamate and is a vitamin for animals because they cannot synthesize either of first two of these components. Interestingly, the sulphonamide group of synthetic antibacterial agents inhibits the growth of sensitive bacteria by competitively inhibiting the incorporation of p-amino benzoate into folic acid. The biologically active forms of folate are based on the reduced form **tetrahydrofolate** (FH_4), produced by two consecutive actions of dihydrofolate reductase (DHFR).

Dihydrofolate reductase is a particularly important enzyme because of the need to regenerate tetrahydrofolate during the synthesis of thymidylate for DNA replication and cell division (see above and Chapter 6). A further modification of folate is the addition of four additional residues of glutamate to the *p*-amino benzoate part of the folate molecule. These are linked α-amino to γ-carboxyl (as in the peptide **glutathione**) and probably serve to retain the folate within cells, as only the mono-glutamyl form can traverse the plasma membrane.

Chapter 10. Deoxyribo-nucleotides are synthesized from ribonucleotides, page 319.
Chapter 6. DNA replication requires a continuous supply of deoxyribonucleotides, page 160.

Tetrahydrofolate derivatives are important in amino acid and nucleotide metabolism

The 'one-carbon' units are mainly derived from the amino acids **serine**, using serine transhydroxymethylase and pyridoxal phosphate and (in mitochondria) from **glycine**, using glycine decarboxylase and NAD^+. The pathways by which this occurs and by which the various folate derivatives are interconverted are outlined in Figure 10.23.

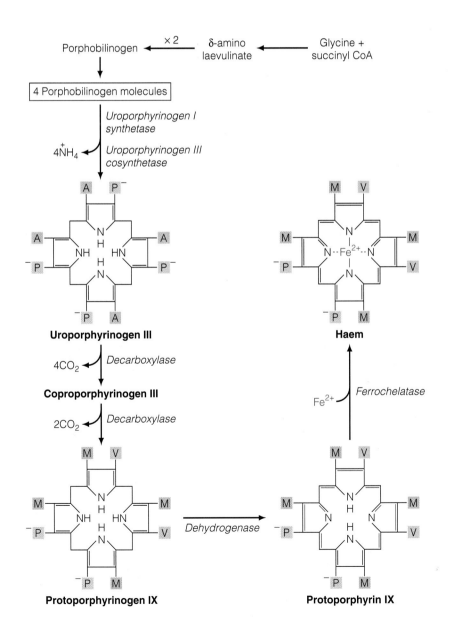

Figure 10.24 An outline of haem biosynthesis. The high-energy bond of succinyl CoA gives it the ability to combine with glycine to form δ-aminolaevulinate, two molecules of which give rise to porphobilinogen by dehydration. Four molecules of porphobilinogen are joined to form a linear tetrapyrrole that is cyclized into uroporphyrinogen III, with its characteristic arrangement of acetyl (A) and propionyl (P) groups. A sequence of decarboxylation and dehydrogenation steps leads to protoporphyrin IX (M = methyl; V = vinyl), into which Fe^{2+} is inserted by ferrochelatase to yield haem.

325

There is a close connection between nucleotide and amino acid metabolism, because formyl units can also be derived from **histidine**, and N^5-methyl-tetrahydrofolate is required for the reconversion of homocysteine to **methionine**. In nucleotide metabolism, purine biosynthesis requires N^{10}-formyl-tetrahydrofolate; N^5,N^{10}-methylene-tetrahydrofolate is the methyl donor in the synthesis of thymidylate. Other folate compounds are involved in amino acid metabolism (see above).

Folate and vitamin B_{12} have a physiological association, in that deficiencies in either can cause anaemia due to interference in the production of red blood cells. This is probably because vitamin B_{12} is required for methionine synthesis; its deficiency leads to an accumulation of N^5-methyl-tetrahydrofolate and a corresponding lack of the folate derivatives required for nucleotide synthesis. The diminished availability of nucleotides impairs the cell division necessary for normal production of red blood cell precursors (see Chapter 12). This represents a *functional* folate deficiency, even though total folate levels appear to be normal.

Chapter 10. Glycine is synthesized from serine, page 307.

Chapter 12. The erythrocyte must survive wear and tear, page 365.

Chapter 7. The mitochondrial respiratory chain comprises four multisubunit assemblies of proteins, page 199.
Chapter 12. The quaternary structure of haemoglobin affects oxygen binding, page 370.

Porphyrin metabolism

All cells need haem derivatives

The need for haem and other porphyrin derivatives in the body is twofold:

1. as prosthetic groups for the various **cytochromes** found in all cells (see Chapter 7);
2. as prosthetic groups for the oxygen-transport proteins **haemoglobin** and **myoglobin** (see Chapter 12).

Haem biosynthesis builds glycine and succinyl CoA into the four-ringed porphyrin molecule

The biosynthesis of haem can be divided into five stages (Figure 10.24).

1. Synthesis of δ-**amino laevulinic acid** (δ-ALA) from glycine and succinyl CoA.
2. Joining of two molecules of δ-ALA to give **porphobilinogen** (PB) with its pyrrole ring.
3. Joining of four molecules of PB to give the ring-like tetrapyrrole **uroporphyrinogen III** (uro III).
4. Decarboxylation and dehydrogenation of uro III to yield **protoporphyrin IX** (pp IX) with its closed ring structure.
5. Addition of a ferrous (Fe^{2+}) ion to pp IX by **ferrochelatase** to form **haem**.

Box 10.7   Clinical jaundice is due to hyperbilirubinaemia

Hyperbilirubinaemia reflects failure to conjugate all of the bilirubin produced, and may arise from several causes.

- Overproduction of bilirubin, due to haemolytic anaemia.

- Inadequate uptake of bilirubin into hepatocytes, due to viral hepatitis or cirrhosis of the liver.

- Failure of the conjugation reaction. This occurs especially in preterm infants because the bilirubin-UDP-glucuronyl transferase begins to be synthesized in the liver *only* near the time of birth. Neonates normally have higher concentrations of bilirubin in the bloodstream than adults, but at excessive concentrations unconjugated bilirubin crosses the blood–brain barrier and is toxic.

- Disorders leading to reduced circulating amounts of albumin, or the unavailability of bilirubin-binding sites on albumin. The latter condition may occur in patients taking sulphonamides, which occupy the bilirubin binding site of albumin.

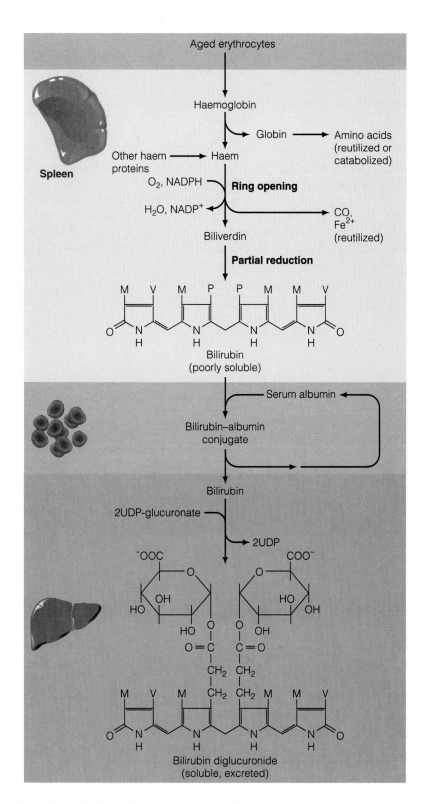

Figure 10.25 An outline of haem degradation. The top part of the figure shows the conversion in the spleen of haem to biliverdin and then bilirubin, with successive release of Fe^{3+} and carbon monoxide. Two glucuronate residues are esterified to the propionyl groups of bilirubin in the liver to form the more soluble bilirubin diglucuronide, secreted into the bile.

Inherited metabolic defects in (3) (and lead poisoning) give rise to various types of porphyrias, in which excessive plasma levels of porphobilinogen cause a variety of symptoms.

Haem degradation leads to the production of bilirubin

The short life of the erythrocyte (about 120 days) and the large amount of haemoglobin it contains means that the body has to dispose of about 1% of its

Chapter 10. The catabolism of purine nucleotides can have important clinical consequences, page 322.
Chapter 12. Binding of small molecules, page 359.

Chapter 13. Detoxification and removal of xenobiotics, page 389.

haem per day. Defects in various steps of this pathway give rise to various types of jaundice. Breakdown of haem (Figure 10.25) leads to production of **bilirubin**, a relatively insoluble molecule with antioxidant properties similar to those of uric acid (see above).

Because of its insolubility, bilirubin is transported through the bloodstream bound to plasma proteins, mostly albumin (see Chapter 12). The bilirubin–albumin complex binds to a receptor on the plasma membrane of a liver cell. The complex dissociates, albumin re-enters the plasma and bilirubin enters the hepatocyte where it is bound tightly to **ligandin**, a protein that constitutes about 6% of the cytoplasmic protein of hepatocytes. The conjugation of bilirubin to mono- or diglucuronates (see Chapter 13) occurs on the surface of the endoplasmic reticulum. The important enzyme in this process is **bilirubin-UDP-glucuronyl transferase** (Figure 10.26). Conjugated bilirubin is secreted as a component of bile (the 'bile pigments') and is degraded by bacteria in the colon which convert it to the water-soluble compound **urobilinogen**, which is excreted in the faeces. After excretion this compound spontaneously oxidizes to the brown-coloured **urobilin**.

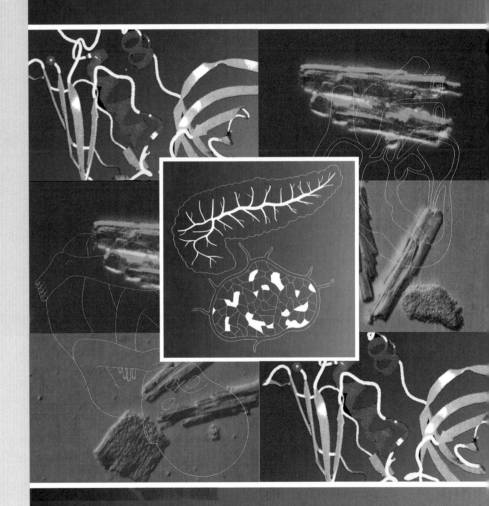

PART D

TISSUES OF THE BODY: THEIR STRUCTURE, FUNCTIONS AND INTERACTIONS

CHAPTER 11
Communications between cells and tissues: hormones, transmitters and growth factors

CHAPTER 12
Blood: molecular functions of cells and plasma

CHAPTER 13
The molecular and metabolic activities of multicellular tissues

CHAPTER 14
From food to energy: molecular aspects of digestion and nutrition

CHAPTER 15
The structural tissues

Communication between cells and tissues: hormones, transmitters and growth factors

Hormones are signal molecules that communicate physiological and nutritional information between cells or tissues. Some hormones are produced by a variety of tissues, while others are produced only by specialized endocrine tissues. Hormones are a chemically diverse range of compounds: these include modified amino acids, small peptides, fatty acid derivatives and steroid derivatives.

Hormones exert their influence by binding to specific receptors, either on the plasma membrane of the target cell or within the cell. Receptors are linked to varied and complex processes within the cell that produce intracellular signals, which in turn regulate metabolic, transcriptional or other events within the cell.

This chapter describes the production and storage of these compounds and the mechanisms through which they act. The chapter concludes with a brief introduction to the properties of growth factors.

This chapter deals predominantly with hormones and other mediators produced in peripheral tissues, with only passing reference to hormones produced in the central nervous system.

Hormones and transmitters provide an information network throughout the body

The coordinated function of multicellular tissues, and of the whole body, depends on a complex system by which the requirements of one cell or tissue are signalled to other cells capable of responding to such signals. The purpose of these communication systems is to ensure coordinated function of the whole animal, and provide mechanisms through which it can respond to environmental or nutritional change.

Hormones, transmitters and growth factors provide the chemical basis for communication or signalling between cells. When these molecules bind to receptors on the plasma membrane or to receptors within the cell they generate intracellular signals responsible for changing patterns of activity within the cell. Traditionally, hormones and transmitters have been regarded as distinct groups of signalling molecules but as our knowledge of the mechanisms of action of these molecules has developed the distinction between the two groups has blurred, as molecules from the two groups share common elements in their mechanisms of action. There is an increasing tendency to classify signalling molecules according to the molecular mechanisms through which they mediate change within cells.

The mechanism of action of extracellular signal molecules is related to their chemical characteristics

Hormones and transmitters are as structurally diverse as their functions. Chemically they may be:

- amino acids or their derivatives
- small protein molecules (peptide hormones)
- derivatives of cholesterol (steroid hormones)
- modified fatty acids (eicosanoids).

Modified amino acids, such as adrenaline, and peptide hormones are water soluble and therefore cannot traverse the lipid membrane of cells. They bind with high affinity to specific proteins called **receptors** on the outer surface of the plasma membrane of target cells. A conformational change induced in the receptor by the binding of these molecules then modifies the behaviour of other membrane proteins, and in turn these modify molecules within the cytoplasm.

Hormones of the thyroid, steroid and eicosanoid classes are lipid soluble and can diffuse across the cell membrane. They must, however, be tightly bound to carrier proteins for transport through the bloodstream to the target tissues. They modify aspects of cell function after binding to receptors in the cytoplasm or nucleus of the cell (Figure 11.1). Some eicosanoids (PGE_1 and PGI_2) are exceptions to this rule because as lipophilic hormones they bind to, and act through, plasma membrane receptors.

Hormones and transmitters are classified according to the relationship between site of production and target site

Hormones may be synthesized continuously and stored in vesicles within cells producing them, ready to be released upon demand. Alternatively, they may be synthesized only in response to a demand.

Once released, hormones travel to the target tissues where they bind to specific receptors. Binding to the receptor induces a diverse range of hormone- and tissue-specific responses some of which are mediated by the production of **second messengers** within the target cells.

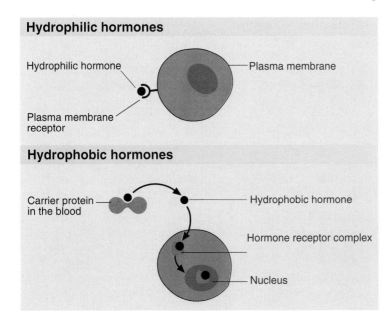

Figure 11.1 Hydrophilic hormones generally bind to receptors on the outer surface of the plasma membrane. Some exert a longer term effect dependent upon the hormone–receptor complex entering the cell by endocytosis. Hydrophobic hormones diffuse across the cell membrane and bind to receptors within the cell.

- **Autocrine** signalling involves a hormone influencing the behaviour of the cell in which it is produced. An example is the eicosanoid group of molecules, which are produced by most cells throughout the body. These molecules are sometimes called **local mediators**, rather than hormones.
- **Paracrine** signalling is local to the site of production of the hormone but between adjacent cells. This mechanism most commonly involves neurohormones or neurotransmitters, for example acetylcholine or noradrenaline, which are released from nerve cells and activate other nerve cells or muscle cells.
- **Endocrine** signalling occurs when signal molecules are synthesized and secreted by cells of specialized endocrine tissue, and transported through the bloodstream to distant target tissues. For example, insulin is produced by β cells of the pancreas, and adrenaline by cells of the adrenal gland, but both influence the metabolism of a variety of other tissues.

Tissues may exhibit both short and long-term responses to hormones

The response of cells to the binding of hormone to receptor varies in rapidity and duration. The response may be short lived, on a timescale of seconds, or may be sustained for periods of hours or days.

- Rapid responses are usually associated with modification of an enzyme, for example by phosphorylation, and are usually expressed as changes in rates of reactions important in regulating metabolism. Rapid response can be terminated by inactivation of the hormone, by inactivation of second messengers or by reversal of the modification of the enzyme.
- Slow and sustained responses are more likely to indicate that the signal molecule alters the pattern of gene expression, perhaps increasing the number of particular enzyme molecules available in the cell. These responses are particularly important in regulation of growth and differentiation.

Some hormones are released in intermittent bursts; others are secreted continuously, with relatively minor variations occurring in the circulating concentration of the hormone. Hormones that alter metabolic activity tend to be released intermittently; those controlling tissue growth and differentiation are constantly circulating.

Chapter 11. Catalytic or enzyme-linked receptors have tyrosine kinase activity, page 337.

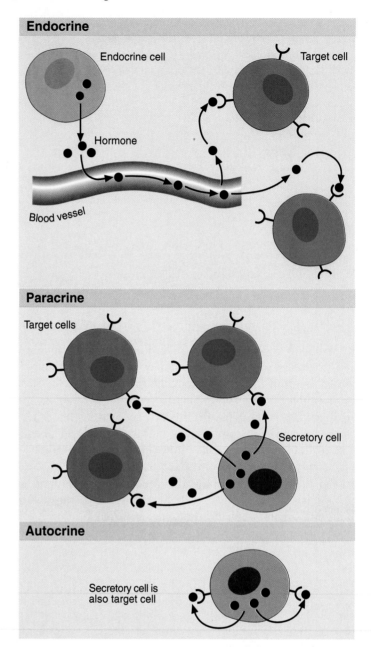

Figure 11.2 This figure illustrates the relationships between the cell producing the hormone and the target cell for autocrine, paracrine and endocrine signalling mechanisms.

Signal molecules control a diverse range of functions

- Cell growth and differentiation
- Electrical and mechanical functions supporting locomotion
- Digestion of foods
- Maintenance of fluid and electrolyte homeostasis
- Wound healing
- Response to infection by viruses or bacteria
- Nutritional support for all tissues

Responsiveness of receptors can be modulated

Variation in receptor number provides a mechanism for regulating the responsiveness of cells to hormone. In the case of the plasma membrane receptors for the peptide hormones insulin, glucagon and the growth factors, endocytosis of the receptor structure (Chapter 3) is the principal mechanism for reducing, or **down-regulating**, the number of receptors available on the plasma membrane surface. The internalized receptor may be recycled to the plasma membrane or may be degraded within lysosomes. Down-regulation may also occur if cells are constantly exposed to abnormally high concentrations of circulating hormone. This mechanism desensitizes the cell, protecting it against being constantly in a state of metabolic 'overdrive'.

Chapter 3. Box 3.1
Endocytosis and exocytosis, page 49.

An alternative mechanism for regulating the responsiveness of a tissue to hormone is through phosphorylation of specific amino acids in the receptor structure. Generally, phosphorylation reduces the affinity of receptor for its hormone, so that a higher hormone concentration is required to produce a response in the cell. An example of this is the phosphorylation of specific threonine and serine residues of β-adrenergic receptors exposed to adrenaline for long periods. Although these phosphorylated receptors still bind adrenaline, the hormone is no longer able to activate the intracellular enzyme to which the hormone is linked.

Mechanisms used by hormones acting through plasma membrane receptors

Plasma membrane receptor molecules are integral proteins, which are believed in most cases to traverse the structure of the phospholipid bilayer. They are therefore able to signal to the interior of the cell membrane the information that the receptor is occupied. The mechanism by which intracellular processes recognize the activation of a receptor on the outer surface of the cell is called **signal transduction**.

Plasma membrane receptors may be classified into three types, depending on the nature of the immediate response when hormone or transmitter binds to the receptor:

- channel linked
- G-protein linked
- catalytic, or enzyme linked.

The intracellular changes induced by both G-protein-linked and enzyme-linked receptors are mediated through kinases, which phosphorylate intracellular proteins. The changes are terminated or reversed by phosphatases, which reverse the phosphorylation. In the case of G-protein-linked processes the kinases are activated by the second messengers generated within the cell and phosphorylate serine or threonine residues of target proteins. For enzyme-linked receptors the cytoplasmic domain of the receptor is often a tyrosine kinase, which phosphorylates tyrosine residues of the cytoplasmic target proteins.

Channel-linked receptors regulate the permeability of membranes to ions

Neurotransmitters such as acetylcholine, γ-amino butyric acid or glutamate bind to specific plasma membrane receptors, which are transmembrane proteins or protein complexes with a receptor-binding domain on the outer surface of the cell and a channel-forming domain passing through the membrane. The channel is ion specific, allowing specific ions to pass through the channel. The ion specificity is determined by the structure of the channel-forming portion of the protein. Binding of transmitter to the receptor opens the channel allowing an influx of the specific ion(s) into the cell. These channels are generally associated with electrical activity in cells. Mechanisms of electrical activity are discussed in Chapter 13.

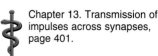

Chapter 13. Transmission of impulses across synapses, page 401.

G-protein-linked receptors mediate a wide variety of processes

When hormone binds to a receptor linked to a G-protein or guanosine triphosphate (GTP)-binding protein, the conformational change induced in the transmembrane receptor protein is communicated to a G-protein located on the inner surface of the plasma membrane. The signal generated by hormone binding to the receptor is then transmitted through the G-protein to membrane-bound enzymes which generate the second messenger molecules within cells. Within any cell a large number of different G-proteins link receptors to a variety of intracellular processes. Currently four groups of G-proteins are recognized (see Chapter 3):

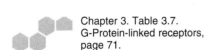

Chapter 3. Table 3.7. G-Protein-linked receptors, page 71.

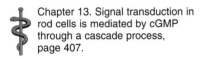

Chapter 13. Signal transduction in rod cells is mediated by cGMP through a cascade process, page 407.

1. The G_s family activate adenylate cyclase to increase the rate of production of cAMP within the cell.

2. The G_i family inhibit adenylate cyclase. Some members of this family interact with ion channels. Transducin, part of the light-detecting mechanism in the eye (Chapter 13) is a member of the G_i family.

3. The G_q family activate phospholipase C.

4. The role of the G_{12} family of G-proteins has not yet been confirmed.

All G-proteins have a common three subunit structure: Gα, Gβ, and Gγ. Each G-protein so far identified has a unique α subunit, of which there are at least 15 types. Some α subunits are specific for a single cell type, others are widely distributed. Several β and γ subunits have been identified. The β and γ subunits probably function other than simply as attachment sites for the α subunit.

G-protein α subunits catalyse GTP hydrolysis

The α subunit has the capacity to bind GDP or GTP and to catalyse the reaction

$$GTP + H_2O \xrightarrow{GTPase} GDP + Pi$$

When hormone binds to the receptor it produces a conformational change in the portion of the receptor exposed to the interior of the cell. G-protein

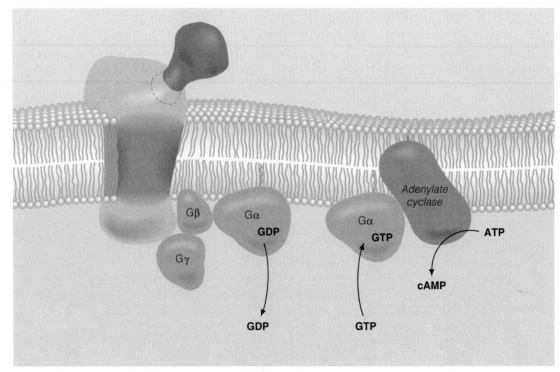

Figure 11.3 The information conveyed when hormones bind to G-protein-linked receptors depends on the α subunit of the G-protein migrating along the membrane to interact with and activate adenylate cyclase, increasing the rate at which it produces cAMP.

associates with the receptor and undergoes a conformational change, which results in the bound GDP being replaced by GTP. The α subunit then dissociates from the β,γ subunit complex, diffuses along the plasma membrane and binds to the effector enzyme (for example, adenylate cyclase), which produces the intracellular second messenger molecules. Activation of effector enzyme by this mechanism is very short lived, and ceases when the GTPase activity of G-protein hydrolyses GTP to GDP. Gα then diffuses back along the membrane and reassociates with the β,γ complex. At this stage production of cAMP returns to basal rates (see Figure 11.3).

G-proteins may be inhibitory (G_i) or stimulatory (G_s), the difference reflecting the presence of stimulatory ($α_s$) or inhibitory ($α_i$) activity of the α subunit in the G-protein complex.

All receptors that interact with G-proteins share several structural features, although their amino acid sequences vary substantially. Each has an extracellular hormone-binding domain, a number (thought most commonly to be seven) of transmembrane α-helices and two hydrophobic regions located on the cytoplasmic face of the membrane, which interact with G-protein. Hormone specificity is conferred by the sequence and conformation of the part of the protein exposed on the outer surface of the plasma membrane.

Catalytic or enzyme-linked receptors have tyrosine kinase activity

Plasma membrane receptors for insulin and growth hormones are multisubunit transmembrane protein complexes with hormone binding domains on the external facing α subunits and tyrosine kinase domains on the β subunits that protrude into the cytoplasm. Binding of the hormone to the receptor stimulates autophosphorylation of the tyrosine kinase domain, with two molecules of ATP used to phosphorylate two tyrosine residues on each of the β chains of the receptor. This phosphorylation activates the capacity of tyrosine kinase to phosphorylate tyrosine residues in other proteins (Chapter 3), and this effect is not lost if insulin dissociates from the receptor site. The exact nature of the target proteins has not been established. The tyrosine kinase activity is terminated by a tyrosine phosphatase, which removes the phosphate groups.

Chapter 3. A family of transmembrane receptors is involved in the G-protein-mediated activation of the adenylate cyclase and phosphoinositide pathways, page 70.

Chapter 3. Growth factors and insulin bind to a family of receptors with intrinsic tyrosine kinase activity, page 72.

Box 11.1 Cholera and whooping cough

Cholera
Cholera, an illness caused by bacterial infection, is characterized by electrolyte loss and dehydration. Untreated it rapidly becomes life threatening. The disease is caused by infection with *Vibrio cholerae*, an organism that thrives in contaminated water supplies and secretes cholera toxin. This toxin catalyses the transfer of ADP-ribose from NAD^+ to the α subunit of G_s subunits of G-proteins, blocking the GTPase activity. G-protein is then permanently in the active state. In cells of the intestinal epithelium the continuous high concentration of cAMP stimulates continuous secretion of chloride and bicarbonate ions and water into the lumen of the gut, causing dehydration.

Whooping cough
Whooping cough is caused by the organism *Bordetella pertussis*, which secretes the causative agent, pertussis toxin. This toxin catalyses the transfer of ADP-ribose from NAD^+ to the G_i unit of G-protein. This blocks the displacement of GDP by GTP and thereby blocks the inhibitory effect of G_i on adenylate cyclase.

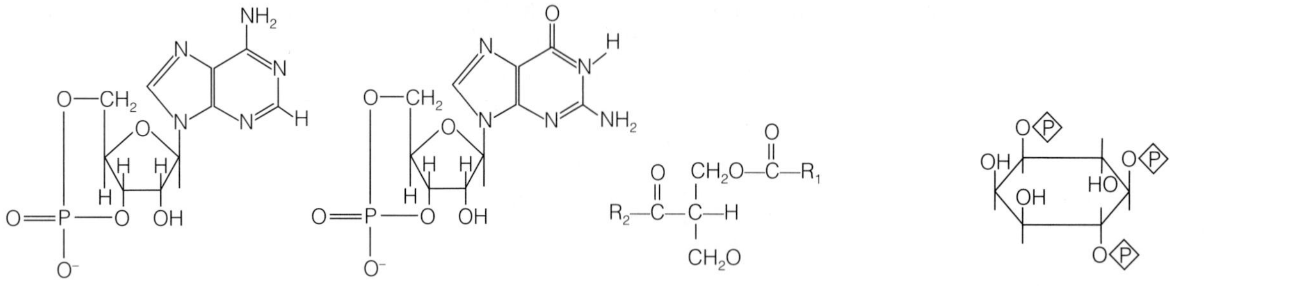

Cyclic AMP (cAMP)   **Cyclic GMP (cGMP)**   *sn*–1,2–Diacylglycerol (DAG)   Inositol 1,4,5–trisphosphate (IP$_3$)

Figure 11.4 Structures of common second-messenger molecules.

Second messengers are small molecules produced within cells in response to hormone binding and signal transduction

The intracellular response to receptor activation is the production of small molecules called second messengers. Common second messengers include

- cAMP
- cGMP
- sn 1,2-diacylglycerol (DAG)
- inositol-1,4,5-trisphosphate (IP$_3$)
- calcium ions.

Structures of common second messenger molecules are shown in Figure 11.4.

The second messengers activate protein kinases that phosphorylate serine or threonine residues of enzymes within the cytoplasm, thus stimulating or inhibiting the activity of these enzymes. Generally, these phosphorylation reactions initiate cascade mechanisms (see Chapter 7) that amplify the hormone-induced processes within cells. The responses may involve processes as diverse as

- release of hormones,
- release of glucose from glycogen stores,
- generation of electrical impulses.

The response to hormone is terminated when specific phosphatases remove the phosphate groups, returning the enzyme activity to basal rates.

The number of second messenger mechanisms is small relative to the diverse range of metabolic and functional responses seen in different tissues. The complexity of responses generated reflects the contributions of three components.

- The capacity of the cell to respond to a particular hormone by expressing receptors specific to that hormone on its plasma membrane.
- The change induced in the receptor by hormone binding.
- The capacity of the enzymes in the cell to respond to the second messenger(s).

In summary, the intracellular response generated by signals that operate through second messengers varies depending on the function of the cell and the enzymes available for modification by the specific second messenger. This means that the same type of hormone–receptor complex may activate different processes in the cells of different tissues.

Chapter 7. Covalent modification of existing enzyme molecules can change their activity, page 195.

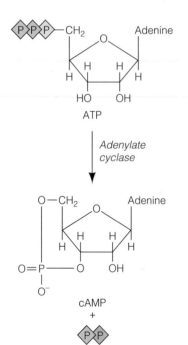

Figure 11.5 Production of cyclic AMP from ATP.

Hormonal responses mediated through adenylate cyclase and the production of cAMP

Many hormones (including adrenaline, glucagon and parathyroid hormone) exert their effects on target cells through G-protein-mediated changes in the activity of adenylate cyclase. The catalytic site of this transmembrane enzyme is located on the inner surface of the plasma membrane and catalyses the production of adenosine cyclic-3',5'-phosphate (cAMP) from ATP (Figure 11.5) within the cell.

Cyclic AMP is a ubiquitous second messenger, but the response of a particular cell type to a change in the intracellular concentration of cAMP depends on the presence of enzymes within the cell that respond to changes in cAMP concentration because they are subject to phosphorylation by cAMP-activated kinases.

Some hormone–receptor complexes operate through G-proteins that stimulate adenylate cyclase; others operate through G-proteins that inhibit adenylate cyclase. For example, adipose tissue has receptors for adrenaline, glucagon and ACTH, all of which operate through G_s proteins to *stimulate* adenylate cyclase, and for prostaglandins and adenosine which operate through G_i to *inhibit* adenylate cyclase. In liver cells the hormones glucagon and adrenaline operate through different receptors, both of which stimulate adenylate cyclase. More than one type of hormone–receptor complex may operate through the same G-protein in a single cell type.

The concentration of cAMP in a cell is reduced by the action of **cAMP-phosphodiesterase**, which hydrolyses the cyclic phosphate bond, producing 5'-AMP (Figure 11.6).

Hormonal responses mediated through inositol phosphates and calcium ions

Receptors for **vasopressin, angiotensin, serotonin** and α-adrenergic receptors are linked to activation of membrane-bound **phospholipase C**. This enzyme catalyses the release of two second messengers, **inositol-1,4,5-trisphosphate (IP₃)** and **diacylglycerol (DAG)**, from the membrane phospholipid **phosphatidylinositol-4,5-bisphosphate (PIP₂)** (Figure 11.7). These products normally function in a synergistic manner, operating through different mechanisms towards a common change of metabolic response. The synthesis of PIP₂ is outlined in Chapter 9 (Figure 9.00).

The second messenger IP₃ is water soluble and passes into the cytoplasm, where it binds to receptors on the endoplasmic reticulum (sarcoplasmic reticulum in smooth muscle). In response, channels in these membranes open, releasing calcium ions into the cytoplasm.

IP₃ is very short-lived, operating on a timescale of seconds before being degraded and inactivated by phosphatases.

DAG is retained within the membrane, where it activates **protein kinase C**. This enzyme is located in the cytoplasm of an unstimulated cell, but and increase in the calcium ion concentration causes it to bind to the plasma membrane, where it is activated by DAG. Protein kinase C phosphorylates serine and threonine residues of target proteins (Figure 11.8).

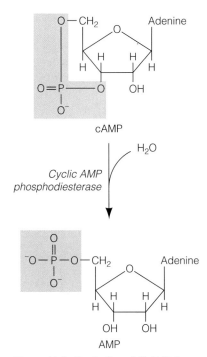

Figure 11.6 Production of 5'-AMP from cAMP.

Chapter 9. Hydrolysis of phospholipids is catalysed by phospholipases with differing specificities for particular bonds, page 284.

Figure 11.7 Production of second messengers from membrane phospholipids

Phosphatidylinositol 4,5-bisphosphate (PIP₂)

+ H₂O

sn-1,2-Diacylglycerol (DAG)

Inositol 1,4,5-trisphosphate (IP₃)

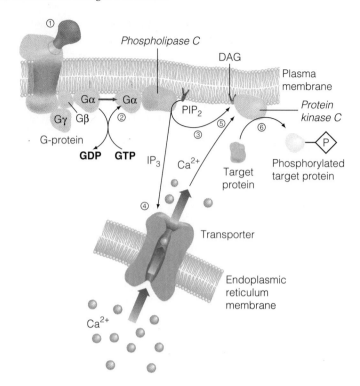

Figure 11.8 Phosphatidylinositol-4,5-bisphosphate, a phospholipid component of the plasma membrane, is the substrate from which these intracellular signal molecules are produced. Details are described in the text. The calcium-mediated part of the pathway depends upon Ca^{2+}-activation of calmodulin (see Figure 11.9).

Chapter 17. Chemicals can participate in both initiation and promotion of tumours, page 532.

Protein kinase C also plays a central role in cell division and proliferation. This is illustrated by the fact that tumour promoters such as the phorbol esters (see Chapter 17) activate the enzyme and transform normal cells into malignant cells, as a result of loss of growth control.

The concentration of calcium ions is a sensitive signal because the normal intracellular calcium ion concentration in very low (<0.2 mM) compared with an extracellular concentration of >2 mM. With such a large gradient across the plasma membrane only a small influx is need to produce a substantial increase in the intracellular concentration. Many cells have intracellular calcium stores and all have calcium extrusion processes.

Cellular responses to increased concentration of calcium in the cytoplasm include

- stimulation of contraction in muscle cells,
- activation of glycogen phosphorylase to release glucose from glycogen stores in liver and muscle cells, and
- exocytosis of some specific granules, notably the insulin-storage granules of pancreatic β cells.

Many calcium-dependent processes operate through the protein calmodulin

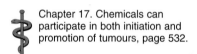
Chapter 12. Oxygen binding by haemoglobin is cooperative, page 371.

Calcium ions bind very tightly to some proteins, forming coordination complexes with both charged and uncharged oxygen atoms. Bound calcium ions can induce substantial changes in protein conformation. Many of the calcium-dependent processes in cells are mediated by the binding of calcium to the protein **calmodulin**. This ubiquitous protein, which may comprise as much as 1% of the cell protein, is rich in aspartate, glutamate and asparagine residues. Calmodulin has four binding sites for calcium, the binding being cooperative in a manner similar to that seen when oxygen binds to haemoglobin (Chapter 12). Calcium ions induce a conformational change in calmodulin, allowing it to associate tightly with a

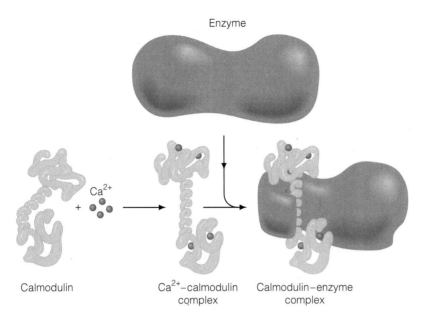

Enzyme

Calmodulin

Ca^{2+}–calmodulin complex

Calmodulin–enzyme complex

Figure 11.9 Calmodulin is a Ca^{2+}-binding protein which responds to changes in calcium ion concentration in a manner which enables it to bind to a variety of other proteins (enzymes) and alter their activity.

variety of other proteins and modify their activities. Among the susceptible enzymes are cAMP phosphodiesterase and glycogen phosphorylase kinase (Chapter 8).

Mechanisms used by hormones acting through intracellular receptors

Lipophilic hormones, such as thyroid hormone and steroid hormones, diffuse across plasma membranes and bind to receptors within the cell. Other hormones (such as insulin, whose short-term actions derive from activation of plasma membrane receptors), have a second, longer term action dependent on their ability to enter the cell by endocytosis. Through mechanisms that are not yet fully understood, insulin acts in the long term to alter rates of protein synthesis.

Mechanism of action of thyroid and steroid hormones

Within the cell these hormones bind tightly to specific receptor molecules located in the cytoplasm or within the nucleus. Binding induces a conformational change in the receptor, increasing its affinity for specific DNA nucleotide sequences. Association of hormone-bound receptor with DNA controls the rate of transcription of the appropriate genes. In many cases the enhancement of transcription appears to operate in two stages: the primary response arises from the binding of hormone-activated receptor to the DNA; a secondary response can arise from the binding to DNA sequences of proteins that have been transcribed as a result of the primary response. This mechanism appears to be complicated by the presence of other tissue-specific gene regulators. The DNA-binding domains of all steroid-activated receptors have similar sequences rich in cysteine, arginine and lysine residues. Thyroid hormone acts through similar mechanisms.

 Box 5.1, Box Table 1. DNA-binding transcription factors, page 118.

Box 11.2   Nitric oxide is a signal molecule

Blood vessels are lined with smooth muscle cells underlying a surface layer of endothelial cells. These endothelial cells produce a relaxing factor (initially termed EDRF: endothelial-derived relaxing factor), and the relaxation of the vessel promoted by this factor is associated with an increase in the amount of cGMP produced in the smooth muscle cells from activation of guanylate cyclase. Cyclic GMP stimulates cGMP-activated protein kinase, which phosphorylates proteins that relax smooth muscle cells and cause vasodilation. The relaxing factor has been identified as nitric oxide (NO), which is produced from arginine by NO synthase, an enzyme found in many mammalian cell types. The several different isoforms of this enzyme are subject to different mechanisms of regulation. In vascular endothelial cells the enzyme is regulated by Ca^{2+}-calmodulin, but other isoforms are Ca^{2+}-independent. The synthesis of NO requires molecular oxygen and the coenzymes tetrahydrobiopterin, FAD and NADPH.

$$\text{Arginine} + O_2 \xrightarrow[\text{NO synthase}]{} \text{Citrulline} + \text{NO}$$

The isoform characteristic of macrophages produces large amounts of NO, which is cytotoxic in these cells. Nitric oxide is known to inhibit iron–sulphur cluster-dependent enzymes and the electron transport chain (Chapter 7).

The production of NO may be stimulated in the cardiovascular system by the shear force on the endothelial cells caused by fluctuations in blood flow, and by a number of agonists. NO counterbalances the vasoconstriction produced by the sympathetic nervous system and by the renin–angiotensin system (Chapter 13).

 Box 7.3. Iron– sulphur (Fe–S) centres in the electron transport chain, page 202.

 Chapter 13. The renin–angiotensin system controls blood pressure and sodium ion concentration in the bloodstream, page 393.

Synthesis, storage, processing, secretion and mode of action of selected hormones

Hormones can be divided into four classes:

1. Modified amino acids,
2. Peptides,
3. Fatty acid derivatives, and
4. Steroids.

Attention will be focused on those hormones with well-characterized mechanisms of action, and those that influence the central metabolic processes.

Modified amino acids

The catecholamines: adrenaline (epinephrine), noradrenaline (norepinephrine) and dopamine

Adrenaline and noradrenaline are synthesized in sympathetic nerve terminals and in cells of the adrenal medulla, where they are stored in specialized granules. These storage granules are released by exocytosis in response to neural signals mediated by the transmitter acetylcholine. Dopamine is generated through the same pathway (Figure 11.10).

Important features of the synthesis of these compounds include

 Table 10.2. Amino acids are precursors of biologically important compounds, page 313.

- The formation of **3,4-dihydroxyphenylalanine (dopa)** from tyrosine, catalysed by **tyrosine hydroxylase**, requires ferrous (Fe^{2+}) ions, vitamin C and the coenzyme tetrahydrobiopterin (see Chapter 10).

Box 11.3   The 'flight or fight' response is mediated by adrenaline

Many tissues of the body are involved in the response to stress or exercise, sometimes described as the 'flight or fight' response. In addition to nutrient mobilization to increase available energy, an increased rate of blood flow is required to transport oxygen and nutrients more rapidly to the peripheral tissue. This is achieved because adrenaline binds to heart muscle cell β adrenergic receptors triggering an increased rate of contraction and increasing blood flow to the peripheral tissues. To ensure that the advantages are centred on those tissues that must respond metabolically to stress or exercise, the simultaneous binding of catecholamines to α adrenergic receptors on smooth muscle cells of arteries supplying the intestinal tract, kidneys and skin causes these vessels to constrict, reducing blood flow to these tissues and thereby diverting blood to skeletal muscle. Constriction of skin vessels further reduces heat loss and increases the rate of metabolism in muscle cells.

- Conversion of dopa to **dopamine**, catalysed by **dopa decarboxylase**, requires the coenzyme pyridoxal phosphate.
- **Dopamine hydroxylase**, which adds a hydroxyl group, converting dopamine to noradrenaline, is a Cu^{2+}-dependent enzyme.
- The methyltransferase which converts noradrenaline to adrenaline does so by adding an 'active' methyl group derived from *S*-adenosyl methionine.

Adrenaline and noradrenaline are stress response hormones. They are released into the bloodstream at times of stress or severe exercise when the peripheral tissues need an increased supply of nutrient to sustain the increased energy requirement. The binding of catecholamines to plasma membrane receptors stimulates

- glycogenolysis in muscle,
- glycogenolysis in the liver, and
- release of fatty acids from triacylglycerols in adipose tissue.

Within seconds these nutrients are supplied through the bloodstream to muscle and other tissues temporarily unable to meet the increased energy demand from their own nutrient reserves.

Box 11.4   The brain produces insufficient dopamine in patients with Parkinson's disease

If dopamine-producing brain cells deteriorate and cease to produce sufficient dopamine, motor function disorders result. This occurs in Parkinson's disease. Administration of dopamine to patients with Parkinson's disease may alleviate these symptoms, but also produces undesirable side-effects outside the central nervous system, affecting cardiac function and the mechanisms controlling blood pressure. Attempts have been made to treat Parkinson's disease by transplanting fetal brain cells into the patient's brain. These appear to be able to establish themselves and produce dopamine. The long-term response to this treatment has not yet been established.

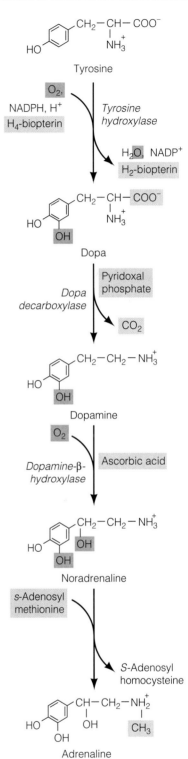

Figure 11.10 Synthesis of adrenaline, noradrenaline and dopamine.

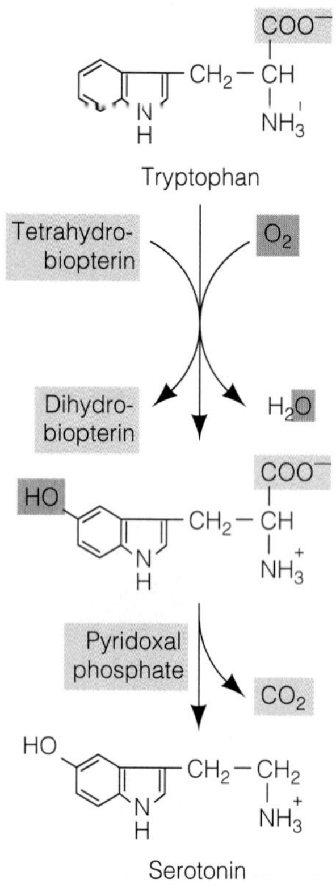

Tryptophan

Tetrahydro-biopterin → O_2

Dihydro-biopterin ← H_2O

Pyridoxal phosphate → CO_2

Serotonin

Figure 11.11 Synthesis of serotonin.

Glutamate

→ CO_2

γ-Aminobutyric acid

→ NH_4^+

Succinic semialdehyde

Succinate

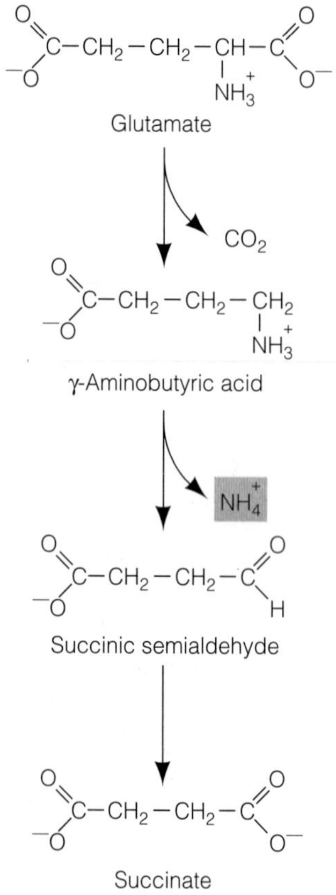

Figure 11.12 Synthesis and inactivation of GABA.

Serotonin

Serotonin (5-hydroxytryptamine) is synthesized from the amino acid tryptophan in a series of reactions requiring pyridoxal phosphate and tetrahydrobiopterin. Serotonin functions as a vasoconstrictor, reducing the blood flow rate and raising blood pressure (Figure 11.11).

γ-Amino butyric acid (GABA)

This is a neurotransmitter that inhibits transmission of impulses from one cell to another in the central nervous system. It is synthesized in the brain by pyridoxal phosphate-dependent decarboxylation of glutamate (Figure 11.12). This transmitter is inactivated by deamination, a reaction which accounts for most of the ammonia produced in the brain. Drugs of the benzodiazepine group, such as Valium, act by increasing the effectiveness of GABA as an inhibitory transmitter.

Histamine

Histamine is produced by decarboxylation of histidine, which is catalysed by **histidine decarboxylase** (Figure 11.13). This hormone is secreted by mast cells as a mechanism for controlling two physiological processes:

1. Stimulation of secretion of acid and pepsin by cells of the gastric mucosa. Histamine analogues, including the drugs cimetidine and ranetidine, inhibit this process and are used in the treatment of peptic ulcers.
2. Control of dilation and constriction of blood vessels in response to inflammation, trauma or allergic reactions.

The effects of excess histamine production are ameliorated by antihistamines, which block the decarboxylase.

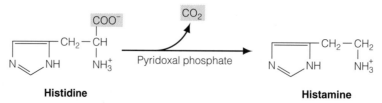

Histidine → CO_2 (Pyridoxal phosphate) → Histamine

Figure 11.13 The production of histamine from histidine is a pyridoxal phosphate-dependent decarboxylation reaction.

Thyroid hormones

The thyroid hormones **tri-iodothyronine** (T_3) and **tetra-iodothyronine** (**Thyroxine, T_4**) are derived by hydrolysis of the glycoprotein **thyroglobulin**, which is synthesized in the thyroid gland. Thyroglobulin contains about 140 tyrosine residues, of which approximately 20% can be iodinated as potential precursors of thyroid hormone.

Iodide is taken up into the thyroid gland through an ATP-dependent pump and is oxidized to iodine by a peroxidase enzyme. The 3 and 5 positions of tyrosine residues in the thyroglobulin molecule are spontaneously iodinated in the presence of iodine to form di-iodotyrosine (DIT). In some tyrosine residues only one of the sites is iodinated, forming mono-iodotyrosine (MIT). Iodinated phenolic groups are transferred so that two iodinated tyrosine residues are coupled to form T_3 (MIT + DIT) or T_4 (DIT + DIT). Both T_3 and T_4 remain bound to thyroglobulin, which is stored in the extracellular space of the thyroid gland.

The thyroid gland is stimulated by **thyroid-stimulating hormone** (**thyrotrophin, TSH**) a peptide hormone synthesized in and secreted by the pituitary gland. This stimulation through adenylate cyclase-linked receptors results in uptake of thyroglobulin into lysosomes where proteolytic digestion of the globulin releases T_3 and T_4. The hormones then pass into the circulation. The increase in circulating thyroid hormone suppresses secretion of TSH through a negative feedback loop to the pituitary gland.

The active form of thyroid hormone is 3,5,3'-tri-iodothyronine (T_3). Circulating T_4 is taken up by the liver or the kidney, where it is converted to T_3 through a NADPH-dependent monodeiodination reaction. A different enzyme converts T_4 to 'reverse T_3' (3,3',5'-tri-iodothyronine) the inactive form. This reaction may provide a mechanism for disposing of excessive amounts of T_4.

The activities of thyroid tissue are under the control of TSH and mediated through the binding of TSH to receptors on cells of the thyroid gland that are linked to adenylate cyclase. Thyroid hormones in circulation control the secretion of TSH from the anterior pituitary. This negative feedback mechanism ensures that a constant plasma concentration of thyroid hormone is maintained.

Thyroid hormones are hydrophobic molecules and most of the circulating hormone is transported bound to proteins, either to a specialized α globulin (thyroid-binding globulin) or to albumin. In the protein-bound form the hormone is inactive and must be released from the protein to enter cells and initiate hormonal action. A small fraction of T_3 and T_4 is free in circulation; this free concentration determines the effectiveness of the hormone.

Within the target cells thyroxine initially binds to a protein in the cytoplasm, where it provides an intracellular pool of the hormone. Thyroxine is then transferred into the nucleus, where it binds to a nuclear receptor protein. The hormone–receptor complex in turn binds to specific nucleotide sequences in DNA, enhancing the transcription of specific proteins through mechanisms similar to those exhibited by steroid hormones.

Thyroid hormone is essential for normal growth and development of young animals. Receptors have been identified on the inner membranes of mitochondria, where the hormone may act to regulate oxygen consumption and ATP synthesis. Thyroid hormone accelerates the rate of oxidative metabolism of carbohydrate and

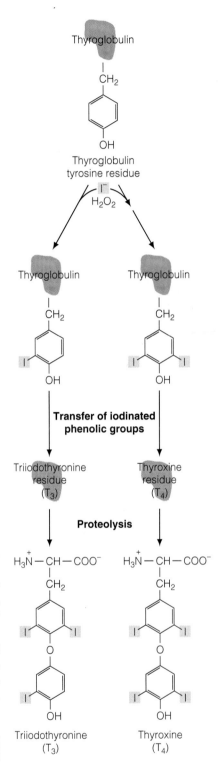

Figure 11.14 Thyroid hormones are produced by iodination of tyrosine residues which make up the protein structure of thyroglobulin. They are stored as part of the protein and released by proteolysis.

> ### Box 11.7   Underactivity of the thyroid
>
> Underactivity of the thyroid, as occurs in hypothyroidism or cretinism, may be due to iodine deficiency during intrauterine development or during the neonatal period. There is a general reduction in growth. Hypothyroidism can also arise from genetic causes. In adults hypothyroidism is called myxoedema, which is characterized by destruction of the thyroid gland by chronic inflammation of cells.

fat, apparently through a mechanism involving increased gene transcription, resulting in enhanced synthesis of the relevant metabolic enzymes.

Peptide hormones

Biosynthesis of peptide hormones

Chapter 5. Targeting of proteins. Page 136.

Peptide hormones are synthesized as precursor polypeptides containing two sequences additional to those required for their hormonal function. One of these is a signal sequence, consisting of about 20 amino acids, providing a predominately hydrophobic region at the *N*-terminal of the polypeptide. This sequence (see Chapter 5) allows for attachment of the signal polypeptide to the endoplasmic reticulum and its insertion into the lumen of the endoplasmic reticulum, where synthesis is completed. The signal peptide is removed by protease action before synthesis of the peptide is completed. The second additional sequence is a 'pro' sequence, *N*-terminal to the functional hormone sequence. As prohormones, these molecules are inactive. Once synthesis is completed prohormone is incorporated into cytoplasmic vesicles, where a trypsin-like enzyme removes the 'pro' sequence, releasing active hormone ready for secretion into the circulation.

Some specific peptide hormones are considered here in detail. For angiotensin and atrial natriuretic peptide the reader is referred to the section on kidney function in Chapter 13. Hormones regulating digestive functions are dealt with in the section on the pancreas, also in Chapter 13.

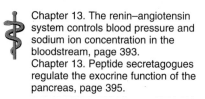

Chapter 13. The renin–angiotensin system controls blood pressure and sodium ion concentration in the bloodstream, page 393.
Chapter 13. Peptide secretagogues regulate the exocrine function of the pancreas, page 395.

Parathyroid hormone

Parathyroid hormone (PTH) is synthesized in, and released from, the parathyroid gland in response to a reduction in the plasma calcium ion concentration; it acts to increase plasma calcium. The secretion of PTH is suppressed when plasma calcium is higher than normal. The prohormone of PTH has 109 amino acid residues, 84 of which remain in the active hormone. Only the 29 *N*-terminal residues are required for hormonal activity, the remainder probably function to bind the hormone to its target site and prolong the life of the circulating hormone by protecting it from proteolytic degradation.

Target tissues possess PTH receptors on their plasma membrane that act through adenylate cyclase; binding of PTH consequently increases the concentration of cAMP. The hormone acts on cells of the kidney, bone and intestine.

- In the kidney PTH decreases tubular reabsorption of phosphate. This eliminates from the body phosphate released from bone, which might otherwise cause generalised calcification of tissues as the solubility product of $Ca_3(PO_4)_2$ is exceeded (Chapter 15). The hormone activates the renal enzyme that hydroxylates 25-hydroxy vitamin D (Chapter 14), increasing the capacity of the kidney to produce the active form of vitamin D (1, 25 dihydroxy vitamin D, or calcitriol).

Chapter 15. Robinson's 'Booster' theory, page 479.
Chapter 14. Vitamin D has a key role in calcium regulation, page 446
Chapter 15. Bone is constantly formed and degraded by specialized cells, page 478.

- In bone PTH increases the number of mature osteoclasts. Osteoclasts produce proteases which degrade connective tissue, releasing bone minerals. Excessive concentrations of PTH can cause erosion of mature well calcified bone. PTH inhibits collagen synthesis by bone osteoblasts (Chapter 15).

- The apparent PTH-related increase in absorption of calcium by intestinal cells is probably secondary to the increase in the amount of active vitamin D produced when PTH activates renal vitamin D hydroxylase.

Calcitonin

Calcitonin is synthesized by and secreted from the thyroid gland. Secretion is regulated solely by the concentration of plasma calcium. The active form of this hormone has 32 amino acids, derived from a 136-amino acid precursor. It produces changes opposite to those produced by PTH. Calcitonin lowers plasma calcium through actions at two sites, bone and kidney.

- In bone, calcitonin directly inhibits bone resorption by osteoclasts.
- In the kidney, calcitonin acts on the membrane calcium pumps responsible for maintaining gradients of calcium across the plasma membrane.

The mechanism of action of calcitonin has not yet been elucidated. This hormone is known to play only a limited role in calcium regulation in humans, but appears to have greater importance in other species.

The pancreas produces three peptide hormones

The α, β and δ cells of the pancreatic islets of Langerhans produce the peptide hormones glucagon, insulin and somatostatin, respectively. Insulin and glucagon have widespread actions; somatostatin acts within the pancreas to inhibit the secretion of insulin and glucagon. The role of the pancreas in the synthesis and secretion of insulin and glucagon, and the relationship between insulin and diabetes, are described in Chapter 13.

Chapter 13. Synthesis and secretion of glucagon, page 397; Synthesis and secretion of insulin, page 398.

Insulin

The mature active form of insulin consists of two peptide chains: an A chain of 21 residues and a B chain of 30 residues. The two chains are covalently linked by two disulphide bonds (Figure 11.15).

Insulin is released from the pancreas in response to high circulating concentrations of glucose and binds to specific receptors in target tissues, particularly muscle and adipose tissue.

Box 4.3. Measurement of enzyme activity and kinetics, page 92.

A chain

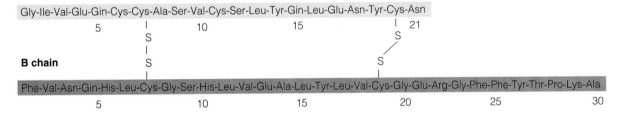

Gly-Ile-Val-Glu-Gin-Cys-Cys-Ala-Ser-Val-Cys-Ser-Leu-Tyr-Gin-Leu-Glu-Asn-Tyr-Cys-Asn

B chain

Phe-Val-Asn-Gin-His-Leu-Cys-Gly-Ser-His-Leu-Val-Glu-Ala-Leu-Tyr-Leu-Val-Cys-Gly-Glu-Arg-Gly-Phe-Phe-Tyr-Thr-Pro-Lys-Ala

Figure 11.15 Proinsulin is a single polypeptide chain. Mature insulin, produced by excision of C-peptide, has two chains linked by disulphide bonds.

Binding of insulin to receptors induces translocation of glucose transporters to the plasma membrane

In the short term insulin reduces plasma glucose concentrations by stimulating its uptake into muscle and adipose tissue cells. Kinetic analysis of the process by which glucose is transported into muscle cells indicates that in the presence of insulin the maximum rate of glucose transport (V_{max}) is increased, but the affinity with which it binds to the transport sites is not altered.

This reflects an insulin-stimulated increase in the number of glucose transporter sites on the plasma membrane of these cells. This increase, which can be 10–20-fold within minutes, is due not to synthesis of new transporter molecules but to the fusion of intracellular vesicles containing stored transporter molecules with the plasma membrane and their recruitment to the outer surface of the plasma membrane. When the insulin concentration falls glucose transporters are withdrawn into the cell by endocytosis and are stored in intracellular vesicles.

The longer term effects of insulin reflect endocytosis of hormone–receptor complex
Insulin hormone–receptor complexes are internalized by endocytosis. The purpose of this may be to bring the tyrosine kinase activity of the receptor protein into proximity with cytoplasmic proteins, which are phosphorylated by the kinase. Within the cell the hormone dissociates from the receptor complex and is degraded. The receptor is eventually recycled to the plasma membrane.

The effects of insulin on metabolism are mediated through enzyme regulation and increased protein synthesis
The long-term metabolic alterations induced by insulin are a reflection of specific phosphorylation of serine and threonine residues of metabolic enzymes, and also of accelerated rates of protein synthesis. Insulin causes

- increased synthesis of glycogen in the liver,
- re-esterification of triacylglycerols in adipose tissue,
- decreased activity of triacylglycerol lipase in adipose tissue, with resultant reduction in the amount of free fatty acid available to muscle,
- increased activity of lipoprotein lipase,
- increased rate of uptake of amino acids into cells,
- increased rates of transcription of many enzymes, notably phosphoenol pyruvate carboxykinase, glucokinase and pyruvate kinase, particularly in liver and muscle.

These long-term changes require exposure to circulating concentrations of insulin 10–100 times greater than those required to accelerate glucose transport. Some tissues, notably erythrocytes and brain, are almost unaffected by insulin.

Glucagon

Glucagon is a single-chain peptide hormone consisting of 29 amino acid residues. It is synthesized by the α cells of the pancreas and is secreted in response to reduction in blood glucose concentration. Glucagon primarily stimulates glycogenolysis and gluconeogenesis in the liver. It binds to specific plasma membrane receptors that act through adenylate cyclase. An increased cellular concentration of cAMP stimulates glycogen phosphorylase and inhibits glycogen synthase (Chapter 8). The amplification of response initiated by activation of adenylate cyclase means that one molecule of glucagon binding to the cell receptor can trigger the release of as many as 3×10^6 molecules of glucose from glycogen stores. Glucagon also stimulates synthesis of glucose from amino acids through gluconeogenesis in the liver (Chapter 8), inhibits acetyl CoA carboxylase through a phosphorylation mechanisms (thereby directing acetyl CoA away from production of malonyl CoA and towards ketogenesis; see Chapter 9) and mobilizes fatty acids from adipose tissue.

Chapter 8. Activation of glycogen phosphorylase and inhibition of glycogen synthase occur in response to a common cascade, page 227.

Chapter 8. Gluconeogenesis: the synthesis of glucose from noncarbohydrate precursors, page 244.

Chapter 9. In the liver accumulating acetyl CoA is converted to ketone bodies, page 271.

Hormones derived from fatty acids

The eicosanoid compounds

Compounds in this family, which includes the **prostaglandins, thromboxanes** and **leukotrienes**, are produced by most cells (Figure 11.16). Their function falls within the class of paracrine signalling; that is they are local hormones influencing the behaviour of immediately adjacent cells. All compounds in this class are very short lived, being degraded almost immediately after entering the circulation. Although eicosanoid compounds are able to influence cell processes

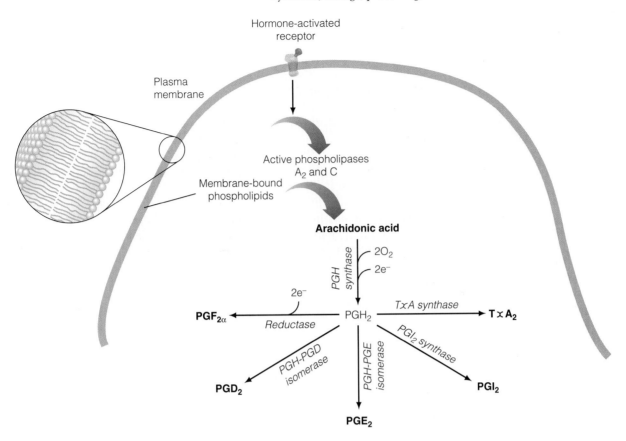

Figure 11.16 Outline of the biosynthetic routes through which arachidonic acid released from membrane phospholipids is converted to the prostaglandins (PG) and to thromboxane (Tx) A$_2$.

directly, a major part of their function appears to involve modulation of tissue responses to other hormones. As for other hormones, the actions of the eicosanoids vary with cell type: for example, prostaglandin E$_1$ (PGE$_1$) inhibits adenylate cyclase in adipose tissue cells but activates this enzyme in other cell types.

The immediate precursors for synthesis of eicosanoids are a series of 20-carbon fatty acids with three, four or five double bonds.

- **Homo-γ-linoleic acid** (C20:3 cis-Δ^8,Δ^{11},Δ^{14}) gives rise to prostaglandins with one double bond.
- **Arachidonic acid** (C20:4 cis-Δ^5,Δ^8,Δ^{11},Δ^{14}) gives rise to prostaglandins with two double bonds.
- **Eicosapentaenoic acid** (C20:5 cis-Δ^5,Δ^8,Δ^{11},Δ^{14},Δ^{17}) gives rise to prostaglandins with three double bonds.

Box 11.8   The benefits of seed oil and fish oil fatty acids

Sunflower and evening primrose seed oils are a particularly good source of γ-linoleic acid. Eicosapentaenoic acid is found in fish oils. The basis of the apparently beneficial effects of the seed oils and fish oil appears to be that they are used to produce prostaglandins which displace the normal mammalian arachidonic acid-derived prostaglandins. Prostaglandins with one or three double bonds have a weaker inflammatory effect than those with two double bonds. There is a growing body of evidence that these oils are beneficial in the treatment of arthritis and psoriasis.

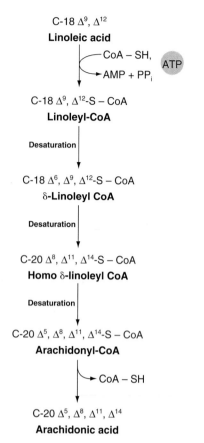

C-18 Δ^9, Δ^{12}
Linoleic acid

CoA – SH,
AMP + PP$_i$ ATP

C-18 Δ^9, Δ^{12}-S – CoA
Linoleyl-CoA

Desaturation

C-18 Δ^6, Δ^9, Δ^{12}-S – CoA
δ-Linoleyl CoA

Desaturation

C-20 Δ^8, Δ^{11}, Δ^{14}-S – CoA
Homo δ-linoleyl CoA

Desaturation

C-20 Δ^5, Δ^8, Δ^{11}, Δ^{14}-S – CoA
Arachidonyl-CoA

CoA – SH

C-20 Δ^5, Δ^8, Δ^{11}, Δ^{14}
Arachidonic acid

Figure 11.17 Synthesis of arachidonic acid from linoleic acid.

Figure 11.18 Structure of arachidonic acid showing point of cyclization.

Figure 11.19 Structures of the major prostaglandins (PGE$_2$ and PGI$_1$).

PGE$_2$

PGI$_2$

Arachidonic acid can be released from membrane phospholipids

Arachidonic acid is an abundant component of membrane phospholipids in plasma and endoplasmic reticulum membranes and is synthesized from linoleic acid, C18:2 cis-Δ^9,Δ^{12}, an essential fatty acid in humans. Linoleic acid is essential because humans lack enzymes capable of incorporating double bonds into fatty acid carbon chains further from the carboxyl end of the molecule than C$_9$ (Figure 11.17).

Large amounts of arachidonic acid (Figure 11.18) are found in the membrane phospholipids of electrically active tissue, including cardiac myocytes. The full significance of this is not yet established, but in cardiac tissue, where phospholipases have degraded phospholipids, large amounts of arachidonic acid are liberated.

The release of arachidonic acid arises from the action of phospholipase A$_2$ on phosphatidylcholine or phosphatidylethanolamine, stimulated by hormones such as adrenaline or bradykinin. Release of arachidonic acid may also be stimulated by the venom injected in a bee sting.

Aspirin acts by inhibiting prostaglandin synthesis

The prostaglandins are 20-carbon fatty acids containing a five-membered ring into which two of the double bonds of arachidonic acid have been incorporated. Examples are shown in Figure 11.19. One of the reactions required for this ring formation is catalysed by a cyclo-oxygenase. Acetylsalicylic acid (aspirin) inhibits this enzyme by acetylation, transferring its acetyl group to the enzyme to form salicylic acid. This inhibition explains the anti-inflammatory action of aspirin. Aspirin also inhibits prostaglandin synthesis in the gastric mucosa and thereby increases gastric acid secretion.

A series of prostaglandin compounds is produced by the action of isomerases and reductases. Thromboxanes are 20-carbon compounds containing a six-membered ether ring (Figure 11.20). They stimulate aggregation of blood platelets as part of the blood clotting mechanism: these compounds are therefore implicated in the mechanism of haemostasis and of vascular disease.

Current medical thinking is that very low doses of aspirin taken regularly may protect against heart attack by reducing thromboxane production. These compounds do not inhibit the synthesis of leukotrienes.

Box 11.9   Eicosanoid therapies

Although the mechanisms of action of the eicosanoids are not fully understood, knowledge of the physiological response to these compounds is sufficiently advanced for useful therapeutic interventions to be implemented. In addition to metabolic effects, mediated through adenylate cyclase, prostaglandins released by uterine cells induce the contractions associated with birth. Other prostaglandins are used to lessen the risk of blood clotting during surgery, as vasodilators in patients with circulatory disorders and, because of their action in inhibiting gastric secretion, in the treatment of stomach ulcers. After coronary artery bypass surgery patients commonly take very small doses, 30–75 mg, of aspirin every day for the remainder of their lives and this appears to be sufficient to inhibit the production of prostaglandins in platelets and thereby to block the thrombotic activity of platelets.

Substantially larger quantities, 1–3 g per day, of aspirin are needed to block the prostaglandin-induced joint inflammation associated with rheumatoid arthritis. Because of the gastric complications associated with long-term use of such high doses of aspirin, current therapy favours the use of steroidal anti-inflammatory drugs in preference to aspirin and related non-steroidal anti-inflammatory compounds.

Ibuprofen, a compound widely used to reduce the pain associated with inflammation and muscle spasm, also inhibits the synthesis of prostaglandins.

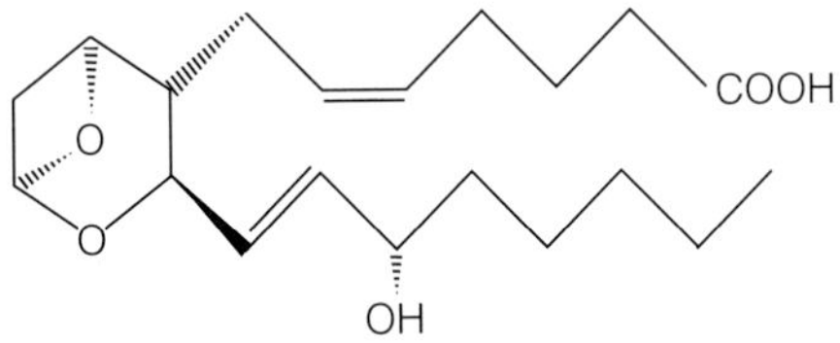

Figure 11.20 Structure of thromboxane (TXA₂).

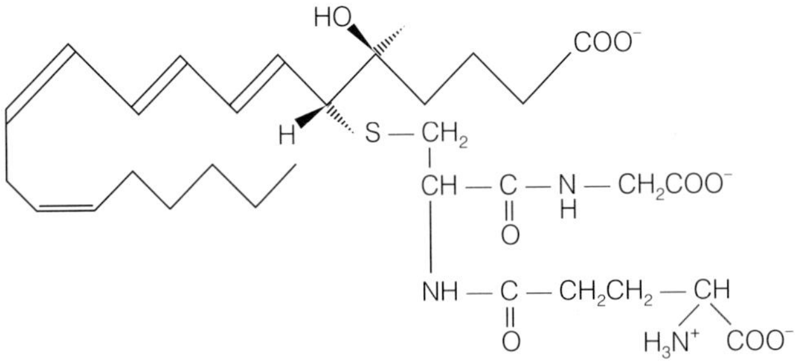

Figure 11.21 Structure of leukotriene C.

Leukotrienes, derivatives of arachidonic acid with three conjugated double bonds but lacking a ring structure (Figure 11.21), are produced by the action of lipo-oxygenases. These compounds, originally found associated with leukocytes, initiate the contraction of smooth muscle cells and may be implicated in the pathogenesis of asthma through constriction of the airways.

Steroid hormones

All steroid hormones are derived from a common synthetic pathway

Steroid hormones are synthesized from cholesterol in steroidogenic tissues (including the adrenal cortex, the gonads and the placenta). These compounds are released immediately into the circulation after synthesis, so that the circulating concentration of hormone is determined by the rate at which it is synthesized. Synthesis occurs only in response to demand, which is signalled by hormones produced in the central nervous system.

The precursor cholesterol may be derived from the circulation, from tissue stores or may be synthesized from acetyl CoA. Synthesis of steroid molecules is regulated by chemical signals from the pituitary in the form of the hormones **adrenocorticotrophic hormone (ACTH)**, **luteinizing hormone (LH)** or **follicle-stimulating hormone (FSH)**, which stimulate adenylate cyclase in the steroid-synthesizing tissues. The cAMP produced stimulates protein kinases which phosphorylate proteins that accelerate the transport of cholesterol into mitochondria.

Conversion of cholesterol to steroid hormones is catalysed by a variety of enzymes located in the mitochondria and on the endoplasmic reticulum. All steroid hormones are synthesized from cholesterol via **pregnenolone**, a compound formed by side-chain cleavage catalysed by a cytochrome P-450-dependent oxidase located in the mitochondrial membrane (see Chapter 7). The enzyme requires molecular oxygen and NADPH, and its synthesis may be induced by hormone action. Synthesis of pregnenolone is stimulated by ACTH. This is the rate-limiting step in the synthesis of all steroid hormone molecules (Figure 11.22).

Box 7.5. Cytochrome *P*-450, page 206.

Chapter 13. The renin–angiotensin system controls blood pressure and sodium ion concentration in the bloodstream, page 393.

Pregnenolone is the precursor of all other steroid hormones

Pregnenolone is converted to **progesterone** (the female sex hormone responsible for the development of the lining of the uterus), and has an essential role in the maintenance of pregnancy. Progesterone is also the precursor of

- the glucocorticoids cortisol and corticosterone, which have a variety of metabolic roles including enhancement of gluconeogenesis and the breakdown of fat and protein,
- the mineralocorticoid aldosterone, which promotes sodium retention and potassium and hydrogen ion loss through the distal tubules of the kidney,
- the female and male sex hormones, which regulate reproduction and determine the sexual characteristics.

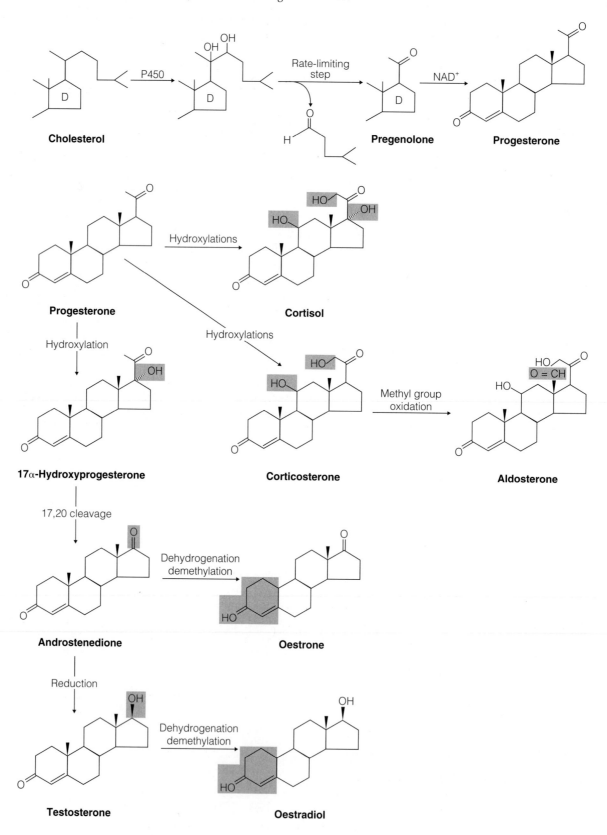

Figure 11.22 Steroid hormones are synthesized from cholesterol through pregnenolone, which is the immediate precursor of adrenacorticoids, mineralocorticoids and sex hormones.

Steroid hormones are transported in plasma bound either to **corticoid binding globulin** or to **sex hormone binding globulin**.

Tissues respond relatively slowly to activation by steroid, and the response is sustained for periods of hours, days or longer. The steroid hormones have roles

Table 11.1 Classes of steroid hormones and their functions.

Functional class	Specific compounds	Principal biological function
Glucocorticoids	Cortisol, corticosterone	Promote gluconeogenesis, suppress inflammatory reactions
Mineralocorticoids	Aldosterone	Regulates cation balance
Progesterone		Regulates changes required to sustain pregnancy
Androgens	Testosterone	Responsible for development and maintenance of male sexual characteristics
Oestrogens	Oestradiol	Responsible for development and maintenance of female sexual characteristics

---

Box 11.10   Hypertension and control of aldosterone production

Both cortisol and aldosterone are synthesized in the adrenal gland. The final reaction for the synthesis of cortisol and corticosterone (occurring mainly in the zona fasiculata) is catalysed by an **11β hydroxylase**, while the final reaction for **aldosterone synthesis** is catalysed by aldosterone synthase, expressed only in the zona glomerulosa. The genes for these enzymes are both located on the long arm of chromosome 8 and there is more than 90% sequence homology between them. The major difference is that expression of the gene for cortisol synthesis is controlled by ACTH while the gene for aldosterone synthesis is controlled by angiotensin II (see Chapter 13: kidney). In a relatively rare mutation the ACTH control sequence is transposed to the aldosterone synthase gene, producing an extra chimaeric gene (expressed throughout the adrenal gland) which expresses the synthase and produces aldosterone in response to increased ACTH.

This disorder can be diagnosed by suppressing the expression of the chimaeric gene with the cortisol analogue **dexamethasone** so that aldosterone secretion becomes independent of angiotensin II in the normal way.

Many of the individuals affected by this mutation are found in the west of Scotland or are descendents of families from that area. The condition is a cause of hypertension, although within families carrying this mutation there are large variations in blood pressure.

---

central to the processes of cell growth and differentiation. All steroid hormones act by changing the rates of gene expression so that rates of transcription of specific genes are altered: where these encode structural proteins the physical characteristics of the tissue may be altered; where they encode enzymes the metabolic activity of the tissue may be altered.

The action of steroid hormones is terminated when the hormones are inactivated by enzymic modification – generally conjugation to glucuronic acid or sulphate, in which form they are water soluble and readily excreted in urine or bile. These inactivation reactions occur in the liver (Chapter 13).

Chapter 13. Detoxification and removal of xenobiotics, page 389.

Growth factors

Most cell types produce peptides that act as growth hormones

Growth hormones may be secreted into the circulation or in saliva and produce a variety of responses. The best understood of these growth factors are **epidermal growth factor (EGF)** and **nerve growth factor (NGF)**, both of which are known to be secreted into saliva from the submaxillary gland but must also be secreted into serum by other tissues, and **platelet-derived growth factor (PDGF)** (Chapter 12).

Target tissues have plasma membrane receptors for growth factors. These receptors, like the insulin receptor, exhibit tyrosine kinase activity, which leads to the phosphorylation of tyrosine residues in a variety of proteins when growth factor binds to the receptor. There is no evidence for intact growth factor in the cytoplasm although apparent degradation products have been identified within cells.

The peptide EGF consists of 53 amino acid residues, while NGF has 113 residues and appears to function as a dimer of two identical peptides. Both of these growth factors are stored in the submaxillary gland as inactive higher molecular weight protein complexes associated with zinc ions.

Differentiation and proliferation of epidermal and epithelial tissue and of liver cells are promoted by EGF, causing a generalized increase in the rates of synthesis of DNA, RNA and protein, in support of an increase in the rates of cell growth and division. The membrane Na+/K+ ATPase activity also increases. Target cells take up more amino acids, purines and pyrimidines, the precursors of proteins and nucleic acids.

The action of NGF is not only to stimulate nerve growth during development but also appears to be essential in maintaining the function and survival of mature nervous tissue. It has been suggested that NGF acts to suppress the expression of genes that code for proteins forming part of the apoptosis pathway (see Chapter 17) through which the normal process of cell death occurs. It also increases the Na+/K+ ATPase activity in these cells and seems to induce the synthesis of selected enzymes of the pathway of catecholamine synthesis.

Platelet-derived growth factor is secreted to promote wound repair in blood vessels. It appears to act as a chemoattractant to induce the migration of fibroblasts and macrophages into the area of damage. It also stimulates proliferation of smooth muscle cells and fibroblasts and the synthesis of extracellular matrix molecules by fibroblasts (Chapter 15).

Chapter 3. The Na+/K+ ion pump of plasma membranes is driven by hydrolysis of ATP, page 64.

Box 17.10. Cell proliferation or cell death?, page 537.

Chapter 15. The extracellular matrix holds together cells of multicellular tissues, page 467.

CHAPTER 12

Blood: molecular functions of cells and plasma

The blood acts as a distribution system throughout the body, delivering nutrients to the tissues and collecting waste products. It also plays a key role in the body's defences against external insult, whether physical or biological in origin. Blood is composed of a solution of small and large molecules – the plasma – in which a variety of cells are suspended: white blood cells (leukocytes), involved in responses to invasion by foreign molecules and organisms (see Chapter 16); red blood cells (erythrocytes), containing haemoglobin that transports oxygen, carbon dioxide and hydrogen ions between lungs and tissues; and platelets, subcellular fragments crucial for the process of blood clotting.

Plasma contains ions, small molecules and proteins. Many plasma proteins bind other molecules, and transport them between tissues. Many of the components of the blood clotting system are plasma proteins that act in a cascade, leading to the production of a clot of fibrin to staunch blood loss.

Erythrocytes are not true cells, but cellular remnants in which a concentrated solution of haemoglobin and other proteins is enclosed within a plasma membrane. Erythrocytes live for 60–120 days and possess a limited range of metabolic activities to maintain themselves and their haemoglobin molecules in a functional state.

Despite its lack of form, the blood is an important tissue or organ of the body and this chapter considers a number of the important biochemical aspects of blood plasma, erythrocytes and platelets. The leukocytes and their roles in the immune and other responses of the body are described more fully in Chapter 16.

Blood contains several types of cellular component

Cellular components circulating in the bloodstream fall into three groups.

Erythrocytes

These are membrane-bound biconcave discs containing high concentrations of **haemoglobin** that acts in the transport of oxygen, carbon dioxide and hydrogen ions between the lungs and the tissues. Erythrocytes are not true cells because they lack a nucleus and intracellular organelles. They are derived from **normoblasts** in the bone marrow by the process of **erythropoiesis** under the control of the glyco-protein hormone **haemopoietin** (see Figure 12.19). Low oxygen tension or a reduction in the number of circulating erythrocytes cause the kidney to secrete haemopoietin into the bloodstream, thus stimulating erythropoiesis.

Leukocytes

Leukocytes are true cells, with a range of roles in combating infection. There are three main types – **granulocytes, monocytes** and **lymphocytes** – made principally in the bone marrow and thymus gland. Circulating leukocytes can stimulate the bone marrow to synthesize more leukocytes by secreting signal peptides into the bloodstream.

Platelets

Platelets are not cells; they are membrane-bound fragments derived from **megakaryocytes** residing in the bone marrow. They play a key role in the process of blood clotting (**haemostasis**).

There are three types of leukocyte

Most leukocytes develop in the bone marrow. The exceptions are the T lympho-cytes, which develop in the thymus, and macrophages, which develop from monocytes in a variety of tissues. Leukocytes circulate in the bloodstream, but their function depends upon their ability to migrate out of the bloodstream and into the tissues. In this sense leukocytes simply use the bloodstream as a route to their site of action. Granulocytes and monocytes combat infection by engulfing or phagocytosing invading organisms and digesting them. In contrast, lymphocytes are involved in the immune response (see Chapter 16). The categories of leukocytes and their principal functions are listed in Table 12.1.

Chapter 16. The basis for the specific or adaptive immune response resides in the lymphocytes, page 485.

Table 12.1 Classification and functions of leukocytes

Leukocyte group	Subgroup	Function
Granulocytes	Neutrophils	Destroy small organisms
	Basophils	Secrete histamine as part of a process mediating inflammatory responses, secrete platelet-activating factor
	Eosinophils	Destroy parasites, modulate allergic inflammatory reactions
Monocytes	Mature into macrophages	Destroy invading organisms
Lymphocytes	B lymphocytes	Synthesize antibodies
	T lymphocytes	Destroy virus-infected cells

Plasma and its proteins

Plasma comprises about half the blood volume (5 l in an adult). Plasma is the aqueous solution in which the 'formed elements of the blood' – erythrocytes, leukocytes and platelets – are suspended. It makes contact with the tissues of the body as blood makes its way round the artery–capillary–vein circuit. Plasma can be obtained from blood by centrifuging down the formed elements, with precautions taken to prevent clotting. The aqueous solution left after blood has been allowed to clot and the clotted material removed is **serum**. The principal difference between plasma and serum is that the latter lacks fibrinogen, the protein that forms the clot (see below). The two terms are often used interchangeably (for example, the major protein of plasma – plasma albumin – is often called serum albumin), but for the purposes of many of the analytical tests that are performed on blood samples it is important whether plasma or serum is used.

The principal constituents of plasma and their normal concentration ranges are shown in Table 12.2 (small molecules and ions) and Tables 12.3 and 12.4 (plasma proteins). The concentrations of many of these constituents are under physiological control and values outside these ranges may well indicate significant dysfunction in the individual: this is the realm of Clinical Biochemistry.

There are many types of plasma proteins

Plasma contains many diverse proteins which vary enormously in their individual concentrations – from **albumin**, which comprises nearly half the total, to many that can be detected only by highly sensitive biological or immunological assays.

As techniques for separating proteins developed, plasma was an obvious biological fluid for analysis, and application of the method of **electrophoresis** revealed four major classes of protein – albumin and the α-, β- and γ-globulins (Figure 12.1). This technique is still used for the analysis of clinical serum samples and is useful where there is a gross change from the normal distribution of the four classes, for example, deficiencies of albumin (see below) or γ-globulin or an excess of γ-globulin. This last condition occurs in **myelomas** or tumours of B lymphocytes, in which large amounts of a single type of antibody are secreted (see Chapter 16).

The four classes show the following general features:

- **albumin** is quantitatively predominant and is a single molecular species (see below);
- **α-globulins** are a large group of different proteins with similar electrophoretic mobility – they can be resolved into two subclasses, the α_1-globulins and α_2-globulins;

Chapter 16. Monoclonal antibodies are produced by cells derived from a single B lymphocyte, page 489.

Table 12.2 The composition of plasma – small molecules and ions

Substance	Concentration
Ions	(meq l^{-1})
Anions	
Chloride	100–110
Bicarbonate	25–30
Phosphate	1.5–2.7
Cations	
Sodium	130–150
Potassium	3.8–5.4
Calcium	4.5–5.6*
Magnesium	1.6–2.2
Small molecules	(mmol l^{-1})
Glucose	70–90
Lactate	1–2
Pyruvate	0.05–0.2
Acetoacetate	0.1–0.3
Urea	3–5
Alanine	0.2–0.6
Glutamine	0.3–0.6
Uric acid	0.1–0.3

Some individual α-globulins and β-globulins are listed in Table 12.3
*Includes approx. 50% protein-bound Ca^{2+}

Table 12.3 Plasma proteins and their properties

Electrophoretic class	Concentration (g l^{-1})	Molecular masses*
Albumin	35–45	66 000
α_1-Globulin	3–6	40–60 000
α_2-Globulin	4–9	100–400 000
β-Globulin	6–11	100–120 000
Fibrinogen†	3	340 000
γ-Globulin	7–15	150 000 (IgG)

*Only albumin and fibrinogen are single proteins – the various globulin fractions are mixtures of many different species of proteins with different functions (see Table 12.4); hence the molecular masses are shown as ranges.
†Fibrinogen is a β-globulin.

Table 12.4 Some plasma proteins and their physiological ligands

Plasma protein	Ligand	Globulin class	Concentration (g l⁻¹)	Molecular mass	Glycosylation (%CHO)
Metal ion binding					
Transferrin	Fe^{3+}	β	2–3	78 000	6
Caeruloplasmin	Cu^{2+}	α_2	0.2–0.6	150 000	7
Albumin	Zn^{2+}, Cu^{2+}, etc.				
Hormone binding					
Thyroid hormone binding protein	Thyroxine	α_1	0.01	58 000	13
Vitamin binding					
Retinol binding protein	Retinol	α_1	0.05	21 000	0
Binding of nonpolar molecules					
Albumin	Fatty acids, bilirubin, drugs, etc.				
Protein binding					
Haptoglobins	Haemoglobin dimer	α_2	1–2	100–400 000	9–19
Proteinase inhibitors					
α_1-Antitrypsin, etc. (see Chapter 17)					
Lipoproteins					
HDL, LDL, VLDL					

Chapter 17. Deficiency in α-antitrypsin causes emphysema and is exacerbated by cigarette smoking, page 527.

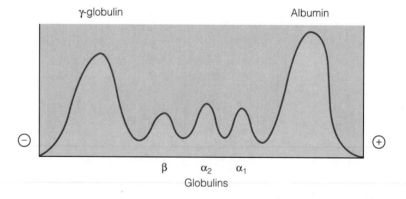

Figure 12.1 Electrophoretic separation of the major groups of plasma proteins. During electrophoresis of plasma at mildly alkaline pH (for example, pH 8.6), albumin and the α- and β-globulins are negatively charged and move to the positive electrode, while γ-globulins carry no charge or a slight positive charge. The figure shows a trace of protein stained with a dye (for example, Ponceau S) after electroporesis on cellulose acetate.

- **β-globulins** are a second large group of different proteins with similar electrophoretic mobility (individual α-globulins and β-globulins are listed in Table 12.4);
- **γ-globulins** form a distinctive group of very similar proteins. These are the soluble antibodies of the immune system (see Chapter 16).

One group of plasma proteins, typified by their low densities, are the **lipoproteins** (see Chapter 9), members of which occur among both the α- and β-globulins.

Chapter 16. Different classes of immunoglobin have related structures, page 492.
Chapter 9. Distinct classes of lipoproteins are responsible for the transport of triacylglycerols, phospholipids and cholesterol, page 287.

Most plasma proteins are made in the liver

With the exception of the γ-globulins, the plasma proteins are synthesized in the liver and secreted into the bloodstream. Apart from albumin they are almost all glycoproteins, typically containing 5–20% sugar residues (Table 12.4), which are added to the polypeptide chain postsynthetically as it passes through the endoplasmic reticulum and Golgi complex to the secretory vesicles of the hepatocyte (see Chapter 3).

It is hard to say just how many different molecular species of plasma protein there are. It is easy to list those that occur in substantial amounts but many proteins are present in very small amounts, and more are being detected as analytical techniques become more sensitive. Although most of the proteins we call 'plasma proteins' are 'intentional' in the sense that they are specifically secreted into the plasma, some are 'accidental' – they are tissue proteins that have 'escaped' into the plasma following tissue damage. Measurement of these proteins, especially when they can be detected by their enzyme activity, is a useful diagnostic indicator of disease (see Chapter 10). This procedure is termed 'clinical enzymology'.

Plasma proteins have specific binding properties

A characteristic of almost all plasma proteins is their ability to bind one or more ligands; indeed, this appears to be the principal function of most of them. The binding of ligands has two principal parameters – *affinity* and *specificity* – and many of the molecules found in serum are bound by two or more proteins, usually with different affinities and specificities. For example, steroid hormones (see Chapter 11) are generally bound in two ways:

1. tightly by a protein specific for a particular steroid or group of related steroids;
2. weakly and nonspecifically by albumin.

A list of plasma proteins and substances bound by them is given in Table 12.4.

Albumin is important in osmoregulation and in binding small molecules

Plasma albumin has two main roles: (1) maintenance of the osmotic balance of the body and (2) binding of a range of physiologically important ligands, both hydrophobic molecules and metal ions. In common with other plasma proteins its structure is stabilized by disulphide bonds, 34 of its 35 cysteine residues participating in 17 disulphide bridges. The single free cysteine can react with heavy metal ions such as Ag^+ and Hg^{2+} and thus limit their toxicity. Albumin is notable among the plasma proteins as being one of the few that are not normally glycosylated.

Osmoregulation
The high concentration and relatively low molecular mass of albumin mean that it makes a large contribution to the osmolarity of plasma, thus preventing the tissues 'waterlogging'. When albumin is severely depleted, as occurs in kidney disease (nephritis) and in protein malnutrition (kwashiorkor) this 'waterlogging' can cause oedema. The size of albumin (66 kDa) is sufficient to ensure that it is not lost through the glomerulus of a healthy kidney.

Binding of small molecules
Hydrophobic molecules
Several sites on the albumin molecule are capable of binding nonpolar molecules (Figure 12.2) and, although this binding is relatively weak and nonspecific, the high concentration of albumin makes it quantitatively important. Important ligands include

- long-chain fatty acids (the 'free fatty acids' of plasma);
- bilirubin;
- steroid hormones;
- nonpolar drug molecules.

Chapter 3. The endoplasmic reticulum and Golgi complex are membrane-bound compartments, page 46.

Box 10.1. Distribution of specific aminotransferase provides diagnostic information when tissues are damaged, page 297.

Chapter 11. Steroid hormones, page 351.

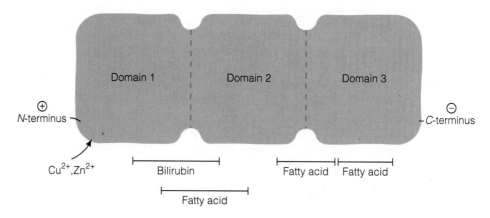

Figure 12.2 Domains in plasma albumin and the binding of physiologically important ligands. Plasma albumin is composed of a single polypeptide chain, the sequence of which is organized into three domains. The figure shows the locations of the various sites that bind fatty acids, bilirubin and other relatively non-polar molecules. Bivalent metal ions, such as Cu²⁺ and Zn²⁺, bind to another site near the *N*-terminus.

Metal ions

The terminal amino group of the albumin polypeptide is important in binding 'trace element' metals such as Cu^{2+} and Zn^{2+}. Again, the high concentration of albumin makes this quantitatively important.

Albumin is made up of a series of **domains** and the amino acid sequence of the protein shows that these have evolved by a process of gene duplication. Different domains are involved in the binding of the various ligands (see Figure 12.2)

Complete deficiency of albumin occurs as a very rare genetic condition, **analbuminaemia**, in which the gene for albumin (a 17 kb region on human chromosome 4, consisting of 15 exons and 14 introns) is defective. In one case, which has been analysed in molecular detail, insertion of a single adenine nucleotide into exon 8 causes a truncated albumin of 273 residues; another showed a splicing defect between exons 6 and 7. Strangely, in these individuals, absence of albumin causes only moderate oedema, probably because the levels of other plasma proteins (including γ-globulins and lipoproteins) increase to compensate. Albumin is considered critical in fetal life and only very rare analbuminaemic individuals survive past the neonatal state.

Blood clotting

When blood is removed from the body and left to stand in a tube, it clots within a few minutes. The clot leaves behind a clear, straw-coloured liquid, the serum, which lacks blood cells and the protein fibrinogen, which is used in clot formation. Blood clotting is an important defence mechanism, limiting the loss of blood from wounds and promoting the healing of injuries to the skin and intestinal lining. Clotting also poses a hazard to the body if not properly controlled; clots must be dissolved and removed when their role in healing is complete.

Clotting is effected by plasma components, as can be shown by the fact that plasma itself can clot even though the erythrocytes and leukocytes have been removed. The essentials of the biochemical mechanisms of these processes are outlined below.

Platelets play an important role in blood clotting

Platelets are cell fragments, produced by 'budding-off' from elongated processes of cells (the **megakaryocytes**) in the bone marrow. As many as 10^4 platelets enter the bloodstream from a single megakaryocyte. Platelets have no nucleus, mitochondria or endoplasmic reticulum, but they enclose dense granules containing several factors involved in blood clotting.

When the endothelial cells that line the blood vessels are damaged the underlying collagen becomes exposed. Platelets adhere specifically to these cells and accumulate and aggregate at these sites of vessel damage. This process is stimulated by **platelet-activating factor**, a plasmalogen (see Chapter 9), released by **basophils** (Figure 12.3).

Aggregation of platelets is accompanied by exocytosis of granule contents, releasing ADP, serotonin, phospholipids, lipoproteins and thromboxane A$_2$ (see Chapter 11) at the site of injury. At the same time a glycoprotein (**von Willebrand factor**) concentrates at the site. This has the dual function of ensuring that the

Chapter 9. Plasmalogens are ether-linked fatty alcohol derivatives, page 283.

Chapter 11. The eicosanoid compounds, page 348.

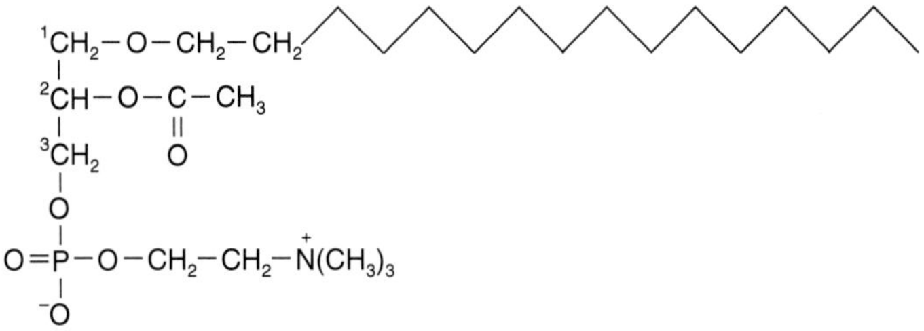

$$^1CH_2-O-CH_2-CH_2$$
$$^2CH-O-C-CH_3$$
$$^3CH_2 \quad\quad O$$
$$O$$
$$O=P-O-CH_2-CH_2-\overset{+}{N}(CH_3)_3$$
$$^-O$$

Figure 12.3 Platelet-activating factor.

Box 12.1 von Willebrand factor – molecular biology and blood clotting

Von Willebrand factor is a multimeric glycoprotein with a central role in blood clotting. Genetic abnormalities of this protein result in von Willebrand's disease, a common human bleeding disorder. The molecular biology of this protein and its gene have been intensively studied and provide a good example of the application of fundamental molecular biology (see Chapter 17) to human disease.

von Willebrand factor is a scarce (10 mg l^{-1}) plasma glycoprotein, which has at least two distinct roles in haemostasis.

1. In the initial adhesion of platelets to collagen exposed after vascular injury, von Willebrand factor binds the platelet to the vessel wall in a very early event in the formation of the initial platelet plug.

2. In the transport and stability of factor VIII in plasma (see Figure 12.3), von Willebrand factor is present in 50-fold molar excess and all factor VIII in plasma is thought to exist as the VIII–von Willebrand factor complex.

The mature protein is large (2050 amino acids) and contains several domains involved in binding to collagen and factor VIII. It is synthesized as a pre-pro-peptide of 2813 amino acids containing a signal peptide and propeptide; these are cleaved off during a complex pattern of maturation. The protein undergoes glycosylation and dimerization in the endoplasmic reticulum, carbohydrate processing and sulphation in the Golgi apparatus and finally multimerization and propeptide cleavage as it is assembled into secretory granules of megakaryocytes (and eventually platelets) or the Weibel–Palade bodies of endothelial cells. The degree of polymerization is high and multimers can contain up to 100 subunits. These large multimers are particularly effective in mediating platelet binding; their levels in plasma rise in response to a number of factors, including adrenaline and vasopressin. The protein has an unusually high content of cysteine residues (8.3%), which are involved in intra- and inter-subunit disulphide bonds (see Box 4.1).

The gene for von Willebrand factor is very large (178 kb), located on human chromosome 12 and illustrates the tissue specificity that can be achieved by the control of gene expression (see Chapter 5), as it is expressed only in endothelial cells and megakaryocytes. The large size of the gene reflects the presence of an extremely large number of introns (51, separating 52 exons). In comparison, after processing, the mature von Willebrand factor mRNA is only 8.7 kb in length.

Box 5.1. Control of transcription in bacteria and humans, page 117.

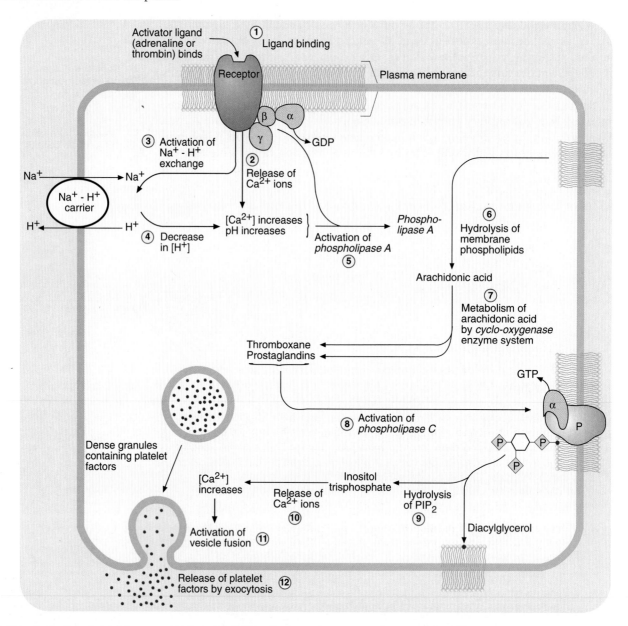

Figure 12.4 The mechanism of platelet activation. Steps 1–12 show the series of signalling events that lead to the release of platelet factors following activation of a platelet by adrenaline or thrombin. Further discussion of individual steps can be found in Chapter 11. In brief: binding of an activating ligand to a GTP-linked receptor raises intracellular pH and Ca^{2+}, leading to release of arachidonic acid by phospholipase A; metabolites of arachidonic acid in turn activate phospholipase C, causing further increase in Ca^{2+} and fusion of platelet factor vesicle with the plasma membrane.

platelets adhere to the damaged vessel surface and binding factor VIII of the blood clotting cascade (see Box 12.1). These released substances help to repair the vessel wall and contribute to haemostasis (Figure 12.4).

The clotting process can be triggered in two ways

Two pathways may trigger the clotting process: (1) the **intrinsic pathway**, in which damaged tissue releases the proteins **kininogen** and **kallikrein** that initiate the clotting process at factor XII (Hageman factor) and (2) the **extrinsic pathway**, in which damaged blood vessels release **tissue factor** and activate factor VII (Figure 12.5).

Specific and sequential proteolytic cleavages give a 'cascade' effect

More than a dozen proteins found in plasma are required to bring about the clotting of blood, and most of these act in a sequential manner leading to the conversion of **prothrombin** to **thrombin** (Figure 12.3). Thrombin, a proteolytic enzyme, acts on **fibrinogen** (see below); the end result of the 'clotting cascade' is

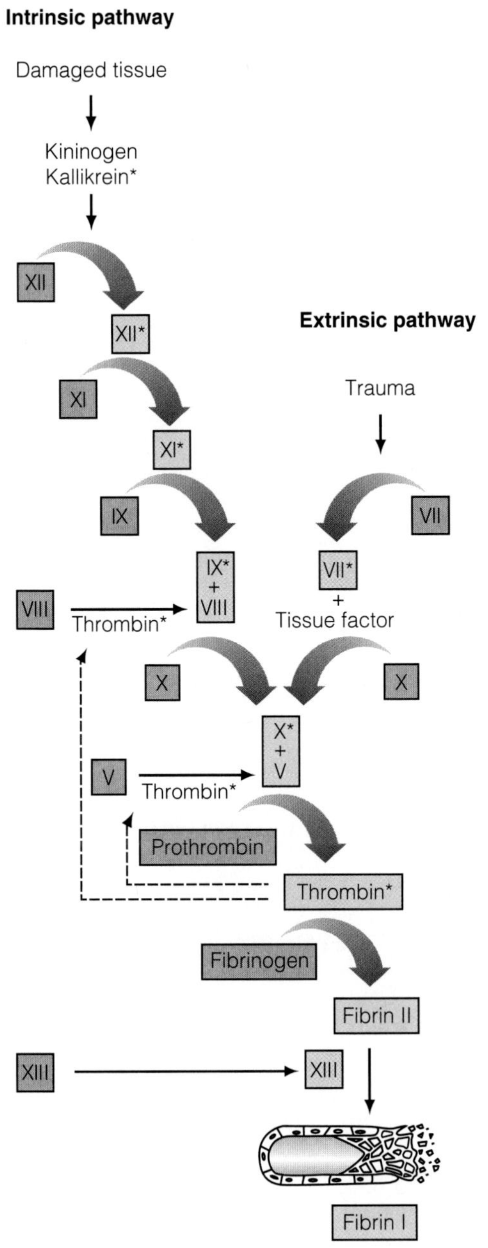

Intrinsic pathway

Damaged tissue

Kininogen
Kallikrein*

XII → XII*

Extrinsic pathway

Trauma

XI → XI*

IX

VII

VIII   IX* + VIII   VII* + Tissue factor
Thrombin*

X   X* + V   X
V   Thrombin*

Prothrombin

Thrombin*

Fibrinogen

Fibrin II

XIII → XIII

Fibrin I

Figure 12.5 Intrinsic and extrinsic initiation of blood clotting. The diagram shows the cascade of interactions amongst the various blood-clotting factors of plasma leading to the production of cross-linked fibrin (fibrin 1).

Factor II: Prothrombin
Factor V: Proaccelerin
Factor VII: Proconvectin
Factor VIII: Antihaemophilic factor
Factor IX: Christmas factor
Factor X: Stuart factor
Factor XI: Thromboplastin precursor
Factor XII: Hageman factor
Factor XIII: Fibrin cross-linking factor

to amplify the initial signal so that enough fibrin is produced for clotting to occur rapidly and prevent excessive blood loss from a wound. The conversion of the inactive prothrombin into active thrombin is effected by specific proteolytic cleavage, as are most of the other steps of the cascade. In classical haemophilia, factor VIII (antihaemophilia factor) is genetically absent, requiring regular injections of material isolated from human blood stocks. Use of infected blood in the preparation of factor VIII has lead to the widespread incidence of AIDS in haemophiliacs (see Chapter 18).

Chapter 18. HIV and AIDS, page 577.

Thrombin catalyses the conversion of fibrinogen to fibrin

Fibrinogen is a highly asymmetric protein composed of six polypeptide chains – α_2, β_2, γ_2 – and its three-dimensional structure is arranged so that the amino termini of these chains are in close proximity in the central area of the molecule. The action of thrombin cleaves off short peptide fragments from four of these termini and exposes

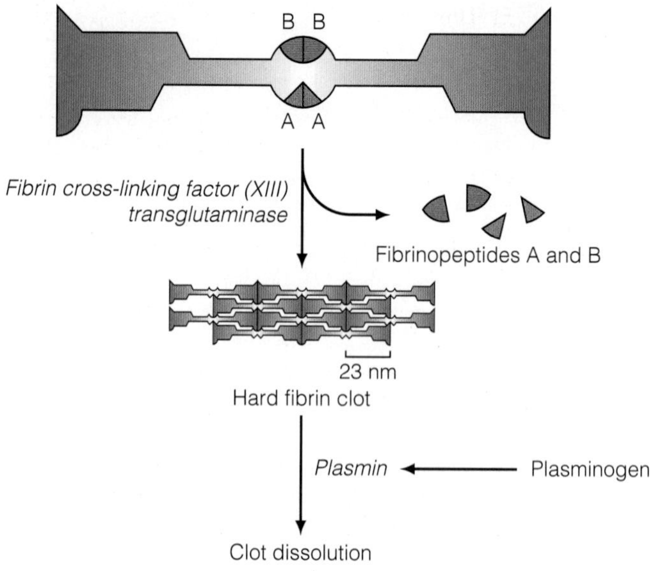

Figure 12.6 Fibrin cross-linking and fibrinolysis. Cleavage of fibrinopeptides from fibrinogen allows fibrin molecules to associate in a soft fibrin clot that is converted to a hard clot by the cross-linking action of transglutaminase. During healing, the hard fibrin clot is digested by the proteolytic action of plasmin, released from its precursor, plasminogen.

a surface in the protein which binds tightly to the end of another fibrinogen molecule. As this process proceeds, more and more fibrinogen molecules interact to form a network of insoluble fibrin and so initiate clot formation (Figure 12.6).

Thrombin is a highly potent enzyme; its uncontrolled production would pose a threat to the body, bringing the risk of undesirable clotting, for example blocking the capillaries. In order to keep thrombin under tight control, plasma contains the protein **antithrombin**, which binds any excess thrombin in plasma.

Cross-linking and fibrinolysis are important in the role of fibrin in clotting and healing

The physical strength of protein-based structures may be increased by the process of covalent cross-linking. The initial 'soft' fibrin clot (fibrin II) is held together only by noncovalent interactions; conversion to a mature ('hard') clot (fibrin I) depends on the action of factor XIII, which has transglutaminase activity (Figure 12.6).

Although a clot needs to be strong enough to seal a wound, if it became a permanent feature of the body wound healing and restoration of the skin to near-pristine condition would be difficult or impossible. The clot and its constituent cross-linked fibrin must be broken down at the appropriate stage in the healing process. The enzyme responsible for this is **plasmin**, which circulates in plasma as the inactive **plasminogen**. Drugs that are plasminogen activators are used clinically, for example in assisting clot dissolution in the treatment of coronary heart disease.

The erythrocyte and oxygen

Oxygen is a crucial component of the biosphere

The vital importance of oxygen is recognized by everyone, but the fundamental reason for this importance may be fully grasped only from a knowledge of biochemistry.

We take oxygen so much for granted – 'the air we breathe' – that it is easy to forget that life on Earth began long before oxygen appeared in substantial amounts in the atmosphere. The first cellular life forms are believed to have appeared by the time the first 1000 million years of Earth history had elapsed. The explosion of multicellular life forms in the Cambrian era occurred about 600 million years ago and oxygen levels in the atmosphere increased from 2500 million years ago. Before that time all life was anaerobic; the atmospheric oxygen was a waste product of the metabolism of some of the primitive unicellular organisms, probably resembling cyanobacteria. The increase in atmospheric oxygen levels must have had a marked effect on the evolution of living things because, even today, oxygen is poisonous to many anaerobic bacteria.

Why does the body need oxygen? The answer is simple: if the energy is to be 'harvested' from food molecules, hydrogen must be removed from them and oxygen is the final acceptor of this hydrogen. Anaerobic organisms adopt many strategies to do the same thing, producing a variety of end products such as ethanol, methane, H_2S, propanoic acid and, indeed, our muscles do this in a short-term fashion when producing lactate from pyruvate.

In molecular terms, oxygen is a direct substrate for only a limited range of reactions within the body – predominantly those catalysed by the group of enzymes called oxidases. Chief among these is **cytochrome oxidase**, the terminal step of the **electron transfer pathway** (see Chapter 7), but a whole range of oxidases use molecular oxygen as their hydrogen acceptor, including **xanthine oxidase** and **cytochrome *P*-450**. Of the tissues of the body, the brain particularly depends on a constant supply of oxygen (see below).

Chapter 7. Cytochrome *c* oxidase (complex IV) reacts with molecular oxygen, page 205.

The transport of oxygen is essential

Small organisms are able to get sufficient oxygen to their cells by simple diffusion, but all larger animals have some kind of oxygen-transport mechanism. A special class of oxygen binding and transporting proteins, the **globins**, appear to have evolved over a billion years ago – almost certainly in response to the increase in atmospheric oxygen. The ancestor of the major human globins – myoglobin and the α- and β-families of haemoglobins – seems to have made its appearance over 500 million years ago.

The characteristic three-dimensional shape of these proteins has been conserved during evolution and is responsible for the noncovalent binding of haem in a distinctive 'cleft'. The more detailed aspects of the structures are finely tuned to the precise role of the individual globin. The human genome contains several families of genes encoding globins and there are specialized globins for embryonic and fetal life (see Figure 12.17 and 12.18).

Oxygen can have toxic effects

Oxygen is potentially hazardous because of its high reactivity. Industrial and other processes carried out in a pure oxygen atmosphere must be carefully controlled. In the body, oxygen can have deleterious effects because of the toxic products that are produced as a result of interactions with biological molecules. These include **hydrogen peroxide** (H_2O_2) and the **superoxide radical** ($\bullet O_2^-$). Their toxicity lies in their powerful oxidizing properties, particularly towards double bonds such as those present in unsaturated fatty acids.

One of the modern views on ageing is that it is largely a result of an accumulation of the products of harmful and uncorrected oxidations of key components of the body. The erythrocyte, which needs to handle large amounts of oxygen, contains systems to protect itself from such damage (see below).

The erythrocyte must survive wear and tear

The erythrocyte is the end product of the erythropoietic system and is derived from a erythroblast by the loss of its nucleus (Figure 12.19). The red blood cell is not a true cell, but rather the remnants of a cell with an extremely high concentration of one protein – **haemoglobin**. The erythrocyte has a finite life in circulation

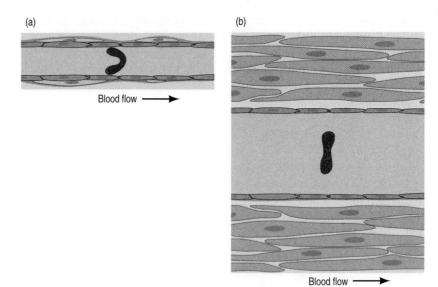

Figure 12.7 (a) Erythrocyte deforming to squeeze through a capillary. (b) The original shape of the erythrocyte is restored when it passes into wider vessels such as arteries.

of 60–120 days, depending on the age of the individual and other factors. In order to survive for this length of time, the erythrocyte possesses a variety of maintenance processes (see below).

Its principal function is the transport of O_2, CO_2 and H^+ between tissues and lungs. The characteristic biconcave disc shape of the human erythrocyte is maintained by internal cytoskeletal elements (see Chapter 3) and provides a high surface to volume ratio. Because the diameter of an erythrocyte is greater than that of the smallest capillaries, it distorts as it passes through these vessels (Figure 12.7), but its original shape is restored once it passes into wider vessels.

The erythrocyte suffers repeated insults during its lifetime. Besides being distorted in the capillaries, it experiences pressure changes and shearing forces as it makes its way round the circulation, especially in the chambers of the heart. To survive this, its membrane must be maintained in a sound condition: if the membrane deteriorates, cell lysis rapidly follows (see below). The heavy traffic in oxygen causes production of superoxide and H_2O_2, which will damage both the cell membrane and the erythrocyte's haemoglobin if they are not disposed of (see below).

Compared with true cells, the erythrocyte is a highly simplified and specialized entity, but its study has taught us many lessons about more complex cells, especially their membranes.

Chapter 3. The cytoskeleton determines cell shape and movement and regulates the organization of internal structures, page 49.

The erythrocyte membrane

The membrane plays a crucial role in the life of the erythrocyte

In one sense, the membrane *is* the red cell, because (given its lack of a nucleus and other internal components of a normal cell) without its membrane the erythrocyte would not exist. Rupture of the erythrocyte membrane (**haemolysis**) leads to loss of haemoglobin and function. Abnormal rigidity of the erythrocyte, such as is found in sickle-cell anaemia (see Chapter 17), results in premature destruction.

The erythrocyte membrane contains phospholipids, sphingolipids and cholesterol. Maintenance of the correct degree of mobility of the fatty acyl chains of the phospholipids and sphingolipids is vital for the proper functioning of the membrane proteins; this is partly achieved by cholesterol, present in roughly equal proportions with protein (see Chapter 3). Some of the lipids, notably the sphingolipids, are glycosylated in a fashion similar to those of the integral membrane proteins.

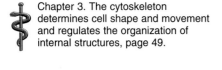

Chapter 17. Sickle-cell anaemia is a genetic disorder whose molecular basis is well understood, page 512.

Chapter 3. Membrane fluidity maintains the appropriate environment for the optimal functioning of membrane proteins and is carefully controlled, page 59.

The external and internal faces of the bilayer have different phospholipid, probably because of the curvature and flexibility of the membrane. The membrane is highly sensitive to certain lipophilic drugs, which can cause distortions of the erythrocyte, such as crenellations of the cell surface.

There are distinctive erythrocyte membrane proteins

For a general discussion of cell membranes and their proteins see Chapter 3.

Integral proteins

Because it is technically easy to isolate pure erythrocyte membranes, the major proteins and glycoproteins are well characterized (Table 12.5) and their study has been important in the development of concepts of membrane proteins in general (see below). This is especially so in the concepts of internal, external and transmembrane **domains**, so clearly seen in the major protein glycophorin (Figure 12.8). A more complex transmembrane structure is seen in the anion channel protein (see below). There are also minor but essential proteins such as the glucose transporter **GLUT 1** and the **Na+/K+-ATPase** (see Chapter 3). The major proteins, and probably most of the minor ones, are glycosylated – sometimes extensively, as in glycophorin, in which 60% of the mass of the glycoprotein is carbohydrate and only 40% is amino acids.

 Chapter 3. Na+/K+-ATPase is asymmetrically organized within the membrane, page 64.

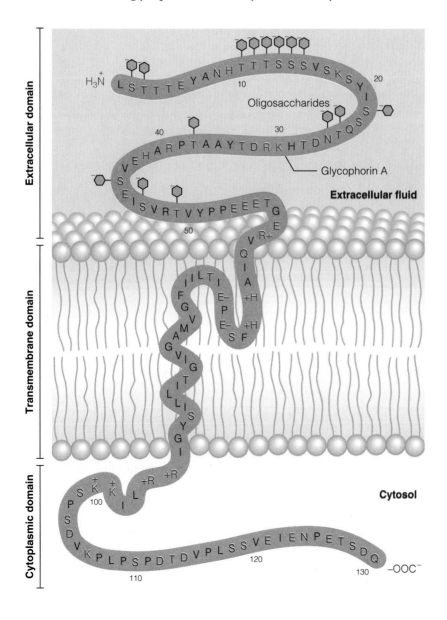

Figure 12.8 Glycophorin.

Table 12.5 Proteins of the erythrocyte membrane

Name	Polypeptide (kDa)	Quaternary structure	Molecules/cell
Integral proteins			
Anion exchange protein	89	Dimer	10^6
Glucose transporter (GLUT 1)	55	Monomer	10^6
Glycophorins	23, 29, 31	Monomer	10^6
Peripheral proteins			
α-Spectrin	260	Tetramer	10^5
β-Spectrin	225	Tetramer	10^6
Ankyrin	215	Monomer	10^5
Actin	43	12–17 oligomer	5×10^5
Glyceraldehyde-3-phosphate dehydrogenase	35	Tetramer	5×10^5

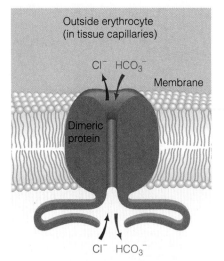

Figure 12.9 The anion channel protein of the erythrocyte.

Peripheral proteins and the cytoskeleton

Some proteins are attached to the internal surface of the erythrocyte membrane and, like all peripheral membrane proteins, these may be removed *in vitro* by chelating agents, high salt concentrations or mild detergents. Some of the cytoskeletal elements of the erythrocyte fall into this category, including spectrin, actin and ankyrin (Table 12.5). Another peripheral protein is the enzyme glyceraldehyde 3-phosphate dehydrogenase; it is not clear why this glycolytic enzyme should be so located, but it may be connected with the importance of glycolysis to the erythrocyte in the production of energy and 2,3-bisphosphoglycerate (see below).

Transport of solutes occurs across the erythrocyte membrane

Bicarbonate and chloride ions

The importance of the erythrocyte in the transport of CO_2 from the tissues to the lungs hinges on the activity of **carbonic anhydrase** within the erythrocyte and the role of this enzyme on the interconversion of CO_2 and bicarbonate.

$$H_2O + CO_2 \underset{\text{Carbonic anyhdrase}}{\rightleftharpoons} H_2CO_3 \rightleftharpoons H^+ + HCO_3^-$$

Bicarbonate must move *out* of the erythrocyte into the tissues to maintain the internal pH of the erythrocyte. This movement of bicarbonate anion is achieved by the **anion channel** (or **anion exchange**) **protein**, whose dimeric structure allows the energy-independent exchange of bicarbonate and chloride ions in the appropriate direction, depending on the location in the body (Figure 12.9).

Glucose

Glucose is the principal source of energy for the erythrocyte and enters from the plasma by facilitated diffusion, which requires the glucose transporter protein GLUT 1. Glucose enters in an energy-independent manner down a concentration gradient (see Chapter 3). This gradient is maintained as glucose is used within the erythrocyte by glycolysis and the pentose phosphate pathway.

Chapter 3. Entry of glucose into cells is mediated by several related tissue-specific transporters, page 68.

Na$^+$ and K$^+$ ions

The balance of these important cations is maintained by the activity of Na$^+$/K$^+$-ATPase, an enzyme present in the plasma membrane of all cells. This is especially significant in the erythrocyte, in which the ATP-dependent maintenance of the difference in Na$^+$/K$^+$ ratios between plasma and erythrocyte is a major energy requirement of the cell. In contrast, the absence of a Ca²⁺-ATPase (see Chapter 3) from the erythrocyte membrane explains the low concentrations of Ca^{2+} found in the cell.

Chapter 3. Muscle sarcoplasmic reticulum contains another type of ion-motive ATPase, page 67.

The surface of the erythrocyte membrane carries blood group substances

As is the case for all plasma membranes, the surface of the erythrocyte is covered in oligosaccharides covalently attached as components of the glycoproteins and glycolipids. However, this glycosylation assumes major significance in the erythrocyte because it characterizes the major blood groups. The **ABO system** is the major blood-grouping system of humans and the four types (A, B, AB and O) can be explained by the presence or absence of two enzymes that add sugar residues to proteins and lipids on the surface of erythrocyte precursor cells. The products of these enzymes – **glycosyl transferases** – are shown in Figure 12.10. Type O individuals lacks both of these enzymes, an AB person possesses both, types A and B have one or the other.

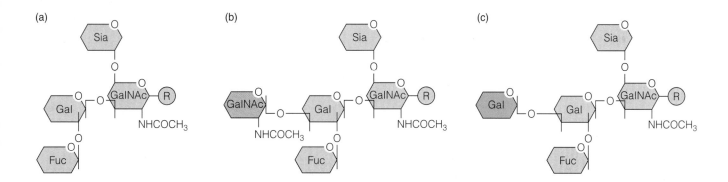

Figure 12.10 The molecular basis of ABO blood groups: (a) type O; (b) type A; (c) type B. All individuals possess the enzymes that add (a) to membrane lipids and proteins. Type A individuals have the transferase that adds *N*-acetyl galactosamine (GalNAc) to (a), producing (b), while type B have the enzyme that adds galactose (Gal) producing (c). AB persons have both enzymes and their membranes contain some molecules carrying (b) and some (c).

The erythrocyte membrane is a model for the plasma membranes of more complex cells

Our knowledge of the plasma membranes of human cells has been greatly advanced by study of the erythrocyte membrane because it possesses several unique features.

1. It is a 'simple' membrane, with far fewer components than the plasma membrane of, for example, a hepatocyte.

2. The erythrocyte membrane is the only membranous component of the human erythrocyte; in contrast, the plasma membrane of a hepatocyte comprises only about 2% of the total membrane of the cell.

3. Erythrocyte membranes may be prepared simply and in large amounts from outdated human blood stocks by spinning down the erythrocytes, lysing them and then spinning down what is already a virtually pure membrane preparation.

Haemoglobin – structure and function

Haemoglobin is understood in great detail

The red cell membrane confines a solution of haemoglobin at a concentration of about 5 mM; only slightly less than that found in crystals of the protein. Haemoglobin is one of the most plentiful and important proteins of the human body, and its structure and function have been studied in great detail. The knowledge obtained from this protein and from myoglobin, the oxygen-binding protein of muscle, have also been of great importance in our understanding of protein structure and function in general, especially at the level of quaternary structure and allosteric effects.

Both proteins undergo an **oxygenation** (as opposed to an oxidation) reaction, in which the iron atom of haem remains in the ferrous (Fe^{2+}) state:

$$Hb(Fe^{2+}) + O_2 \longrightarrow Hb(Fe^{2+}) — O_2$$
$$\text{Deoxyhaemoglobin} \qquad\qquad \text{Oxyhaemoglobin}$$

Globins have distinctive three-dimensional structures

Haemoglobin is a protein of four polypeptide subunits, two chains each of α-globin and β-globin forming an $\alpha_2\beta_2$ tetrameric structure. This is in contrast to the monomeric structure of myoglobin (Figure 12.11). Each globin chain binds one molecule of haem in a precise but noncovalent fashion. Contrast this to cytochrome *c* (see Chapter 7), in which the same type of porphyrin is bound covalently. The three-dimensional structures of α-globin, β-globin, myoglobin (and indeed of all globins) are very similar, with an unusually high (80%) α-helical content arranged in eight separate helices (named A to H from *N*- to *C*-termini). One of the most

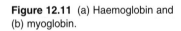

Chapter 7. Ubiquinone-cytochrome *c* reductase (complex III) contains cytochromes *b* and c_1, page 202.

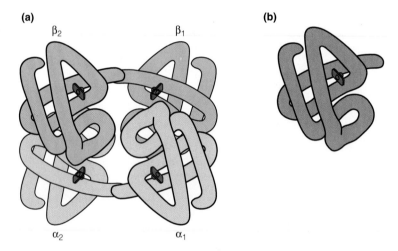

Figure 12.11 (a) Haemoglobin and (b) myoglobin.

characteristic features of the folding of the globin chain is the way in which several helices (notably E, F and H) are arranged to form a 'hydrophobic pocket' in which the haem is bound (Figure 12.12), predominantly by hydrophobic interactions with specific side-chain residues such as phenylalanine CD1 and leucine F4. These amino acids are completely conserved in all globins, showing their importance. Also conserved are the two histidines (E7 and F8) that bind the Fe^{2+} (Figure 12.14).

The quaternary structure of haemoglobin affects oxygen binding

The $\alpha_2\beta_2$ tetrameric nature of haemoglobin was revealed by the determination of its three-dimensional structure by X-ray crystallography and comparison with that of myoglobin. Dilution of a solution of haemoglobin causes the tetramer to dissociate to two αβ dimers, showing that α–β interactions are stronger than α–α or β–β. The contacts between subunits involve weak interactions, such as the electrostatic bond between aspartate 126 and arginine 141, and are of great importance in the allosteric properties of haemoglobin in oxygen binding. An essential

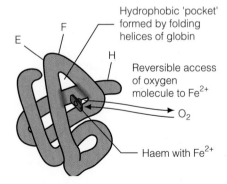

Hydrophobic 'pocket' formed by folding helices of globin

Reversible access of oxygen molecule to Fe^{2+}

O_2

Haem with Fe^{2+}

Figure 12.12 The hydrophobic 'pocket' in globins, showing helices E, F and H.

feature of this behaviour is that small but significant changes in the structure of one subunit caused by the binding of oxygen are transmitted to other subunits and change their oxygen-binding properties. This is the reason for the cooperative binding of oxygen by haemoglobin (see below). The overall effect of oxygen binding is to reduce the size of the central cavity of the tetramer.

There are physiologically important ligands of haemoglobin

The principal ligand of haemoglobin is oxygen, but other physiologically important ligands are H^+, CO_2 and 2,3-bisphosphoglycerate (2,3BPG), all of which bind at sites other than the oxygen-binding site (Figure 12.13). An important alternative ligand of the oxygen site is carbon monoxide. Carbon monoxide binds to haemoglobin in a similar manner to oxygen, but does so about 500 times more tightly, thus contributing to its toxicity. Carbon monoxide is a common atmospheric pollutant, but it also arises in the body during haem degradation (see below).

H^+, CO_2 and 2,3BPG all bind to haemoglobin to favour the *deoxy*haemoglobin form, to *promote the release of oxygen*. There is a clear physiological need for this behaviour, because oxygen is required at the tissues, where the concentrations of H^+ and CO_2 are highest. A haemoglobin that bound its oxygen too tightly would fail to release on adequate supply of oxygen to the tissues.

Oxygen binding by haemoglobin is cooperative

The binding of oxygen by haemoglobin is **cooperative**, as reflected in the sigmoidal shape of the oxygen binding curve and in contrast to the hyperbolic shape of the curve for myoglobin (Figure 12.14). Haemoglobin isolated from

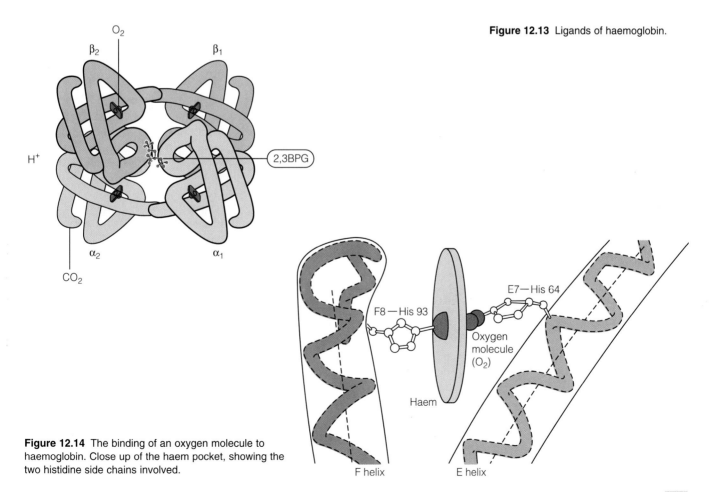

Figure 12.13 Ligands of haemoglobin.

Figure 12.14 The binding of an oxygen molecule to haemoglobin. Close up of the haem pocket, showing the two histidine side chains involved.

erythrocytes has a much higher affinity for oxygen than that present in the cell, which posed a puzzle for many years until it was recognized that haemoglobin occurs in the erythrocyte predominantly in the form of a complex with 2,3BPG. This complex binds oxygen 26 times less tightly because the presence of this small molecule (carrying five negative charges) in the central cavity of the protein favours the *deoxy* form by linking the β-subunits. One reason why fetal haemoglobin (see below) has a higher affinity for oxygen than the adult form is that it binds 2,3BPG less strongly. Conditions such as pulmonary disease or life at high altitudes elevate the concentration of 2,3BPG in the erythrocyte from the normal 3 mM to about 5 mM. Higher concentrations of 2,3BPG favour the dissociation of oxyhaemoglobin when erythrocytes pass through the capillaries of peripheral tissues, allowing delivery of oxygen. The concentrations of H^+ and CO_2 vary as the erythrocyte moves between the lungs and other tissues. The effects of both favour the release of oxygen at the tissues (Figure 12.15). The **Bohr effect** shows that decreasing pH shifts the oxygen-binding curve to a higher O_2 concentration; this is because deoxyhaemoglobin is a stronger acid (protonation of haemoglobin as it loses its O_2 is responsible for about half of the buffering capacity of the blood). The effect of CO_2 is similar and is effected by the reversible addition of CO_2 to the *N*-termini of the globin chains; this **carbaminohaemoglobin** is also an important part of the transport of CO_2 from the tissues to the lungs, accounting for about 15% of the total. The structure and function of haemoglobin are finely tuned to its movement between the oxygen tensions of lung and peripheral tissues (Figure 12.15a).

The anion exchange protein participates in carbon dioxide metabolism

When CO_2 is produced in the tissues, it diffuses into the erythrocyte where it is hydrated by **carbonic anhydrase**, a zinc metalloenzyme which catalyses the reaction

$$CO_2 + H_2O \xrightarrow[\textit{Carbonic anhydrase}]{} H_2CO_3$$

This reaction occurs spontaneously, but at a rate too slow for adequate gas exchange. Its importance lies in the subsequent ionization of carbonic acid to bicarbonate ion (HCO_3^-) and H^+. The latter protonates haemoglobin; bicarbonate passes out of the erythrocyte through the anion-exchange protein (see above) by exchanging with chloride ions (Figure 12.16). Thus, in the erythrocyte the net conversion of carbon dioxide to bicarbonate is achieved as the erythrocyte passes through the microcapillaries of peripheral tissues. The opposite sequence of events occurs in the lungs, leading to the production and eventual expiration of CO_2. By combining with H^+ ions as it releases its oxygen to the tissues, haemoglobin allows CO_2 to be transported in the form of HCO_3^-, producing only a small change in blood pH (Figure 12.16).

Other forms of haemoglobin occur during fetal development

The principal form of haemoglobin in normal adults is $\alpha_2\beta_2$, although other forms are found in certain individuals (see Chapter 17). In the neonate and fetus, however, forms of haemoglobin are found with different structures and oxygen-binding properties, reflecting the fact that the fetus has to obtain its oxygen from maternal oxyhaemoglobin across the placenta.

In the first 1–3 months of human fetal life the predominant form of haemoglobin is $\zeta_2\varepsilon_2$, where the ζ subunit is equivalent to α and the ε to β. During this period, α subunits progressively replace ζ and ε is replaced by γ, to give the $\alpha_2\gamma_2$ **fetal haemoglobin**, which in turn gives way to adult haemoglobin from 1 month before birth until the age of 3 months. These fetal forms have higher affinities for oxygen

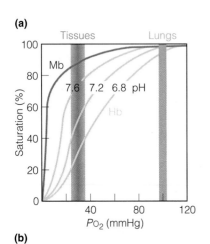

(a)

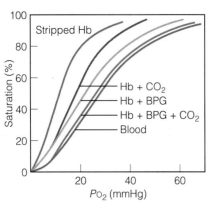

(b)

Figure 12.15 Effects of hydrogen ions (a) and carbon dioxide (b) on binding of oxygen by haemoglobin.

Chapter 17. Study of mutant haemoglobins has increased understanding both of mutations and of protein structure and function, page 513.

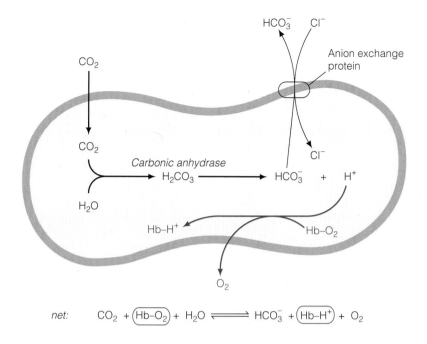

$$net: \quad CO_2 + \boxed{Hb\text{-}O_2} + H_2O \rightleftharpoons HCO_3^- + \boxed{Hb\text{-}H^+} + O_2$$

Figure 12.16 Carbon dioxide, bicarbonate and the release of oxygen from the erythrocyte.

than the maternal $\alpha_2\beta_2$ haemoglobin, which allow fetal haemoglobin to draw oxygen from maternal (Figure 12.17). It is apparently essential that the fetal blood operates with a different haemoglobin because the pH of fetal blood is significantly lower than that of maternal. Maternal-type $\alpha_2\beta_2$ haemoglobin in the fetal circulation would have a lower affinity for oxygen, because of the Bohr effect, and would therefore be ineffective in binding oxygen.

Globin genes and their expression have been intensively studied

Comparison of the amino acid sequences of globins reveals that they belong to two families, typified by α- and β-globin. The genes of the α-family reside on human chromosome 16 and those of the β-family on chromosome 11. Both are accompanied by several globin 'pseudogenes' that are not expressed (Figure 12.18). The myoglobin gene lies on chromosome 22. Both α- and β-globin genes are composed of three exons: exon 1 encoding helices A and B; exon 2 helices C to F; and exon 3 helices G and H. The organization of the start of an α-globin gene is shown in Figure 5.13.

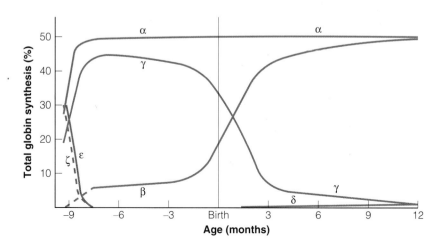

Figure 12.17 Synthesis of various globins in the perinatal period.

373

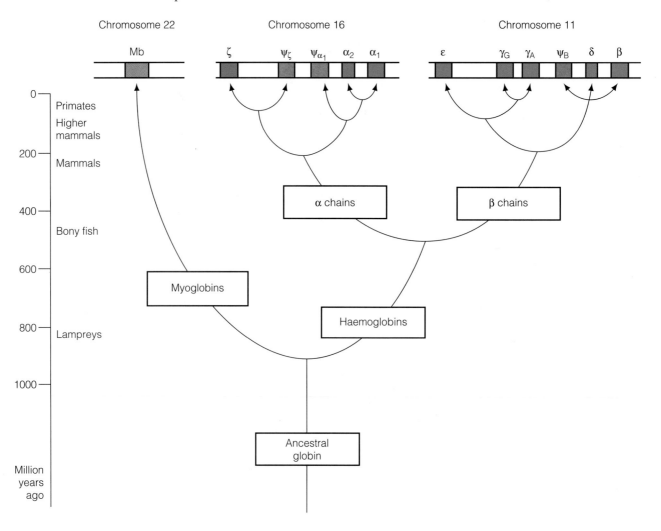

Figure 12.18 Evolution of genes for globins. Alignment and comparison of globin sequences allows construction of this molecular phylogenetic tree. The genes shown in grey are pseudogenes, related to globins, but not expressed.

Haemoglobin is synthesized in the reticulocyte

Synthesis of haemoglobin occurs in the erythroblast and reticulocyte stages of erythrocyte development (Figure 12.19) and accounts for much of the protein synthesis of that cell. It is accompanied by haem synthesis and is tightly controlled by the available concentration of haem.

Haem synthesis must occur in all cell types, because, although haemoglobin is made only in the reticulocyte, haemoproteins are required by every cell, especially for the various cytochromes. The complex pathway is shown in outline in Figure 10.20, starting from succinyl CoA (derived from the citric acid cycle) and the amino acid glycine, and leading eventually to protoporphyrin IX and haem. A significant feature of the pathway is the achievement of the arrangement of the acetyl and propionyl substituents in uroporphyrinogen III, which is necessary to bring the two final propionyl groups on the same side of the molecule so that they can protrude from the 'haem pocket' of haemoglobin into the aqueous environment.

There is a mechanism for the degradation of haemoglobin after erythrocyte haemolysis

When the erythrocyte has reached the end of its life span, changes in its cell surface are recognized and it is normally degraded in the spleen. Cells that are prematurely

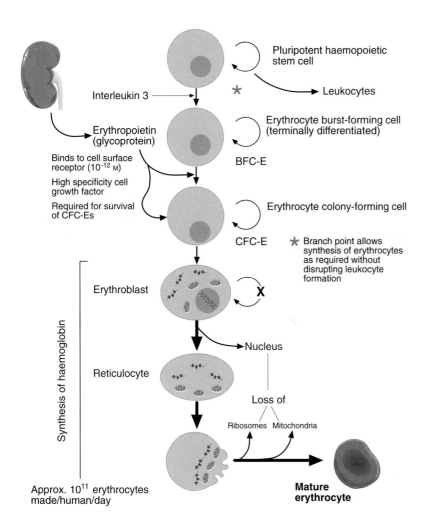

Pluripotent haemopoietic stem cell

Interleukin 3

Leukocytes

*

Erythropoietin (glycoprotein)

Erythrocyte burst-forming cell (terminally differentiated)

Binds to cell surface receptor (10^{-12} M)

High specificity cell growth factor

Required for survival of CFC-Es

BFC-E

Erythrocyte colony-forming cell

CFC-E

★ Branch point allows synthesis of erythrocytes as required without disrupting leukocyte formation

Synthesis of haemoglobin

Erythroblast

X

Nucleus

Reticulocyte

Loss of

Ribosomes Mitochondria

Approx. 10^{11} erythrocytes made/human/day

Mature erythrocyte

Figure 12.19 Formation of erythrocytes.

haemolysed will release their haemoglobin into the plasma where it dissociates into dimers, which are then bound by the plasma protein **haptoglobin**.

Once haem has been removed from globin, it is spontaneously oxidized to the ferric (Fe^{3+}) state and acted on by **haem oxygenase**, which uses molecular oxygen and NADPH to open the ring, producing **biliverdin** and carbon monoxide. The iron released is recovered for reuse (see Chapter 14) as the body can not afford to waste this valuable element. **Biliverdin reductase** uses NADPH to reduce biliverdin to **bilirubin**, the principal degradation product of haem. Bilirubin is a relatively insoluble molecule and is transported in the plasma (mostly bound to albumin; see Chapter 10) to the liver, where it is rendered water soluble by the addition of two glucuronide residues (producing **bilirubin diglucuronide**) and excreted in the bile as the bile pigments (Figure 10.21). The conjugated bilirubin is subsequently acted upon by the gut flora and the products are the principal colorants of faeces. The metabolism of bilirubin and its disturbances are of great importance in liver disease and its diagnosis (see below).

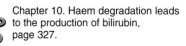
Chapter 14. Iron, page 453.

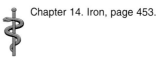
Chapter 10. Haem degradation leads to the production of bilirubin, page 327.

Metabolism in the erythrocyte

The erythrocyte is not a true cell in the normal meaning of the word, as it lacks most of the capabilities of cells (for example, the abilities to make proteins and to multiply by cell division). In addition, the erythrocytes of mammals lack all subcellular organelles, particularly mitochondria (avian erythrocytes *do* contain a nucleus, but this is functionally 'shut down'). Because of the central role of the mitochondrion in aerobic cellular metabolism, the erythrocyte must use a restricted range of pathways during its life span. Fatty acids and ketone bodies can not be metabolized.

The erythrocyte uses blood glucose for its sustenance

The erythrocyte is bathed in a solution of glucose which is maintained close to 5 mM (see Chapter 14). Glucose is taken through the plasma membrane into the erythrocyte by the glucose transporter GLUT 1 (see Chapter 3) and phosphorylated to G6P by hexokinase. The erythrocyte may then use glucose in glycolysis or the pentose phosphate pathway.

Glycolysis in the erythrocyte produces lactate and 2,3BPG

It is ironic that amid the heavy traffic in oxygen that goes on in the erythrocyte, the cell operates strictly anaerobic metabolism of lactate and its diffusion, leading to the production and excretion of lactate into plasma (Figure 12.20). This provides the ATP required by the cell via substrate-level phosphorylation by the phosphoglycerate kinase from 1,3-bisphosphoglycerate and pyruvate kinase from phosphoenol pyruvate. The process is maintained by reoxidation by lactate dehydrogenase of the NADH produced by glyceraldehyde 3-phosphate dehydrogenase. Glyceraldehyde 3-phosphate dehydrogenase is a peripheral protein on the internal surface of the erythrocyte membrane, reflecting its key role in the production of 1,3-bisphosphoglycerate. This molecule is converted to 2,3-bisphosphoglycerate (2,3BPG) by **bisphosphoglycerate isomerase** in a reaction that uses the energy in the 1-position to transfer the phosphate to the 2-position. Note that 2,3BPG does not contain a 'high-energy' bond.

$$1,3BPG \xrightarrow[\textit{Bisphosphoglycerate isomerase}]{} 2,3BPG$$

2,3BPG plays a key role in modulating the binding of oxygen by haemoglobin (see above).

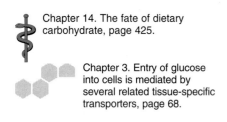
Chapter 14. The fate of dietary carbohydrate, page 425.

Chapter 3. Entry of glucose into cells is mediated by several related tissue-specific transporters, page 68.

Figure 12.20 Metabolic activities of the erythrocyte.

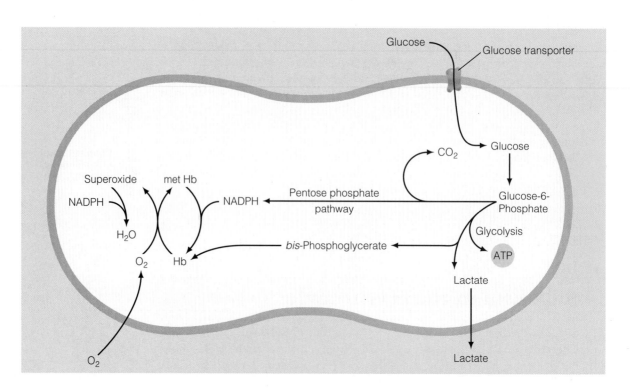

The erythrocyte has an active pentose phosphate pathway for generation of NADPH

This pathway (see Chapter 8) has two phases – oxidative and sugar phosphate inter-conversion. The two dehydrogenases of the oxidative phase provide a mechanism for the regeneration of the NADPH that is required continuously for the maintenance of the erythrocyte (see below). Approximately 15% of the G6P metabolized by the erythrocyte follows this pathway. The importance of the dehydrogenases is shown by the human genetic defect leading to **haemolytic anaemia**, in which the afflicted individuals lack G6P dehydrogenase and have excessively fragile erythrocytes.

Chapter 8. The pentose phosphate pathway, page 258.

The erythrocyte needs to maintain its adenine nucleotides

Erythrocyte ATP can be regenerated from ADP and ATP by substrate-level phosphorylation in glycolysis and by the action of adenylate kinase. The amino group of adenine nucleotides may also be lost to form inosine nucleotides. The erythrocyte contains the enzyme **IMP aminotransferase**, which catalyses the reaction

$$\text{IMP} + \text{glutamine} \xrightarrow[\textit{IMP aminotransferase}]{} \text{AMP} + \text{glutamine}$$

This route is exploited by the technique used for stabilizing blood for transfusion, in which inosine is added to the stored blood in order to maintain adequate levels of purine nucleotides during storage and before transfusion.

Maintenance of the erythrocyte

Because the erythrocyte is not capable of protein synthesis, it can not replace any protein or membrane components that become nonfunctional. It is therefore essential that, during the 60–120 days of its life span, it can prevent or repair damage to essential molecules such as haemoglobin and membrane lipids and proteins.

An alternative interaction between haemoglobin and oxygen is harmful

As discussed above, oxygen is a highly reactive and potentially hazardous molecule and the high flux of oxygen in and out of the erythrocyte makes it vulnerable to oxidation, especially in view of the high concentration of haemoglobin within the cell.

Besides the normal oxygenation of haemoglobin (see above), interactions between these molecules can also lead to oxidation in the following reaction:

$$\begin{array}{ccc} \text{Hb}(\text{Fe}^{2+}) + \text{O}_2 & \longrightarrow & \text{Hb}(\text{Fe}^{3+}) + \bullet\text{O}_2^- \\ \text{Deoxyhaemoglobin} & & \text{Methaemoglobin} \end{array}$$

There are two serious consequences

- methaemoglobin is not capable of binding oxygen;
- the superoxide radical is a highly reactive species which can oxidize many biological molecules, especially the double bonds of the fatty acyl chains of membrane lipids.

If not corrected, this reaction and the effects of its products will progressively convert haemoglobin to methaemoglobin, damage the erythrocyte membrane and eventually cause premature cell lysis. In order to prevent this, the erythrocyte contains enzymes that reduce methaemoglobin and detoxify superoxide. As these require a regular supply of NADPH, the pentose phosphate pathway is important for survival of erythrocytes (see above).

The reduction of methaemoglobin requires NADPH

This is achieved by **methaemoglobin reductase**:

$$Hb(Fe^{3+}) \longrightarrow Hb(Fe^{2+})$$
$$NADPH + H^+ \qquad NADP$$

Methaemoglobin is a normal component of the erythrocyte, typically comprising about 0.5% of the haemoglobin of the cell. The presence of methaemoglobin impairs oxygen transport only when the proportion is substantially raised above this value as, for example, by the action in some individuals of drugs such as the antimalarial agent primaquine. A high content of methaemoglobin is clearly deleterious because less oxygen can be transported. This situation is aggravated by the fact that significant levels of methaemoglobin increase the affinity of the remaining haemoglobin for oxygen and thus release it less readily to the tissues.

Detoxification of superoxide needs NADPH and the peptide glutathione

The detoxification of superoxide by the erythrocyte is a more complex process, involving a number of components. It also depends on a supply of NADPH as the ultimate reducing agent (Figure 12.21).

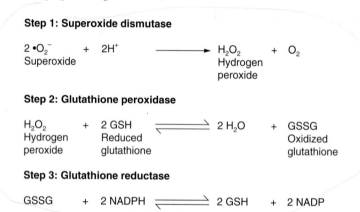

Step 1: Superoxide dismutase

$$2 \cdot O_2^- + 2H^+ \longrightarrow H_2O_2 + O_2$$

Superoxide → Hydrogen peroxide

Step 2: Glutathione peroxidase

$$H_2O_2 + 2\,GSH \rightleftharpoons 2\,H_2O + GSSG$$

Hydrogen peroxide, Reduced glutathione → Oxidized glutathione

Step 3: Glutathione reductase

$$GSSG + 2\,NADPH \rightleftharpoons 2\,GSH + 2\,NADP$$

Figure 12.21 Detoxification of superoxide.

The net reaction emphasizes the importance of an adequate supply of NADPH for the erythrocyte. For each molecule of haemoglobin oxidized, two molecules of NADPH are required to remove both the methaemoglobin and superoxide products.

Superoxide dismutase is a very widespread enzyme in nature and has a three-dimensional structure resembling that of the immunoglobulin domain (see Chapter 16), suggesting a link between this structure and 'protective functions'. **Glutathione peroxidase** is unusual in containing the trace element selenium; the importance of this enzyme for mammals is seen in the veterinary problem of sheep grazed in areas of selenium deficiency in soils suffering from a haemolytic anaemia (see above, human G6P dehydrogenase deficiency). Selenium is also being marketed to humans as an 'anti-ageing' product.

Glutathione is a tripeptide (γ-glutamyl-cysteinyl-glycine) which is ubiquitous in mammalian cells (see Chapter 10), occurring at a level of around 5 mM. Within cells, the sulphydryl on the cysteine residue is about 95% reduced (GSH) and helps to maintain the reduced state of intracellular components, especially proteins. Two molecules of glutathione can be oxidized to form the disulphide bond of oxidized glutathione (GSSG), an oxidation which is reversed by **glutathione reductase**. The reduced GSH of the erythrocyte also protects the sulphydryl groups of haemoglobin from oxidation.

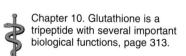

Chapter 16. Immunoglobulins have a multidomain organization, page 490.

Chapter 10. Glutathione is a tripeptide with several important biological functions, page 313.

CHAPTER 13

The molecular and metabolic activities of multicellular tissues

In this chapter we examine the functional characteristics of individual multicellular mammalian tissues. Connective tissue and the contractile aspects of muscle are dealt with in Chapter 15.

Adipose tissue

Adipose tissue is classified as white or brown. White adipose tissue is purely a store of triacylglycerol, and the metabolic functions of white adipocytes are restricted to the reactions required for synthesis and degradation of triacylglycerols. Brown adipose tissue is metabolically more complex than white. The adipocytes of brown adipose tissue store triacylglycerol and have the capacity to metabolize fatty acids as a mechanism for generating heat and raising body temperature.

Muscle

Myocytes, the contractile cells of muscle, have the capacity to store both glycogen and triacylglycerol. These cells come in two types: red and white. The difference reflects different patterns of metabolism and determines the mechanical characteristics of the muscle. Most of the metabolic activity is directed towards providing energy for contraction. Myocytes make little metabolic contribution towards the activity of other tissues, except in starvation or protein malnutrition when myocyte structural proteins may be degraded to meet the energy needs of other tissues.

Liver

Hepatocytes are the distinctive cell type of the liver. Most of their metabolic activity is directed towards sustaining other tissues, through the provision of glucose or ketone bodies to other tissues. Hepatocytes also synthesize most of the plasma proteins and convert many compounds to soluble nontoxic forms so that they may be excreted.

Kidney

Renal tissue is composed of many specialized cells which cooperate to regulate the fluid volume and ionic composition of the body. The kidney also has specialized hormone-dependent activities.

Pancreas

The pancreas is made up of many specialized cell types. In this chapter we describe both the exocrine activity of the pancreas (which enables it to synthesize, store and secrete enzymes into the digestive tract) and the endocrine activity of the pancreas (involving the production, storage and secretion of the peptide hormones insulin and glucagon). The molecular basis of diabetes is described.

Electrically active tissue

Electrically active cells are found in a diverse range of tissues: central and peripheral neural tissue, myocytes and light-detecting cells of the eye. The molecular mechanisms underlying electrical activity of tissues are described.

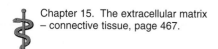

Chapter 15. The extracellular matrix – connective tissue, page 467.

The cells of mammalian tissues lie in a complex environment of extracellular molecules that make up the connective tissue holding the cells of each tissue together. **Connective tissue** molecules provide adhesion between cells (see Chapter 15). Small blood vessels, **microcapillaries**, pass through the connective tissue to bring oxygen, hormones and nutrients to the cells and remove waste molecules. The capillaries are lined with **endothelial** and **smooth muscle** cells, which often exhibit metabolic activities distinct from those of the major cell type in the tissue, and may make a major and distinctive contribution to the function of the tissue. In the liver only about 60% of the cells are hepatocytes, about 30% being endothelial cells. However, endothelial cells are small compared with hepatocytes and minute compared with cardiac myocytes so that, despite their large numerical proportion, they make up only a small fraction of the mass of most tissues.

For the purposes of this chapter we assume that students have acquainted themselves with the central metabolic pathways presented in Chapters 7, 8, 9 and 10.

The metabolic functions of adipose tissue

Animals possess two different types of adipose tissue. **White adipose tissue** cells serve solely as storage sites for triacylglycerol. This is the tissue in which nutrient in excess of that immediately required for energy production is stored as triacylglycerol – to be mobilized and used when there is no nutrient input from the

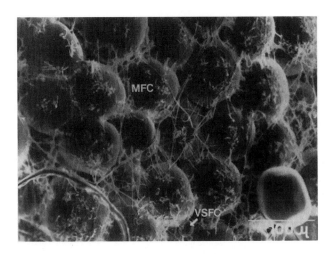

Figure 13.1 Human adipose tissue, showing the two sizes of adipocyte: MFC (mature fat cell) and VSFC (very small fat cell) seen in all animal adipose tissue. The two cell types may have different functions, for in obesity both MFCs and VSFCs undergo hyperplasia (increase in cell numbers) but only MFCs undergo hypertrophy (increase in cell size). the tissue is interspersed with collagen fibres.

digestive system. The amount of triacylglycerol stored in humans is theoretically sufficient to meet the normal energy needs for about 40 days.

Brown adipose tissue is both structurally and functionally more complex than white tissue. It stores triacylglycerols, which can be metabolized within the adipocytes to generate energy in the form of heat. The brown colour is due to the cytochromes present in the many mitochondria present in these cells.

White adipocytes are sites for storage of triacylglycerol

Most white adipose tissue is found under the skin where it provides insulation, but it is also located around internal organs to provide mechanical protection. White adipocytes are occupied almost entirely by a single droplet of triacylglycerol (Figure 13.1). A small volume of cytoplasm at the periphery of the cell contains a nucleus, and the enzymes of glycolysis and gluconeogenesis (but no glucose-6-phosphatase) and the enzymes required to assemble and hydrolyse triacylglycerols.

Triacylglycerol from the diet is directed towards adipose tissue

Triacylglycerols enter the bloodstream from the digestive tract as **chylomicrons**. Chylomicrons have a half-life in circulation of about 10 minutes, reflecting the rapidity with which triacylglycerols are taken up into tissues. Most of the circulating triacylglycerol is extracted by adipose tissue, although small amounts enter liver cells and heart muscle cells. The relatively small capacity of other tissues to take up triacylglycerol ensures that, at times when glucose is available, most of the dietary triacylglycerols are stored in adipocytes rather than being metabolized.

The capillary endothelial cells of adipose tissue are rich in the enzyme **lipoprotein lipase** (sometimes called **clearing factor lipase**). This enzyme, located on the outer plasma membrane of endothelial cells, catalyses the hydrolysis of triacylglycerols. The fatty acids produced are taken up by the endothelial cells and passed to the adipocytes for storage as triacylglycerol. The glycerol produced is transported to the liver.

Adipose tissue lipoprotein lipase controls the storage of triacylglycerol

The amount of lipoprotein lipase present on the surface of these endothelial cells reflects the nutritional state of the body: in starvation the number of molecules is

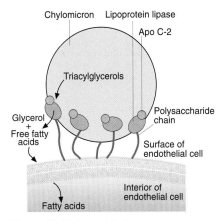

Figure 13.2 Lipoprotein lipase on endothelial cell plasma membranes facilitating the uptake of fatty acids into adipocytes

381

reduced; on re-feeding it increases. This control means that the maximum amount of triacylglycerol-derived fatty acid is taken up and stored in adipose tissue immediately after a meal and that between meals, when triacylglycerol is required by other tissues, not all the circulating triacylglycerol is extracted by adipose tissue. Lipoprotein lipase is activated when insulin secretion is increased and inhibited when adrenaline secretion is increased (see Chapter 9).

Chapter 9. Table 9.1, Hormones that activate hydrolysis of triacylglycerol in adipose tissues, page 263.

Fatty acids are re-esterified for storage within adipocytes

Re-esterification of fatty acids within adipocytes depends on the availability of glycerol-3-phosphate. Adipose tissue lacks glycerol kinase, and must derive its glycerol-3-phosphate either through glycolysis from glucose or through gluconeogenesis from pyruvate (see Figure 13.3).

In adipose tissue the ratio of AMP to ATP is substantially higher than in most other tissues. For this reason, adenine nucleotides play a less significant role in regulating phosphofructokinase and the rate of glycolysis in adipose tissue (see Chapter 8).

Chapter 8. Phosphofructo-kinase is the key regulatory enzyme of glycolysis, page 239.

Fatty acids released from triacylglycerols in adipose tissue are continuously re-esterified

Adipose tissue has three lipases: tri-, di- and monoacylglycerol lipase. Fatty acids are released from triacylglycerols by progressive hydrolysis (Figure 13.4), producing sequentially diglycerides, monoglycerides and free fatty acids plus glycerol, all of which diffuse into the bloodstream.

Provided that glucose is available in the adipocyte, fatty acids released by lipolysis are continuously re-esterified, the balance between re-esterification and release of free fatty acids into the bloodstream being controlled by insulin. Insulin is a potent inhibitor of lipolysis, because the adipocyte glucose transporter, GLUT 4, is insulin-dependent (see Chapter 3). When blood glucose is low, circulating insulin concentration is also low and there are few GLUT 4 transporters in the plasma membrane. The amount of glucose that enters the cell is therefore insufficient to produce the glycerol-3-phosphate required for re-esterification of fatty acids. Thus, in a situation when fatty acids are required by other tissues as nutrients, they will be released from adipocytes rather than re-esterified. After a meal

Chapter 3. Entry of glucose into cells is mediated by several related tissue-specific transporters, page 68.

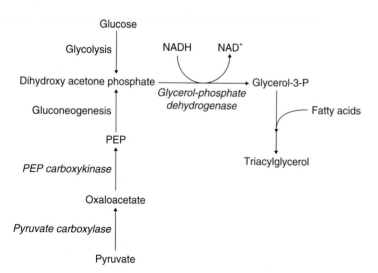

Figure 13.3 Synthesis of glycerol-3-phosphate and triacylglycerols in adipocytes.

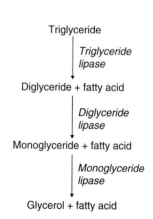

Figure 13.4 Fatty acids are sequentially hydrolysed from triacylglycerols by the action of three lipases.

the blood glucose concentration rises, more insulin is secreted into the blood-stream, GLUT 4 transporters move into the plasma membrane from endosomes and fatty acids released from triacylglycerols in adipose tissue are re-esterified. The lipolysis and esterification reactions involved are described in Chapter 9.

Fatty acids released from adipose tissue are transported in the bloodstream bound to serum albumin. The glycerol produced returns to the liver, where it is a substrate for gluconeogenesis (see Chapter 8).

Triacylglycerol lipase is hormone sensitive, being activated by adrenaline, noradrenaline, ACTH and glucagon. Binding of hormone to adipocyte receptors activates an adenylate cyclase-mediated amplification **cascade** (see Chapter 11), through which triacylglycerol lipase is phosphorylated. The di- and monoacyl-glycerol lipases are not hormone sensitive. Lipolysis in adipose tissue is also stimulated by growth hormone, by a mechanism operating on a longer timescale, through changes in transcription rates of certain genes.

Chapter 8. Adipose tissue and liver cooperate to use glycerol, page 252.
Chapter 9. Fatty acids are released from triacylglycerol stores by the action of a hormone-sensitive lipase, page 263.

Brown adipose tissue has a thermogenic role

Brown adipose tissue is located mostly in the dorsal thorax, where its proximity to the heart may be important. It is also found around the kidneys and adrenal glands. Brown adipocytes are of particular importance in the young and in hibernating animals because of their capacity to generate heat.

Brown adipose tissue is highly vascularized and innervated with noradrenergic fibres. The cells are smaller and more complex than those of white adipose tissue. Each cell contains many small lipid droplets and many mitochondria. The presence of mitochondria provides the clue to the main difference between white and brown adipose tissue: brown adipocytes can oxidize the fat they store.

Thermogenesis is stimulated by noradrenaline

Shivering, the physiological response to cold, stimulates release of noradrenaline from the nerve fibres within brown adipose tissue. Noradrenaline binds to plasma membrane receptors, which are linked through G-proteins to activate adenylate cyclase (Chapter 11). Cyclic AMP, the second messenger, activates triacylglycerol lipase to hydrolyse triacylglycerols (Figure 13.5). The released fatty acids are oxidized within the adipocyte through β-oxidation and the citric acid cycle.

Chapter 11. Homonal responses mediated through adenylate cyclase and the production of cAMP, page 338.

Thermogenin dissipates mitochondrial proton gradients

Mitochondrial function in brown adipocytes differs from that in liver, muscle or other tissues in that electron transport and oxidative phosphorylation may be uncoupled.

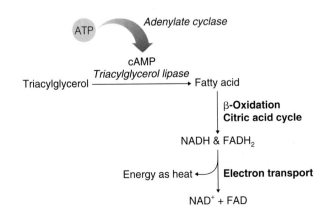

Figure 13.5 When fatty acids are oxidized in uncoupled brown adipose tissue the energy generated by the electron transport chain is not coupled to phosphorylation of ADP but is instead released as heat.

Chapter 7. Transfer of reducing equivalents down the electron transport pathway generates a proton gradient across the inner membrane, page 206.

The inner membranes of brown adipocyte mitochondria contain **thermogenin**, a protein that forms channels through the membrane. These channels increase the permeability of the inner mitochondrial membrane to protons, dissipating the proton gradient generated by the **electron transport chain** (Chapter 7). Movement of protons through these channels is inhibited by physiological concentrations of adenine and guanine di- and trinucleotides, ensuring that thermogenesis is not continuous. The inhibitory effect of these nucleotides is overcome by the long-chain fatty acids released from triacylglycerols. Protons then flow through the channels. This uncoupling ensures that the energy released from the oxidation of NADH is converted to heat rather than being coupled to the phosphorylation of ADP.

The metabolic functions of muscle

The metabolic activity of **myocytes** reflects both the contractile activity of the muscle and the nutritional state of the body. Cardiac myocytes work constantly, with relatively small variations in the mechanical demands being made on them. However, the metabolic patterns in cardiac myocytes *do* change depending on the nutritional state of the whole body and changes in metabolic demand associated with changes in the rate or force of contraction. The contractile properties of muscle are described in Chapter 15.

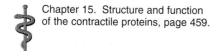

Chapter 15. Structure and function of the contractile proteins, page 459.

The energy consumed by muscle reflects contractile activity

Contraction of muscle fibres depends on the conversion of chemical energy (as ATP) into mechanical energy – the shortening of muscle fibres. During exercise the requirement for ATP can be as high as 200 µmol ATP per gram of muscle per minute. The amount of glycogen normally stored in muscle is equivalent to about 20 µmol glucose per gram of muscle, although training can increase this. Complete oxidation of muscle glycogen produces about 700 µmol ATP per gram of muscle, sufficient to support contractile activity for about 3.5 minutes. If the metabolism of glycogen proceeds only to the formation of lactate, the ATP produced will support contraction for only about 20 s.

Muscle contraction is sustained only while the rate of hydrolysis of the γ-phosphate of ATP is matched by the rate of rephosphorylation of ADP. To provide the energy to support vigorous or prolonged exercise, muscle cells metabolize fatty acids derived from triacylglycerols stored within muscle cells as small lipid droplets and fatty acids released from adipose tissue.

Muscle cells have an additional energy source: creatine phosphate

Creatine phosphate is a 'high-energy' compound found in muscle and brain. The standard free-energy change associated with the hydrolysis of the phosphate bond of creatine phosphate is almost identical to that obtained from the hydrolysis of the terminal phosphate of ATP. This means that the phosphate of creatine phosphate is readily exchangeable with the γ-phosphate of ATP through a reaction catalysed by **creatine kinase**.

This reaction allows muscle cells to store additional energy as creatine phosphate at times when the production of ATP from metabolism exceeds the energy requirement of the cell. Conversely, it provides an energy buffer at times when the energy requirement exceeds the rate at which ADP is phosphorylated through metabolism. In a myocyte at rest the amount of creatine phosphate is about the same as the amount of ATP, so during exercise it will be expended very rapidly if there is insufficient oxygen to support the rephosphorylation of ADP through oxidative phosphorylation.

Adenylate kinase provides energy by transferring phosphate groups between adenine nucleotides

The α-β phosphoanhydride bond of ADP is a high-energy bond. The enzyme **adenylate kinase** transfers a phosphate group from one molecule of ADP to another, producing one molecule each of ATP and AMP:

$$ADP + ADP \leftrightharpoons ATP + AMP$$

This is an extremely important reaction, because the AMP produced signals that the energy charge in muscle cells is low and activates glycogenolysis and glycolysis through allosteric effects on glycogen phosphorylase and phosphofructokinase (Chapter 8). A disadvantage of this reaction is that if AMP accumulates in muscle cells it will be deaminated by AMP deaminase, releasing ammonium ions and IMP, thus:

$$AMP \rightarrow IMP + NH_4^+$$

Chapter 8. Glycogen phosphorylase and glycogen synthase are both subject to allosteric regulation, page 226. Chapter 8. Phosphofructo-kinase is the key regulatory enzyme of glycolysis, page 239.

Creatine is synthesized in the liver

Creatine is synthesized in the liver from glycine, arginine and methionine. Arginine and methionine are **essential amino acids** (Chapter 10) – individuals with a nutritional deficit of these amino acids, or suffering from liver failure, are unable to synthesize sufficient creatine. Creatine is transported from liver to both muscle and brain cells, where it is phosphorylated.

Chapter 10. Biosynthesis of amino acids, page 306. Chapter 14. Table 14.4, Essential and nonessential amino acids, page 429.

The metabolic profile of muscle cells determines their contractile function

The microscopic appearance of a muscle cell provides the clues to its mechanical function. Muscle fibres are categorized as **red muscle cells** (type I) or **white muscle cells** (type II). It is important to remember that individual muscles are not composed entirely of white or red cells; most contain both cell types in varying proportions.

Box 13.1 Creatinine production can be used to measure the amount of muscle tissue

Creatine and creatine phosphate are degraded spontaneously to creatinine. The constant need to replace creatine is explained by the fact that both creatine and creatine phosphate are spontaneously cyclized, producing the compound **creatinine**, which is then excreted in the urine. Creatinine is a useful diagnostic marker; the amount excreted in a 24-h period is proportional to the muscle mass of each individual. Approximately 2% of the total pool of creatine and creatine phosphate is converted to creatinine every day, corresponding to 400 mg per 100 g wet weight of muscle tissue (8 mg creatinine would be formed each day per 100 g of muscle). By collecting a 24-h urine sample and measuring the amount of creatinine, it is possible to determine the muscle mass of an individual. For example, if the total urinary creatinine output is 1 g in 24 h, the individual has 12.5 kg of muscle. If muscles are damaged, by traumatic injury or degenerative disease (such as muscular dystrophy), there might be an increase in the amount of creatinine excreted disproportionate to the muscle mass.

Red muscles have a high oxidative capacity

Red muscles derive their colour from the haem group of proteins that they contain in high concentrations, the oxygen-binding protein **myoglobin** and the cytochromes of the **electron transport chain**. A relatively large proportion of the volume of red muscle cells is occupied by mitochondria, which contain the enzymes of oxidative metabolism and the electron transport chain. Red muscle cells are able to extract all of the nutrient value from carbohydrates and fats but have a high demand for oxygen.

Fat droplets containing triacylglycerol are often seen within red muscle cells. The oxidative capacity of these cells means that they are able to metabolize fatty acids rapidly.

Metabolism in white muscle cells is designed for short-term bursts of energy

White myocytes contain less myoglobin and cytochrome and fewer mitochondria than red myocytes. All mammalian muscle cells possess some mitochondria and therefore the capacity for some oxidative energy production. A white muscle at rest contains large numbers of glycogen granules between the contractile filament bundles. The enzymes that degrade glycogen (as well as the enzymes that control the degradation of glycogen) are stored within these granules, ensuring efficient mobilization of G1P from glycogen (Chapter 8). The energy yield of ATP produced from each molecule of glucose is relatively low if pyruvate is converted to lactate rather than entering the citric acid cycle. However, white muscle cells have high concentrations of the enzymes of glycolysis and can convert glucose to pyruvate or lactate very rapidly. The timescale on which they can sustain this energy production reflects the amount of glycogen stored in the cell and the rate at which they take up glucose from the bloodstream. The smaller oxidative capacity of white myocytes is sufficient to enable them to derive some energy from the oxidation of fatty acids.

White muscle cells contain less hexokinase than red skeletal or heart muscle cells. However, this is not disadvantageous because white muscle cells derive most of their G6P from their glycogen stores rather than from circulating glucose. No muscle cells contain glucose-6-phosphatase, so they are unable to release glucose to the circulation, and G6P accumulating in the cell will either be oxidized through glycolysis or the pentose phosphate pathway or be converted to glycogen for storage.

Chapter 8. When the energy status of a cell is low monosaccharides will be released from glycogen stores, page 225.

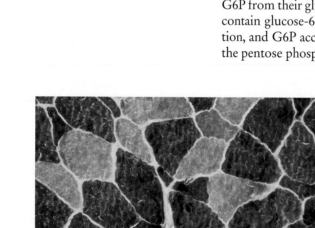
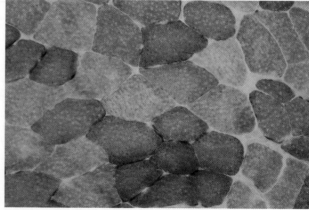

(a) (b)

Figure 13.6 Histochemical demonstration of differences between red and white muscle fibres in a single human skeletal muscle. The tissue is sectioned across the fibre, showing individual fibrils. In (a) the tissue is stained for myofibrillar ATPase activity. The darker staining indicates greater rates of ATPase characteristic of white muscle cells. In (b) the tissue is stained for cytochrome oxidase, and darker staining is seen in those fibres with greater density of mitochondrial cristae: the red muscle fibres. Figures courtesy of Dr W.M.H. Bevan, University of Glasgow Department of Histopathology.

Box 13.2   Lactate dehydrogenase exists as several isoenzymes

Lactate dehydrogenase is found in many cell types, and therefore in environments where the metabolic requirements differ. For example, skeletal muscle needs to produce lactate but in the liver lactate is converted to pyruvate to make it available as a substrate for gluconeogenesis. Lactate dehydrogenase is able to accommodate these opposing requirements because it occurs as tissue-specific **isoenzymes**.

This four-subunit enzyme is made of varying proportions of two types of subunit: M and H, making five variations of the tetramer possible: M_4 (type 5), M_3H, M_2H_2, MH_3 and H_4 (type 1). The H subunit predominates in the heart, the M subunit predominates in skeletal muscle and the liver. The functional difference between M and H subunits is that the M subunit has higher affinity for substrate than the H subunit. In addition, the H subunit is allosterically inhibited by high concentrations of pyruvate.

The isoenzymes of lactate dehydrogenase can be separated by electrophoresis, which means that serum lactate dehydrogenase can be used as a marker of tissue-specific damage. For example, after a myocardial infarction damaged heart myocytes release H_4 into the plasma.

Sustaining energy production in white muscle fibres depends on coupled dehydrogenase reactions

In order to sustain the rate of energy production through glycolysis white muscle cells need to reoxidize NADH to maintain a supply of NAD^+ for glyceraldehyde-3-phosphate dehydrogenase. As they contain relatively few mitochondria these cells can not oxidize NADH through the electron transport chain fast enough to sustain glycolysis if energy demand is high. As an alternative, muscle cells use **lactate dehydrogenase** to convert NADH to NAD^+ in the following reaction.

$$\text{Pyruvate} + \text{NADH} + H^+ \rightarrow \text{Lactate} + NAD^+$$

Muscle contains large amounts of lactate dehydrogenase and this reaction permits muscle cells to shift some of the metabolic burden to the liver. The lactate produced in muscle is transported to the liver where it is used as a substrate for gluconeogenesis and is converted to glucose, which then may be returned to muscle or sent to other peripheral tissues. This is the **Cori cycle** (Chapter 8).

Chapter 8. Liver and muscle co-operate to conserve energy through the Cori cycle and the alanine cycle, page 251.

There is a close relationship between the mechanical activity of muscle cells and processes through which they derive their energy

Red muscle fibres, because of their high oxidative capacity, are able to maintain contractile tension for long periods of time. These are the 'posture' muscles and have a slow 'twitch' speed reflecting a relatively slow rate of hydrolysis of ATP by the myosin molecules (this is described more fully in Chapter 15).

White muscle fibres produce large amounts of energy but only for relatively short periods of time, being largely dependent on endogenous glycogen stores. These are the 'sprinting' muscles. They have fast 'twitch' speeds and their myosin molecules hydrolyse ATP at a correspondingly fast rate.

The average diameter of fibres also differs, red fibres being smaller in diameter than white fibres. Diffusion of solutes into the interior of a fibre is more efficient if its diameter is small: it is logical therefore that fibres dependent on diffusion of oxygen for metabolism should have a smaller diameter. In addition, capillary density in red muscle is higher, enabling oxygen to be provided more efficiently.

Chapter 15. Structure and function of the contractile proteins, page 459.

Metabolic functions of the liver

The biochemical functions of the liver are so many and varied that a description of its activities might require a comprehensive textbook of biochemistry. In part, this reflects the fact that the liver is the first tissue through which the contents of the hepatic portal vein pass, carrying molecules absorbed from the digestive system. Many of the metabolic functions of the liver are described in detail elsewhere and, in particular, students are referred to the chapters on metabolism (Chapters 7–10) and nutrition (Chapter 14).

Many of these metabolic processes provide nutrient to other tissues, but the liver must also meet its own energy needs. The liver needs

- ATP for tissue maintenance and for synthesis of macromolecules including glycogen, triacylglycerols and proteins.
- Reduced nucleotides, NADH, NADPH and $FADH_2$, for synthetic processes including gluconeogenesis and the synthesis of fatty acids, cholesterol and other compounds.

Utilization of glucose in the liver

Chapter 8. Glucokinase in the liver helps to regulate the distribution of glucose between different tissues, page 223.

The liver takes up glucose from the bloodstream only if all other tissues are adequately supplied. Entry of glucose into hepatocytes is independent of insulin, and glucokinase (the enzyme that phosphorylates glucose in the liver) has a much lower affinity for glucose than does the hexokinase characteristic of other tissues. Liver glucokinase is regulated by the binding of an inhibitory protein, a relationship mediated by F1P and F6P (Chapter 8).

When the blood glucose concentration is high, glucose is stored in the liver as glycogen. About 30% of the glucose metabolized in the liver goes through the pentose phosphate pathway (Chapter 8), ensuring that hepatocytes have sufficient NADPH for the reductive biosynthetic pathways, including lipogenesis (Chapter 9).

Chapter 8. The pentose phosphate pathway, page 258.
Chapter 9. The synthesis of fatty acids begins with the attachment of acetyl CoA and malonyl CoA to the acyl carrier region, page 277.

Gluconeogenesis

Chapter 8. Glucose-6-phosphatase, page 249.

Although partial reversal of glycolysis – the conversion of pyruvate to G6P – occurs in all tissues, hepatocytes are the only cells with **glucose-6-phosphatase**, which enables the liver to produce glucose and export it into the circulation (see Chapter 8). The presence of glucose-6-phosphatase gives the liver a central role in the distribution of nutrient between tissues. Among the substrates for gluconeogenesis the liver uses lactate and alanine derived from muscle and glycerol derived from adipose tissue.

Chapter 10. Ribose is activated for nucleotide biosynthesis as PRPP, page 315.

Box 13.3   Glucose-6-phosphatase deficiency and gout

Glucose-6-phosphatase deficiency is one of the predisposing factors for the development of **gout**, a condition in which crystals of sodium urate are deposited in the joints. This condition leaves sufferers with extremely painful swollen joints. If glucose-6-phosphatase is deficient in the liver, even short-term starvation will lead to an increase in the concentration of G6P in hepatocytes. The rate of utilization of G6P through the pentose phosphate pathway will therefore increase, producing increased amounts of ribose-5-phosphate and PRPP and causing overproduction of purine nucleotides (Chapter 10). The catabolism of these excess purine nucleotides leads to the production of hypoxanthine, which is oxidized to xanthine and uric acid by xanthine oxidase. Gout is treated with allopurinol, which is not an inhibitor of xanthine oxidase but a substrate that binds to the enzyme with greater affinity than the natural substrates, ensuring that more hypoxanthine and xanthine are excreted in urine and that less uric acid accumulates.

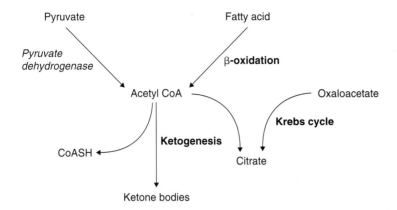

Figure 13.7 The generation of acetyl CoA and ketone bodies in hepatocytes. Details of these reactions are given in Chapter 9.

Ketogenesis

Significant ketone body production occurs *only* in the liver. Ketone bodies are produced when the rate of production of acetyl CoA exceeds the capacity of the citric acid cycle to oxidize them, that is, when the rate of β-oxidation of fatty acids is increased and the rate of the citric acid cycle is depressed. If acetyl CoA were to accumulate in the liver a major fraction of coenzyme A would be sequestered in this form, rendering it unavailable for other processes. The conversion of acetyl CoA to ketone bodies releases the coenzyme A, and generates soluble nutrient that can be transported to other tissues (Figure 13.7). Ketone bodies are released into the bloodstream, providing nutrient to other tissues, but they can not be used in the liver because the liver lacks **succinyl CoA: 3-oxoacyl CoA transferase**, the enzyme that catalyses the first reaction in the utilization of ketone bodies (Chapter 9).

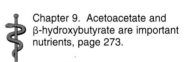

Chapter 9. Acetoacetate and β-hydroxybutyrate are important nutrients, page 273.

Detoxification and removal of xenobiotics

Lipophilic compounds are not readily excreted from the body. The liver facilitates excretion of such molecules by conjugating them with compounds that increase their hydrophilic character. Compounds that need to be solubilized in this way may be natural or artificial substances in the general category of **xenobiotics** and include metabolites of steroids, bacterial toxins produced by intestinal flora, pharmaceutical compounds, artificial sweeteners, natural pigments in food, exogenous food colorants and preservatives and other materials that we ingest involuntarily. The most common conjugation processes include addition of sulphates or glucuronates. All conjugation processes use energy.

Conjugation with glucuronate
UDP-glucuronate is the immediate donor of glucuronate groups. It is produced by reacting G1P with UTP to form UDP-glucose, in which form the glucose moiety is oxidized to glucuronate in a reaction catalysed by **UDP-glucose dehydrogenase**. The product is a high-energy compound that can contribute its energy to the transfer of the glucuronate to the substrate being solubilized (Figure 13.8).

Conjugation with sulphate
The sulphate donor in these reactions is the high-energy compound 3'-phospho-adenosine-5'-phosphosulphate, produced by reaction of sulphate ions and ATP and catalysed by an ATP sulphurylase.

Hydroxylation
The addition of the hydroxyl group is sometimes a necessary precursor for addition of glucuronate or sulphate groups and a variety of **hydroxylases** in the

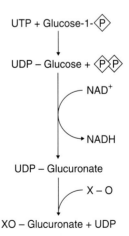

Figure 13.8 Glucuronation as a detoxification mechanism. The xenobiotic X-O is coupled to a solubilizing residue donated by the high-energy compound UDP-glucuronate.

Chapter 7. Box 7.5, Cytochrome *P*-450, page 206.

liver are capable of doing this. Hydroxylation reactions require molecular oxygen and NADPH and the activity of a cytochrome P-450 system (see Box 7.5). One of the most interesting aspects of this system is that some cytochrome P-450 systems are inducible; continued exposure to xenobiotics increases the liver's capacity to deal with them.

Other liver-specific metabolic processes

Many metabolic processes are specific to, or occur mainly in, the liver but have been described in other chapters of this text. In particular the reader is reminded of the importance of the liver in

- synthesis of plasma proteins (Chapter 12);
- production of bile (Chapter 14);
- hydroxylation of vitamin D (Chapter 14);
- synthesis of creatine (Chapter 15);
- synthesis of urea (Chapter 10);
- pyrophosphorylation of thiamine (Chapter 14);
- storage of iron and of vitamins A, B_{12} and folate (Chapter 14); and
- metabolism of fructose and galactose (Chapter 14).

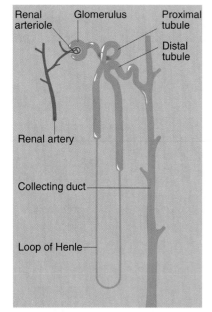

Figure 13.9 Diagram of a nephron. The proximal tubule is the main site of reabsorption of sodium, chloride and water and the main site at which ammonia enters the tubule. Glucose is reabsorbed largely from the proximal tubule. Bicarbonate is reabsorbed from the distal tubule, which also functions in sodium and potassium homeostasis.

Chapter 15. Loose connective tissue provides adhesive material between cells of soft tissues, page 467.

Metabolic functions of the kidney

In this section some of the molecular aspects of renal function will be presented. No attempt is made to explain the details of tubular functions through which the kidney controls the ionic composition of body fluids: for this information you should refer to a suitable textbook of physiology.

The renal glomerulus functions as a filter

The function of the kidney is to filter blood plasma, extracting from it small water-soluble molecules that can be reabsorbed from the renal tubule or excreted in the urine (Figure 13.9). Small molecules pass through the **glomerulus** non-selectively, through pores in the **basement membrane**. The glomerular filtrate then contains amino acids, sugars, vitamins, urea and ions at concentrations corresponding to those found in blood plasma. Larger molecules, especially proteins, to which the basement membrane of the glomerulus is impermeable, are normally retained in the plasma. In addition to the filtered molecules and salts, substances are secreted into the tubular filtrate from the interstitial fluids of the kidney.

The basement membrane of the glomerulus (Figure 13.10) is made up of a network of macromolecules including type IV **collagen**, the glycoproteins **laminin** and **entactin** and proteoglycans including **heparan sulphate** (see Chapter 15). The glycoproteins of the basement membrane have many strongly negatively charged carboxyl and sulphonate groups, forming a layer that inhibits the passage, and therefore loss to the urine, of negatively charged molecules. Alterations in the structure of the basement membrane impair renal function and are believed to occur in a variety of disorders, including diabetes. The appearance of large molecules or abnormal concentrations of nutrients in urine may suggest a metabolic abnormality or destruction of structural molecules of the glomerulus or tubules.

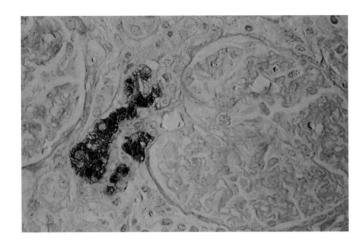

Figure 13.10 A human renal glomerulus and the renin-secreting cells of the juxtaglomerular apparatus (JGA). These cells are stained brown by an immunoperoxidase technique that uses renin-specific antibodies to identify the renin-secreting cells. In this case the renal artery is narrowed as the number and size of the renin-secreting cells has increased compared with a normal healthy kidney. Figure reproduced by permission of the Editor of *Muir's Textbook of Pathology*.

Homeostasis is maintained by reabsorption of some molecules and ions through the renal tubules

Many of the small molecules entering the glomerular filtrate are essential to body function and some are reabsorbed through the renal tubules and restored to the bloodstream. Glucose is reabsorbed by a saturable active-transport process indirectly dependent on ATP. It is cotransported with Na^+, and the Na^+ concentration within the tubule cells is regulated by the Na^+/K^+-ATPase pump (see Chapter 3). The maximum rate of reabsorption of glucose is achieved with a plasma or tubular glucose concentration of about 10 mM (this is called the **threshold concentration**). If the concentration of glucose is greater than the threshold concentration the transport process becomes saturated and glucose appears in the urine (glycosuria).

The limited capacity of renal tubules to reabsorb glucose is used as a diagnostic test for diabetes – the **glucose tolerance test**. In type I diabetics (see below) the amount of insulin secreted is insufficient to stimulate glucose uptake from the bloodstream into tissues after a meal. If this happens the glucose concentration in the bloodstream is likely to exceed the renal threshold concentration and glucose will be excreted in the urine. An oral test dose of 75 g glucose should not cause glucosuria in normal individuals, but will if diabetes is not properly controlled. Additional criteria such as measurements of fasting blood glucose are needed for the diagnosis of diabetes, because a failure to reabsorb glucose may also be caused by impaired renal function.

Chapter 3. The Na^+/K^+ ion pump of plasma membranes is driven by hydrolysis of ATP, page 64.

The kidney exhibits a broad range of metabolic activity

Energy is provided in the kidney from oxidation of glucose or fatty acids and used for metabolic processes, active transport, maintenance of ion gradients and tissue repair. Fatty acids are the major nutrient for kidney tissue. The fatty acids are derived in part from triacylglycerols stored in renal tissue, but most come from adipose tissue. Renal tissue is capable of both lipogenesis and a limited amount of gluconeogenesis.

The kidney regulates acid–base balance by altering the amount of acid or bicarbonate it excretes

Acid is produced constantly by metabolic processes. The kidney has a major role in regulating acid–base balance, which it does by limiting the excretion of bicarbonate or by increasing the excretion of acid. As much as 50 millimoles of free H^+

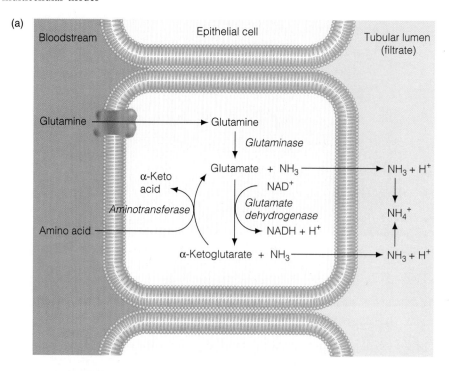

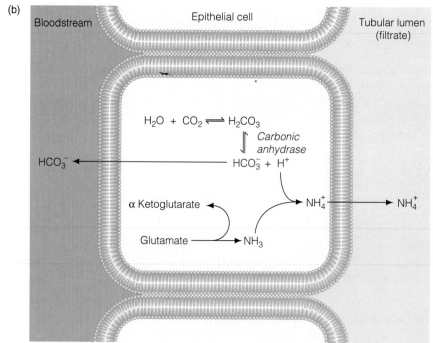

Figure 13.11 Excretion of acid and ammonia by the kidney. In (a) amino groups released from glutamine and glutamate within tubular epithelial cells combine with hydrogen ions in the tubular lumen and body acidity is reduced by the excretion of ammonium ions (NH_4^+). In (b) CO_2 and H_2O produced from metabolism within the cell are substrates for the carbonic anhydrase reaction. Both the reabsorption of HCO_3^- and the excretion of H^+ contribute to a reduction of body acidity.

Chapter 10. Glutamate acts as the major common pool for amino groups, page 296.

may be excreted per day, through primary active proton transport or as NH_4^+ and $H_2PO_4^-$ formed when NH_3 and HPO_4^{2-} pick up protons available from H^+/Na^+ exchange. Ammonia is released from glutamine or other amino acids in the tubular epithelium through the reactions catalysed by **glutaminase** and **glutamate dehydrogenase** (see Chapter 10), and ammonium ions carry 60–70% of the excreted protons. (Figure 13.11) The α-ketoglutarate produced augments the citric acid cycle metabolite pool in the renal tubule cells.

Tubular cells contain carbonic anhydrase, which enables the kidney to use CO_2 produced from metabolic processes to generate bicarbonate ions and protons, thus:

$$CO_2 + H_2O \leftrightharpoons H_2CO_3 \leftrightharpoons HCO_3^- + H^+$$

The protons are excreted as described above, and reabsorption of the bicarbonate ions will further decrease the acidity of body fluids. **Titratable acidity** of urine is defined as the pH (or H^+ concentration) difference, between plasma and urine. Total acidity is the sum of titratable acidity H^+ and NH_4^+.

Antidiuretic hormone controls water reabsorption in the kidney

The kidney conserves body water by concentrating the solutes it excretes. The osmolarity of urine varies widely, reflecting fluid intake, and allows the kidney to excrete excess solutes with minimal loss of water, consistent with maintaining blood volume. **Antidiuretic hormone (ADH)**, previously called **vasopressin**, is a small peptide synthesized in the hypothalamus and secreted from the posterior pituitary that monitors the osmolarity of blood. ADH acts to increase the renal reabsorption of water. The mechanism is mediated through collecting duct cells in the kidney with ADH-specific receptors linked to the activation of adenylate cyclase. The increased concentration of cAMP in collecting duct cells promotes protein kinase-dependent phosphorylation of membrane proteins thereby increasing the permeability of the collecting duct to water. Secretion of ADH is inhibited by alcohol, which explains the excessive loss of water and dehydration associated with alcohol intake. Tea, coffee and cocoa all act as **diuretics**, increasing water loss from the body, because they contain respectively theophylline, caffeine and theobromine, all of which inhibit cAMP phosphodiesterase and thereby block the breakdown of cAMP (Chapter 11). The continuing presence of cAMP maintains the permeability of the collecting ducts to water.

Chapter 11. Hormonal responses mediated through adenylate cyclase and the production of cAMP, page 338.

The renin–angiotensin system controls blood pressure and sodium ion concentration in the bloodstream

A specialized group of cells, called the **juxtaglomerular apparatus (JGA)**, is located at the base of the glomerulus, where the arterioles enter and leave the glomerular structure. Cells of the JGA are sensitive to changes in blood pressure and plasma Na^+ concentration. In response to reduced blood pressure or reduced Na^+ concentration cells of the renal arterioles in the JGA release **renin**, a peptide with proteolytic activity, into the bloodstream.

Angiotensinogen is a peptide synthesized by and secreted from the liver. In the bloodstream renin catalyses the conversion of angiotensinogen to **angiotensin I**, which is then cleaved by the endothelial cell protease **angiotensin converting enzyme (ACE)**, releasing the active hormone **angiotensin II** (Figure 13.12).

Angiotensin II acts immediately as a vasoconstrictor and stimulates the adrenal gland to secrete the steroid hormone **aldosterone**. Aldosterone acts on the distal tubule of the kidney to increase the reabsorption of Na^+ and water. The mechanism is thought to be mediated by the binding of an aldosterone-receptor complex to DNA sequences, promoting the synthesis of a specific protein which has yet to be characterized. The result is that blood volume increases and blood pressure rises (Figure 13.13). Drugs that inhibit ACE (such as captopril) are used in the treatment of hypertension (high blood pressure), and their use in a wide variety of other clinical situations is currently under investigation.

This renin–angiotensin–aldosterone system is opposed by a peptide secreted from the atrial muscle cells of the heart. **Atrial natriuretic peptide (ANP)** is released in response to increased blood pressure or increased blood volume, both of which induce stretching of atrial tissue. ANP inhibits the release of renin from the cells of the JGA, thus reducing the secretion of aldosterone and Na^+ reabsorption, leading to a reduction in blood volume and blood pressure.

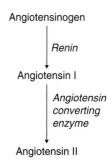

Figure 13.12 Activation of angiotensin occur in the bloodstream.

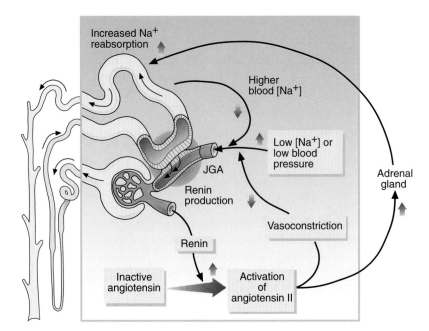

Figure 13.13 Sodium and water regulation by the renin–angiotensin–aldosterone system in the kidney.

Functions of the pancreas

The pancreas has two major functions:

1. It is an **exocrine** organ, secreting digestive enzymes from the acinar cells through the pancreatic ducts into the lumen of the upper small intestine.
2. It is an **endocrine** organ, secreting into the bloodstream peptide hormones that are produced in the cells of the Islets of Langerhans and control metabolism throughout the body.

Exocrine activity of the pancreas

Many of the digestive enzymes acting in the intestine are produced in the pancreas and secreted into the intestinal lumen in response to a complex control mechanism mediated by hormones synthesized and stored in the epithelial cells at the lumenal surface of the intestine.

The digestive enzymes are synthesized as proenzymes on the rough endoplasmic reticulum of pancreatic exocrine cells (the **acinar cells**). The proenzymes are then transferred to the Golgi complex, where glycosylation reactions produce the completed glycoprotein proenzymes, or **zymogens**. These are then concentrated into membrane-bound secretory vesicles (**zymogen granules**), which bud off from the Golgi and are stored near the apical surface of the cell, adjacent to the pancreatic duct. The apical and basolateral membranes of the acinar cells have distinctive molecular compositions, which ensure that zymogen granules can fuse *only* with apical membranes. Stimulation of the exocrine cell, electrically or by hormones, causes the zymogen granule membrane to fuse with the plasma membrane and release the digestive enzymes into the pancreatic duct, where they enter the lumen of the upper small intestine (Figure 13.14).

Chapter 3. The endoplasmic reticulum and Golgi complex are membrane-bound compartments, page 46.

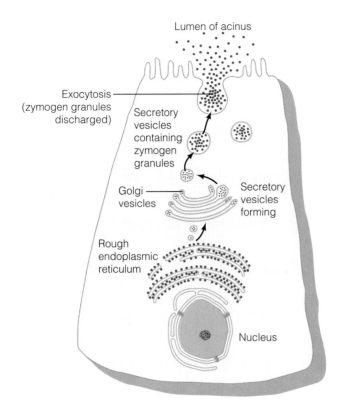

Figure 13.14 Newly synthesized secretory proteins are transferred from the rough endoplasmic reticulum to Golgi vesicles, where they accumulate in zymogen granules that bud from the Golgi. The granules are stored adjacent to the apical surface of the plasma membrane, ready for exocytosis into the pancreatic duct and intestinal lumen.

Peptide secretagogues regulate the exocrine function of the pancreas

The hormonal stimulus for secretion of the digestive enzymes from the pancreas comes from the binding of compounds called **secretagogues** to specific receptors on the exocrine cells. The secretagogues for pancreatic exocrine function include acetylcholine and the peptide hormones **gastrin, secretin** and **pancreozymin**, which are synthesized and stored within granules of epithelial cells of the gastrointestinal tract. Each peptide is produced by a specific epithelial cell type.

These peptide hormones are secreted into the bloodstream in response to dietary peptides in the digestive tract. Amino acids and di- and tripeptides produced when pepsin digests proteins in the stomach pass into the upper small intestine and are taken up by intestinal epithelial cells, where they stimulate release of secretagogues into the circulation. These hormonal peptides bind to specific receptors on the outer surface of exocrine cells of the pancreas, causing it to release digestive enzymes. Thus the products of the action of pepsin signal to the intestine the requirement for further digestive activity. The secretagogue hormones provide a mechanism of constant signalling between the epithelial cells of the intestine that monitor the need for digestive enzymes and the pancreatic exocrine cells that produce and secrete these digestive enzymes into the intestinal lumen. The main pancreatic ducts fuse with the lumen of the small intestine, allowing the digestive enzymes to be emptied directly into the intestinal lumen (Figure 13.15).

See Table 14.2 for a list of peptide secretagogues produced by the intestinal epithelium that stimulate secretion of pancreatic enzymes

Chapter 14. Table 14.3, Peptide hormones (secretagogues) controlling the digestive system, page 414.

Secretagogues express their activity through intracellular signalling processes

Secretagogues (peptides or neurotransmitters) bind to specific pancreatic plasma membrane receptors and initiate a signalling sequence within the pancreatic cells.

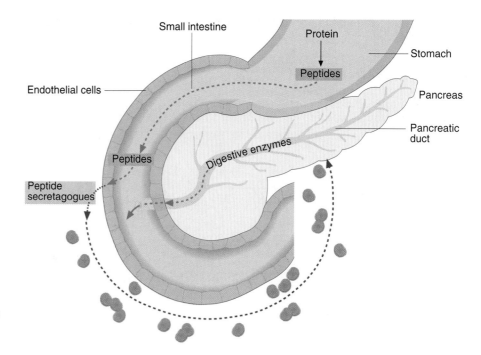

Figure 13.15 Interactions between exocrine cells of the pancreas and epithelial cells of the intestinal wall. Pancreatic enzymes pass directly from the pancreatic duct to the lumen of the intestinal tract. The release of these digestive enzymes is in response to signals from hormones produced in the epithelial cells of the tract.

Chapter 11. The mechanism of action of extracellular signal molecules is related to their chemical characteristics, page 332.

The end response is the release of digestive enzymes. The intracellular signalling sequence is a series of chemical changes analogous to those occurring in response to the binding of hormones such as glucagon and adrenaline in other tissues (Chapter 11). Signalling may be through activation of either

- adenylate cyclase and the production of cyclic AMP within the cell, or
- phospholipase C, the production of IP$_3$ and DAG triggering the release of calcium into the cytoplasm and the activation of protein kinase C.

The acetylcholine and pancreozymin receptors of the exocrine cell operate through activation of phospholipase C but the secretin receptor is linked to the activation of adenylate cyclase.

The pancreas normally produces digestive enzymes in amounts ten times greater than required to ensure adequate digestion. A very substantial reduction in the amount of enzyme synthesized would therefore be required for any discernible effect on the efficiency of digestion. Nevertheless, this *can* occur – in a condition called **pancreatic exocrine insufficiency**.

Endocrine activity of the pancreas

The Islets of Langerhans are scattered throughout the pancreas (Figure 13.16) and contain three cell types, each of which synthesizes a different peptide hormone: the α **cells** produce glucagon, the β **cells** insulin and the γ **cells** somatostatin. These peptides are stored in secretory granules which fuse with the plasma membrane in response to the stimulus to release hormone into the bloodstream. All have short half-lives of a few minutes in the circulation. Glucagon and insulin control metabolism in many different cell types, having broadly opposite effects. **Somatostatin** is a 14-amino acid peptide which inhibits the secretion of both insulin and glucagon from the pancreas. It also has growth factor properties, which influence both the endocrine and the exocrine cells of the pancreas.

The mechanisms of action and metabolic effects of hormones produced by the pancreas are described in Chapter 11.

Chapter 11. The pancreas produces three peptide hormones, page 347.

Synthesis and secretion of glucagon

Glucagon is synthesized as a larger precursor, which is then broken down to the active 29-amino acid peptide. Relatively little is known about events regulating the

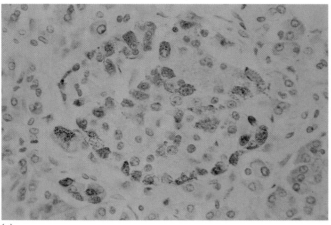

(a)

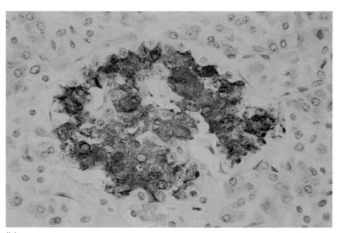

(b)

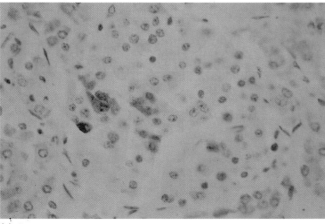

(c)

Figure 13.16 A normal human islet of Langerhans has been immunostained (brown colour) with antibodies recognizing (a) glucagon-producing cells, (b) insulin-producing cells and (c) cells producing somatostatin. These are serial sections cut from the same islet cell. Comparing them, it is clear that no individual cell produces more than one of the three hormones. Figure courtesy of Dr A.K. Foulis, University of Glasgow Department of Pathology.

release of glucagon from the pancreas compared with our current knowledge of events regulating insulin secretion. We know that the pancreas secretes glucagon in response to low concentrations of glucose and insulin in the bloodstream.

Synthesis and secretion of insulin

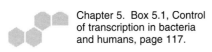

Chapter 5. Box 5.1, Control of transcription in bacteria and humans, page 117.

Expression of the insulin gene in β islet cells is regulated by DNA sequence elements upstream from the insulin gene (see Chapter 5). Both positive and negative gene regulatory proteins probably exist, but these have not yet been fully characterized.

Insulin is synthesized as a single polypeptide **prohormone** of 86 amino acids, with an additional *N*-terminal signal sequence. The prohormone is stored as zinc crystals in secretory granules of the β cells. Before it is secreted into the circulation a central **C-peptide** (31 amino acids) and four additional amino acids are removed by proteolytic cleavage – the mature peptide is composed of two polypeptide chains of 21 and 30 amino acids held together by two interchain disulphide bonds (Figure 13.17). Secretion of insulin is very rapid – as many as 1.8×10^8 granules may be released per second from the pancreas.

Box 13.4 C-peptide and insulin overdose

C-peptide is released into the bloodstream in amounts equimolar with mature insulin. Commercial insulin used in the treatment of diabetes is the mature form and does not include C-peptide. The fact that endogenous C-peptide is secreted into the bloodstream and can be quantified provides a means of assessing whether excessive concentrations of insulin in the bloodstream arise from overproduction by the pancreas or from overdosing on insulin. If the amount of mature insulin exceeds the amount of C-peptide in the circulation the individual has probably injected commercial insulin.

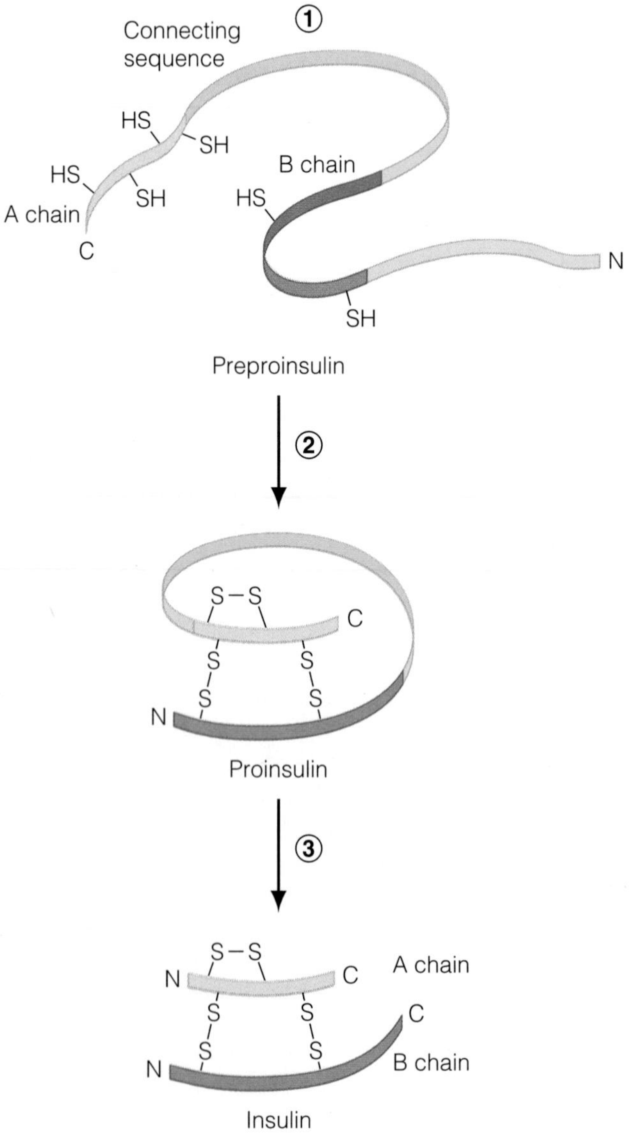

Figure 13.17 Structure and processing of insulin to produce the active hormone. ① Newly synthesized pre-prinsulin. ② Leader sequences removed and disulphide bonds formed. ③ C-peptide excised forming mature active insulin.

Secretion of insulin is regulated by plasma glucose concentration, mediated through islet cell glucokinase

The primary stimulus for secretion of insulin is plasma glucose concentration and appears to be mediated by influx of calcium ions. The pancreas is the one tissue in addition to the liver that possesses glucokinase, a glucose-specific enzyme with a low affinity (high K_m) for substrate. The pancreas also contains hexokinase, an enzyme with a high affinity (low K_m) for glucose. Hexokinase provides G6P to meet the energy needs of pancreatic cells, while glucokinase operates as a signal mechanism, causing a more rapid reaction at high blood glucose concentration and initiating the sequence that stimulates secretion of insulin from β cells. Pancreatic glucokinase is a monomer, free of allosteric or other regulatory controls; its activity is determined solely by the concentration of glucose.

Pancreatic β cells recognize the accelerated metabolic flux rate through glycolysis, the citric acid cycle and oxidative phosphorylation that is initiated by glucokinase as a signal for secretion of insulin. The accelerated ATP production, and the influence of ATP on membrane ion channels, is probably the immediate signal for secretion of insulin. Secretion of insulin is inhibited if glucokinase is inhibited.

Hormonal systems must respond not only to the short-term changes in nutrient availability that reflect normal intervals between meals but also to longer-term changes in nutritional state. The long-term response generally depends on induction of enzymes. During starvation the activity of liver and pancreatic glucokinase decreases; refeeding induces glucokinase. Insulin prevents the drop in glucokinase activity usually seen in starvation, despite low glucose concentrations. This observation establishes that the nutritionally induced changes in the expression of glucokinase reflect changes in circulating concentrations of insulin rather than changes in glucose concentration.

Diabetes

Diabetes is a disorder arising from

- inadequate production of insulin by β cells of the pancreas (juvenile onset diabetes, or type I insulin-dependent diabetes);
- failure of insulin target tissue to recognize the presence of insulin (maturity onset diabetes, or type II insulin-independent diabetes);
- failure of the glucokinase-mediated pancreatic signalling system (maturity onset diabetes of the young).

Juvenile diabetes

Juvenile diabetes is caused by destruction of pancreatic islet cells, most commonly by autoimmune disease or viral infection. A few cases have a genetic component (see Chapter 16), in which, for example, the structure of insulin or the structure of the insulin receptor is abnormal so that the receptor fails to recognize the hormone. Many of these patients also have a deficiency of liver glucokinase, so the liver is less efficient at buffering blood glucose concentration. In all cases, except those with a genetic abnormality in the receptor structure, the condition may be ameliorated by administration of insulin. However, as insulin is a peptide and is hydrolysed in the digestive system, it can not be given orally but must be administered by injection.

Chapter 16. Box 16.8, HLA and disease, page 503.

Under normal physiological control there is a 7–10-fold difference between the concentrations of insulin between meals and following a meal. These changes are difficult to mimic if insulin is provided by injection, and presents a major difficulty in controlling the insulin dose for diabetics.

If plasma concentrations of insulin are persistently low, the resulting hyperglycaemia causes inappropriate glycosylation of molecules, particularly blood and tissue proteins. If excessive amounts of insulin are administered, glucose is taken into muscle and adipose tissue to an extent which depletes the blood glucose concentration. The resulting hypoglycaemia imperils energy provision to the brain. Clinical hypoglycaemia is defined as a blood glucose concentration of less than 2.2 mM.

Maturity onset diabetes

Maturity onset diabetes is most commonly associated with obesity in middle-aged people. It is due to a reduction in the number or affinity of insulin receptors on the plasma membranes of cells of target tissues or to abnormal binding of insulin to the receptors. The relationship with obesity arises because persistent dietary excess leads to hyperinsulinemia, caused by a constantly high level of insulin secretion. This leads to a reduction in the number of insulin receptors. In its early stages this form of diabetes can be reversed by modifying the diet. If the patient adopts a weight-loss regime the down-regulation of receptors can be at least partially reversed.

Maturity onset diabetes of the young

Maturity onset diabetes of the young (MODY) is characterized by a genetic abnormality in the pancreatic glucokinase gene, so that the β cells do not respond normally to changes in circulating glucose concentration. These patients may have a normal capacity to synthesize, store and secrete insulin, but the signalling system that initiates its secretion does not respond within the appropriate glucose concentration range.

Diabetes is associated with a broad spectrum of metabolic change

Chapter 14. Distribution of nutrients in starvation, page 437.

Chapter 9. In the liver accumulating acetyl CoA is converted to ketone bodies, page 271.

Chapter 13. The kidney regulates acid–base balance by altering the amount of acid or bicarbonate it excretes, page 391.

Chapter 3. Lysosomes digest ingested macromolecules and participate in the turnover of intracellular components, page 48.

In diabetes of any origin the body responds by maintaining itself in the metabolic state appropriate to starvation, even immediately following a meal. The lack of insulin, or lack of insulin responsiveness, is accompanied by increased secretion of glucagon, even when the blood glucose concentration is high. Tissues that are normally insulin-responsive do not take up glucose. After a meal the glucose concentration in the bloodstream remains high, and if its concentration exceeds the renal threshold of about 10 mM glucose is excreted in the urine.

Insulin normally suppresses lipolysis in adipose tissue. In untreated diabetes this inhibition is relieved by the low insulin concentration and lipolysis is accelerated, increasing the rate at which glycerol and fatty acids are released from triacylglycerols. Uptake of glucose into adipose tissue is insulin-dependent, so in diabetes adipocytes are short of glucose and therefore produce little glycerol-3-phosphate. The fatty acids are not re-esterified; they are instead released from adipose tissue into the bloodstream and transported to the liver.

In these metabolic conditions fatty acids are oxidized to acetyl CoA in the liver and then converted to ketone bodies, which are released into the bloodstream. The concentration of ketone bodies in circulation (normally 1–2 mM) may reach 20 mM and, as these are carboxylic acids, the pH of blood may fall from 7.4 to 6.8. Ketone bodies are excreted in the urine, the pH of which may fall to 4.5. Even in this relatively acidic environment ketone bodies are largely ionized and their charge is balanced by the loss of cations (mostly Na^+) except to the extent that the kidney can substitute NH_4^+ for Na^+. Thus the loss of glucose in urine is accompanied by water and electrolyte loss, the individual becomes dehydrated and suffers ion imbalances. Weight loss and muscle weakness may occur because the main gluconeogenic substrates available to tissues under these conditions are amino acids derived from the breakdown of muscle protein.

In all types of diabetes, internalization of insulin/receptor complexes does not occur to the normal extent, either because insulin is not available or because receptor numbers are subnormal. Therefore the normal intracellular effects of insulin on the metabolic enzymes of glycogenesis and glycolysis are reduced; a general failure of muscle and liver and adipose tissue to respond to the nutritional state ensues.

Diabetic ketosis and starvation ketosis have different consequences

The contrast between diabetic ketosis and physiological or starvation ketosis, which may develop if there is an extended interval between meals, is that in the latter cases insulin responsiveness is retained. This difference protects many metabolic responses:

- lipolysis is less severe, limiting the amount of fatty acid available to the liver;
- ketogenesis remains within normal limits so that plasma ketone body concentrations seldom exceed 2–3 mM;
- glucose continues to be taken up by cells of peripheral tissues;
- liver gluconeogenesis achieves its objective of providing glucose to these tissues; and
- glucose is not lost in the urine and dehydration and ion losses do not develop.

The molecular basis of electrical activity in tissues

Many body processes depend on the electrical signals generated by movement of ions across membranes that alter ion gradients. The following discussion is restricted to the well characterized biochemical aspects of transmission of impulses, specifically at neuromuscular junctions, in the brain and in rods cells of the eye. In what follows it is assumed that students will be familiar with the properties of membranes and ion pumps (see Chapter 3).

Chapter 3. Transport of molecules and ions across membranes, page 62.

Transmission of impulses across synapses

A **synapse** is a specialized region between two adjacent cells involving a **synaptic cleft** (gap) across which cells communicate. The communication is unidirectional: on one side of the gap is the **presynaptic** cell, from which a signal is sent across the gap to the **postsynaptic** cell. The membrane of the postsynaptic cell, adjacent to the gap, has receptors capable of recognizing the signal, which may be electrical or chemical. An electrical impulse may pass directly from the presynaptic membrane to the postsynaptic membrane but the more common form of transmission is across a *chemical* synapse.

In a chemical synapse (Figure 13.18) the presynaptic cell contains a chemical transmitter substance in vesicles immediately adjacent to the gap. The transmitter is released into the gap by exocytosis of the storage vesicles and binds to receptors on the postsynaptic membrane. A feature of chemical transmitter molecules is that they are degraded, inactivated or reabsorbed very rapidly after they are released into the synaptic cleft.

Membrane potentials and the generation of action potentials

Because negatively charged macromolecules are restricted to the interior of cells (described as the **fixed negative charge**) it is impossible to achieve both electrical

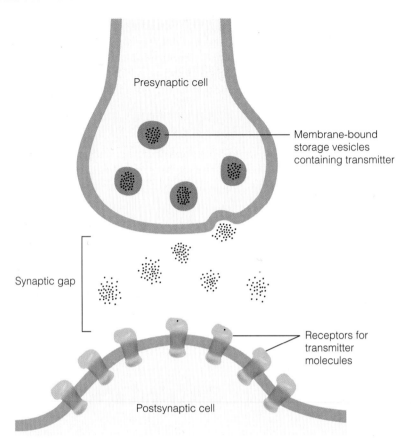

Figure 13.18 Diagram of a chemical synapse. Chemical transmitter molecules are released from the presynaptic cell, diffuse across the gap and bind to receptors on the postsynaptic cell.

Presynaptic cell

Membrane-bound storage vesicles containing transmitter

Synaptic gap

Receptors for transmitter molecules

Postsynaptic cell

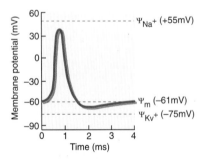

Figure 13.19 As an action potential is generated, the membrane becomes transiently permeable to sodium ions, which move into the cell. This is followed by an efflux of potassium ions. (a) Shows the conductance changes due to changes in permeability of the membrane and (b) the changes in membrane potential. The potential becomes positive as Na⁺ enters the cell, decreases and overshoots as a negative potential as K⁺ leave, but then returns to the normal resting potential.

and chemical neutrality across plasma membranes. Cations, K^+ and, to a lesser extent, Na^+ enter the cell to balance the fixed negative charge, but in the process set up chemical gradients across the membrane. The membrane is more permeable to potassium ions than to sodium ions so the potassium ion concentration is greater within the cell and the sodium ion concentration is greater in the extracellular fluid. This distribution gives rise to an electrical potential, described as the **resting potential**, across the plasma membrane. The magnitude of the resting potential is characteristic of different cell types but in many it is of the order of 70 mV, the inside of the cell being negative with respect to the outside.

These ion gradients, which provide a source of available energy for the cell, are sustained by Na^+/K^+-ATPase pumps in the plasma membrane. While the resting potential is maintained the net flow of charged species across the membrane is zero, so that the diffusion of ions against their concentration gradients is matched by active transport through the ATPase pump.

An **action potential** is generated by stimulating the plasma membrane of an excitable cell (Figure 13.19). The stimulus may be electrical or chemical. The feature distinguishing plasma membranes of excitable tissues from those of other tissues is that excitable membranes possess sodium channels composed of proteins that respond to a stimulus with a conformational change, which allows rapid transient influx of Na^+ ions. As a result, the voltage difference across the membrane is diminished and may even be reversed, so that the charge within the cell becomes positive with respect to the outside. This process is described as **depolarization** of the membrane. The membrane is repolarized, and the resting potential restored, as the Na^+-pump extrudes Na^+. Action potentials are propagated along the plasma membrane of a cell by controlled sequential changes in membrane permeability to cations in adjacent regions.

Neurotransmitters and their mechanism of action

Neurotransmitters alter the ion permeability of cell membranes. They bind to specific receptors associated with ion channels, thereby changing the conformation of channel proteins and permitting movement of ions through the channel. Many chemical substances are able to act as neurotransmitters. With the exception of **acetylcholine**, most natural transmitters are either small peptides or derivatives of amino acids. Some receptors are structurally part of an ion channel, so that binding of the transmitter to the receptor immediately triggers opening of the channel and movement of ions across the membrane. Other receptors are indirectly linked to ion channels, probably through G-proteins, to adenylate cyclase or phospholipase C, in which case the channel opens in response to changes in cAMP or Ca^{2+} concentrations (see Chapter 11 for descriptions of these mechanisms). Transmitters may be **excitatory** or **inhibitory**.

Chapter 11. Channel-linked receptors regulate the permeability of membranes to ions, page 335.

Excitatory transmitters

Excitatory transmitters bind to receptors linked to cation channels, causing a reduction in the membrane potential – a **depolarization**. Sodium ions flow into cells and potassium ions flow out, each moving down a concentration gradient, dissipating the normal resting potential.

Inhibitory transmitters

Inhibitory transmitters bind to receptors linked to anion channels admitting negatively charged ions, mostly Cl^-, into the cells and causing **hyperpolarization**, an increase in the membrane potential. The effect is to make subsequent depolarization of the membrane more difficult.

Transmission of impulses across neuromuscular junctions – the mechanism of action of acetylcholine

Acetylcholine is stored in membrane-bound vesicles located adjacent to the presynaptic membrane at neuromuscular junctions and anchored to the cytoskeletal matrix of the cell. The adjacent part of the cell membrane has a high density of calcium channels. When the presynaptic cell is stimulated the permeability of these calcium channels changes, permitting influx of Ca^{2+} and facilitating the detachment of the vesicles from the cytoskeletal matrix and their fusion with the plasma membrane (Figure 13.20). The fusion of the vesicle and plasma membranes appears

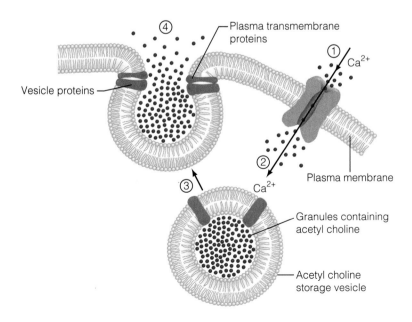

Figure 13.20 Each granule within the storage vesicle contains perhaps thousands of molecules of acetylcholine. Fusion of storage vesicles with the plasma membrane preceeds release of acetylcholine into the synapse. Stimulation of presynaptic cells results in an influx of calcium ions ①, binding of calcium ions to the vesicles ②, which then fuse with the plasma membrane ③, forming a channel through which acetylcholine is released ④.

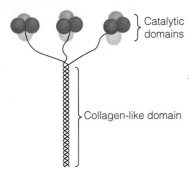

Figure 13.21 Diagrammatic representation of the structure of acetylcholinesterase.

Chapter 15. Collagen, page 470.

to generate a channel, probably through the recognition and binding of vesicle and plasma membrane proteins. Acetylcholine is then released through these channels into the synaptic cleft.

Acetylcholine diffuses across the cleft and binds to receptors on the postsynaptic membrane, where it induces a change in permeability to ions and depolarizes the membrane. The acetylcholine receptor is composed of four different subunits ($\alpha_2\beta\gamma\delta$), each of which is believed to contain four or five transmembrane helices. Each subunit contributes negatively charged aspartate or glutamate residues to form rings of negative charge at the core of the channel. This aids the movement of positively charged ions through the channel.

After release of the transmitter, the empty vesicles are reabsorbed into the presynaptic cell by endocytosis. The acetylcholine is hydrolysed by **acetylcholinesterase**, which is located in the synaptic gap bound to a network of collagen fibres. This terminates the signal, and restores the excitable state of the membrane.

$$H_3C-\overset{\overset{\displaystyle O}{\|}}{C}-O-CH_2CH_2\overset{+}{N}(CH_3)_3 + H_2O \xrightarrow{\textit{Acetylcholinesterase}} H_3C-\overset{\overset{\displaystyle O}{\|}}{C}-O^- + HO-CH_2CH_2\overset{+}{N}(CH_3)_3 + H^+$$

Acetylcholine **Acetate** **Choline**

Acetylcholinesterase is made up of three polypeptide chains. Each polypeptide contains a globular catalytic head at one end of the molecule and extends into a fibrous tail twisted into a single collagen-like domain at the other (Figure 13.21). This feature is thought to aid association of the enzyme with the structural collagen molecules of the basal lamina in the synaptic gap.

Acetylcholine is resynthesized in the presynaptic cell from choline and acetyl CoA in a reaction catalysed by **choline acetyltransferase.**. Acetyl CoA is derived from the metabolism of glucose within the nerve cells, but choline is cotransported into the cell with Na$^+$; a process that can be rate-limiting for the synthesis of the transmitter. Alternatively, choline may be synthesized within the cell from serine.

$$H_3C-\overset{\overset{\displaystyle O}{\|}}{C}-S-CoA + HO-CH_2CH_2\overset{+}{N}(CH_3)_3 \xrightarrow[\textit{Choline acetyltransferase}]{HS\text{-}CoA} H_3C-\overset{\overset{\displaystyle O}{\|}}{C}-O-CH_2CH_2\overset{+}{N}(CH_3)_3$$

Acetyl-CoA **Choline** **Acetylcholine**

Transmission of impulses in the brain

The central nervous system has both excitatory and inhibitory transmitters.

- The major excitatory transmitters are acetylcholine, glutamate and the catecholamines – dopamine, adrenaline and noradrenaline.
- The major inhibitory transmitters are γ-aminobutyric acid (GABA), glycine and taurine.

Box 13.5 Suxamethonium blocks the binding of acetylcholine

Suxamethonium is used during surgery to relax muscles so that the surgeon can more easily cut through them. This molecule, which is a head-to-head dimer of acetylcholine, binds at the acetylcholine receptor, blocking access to the transmitter.

$$(CH_3)_3\overset{+}{N}.(CH_2)_2.OCOCH_2.CH_2COO.(CH_2)_2.\overset{+}{N}(CH_3)_3$$

$$\alpha\text{-ketoglutarate} + NH_4^+$$

Glutamate
dehydrogenase

Glutamate

Glutamate
decarboxylase

$$\gamma\text{-aminobutyric acid} + CO_2$$

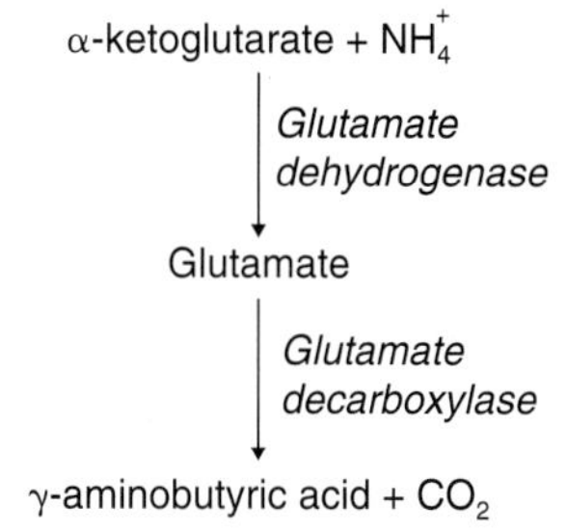

Figure 13.22 Synthesis of GABA.

GABA is the major inhibitory transmitter in brain cells

The concentration of GABA in the brain is about 500 times greater than that of other transmitters. Receptors for GABA are linked to ion channels in which positively charged arginine and lysine residues exposed in the core of the channel participate in the movement of anions across the membrane.

GABA is synthesized in the brain via glutamate from α-ketoglutarate derived from the citric acid cycle (Figure 13.22). The structure of GABA is shown in Chapter 11. The large amounts of this neurotransmitter being synthesized, and the associated drain on the citric acid cycle metabolite pool, accounts in part for the high glucose demand of the brain. An aminotransferase converts α-ketoglutarate to glutamate. Glutamate, which is itself an excitatory transmitter in the brain, is then converted to GABA by glutamate decarboxylase in a pyridoxal phosphate-dependent reaction.

Inactivation of GABA occurs through oxidative deamination, a two-step reaction producing succinate, which re-enters the citric acid cycle pool (Figure 13.23). This reaction also requires pyridoxal phosphate. Drugs that bind to the aldehyde form of pyridoxal phosphate block its participation in catalytic activities, reducing the rate of synthesis of GABA. However, they have no influence on the rate of inactivation of GABA, so overall there is a reduction in the availability of the inhibitory transmitter in the presence of drugs of this type.

The inactivation of GABA gives rise to large quantities of ammonia in the brain (Chapter 10). This ammonia is incorporated into glutamine for transport to the liver.

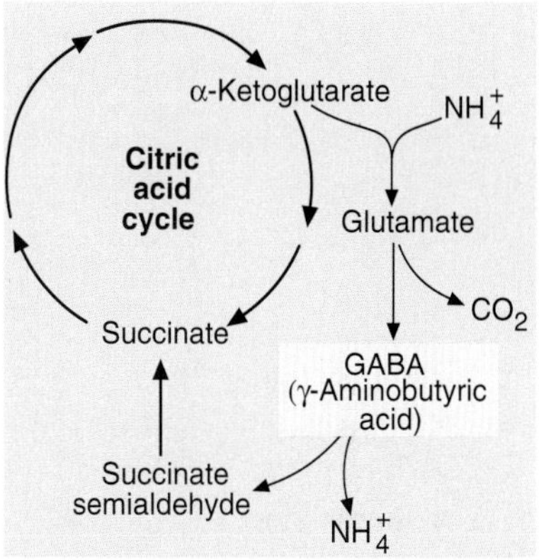

Figure 13.23 Although the degradation of GABA potentially returns succinate to the citric acid cycle metabolite pool, this bypasses two energy-producing reactions of the cycle: α-ketoglutarate dehydrogenase, which produces NADH as a substrate for oxidative phosphorylation, and succinyl CoA synthetase, which involves substrate-level phosphorylation of GDP and produces GTP.

Chapter 10. Most of the ammonia in the brain is derived from inactivation of neurotransmitters, page 298.

Inactivation of catecholamines also adds to the production of ammonia

The catecholamines adrenaline, noradrenaline and dopamine are inactivated by the action of **catecholamine O-methyltransferase**, through a methylation reaction, and by **monoamine oxidase**, which catalyses the conversion of the amines to aldehydes

$$RCH_2NH_2 + O_2 \rightarrow RCHO + NH_3 + H_2O_2$$

a reaction that adds to the amount of ammonia generated in the brain.

The pathway for synthesis of adrenaline, noradrenaline and dopamine is described in Chapter 11.

Chapter 11. Synthesis, storage, processing, secretion and mode of action of selected hormones, page 342.

Energy production in the brain

Metabolism in the brain is directed towards

- generation of ATP to support the electrical activity of the brain through maintaining transmembrane Na^+/K^+ gradients;

- synthesis and inactivation of neurotransmitters; and
- maintenance of cell structure, particularly cell membranes.

The major nutrient for the brain is glucose, but these cells are also able to metabolize the ketone bodies, acetoacetate and β-hydroxybutyrate that become available during fasting and starvation.

It is important to remember that, although ketone bodies can replace glucose in the brain as a substrate for ATP production, they can not replace glucose as a source of intermediates from which neurotransmitters are synthesized. At the same time the brain must have enough acetyl CoA for the synthesis of cholesterol to maintain membrane structure. Ketone bodies are a useful source of acetyl CoA for this purpose.

The brain contains a relatively low concentration of glucose, and does not store glycogen. Brain cells therefore depend on glucose entering the brain from the circulation – the brain extracts about 70% of the glucose circulating in the bloodstream. It also consumes 20% of the oxygen carried by blood.

The brain does have one energy reserve in addition to ATP: the high-energy phosphate compound **creatine phosphate**. The ability of creatine kinase to transfer the high-energy phosphate between adenine nucleotides and creatine

$$\text{Creatine} + \text{ATP} \underset{\substack{\textit{Creatine} \\ \textit{kinase}}}{\rightleftharpoons} \text{Creatine} - \text{P} + \text{ADP}$$

provides the brain cells with a short-term buffer against energy depletion.

The brain contains complex lipids which function as structural components and control electrical conductance

Myelin is composed of lipid, lipoproteins and basic proteins. It forms an insulating layer around cells of the **white matter** of the brain and facilitates conduction of impulses along the cell. The major lipids of myelin are cholesterol, sphingomyelins, cerebrosides, phosphatidylcholine, phosphatidylethanolamine and plasmalogens and are synthesized in the brain. The major lipoprotein has a high percentage of polar amino acids.

Failure to establish adequate myelination or subsequent demyelination of nerve cells has serious clinical consequences. The most common demyelination disease is **multiple sclerosis**. The demyelination characteristic of this disease appears to be irreversible, and the resultant loss of control of conductance gives rise to a variety of neurological symptoms.

Transmission of light impulses by rod cells in the eye

The **rod cells** of the **retina** of the eye possess excitable receptors that are activated by light. These receptors convert light into electrical impulses, which are transmitted to the brain through bipolar neurones of the optic nerve.

The outer segment of rod cells is made up of stacks of membranous discs packed with the photoreceptor molecule **rhodopsin**. Rhodopsin is made up of the protein **opsin**, which does not respond to light, and a prosthetic group **11-*cis*-retinal**, the chromophore which absorbs light in the 400–600 nm wavelength range. **Vitamin A**, an essential component of the diet, is the precursor of this chromophore. Rhodopsin is embedded in the disc membrane and forms a structure with seven transmembrane helices reminiscent of those characterizing the β-adrenergic receptor.

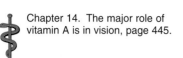

Chapter 14. The major role of vitamin A is in vision, page 445.

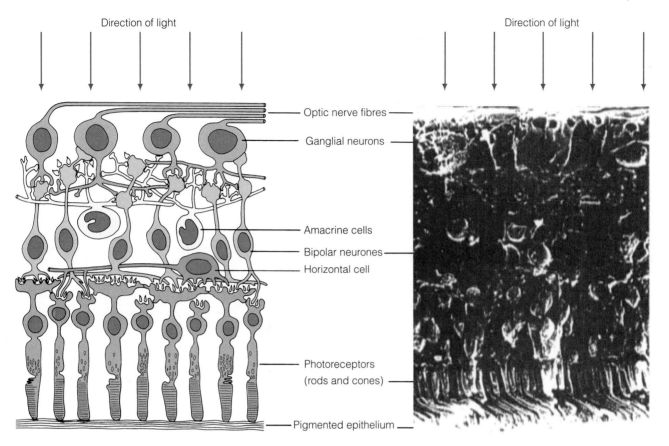

Direction of light

Direction of light

- Optic nerve fibres
- Ganglial neurons
- Amacrine cells
- Bipolar neurones
- Horizontal cell
- Photoreceptors (rods and cones)
- Pigmented epithelium

Figure 13.24 Diagrammatic and electron-microscopic views of the three layers of neural cells in the human retina. The rod cells detect low levels of light and the cone cells brighter light. Horizontal and amacrine cells integrate signals between the photoreceptor cells and the bipolar cells.

The immediate response to light is the isomerization of retinal

The absorption of light triggers the isomerization of the chromophore, converting it to **all-*trans*-retinal**. The complex goes through a series of intermediates, forming **metarhodopsin**. The conformational change induced by this isomerization produces an unstable relationship between the protein and the chromophore, which immediately dissociate. In the dissociated state the isomerization of the chromophore is reversed by **retinal isomerase** and the 11-*cis*-retinal binds to opsin once more (Figure 13.25).

The final event in the light-detection sequence is the closure of sodium channels

In the dark, the cation-specific channels in the plasma membrane of rod cells are open, allowing constant influx of sodium ions. As a result, the resting membrane potential is only about –30 mV. The rod cell responds to a pulse of light by a sequence of changes that result in these cation channels closing and the membrane potential increasing, causing hyperpolarization of the membrane.

Signal transduction in rod cells is mediated by cGMP through a cascade process

The relationship between activation of rhodopsin and closing of the sodium channels is mediated through a **cascade** process involving the protein **transducin** and **cGMP phosphodiesterase**. **Guanylate cyclase** catalyses the production of cGMP from GTP. In rod cells the activity of this enzyme does not vary with

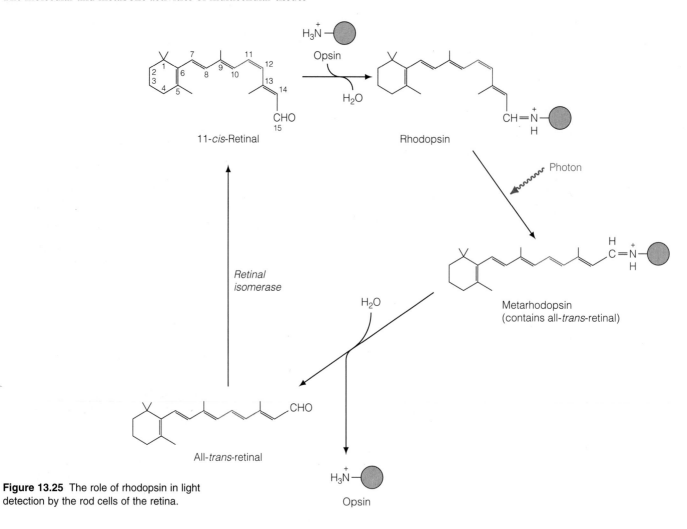

Figure 13.25 The role of rhodopsin in light detection by the rod cells of the retina.

exposure to light. By contrast, the enzyme that hydrolyses cGMP, **cGMP phosphodiesterase**, is active on exposure to light but inactivated in the dark .

$$GTP \xrightarrow{\text{Guanylate cyclase}} cGMP + PPi$$

$$\downarrow \text{cGMP Phosphodiesterase}$$

$$GMP$$

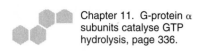

Chapter 11. G-protein α subunits catalyse GTP hydrolysis, page 336.

Transducin is a signal-coupling protein unique to the visual system, but related to the G-protein family. It is composed of α, β and γ subunits, the α subunit having GTPase activity and ability to exchange GDP for GTP, as do other G-proteins (Chapter 11). Photoactivated rhodopsin binds to transducin, activating it and allowing GDP to be exchanged for GTP. The two proteins then dissociate and the rhodopsin is free to activate other transducin molecules (one molecule of activated rhodopsin can activate about 500 molecules of transducin before dissociation of the chromophore terminates the process). Each GTP-activated α-subunit of transducin dissociates from its β–γ subunit complex and associates with cyclic GMP phosphodiesterase. This association displaces an inhibitory subunit of phosphodiesterase, activating phosphodiesterase and accelerating the hydrolysis of cGMP, so that the concentration of cGMP falls. This is a second amplification step in the cascade (Figure 13.26).

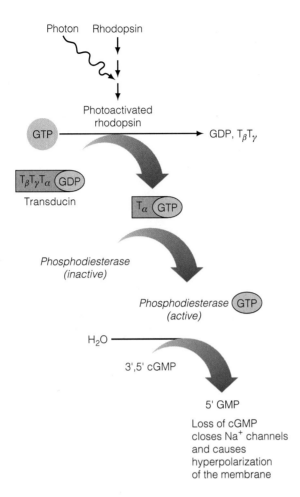

Figure 13.26 Transducin is a specialized G-protein, activated by photoactivated rhodopsin to trigger a cascade. This cascade accelerates the hydrolysis of cGMP, closing the Na^+ channels in the light-detecting rod cells, which then transmit the signal to the optic nerve.

The sodium channel of rod cells is a multimeric protein which binds cGMP cooperatively, making it extremely sensitive to small changes in cGMP concentration. At least three molecules of cGMP must bind to open the channel fully. In the dark the concentration of cGMP is enough to ensure that the channels remain open, maintaining the flux of Na^+ ions into the rod cell and maintaining a relatively low negative potential across the membrane. The light-induced hydrolysis of cGMP reduces its concentration to a level insufficient to keep the channel open, diminishing the influx of Na^+ and increasing the negative membrane potential, thereby activating the bipolar neurone that transmits the information to the optic nerve.

CHAPTER 14

From food to energy: molecular aspects of digestion and nutrition

This chapter describes the mechanisms of digestion and absorption of nutrients and the ways in which different tissues interact to balance nutritional requirements of the whole body. We also introduce vitamin and mineral nutrition.

The key to survival of animals is metabolic cooperation between different tissues. Some tissues are specialized for storing and/or distributing nutrients to other tissues. Others have little or no nutrient storage capacity. We examine the ways in which foodstuffs absorbed from the digestive tract are distributed throughout the body for the purposes of immediate energy production or temporary storage of nutrient. We consider how nutrient distribution changes with nutritional status, how structural tissues of the body can be converted to nutrient in starvation and briefly consider nutritional and metabolic aspects of obesity and temperature regulation.

Vitamins are essential to a healthy diet. These small organic molecules, or their coenzyme derivatives, participate in a wide range of enzyme-catalysed reactions. Minerals, including the trace elements, are also vital to survival. Attention is focused on those ions whose absorption and function are well understood.

Digestion and absorption of nutrients

Chapter 13. Exocrine activity of the pancreas, page 394.

Chapter 3. Transport of molecules and ions across membranes, page 62.

In this section we focus on the processes by which food is degraded in the intestine and the fragments absorbed into the bloodstream. The role of the **pancreas** in secreting digestive enzymes has been described in Chapter 13. Students should be familiar with the processes by which molecules or particles cross cell membranes (described in Chapter 3) and it is assumed that students will have read Chapters 7–10, which cover details of the metabolic processes. Where metabolic pathways are referred to in this chapter they will be presented only in outline.

Foodstuffs must be degraded to small fragments in the digestive tract

The digestive processes by which macromolecules are degraded include

- mechanical homogenization and hydration;
- denaturation and hydrolysis by hydrochloric acid;
- degradation by enzymes;
- physical dispersion and emulsification.

Food consists mainly of macromolecules (proteins, carbohydrates, fats and nucleic acids), which can not be absorbed directly from the intestinal tract but must first be degraded to smaller fragments.

- Carbohydrates are absorbed as mono- and disaccharides.
- Proteins are absorbed as amino acids or dipeptides.
- Lipids are absorbed as fatty acids and monoacylglycerols.

The digestive enzymes are hydrolases, breaking covalent bonds in the nutrient molecules by the addition of a molecule of water across the bond and include

- amylase (produced by the salivary glands);
- pepsin (produced in the stomach);
- enzymes of the gastrointestinal tract (produced by epithelial cells of the tract);
- pancreatic enzymes (which are secreted into the lumen of the gut).

The content and function of digestive juices are listed in Table 14.1. Most enzymic digestion occurs in the lumen of the duodenum, where macromolecules are degraded into small fragments (see Table 14.2). These fragments are often not small

Table 14.1 Production and function of digestive secretions

Digestive secretion	Organ	Major constituents	Digestive function
Saliva	Salivary glands in the mouth	Amylase, proteins, glycoproteins, proteoglycans, electrolytes	Digestion of starch to maltose, lubrication, buffering against bacterial acids, physical protection of teeth
Gastric juice	Stomach	HCl	Denaturation of proteins
		Pepsinogen	Digestion of proteins
		Lipase, glycoproteins	Absorption of vitamin B_{12}
		Intrinsic factor	
Bile	Produced in the liver, concentrated and stored in the gall bladder	Glycocholic acid and taurocholic acid	Excretion of bile pigments, cholesterol and drugs, emulsification of fats
		Bicarbonate	Neutralizes acidity of gastric juice
Pancreatic juice	Produced by the pancreas, secreted with bile into intestinal tract	Na^+, K^+, Ca^{2+}, HCO_3^-, numerous digestive enzymes (see Table 14.2)	Enzymic digestion of the majority of foodstuffs

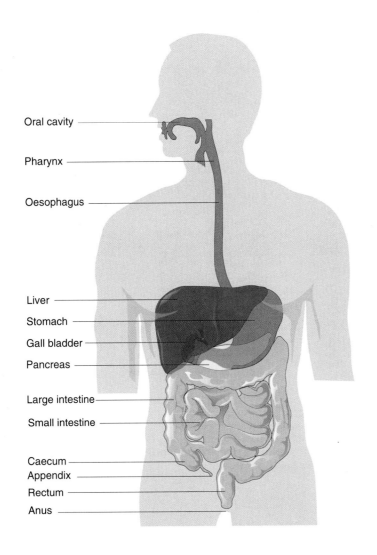

Oral cavity

Pharynx

Oesophagus

Liver

Stomach

Gall bladder

Pancreas

Large intestine

Small intestine

Caecum

Appendix

Rectum

Anus

Box 14.1   Stomach pH is maintained by active secretion of HCl

Secretion of hydrochloric acid into the stomach depends upon an ATP-dependent K^+/H^+ antiport located in the lumenal surface of the plasma membrane of oxyntic (parietal) cells lining the stomach cavity. This enzyme, which pumps K^+ into the cell and H^+ into the lumen of the stomach against a H^+ gradient, has not been found in other cells. The chloride ions are derived from the bloodstream and exchange for HCO_3^- ions crossing the basolateral surface of the cell. H^+ and HCO_3^- ions are generated in the cell from metabolic CO_2 and H_2O through the carbonic anhydrase reaction. The overall process maintains normal pH within the parietal cell, secretes HCl into the stomach and bicarbonate into the bloodstream.

Figure 14.1 The human digestive system.

Table 14.2 The digestive enzymes and their functions

Nutrient substrate	Enzymes	Specific role	Digestive secretion
Carbohydrate	Salivary amylase	Digestion of starch to maltose and isomaltose	Saliva
	Pancreatic amylase		Pancreatic juice
	Isomaltase	Initiates digestion of limit dextrin	Pancreatic juice
Protein	Pepsin	Endopeptidase	Gastric juice
	Trypsin	Activates proenzymes in the digestive tract; endopeptidase	Pancreatic juice
	Chymotrypsin	Endopeptidase	Pancreatic juice
	Elastase	Endopeptidase	Pancreatic juice
	Carboxypeptidases	Exopeptidases	Pancreatic juice
	Aminopeptidases	Exopeptidase	Pancreatic juice
Lipid	Lipase	Hydrolyses mono-, di- and triacylglycerols	Pancreatic juice
	Phospholipase	Hydrolyses phospholipids	Pancreatic juice
	Cholesterolesterase	Hydrolyses cholesterol esters	Pancreatic juice
Nucleic acids	Ribonuclease*	Hydrolyses phosphodiester bonds of nucleic acids	Pancreatic juice
	Deoxyribonuclease*		Pancreatic juice

*These are more significant in ruminants than in humans

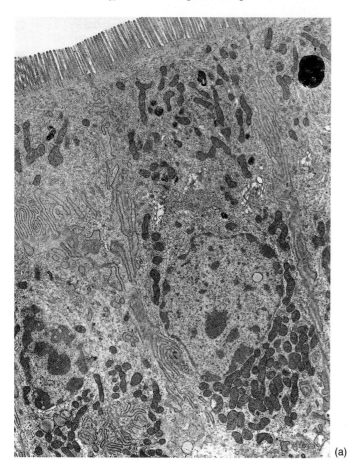

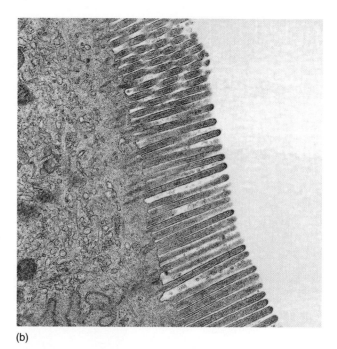

(b)

Figure 14.2 Electron microscopic view of an intestinal epithelial cell. Note the numerous microvilli, which provide sites for absorption of nutrient from the lumen of the gastrointestinal tract. The extensive volume of epithelial cell (a) from which microvilli extend is rich in mitochondria, providing energy for absorption and transfer of nutrients to the bloodstream and Golgi. (b) A detailed view of the structure of microvilli bordering the lumen of the intestinal tract.

(a)

Table 14.3 Peptide hormones (secretagogues) controlling the digestive system

Hormone	Secreted from cell type	Secreted in response to	Functions
Gastrin	Mucosal cells of stomach, duodenum and jejunum	Food in the stomach	Stimulates secretion of acid and pepsin and increases motility
Secretin	Mucosal cells of duodenum and jejunum	Presence of acid (pH less than 5) in duodenum and jejunum	Stimulates secretion by pancreas of enzymes, water and electrolytes Potentiates effects of cholecystokinin in pancreas Increases secretion of bicarbonate by the liver Delays gastric emptying
Pancreozymin (cholecystokinin)	Mucosal cells of duodenum and jejunum	Distension of intestine Presence of amino acids, peptides, fatty acids and hydrogen ions in intestinal epithelial cells	Stimulates enzyme secretion from the pancreas Causes contraction of the gall bladder
Enteroglucagon	Mucosal cells of stomach, small intestine and colon		Delays passage of food from the stomach to the intestine Stimulates release of insulin from the pancreas
Gastric inhibitory peptide	Mucosal cells of the duodenum and jejunum	Action of digested products reaching the duodenum	Inhibits gastric secretions Inhibits gastric emptying Stimulates release of insulin from the pancreas

enough to be taken into the bloodstream and the final stage of digestion, by which these fragments are converted to monomers such as amino acids and monosaccharides, occurs as or after the fragments are absorbed into the epithelial cells at the lumenal surface of the digestive tract. These epithelial cells possess **disaccharidases** and **dipeptidases** which complete the digestive process. These enzymes are glycoproteins, and it is believed that their sugar residues protect them against attack by enzymes in the lumen.

The lumen of the digestive tract is lined with specialized epithelial cells

Absorption of nutrients is controlled by the epithelial cells lining the lumen of the digestive tract. These cells take up the products of digestion and pass them across the basolateral membrane into the bloodstream. The lumenal surface, or **brush border**, consists of a large number of finger-like extensions of the cell surface, called **microvilli** (Figure 14.2). These microvilli increase the surface area across which absorption occurs, and thereby increase the rate of absorption of nutrient molecules. The plasma membrane contains specific transport pores through which nutrients enter the epithelial cells.

Digestive secretions are released in response to hormonal signals

Cells of the gastrointestinal tract produce a group of polypeptide hormones that regulate the release of digestive secretions. As food enters the digestive tract these hormones are released from cells lining the tract, absorbed and passed into the bloodstream. The individual functions and the stimuli promoting the secretion of these hormones are set out in Table 14.3. The exocrine cells of the pancreas are their prime target (see Figure 13.15).

Chapter 13, Peptide secretogogues regulate the exocrine function of the pancreas, page 395.

Digestion and absorption of protein

Undigested proteins can not be absorbed from the gut except in the neonate. During the first few days of life some proteins, notably the antibodies in maternal milk (colostrum), are absorbed directly into the infant's bloodstream by pinocytosis. This process provides the infant with passive immunity to infection during a period when its own immune system is not fully developed.

The minimum daily amount of protein for a healthy adult diet is about 40 g (this is the **exogenous protein** entering the digestive system). The amount of protein actually digested per day may be twice this amount because the digestive tract secretes about 30 g of protein, largely in the form of enzymes. In addition, some of the cells lining the tract are sloughed off, releasing additional protein to be digested. These are the **endogenous** sources of protein.

Synthesis and secretion of proteolytic enzymes

Many of the digestive enzymes are secreted as inactive proenzymes, **zymogens**, which are synthesized on the endoplasmic reticulum of cells of the pancreas, the salivary glands or the gastric mucosa and are then transferred to the Golgi complex where they are processed to glycoproteins and packaged into zymogen granules. In response to the stimulus for enzyme secretion the granules fuse with the plasma membrane and empty their contents into the lumen, where the enzymes are activated. Amylase, lipase and enzymes of the lumenal epithelial cells are not secreted as proenzymes.

Digestive enzymes hydrolyse different types of peptide bond

The enzymes that digest proteins are either **endopeptidases**, which hydrolyse internal bonds of the protein, or **exopeptidases**, which hydrolyse terminal peptide bonds (the latter are called **carboxypeptidases** if they remove amino acids from the C-terminus of the protein or **aminopeptidases** if they remove amino acids from the N-terminus). Endopeptidases break proteins into smaller fragments, increasing the number of C- and N-terminal residues available for hydrolysis by the exopeptidases, and thereby increase the rate at which proteins are digested.

Endopeptidases are specific for the amino acid adjacent to the peptide bonds they hydrolyse. For example, **pepsin** hydrolyses peptide bonds with aromatic amino acids (tyrosine, tryptophan or phenylalanine) on the N-terminal side of the bond, **trypsin** hydrolyses peptide bonds with positively charged residues (arginine or lysine) on the C-terminal side of this residue.

Proteolysis in the stomach

Pepsinogen, released from the mucosal cells of the stomach, is activated by HCl from the oxyntic cells and by autocatalysis. Activation involves release of a 44-amino acid peptide from the N-terminus of the zymogen to produce the active enzyme, **pepsin**. The released peptide remains bound to pepsin and acts as an inhibitor if the pH is greater than 2, dissociating and allowing full expression of the activity of pepsin when the pH is below 2. Pepsin is maximally active at pH 1.5–2 and digests most proteins in the normal environment of the stomach. The acidity of the stomach serves both to destroy bacteria and to denature proteins, increasing their susceptibility to digestion.

The human stomach secretes small amounts of several genetically distinct isozymes of pepsinogen. All have similar specificities, and the significance of the different forms is not yet known.

Proteolysis in the intestine

Pancreatic enzymes are activated by proteolysis, catalysed initially by **enteropeptidase** (previously called **enterokinase**), which is produced in, and secreted from, the epithelial cells of the duodenum. Efficient digestion in the duodenum depends on the fact that partly digested food passing from the acidic environment of the stomach is neutralized by bicarbonate in the bile, because the pancreatic enzymes are active at neutral pH and would be inactivated in the

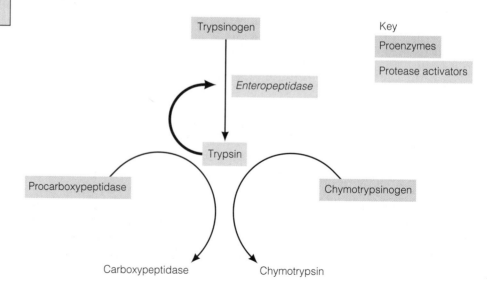

Figure 14.3 In the digestive tract pancreatic zymogens (proenzymes) are activated by proteolysis, which is initiated when enteropeptidase activates trypsinogen, converting it to trypsin.

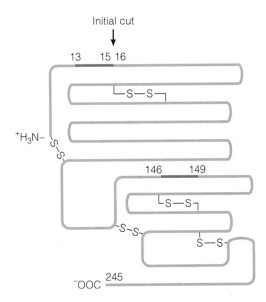

Figure 14.4 The production of α-chymotrypsin (the active form) from chymotrypsinogen involves excision of two internal peptides (black sections). Chymotrypsin is held together by disulphide bonds.

acidic conditions prevailing in the stomach. Pancreatic juice, which mixes with the bile before entering the intestine, also contains bicarbonate. The enzyme and electrolyte components of pancreatic juice are released under different hormonal controls (see Table 14.2), ensuring a balanced environment for digestion.

Trypsinogen, the inactive precursor of **trypsin**, is activated initially by **enteropeptidase**, which removes six amino acids from its *N*-terminus. Further molecules of trypsinogen are then activated by autocatalysis. Pancreatic juice also contains a **trypsin inhibitor**, a small protein that binds tightly to trypsin and prevents the premature conversion of trypsinogen to trypsin.

Trypsin also activates other enzymes released in the pancreatic juice, including **chymotrypsinogen**, **proelastase** and **procarboxypeptidase**. Conversion of chymotrypsinogen to **chymotrypsin** is an example of activation through cleavage of internal peptide bonds (Figure 14.4).

Absorption of amino acids

Amino acids liberated by digestion are absorbed by the epithelial cells of the gut. Dipeptides are hydrolysed by epithelial cell dipeptidases. There appear to be several amino acid transporting systems specific for different classes of amino acid. It is probable that absorption processes analogous to the γ-glutamyl cycle of the kidney (see Chapter 10) exist in the intestinal wall.

 Chapter 10. Glutathione drives amino acid transport through the γ-glutamyl cycle, page 314.

Digestion and absorption of carbohydrate

Starch (amylose and amylopectin – see Chapter 2), the major carbohydrate in the human diet, is hydrolysed to isomaltose and limit dextrin by **salivary amylase** and **pancreatic amylase**.

These enzymes hydrolyse the α,1–4 glycosidic bonds but are unable to hydrolyse the α,1–6 bonds at branch points of amylopectin. The effectiveness of salivary amylase depends on the time the food remains in the mouth and the extent to which it is homogenized by chewing. Salivary amylase is inactivated in the acidic environment of the stomach, although some enzyme molecules survive within the food bolus to continue the digestion process. Amylase and **isomaltase** (which hydrolyses α,1–6 glycosidic bonds) in the pancreatic secretions complete the digestion of starch to disaccharides. **Disaccharidases** in the brush border cells convert

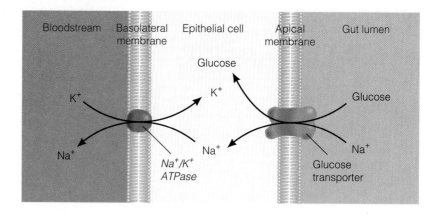

Figure 14.5 Glucose enters epithelial cells of the intestinal lumen by cotransport with Na⁺ through a symport in the apical membrane. It passes into the bloodstream by facilitated diffusion through a low-affinity transporter located in the basolateral membrane, while the Na⁺ concentration within the cell is maintained by the Na⁺/K⁺-ATPase in the basolateral membrane.

Chapter 3. Active transport mechanisms can be driven by the simultaneous movement of ions down a concentration gradient, page 68.

Box 11.1. Cholera and whooping cough, page 337.

these to monosaccharides, which are then absorbed into the bloodstream. Specific disaccharidases hydrolyse lactose (**lactase**) and sucrose (**sucrase**), thereby facilitating the absorption of these molecules.

Transport of glucose from the lumen into the epithelial cell is tightly coupled to the uptake of Na⁺, but glucose then passes to the bloodstream by facilitated diffusion (Figure 14.5; see Chapter 3). During this process the Na⁺ gradient across the epithelial cell plasma membrane is maintained by a Na⁺/K⁺-ATPase pump on the basolateral membrane, which moves Na⁺ into the bloodstream in exchange for K⁺.

Our understanding of this mechanism is exploited in the treatment of cholera (Box 11.1) and other conditions in which severe dehydration and the associated ionic imbalances must be reversed. Rehydration fluids need to contain glucose as well as sodium because glucose aids the absorption of sodium ions.

Digestion and absorption of lipids

About 40–50% of our energy requirement is normally provided as lipid. This can be as much as 150 g fat per day. Most dietary lipid is triacylglycerol, which is hydrolysed to a mixture of fatty acids and monoacylglycerols, partly by lipase in the stomach but mostly by **pancreatic lipase** in the duodenum. Pancreatic juice also contains **phospholipase**, which degrades phospholipids, and **cholesterolesterase**, which degrades cholesterol esters.

The hydrophobic character of lipids complicates their digestion and absorption

Lipids are not miscible with water, which complicates their digestion and absorption. Enzymes that digest lipids are water soluble and mix poorly with their substrates. Even partially digested fats aggregate into complexes that can not readily be absorbed through cell membranes. To overcome the poor water solubility of these compounds, and to aid their digestion and absorption, fats must be dispersed and **emulsified**. This is aided by the **bile salts**.

Gastric and pancreatic lipases convert triacylglycerols to mono- or diacylglycerols and free fatty acids. The exposure of free carboxyl groups through these reactions increases the tendency of the fatty acids to interact with the aqueous environment and form smaller lipid droplets: this is the process of emulsification. Pancreatic lipase specifically hydrolyses esters in the C-1 position on the glycerol backbone. To aid the interaction of enzyme with substrate a small protein (**colipase**) is bound to pancreatic lipase at the water–lipid interface, effectively anchoring the enzyme to the lipid globule and at the same time activating it.

Pancreatic cholesterolesterase and **phospholipase A₂** are both secreted from the pancreas as proenzymes. Both require to be activated by trypsin and require bile salts for activity. The interaction of the two enzymes with substrate is aided by bile salts which act as emulsifying agents.

Bile salts facilitate the digestion and absorption of lipids

Bile acids (or bile salts) are biological detergents synthesized by the liver, concentrated and stored by the gall bladder and secreted into the duodenum as a component of bile. Bile salts, the bulk of which are derivatives of **cholic acid** and **chenodeoxycholic acid**, are synthesized from cholesterol. Their production represents the major metabolic fate of cholesterol in the body.

Primary bile salts are 24-carbon compounds retaining the ring structure characteristic of cholesterol, and having two or three hydroxyl groups and a side chain that terminates in a carboxyl group. They are anions with detergent properties. At concentrations above 2–5 mM the primary bile salts form **micelles**, aggregates of molecules in which the hydrophobic portions interact at the centre of the aggregate, leaving the hydrophilic portions in contact with the aqueous environment. Individual molecules of the micelle are in equilibrium with those remaining free in solution. The hydrophobic component of a bile salt molecule is the surface of the ring structure; the hydrophilic moieties are the carboxyl, hydroxyl and sulphonate residues.

Bile acids are conjugated in the liver with glycine (forming **glycocholate**) or with taurine (forming **taurocholate**) (Figure 14.7). These components increase the polarity of the bile salt side-chains.

Conjugated bile salts are secreted into the intestinal tract where they interact with triacylglycerols, fatty acids and phospholipids by emulsifying the lipid molecules. The bile salts remain at the lipid–water interface, increasing the contact between lipid molecules and lipase, facilitating the hydrolysis of triacylglycerols and the release of fatty acids and monoacylglycerols. The hydrolytic products are released from the larger lipid droplets and incorporated into smaller bile salt micelles. The solubilizing effect of bile salts facilitates the absorption of the digestion products of these small micelles into mucosal cells of the lumen (Figure 14.6).

Bile salts are constantly recycled through the enterohepatic circulation

The amount of bile salts secreted into the intestine varies between 16 and 70 g per day, yet the total amount of bile salt in the body may be as little as 3–4 g. This is

Figure 14.6 Bile salts associate with triacylglycerols, and these complexes aggregate into micelles in the intestinal tract. Pancreatic lipase binds to the polar external surface of the micelle, liberating fatty acids into smaller micelles, which can be absorbed through the intestinal epithelium.

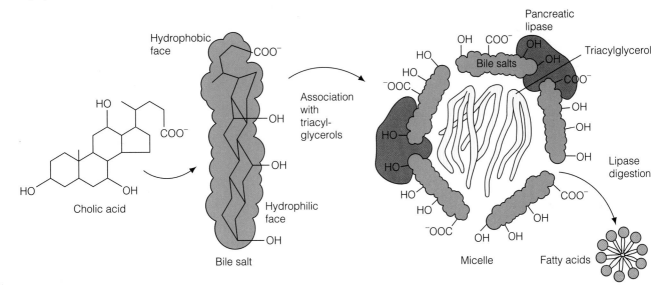

419

Figure 14.7 Synthesis of glycocholic and taurocholic acids.

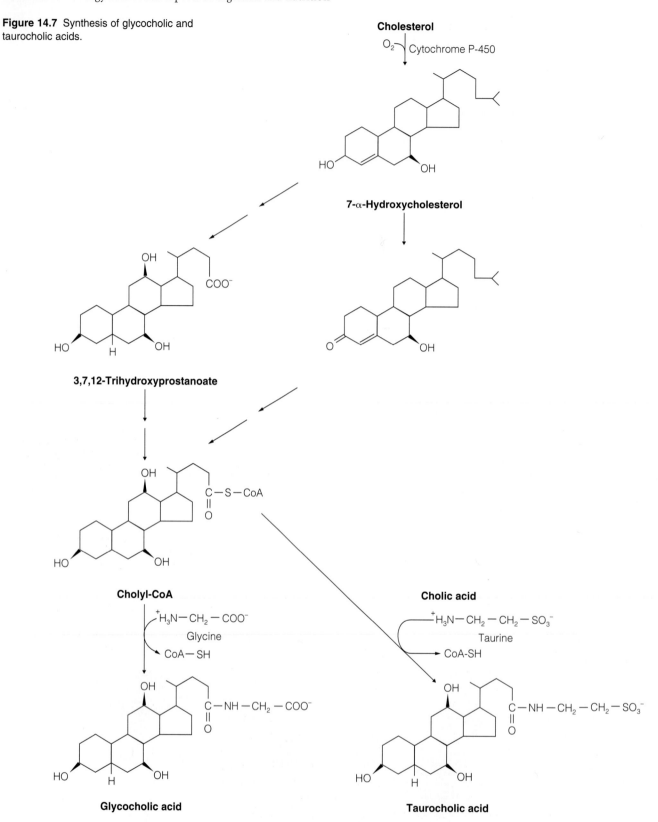

explained by the constant recycling of bile components between the intestine and the liver: the **enterohepatic circulation** (Figure 14.8). Conjugated bile salts secreted into the intestine are hydrolysed by bacterial enzymes, which remove the taurine or glycine residues and one of the hydroxyl groups. The unconjugated bile

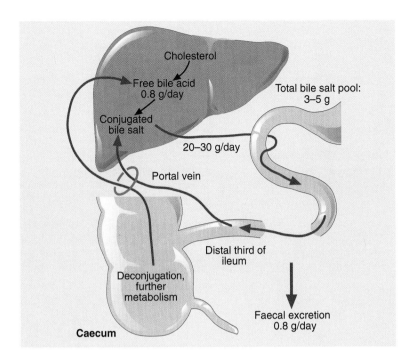

Figure 14.8 Enterohepatic circulation through which bile is recycled between the liver and the digestive tract.

salts are reabsorbed by active transport through the epithelium of the lower small intestine into the portal blood, whence they are taken up by the liver, reconjugated with taurine or glycine, and recycled through the intestine.

Those bile salt molecules that are not reabsorbed through the enterohepatic circulation are excreted in the faeces along with **bile pigments** (the degradation products of haemoglobin). Approximately 350 mg of cholesterol is converted to bile acids, and about twice as much is excreted, each day.

Bile salts also facilitate the absorption of fat-soluble vitamins, particularly vitamin D. Obstructive jaundice brings a risk of vitamin K deficiency, with consequences for blood clotting processes (Chapter 12), which is particularly important if surgery is required to relieve the obstruction.

Chapter 12, There is a mechanism for degradation of haemoglobin after red cell haemolysis, page 374.

Chapter 12. Specific and sequential proteolytic cleavage give a 'cascade' effect, page 362.

Triacylglycerols hydrolysed in the gut are reassembled within epithelial cells

Free fatty acids released from dietary triacylglycerols by the action of lipases are taken up by the epithelial cells of the intestine. About 10% of free fatty acids are absorbed directly into the portal bloodstream and transported to the liver bound to serum albumin. Most of these fatty acids have short chains, with fewer than 12 carbon atoms. The remaining 90% of the fatty acids are bound to a **fatty acid binding protein** in the cytoplasm of the intestinal cell and are then transported into the smooth endoplasmic reticulum for reassembly into triacylglycerols. The triacylglycerols, together with phospholipids and cholesterol, acquire a protein coat, forming lipoprotein globules (**chylomicrons**) (Figure 14.9). Chylomicrons are involved in the dispersion and partial solubilization of fat for transport. Chylomicrons are secreted into the intestinal lymph vessels, whence they are introduced into the bloodstream through the thoracic duct. This ensures that muscle and adipose tissue (rather than the liver) have first access to these triacylglycerols.

This resynthesis of triacylglycerols balances the usage and storage of fatty acids and glucose absorbed from a meal. If all of the dietary triacylglycerol entered the circulation as free fatty acids, the metabolism of free fatty acids by tissues would increase, which would reduce the amount of triacylglycerol entering the lipid

Chapter 13, Homeostasis is maintained by reabsorption of some molecules and ions through the renal tubules, page 391.

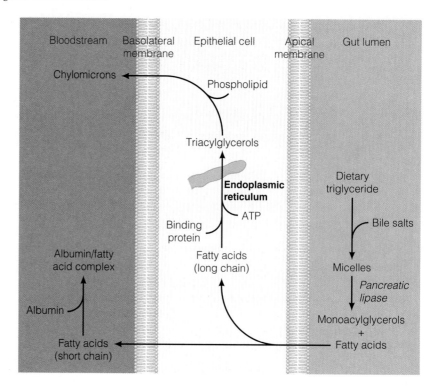

Figure 14.9 Monoacylglycerols and fatty acids are absorbed through the lumenal membrane of epithelial cells. Short-chain fatty acids diffuse into the bloodstream, but long-chain fatty acids are reassembled into triacylglycerols in the endoplasmic reticulum. These are then complexed with phospholipids to form chylomicrons.

storage cells of adipose tissue in the immediate aftermath of a meal. At the same time, less of the dietary glucose would be taken into cells. Blood glucose would remain high enough to exceed the renal threshold and glucose would be excreted, wastefully, by the kidneys.

Nutritional aspects of metabolism

Animals that eat only intermittently need mechanisms to sustain energy provision between meals. In the aftermath of a meal the energy for tissue functions is derived from the metabolism of nutrients absorbed from the digestive tract. Between meals, even when the animal is asleep, all tissues continue to consume energy and it is therefore essential that some of the nutrient absorbed after each meal is converted to an energy reserve that can be called on between meals. The human body has the capacity to store carbohydrate, particularly in muscle and liver, and fat, mainly in adipose tissue. Although carbohydrate provides an important short-term nutrient supply, the amount of carbohydrate that humans store is small compared to the amount of fat: glycogen stores are limited but fat stores provide a huge reservoir of nutrient energy. After only a few hours without food, stored fat becomes the main nutrient.

Tissues vary in their patterns of nutrient consumption

The amount of energy required to sustain function varies between tissues. At one extreme, the brain uses a constant amount of energy at all times – whether we are awake, asleep or trying to resolve a difficult intellectual problem, the brain requires the same amount of energy. The amount of energy required by muscle cells varies greatly depending on the extent of physical activity. At rest, muscles (and indeed all cells) need some energy for repair and maintenance of cell structure and to preserve Na⁺ and K⁺ gradients across the cell membrane. (It has been estimated that up to 40% of all the energy consumed by the body at rest is used to maintain electrolyte gradients across cell membranes.) With increased muscular activity, more energy is required to support contractility, in amounts reflecting the extent of physical exertion.

Fat in the diet

Fat is the most concentrated form in which nutritional energy is available, contributing about 39 kJ g⁻¹, more than twice the energy per gram available from carbohydrates or proteins. In a typical Western European or North American diet 50% of the energy content is likely to be lipid, but for a healthy diet we require only about 5% of our total energy content to be lipid. About 90% of our lipid intake is triacylglycerols, containing mostly saturated or monounsaturated fatty acids, cholesterol, cholesterol esters and phospholipids. Individuals taking in more plant-derived and less animal-derived fat will consume more polyunsaturated fatty acids.

Controversy continues among nutritionists and clinicians about the potential health risks of a diet containing too much lipid, about the relative health hazards associated with saturated and polyunsaturated fatty acids and (more recently) about the advantages of unsaturated fatty acids containing *cis* double bonds compared with those containing *trans* double bonds. However, it is generally agreed that a diet containing excessive amounts of lipid will lead to deposits of cholesterol and cholesterol esters along arterial walls – with the result that blood flow may become impaired. At the other extreme, a diet deficient in the **essential fatty acids** (Chapters 9 and 11) can be a cause of malnutrition.

Chapter 2, Unsaturated fatty acids have *cis* double bonds in specific locations, page 24.

Chapter 9. Longer chain and unsaturated fatty acids are synthesized from palmityl CoA, page 279.

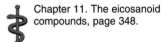

Chapter 11. The eicosanoid compounds, page 348.

Some lipids are necessary in the diet to sustain health

Animals are able to synthesize most of the lipids they need from other dietary constituents, including carbohydrates and amino acids. However, some must be provided by the diet:

- to provide adequate amounts of the essential fatty acids;
- to aid the absorption of fat-soluble vitamins across the intestinal mucosa;
- to ensure the palatability of the diet (a diet devoid of fat is not tolerated).

Synthesis of linoleic acid, linolenic acid and arachidonic acid either from the corresponding saturated fatty acids or from acetyl CoA requires enzymes capable of introducing double bonds near to the methyl group of saturated fatty acids (Figure 14.10). Humans do not possess these enzymes and thus can not synthesize linoleic (18:2 *cis*-Δ^9,Δ^{12}) and linolenic (18:3 *cis*-$\Delta^9,\Delta^{12},\Delta^{15}$) acids, which are nutritionally essential. We therefore also lack the substrate (linoleic acid) from which to synthesize arachidonic acid, an essential precursor of the eicosanoid hormones (see Chapter 11).

Chapter 11. The eicosanoid compounds, page 348.

Dietary lipids are used for a variety of purposes

Triacylglycerols in the diet may be used as an immediate energy source or may be stored. Most cells have the capacity to synthesize and store some triacylglycerol, but adipose tissue is the main storage site. Most of the triacylglycerol produced in the liver is used in the synthesis of lipoproteins to be exported to the bloodstream.

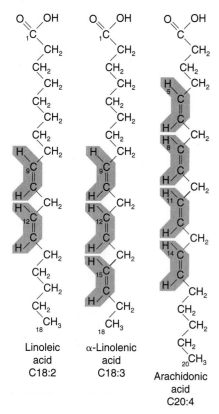

Figure 14.10 The structures of linoleic, linolenic and arachidonic acids.

Most of the long-chain (C_{20} or larger) polyunsaturated fatty acids are incorporated into membrane phospholipids where they influence the conformation and fluidity of the membrane bilayer (see Chapter 3). If the diet is deficient in these fatty acids, membrane composition will change and membrane functions may be impaired.

Triacylglycerols and fatty acids are not soluble in the blood, and special transport strategies are required to transport fats absorbed from the diet and between tissues. These include

- incorporating triacylglycerols into chylomicrons or VLDL;
- transferring triacylglycerols to lipoproteins;
- binding fatty acids to albumin.

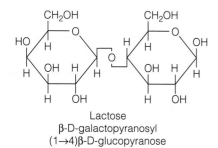

Lactose
β-D-galactopyranosyl
(1→4)β-D-glucopyranose

Sucrose
α-D-glucopyranosyl
(1→2)β-D-fructofuranoside

Figure 14.11 The structures of sucrose and lactose.

Carbohydrate in the diet

Dietary sources of carbohydrate

About 60% of our dietary carbohydrate is the plant storage polysaccharide starch. We may also consume small amounts (less than 1%) of glycogen, the polysaccharide derived from animal muscle and liver. In Western diets about 25% of the dietary carbohydrate is **sucrose** ('sugar'), a disaccharide of glucose and fructose. Milk, the main source of food available to the very young, contains **lactose**, a disaccharide of **galactose** and **glucose** (Figure 14.11).

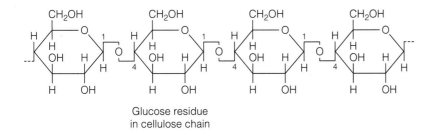

Figure 14.12 α,1–4 linked glucose units found in glycogen.

Glucose residue at the nonreducing end of the glycogen chain

Glycogen chain

Polysaccharides of glucose may take a variety of forms

Glycogen (Figure 14.12) is a large, highly branched polymer of glucose in which linear chains of glucose residues are linked by α,1–4 glycosidic bonds and branches are created through the formation of α,1–6 glycosidic linkages. A branch will occur every 8–12 glucose residues on average (Chapter 2).

Starch is a glucose polymer that is present as a mixture of two forms: **amylose** is an unbranched α,1–4 glucose polymer; **amylopectin** is an α,1–4 linked polymer with branches formed by α,1–6 linkages. Amylopectin has fewer branches than glycogen, but they may be longer – up to 30 residues. Both forms of starch can be used by animals for nutrition.

Plants produce another polymer of glucose, **cellulose**, an unbranched polymer of β,1–4–linked glucose units (Figure 14.13). This configuration forces the polymer into long straight chains and confers properties consistent with its role as a structural molecule in plants. Cellulose is not a nutritionally useful molecule in the human diet because we do not possess the **cellulase** enzymes that degrade it. However, cellulose forms the fibre content of the human diet, and has an important role in maintaining digestive functions. Ruminants and other herbivores have cellulase-secreting bacteria in their digestive system that are able to degrade cellulose to short-chain oligosaccharides, which the animal is able to absorb and use.

Chapter 2. Major carbohydrate fuel reserves are polymers of glucose linked by *O*-glycosidic bonds, page 19.

Glucose residue in cellulose chain

Figure 14.13 β,1–4 linked glucose units characteristic of cellulose.

The fate of dietary carbohydrate

After a meal the concentration of glucose in the circulation typically increases from the resting value of 4–5 mM to 8–9 mM. The pancreas responds to the increased glucose by reducing the secretion of glucagon and increasing the secretion of insulin (Chapter 13). These hormonal changes accelerate the rate of glucose uptake into tissues, particularly muscle and adipose tissue. Liver cells take up glucose independently of changes in circulating insulin, responding directly to the increased blood concentration. If more glucose is entering cells than is required immediately, it is

Chapter 13. Secretion of insulin is regulated by plasma glucose concentration, mediated through islet cell glucokinase, page 399.

Chapter 8. When glucose is plentiful it is stored as glycogen, page 224.

Chapter 9. Synthesis of triacylglycerols and membrane lipids, page 279.

Box 14.4 Glycogen stores in the neonate

Late in fetal development, amounts of glycogen 3–4 times greater than those found in adult tissues are stored in liver and muscle. This glycogen is consumed immediately after birth and is thought to provide an energy buffer for the infant in the first hours of independent existence when it is required to assume the unaccustomed demands of breathing, maintaining body temperature and moving. At the same, time the infant must adapt to a diet containing lactose. Metabolism of the galactose derived from lactose requires that suitable enzymes are induced in the liver. During this phase the infant is able to sustain its energy requirements from these large stores of glycogen.

Box 8.2. Glycogen storage diseases, page 230.

stored as glycogen (in liver and muscle cells – see Chapter 8) or used for the synthesis of triacylglycerol (in adipose tissue and liver – see Chapter 9). The blood glucose concentration returns to its pre-prandial value over a period of about 2 h.

Most of the glucose taken into the liver in a normal individual is stored as glycogen, which can be degraded later and its glucose released into the circulation for the benefit of other tissues. In this way the liver buffers the blood glucose concentration, taking up glucose when large quantities are being absorbed from the digestive system and releasing it to other tissues when the blood glucose concentration is low.

Why does the body store carbohydrate as an energy source?

Although carbohydrate provides less energy per gram than fat, carbohydrate has two major advantages over fat as an energy store.

1. It can be oxidized to produce ATP through glycolysis in tissues that are starved of oxygen – fat can *not* be oxidized without oxygen. During periods of strenuous exercise, muscle cells may need to generate energy at a rate greater than can be sustained by the rate at which oxygen is made available to muscle from circulating red blood cells. Glycogen is therefore an important store of energy for muscle.
2. Animals are unable to convert fat to glucose and so, because glucose is an essential nutrient for brain, it must be obtained from glycogen stores or gluconeogenesis.

The storage of carbohydrate as a branched polysaccharide has several advantages over storage of monosaccharides

1. The formation of polysaccharides from monosaccharides reduces the number of molecules within the cell and therefore reduces the overall osmotic effect.
2. Its branched structure ensures that glycogen can be stored compactly, occupying less space in the cell.
3. The high density of branches in glycogen increases the number of points from which degradation and release of monosaccharides can be initiated. This means that glucose can be released more rapidly from glycogen than from a less highly branched polymer.
4. The branched structure ensures that a core limit structure will be maintained on to which extensive branching may be re-established when glucose becomes available from the diet.

Glycogen is stored as cytoplasmic granules, which can be seen clearly under the electron microscope in sections of muscle and liver tissues. The processes of storage and degradation of glycogen are made more efficient by the fact that these granules also contain the enzymes responsible for glycogen synthesis and degradation – and some of the enzymes that regulate these processes.

The amount of free glucose in the body is equivalent to about 120–160 kJ of energy, whereas the total amount of glycogen may be equivalent to as much as 2400 kJ of energy. The tissue with the greatest capacity (per gram of tissue) to store glycogen is the liver, but the amount of muscle in the body is so much greater than the amount of liver that most of the body glycogen is found in muscle cells. The metabolic basis of glycogen-storage diseases is described in Chapter 8.

Lactose provides an important source of carbohydrate

The rate at which lactose is absorbed from the gut increases in the presence of high Ca^{2+} concentrations. The mechanism of this **cooperative** effect is not clear, but it is important in infants dependent on calcium for bone growth and lactose as an energy source. Epithelial cells of the gut contain **lactase**, which hydrolyses lactose to the monosaccharides galactose and glucose. Some individuals lack lactase and accumulate lactose in the gut, where it is degraded by intestinal bacteria. The products (CO_2, H_2, and organic acids) cause digestive problems including diarrhoea and flatulence. This condition is termed **lactose intolerance**. If this condition persists it will lead to long-term intestinal problems and commercial sources of low-lactose milk (lactase-treated) are available to overcome the problems. Many adult cats suffer the same condition. Lactase-deficient individuals are able to tolerate yoghurt because lactose is hydrolysed in the fermentation process by which yoghurt is produced.

Normally the liver metabolizes both fructose and galactose, converting them to products that enter glycolysis (Figure 14.14).

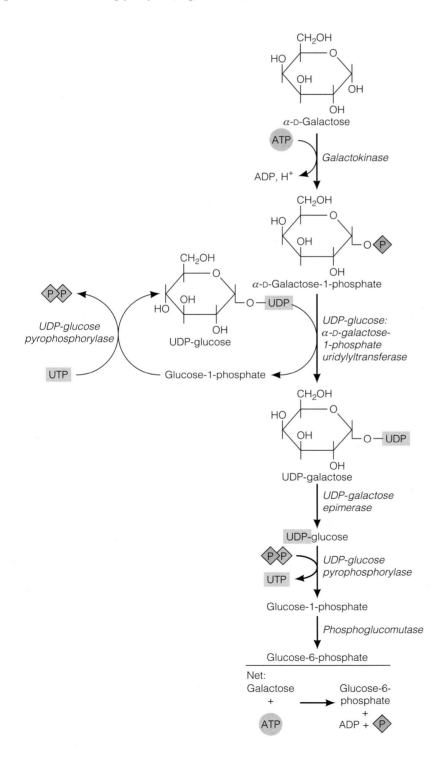

Figure 14.14 Galactose, derived from hydrolysis of the milk disaccharide lactose, is converted to G6P through a complex series of reactions involving UDP-derivatives of the sugars. Deficiencies of all of these enzymes have been identified and give rise to galactosaemia, accompanied by diarrhoea and enlargement of the liver and other organs. Treatment requires that milk and milk products are eliminated from the diet. This pathway operates in reverse for the synthesis of the galactose component of lactose in the mammary gland.

427

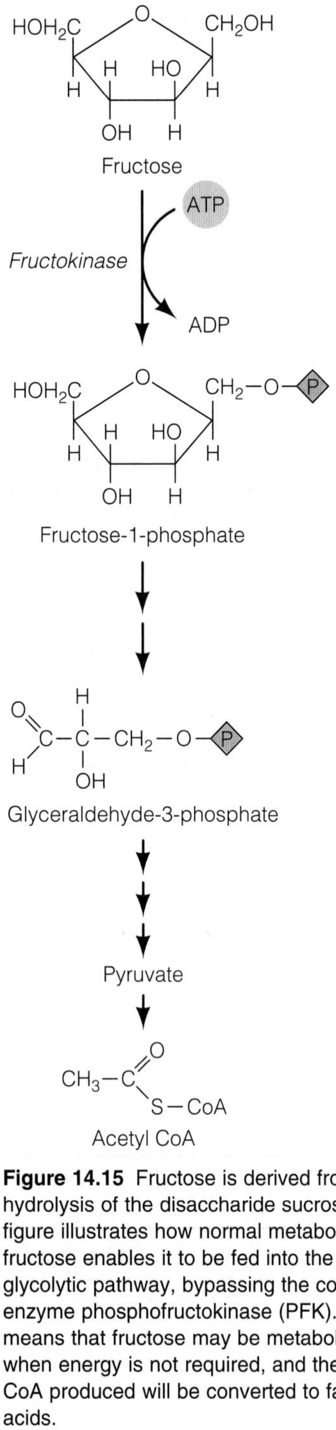

Figure 14.15 Fructose is derived from hydrolysis of the disaccharide sucrose. The figure illustrates how normal metabolism of fructose enables it to be fed into the glycolytic pathway, bypassing the control enzyme phosphofructokinase (PFK). This means that fructose may be metabolized when energy is not required, and the acetyl CoA produced will be converted to fatty acids.

Deficiencies of all enzymes of this pathway have been identified. The most common is a deficiency of UDP glycose: galactose-1–phosphate uridylyl transferase, which leads to severe diarrhoea in infants. This condition is termed **galactosaemia**.

Sucrose is a major source of carbohydrate in Western diets

Sucrose, glucose-(α1,β2)-fructose, is hydrolysed by **sucrase** in intestinal epithelial cells.

The resultant fructose is converted through F1P to glyceraldehyde-3–phosphate, bypassing phosphofructokinase which usually regulates the flux of metabolites through glycolysis. This means that fructose can be converted to acetyl CoA and fatty acids, unrestrained by the metabolic controls normally regulating the production of fatty acids from glucose (Figure 14.15).

Although the idea remains controversial, it is possible that this phenomenon accounts for the high concentration of low-density lipoproteins in the bloodstream of individuals on a high-sucrose Western European or North American diet.

Protein in the diet

Adults require at least 7–10% of the caloric value of their diet as protein. More is required in pregnancy and lactation. In Western diets protein provides closer to 15% of the energy intake.

Most of the amino acids in the body are part of the structural and other proteins. Any additional amino acids needed for protein synthesis must be found from dietary protein or from degradation of body proteins. The protein nitrogen excreted per day, if there is no protein intake, represents 0.5 g protein kg^{-1} body weight (about 33 g for the average person). This must be replaced each day in order to maintain **nitrogen equilibrium**.

Protein is the most complex nutrient macromolecule

Proteins are polymers of about 20 different amino acids; they are degraded in the digestive tract and absorbed as amino acids or small peptides (see above). Once absorbed, these amino acids provide the building blocks for growth and repair of the structural components of tissues and replacement of metabolic enzymes and peptide hormones. Amino acids are also precursors for the synthesis of many essential nonprotein components of the body: the haem of haemoglobin, the nitrogenous bases of nucleic acids, the creatine for muscle and other tissues.

The amount of protein needed in the diet is a reflection of

- the requirement for synthesis of new protein molecules; and
- the requirement for synthesis of other nitrogenous compounds.

The composition of ingested protein is important because the proportions of amino acids in the dietary proteins are not the same as those we require to synthesize proteins. Animal protein contains most of the amino acids we require; vegetable sources of protein contain a less balanced range. In particular, cereal proteins are usually deficient in methionine or lysine, and may provide none of these amino acids. This means that vegetarians have to select foods more carefully than meat eaters to ensure a balanced intake of essential amino acids. As all diets, even in nutritionally deprived communities, contain several different sources of protein, each with a distinctive amino acid composition, the composition of individual foods is less important than the overall composition of the diet.

Consumption of a diet with an unnecessarily high protein content means that most of the amino acid in dietary protein is not required for protein synthesis, and is metabolized. The amino groups are removed and either reused or excreted after being converted to urea (Chapter 10). The carbon skeletons are fed into the central metabolic pathways to provide energy or substrates for the synthesis of other molecules.

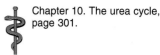

Chapter 10. The urea cycle, page 301.

Essential amino acids are those that can not be synthesized in the body

The terms **essential** and **nonessential** applied to amino acids have exactly the same meaning as when applied to fatty acids. Essential amino acids are those that must be provided in our diet because we do not have the enzymes to synthesize them, at least in the quantities required. We are able to produce some of the amino acids required – these are the nonessential amino acids.

The distinction between essential and nonessential amino acids is a reflection of whether cells have the enzymes to synthesise the necessary carbon skeleton.

Whether an amino acid is essential or not varies between species. Simpler organisms are generally able to synthesize more amino acids than animals are. As with fatty acids, those amino acids with a complex carbon skeleton are more likely to be essential. Study of the synthesis by simpler organisms of amino acids that are essential for humans reveals that it requires several (often as many as six or seven) different enzymes. It could therefore be argued that the fact that we no longer have the capacity to make these amino acids saves us from expending energy on complex syntheses – but we have made ourselves more vulnerable and more dependent on dietary sources.

Table 14.4 Essential and nonessential amino acids

Essential	Nonessential
Arginine	Alanine
Histidine	Aparagine
Isoleucine	Apsartate
Leucine	Cysteine
Lysine	Glutamate
Methionine	Glutamine
Phenylalanine	Glycine
Threonine	Proline
Tryptophan	Serine
Valine	Tyrosine

The distinction between essential and nonessential amino acids is relative rather than absolute

The definition of the nutritional status of an amino acid may be different in younger and more mature individuals. This is because the demand for amino acids is usually greater during growth, and the rate of synthesis of some amino acids may not meet the demand. Particular examples are arginine and histidine. Humans are capable of synthesizing these amino acids but at a slow rate – too slow to meet the needs of growing children and sometimes too slow to meet the demand in adults.

If a nonessential amino acid is synthesized from an essential one, then the amount of the essential acid in the diet will influence the rate at which the nonessential one can be produced. For example, tyrosine (nonessential) is produced from phenylalanine (essential). Cysteine (nonessential) is synthesized from methionine, which is essential, and serine, which is nonessential (see Chapter 10).

Chapter 10. Cysteine is produced from methionine and serine, page 307.

There is a nutritional advantage in providing all amino acids in the diet

Using one amino acid to produce another depletes the total pool available for protein synthesis. It is therefore true that provision of *all* amino acids in the diet, both the essential and the nonessential, is preferable to dependence on their synthesis. In addition, synthesis of amino acids consumes energy, and in some instances depends on induction of the appropriate enzymes.

Synthesis of nonessential amino acids will occur only if the necessary carbon skeletons are available (or can be synthesized) and the necessary coenzymes are available. If the carbon skeletons that might be used for production of nonessential amino acids are required as an energy source they may not be available for amino acid synthesis.

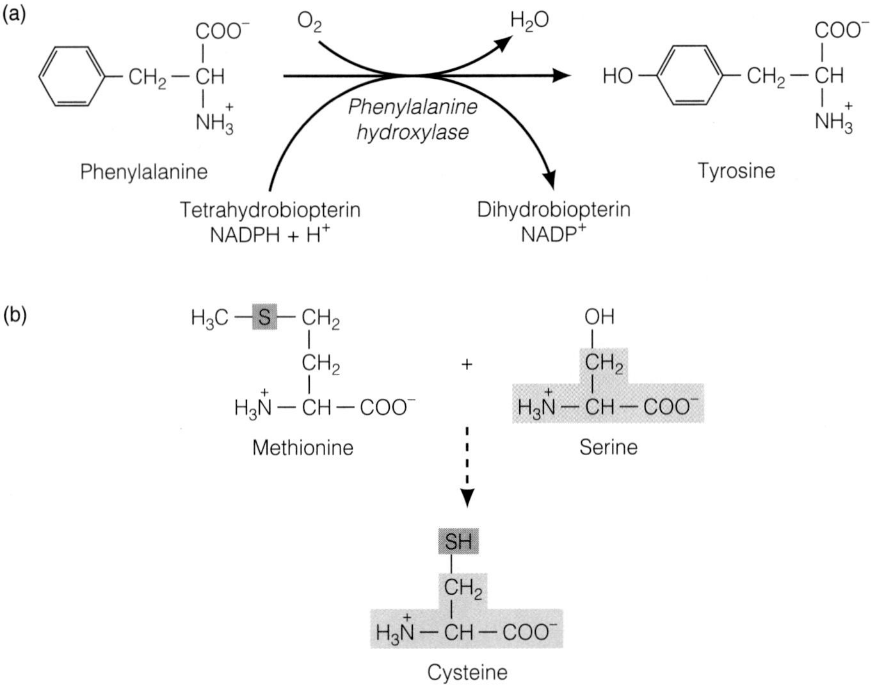

Figure 14.16 Examples of the synthesis of nonessential amino acids from essential amino acids. (a) Synthesis of tyrosine from phenylalanine. (b) Synthesis of cysteine from methionine and serine.

Healthy individuals are in nitrogen equilibrium – excreting nitrogen to balance the amount ingested

Proteins within cells are constantly being degraded and resynthesized at varying rates. This **protein turnover** reflects the continuous repair processes operating within cells. A person ingesting enough of the essential amino acids should be able to replace all the degraded proteins by resynthesis at the same rate at which they are broken down, thus keeping the amount of body protein constant. Most of the body nitrogen is in the form of protein, so any change in the total amount of nitrogen provides an indication of change in the amount of protein. A healthy adult will be in **nitrogen equilibrium**, with the amount of nitrogen ingested balanced by that excreted. The **nitrogen balance** is the difference between the amount of nitrogen excreted and the amount taken in.

How would we determine the nitrogen balance of an individual? This requires that the amount of nitrogen ingested and the amount of nitrogen excreted are both measured. Determining the amount of ingested nitrogen is much easier if the subject is fed a 'synthetic' diet – taking in amino acids rather than complex protein in the form of conventional food. Nitrogen excreted is measured in urine and faeces.

- A *positive* nitrogen balance means that less nitrogen is excreted than is taken in. This is normal in the young, in those undergoing physical training to increase muscle mass or in patients recovering from traumatic injury, surgery, or subject to lengthy periods of immobilization.

- If the nitrogen balance is *negative* more nitrogen is excreted than is taken in. This occurs in elderly people, in patients with degenerative diseases, in individuals on a protein intake that does not contain enough of all of the essential amino acids and during starvation.

Measurements of nitrogen balance provide a method for establishing whether a particular amino acid is essential. If one essential amino acid is omitted from the diet, protein synthesis will not proceed normally and a negative nitrogen balance will develop. If the omitted amino acid is nonessential, it will be synthesized from other molecules and protein synthesis will not be limited. In this case nitrogen equilibrium will be maintained.

Metabolic cooperation between tissues

The bloodstream provides a transport route through the body.

- Dietary nutrients absorbed from the gut are transported to tissues for metabolism or storage.
- Nutrients liberated from endogenous energy stores such as adipose tissue are transported to other tissues that need energy.
- Oxygen is transported by erythrocytes (see Chapter 12) from the lungs to peripheral tissues where it is required to sustain oxidative metabolism.
- The carbon dioxide produced by tissue metabolism is carried to the lungs.
- Hormones and other molecules travel from one tissue to another and influence the metabolic state of tissues to ensure their functional survival.
- Metabolites produced in peripheral tissues, which can safely be disposed of only after further metabolism or modification in the liver, must be transported from their sites of production.

Intertissue metabolic cooperation is mediated through these transport processes.

Chapter 12. Oxygen binding by haemoglobin is cooperative, page 371.

Tissues have different capacities to store energy

The capacity of different tissues to store energy is enormously variable – from the brain (which has essentially no energy storage capacity) to white adipose tissue which functions purely as a storage site for triacylglycerols.

Adipose tissue is the ultimate example of a specialized energy storage tissue. At times when nutrient intake exceeds energy requirements the excess is stored as triacylglycerols in adipocytes. When energy demand exceeds supply, these triacylglycerols are mobilized and the fatty acids are released into the bloodstream to be taken up by other tissues able to metabolize them as a source of energy.

Chapter 13. White adipocytes are sites for storage of triacylglycerol, page 381.

The liver is metabolically much more sophisticated. It has the capacity to store and metabolize both carbohydrates (glycogen) and fats (triacylglycerol). Most of the carbohydrate stored as glycogen in the liver is produced through gluconeogenesis rather than provided directly from dietary glucose. Even immediately after a meal, only a small amount of dietary glucose is taken up by the liver.

Chapter 13. Utilization of glucose in the liver, page 388.

Muscle cells also have the capacity to store both glycogen and small amounts of triacylglycerol. Most of the glycogen stored in muscle is derived directly from dietary glucose, and is eventually metabolized to provide energy within the muscle cells.

Chapter 13. Metabolism in white muscle cells is designed for short-term bursts of energy, page 386.

Tissues demonstrate a high degree of metabolic cooperation

The impact and importance of metabolic cooperation between tissues can not be overemphasized if we are to understand how metabolism is controlled and body nutrition coordinated. The key to this cooperation is the role of hormones and other signalling molecules, many of which are released in response to nutrient changes detected in the bloodstream by endocrine tissues.

Within a single cell, metabolic controls, mostly at the level of regulating enzyme activity, ensure that energy is provided in the amount needed. Within the whole body, hormonal controls operate to ensure that those tissues that are unable to store energy, such as the brain, are provided with nutrient by other tissues that are capable of exporting energy, notably the liver.

Liver is able not only to store and release energy but also to convert the metabolites released from muscle or adipose tissue into others useful to more glucose-dependent tissues such as the brain or erythrocytes. A metabolic imbalance in one tissue may prompt the release of nutrients useful and usable by another with a different metabolite balance or a different capacity to metabolize specific intermediates. A typical example is the production of ketone bodies in the liver and their export to brain and muscle.

Distribution of nutrients in the well fed state

The distribution of nutrients between tissues reflects the interval between meals. Immediately after a meal high concentrations of glucose, amino acids and triacylglycerols enter the bloodstream. In a well fed state most of the nutrient required is obtained directly from the diet.

In a well fed state energy is stored in adipose tissue

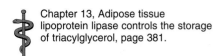

Chapter 13, Adipose tissue lipoprotein lipase controls the storage of triacylglycerol, page 381.

Most of the triacylglycerol absorbed from the digestive tract enters the bloodstream as chylomicrons (via the lymphatics) through the thoracic duct. This route avoids the liver and ensures that adipose tissue and muscle have first call on dietary triacylglycerols. Circulating triacylglycerols are hydrolysed by lipoprotein lipase

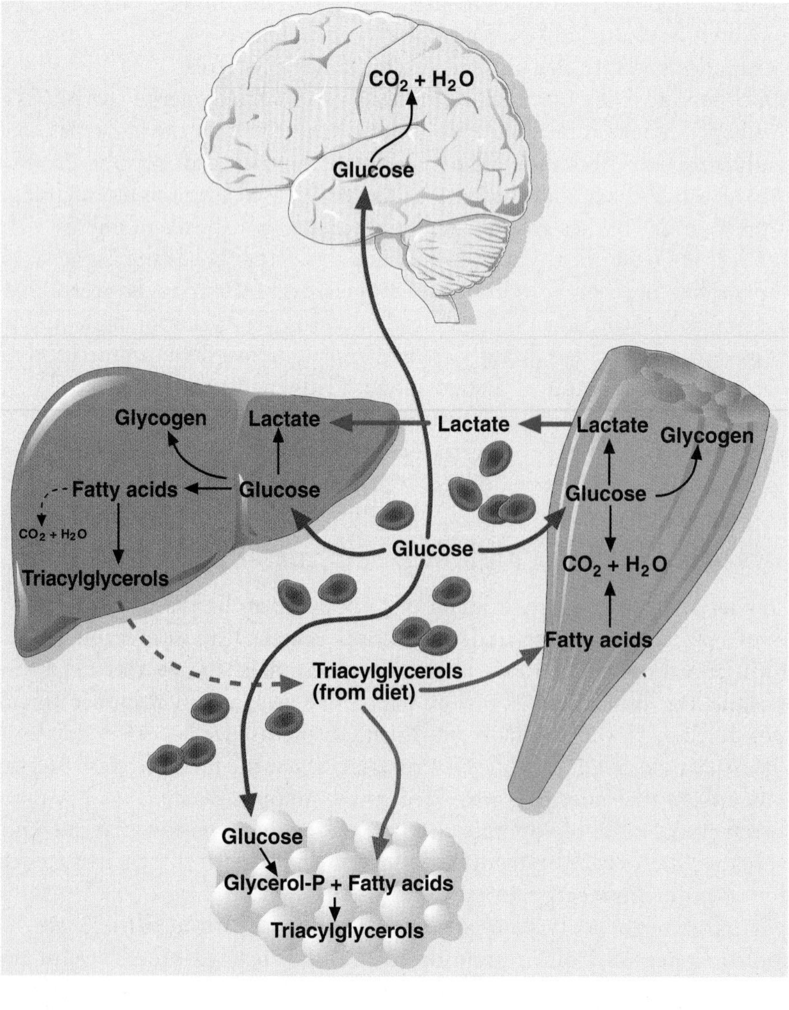

Figure 14.17 After a meal, blood glucose concentration rises to 8–9 mM within an hour. Glucose will be taken up by all tissues. Some will be converted to glycogen for storage in muscle and to some extent in the liver, and some will be oxidized as an immediate source of energy. Most of the circulating fatty acid is in the form of triacylglycerols, absorbed from the gut and en route to adipose tissue for storage. Some of the triacylglycerol is produced in the liver, which, recognizing the caloric excess, converts glucose to fat and exports it to adipose tissue. The liver meets its energy needs from oxidation of amino acids in excess of those required for protein synthesis and to some extent from the oxidation of glucose. Muscle oxidizes both glucose and fatty acids and the brain utilizes only glucose.

on the plasma membrane of capillary endothelial cells of adipose tissue. The fatty acids released are absorbed into adipose tissue and reassembled into triacylglycerols for storage.

In a well fed state the liver replenishes its own energy reserves

In a well fed state or a high-protein diet, liver cells preferentially metabolize the amino acids and keto acids produced by the breakdown of excess dietary protein for their energy needs. This ensures that most of the glucose entering the bloodstream from the digestive system is directed towards glucose-dependent tissues. Nevertheless, the liver continues to derive some energy from the oxidation of glucose. When the blood glucose concentration is high, the liver converts glucose to glycogen and metabolizes glucose to provide ATP and reduced nucleotides for its anabolic functions.

ATP is needed for general maintenance of cells and for a variety of processes including protein synthesis. Reduced nucleotides are required by the liver as coenzymes for fatty acid synthesis, cholesterol synthesis and gluconeogenesis.

Pyruvate and lactate returned to the liver from peripheral tissues are used for lipogenesis. The K_m values of the enzymes responsible for catabolism of amino acids are much greater than those of the aminoacyl tRNA synthetases involved in protein synthesis (Chapter 5) so the synthesis of plasma proteins takes precedence over catabolism unless amino acids accumulate in the liver at high concentrations.

Chapter 5. tRNAs are 'loaded' with amino acids by specific aminoacyl-tRNA synthetases, page 130.

In a well fed state muscle takes glucose and fatty acids from the bloodstream

Skeletal muscle at rest uses relatively little energy compared with its consumption during exercise. At rest, muscles require only enough energy to carry out processes such as maintaining ion gradients, replenishing nutrient stores and sustaining a relatively slow rate of protein synthesis. In a well nourished person, resting muscles take up glucose and fatty acids from the bloodstream. Most of the glucose is stored as glycogen, the fatty acids being used as the immediate source of energy. In a resting muscle, whether red or white, the oxygen supply and the catalytic capacity of mitochondria will support the complete oxidation of carbohydrate or fat within the muscle cell.

During sustained periods of contractile activity the pattern of metabolic activity changes, and differences in the metabolic capacity of red and white muscle fibres become apparent (Chapter 13).

Muscle uses its energy stores to sustain its own needs. The glucose released from glycogen and the fatty acids released from triacylglycerols are broken down to provide energy. Additional fatty acids may be provided from the circulation. Thus, within the normal range of feeding intervals, muscle cells lead a metabolically selfish existence, harvesting from the circulation sufficient energy for their own needs but contributing little to the metabolic needs of other tissues.

Chapter 13. Sustaining energy production in white muscle fibres depends on coupled dehydrogenase reactions, page 387.

White muscle cells store much more glycogen than red muscle cells. They have a much greater capacity for glycolytic metabolism but many fewer mitochondria. If the rate of production of pyruvate through glycolysis exceeds the capacity of the mitochondrial enzymes to convert pyruvate to acetyl CoA and to metabolize acetyl CoA through the citric acid cycle, lactate dehydrogenase will convert pyruvate to lactate.

Lactate is not wasted as an energy source, for the liver takes up lactate released from muscle and converts it to glucose. The glucose is then released into the bloodstream and is available to meet the continuing contractile energy demands of muscle (see Figure 14.21).

Distribution of nutrient between meals

The interval between meals can vary substantially. Metabolic cooperation between different tissues becomes increasingly important as the interval between meals increases. At all times keeping the level of glucose in the plasma high enough to sustain brain and other glucose-dependent tissues is the prime metabolic drive of the body. In a metabolic sense the needs of all other tissues are subordinate to those of the brain.

Between meals the metabolic activity of the liver is directed towards sustaining blood glucose concentration

Between meals, once glucose is no longer entering the bloodstream directly from the digestive system, blood glucose is maintained by the liver, initially by degrading endogenous glycogen stores. The hormonal responses triggered by changes in blood glucose concentration ensure that other metabolites become available before liver glycogen has been fully depleted, partially protecting the available carbohydrate for the carbohydrate-dependent tissues.

Glycogenolysis is regulated by the pancreatic hormones (see Chapter 13). As glucose is withdrawn from the bloodstream, the pancreas responds to the decreasing concentration by reducing the secretion of insulin so that less glucose is taken up by muscle and adipose tissue. At the same time the secretion of glucagon is

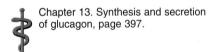
Chapter 13. Synthesis and secretion of glucagon, page 397.

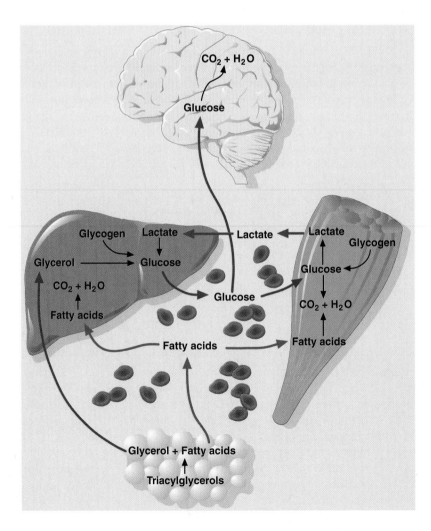

Figure 14.18 Between meals the circulating concentration of glucose would drop rapidly if it were not maintained by the liver. Glucose is provided initially from liver glycogen reserves, but also through gluconeogenesis (using glycerol from adipose tissue and lactate from muscle and erythrocytes). By several hours after a meal, the liver is deriving its own energy requirement from oxidation of fatty acids. Muscles are using their glycogen and fatty acids released from adipose tissue. The energy needs of the brain are still met entirely by glucose.

increased and glucagon binds to liver receptors, accelerating the degradation of glycogen to glucose. Glucagon also inhibits the synthesis of glycogen (glycogenesis) in the liver and stimulates gluconeogenesis. All of these processes contribute to an increase in the amount of glucose generated by liver cells. The total amount of free glucose in the body is only about 0.1–0.2 mol, but a normal human liver has the capacity to store about 0.5 mol as glycogen.

The gluconeogenic activity of the liver enables it to act as an efficient scavenger of metabolites released from other tissues

The major gluconeogenic substrates are

- Glycerol, released by the degradation of triacylglycerols in adipose tissue.
- Lactate and pyruvate, produced from glycolytic metabolism in peripheral tissues (muscle and erythrocytes).
- Glucogenic amino acids, or their carbon skeletons, released by the degradation of muscle proteins.

As liver glycogen is depleted, these metabolites are returned to the liver (Figure 14.19) and used as substrates for gluconeogenesis, rather than lipogenesis (which predominates in the well fed state).

Both the alanine cycle, by which alanine is returned to the liver from muscle (Figure 14.20), and the Cori cycle, by which lactate is returned to liver from muscle, assume greater importance between meals. These cycles ensure that carbon atoms released from peripheral tissues are converted in the liver to glucose, which is then returned to the peripheral tissues. There is no overall increase in the carbon pool from these processes. When new carbon sources are needed, amino acids (derived from protein) and glycerol (derived from adipose tissue triacylglycerol) meet most of this need.

As the interval between meals increases, controls within the liver ensure that glucose is exported rather than metabolized in the liver. However, the liver has a continuing energy demand to sustain its other functions. To meet these energy needs the liver oxidizes fatty acids, generating acetyl CoA and thereafter ketone bodies. Acetyl CoA in liver cells is maintained at a concentration high enough to regulate the three key enzymes at the interface between glycolysis, gluconeogenesis and the citric acid cycle. Acetyl CoA inhibits the **pyruvate dehydrogenase**

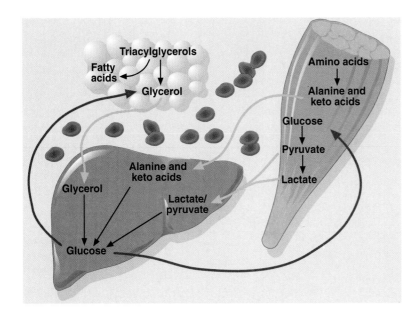

Figure 14.19 Glucogenic substrates returning to the liver from adipose tissue and muscle.

435

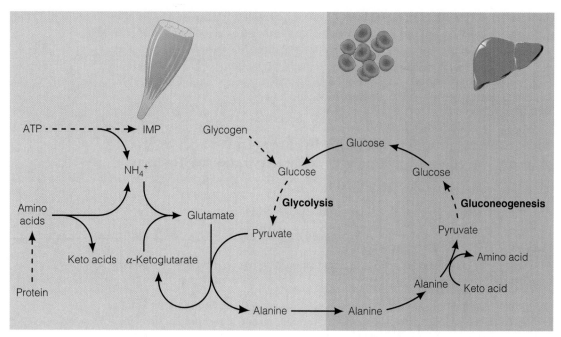

Figure 14.20 The alanine cycle depicts the interactions between liver and muscle cells which ensure that ammonia and pyruvate produced in muscle cells are transported to the liver as alanine. Alanine is used in the liver as a substrate for gluconeogenesis, producing glucose, which is then exported to glucose-dependent tissues.

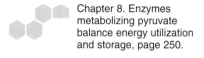

Chapter 8. Enzymes metabolizing pyruvate balance energy utilization and storage, page 250.

complex, which directs metabolites of glucose into the citric acid cycle, and activates **pyruvate carboxylase**, which diverts pyruvate into gluconeogenesis through oxaloacetate and phosphoenol pyruvate. Acetyl CoA also inhibits **pyruvate kinase**, thereby facilitating gluconeogenesis (see Figure 8.22).

A consequence of these regulatory mechanisms is that, by maintaining acetyl CoA production, the oxidation of fatty acids inhibits entry of pyruvate into the citric acid cycle and directs it into gluconeogenesis.

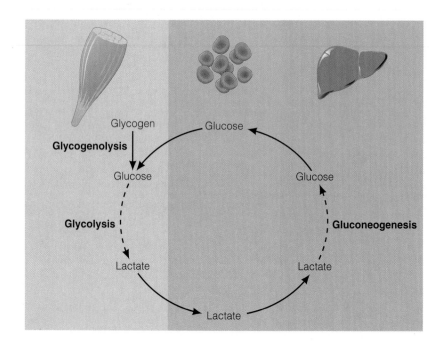

Figure 14.21 The Cori cycle depicts the process by which the liver conserves the carbon source of lactate through gluconeogenesis. The glucose produced is exported from the liver to glucose-dependent tissues.

Between meals adipose tissue provides fatty acids and glycerol to other tissues

In the early fasting state that develops once nutrients from a meal are no longer being absorbed into the bloodstream, adipose tissue provides fatty acids to other tissues. Reduced blood glucose concentration suppresses the secretion of insulin and thus relieves the inhibitory effect of insulin on net hydrolysis of triacylglycerol in adipose tissue. Hormone-sensitive lipase of adipose tissue cells hydrolyses triacylglycerols, releasing both fatty acids and glycerol into the bloodstream. Fatty acids are metabolized by most tissues other than the brain and red blood cells. By 8–10 hours after a meal about half of the energy requirement of muscle tissue is being met by the oxidation of fatty acids. In muscle and heart the oxidation of fatty acids inhibits glycolysis, reducing the demand these tissues make on glucose from the bloodstream. The glycerol released during the hydrolysis of triacylglycerols is taken up by the liver, where it is used as a substrate for gluconeogenesis.

Distribution of nutrients in starvation

There is no absolute definition of starvation; a spectrum of tissue-specific metabolic change develops as the interval between meals increases. The time course of the development of starvation reflects the amount of energy previously stored by the individual.

The glucose/fatty acid cycle allows glucose to be conserved at the expense of fatty acids

Independent of hormone-mediated responses, the uptake and metabolism of fatty acids and glucose by tissues such as muscle are interdependent. This relationship ensures that blood glucose is maintained, so that it can be directed towards glucose-dependent tissues, while other tissues use fatty acids. If plasma fatty acid concentrations rise, glucose use in muscle falls. Normal variations in blood glucose are small compared with normal variations in the plasma concentration of fatty acids, which may be as great as five-fold under normal dietary circumstances. The effects of the glucose/fatty acid cycle are most prominent in a nutrient-depleted state or in starvation when the mobilization of fatty acids from adipose tissue leads to inhibition of glucose metabolism in muscle.

When the blood glucose concentration falls, fatty acids are mobilized from adipose tissue and are taken up and oxidized by muscle. This reduces the demand on blood glucose, the concentration of which will be maintained by glycogenolysis and gluconeogenesis in the liver, so glucose is taken up into adipose tissue and fatty acid synthesis increases. There is a reciprocal inhibition of fatty acid mobilization in adipose tissue, with a resultant reduction in plasma fatty acid concentration. The amount of fatty acid oxidized by muscle cells drops and the amount of glucose metabolized increases.

These changes are mediated by appropriate hormonal responses.

In starvation ketone bodies are a major nutrient

Adults are able to survive for several days without food before ketogenesis becomes a dominant metabolic pattern. Although glycogen stores are effectively depleted within hours, the liver uses gluconeogenic substrates to maintain blood glucose and hormonal changes ensure a shift away from carbohydrate metabolism towards mobilization of triacylglycerols and oxidation of fatty acids. Only after several days of fasting does ketogenesis become noticeable.

 Chapter 9. In the liver accumulating acetyl CoA is converted to ketone bodies, page 271.

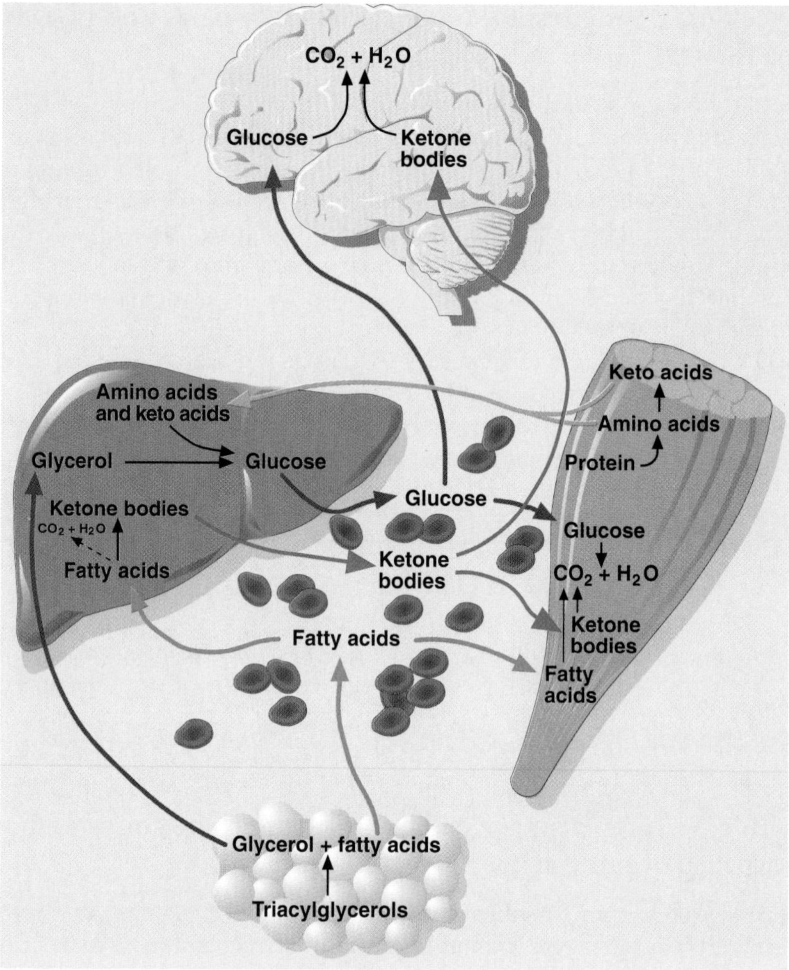

Figure 14.22 During starvation, most of the circulating fat is free fatty acids, released from adipose tissue. Gluconeogenic substrates are glycerol from adipose tissue and amino and keto acids from muscle protein. After several days the circulating concentration of ketone bodies increases, reflecting the imbalance in the liver between carbohydrate and fatty acids. Muscle oxidizes both ketone bodies and fatty acids and exports to the liver glycogenic substrates derived from the degradation of the structural proteins. As the blood glucose concentration drops to the hypoglycaemic level of about 2 mM, the brain adapts to utilizing increasing amounts of ketone bodies, while also consuming almost all the glucose that the liver can produce.

Chapter 9. In the liver accumulating acetyl CoA is converted to ketone bodies, page 271.

Infants respond differently, developing ketosis after much shorter intervals without food. This difference appears to reflect the higher basal metabolic rate of young children and the importance of mechanisms designed to sustain growth in immature animals. The increase in tissue mass during growth and development requires that protein synthesis is maintained. Through mechanisms which are not yet well defined, muscle tissue in young animals is less likely to be mobilized to support gluconeogenesis than in adults.

When blood glucose is low, the liver retains very little of the glucose derived from the diet or from gluconeogenesis. Acetyl CoA is produced from the oxidation of fatty acids within the liver more rapidly than it can be incorporated into the citric acid cycle, as reflected in the production and release from the liver of ketone bodies. The liver is the only tissue able to produce significant amounts of ketone bodies (Chapter 9). The importance to the liver of ketone body production under these conditions is that coenzyme A is prevented from accumulating as acetyl CoA, freeing it for participation in other reactions. The advantage to other tissues is that ketone bodies are water soluble and can be distributed in the bloodstream.

The brain is the major user of ketone bodies in starvation

If blood glucose is low but ketone bodies are available, the brain may derive up to 50–60% of its energy requirement from the oxidation of ketone bodies because

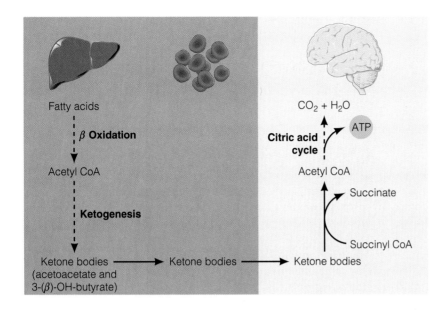

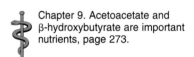

Figure 14.23 Ketone bodies synthesized and exported from the liver are used in other tissues, for example the brain, where they provide a valuable source of energy.

brain cells continue to be provided with some glucose by the liver and continue to produce oxaloacetate. Acetoacetate and β-hydroxybutyrate entering brain cells can be converted to acetyl CoA, which then enters the citric acid cycle. Thus ketone bodies are produced in the liver to be transported to other tissues such as the brain, which have the capacity to use them as an energy source (Figure 14.23). The brain may consume as much as 80% of the ketone bodies released from the liver, the oxidation of ketone bodies reflecting the availability of oxaloacetate produced from glucose. Oxidation of ketone bodies by the brain also has the advantage of conserving body protein, for less protein is required for gluconeogenesis as the brain adapts to derive a large fraction of its energy from ketone bodies.

Cardiac muscle also has a substantial capacity to oxidize ketone bodies.

Chapter 9. Acetoacetate and β-hydroxybutyrate are important nutrients, page 273.

In starvation muscles use fatty acids and degrade their structural proteins

After prolonged periods without food, muscle glycogen will have been used up and the nutrient available to muscle cells are the fatty acids and ketone bodies from adipose tissue and the liver. In addition, the demand for glucogenic substrate within muscle cells and other body tissues leads to use of the amino acids released from degradation of muscle proteins.

Maintenance of blood glucose is essential to the function of the brain

In starvation brain cells use about 70% of the circulating glucose to produce ATP to maintain ion gradients essential for transmission of nerve impulses and for the synthesis of neurotransmitters. Brain cells have little capacity to store any nutrients.

If an individual becomes hypoglycaemic the rates at which glucose crosses the blood–brain barrier and enters the brain cells are both reduced. The rate at which ATP is produced within brain cells decreases, impairing their electrical activity and eventually leading to coma.

Degradation of cellular macromolecules is part of a process of repair and maintenance

Maintenance of tissue structures depends on continuous degradation and resynthesis of RNA and proteins. The degradation of macromolecules is carried out by enzymes in the cytosol and in lysosomes. Uncontrolled activity of degradative processes would damage molecules in a manner harmful to the survival of the cell, so many 'scavenging' enzymes are stored in membrane-bound structures – the lysosomes. Lysosomal enzymes operate at maximal rates at about pH 5, which is maintained by ATP-dependent hydrogen ion pumps in the membranes of these organelles.

We shall concentrate on the processes by which cell proteins are degraded, but it is important to remember that lysosomes also contain a battery of enzymes capable of degrading nucleic acids, polysaccharides and complex lipids. Quite large fragments of other subcellular organelles have been seen within lysosomes.

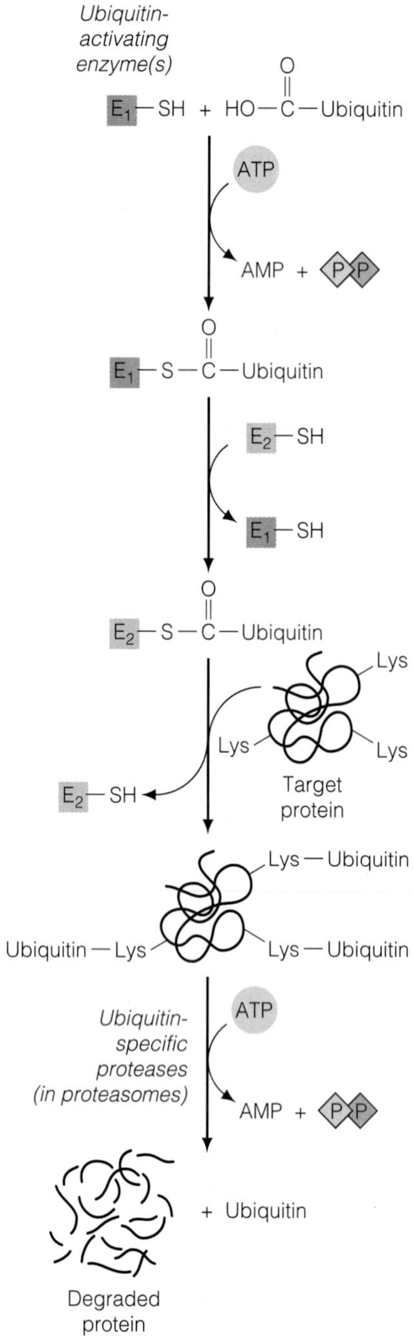

Protein turnover is the process by which the structure and function of tissue is maintained

Protein turnover is a normal process by which protein is degraded, releasing amino acids, and new protein molecules are synthesized. In a normal cell under conditions of nutrient balance the rate of degradation of each protein should be exactly balanced by the rate at which new molecules are synthesized. When there is a demand for amino acids as gluconeogenic substrates, protein degradation will outstrip synthesis and the amount of tissue protein will progressively decline.

The normal rate of turnover varies enormously between different proteins. It is expressed as the half-life of the protein: the time over which half of the molecules of that protein will be degraded (and replaced in a healthy mature tissue). Plasma proteins and proteins of the liver are among those with the shortest half-lives. Very long half-lives, of the order of 10–30 days, are exhibited by structural proteins of muscle. Connective tissue proteins, such as collagen, are very stable – the half-life of collagen is measured in weeks. This is perhaps unsurprising because mature collagen is located in the extracellular space of connective tissue.

Some proteins are degraded within the lysosomes after protein is engulfed into lysosomes. If lysosomal membranes are disrupted enzymes will be released into the cytoplasm, causing excessive destruction of cellular molecules. However, a very similar process determines the natural destruction of cells as part of the process of programmed cell death (apoptosis; see Chapter 17). The death of specific cell types is a vital part of developmental processes – some cells have functions that are relevant only during very early stages of development.

Other proteins are degraded by proteases that are normally found in the cytoplasm. These proteases are usually inactive and require specific signals to trigger them. The best understood of these signals is that generated when the 76-residue peptide **ubiquitin** is covalently attached to the protein to be digested (Figure 14.24). Ubiquitin is activated when it is coupled through a thiol group to a specific ubiquitin-activating enzyme. This enzyme transfers ubiquitin to a second enzyme, which appears to recognize proteins destined for proteolysis and transfers ubiquitin to ε-NH_2 groups of lysine residues in these proteins. Specific cytoplasmic proteases, in particles called **proteasomes**, recognize the attachment of ubiquitin as a signal to digest a protein.

Figure 14.24 The enzymes labelled E1 and E2 are involved in transferring ubiquitin to proteins targeted for programmed destruction.

If other nutrients are in short supply, resynthesis of protein is inhibited and amino acids are metabolized

If the total nutrient intake is insufficient, amino acids are mobilized from body protein and their carbon skeletons used as a source of nutrient. Several factors determine the proteins that will be used to provide energy and precursors for synthesis of other molecules if dietary nutrient is inadequate.

- The body appears to have mechanisms, not yet fully understood, by which it distinguishes **dispensable proteins** from **indispensable proteins**. The resynthesis of indispensable proteins takes priority over the resynthesis of dispensable proteins. In general those proteins that control physiological function are preserved at the expense of structural proteins.

- Proteins with the fastest normal turnover are most vulnerable because they are most likely to be used as an energy source during starvation. This means that plasma proteins, particularly albumin, and the proteins of tissues such as the liver and pancreas provide energy at an earlier stage of starvation than the more slowly metabolized structural proteins of muscle.

These two factors do not always operate to the same end. For example haemoglobin, although clearly an important (indispensable) protein, is one of the most vulnerable because it has a rapid turnover rate, approximately 1% of the total body haemoglobin being degraded each day as the normal life of red blood cells (120 days) expires. A lack of amino acids may reduce the rate of resynthesis of haemoglobin, resulting in a reduced capacity of erythrocytes to carry oxygen to the peripheral tissues and a reduced rate of oxidative metabolism.

Chapter 12. There is a mechanism for the degradation of haemoglobin after erythrocyte haemolysis, page 374.

The nonmuscle proteins represent a relatively small fraction of the protein mass of the body so that, despite the greater vulnerability of plasma proteins, loss of muscle will occur relatively early in starvation. Dependence on protein as an energy supply will be manifest in degradation of muscle proteins, with loss of muscle mass, loss of muscle power and general emaciation.

In starvation muscle protein is degraded in a controlled manner to support the nutritional needs of other tissues

A complex mechanism operates within muscle cells, the purpose of which appears to be to ensure efficient use of the amino acids derived from muscle protein breakdown. The amino acids produced by proteolysis in muscle are not released directly into the circulation (this is apparent from the observation that alanine makes up about 30% of the amino acid released, yet comprises less than 10% of muscle protein). Branched-chain amino acids (leucine, isoleucine and valine) released from proteins appear to be used as substrates for aminotransferase reactions in which pyruvate is converted to alanine, with the production of branched-chain keto acids. The alanine is transported to the liver, where it is converted to pyruvate and then to glucose through gluconeogenesis (Figure 14.25).

This mechanism offers an economy in the liver: in an environment in which glycogenic intermediates are needed the liver does not have to metabolize nonglycogenic amino acids and does not need to sustain a full battery of enzymes capable of converting all of the glycogenic amino acids to substrates for gluconeogenesis. In muscle cells the branched-chain keto acids formed may be metabolized to provide energy within muscle cells, although most are transported to the liver.

This is a striking example of intertissue cooperation for, whereas muscle cells have a high capacity to transaminate branched-chain amino acids, they have a much lower capacity to oxidize the corresponding keto acids. By contrast, the liver has low levels of the branch-chain aminotransferase enzymes but oxidizes the keto acids efficiently.

Chapter 10. Ammonia is transported from skeletal muscle to the liver as alanine, page 299.

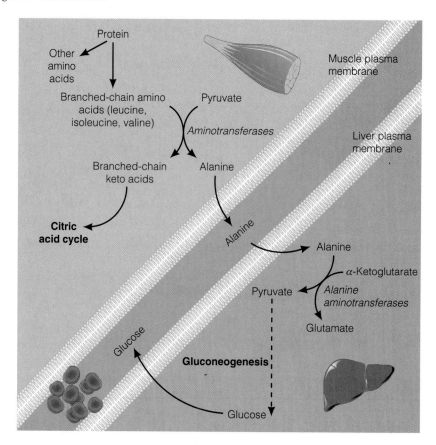

Figure 14.25 In starvation, muscle protein provides an energy source. The aminotransferase reactions and distribution of carbon skeletons between muscle and liver cells are designed to ensure that this breakdown of body tissue is carefully regulated to ensure efficient utilization of the energy available.

Survival clearly depends on maintaining the structure of the body, so there is a limit to the amount of protein that can be degraded without causing a life-threatening situation. Protective mechanisms are seen in the progressive reduction in nitrogen excretion as the period of food deprivation increases. Most tissues then derive energy from the oxidation of fatty acids or ketone bodies in an attempt to minimize the degradation of muscle protein.

Obesity

If the daily caloric intake is in perfect balance with the energy requirements, mature healthy individuals maintain a constant body weight. If the number of calories stored immediately after a meal is persistently greater than that needed to sustain energy requirements between meals, weight gain and eventual **obesity** will occur. Both caloric intake and energy expenditure contribute to the overall balance, and a change in either will alter the amount of fat stored.

Assessment of obesity and the development of sensible weight-loss regimes are complicated by the fact that a large fraction (40–70%) of body weight is water. Reversal of obesity requires a reduction in body fat, but some weight loss regimes achieve substantial reduction in the amount of body water with relatively little loss of body fat. Low carbohydrate diets may produce more dramatic weight loss but, because the glycogen lost in these regimes is hydrophilic, a substantial fraction of the weight lost may be body water. One gram of glycogen binds approximately 2 g water, so a substantial weight loss may involve very little loss of stored energy. A low-carbohydrate diet that is ketogenic may bring other metabolic disadvantages, notably the mobilization of body protein to produce amino acids for gluconeogenesis. A diet balanced with respect to intake of carbohydrate, fat and protein, but with a reduced overall caloric value, is more effective in reducing obesity than an unbalanced diet.

Many factors contribute to the wide variation in the extent to which fat is stored by individuals on identical diets. Caloric intake is not the only factor, for most individuals increase the amount of fat they store on a high-fat diet compared with a low-fat diet even if the total caloric value of the diets is identical. A great deal remains to be clarified in this area, and only some of the better understood aspects will be considered here.

Chapter 13, White adipocytes are sites for storage of triacylglycerol, page 381.

The capacity of the body to store fat depends on the number of adipocytes present, and it has been suggested that infants consistently fed a caloric excess will develop more adipocytes, thus increasing the ease with which caloric excess may be stored as fat throughout the life-time. In addition, genetic factors may contribute to the number of adipocytes.

Obese persons often have an increased circulating concentration of insulin. It appears that persistent excessive carbohydrate intake leads to overproduction of insulin and both hypertrophy (increase in the size of cells) and hyperplasia (increase in the number of cells) of the endocrine cells of the pancreas, which exacerbates the overproduction of insulin. Insulin stimulates the conversion of carbohydrate to fatty acids in the liver and the storage of triacylglycerol in adipose tissue. In the long term, hyperinsulinaemia causes a reduction in the number of insulin receptors so that tissues become less responsive to insulin. The inhibitory effect of insulin on lipolysis is relieved, and the circulating concentration of free fatty acids increases.

Chapter 13, Synthesis and secretion of insulin, page 398.

Thermogenesis

Thermogenesis, the process by which body temperature is maintained, requires energy. Both excess dietary intake and exercise may stimulate heat production. Small variations in the capacity to respond to thermogenic stimuli could substantially alter the amount of energy available for storage. In most animals thermogenesis is the responsibility of brown adipose tissue but, in humans, muscle also has a role. In thermogenic cells oxidative phosphorylation is partially uncoupled so that the energy is released as heat. The mechanism is described in Chapter 13.

It is an interesting anomaly that lean rats are better able than obese rats to survive low temperatures. Whether lean individuals have more brown adipose tissue than obese individuals, or whether other factors determine the greater efficiency with which body temperature is maintained in lean individuals, remains to be clarified. Whatever the reason, thermogenesis is more efficient in lean than in obese rats, and there is some evidence that the same applies to humans.

Chapter 13. Thermogenesis is stimulated by noradrenaline, page 383.

The role of vitamins in metabolism

Vitamins are small organic molecules, many of which are precursors of coenzymes (Figure 14.27). In most cases the dietary vitamin must be chemically modified within the body to the coenzyme form that participates in metabolic reactions. The modification is sometimes as simple as the addition of a phosphate group, but may be a great deal more complex.

Vitamins are essential in our diet because we lack the enzymes required for their synthesis. A diet containing adequate amino acids and fatty acids will not support

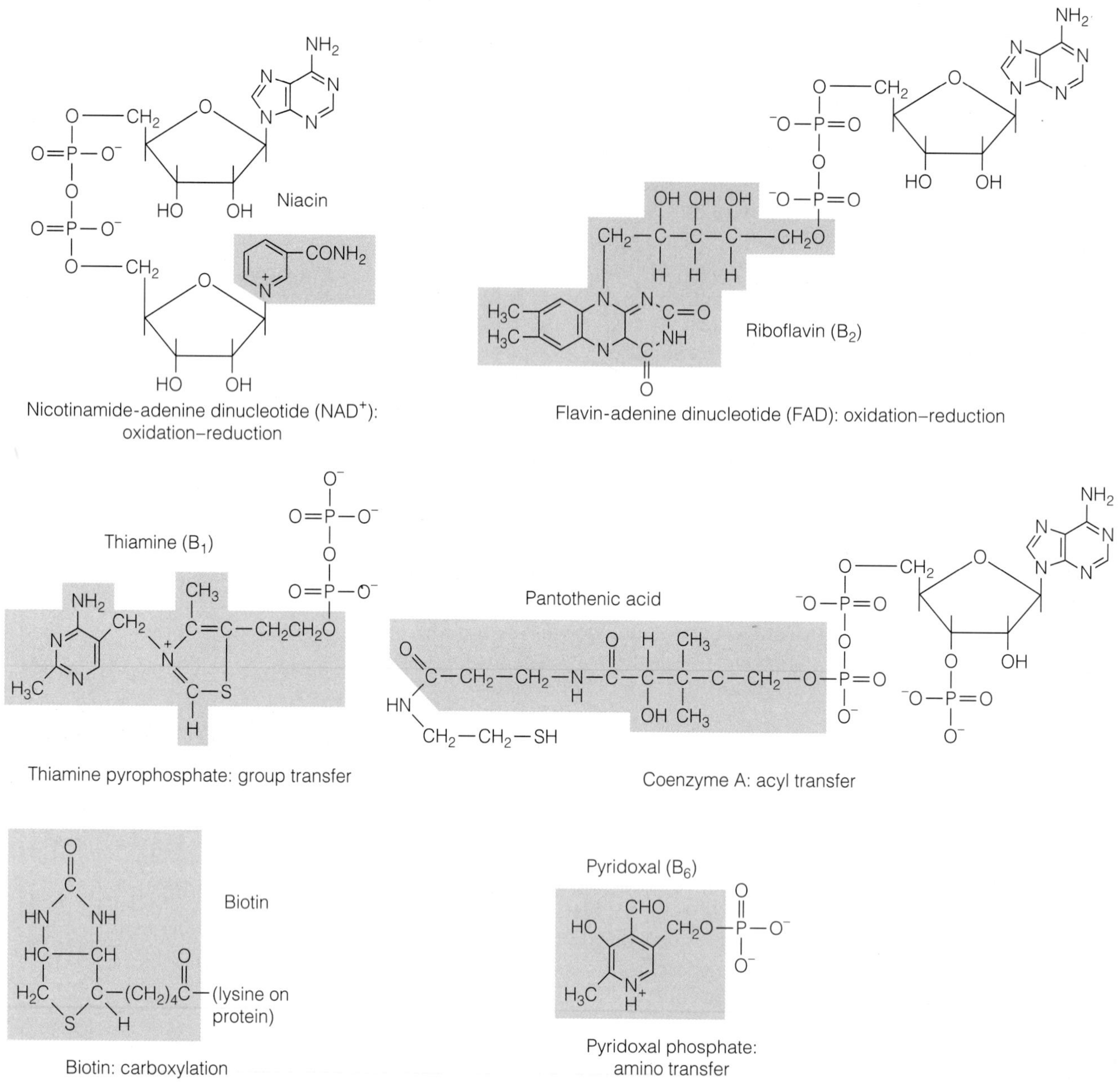

Figure 14.26 Structures of some coenzymes and their relationship to vitamin structure. The dietary vitamin component of the co-enzyme is shown on a blue background.

growth or maintain health unless it also contains appropriate quantities of all the vitamins. Vitamin deficiency in the young stunts growth, and at any age causes characteristic 'deficiency' diseases and consequent metabolic disorders. Many of the disease manifestations of vitamin deficiency are rapidly reversed by increasing the vitamin content of the diet. Vitamins are required as a daily component of the diet but, with the exception of vitamin C, only in very small quantities.

The primary role of most vitamins or their derivatives is as coenzymes or prosthetic groups in metabolic reactions, although some appear to have a broader regulatory role. Vitamins are classified into two groups: the **fat-soluble** and the **water-soluble** vitamins. The latter tend to be precursors of coenzymes, the former tend to exhibit broader regulatory functions. The body storage capacity for vitamins varies, but the fat-soluble vitamins are stored more efficiently – by being retained within body fat deposits – than the water-soluble vitamins, which are readily lost in the urine.

Fat-soluble vitamins

There are four fat-soluble vitamins: **A, D, E** and **K**. Vitamins A and D are stored very efficiently in the liver, and produce toxic effects if present in excessive amounts. Fat-soluble vitamins are absorbed with bile through the upper small intestine in association with dietary fat.

The major role of vitamin A is in vision

Vitamin A, **retinol**, can be stored in high concentrations in the liver. β-Carotene is a related compound, essentially two molecules of retinol condensed tail-to-tail, which is found in plants and is responsible for the red colour of carrots. Retinol is essential for detection of light by the eye, and its role was first recognized when it was discovered to be effective against night-blindness, the loss of visual acuity in dim light. The first evidence of its toxicity when taken in excessive amounts came from observations of Eskimos consuming large amounts of polar bear liver, the oil of which is particularly rich in vitamin A.

The conversion of β-carotene to retinol, which occurs in the mucosal cells of the intestine, is inefficient, so that β-carotene must be consumed in much greater amounts than the molar equivalent as retinol (Figure 14.27). Retinol is stored in the liver and transported to other tissues bound to **retinol binding protein** in the plasma. The toxicity of vitamin A appears to result from its ability to destabilize lysosomal membranes, in which it is soluble, leading to the release of proteolytic enzymes and generalized cell damage, particularly in the liver.

In addition to its role in vision, which is discussed in Chapter 13, retinol and related derivatives, including β-carotene, contribute to a variety of other functions:

1. Maintenance of epithelial tissue through
   - synthesis of membrane glycoproteins and mucopolysaccharides;
   - inhibition of the synthesis of high-molecular weight forms of keratin which lead to abnormal keratinization of epithelial surfaces and the cornea of the eye.

Chapter 13. Transmission of light impulses by rod cells in the eye, page 406.

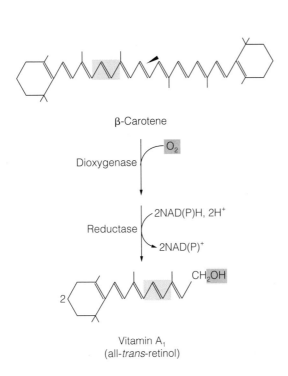

β-Carotene

Dioxygenase — O_2

2NAD(P)H, 2H⁺

Reductase

2NAD(P)⁺

CH_2OH

2

Vitamin A₁
(all-*trans*-retinol)

Figure 14.27 Conversion of β-carotene to vitamin A. The position of the arrow indicates the bond which is cleaved.

2. Regulation of cell growth and differentiation through binding to specific receptor proteins that bind to DNA and influence gene expression.
3. Synthesis of the iron transport protein, transferrin.
4. Antioxidant properties, important in protecting tissues against free radical damage.

Vitamin D has a key role in calcium regulation

Chapter 11. Parathyroid hormone, page 346.

Vitamin D is classified as a vitamin, but there is no necessity for it to be provided in the diet unless there is little exposure to daylight. Dairy products are a common dietary source of vitamin D. Vitamin D functions in close coordination with the peptide hormones **calcitonin** and **parathyroid hormone**, which are dealt with in Chapter 11. It has a major role in controlling calcium homeostasis in the body, increasing the serum calcium concentration by increasing

- intestinal uptake of calcium and phosphate;
- renal reabsorption of calcium and phosphate;
- mobilization of calcium and phosphate from bone.

Vitamin D deficiency is seen in the development of rickets in children and osteomalacia in adults; both are disorders of bone mineralization.

Synthesis of the active vitamin D derivative (calcitriol)

Vitamin D is a derivative of cholesterol. The active form is synthesized through a sequence of reactions occurring in different tissues. Provitamin D, 7–dehydrocholesterol, arises in the skin from the oxidation of cholesterol. It is subsequently converted to cholecalciferol by the action of ultraviolet light (Figure 14.28). The reaction involves opening the B ring of the provitamin.

Alternatively, vitamin D can be absorbed from the diet. In areas of low sunlight this source assumes greater importance.

Cholecalciferol is transported first to the liver, where a mitochondrial hydroxylase introduces an -OH group at position 25. The reaction requires molecular oxygen and NADPH and the product may be stored in the liver if there is already sufficient of the active form of vitamin D in the body. When more is required the 25-OH compound is transported to the kidney where it induces the synthesis of the hydroxylase which introduces another -OH group at position 1, converting it to 1,25-dihydroxycholecalciferol (calcitriol) (Figure 14.29).

Evidence for induction of this enzyme is provided by the demonstration that the addition of the –OH group at position 1 does not occur in the presence of actinomycin D or cycloheximide – inhibitors of protein synthesis. This 1-hydroxylase activity, which also requires molecular oxygen and NADPH, is controlled by parathyroid hormone, by calcium and phosphate concentrations in plasma and by negative feedback from the active vitamin.

The biologically active form of vitamin D is then transported to the mucosal cells of the intestine, its major site of action, where it stimulates calcium absorption. These cells also contain a cytoplasmic calcium-binding protein, synthesis of

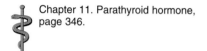

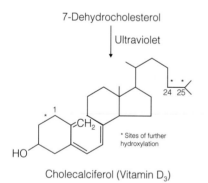

7-Dehydrocholesterol

Ultraviolet

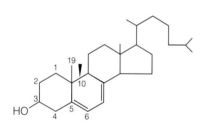

Cholecalciferol (Vitamin D₃)

Figure 14.28 Formation of cholecalciferol.

Figure 14.29 Structure of 1,25-dihydroxycholecalciferol.

which is induced by calcitriol and which probably acts in calcium storage rather than calcium transport. This protein contains a large number of amino acids with carboxyl side-chains, allowing it to bind calcium ions electrostatically and perhaps also preventing the uptake of calcium into the mitochondria. The same protein has been found in kidney tubule cells, where it may be responsible for limiting the urinary losses of calcium.

Vitamin E protects tissues through its antioxidant properties

Vitamin E is a name for a group of compounds also called **tocopherols** (α, β, γ and δ), of which α is the most potent. These compounds are found in plant oils and are highly lipophilic so that they accumulate in fat deposits, cell membranes and circulating lipoproteins.

The major role of tocopherols is as antioxidants. The methylene group between two double bonds of a polyunsaturated fatty acid is susceptible to hydrogen abstraction, forming a free radical. Polyunsaturated fatty acids incorporated into complex lipids of membrane structures or fat stores of both living cells and foods are susceptible to these reactions. Peroxidation reactions are initiated by radicals such as OH• (the hydroxyl radical) and can generate chain reactions in the presence of oxygen, producing large amounts of hydroperoxides. The reactions are not enzymic, but are initiated by ionizing and ultraviolet radiations or catalysed by metal ions such as iron or copper or by haemoglobin. Formation of hydroperoxides results in a loss of unsaturated fatty acids as well as production of powerful oxidizing agents which react with, and damage, DNA, proteins and membranes. Hydroperoxides are unstable and break down to produce dialdehydes, the most common of which is malondialdehyde, and ketones.

In foods, peroxidation can lead to depletion of essential fatty acids, reducing the nutrient quality.

In cells, peroxidation of membrane lipids eventually results in a loss of the structural integrity of the membrane. The membranes of erythrocytes, lysosomes and the endoplasmic reticulum are particularly susceptible to peroxidation damage.

Malondialdehyde can form cross-links between various molecules, which may cause

- cytotoxicity;
- mutagenesis;
- membrane disruption;
- modification of enzymes.

If these complexes of dialdehydes with lipids and proteins persist in tissues they polymerize, forming **lipofuscin**, an insoluble pigment which accumulates in ageing tissue.

Vitamin E protects against peroxidation, blocking the autooxidation by interacting with the free radical to terminate the chain reaction. Some membranes have large amounts of vitamin E embedded in their structure, providing immediate protection against the generation of these chain reactions.

Vitamin K makes a major contribution to the blood-clotting process

Vitamin K is a quinone derivative found in green vegetables; it is also synthesized by intestinal bacteria. Deficiency of vitamin K is seldom seen except in patients with severe fat malabsorption problems or on antibiotic therapies that deplete the normal intestinal flora. Vitamin K is essential for the carboxylation of specific glutamate residues of prothrombin which are converted to γ-carboxyglutamic acids as a

> **Box 14.5 Absorption and transport of vitamin E depend on circulating lipoproteins**
>
> A deficiency of vitamin E occurs in individuals with a genetic deficiency of apolipoprotein B, a condition called **abetalipoproteinaemia**. Absorption and transport of vitamin E are compromised because chylomicrons and VLDL are not present in the bloodstream at normal concentrations. The condition leads to the development of retinopathies and neuropathies, diseases of the retina and nervous system, many of the symptoms of which can be alleviated by the administration of α-tocopherol.

Chapter 12. Blood clotting, page 360.

postsynthetic modification, essential for its function in blood clotting (Chapter 12). These carboxylated residues chelate calcium ions as part of the initiation of blood coagulation. Carboxylated glutamic acid residues also occur in many other tissues and the role of vitamin K may be more widespread than is currently recognized.

Water-soluble vitamins

The water-soluble vitamins are mainly required in very small amounts, and deficiencies are rarely seen in modern Western diets except among alcoholics. Alcoholics seldom consume a balanced diet, and excessive intake of alcohol produces changes in the gastrointestinal tract that may impair absorption of vitamins from the diet. It also reduces the capacity of the liver to convert vitamins to their corresponding active coenzymes (see Figure 14.26).

Thiamine: vitamin B_1

Deficiency of thiamine was first recognized as the disease **beriberi** among people consuming refined rice. Beriberi is characterized by neuromuscular weakness and cardiac failure. When rice husks were included in the diet, the disease regressed. Meat is also a good source of vitamin B_1. The active form of the vitamin is **thiamine pyrophosphate**, the addition of pyrophosphate being catalysed by a liver enzyme. In alcoholics the capacity of the liver to catalyse this conversion may be compromised to the extent that the individual appears to be thiamine deficient. Thiamine pyrophosphate is one of the coenzymes for the reactions catalysed by the pyruvate dehydrogenase and α-ketoglutarate dehydrogenase complexes, and the transketolase enzymes of the pentose phosphate pathway. Deficiency of this vitamin therefore leads to a failure of energy generation. In addition vitamin B_1 has a role in the synthesis of acetylcholine in nervous tissue.

Riboflavin: vitamin B_2

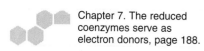

Chapter 7. The reduced coenzymes serve as electron donors, page 188.

Riboflavin is a component of the flavin coenzymes FMN and FAD, which are tightly bound to flavin-dependent dehydrogenase enzymes such as succinate dehydrogenase. These coenzymes act as carriers of hydrogen atoms in metabolic oxidation–reduction reactions (see Chapter 7). Dairy and cereal products are rich in riboflavin.

Nicotinic acid (niacin)

Nicotinic acid is classed as a vitamin, although humans do possess the enzymes to synthesize it, albeit in insufficient amounts, from tryptophan. It is present in abundance in meat, milk and eggs. This vitamin forms part of the oxidation–reduction coenzymes NAD^+ and $NADP^+$, which are carriers of hydrogen atoms. Deficiency leads to the development of **pellagra**, characterized by neurological degeneration.

Pyridoxine: vitamin B_6

Pyridoxine is found in meat, eggs, vegetables and cereals. This vitamin, which exists also as pyridoxamine and pyridoxal, is converted to the coenzyme form by addition of a phosphate group. This coenzyme is associated with enzymes involved in aminotransferase reactions, amino acid decarboxylations and other amino acid modifying reactions including the synthesis of the physiologically important compounds serotonin, noradrenaline, histamine and haem groups. The requirement for pyridoxine depends to some extent on the protein content of the diet. It is also required for the degradation of glycogen, as the coenzyme for glycogen phosphorylase.

Pyridoxal phosphate-dependent reactions involve the formation of a Schiff base between an amino acid and the aldehyde group of the coenzyme. In the absence of substrate the coenzyme is covalently bound to the enzymes. As the coenzyme is tightly bound, nutritional deficiencies are rarely observed. However, some of the early drugs used to treat tuberculosis had toxic side-effects that were traced to the fact that they reacted with the aldehyde form of pyridoxal phosphate, inhibiting its interaction with these enzymes and giving rise to symptoms similar to those of vitamin B_6 deficiency.

Chapter 10. Pyridoxal phosphate is the essential coenzyme for aminotransferases, page 296.

Pantothenic acid

Pantothenic acid is a component of coenzyme A, which is a carrier of acyl groups in metabolism. It is required for the metabolism of fat, carbohydrate and protein. This vitamin occurs widely in food and deficiencies are rare.

Biotin

Biotin is found covalently attached to the ε-amino side chain of lysine residues at the active site of enzymes catalysing carboxylation reactions. It is a carrier of carboxyl groups in reactions generally also requiring ATP, for example pyruvate carboxylase, a key enzyme in gluconeogenesis. Biotin is synthesized by intestinal bacteria, but is also found in high quantities in eggs where it is bound to a protein called **avidin**. Cooking eggs releases biotin from avidin, but for those on a diet of raw eggs the presence of native avidin will prevent biotin being absorbed from the gut.

Folic acid

Folic acid and its derivatives are present in several foods, notably leaf vegetables but also in meat. Folate deficiency results in one form of anaemia. The molecule consists of the pteridine ring with a *p*-aminobenzoic acid residue linked to at least one, and at most seven, glutamyl residues (see Figure 10.19). The precise number of glutamyl residues varies with the source but these residues must be removed in the mucosal cells for efficient absorption of folate because of the negative charge associated with them. Once absorbed folate is converted to dihydrofolate and tetrahydrofolate by dihydrofolate reductase, and the polyglutamate residues are added. The negative charges have the advantage of trapping the molecule within cells. Tetrahydrofolate derivatives are intermediate carriers of one-carbon groups, which may be in the form of methyl ($-CH_3$), methylene ($-CH_2-$), methenyl ($-CH_2=$), formyl ($-CHO$) or formimino ($-CH=NH_2$) groups. Reactions of these types are required for the synthesis of choline, methionine, glycine, serine, purines and dTMP (see Chapter 10). Folates are therefore important for cell division and for protein synthesis. Folic acid deficiency is often found in alcoholics who do not eat a balanced diet; extra supplies are required in pregnancy and lactation.

Chapter 10. Tetrahydrofolate derivatives are important in amino acid and nucleotide metabolism, page 325.

Folic acid deficiency inhibits DNA synthesis. The characteristic symptom is a reduction in the rate of maturation of, and production of abnormally large, erythrocytes; a condition called **macrocytic anaemia**.

Cobalamin: vitamin B_{12}

Cobalamin has a complex tetrapyrrole ring structure (the **corrin ring**), to which cobalt is complexed. It is not synthesized by animals or plants, only by some microorganisms. It acts as a coenzyme in reactions in which a methyl group is shifted from one carbon to an adjacent carbon. Deficiency, associated with the development of **pernicious (megaloblastic) anaemia**, is probably due to failure to absorb the vitamin, a process dependent on production in the stomach of a glycoprotein called **intrinsic factor**. This protein binds the vitamin to receptors for absorption, but may also protect the vitamin from uptake by intestinal bacteria. Vitamin B_{12} is transported in plasma bound to a specific binding protein. Its metabolic role in mammals is not fully understood but, among other reactions, it is required for the conversion of methylmalonyl CoA to succinyl CoA (which may be used as an energy source or as a precursor for porphyrin synthesis) and also for the conversion of homocysteine to methionine. It is important that vitamin B_{12} deficiency is recognized early or permanent neurological damage may result. The metabolism and function of vitamin B_{12} is closely connected to that of folic acid.

Vitamin C: ascorbic acid

Vitamin C is structurally related to glucose, and in most nonprimate mammals is synthesized from glucose. Some species, including humans, apes and guinea-pigs, are unable to synthesize vitamin C, which is therefore required in the diet. Citrus fruits are a rich source of this vitamin. Vitamin C is a reducing agent in hydroxylation reactions and has a key role in maintaining the reduced (ferrous) state of iron essential for its absorption.

There is also evidence that vitamin C acts through nonenzymic mechanisms to aid the absorption of iron from the stomach by maintaining it in the reduced (ferrous) state, and to protect several of the other vitamins, including vitamin E, from oxidation as part of a general antioxidant role.

Chapter 15. Synthesis of procollagen is completed by postsynthetic modifications, page 473.

Box 14.6   Scurvy

Vitamin C is required for the hydroxylation of lysine and proline residues in collagen molecules (Chapter 15). In vitamin C-deficient individuals collagen is not adequately hydroxylated, which leads to

- failure to align polypeptides into stable triple helical structures;
- failure of glycosylation reactions;
- inadequate cross-linking of fibres.

The resulting condition is called **scurvy**. This disorder, which took a heavy toll of the lives of sailors in the sixteenth, seventeenth and eighteenth centuries, is characterized by

- anaemia;
- poor wound healing;
- increased fragility of capillaries resulting in bleeding into the joints and under the skin;
- inadequate bone growth in the young;
- weakening of bones and joints and loosening of teeth;
- Increased vulnerability to infection.

The symptoms are quite rapidly reversed by administration of vitamin C.

The role of inorganic minerals

Small amounts of many minerals are absorbed as part of the diet. Some of these are necessary for health, others have no recognized function. If some minerals are absorbed in too large amounts, toxicity can occur.

The major cations are sodium (Na^+) potassium (K^+), calcium (Ca^{2+}) and magnesium (Mg^{2+}). The major anions are chloride (Cl^-), iodide (I^-), fluoride (F^-) and phosphate (PO_4^{2-}). The essential trace element cations, those for which there is a recognized function, are iron, copper, manganese, cobalt, zinc, molybdenum, selenium and chromium. All of the trace elements function as ions. For those with more than one ionization state, the transition between charged states may be part of the mechanism of action; others are functional in only one ionization state. Some of the functions of trace elements are listed in Table 14.5.

Sodium is the major extracellular cation, **potassium** is the major intracellular cation. The distribution of these ions across plasma membranes creates ion gradients, which are essential for the generation of action potentials in muscle and nerve cells. These processes are described in Chapter 13. Significant amounts of both sodium and potassium are also found in bone.

Chapter 13. Membrane potentials and the generation of action potentials, page 401.

Table 14.5 Trace elements and their functions

Element	Dependent processes or molecules	Comments
Manganese	Isocitrate dehydrogenase Pyruvate carboxylase Acetyl CoA carboxylase Arginase Serine hydroxymethylase Synthesis of glycoproteins	Redox reactions
Zinc	RNA polymerase DNA polymerase Gustin Carbonic anhydrase Alcohol dehydrogenase Carboxypeptidase Alkaline phosphatase Wound-healing processes Salivary peptide	
Cobalt	Constituent of vitamin B_{12}	Essential for methionine biosynthesis and propionate metabolism
Copper	Cytochrome oxidase Lysyl oxidase Lysyl and prolyl hydroxylase Superoxide dismutase $C_{18} \Delta^9$ desaturase	In Wilson's disease copper accumulates, causing cirrhosis of the liver and neurological disorders
Iodine	Thyroid hormone	Deficiency leads to goitre and cretinism
Chromium		Potentiates the action of insulin in accelerating glucose uptake into tissues
Selenium	Glutathione peroxidase	Essential for neutralizing peroxides and free radicals and deiodination of thyroxine
Molybdenum	Xanthine oxidase Aldehyde dehydrogenase	Oxidation/reduction reactions

Chapter 15. Bone is constantly formed and degraded by specialized cells, page 478.

Calcium

The daily dietary requirement for calcium is 600 mg, an amount needed to replace large losses through kidney, intestine and bile. Greater amounts of calcium are required by the young to support bone growth. Absorption of calcium from the intestinal tract is increased by the presence of lactose or basic amino acids and of vitamin D_3 in intestinal cells. Absorption of calcium is inhibited by the presence of phytates (inositol hexaphosphate) in the diet. These compounds, which are found in cereals and are mostly excreted in the faeces, have a high affinity for calcium.

Many processes depend on the concentration of calcium in plasma being maintained within narrow limits

Most of the body calcium is located in bone, but the relatively smaller amount in plasma controls many vital functions. Calcium is normally present in plasma at 2.5 mM, a much higher concentration than in the cytoplasm. About half of the circulating calcium is bound to plasma proteins, mostly albumin; the rest exists as free ions. It is vital that the circulating concentration of calcium remains within narrow limits (2.1–2.6 mM) to support a variety of essential functions, including

- contractility of some muscles, particularly in the heart;
- plasma membrane ion channel activities;
- transmission of impulses in nervous tissue;
- enzyme activities;
- maintenance of bones and teeth;
- blood clotting.

Changes in plasma calcium concentration can be life-threatening

The plasma calcium concentration is controlled by two peptide hormones: parathyroid hormone and (to a lesser extent in humans) calcitonin. The function of parathyroid hormone is described in Chapter 11. The plasma calcium concentration must be maintained within very fine limits to avoid life-threatening loss of membrane excitability and contractility in heart muscle. This reflects the fact that the excitability of muscle (including heart muscle) and nerve cells depends on movement of calcium ions across cell membranes through calcium ion channels which are sensitive to small changes in plasma calcium ion concentrations. Although skeletal muscle stores most of the calcium it needs to regulate contraction of sarcomeres within the sarcoplasmic reticulum, cardiac muscle cells store less calcium. The calcium required to initiate contraction of cardiac muscle filaments is derived mostly from calcium crossing plasma membranes.

The consequences of **hypocalcaemia** (abnormally low blood calcium) include increases in membrane permeability, which has particularly profound effects on nervous tissue. It can cause increased membrane excitability, spontaneous discharges of nerve impulses and tetany.

The consequences of **hypercalcaemia** (abnormally high blood calcium) include depression of the nervous system, sluggish reflexes and muscle weakness. Failure of cardiac contractility is a particular hazard.

Chapter 11. Parathyroid hormone, page 346.

Box 14.7 Renal osteodystrophy

Renal osteodystrophy arises from lack of renal 1-hydroxylase activity, which normally converts 25-hydroxycholecalciferol to 1,25-di-hydroxycholecalciferol (see above). With insufficient vitamin D_3, calcium absorption from the gut is impaired and, in order to maintain serum calcium, calcium is mobilized from bone. The kidneys retain phosphate and soft tissues become calcified, further reducing serum calcium. In response, secretion of parathyroid hormone increases and progressive bone demineralization occurs. Treatment of this condition requires intravenous administration of 1,25-dihydroxy-cholecalciferol.

Magnesium

Magnesium is required by all ATP-dependent processes, because ATP (and other nucleotides such as GTP) functions as a Mg^{2+} complex. It is also essential for neuromuscular transmission and is a cofactor for many enzyme-catalysed reaction.

Selenium

Trace amounts of selenium in the diet protect against autooxidation. Selenium protects through a mechanism different from that of vitamin E. This ion is part of the active site of the enzyme **glutathione peroxidase,** which destroys lipid peroxides after they are formed, converting them to metabolizable hydroxy acids, thereby protecting tissue molecules from hydroperoxide-induced damage (Figure 14.31). The protective reaction is coupled to oxidation of **glutathione** (GSH), which is subsequently regenerated in an NADP-dependent reaction. This coupled mechanism is similar to that by which the superoxide in erythrocytes is rendered harmless (Chapter 12).

Selenium is also required for the conversion of thyroxine (T_4) to T_3, the active form of this hormone.

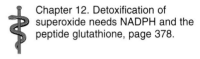

Chapter 12. Detoxification of superoxide needs NADPH and the peptide glutathione, page 378.

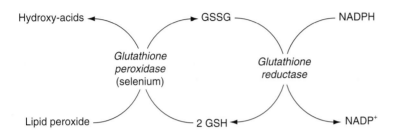

Figure 14.30 Coupled mechanism of the selenium-dependent oxidation of glutathione.

Iron

Iron is essential for a number of metabolic and physiological processes. Iron-containing proteins are of the haem or nonhaem types (see Box 7.3), depending on whether the iron is bound to protoporphyrin ring structures. Most proteins bind the ferric (Fe^{3+}) form of iron with higher affinity than the ferrous (Fe^{2+}) form. The iron content of the body is 3–4 g, of which as much as 2.5 g is found in haemoglobin. The haem proteins include haemoglobin, myoglobin and the cytochromes of the electron transport chain. The bulk of nonhaem iron-containing proteins are involved in transport or storage of iron. Other iron-dependent enzymes, including lysyl and prolyl hydroxylase, are involved in oxidation reactions.

Box 7.3. Iron–sulphur (Fe–S) centres in the electron transport chain, page 202.

The availability of iron is controlled by specific proteins

It is perhaps because of its many roles crucial to the survival of animals that more is known about how the body handles iron than about most other trace elements. Iron is found in many foods. Ferric ions in the diet are reduced to ferrous ions before being absorbed. Ascorbic acid facilitates the process, maintaining iron in the reduced state. In the intestinal mucosal cells iron is bound by the iron-storage protein **ferritin.** Ferritin is an apoprotein, a multisubunit complex with the capacity to bind as many as 4500 Fe^{3+} ions in a core at the centre of the complex. This interaction regulates the amount of iron entering the bloodstream because the amount of apoferritin in these cells is inversely proportional to the amount of iron in the body – apoferritin thus serves as an iron trap, preventing excessive amounts of iron being absorbed into the bloodstream. Ferrous ions are oxidized to the ferric state in the ferritin complex. The synthesis of ferritin is controlled at the level of translation, by a protein that binds to the mRNA, thereby stabilizing it against destruction and extending the period over which each molecule of mRNA is available for translation.

Box 5.1. Control of transcription in bacteria and humans, page 117.

Iron bound to apoferritin can not pass from the mucosal cells into the capillary blood supply. Iron that does cross into the capillaries is transported bound to the plasma protein **transferrin.** Only ferric ions bind to transferrin, and their binding is dependent on the simultaneous binding of a carbonate ion. Transferrin is a β-glycoprotein, synthesized in the liver, with two ferric ion binding sites. Iron enters cells after iron-loaded transferrin binds to specific plasma membrane receptors. The transferrin–receptor complex is internalized and the iron released in lysosomes. The receptor–transporter complex then fuses with the plasma membrane and transferrin is released into the plasma by exocytosis. As iron is constantly bound to protein from the moment it crosses the mucosal wall, very little is lost from the body. This binding is also important in protection against the toxic effects of free iron ions.

When erythrocytes are degraded in the reticuloendothelial system at the end of their 120-day life, iron is released from the haemoglobin, bound to transferrin and returned to the bone marrow for reincorporation into new haem molecules.

Iron deficiency results in **microcytic anaemia**. This is a common problem for women.

Lactoferrin is an iron-binding protein in milk

The role of this protein is not certain, but it may be essential for the absorption of iron in infants. The tight binding of iron to this protein prevents the growth of iron-dependent microorganisms in milk and in the gut of infants. Some microbes have evolved a mechanism allowing them to defeat this threat to their proliferation by secreting competing iron-chelating molecules, which abstract iron from lactoferrin.

Copper

Copper is a crucial component of several enzymes.

Chapter 7. Cytochrome c oxidase (complex IV) reacts with molecular oxygen, page 205.
Chapter 15. Collagen fibres are stabilized and strengthened by the formation of covalent cross-links between molecules, page 475.
Chapter 12. Detoxification of superoxide needs NADPH and the peptide glutathione, page 378.
Chapter 9. Longer chain and unsaturated fatty acids are synthesized from palmityl CoA, page 279.

1. **Cytochrome oxidase**, which catalyses the terminal reaction of the electron transport chain (see Chapter 7).
2. **Lysyl oxidase**, which oxidizes hydroxylysine residues incorporated into the polypeptide chains of collagen and elastin molecules (see Chapter 15).
3. **Superoxide dismutase**, which destroys free radicals thereby protecting cells from radical induced physical damage (see Chapter 12).
4. C_{18} Δ^{-9} **desaturase**, which is responsible for introducing double bonds into C_{16} and C_{18} saturated fatty acids (see Chapter 9).

Copper competes with zinc for absorption from the gut. In the circulation, copper is transported bound to **ceruloplasmin** or albumin.

Menke's disease is a fatal sex-linked disease characterized by low serum copper and low ceruloplasmin concentrations.

Wilson's disease is characterized by low ceruloplasmin levels but high serum copper, which results in deposition of copper in soft tissues and resulting complications in kidney, liver and brain.

Phosphate

Chapter 15. The mineral composition of bone, page 478.

Many processes are dependent on the concentration of phosphate in plasma or the cytoplasm. Most of the body phosphate is found in calcified tissues as a component of the **hydroxyapatite** crystal, from which it is mobilized, with calcium, in response to reduced plasma calcium ion concentrations. Inorganic phosphate is required within cells to sustain metabolic reactions. Phosphorylated intermediates play a significant role in metabolism, structure and function of all cells.

The structural tissues

This chapter describes the two major structural tissues of the body: muscle and connective tissue.

Muscle

Muscle cells are made up of overlapping arrays of actin and myosin filaments. Contraction occurs when the energy of ATP is utilized for inducing conformational change in the structure of the myosin molecule. The energy is then converted to movement, a sliding of the filaments. Contractility is controlled by the calcium ion concentration because thin filaments of skeletal muscle have a calcium receptor that activates the filaments.

Connective tissue

Connective tissue provides adhesion between cells, strength in tendons, shock-absorbing properties in joints, flexibility to the layers of skin and forms the organic matrix of bone. The different properties of these tissues reflect differences in the types of glycoproteins and proteoglycans present, and differences in the ways in which these molecules assemble to form the tissue.

The structure and synthesis of collagen and some disorders of collagen structure are described. Elastin is described briefly. Bones and teeth, in which connective tissue becomes calcified, make up a major fraction of connective tissue. Current theories of how calcification may occur and the processes that control calcification are described.

Chapter 13. The metabolic functions of muscle, page 384.

The contractile cells, or fibres, of muscle are called **myocytes**. Muscles also contain other cell types, notably the **endothelial** and **smooth muscle** cells of the **microvascular** (capillary blood supply) system. These cells are extremely small compared with myocytes and occupy a small fraction of the muscle tissue. However, some of the metabolic activities associated with muscle are confined almost exclusively to cell types other than myocytes. This section deals with the structure and contractile function of myocytes. Metabolic aspects of muscle function are described in Chapter 13.

The structure of muscle cells

Muscle cells are cylindrical in shape, up to 40 mm long and 50 μm in diameter. Because of this large size, each myocyte needs more than one nucleus to provide information for protein synthesis and cellular repair. Myocytes are multinucleate, having as many as 100 nuclei per cell. Cardiac myocytes are shorter (100–150 μm) and most are binucleate.

Muscle cells have specialized membrane systems

The plasma membrane of muscle cells is called the **sarcolemmal membrane** or the **sarcolemma**. As in other cell types, this membrane possesses transport sites for glucose and amino acids, contains the Na^+/K^+-ATPase pumps that maintain the cell's electrical potential and receptors for hormones such as insulin and adrenaline (Figure 15.1).

The **transverse**, or **T-tubule**, membrane system crosses the fibre, passing between the fibrils and providing continuity between the plasma membrane and the **sarcoplasmic reticulum**. T-tubules provide a communication system, allowing rapid transfer of nutrients and ions from the extracellular space to the interior of the fibre, and facilitating the transmission of electrical signals from the plasma membrane to the sarcoplasmic reticulum.

The sarcoplasmic reticulum is a lace-like structure, wrapped around each of the fibrils making up the muscle fibre. Although the microscopic appearance of this organelle is reminiscent of the endoplasmic reticulum, the sarcoplasmic reticulum is specialized for the storage of calcium and regulates the changes in cytoplasmic

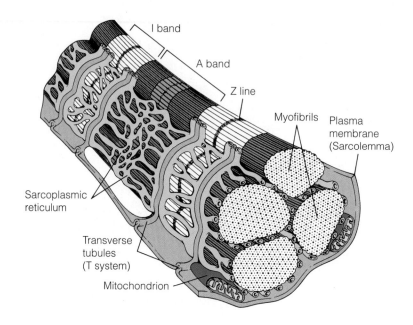

Figure 15.1 Diagram of the membrane systems of striated muscle. Four fibrils of a muscle cell. Each fibril is made up of contractile filaments and is wrapped with a network of sarcoplasmic reticulum. Mitochondria lie in rows between the fibrils, and the cell is bounded by the sarcolemmal (plasma) membrane.

calcium ion concentration that control contraction and relaxation of muscle fibrils. The amount of sarcoplasmic reticulum varies between muscle types in a way that reflects the diameter of muscle fibres. Cardiac muscle has a sparse sarcoplasmic reticulum and red skeletal fibres have less than white skeletal fibres. The reason for this difference is that larger diameter fibres have greater diffusion distances between the extracellular space and the centre of the fibre and are therefore more dependent on an intracellular supply of calcium. Fibrils within small-diameter cells are provided with calcium more rapidly from the extracellular space.

The contractile elements of skeletal muscle are composed of protein filaments arranged into fibrils

Myocytes are made up of a large number of **fibrils** extending along the length of the cell. Within each fibril an alternating pattern of light bands (**I bands**) and dark bands (**A bands**) creates a characteristic striated pattern along the length of the fibre, a pattern that is clearly visible under the light microscope (Figure 15.2). The A bands appear dark because of their ability to polarize light; they are **anisotropic** (A). This property reflects the organization of molecules into filaments. The light bands do not polarize light; they are **isotropic** (I).

Fibrils are composed of arrays of thick and thin **filaments** (Figure 15.3). Thin filaments lie in the light band region and are composed of aggregated actin molecules. Thick filaments, located in the dark bands, are composed of aggregated myosin molecules. In a relaxed muscle the sarcomere length is 2.4 µm and thin and thick filaments partially overlap in the A band. In a fully contracted muscle the sarcomere length is 1.6 µm and overlap between the filaments occurs throughout the length of the sarcomere so that the I band is no longer visible in tissue sections.

The thin filaments are held in position by a structure called the **Z line** (**Z disc** or **Z band**), which lies at the centre of the I band. Thin filaments project in both directions from the Z line. The contractile unit of muscle, the **sarcomere**, is defined as the material lying between two adjacent Z lines. Thick filaments are stabilized at

Figure 15.2 Organization of skeletal (striated) muscle. Cells (fibres) run longitudinally along the muscle. Each is made up of many myofibrils and each myofibril is constructed of overlapping arrays of thick and thin filaments.

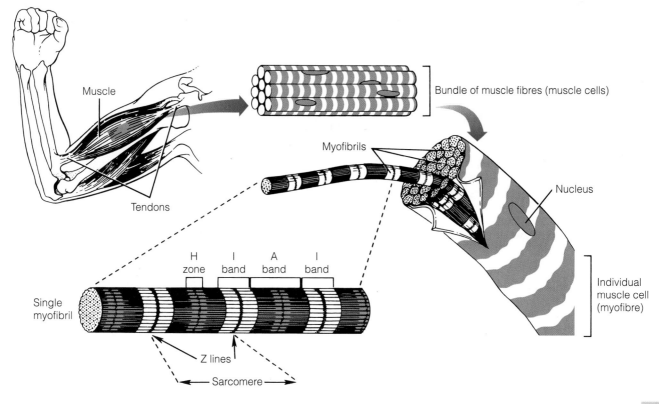

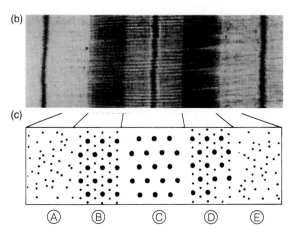

(a)

Sarcomere

Myosin thick filaments reverse polarity
at midline of sarcomere (the M line)

Z line   Myosin (thick)   Actin (thin)   Z line
          filaments   filaments

(b)

(c)

Ⓐ Thin filaments only
Ⓑ Thick and thin filaments
Ⓒ Thick filaments only
Ⓓ Thick and thin filaments
Ⓔ Thin filaments only

Figure 15.3 Detail of the structure of a sarcomere at rest length. Each sarcomere is made up of overlapping thick and thin filaments (a) and is bounded by Z-line structures at each end. Thin filaments are attached to the Z lines while thick filaments appear to be floating at the centre of the sarcomere. We now know that thick filaments are also attached to the Z lines through a thin elastic protein called titin. (b) An electron micrograph of the same structure. (c) Shows the arrangement of thick and thin filaments in cross-section at various points along the sarcomere.

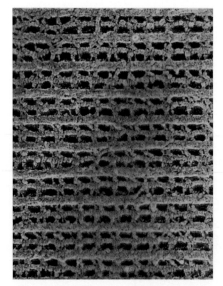

Figure 15.4 When a muscle cell is depleted of ATP, cross-bridges form between the thick and thin filaments, making the fibre inextensible. These are seen in this high-resolution electron micrograph.

Chapter 3. Phospholipids spontaneously form lipid bilayers in water, page 52.

the centre of the sarcomere by the **M line** structure. The three-dimensional arrangement of filaments is such that, when viewed in cross-section, the thick filaments form trigonal arrays while the thin filaments form hexagonal arrays. These patterns are clearly seen in transverse sections of muscle tissue viewed under the electron microscope (Figure 15.3(c)).

When thin-sectioning techniques were developed for electron microscopy, it became clear that structural projections from the thick filaments attach to the thin filaments. In a muscle depleted of ATP these connections are fixed, but they are broken if the muscle is perfused with ATP. These projections are called **cross-bridges** and have a central role in the mechanism of contraction (Figure 15.4).

Sarcomere shortening is explained by the 'Sliding Filament Theory'

Sarcomere shortening occurs through a progressive interdigitation of thin and thick filaments. During this process the lengths of individual filaments do not change. Thick filaments are unable to pass through the Z line structures, so when the sarcomere length is equal to the length of the thick filaments the fibre is fully contracted (Figure 15.5).

Rest length of the sarcomere is determined by the thermodynamic stability of the plasma membrane and interactions between membrane molecules and the cytoskeletal proteins anchoring the filaments to the membranes. As a muscle cell shortens the lipid bilayer of the plasma membrane becomes distorted, disrupting the intrinsic stability of the phospholipid bilayer (see Chapter 3). The self-healing properties of membranes ensure that restoration of a stable bilayer limits the relaxation of the sarcomeres. Cytoskeletal proteins have a vital role in determining the stable shape of the cell and in maintaining the organization of myofilaments.

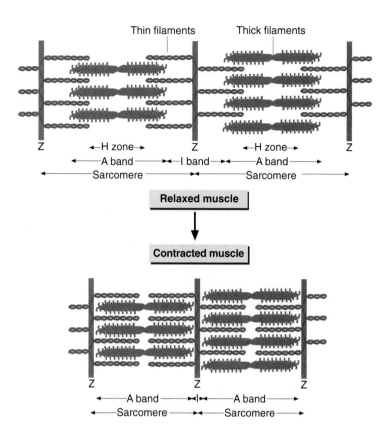

Figure 15.5 Relaxed and contracted sarcomeres. The progressive overlap of thick and thin filaments leads to the shortening of sarcomere length (contraction). The length of the fully relaxed sarcomere is 2.4 µm; the fully contracted sarcomere is 1.6 µm in length. In fully contracted muscles the sarcomere length is identical to the length of the thick filaments.

Structure and function of the contractile proteins

Thick filaments are composed of aggregates of myosin molecules

Myosin has both structural and catalytic properties. It provides

- the structure of the thick filament;
- the cross-bridges, which project from the thick filament to link adjacent thick and thin filaments;
- the catalytic site that releases the chemical energy of ATP, making it available as mechanical energy for sarcomere shortening.

Myosin molecules (Figure 15.6) are made up of six polypeptide chains, two heavy chains with mass of about 200 000 Da, and four light chains with molecular weights of 15 000–25 000 Da.

The **tail portions** (*N*-terminal ends) of the two heavy chains are α-helices, which intertwine to form a coiled-coil structure, 150 nm long and 2 nm in diameter. The structure is stabilized by hydrophobic interactions. At the *C*-terminal end, each of the two heavy chains assumes a globular form; these are the **myosin heads**. These heads contain both the actin-binding sites and the catalytic sites (each head is capable of both functions). Two light chains bind to each of the heads.

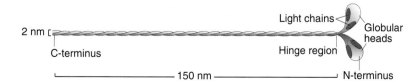

Figure 15.6 Diagrammatic structure of a myosin molecule.

Myosin molecules aggregate into thick filaments

At the ion concentrations found in muscle cells, myosin molecules are insoluble and assemble spontaneously into thick filament structures, bipolar aggregates of 300–400 myosin molecules. Within the thick filaments myosin molecules form tail-to-tail arrays, with heads projecting from both ends, leaving a bare zone at the centre of the filament. Most of the length of the tail aggregates with other myosin molecules to form the core of the filament, but a small part of the tail and the heads project from the filament. Myosin tails pack around a six-fold axis, so that in cross-section heads project from the filament in a regular pattern, each head being displaced from the adjacent heads by 60° around the filament core. Within a fibril this arrangement places every myosin head directly opposite a thin filament (Figure 15.7).

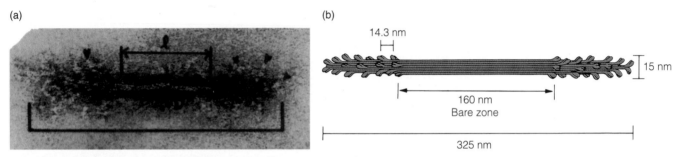

Figure 15.7 Aggregation of myosin molecules into thick filaments. (a) Electron micrograph of purified myosin molecules aggregated into a synthetic thick filament, a structure analogous to that formed *in vivo*. (b) Diagrammatic details of the structure. Note the bare zone at the centre of the filament caused by the bipolar tail-to-tail aggregation of molecules.

Myosin molecules have 'hinge' regions that allow cross-bridges to form between thin and thick filaments

The section of myosin tail adjacent to the heads projects from the filament core because a local irregularity in the α-helices introduces a short segment with a flexible structure. This is called the **hinge segment** and it marks the junction between the core tail and the cross-bridge. A second hinge point is located at the base of each head of the molecule, where the extended helical structure of the tail changes to the globular structure of the head. The portions of the molecule defined by these hinge regions can be separated after proteolytic digestion. In the test tube, enzymes such as **chymotrypsin** can digest the hinge regions but are unable to attack the regular helix. Digestion releases several distinct fragments, each with an identifiable function (Figure 15.8).

The portion of the molecule most sensitive to digestion is the cross-bridge hinge point. The portion of the tail normally comprising the filament structure is called **light meromyosin**; the remainder of the molecule – the upper portion of the tail with two heads attached – is called **heavy meromyosin**. 'Heavy' and 'light' refer to the relative molecular weights of the two fragments. Continued digestion results in release of the heads from the cross-bridge portion of the tail. Isolated myosin heads retain their ability to hydrolyse ATP and the capacity to bind to actin.

Myosin light chains influence the rate of hydrolysis of ATP

The light chains of the myosin molecule do not contribute to the structure of the catalytic site, yet they affect the rate at which ATP is hydrolysed by myosin. Different classes of striated muscle have characteristic types of myosin light chains, characterized by molecular size and amino acid sequence. The light chains appear to regulate the rate of ATP hydrolysis and the twitch speed of the muscle.

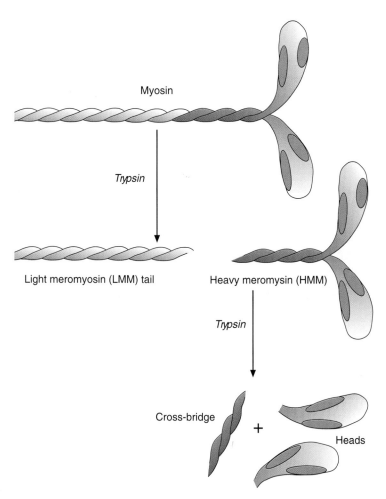

Figure 15.8 Proteolytic digestion products of myosin. Digestion of the myosin tail by trypsin at a specific point in the structure indicates the absence of regular helix at this point in the tail. Further digestion releases two heads from the molecule. Each head retains actin-binding and ATPase activities.

Figure 15.9 Actin molecules polymerized into a double-stranded filament.

Actin polymerizes to form thin filaments

Thin filaments have three major protein constituents. The core of the filament is made of **globular** actin (**g actin**) molecules polymerized into long strands, two of which intertwine to form filament structures (**f actin**) (Figure 15.9).

Actin molecules have a molecular mass of about 42 000 Da. Actin is ubiquitous in nature, being found in organisms as diverse as mammals and fungi. The actin produced in muscle and nonmuscle cells is derived from different genes, but the properties of actin from different organisms are only slightly different and exhibit only minor differences in amino acid sequence. In mature skeletal muscle cells the actin filaments do not disaggregate after the initial polymerization associated with synthesis of new filaments, but in other mammalian cell types the function of actin depends on its capacity to polymerize and depolymerize in response to specific signals. Actin is a major component of the cytoskeletal complex of cells (see Chapter 3). Muscle cell cytoskeletal actin molecules are constantly polymerizing and depolymerizing to accommodate the changing shape of myocytes during the cycle of contraction and relaxation.

Each molecule of actin has a purine nucleotide binding site, and in thin filament actin of mature muscle cells this site is occupied by a molecule of ADP. Although it seems likely that this ADP contributes to the nucleotide pool of the cell in some

 Chapter 3. The cytoskeleton determines cell shape and movement and regulates the organization of internal structures, page 49.

way, its precise role has not been established. As new muscle cells develop, a molecule of ATP is bound to each actin molecule before it polymerizes. During polymerization the terminal phosphate of ATP is lost as the energy is consumed to facilitate the polymerization, leaving ADP.

Tropomyosin and troponin regulate the interactions between actin and myosin in skeletal and cardiac muscles

Tropomyosin is a fibrous protein with two α-helical polypeptide chains wound into a coiled coil. The protein has a molecular mass of about 35 000 Da and forms a 40-nm long rod that lies along the groove formed between the two strands of actin making up the core of the thin filament (Figure 15.10). Tropomyosin molecules polymerize head-to-tail, each molecule extending over seven actin molecules in the filament. The interaction between actin and tropomyosin is not fixed, but allows for a shift in the position of tropomyosin. It may be bound to actin at either peripheral positions or central to the actin groove (Figure 15.10). Depending on the position it occupies, tropomyosin either obstructs or facilitates the interaction between the actin and myosin heads: positioned on the outer surface of actin molecules, tropomyosin blocks the myosin-binding site. With tropomyosin in the central position the myosin binding site is accessible. In a resting muscle the interaction between actin and myosin is blocked.

Troponin is a globular protein with three subunits, designated TnT, TnI and TnC. Each subunit has a different function. One molecule of troponin is bound to each molecule of tropomyosin. TnT, the tropomyosin-binding subunit, binds at a site near the *C*-terminal of tropomyosin anchoring both TnI and, through that subunit, TnC to tropomyosin. TnI, the inhibitory subunit, is linked directly to actin as well as tropomyosin. Troponin responds to changes in the concentration of Ca^{2+} in the cytoplasm through TnC, the calcium-binding subunit of troponin.

When the cytoplasmic calcium ion concentration increases from the resting concentration of approximately 10^{-7} M to the activating concentration of approximately 10^{-5} M, the following sequence is initiated (see Figure 15.11).

Table 3.2 Cytoskeletal elements, page 50.

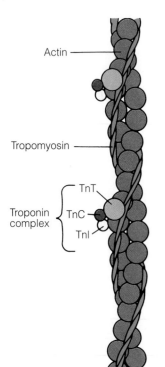

Figure 15.10 The structure of a thin filament.

Figure 15.11 Response of thin filament proteins to changes in cytoplasmic calcium ion concentration. A single myosin head is shown projecting from a thick filament. (a) The arrangement of tropomyosin and troponin subunits blocks the interaction of the myosin head with an actin molecule in the relaxed state. In (b) an indication is given of how, when calcium binds to TN-C, the tropomyosin strand is dragged into the central position of the actin double strand, exposing the binding site of actin to the myosin head.

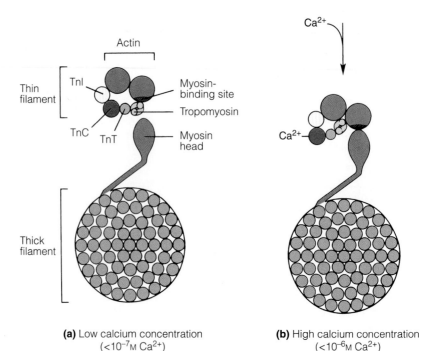

(a) Low calcium concentration ($<10^{-7}$ M Ca^{2+})

(b) High calcium concentration ($<10^{-6}$ M Ca^{2+})

1. Calcium ions bind to specific sites on TnC, changing the protein conformation.
2. The change in TnC is transmitted to TnI, releasing the inhibitory interaction between TnI and tropomyosin.
3. Tropomyosin moves to the central binding position on actin.
4. The myosin-binding site on the actin molecule is exposed, allowing the interaction between actin and myosin,
5. The sarcomere shortens using energy released by the hydrolysis of ATP.

TnC has four Ca^{2+} binding sites: two high-affinity sites, binding Ca^{2+} with an association constant (K_a) of 2×10^{-6} M, and two low-affinity binding sites, K_a 4×10^{-4} M. It appears that two of these sites are constantly occupied but development of full contractile tension requires that all four sites bind calcium ions.

When the cytoplasmic calcium ion concentration falls to the resting value of 10^{-7} M, the sequence of events described above is reversed, blocking further interactions between actin and myosin. Muscle enters the resting state and the filaments slide apart passively, restoring the sarcomere to rest length.

The organization of the smooth muscle contractile system is different

Vertebrate smooth muscle does not have the sarcomere structure characteristic of skeletal muscle. Instead, myosin molecules in these cells polymerize into bipolar bundles or ribbons and actin molecules polymerize into filaments in response to signals that stimulate a contractile response. In the absence of appropriate signals these polymers dissociate into single molecules. Smooth muscles have only a sparse sarcoplasmic reticulum network and depend on calcium entering the cell from the extracellular space. For these reasons development of contractile tension is slow in smooth muscle relative to that in skeletal muscle.

Calcium sensitivity of smooth muscle cells is mediated through myosin light chains

Smooth muscle contains tropomyosin but not troponin, but the contractile response of smooth muscle is similarly triggered by increases in the cytoplasmic calcium ion concentration. At low calcium ion concentrations, the myosin light chains obstruct the binding of myosin heads to actin. The response of smooth muscle cells to increases in calcium ion concentration is mediated by the calcium-binding protein **calmodulin**. When the calcium concentration increases and saturates the binding sites of calmodulin it binds to, and activates, **myosin light chain kinase** (MLC) (Figure 15.12), which phosphorylates the light chains of the myosin molecule. This

 Chapter 8, Activation of glycogen phosphorylase and inhibition of glycogen synthase occur in response to a common cascade, page 227.

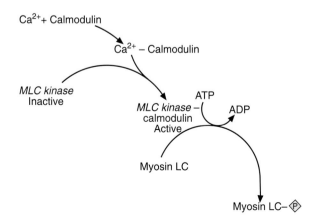

Figure 15.12 Activation of smooth muscle myosin.

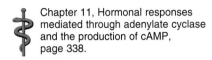

Chapter 11, Hormonal responses mediated through adenylate cyclase and the production of cAMP, page 338.

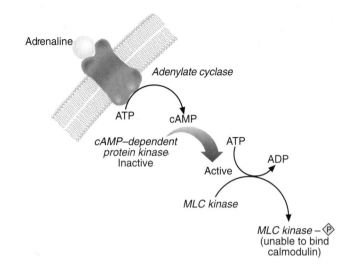

Figure 15.13 Inhibition of smooth muscle contractility by adrenaline.

phosphorylation is a prerequisite for the polymerization of myosin molecules into bundles and for the conformational changes in the myosin molecule that permit the interaction of myosin and actin and the development of contractile tension.

Contraction of some smooth muscle types is inhibited by adrenaline (Figure 15.13). The mechanism of inhibition involves activation of cAMP protein kinase, which phosphorylates myosin light chain kinase. This reduces the affinity of the kinase for calmodulin and thereby blocks the phosphorylation of myosin light chain kinase, maintaining the relaxed state of the muscle cell.

Chemical energy is converted to mechanical energy during contraction

The progressive interdigitation of thin and thick filaments by which the sarcomere shortens is a mechanical process powered by the hydrolysis of ATP. We will look first at the mechanical sequence, and then describe how chemical energy is provided for these events.

The mechanical events of the contractile cycle

The contractile cycle is initiated when myosin cross-bridges reach across the inter-filament space and myosin heads attach to specific sites on the actin molecules of the thin filament. This cross-bridge formation is facilitated by the flexible hinge region between light and heavy meromyosin. Attachment of myosin heads to actin is followed by rotation of the myosin head around the hinge at the base of the head, pulling thin filaments towards the centre of the sarcomere. The cross-bridge detaches from the thin filament and the head is restored to its initial orientation through changes at the two hinge regions. The myosin head is now positioned to repeat the cycle by attaching to an actin molecule further along the thin filament (Figure 15.14). Each myosin molecule has two heads and both heads have the capacity to bind actin. Whether the two heads cooperate in some way is not yet established.

ATP is hydrolysed to provide energy to power the mechanical cycle

The mechanical cycle is provided with energy from the hydrolysis of the terminal phosphate of ATP. The relationship between the catalytic process and the mechanical cycle must accommodate the following observations.

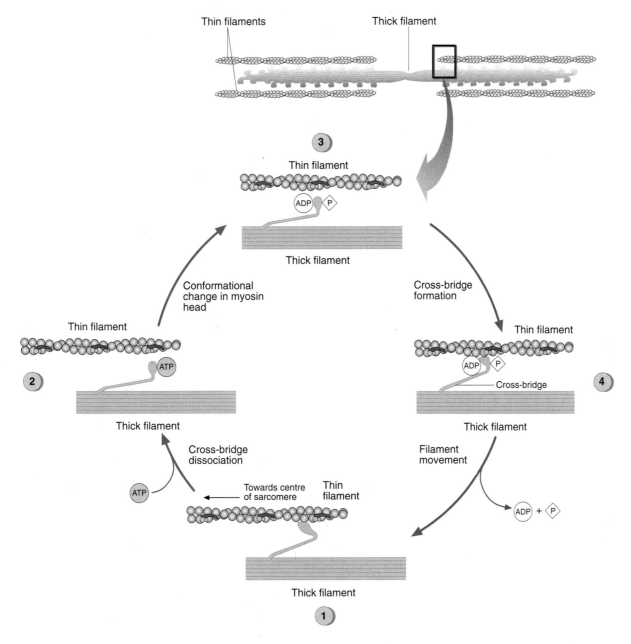

Figure 15.14 The mechanical cycle by which sarcomeres shorten. The myosin molecule can rotate about both 'hinge points', allowing both a change in the orientation of the cross-bridge portion of the tail and rotation of the heads. With the completion of each cycle the thin filament has been dragged further towards the centre of the sarcomere. ① Actomysin in 'rigor'-like complex which is broken, ②, as ATP dissociates actin from myosin. ③ High energy myosin head with ADP and Pi bound. ④ Cross-bridge forms prior to release of ADP and Pi, and completing the cycle.

- ATP is hydrolysed by myosin.
- Actin has no catalytic activity and the actin-binding and ATPase sites of the myosin head are distinct.
- In the presence of actin, ATP hydrolysis by myosin is accelerated about ten-fold.
- In muscle depleted of ATP, **rigor complexes** are formed in which myosin cross-bridges attach to actin, making the muscle inextensible.
- The addition of ATP to a muscle in rigor leads to the dissociation of myosin and actin and breaking of the cross-bridge links.
- If ATP is added to a myosin–actin complex the proteins dissociate before ATP is hydrolysed because myosin has a higher affinity for ATP than for actin.
- In the presence of actin, the products of this reaction are released from the catalytic site much more readily than in the absence of actin.
- After the hydrolysis of ATP, the myosin•ADP•Pi complex reassociates with actin. ADP and Pi then dissociate because myosin heads have a higher affinity for actin than for ADP and Pi.

These requirements are satisfied only if ATP is hydrolysed after the cross-bridge dissociates. This implies that the energy-generating step, ATP hydrolysis, occurs when actin is *not* attached to myosin. This must be reconciled with the fact that the mechanical event, the progressive interdigitation of filaments, requires the interaction of actin and myosin.

The energy of ATP is converted first into a high-energy protein conformation and then into mechanical energy

If ATP is added to a solution of myosin, its hydrolysis is accompanied by a substantial change in the conformation of myosin. If, instead, ADP and Pi are added to myosin, both bind at the catalytic site but there is no change in the protein conformation. This shows that when ATP is hydrolysed by myosin the energy released is transferred into an energized protein conformation. The myosin head assumes a high-energy form. After this high-energy myosin–product complex reassociates with actin, ADP and Pi dissociate, releasing the energy stored in the myosin head. The conformational energy is converted to mechanical energy (rotation of the myosin head), which pulls the thin filament towards the centre of the sarcomere.

The many myosin heads of an individual filament do not enter the mechanical cycle in a synchronous manner. The sliding of filaments proceeds in a smooth continuous manner because the individual heads are at different phases of the cycle at any instant.

Control of the cytoplasmic calcium ion concentration

The sarcoplasmic reticulum has mechanisms for the uptake, storage and release of calcium

Contraction of muscle fibres is initiated when the cytoplasmic calcium ion concentration increases sufficiently to saturate the calcium-binding sites of troponin. In resting muscle calcium is bound to specific proteins within the sarcoplasmic reticulum. Depolarization of the plasma membrane transmits a signal through the T-tubule system which reaches the sarcoplasmic reticulum within milliseconds. Ion permeability changes result in the opening of calcium channels, allowing efflux of the stored Ca^{2+} ions into the cytoplasm. The process is so rapid that all sarcomeres within a single fibre contract simultaneously.

When the sarcolemmal membrane repolarizes, calcium channels close and a Ca^{2+}-activated ATPase in the sarcoplasmic reticulum is activated, pumping Ca^{2+} into the lumen of the reticulum. Activation of the ATPase occurs through ATP-dependent phosphorylation of the pump. In the activated state the pump hydrolyses ATP, transferring two Ca^{2+} from the cytoplasm for every molecule of ATP expended. Once the cytoplasmic Ca^{2+} concentration is reduced to levels that ensure relaxation of the contractile filaments (dissociation of Ca^{2+} from troponin), the pump is inactivated to protect the cytoplasm against excessive depletion of Ca^{2+} ions, which are also required for other processes. The ATPase is inactivated through a phosphatase, which removes the phosphate residues that initially activated the pump.

Calcium taken into the sarcoplasmic reticulum is stored bound to two proteins. Both of these proteins have large numbers of aspartate and glutamate residues with free carboxyl groups, providing sites to which calcium ions bind electrostatically. **Calsequestrin** is a protein with a relatively low affinity for calcium (K_d 4×10^{-5} M), but the capacity to bind large numbers of calcium ions. **High-affinity calcium binding protein**, as its name suggests, binds calcium with a high affinity (K_d 3×10^{-6} M) but has a relatively small number of binding sites. Working together, these two proteins ensure that the cytoplasmic Ca^{2+} concentration is finely controlled.

The extracellular matrix – connective tissue

Connective tissue has many functions, including strengthening and supporting tissues, providing an adhesive link between the cells of a tissue and providing the matrix for crystallization of inorganic salts during the formation of bones and teeth. There are two major components of connective tissue: the **ground substance** and the **protein fibre**. The ground substance is made up of large complex polysaccharides and proteoglycans that attract water and ions, forming a viscous matrix in the extracellular space. Connective tissue proteins are secreted into this matrix from the cells in which they are synthesized. **Collagen** is the most abundant connective tissue protein. **Elastin** is found particularly in those connective tissues that require elastic properties for their function. Connective tissue is described as **loose** or **dense** depending on the organization and content of protein fibre. Each type serves different functions.

Loose connective tissue provides adhesive material between cells of soft tissues

Cells of many soft tissues are not in continuous contact with one another, but lie in a bed of loose connective tissue composed of collagen fibres surrounded by a gel of hydrated proteoglycans. Loose connective tissue has several identifiable layers: adjacent to the cells is the **basement membrane**, which is itself multilayered, with the **basal lamina** being immediately adjacent to the plasma membrane. The basement membrane is composed of type IV collagen molecules, the glycoproteins **laminin** and **entactin** in a proteoglycan ground substance composed largely of heparan sulphates. The collagen fibres of the basal lamina form a two-dimensional reticulum quite different from the organized packing of collagen fibres in structural (dense) connective tissue such as tendons.

The basement membrane has a diverse range of functions. In the renal glomerulus it acts as a filter; at neuromuscular junctions it has specialized structures apparently related to the responsive nature of these surfaces. It provides a framework around which tissue can regenerate and probably contributes to morphogenesis and cell proliferation.

Beyond the basement membrane is a more extensive region of loose connective tissue. Small blood vessels pass through this tissue, bringing oxygen, nutrients, hormones and ions as well as fibroblasts and other blood-borne cells. The proteoglycan gel permits free diffusion of molecules and cells from the bloodstream.

The extracellular matrix holds together cells of multicellular tissues

Proteins of the extracellular matrix are linked through plasma membrane proteins to the cytoskeletal proteins (Figure 15.15). A family of receptor proteins (the **integrins**) are integral proteins of the plasma membrane. They possess binding sites on the extracellular face of the membrane for a group of **adhesion proteins** including laminin and **fibronectin**. These adhesion proteins also bind to the proteoglycan molecules, so that the entire extracellular matrix is anchored to the plasma membrane.

Within the cell, the portion of the integrin molecule exposed on the internal face of the plasma membrane interacts with the filaments of cytoskeletal proteins lining the interior of the plasma membrane, maintaining the shape and rigidity of cells. In mammalian cells the cytoskeletal microfilaments are composed mostly of polymerized actin molecules cross-linked into a matrix by molecules of **filamin** or **fodrin**. Filamin cross-links actin polymers into perpendicular arrays; fodrin cross-links them into parallel arrays.

Chapter 3. Membrane-associated proteins are categorized by the nature of their interaction with the lipid bilayer, page 53.

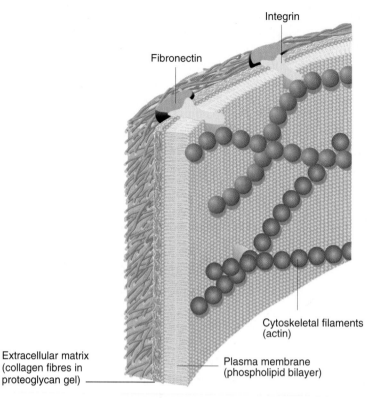

Figure 15.15 The structural relationship between cytoskeleton, plasma membrane and basement membrane. In multicellular tissues individual cells are held together through a complex network of proteins and glycoproteins that link the cytoskeletal proteins (lying inside the plasma membrane) through membrane integral proteins (integrins) to fibronectin and the collagen fibres of the extracellular space.

Dense connective tissue is composed mostly of protein fibres

The more organized type of connective tissue found in skin, tendon, cartilage and bone is known as **dense connective tissue**. Most of this tissue is occupied by large, highly organized protein fibres, with few cells and a smaller proportion of the space occupied by proteoglycan gel compared with loose connective tissue.

Mucopolysaccharides make up the carbohydrate content of the ground substance

Mucopolysaccharides, or **glycosaminoglycans**, are unbranched polymers made up of repeating disaccharide units. One of the sugar units is usually an *N*-acetylated hexosamine. Uronic acids occur commonly, the negatively charged carboxyl groups making these molecules polyanions. Repulsion between negative charges encourages the molecules to form elongated structures occupying large volumes, and also to

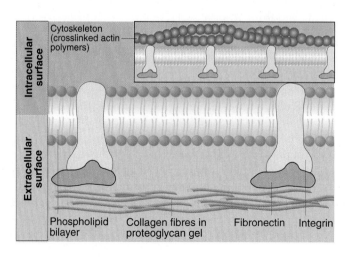

Figure 15.16 Cross-sectional view of the links between cytoskeletal (intracellular), integral membrane and loose connective tissue (extracellular) proteins in a multicellular tissue.

attract cations (Na⁺, K⁺ and Ca²⁺) to balance the negative charge. The osmotic effect of the accumulating ions draws water into the structure, creating a highly hydrated and viscous gel. One gram of glycosaminoglycan can absorb up to one litre of water.

The major mucopolysaccharides (Figure 15.17) are

- **hyaluronic acid**, a β,1–4 polysaccharide of β,1–3-linked D-glucuronic acid and N-acetyl-D-glucosamine, found throughout the body and the commonest mucopolysaccharide;
- **chondroitin sulphate**, a polysaccharide of D-glucuronic acid and N-acetyl-D-galactosamine-6-sulphate, the major mucopolysaccharide of cartilage;
- **dermatan sulphate**, a polymer of α-L-iduronate and N-acetyl-D-galactosamine-4-sulphate, found mostly in the skin;
- **keratan sulphate**, a polysaccharide of galactose and N-acetyl-D-galactosamine-6–sulphate, has an important role in attracting water into the cornea of the eye.

Box 15.1 Hyaluronidase

Hyaluronidase is an enzyme secreted by pathogenic bacteria, providing them with a mechanism for invasion of the host tissue. The enzyme hydrolyses β,1–4 linkages. Infection by these organisms can result in disintegration of the connective tissues.

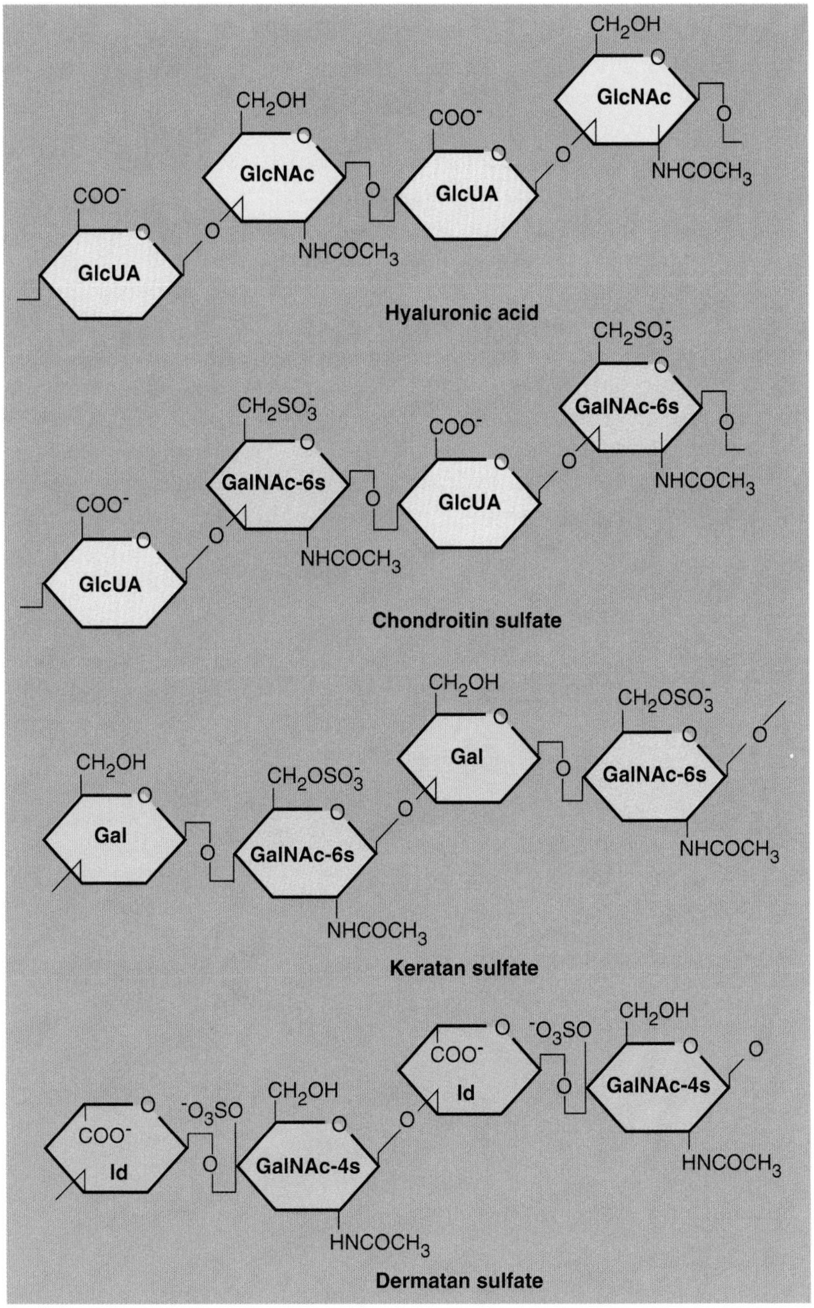

Figure 15.17 Structures of disaccharides comprising the major mucopolysaccharides of connective tissue. GlcUA, glucuronic acid; GlcNAc, N-acetyl-D-glucosamine; Gal, galactose; GalNAc-6s, N-acetyl-D-galactosamine-6-sulphate; Id, iduronate.

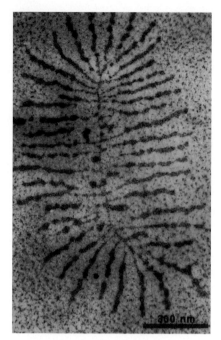

Figure 15.18 Proteoglycan structure of cartilage. Electron micrograph.

Most of the mucopolysaccharides, with the exception of hyaluronic acid, occur in the form of proteoglycan, which has a small fraction (approximately 5% by weight) of protein covalently attached to the carbohydrate chains.

Proteoglycans are vast molecules, each of which may have a molecular weight of greater than a million, made up of a core protein attached to molecules of hyaluronic acid. In addition, smaller mucopolysaccharides, keratan sulphate, heparan sulphate, chondroitin sulphate or dermatan sulphate, are attached to the core protein. The whole structure forms a vast extended network (Figure 15.18).

The viscosity of the ground substance matrix allows joint tissue to resist deformations and absorb shock. Interwoven among this proteoglycan are fibres of collagen and elastin, which strengthen the structure.

Collagen

Collagen is found in all mammalian tissues, and is the single most abundant protein in the animal kingdom. In humans it comprises 25–33% of total protein and about 6% of total body weight. The structure of individual molecules and the manner in which molecules are packed into fibres varies according to the role collagen fills in each tissue. The amount of collagen varies enormously from one tissue to another (Table 15.1).

Collagen is a fibrous protein, made up of three polypeptide chains, each with a molecular weight of about 100 000. The strength of collagen arises from several levels of organization (Figure 15.19). First, each polypeptide chain is twisted into a tight left-handed helix with only three amino acid residues per turn. Three chains are then twisted together in the opposite direction, forming a right-handed superhelix. The triple helix is stabilized by interchain hydrogen bonds from the hydroxyl groups of hydroxyproline and from the peptidyl amino and carbonyl groups of glycine residues. Approximately 90% of the proline residues need to be hydroxylated if a stable triple helix is to be maintained.

Table 15.1 Collagen content of tissues

Tissue	Percentage by weight of collagen
Liver	4
Lungs	10
Aorta	12–24
Cartilage	50
Cornea	68
Skin	72

Figure 15.19 Each tropocollagen molecule comprises three left-handed polypeptide helices. The three then associate to form a right-handed superhelix. (a) The structure of the single-stranded helix, in which all of the glycine residues are aligned along one edge of the helix. (b, c) The triple helix, in which all three chains have glycine residues oriented towards the core of the superhelix.

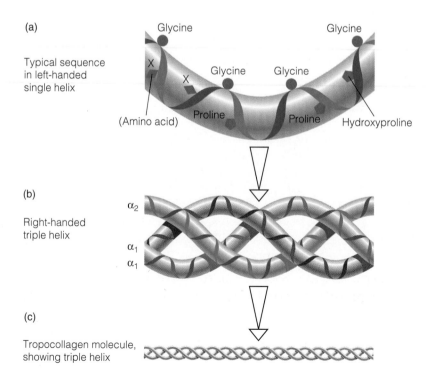

(a)

Typical sequence in left-handed single helix

(b)

Right-handed triple helix

(c)

Tropocollagen molecule, showing triple helix

The structure and properties of collagen molecules vary to meet the functional differences of tissues

Seven genetically distinct chains have been recognized as components of collagen molecules. To date, 13 types of molecule have been identified but only those best understood will be described here. Type I makes up 90% of the collagen in most animals. Types I, II, and III all form triple helical molecules approximately 300 nm in length and 1.5 nm in diameter, and assemble into fibrils in which the molecules are packed in similar arrays. The fibrils vary in size in different tissues.

- **Type I** is found in tendons, bones, skin and cornea. Each molecule is made up of two α_1 and one α_2 chains. The molecules are packed side-by-side in parallel bundles, creating fibrils 50 nm in diameter and several millimetres in length.
- **Type II**, which is the major form of collagen in cartilage, has three identical polypeptide chains, all α_1. It forms fibrils of smaller diameter than type I with fewer collagen molecules per fibril.
- **Type III**, in which all chains are α_1, is found in blood vessels, scar tissue, intestinal and uterine walls.
- **Type IV** is composed of two α_1 chains and one α_2 chain. It is found in the basal lamina and retains globular sections at the ends of the molecule. A flexible region in the centre of the molecule enables it to form network arrays.
- **Type V** is made up of one each of the α_1, α_2 and α_3 chains, and generally forms smaller aggregates. These molecules are frequently associated with type I collagen and are thought to regulate the diameter of the fibril.
- **Type VI**, which is a much shorter triple helix with globular zones at both ends of the molecule, forms fibrils linking the type I fibrils in some tissues.
- **Type IX** tends to be associated with type II fibrils and has a flexible joint disrupting the regularity of the helix. Chondroitin sulphate is attached at this joint and protrudes from the fibril in a manner that probably determines interactions between adjacent fibrils, and which may be important for anchoring the fibril in the ground substance.

The assembly of collagen molecules into fibres reflects the functional requirements of the tissue

In the **cornea** of the eye collagen is packed into planar sheets. In each sheet the fibres lie parallel, but adjacent sheets lie perpendicular to one another. The sheets are stabilized by keratan sulphate interspersed between them. This array is important in maintaining the transparency of the cornea.

In **tendons**, which connect muscles to bones, collagen is arranged into parallel bundles, with the fibrils oriented in the direction in which stress will be applied, which confers **tensile** strength on the tendon.

Collagen molecules in **cartilage**, which has a load-bearing role, are randomly oriented rather than being organized in any regular pattern. This allows cartilage to resist **compressive** forces.

In **skin**, the necessary but limited extensibility is conferred by layering successive sheets of collagen at a variety of different angles.

The properties of connective tissue reflect the molecular composition of the ground substance as well as the type of collagen

The different properties of connective tissues can be explained by the following variable factors.

Box 15.2 Destruction of connective tissue

Excretion of hydroxyproline or hydroxylysine in urine is diagnostic of the breakdown of connective tissue. Because connective tissue protein usually is very stable and not readily degraded, even in individuals suffering severe deficiency of dietary protein, the appearance of these amino acids in the urine provides an unequivocal indicator of abnormal degeneration of connective tissue.

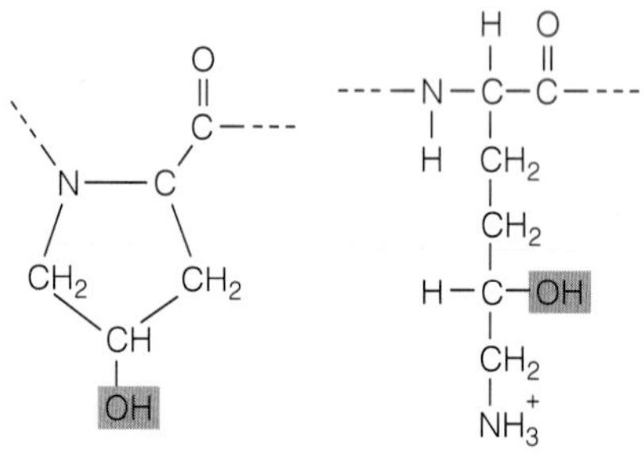

Figure 15.20 Structures of hydroxyproline and hydroxylysine.

- Tissues composed predominantly of types I, II or III collagen may also contain small amounts of the minor forms of collagen, which differ in structure and confer different properties on the tissue.
- Collagen molecules contain different amounts of covalently bound carbohydrate.
- The composition of the ground substance has a central role in determining the properties of connective tissue.

Collagen molecules have a distinctive amino acid composition

Every third amino acid in the primary sequence of collagen is glycine, and 11% of the residues are alanine. As there are three amino acids per turn of the left-handed helix, all of the glycine residues are positioned along the same side of the helix. When the three chains intertwine to form the superhelix, these glycine-rich faces interact at the centre of the triple helix. Glycine is the only amino acid small enough to be accommodated at the centre of the triple-stranded helix. If one glycine residue is replaced by a different amino acid, the stability of the helix is disrupted and the collagen molecule will not be incorporated into a fibre. In collagen type I the two sequences –gly–pro–*y*– and –gly–*x*–OHpro– (where *x* and *y* may be any other amino acid) occur more than 100 times each, accounting for about 60% of the molecule.

Collagen contains two hydroxylated amino acids, hydroxylysine and hydroxyproline, which are not found in proteins other than connective tissue. Of the total residues 15–25% are either proline or hydroxyproline (Figure 15.20). These highly inflexible residues determine the pitch of the helix and add to the rigidity of the structure. Hydroxyproline residues also contribute to hydrogen bonding, possibly through bridging water molecules. The proportion of proline and hydroxyproline residues is tissue specific and confers on the collagen properties essential for tissue function.

The number of hydroxylated lysine residues also varies from one collagen type to another: type I generally has fewer than ten hydroxylysine residues per chain, type II has more than 20 per chain and type IV may have more than 40 per chain. These residues determine the potential sites for addition of carbohydrate units and the potential sites for the formation of interchain cross-links.

Synthesis of procollagen polypeptide chains

Most collagen is synthesized by fibroblasts, but epithelial cells also produce collagen. Synthesis occurs on the rough endoplasmic reticulum and glycosylation occurs in the Golgi complex.

Each polypeptide chain is synthesized with additional 'pro-' peptides at the *N*- and *C*-termini (Figure 15.21). These sequences contain large numbers of cysteine but few glycine residues, so they are unable to form the left-handed collagen helix.

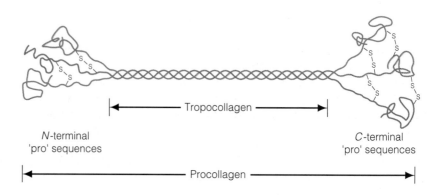

Figure 15.21 The structure of procollagen molecules.

Cysteine residues have a key role in aligning the three chains. They form interchain disulphide bonds, which lock the chains together in the correct alignment as they twist into the triple helix. The 'pro-' sequences, amounting to about 20% of each polypeptide, are removed from most types of collagen before assembly into fibres.

The importance of the pro- sequences has been illustrated in experiments in which mature collagen is denatured by heating. Heated mature collagen unwinds into a globular conformation, forming the substance we call **gelatin**. It is not possible to renature this protein into the fibrous form. In contrast, denatured procollagen *will* renature, the pro- sequences directing the alignment of polypeptide chains into a triple helix.

Synthesis of procollagen is completed by postsynthetic modifications

Hydroxylation reactions

Hydroxylation of proline and lysine residues occurs only after protein synthesis is completed. Hydroxylated amino acids can not be incorporated into collagen. Enzymes that hydroxylate lysine or proline – **lysyl hydroxylase** and **prolyl hydroxylase** – are located on the rough endoplasmic reticulum. These enzymes require ascorbate (vitamin C) to maintain the reduced state of iron, which is an essential cofactor for these reactions.

$$\text{α-Ketoglutarate + proline (lysine) + O}_2 \xrightarrow{\textit{Fe}^{2+}\textit{ascorbate}} \text{Succinate + OH-proline (OH-lysine) + CO}_2$$

Only those lysine and proline residues in the sequences

$$-x-pro-gly-\quad \text{or}\quad -x-lys-gly-$$

are hydroxylated. Hydroxylation of lysine residues is the prerequisite for the subsequent glycosylation and provides the sites for the cross-linking reactions that give rise to the ageing process of collagen fibres.

Glycosylation reactions

Sugar residues are covalently attached to the hydroxyl groups of hydroxylysine residues. Each monosaccharide residue is added from an activated substrate, a uridine diphosphate (UDP)–sugar, in a reaction catalysed by a transferase specific for the sugar being added. Generally, the first residue added is galactose and the second is glucose. These transferases are Mn^{2+}-dependent; an example of the need for manganese as a trace element in the diet.

Chapter 8. When glucose is plentiful it is stored as glycogen, page 224.

$$\text{UDP-Galactose+} \atop \text{~OH-lysine ~} \xrightarrow{\textit{Galactosyl transferase}} \text{~Galactosyl-OH-lys~ + UDP}$$

Sugars are not added to all hydroxylysine residues in procollagen. The pattern of glycosylation for any procollagen molecule is determined principally by the protein sequence and structure. There is no template determining the order in which sugars are added. Instead, sugars are added to hydroxylysine residues in a suitable three-dimensional environment and reflecting available sugars and transferases. Each transferase is specific not only for the sugar but also for hydroxylysine and the adjoining protein structure.

The total number of sugar branches attached to hydroxylysine residues in collagen varies between tissues, although generally each unit is no larger than a disaccharide. As even these small sugar branches are bulky and attract water, their number and orientation about the helix has an impact on the packing of tropocollagen molecules into collagen fibres. Although tendon collagen possesses six carbohydrate units per molecule, collagen molecules in the lens of the eye may have as many as 110 carbohydrate units.

'Pro' sequences are cleaved in the extracellular space

Tropocollagen molecules are secreted from fibroblasts by a mechanism which recognizes only those molecules for which the hydroxylation and glycosylation reactions have been completed. The 'pro-' portions of the molecule are then cleaved by **procollagen carboxypeptidases** and **aminopeptidases**, enzymes that are strongly inhibited by sequences in which every third amino acid residue is glycine. Thus, only the 'pro-' sequences are removed, leaving the completed tropocollagen molecule. The nonhelical portions of types IV and V collagen are not removed after secretion, accounting for the fact that these types do not assemble into fibrils.

Tropocollagen molecules aggregate into collagen fibres

Collagen fibrils of types I, II and III are formed by a head-to-tail alignment of tropocollagen molecules. This alignment includes a 40 nm gap between the head of one molecule and the tail of the next.

The importance of this gap appears to be two-fold: first, as fibres age, covalent cross-links are formed between molecules and this gap allows access of the enzyme **lysyl oxidase**, which catalyses the formation of cross-links; secondly, for bone formation, the gap between collagen molecules in these rows may be important in providing the nucleation sites for mineral deposits.

Fibres are formed from collagen molecules layered into parallel arrays, with individual molecules in one head-to-tail row shifted by one-quarter of their length (67 nm) from those in the next row. This alignment, described as the **quarter stagger array** (Figure 15.22), gives collagen fibres the characteristic banding pattern seen under the electron microscope. Interactions between molecules in adjacent rows are noncovalent in newly assembled fibres, and depend on the clustering of charged and uncharged amino acid residues at intervals along the triple-helical molecules. As the fibre ages, covalent cross-links are formed to stabilize and strengthen the fibres, which become increasingly insoluble and thicken progressively. Mature fibres may vary in diameter from 10 to 200 nm, depending on the tissue.

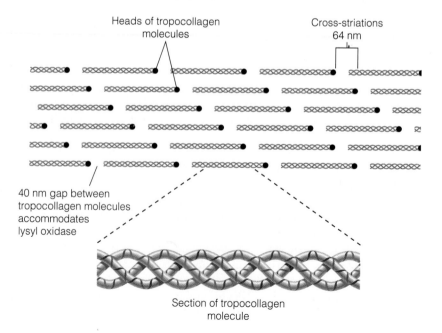

Figure 15.22 The quarter-stagger array of collagen fibres. Collagen fibres are formed when tropocollagen molecules line up in head-to-tail arrays with a 40 nm gap between molecules. In adjacent rows the molecules are staggered by one quarter of the head-to-head length. This gives rise to a characteristic banded pattern seen in the electron microscope.

Box 15.3 Clinical disorders arise from errors in collagen synthesis

The causes of many clinical disorders primarily affecting connective tissue are now understood in molecular detail. Some are genetically transmitted disorders, but others occur as spontaneous mutations within an individual. In some cases a single name is used to describe a syndrome, the clinical expression of which can be highly variable. This is at least partly due to the fact that there may be more than one type of molecular change contributing to a similar clinical condition.

Osteogenesis imperfecta

This is a group of genetic disorders caused by the synthesis of abnormally short α_1 chains or by substitution of a glycine residue. In the first case, when abnormally short chains associate with α_2 chains of normal length, a stable helix can not be produced and these molecules are not assimilated into a fibre. The replacement of a glycine by a cysteine or an arginine residue blocks the formation of a stable triple helix. Osteogenesis imperfecta is manifest as brittle bones, children suffering repeated fractures and bone deformities.

Ehlers–Danlos syndrome

This group of disorders arises from structural weakness of connective tissue. Individuals with this syndrome may have hyper-extensible skin and suffer recurrent joint dislocations. The problem may be an inherited deficiency of the enzyme lysyl hydroxylase. The collagen produced has fewer hydroxylated lysine residues, which leads to deficient cross-linking of molecules during ageing. In some individuals procollagen peptidase activity is abnormally low, so the conversion of procollagen to tropocollagen is not complete. Molecules retain some or all of the 'pro-' sequences and are unable to align into head-to-tail arrays that initiate the assembly of fibres.

Marfan's syndrome

Marfan's syndrome is an inherited disorder, manifest as an apparent defect in cross-linking collagen molecules in the fibre. In patients suffering from this disorder the amount and activities of lysyl hydroxylase and oxidase are normal; the problem is believed to arise from a mutation leading to the insertion of extra amino acids near the C-termini of α_2 chains. Because of this, lysine and hydroxylysine residues are not in an alignment that permits the formation of cross-links. It has recently been established that a small glycoprotein, **fibrillin**, which is important as an elastic element of the extracellular matrix, is present in reduced amounts or is defective in tissue affected by Marfan's syndrome. The syndrome affects the strength of many tissues, primarily in the cardiovascular system, where weakening of the aortic wall may result in aortic rupture. The ocular and skeletal systems, lungs and nervous system may also be affected.

Collagen fibres are stabilized and strengthened by the formation of covalent cross-links between molecules

Covalent linkages between the component chains of a single tropocollagen molecule and between adjacent molecules are formed by lysyl oxidase. Lysyl oxidase is a copper-dependent enzyme which oxidizes the ϵ-amino groups of lysine residues to aldehyde groups. Subsequent nonenzymic condensation of aldehyde groups on different chains gives rise to the covalent cross-link that secures the component molecules within the fibre (Figure 15.23). These cross-links connect sites near the N-terminus of one molecule to sites near the C-terminus of another. The extent of cross-linking varies between tissues (Figure 15.24).

Collagen is a metabolically inert protein

Even in individuals suffering a severe nutritional deficit of amino acids, collagen is unlikely to be degraded. However, if limbs are immobilized for very long periods, collagen may begin to disappear from bones.

Collagen is also degraded by bacterial or mammalian collagenases, enzymes that specifically hydrolyse gly–pro peptide bonds. The bacterium *Clostridium histolyticum* secretes a battery of collagenases as a mechanism for invading connective tissue and some mammalian collagenases are important for the growth and remodelling of tissue. The mammalian enzymes have essential roles in the

(a)

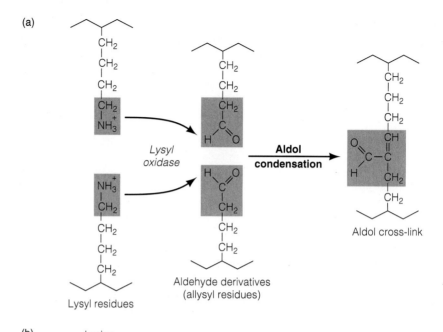

Lysyl residues

Aldehyde derivatives
(allysyl residues)

Aldol cross-link

(b)

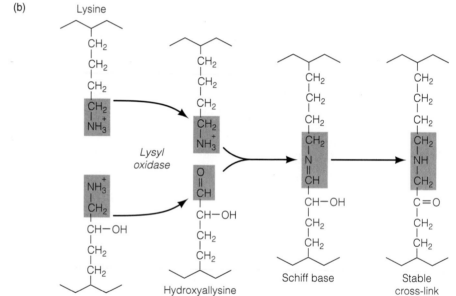

Lysine

Hydroxylysine
residue

Hydroxyallysine
residue

Schiff base

Stable
cross-link

Figure 15.23 Formation of collagen cross-links. Where the ε-amino groups of two lysyl or hydroxylysyl residues are oxidized to aldehydes by lysyl oxidase (a) a spontaneous aldol condensation between the two allysyl (or hydroxyallysyl) residues produces an aldol cross-link. If only one of the residues is oxidized (b) a nucleophilic addition reaction occurs between one allysyl residue and an unmodified residue.

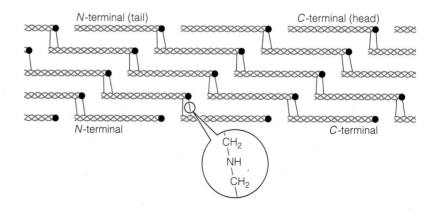

Figure 15.24 Cross-linked collagen molecules. The 'ageing' of collagen fibres involves the formation of covalent cross-links between tropocollagen molecules in adjacent rows. Cross-links are formed near the ends of molecules by lysyl oxidase, which occupies the 40 nm gaps between molecules.

formation and healing of scar tissue and in the remodelling of the walls of the uterus after birth. The activity of these mammalian collagenases is carefully controlled by specific inhibitory proteins.

Elastin

Elastin is a globular protein found in connective tissues that require elasticity in addition to tensile strength. Most elastin is found in the walls of blood vessels and in the lungs, skin and ligaments. This protein, like collagen, is one-third glycine residues. It is also rich in proline, but has very little hydroxyproline and no hydroxylysine. Individual molecules are probably formed of helical arrays of β-turns (this is called a β spiral).

In elastic tissue the globular units of elastin are cross-linked into three-dimensional arrays by the formation of covalent bonds between lysine residues. Generally these cross-links are formed between four lysine residues in which ε-amino groups have been oxidized to aldehydes (Figure 15.25). The enzyme catalysing this reaction is Cu^{2+}-dependent. In individuals lacking the required dietary trace of copper the tensile strength of blood vessels is diminished and the aorta does not recoil properly from the stretch imposed as blood is expelled from the heart.

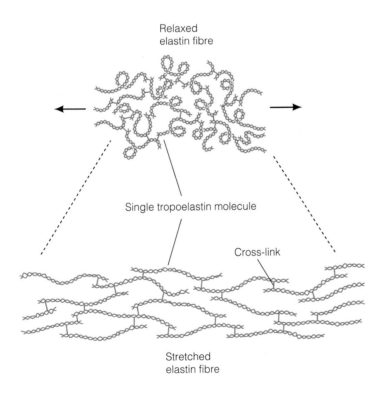

Relaxed elastin fibre

Single tropoelastin molecule

Cross-link

Stretched elastin fibre

Figure 15.25 The structure of elastin fibres.

Individual elastin molecules can be stretched, elongating the protein, but recoil to a tight globular form. Intermolecular cross-links are probably important in allowing the tissue to return to its original shape after stretching. This property is important in damping out the pulsations as blood is ejected from the heart and smoothing the flow of blood through the vessels.

Elastic tissue generally contains mixtures of elastin and collagen. The collagen fibres act to limit the extent of stretching, thereby preventing tearing of the tissue.

The mineral composition of bone

About 99% of the body's calcium is found in bones and teeth, with only 1% in the soluble phase of cells and blood plasma. Calcification of bones and teeth occurs when crystals of **hydroxyapatite** (a complex of calcium, hydroxyl and phosphate ions, with the general form $Ca_{10}(PO_4)_6(OH)_2$) are laid down between the collagen fibres of these specialized forms of connective tissue. The calcium and phosphate ions within calcified tissue are in a dynamic equilibrium with those in plasma. Bone is the major repository for both calcium and phosphate, whence it may be mobilized to meet the demands of other tissues. This equilibrium is regulated by vitamin D (Chapter 14) and by two peptide hormones, **parathyroid hormone** and **calcitonin** (Chapter 11), the latter being of relatively minor importance in humans. Bone also contains 70% of the citrate, 60% of the magnesium and 50% of the sodium found in the body.

Chapter 14. Vitamin D has a key role in calcium regulation, page 446.
Chapter 11. Parathyroid hormone, page 346.

Bone is constantly formed and degraded by specialized cells

The constant turnover of bone crystal is important in providing a mechanism for growth and repair after accidental damage to the skeleton. Three types of bone cells participate in the formation and maintenance of bone.

- **Osteoblasts** are migratory cells responsible for the synthesis and secretion of the collagen fibres and proteoglycans that form the matrix, which is subsequently calcified. These cells also produce a small calcium-binding protein called **osteocalcin**, which may have a role in the mineralization of bone.
- **Osteocytes** lie within the bone structure, developing long filamentous projections of their plasma membrane which enable them to interlace the crystals.
- **Osteoclasts** are found on the surface of bone crystals and appear to be important in the mechanism of bone resorption by which serum concentrations of calcium and phosphate ions are maintained.

Osteoblasts and osteoclasts are in physical contact through extensions of their plasma membranes. These contacts allow osteoblasts to signal to osteoclasts in a complex mechanism by which bone calcium is mobilized. The primary messenger for this mechanism is **parathyroid hormone**, which binds to receptors on the osteoblast membranes, stimulating them to release activators such as interleukin-1 and prostaglandins. These activators stimulate the osteoclasts which, over a period of 2–3 hours, increase in size, divide and increase their hydrolytic activity. The outcome is the breakdown of connective tissue and release of calcium and phosphate into the circulation. Parathyroid hormone may also act directly on osteoclasts, through a rapid mechanism which stimulates them to release calcium from their mitochondrial stores.

Osteocytes also have a central role in the mineralization process. As hydroxyapatite crystals are laid down among the collagen fibres, bone-forming cells secrete hydrolytic enzymes that degrade the proteoglycans, which are progressively displaced by the mineral crystals. Fully mineralized bone has little ground substance and is essentially water free.

Abnormal calcification of tissues may occur when hydroxyapatite crystals are deposited, particularly in the kidney or along artery walls. Until a great deal more is understood about the normal mechanism of calcification, it will remain difficult to understand or prevent the occurrence of these abnormal crystal deposits.

Theories of calcification

We do not yet fully understand the signals that determine which types of connective tissue will become calcified. There are, however, a number of plausible theories, any or all of which may be applicable.

Box 15.4 Calcification of arterial walls in atherosclerosis

Calcification of atherosclerotic plaque is not simply a passive precipitation of calcium salts, but the deposition of hydroxyapatite in the same crystalline form as that found in the structure of bone. The cells of arterial walls contain several bone-related proteins, including **osteopontin** (which has a role in crystal formation), **osteonectin** (a bone glycoprotein), **osteocalcin** (a protein that has a role in bone remodelling) and a potent **bone differentiation** protein. This suggests that calcification of arterial walls may be due to osteogenic differentiation of cells of the artery wall. It has been suggested that osteogenesis is stimulated by an unidentified factor within the atherosclerotic plaque.

Box 15.5   Disorders of bone metabolism

Abnormal bone loss

In conditions where plasma calcium levels are persistently low, secretion of parathyroid hormone is increased and calcium is mobilized from bone. This can lead to progressive demineralization of bone. There are several recognized causes of this. The most common is **osteoporosis**, which commonly occurs in postmenopausal women as the amount of circulating oestrogen decreases. This problem can be corrected by treatment with oestrogen. It also occurs in patients with renal disease and in patients whose limbs are immobilized for long periods of time.

Failure to form bone

Rickets is a condition in which bones and teeth are poorly calcified. It is due to a deficiency of vitamin D (see Chapter 14). The condition was widespread among children as recently as the 1930s and 1940s. The observation that it did not occur among groups of people consuming a diet rich in fish or fish oils lead to the identification of vitamin D deficiency as the central cause.

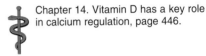

Chapter 14. Vitamin D has a key role in calcium regulation, page 446.

Robison's 'Booster' theory

As is the case for all crystalline materials, calcium phosphates will crystallize if the product of concentrations of the contributing ions exceeds the solubility product.

For the reaction

$$3Ca^{2+} + 2PO_4^{3-} \leftrightharpoons Ca_3(PO_4)_2$$

the solubility constant is expressed as

$$K_{sp} = [Ca^{2+}].[PO_4^{3-}]$$

where the K_{sp} for this reaction is 25 at pH 7.0.

In a biological system, crystallization processes can not be looked at in such simple terms, because a substantial fraction of the ions present may be bound to membrane molecules or to other proteins. Bone contains relatively high concentrations of the enzyme alkaline phosphatase, which has the capacity to remove phosphate groups from phosphate esters such as glucose-phosphate, thereby increasing the local concentration of free phosphate groups. Robison's theory predicts that the presence of this enzyme in bone-forming regions might result in the solubility product for $Ca_3(PO_4)_2$ being exceeded, with resultant initiation of mineralization. The theory has been criticized because other tissues that do not become mineralized also have high concentrations of alkaline phosphatase.

Nucleation mechanism

This theory arises from the observation that crystallization occurs most efficiently in an environment seeded with crystals, or in the presence of a matrix surface with charge distribution or geometry similar to that which would be formed by the crystals. It is thought that the collagen fibres might present such a matrix, but it is difficult to understand why all connective tissues are not then subject to calcification.

Role of mitochondria

Mitochondria are able to accumulate large amounts of calcium phosphate, which leads to the development of granules within them. If these packaged granules were to leave the cell by exocytosis and be deposited among the collagen fibres they might form mineralization centres.

PART E

MOLECULAR MEDICINE: IMMUNOLOGY, GENETICS AND MECHANISMS OF INFECTION

CHAPTER 16
Immunoglobins and defence
mechanisms of the body

CHAPTER 17
The molecular genetic basis of
disease

CHAPTER 18
Invaders of the body

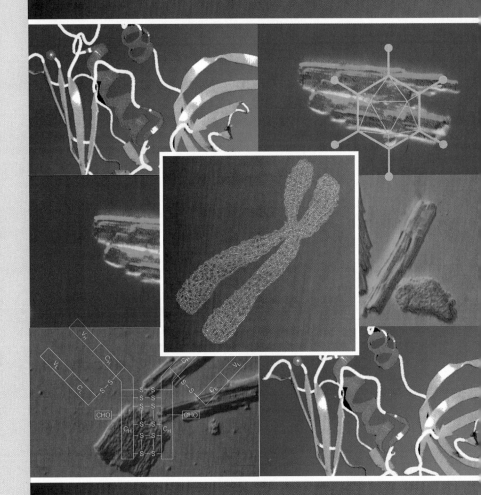

Immunoglobulins and defence mechanisms of the body

Animals respond by producing antibodies (immunoglobulins) to foreign macromolecules (antigens) that enter the body as soluble proteins, polysaccharides, DNA or, more usually, in the form of invading viruses, bacteria or parasites. Antibodies exhibit high affinity for the antigen(s) that induce their formation.

Antibody specificity is achieved by exposing pre-existing clones of B lymphocytes to antigen (clonal selection). The B lymphocytes proliferate and differentiate into plasma cells that secrete large quantities of antibody. Five major classes of antibodies exist: IgG, IgA, IgM, IgD and IgE. These have distinctive distributions in body fluids and separate physiological roles.

Antibodies react with antigens on the surfaces of invading microorganisms and trigger the complement cascade, which forms a multiprotein membrane attack complex that perforates the membrane of the infected cell, causing lysis and cell death.

The diversity of the antibody population reflects the presence of variable (V) genes for heavy (H) and light (L) chains, which are joined during lymphocyte differentiation to a limited number of constant region (C) genes.

Cellular immune responses are mediated primarily by T lymphocytes, which are triggered by peptide fragments presented on the surface of infected cells by the major histocompatibility complexes (MHCs).

The nature of the immune response

The major pathogenic (disease-causing) organisms of humans include viruses, bacteria and parasites. Humans and other mammals have evolved monitoring processes to enable the body to recognize invasion by such organisms, as well as responses to enable the body to combat the resultant infections. These processes collectively make up the **immune response**.

The immune response may be grouped into two functional 'arms': the innate (or 'nonspecific') immune system and the adaptive (or 'specific') immune system. The innate immune system exists to *prevent* infection of the host; the adaptive immune system serves to *eradicate* infection and to generate specific immunological memory to prevent *reinfection*. The nonspecific immune system shows no memory, has no power to discriminate between different categories of pathogen and always mounts a response of similar magnitude. The specific immune response is shaped to generate the response that will eradicate the pathogen quickly and effectively. A memory of the specific immune response is stored so that the response will be greater and faster if a particular pathogen is encountered again. The capacity of the specific immune system to display memory forms the basis for vaccination strategies.

The innate immune response is a collection of barriers to infection

A pathogen must pass the barriers of the innate immune response before it can successfully colonize the host. These barriers begin at the skin, which acts as a physical barrier and prevents a pathogen gaining access to sites within the host where it can survive and replicate. A number of chemical and biochemical barriers at various anatomical locations serve to eradicate pathogens:

- A thin film of the long-chain fatty acid oleate from sweat covers most skin surfaces. It is toxic to bacteria and any enveloped viruses that may come into contact with it.
- The pH of the stomach is low due to the presence of HCl; this chemical barrier eliminates many of the pathogens ingested with food.
- Lysozyme, an enzyme in found tears, is bacteriostatic.
- Amylase, found in saliva, affords protection against some ingested pathogens.
- Some of the components of the plasma complement response spontaneously attack bacterial cell walls. Complement can also be harnessed by antibodies to provide a massive cytotoxic (cell-killing) impact on invading organisms (see later).
- A cellular barrier exists in the bloodstream and in the tissues in the form of cells designed to ingest and destroy foreign particles; these phagocytic cells include monocytes (which include macrophages) and polymorphonuclear leukocytes (eosinophils, basophils and neutrophils, known as 'polymorphs'). In a manner similar to that of the lytic proteins of the complement cascade, polymorphs can also be specifically directed by antibody to attack a single target, thus focusing their cytotoxic potential.
- The 'senses' can also be regarded as a nonspecific immune response. Thus, an individual who is allergic to pollen (the allergy itself is an immunological disorder) is likely to avoid locations where flowering plants are abundant.

If the innate immune system is breached, the adaptive immune system responds

The barriers of the innate immune system protect the host from most (probably more than 99%) of the pathogens with which it comes into contact. However, if the barriers of the nonspecific system are breached, for example by a cut or a burn, and a pathogen gains access to the host tissues, the adaptive immune system responds immediately. The key aspects of the adaptive immune system are **specificity, adaptivity** and **memory**. When a foreign organism enters the host, the adaptive response is specific to that pathogen. Thus, infection with measles virus will not induce antibodies to polio virus or vice versa. Moreover, the type of response is that which will eradicate the organism most efficiently. For example, antibodies are useful for eradication of blood-borne bacterial diseases, but cytotoxic T cells are most suited to eradicate intracellular pathogens such as viruses. Finally, the host retains an immunological memory of the infection so that if it encounters that pathogen again a more rapid response is mounted and clinical symptoms of the disease either do not develop or are reduced or minimized.

The basis for the specific or adaptive immune response resides in the lymphocytes

Lymphocytes are of two types: T lymphocytes, which mature in the *t*hymus, and B lymphocytes, which arise in the *b*one marrow. The thymus and the bone marrow are the primary immune tissues; the sites of storage of lymphocytes (the spleen, lymph nodes, tonsils, blood) are the secondary lymphoid organs, sometimes referred to collectively as peripheral lymphoid tissue.

The adaptive immune system possesses two principal components.

- The **humoral** immune response is primarily directed against soluble antigens. The B lymphocytes are stimulated to form plasma cells which synthesize specific immunoglobulin molecules (antibodies) that are able to bind with high selectivity to the foreign substance, either precipitating it directly or marking it for digestion by scavenging cells, the macrophages.

- The **cellular** immune response is primarily directed against foreign organisms or cells infected by viruses, which signal their presence to the immune system by having foreign motifs on their surface (Table 16.1). The T lymphocytes are responsible for recognizing and killing these invaders. They also stimulate the humoral immune response by interacting with B cells, the precursors of plasma cells.

Antibodies are secreted in large amounts by plasma cells

When a foreign particle (the immunogen) is detected the immune system responds by producing proteins, called **antibodies** or **immunoglobulins**, each with a defined specificity for the immunogen that elicited its synthesis.

Table 16.1 Cells of the immune response

Cell type	Location	Function
B Lymphocytes	Blood, lymphoid organs	Production of antibodies
T Lymphocytes (CD4$^+$)	Blood, lymphoid organs	Regulation of immune responses, production of cytokines
T Lymphocytes (CD8$^+$)	Blood, lymphoid organs	Cytotoxic killing of virus-infected and cancerous cells
Macrophages	Tissues, lymphoid organs	Antigen presentation, phagocytosis and intracellular killing of bacteria
Monocytes	Blood	Presentation of antigen, intracellular killing of bacteria
Neutrophils	Blood	Phagocytosis and intracellular killing of bacteria
Eosinophils	Blood	Phagocytosis and intracellular killing of bacteria
Basophils	Blood	Eradication of parasites, phagocytosis and intracellular killing of bacteria
Dendritic cells	Lymphoid organs	Presentation of antigen

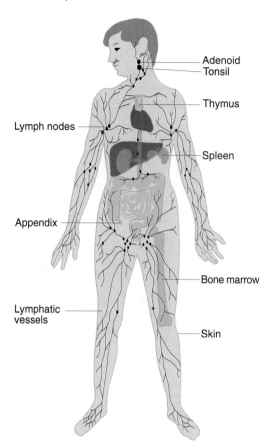

Figure 16.1 Major organs of the human immune defence system

Although the immune system has the ability to mount a response to any foreign substance, not all agents induce this response. A foreign substance that generates an antibody response is called an **immunogen**; any agent to which antibody can bind is termed an **antigen**. Thus, some molecules are poorly immunogenic (for example nucleic acids) but antibodies bind to them. The delineation is size: small molecules ($M_r < 10\ 000$) tend to be very poor immunogens in their own right. However, if small molecules (which are referred to as **haptens**) are linked to a larger molecule, usually a protein, they are rendered potently immunogenic and an antibody response is induced; the antibodies so formed readily bind to the free hapten. Haptens are therefore poor immunogens but good antigens. This distinction is not only academic, as certain drug allergies with clinical consequences can be traced to this mechanism (see Box 16.2)

The primary immune response is characterized by IgM production and the secondary response by IgG production

If an antigen enters the bloodstream, antibody specific for that antigen appears in the bloodstream within a few days. The antibodies secreted during this **primary response** are principally of the immunoglobulin M (IgM) class (see below). After reaching a peak 7–10 days later IgM levels start to decline, being replaced by antibodies of the immunoglobulin G (IgG) class. Peak levels of IgG are achieved about 3 weeks after infection. This pattern of antibody production is characteristic of the primary response, but also indicates that immunological memory has been established. At this stage, if a booster injection of antigen is given, a rapid and massive increase in the amount of IgG occurs over the next 48 hours and high levels persist in the circulation for extended periods. This accelerated IgG production is characteristic of the **secondary response**, sometimes referred to as the **challenge,**

Figure 16.2 Time course of antibody production and isotype switching after exposure to a foreign antigen in rabbits. Kinetics of appearance of antibody classes IgM and IgG following injection of antigen.

Box 16.1 Congenital immunodeficiency diseases

A number of inherited diseases have shaped our understanding of how the immune system works. A number of these conditions are X-linked and so are found predominantly in males. The most lethal of these is **severe combined immunodeficiency** (SCID), in which the sufferer is unable to produce either T or B lymphocytes. Such individuals develop severe and repeated bacterial and viral infections and the prognosis is extremely bleak. Few survive beyond childhood unless they are successfully treated by bone marrow transplantation or are kept in sterile, germ-free environments.

A marginally less severe condition is **DiGeorge syndrome** or **thymic aplasia**, in which the individual has no thymus. In this autosomal recessive trait, the sufferer is unable to produce T lymphocytes. Normal numbers of circulating B cells are present but they make only IgM type antibodies due to a lack of T-cell help (see main text). These children suffer repeated viral and fungal infections, but are able to resist bacterial diseases. DiGeorge syndrome exemplifies the importance of T cells in eradicating intracellular infections. It is often fatal.

The converse situation, in which individuals have no B cells but normal levels of T cells, is manifest in **Bruton's agammaglobulinaemia**, an X-linked condition. The lack of B cells and circulating antibody leaves the child vulnerable to blood-borne bacterial infections but, because numbers of T cells are normal, able to resist viral infections. This disease was fatal as recently as 30 years ago but is now readily treated by administration of antibiotics.

The importance of the nonspecific system is illustrated by another X-linked disease in which the phagocytes are able to take up foreign organisms, but fail to destroy them. **Chronic granulomatous disease** (CGD) results from a defect in the gene for cytochrome b_{245}, which renders the individual unable to generate the oxidative burst necessary for killing pathogens within the phagocyte.

---

memory or **recall response**. Not only is more IgG produced in a shorter time period, but the overall affinity of the antibodies for antigen produced in the secondary response is much higher than that observed in the primary response (Figure 16.2). In some cases the injected antigen will induce IgA and IgE class antibodies in the secondary response as well as IgG. A similar time course of antibody response is noted in response to any injected antigen, for example in vaccination with tetanus toxoid or immunization with polio virus vaccines (see Box 16.3).

Antibodies are formed by selection of pre-existing B-cell clones

Proteins such as haemoglobin have required millions of years of evolution to produce a polypeptide that fulfils its role with optimal efficiency. In contrast, the immune defences of the body ensure that immunoglobulins with high specificity

Box 16.2 Penicillin: hapten and immunogen

The active component of the antibiotic penicillin is the β-lactam ring, which interferes with bacterial cell wall biosynthesis and so kills them (Chapter 18). Most individuals exposed to penicillin suffer no ill-effects but some patients develop a severe allergy to the drug, which can be fatal.

Penicillin is small (M_r < 1000) and can act as a hapten. It is, however, a poor immunogen and generally does not induce an antibody response. In some individuals, the drug binds to platelet membranes; the platelet then acts as a carrier and the β-lactam ring is rendered immunogenic. On first administration, the patient develops a short-lived IgM-type antibody response, which usually passes undetected; however, immunological memory has been laid down. A second treatment with penicillin will induce a rapid and substantial IgG antibody response. The antibodies attach to the β-lactam ring bound at the platelet surface, targeting complement and phagocytes to the platelet and ultimately destroying it. Because platelets are central to the blood clotting response (see Chapter 12), the outcome can be fatal. Once sensitized to penicillin, an individual must not be treated with the drug as a potentially lethal allergic reaction will be provoked.

Box 18.3. The modes of action of some antibiotics that inhibit bacterial protein synthesis, page 558.

Chapter 12. Platelets play an important role in blood clotting, page 360.

Box 16.3 Polio virus vaccines

Polio virus is transmitted on water droplets and infection usually occurs following inhalation of the virus and viral entry to epithelial cells of the upper respiratory tract. The consequences of infection can be severe, although effective vaccination has made the disease rare. As polio virus infects via the upper respiratory tract, it follows that the most useful antibodies will be found at that location. The principal protective antibodies in the gut and airways are of the secretory IgA (sIgA) type.

Vaccination trials have shown that disabled live viruses ('attenuated' viruses) usually induce the best response. Three types of vaccine have been used for polio virus: a killed virus preparation given intramuscularly; the same preparation given as a spray into the nose and mouth; an attenuated virus administered as an intranasal inoculation. Intramuscular injection results in high titres of IgG in the bloodstream but little or no sIgA in the mouth or nasal secretions; it affords essentially no protection against the virus. Killed virus as a nasal spray produces a transient but small increase in sIgA levels and a similarly brief period of protection against the live infectious poliovirus. The attenuated virus vaccine given intranasally results in rapid elevation of anti-polio virus antibodies of the sIgA class, which persist for several years and fully protect against the live virus.

These studies demonstrate that live viral vaccines are often better than killed viruses or subunit preparations and that the route of administration is important. The polio virus vaccine available today is administered as a few drops of attenuated live virus in saline directly into the mouth (younger patients usually receive it on a sugar cube).

for a particular antigen are present in high concentrations within a few days of first encountering the antigen.

The **Clonal Selection Theory**, which explains this remarkable phenomenon, requires that a vast range of antibody-producing cells must exist in the human body, independent of exposure to any specific antigen. The production of a particular antibody or set of antibodies is stimulated by the presence of a given antigen (Figure 16.3).

Figure 16.3 Mechanism of the immune response – the clonal selection theory. Precursor stem cells, produced in bone marrow, differentiate and migrate to lymphoid tissue. Each individual cell presents a specific antibody on its cell surface. The cell is stimulated into rapid proliferation if it encounters the appropriate antigen and secretes large amounts of soluble antibodies.

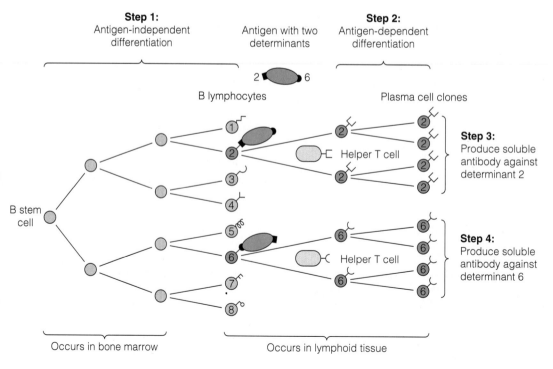

Antibodies raised to a single hapten or epitope are normally heterogeneous

The normal antibody response to a polypeptide antigen is **polyclonal** in nature, involving stimulation of several hundreds to several thousands of B-cell clones. Although every clone in a polyclonal response reacts to the same immunogen, each is specific for a precise **epitope** or **antigenic determinant** on that antigen. A specific epitope may be only a tiny portion of the whole molecule or cell that is the antigen. A repertoire of antibodies is thus produced, each of which differs in specificity and affinity although they act in concert to neutralize the one antigen.

Antibody responses produced by conventional techniques of immunization are always polyclonal in nature, with many different B cells responding to the immunogen. Every B cell proliferates into a family of plasma cells, each of which will secrete identical antibodies to the same epitope.

Monoclonal antibodies are produced by cells derived from a single B lymphocyte

An important advance in understanding clonal selection was the isolation of B-cell clones from patients with tumours of certain leukocytes (plasma cells) called plasmacytomas (myelomas). These B cells secrete large amounts of immunoglobulin of a single epitope specificity. Multiple myeloma is a form of leukaemia in which a single B lymphocyte or stem cell has been transformed to a malignant form such that this individual clone divides uncontrollably and continues to produce a single type (i.e. monoclonal) of antibody. Human myeloma cells may be cultured *in vitro*, producing large quantities of monoclonal antibody for study.

The clonal selection theory is now understood in molecular and cellular terms

Precursors of mature B cells in the bone marrow, termed stem cells, differentiate to become B lymphocytes. Each B cell has molecules of a single type of antibody anchored to the outer surface of its plasma membrane through a hydrophobic tail. This surface antibody confers on each B cell the ability to recognize one specific antigen. When antigen binds to these surface antibodies, the B-cell clone is stimulated to replicate, generating many daughter cells (Figure 16.3). Some of these, the **effector** or **plasma cells**, secrete large amounts of immunoglobulin with the same antibody-combining site as the original membrane-bound form but lacking the hydrophobic tail. Some of the cells in the clone become **memory cells**, persisting in the blood, and particularly in peripheral lymphoid tissue, for extended periods. The presence of memory B cells permits the body to respond rapidly and vigorously if subsequently exposed to the same antigen.

Chapter 3. Membrane-associated proteins are categorized by the nature of their interaction with the lipid bilayer, page 53.

Proliferation of B cells is enhanced by a specific subset of T cells, the **helper T cells**. Helper T cells interact directly with B cells that display the appropriate foreign epitopes on their cell surface (see below). This interaction enhances proliferation of B cells through the provision of B cell growth factors secreted by activated helper T cells.

Chapter 16. T-cell receptors interact with foreign peptides presented on the surface of infected cells, page 500.

Immature B cells exposed to 'self' antigens are destroyed

B cells in bone marrow are an immature form. If they bind antigen at this stage they undergo **apoptosis** (programmed cell death). Mature B cells exposed to antigen undergo growth and cell division. Maturation of B cells involves changes in their signalling pathways (see Chapter 11), but how these changes determine the difference in the way that B cells respond to antigen binding has not yet been determined. Destruction of immature B cells that have encountered and responded to

Chapter 11, Hormones and transmitters are classified according to the relationship between site of production and target site, page 332.

antigen in the bone marrow is an important protective mechanism because antigens available in bone marrow are likely to be 'self' antigens. The survival of these B cells would create the potential for **autoimmunity**, reflecting the survival of B cells programmed to destroy the host tissue.

An understanding of immunological defence mechanisms has stimulated many areas of research

Antibodies are important research tools for the scientist. They also serve as specific probes in the diagnosis of infection, tissue damage, cancer and in the routine measurement of many blood and tissue components. The molecules and genes involved in the immune response are currently being identified and a better understanding developed of the cell–cell interactions and control processes.

The structure of immunoglobulins

Immunoglobulins have a multidomain organization

The five major classes of antibody molecules (IgA, IgD, IgE, IgG and IgM) perform a variety of functions that contribute to the repertoire of the immune response. All of these antibodies are variations on the basic immunoglobulin unit (Figure 16.4).

Immunoglobulins are multimeric proteins composed of two heavy (M_r 53 000) and two light (M_r 25 000) chains linked by disulphide bonds and forming a four-chain Y-shaped molecule. Both chains possess **variable** (V) and **hypervariable** regions located in the N-terminal regions. These are associated with the antigen-binding site. Both chains also possess **constant** (C) regions, which have essentially the same sequence in all antibodies of a given class and are involved in functions other than antigen recognition. Amino acids of both the heavy and the light chains contribute to the antigen-binding sites. Each immunoglobulin is bivalent, having two identical antigen-binding sites (Figure 16.4).

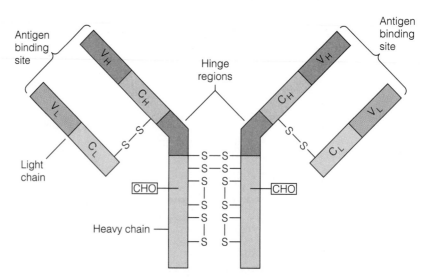

Figure 16.4 General representation of an antibody (IgG) molecule. Antibodies (immunoglobins) are composed of two heavy and two light chains linked by disulphide bonds. Both types of chains contain variable (V) regions, which are involved in determining the specificity for a given antigen, and constant (C) regions, which are the same in all antibodies of a given class. CHO represents carbohydrate components of the molecule.

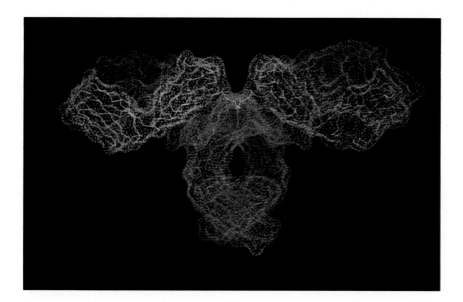

Figure 16.5 Three-dimensional compu graphics representation of the Y-shaped immunoglobulin molecule.

Antigen-recognition sites are located on the two individual 'arms' of the antibody molecule

The original evidence for the Y-shaped organization of the antibody molecule (Figure 16.6) came from electron microscopy. These studies also confirmed that antigens attach near the N-terminal regions of the 'arms' of the molecule.

Controlled proteolysis of immunoglobulin molecules generates monovalent antigen-binding fragments; the 'arms' of the Y-shaped molecule are F_{ab} fragments (Figure 16.6). These 'arms' bind antigen but the remaining 'stalk' of the Y-shaped molecule, designated F_c, does not.

The F_{ab} regions are joined to the F_c region by a flexible 'hinge', which permits wide variation in the angle between adjacent F_{ab} units. This conformational mobility, termed **segmental flexibility**, is an important property of antibodies for several reasons.

- If the distance between the two F_{ab} units is adjusted both sites can bind antigen so that, for example, adjacent molecules on a virus particle could be bound to the two sites of an immunoglobulin molecule.

- The affinity of an antibody for its antigen is increased 1000–10, 000-fold when both antigen-combining sites on the antibody are occupied.

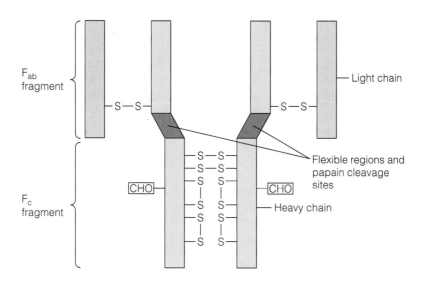

Figure 16.6 Generation of monovalent F_{ab} fragments and F_c fragment by proteolytic cleavage of IgG.

491

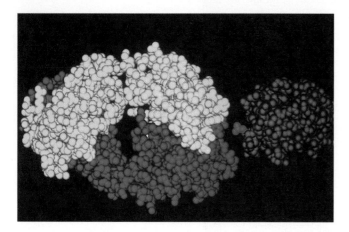
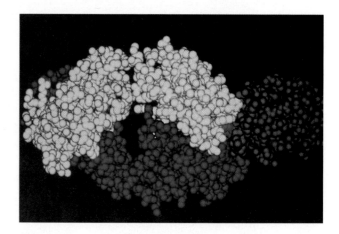

Figure 16.7 Interaction of lysozyme with the F_{ab} fragment of an antibody. The heavy and light chain fragments of the F_{ab} fragments are shown in blue and yellow, respectively. Part of the surface of lysozyme fits closely into the antigen-binding site formed by the light and heavy chains. Glu 121 (red) inserts into a cleft in the antibody and a number of surface contacts are made via hydrogen bonds.

- The transmission of subtle conformational alterations to the F_c region, once an antibody–antigen complex has been formed, stimulates the effector functions of this region, for example in complement fixation (see later in this chapter).

The detailed structures of several antigen–antibody complexes have been determined by X-ray diffraction. For example, the interaction of the F_{ab} fragment of an antibody to lysozyme with lysozyme is through close contacts between 16 residues on lysozyme and ten heavy-chain and seven light-chain amino acids on the antigen recognition site (Figure 16.7). The surfaces of antigen and antibody fit together in a highly complementary fashion, making multiple contacts over a large surface area in the two proteins. These contacts are supplemented by a smaller number of 'lock and key' type contacts similar to those seen in enzyme–substrate interactions.

Different classes of immunoglobulin have related structures

The major classes of immunoglobulin have characteristic distribution patterns in body fluids, reflecting their distinctive biological functions in different aspects of immune defence. The constant regions of the heavy chains are specific for the class of antibodies. In the most abundant immunoglobulin class, IgG, the heavy chains are designated γ; those in A, D, E and M are termed α, δ, ε and μ, respectively. Two kinds of light chain, termed λ or κ, are present in the various immunoglobulins.

Box 16.4 IgG: passive protection and damage

IgG molecules have the ability, unique among antibodies, to pass from the maternal circulation across the placenta and into the fetal bloodstream. This is advantageous for the neonate for two reasons:

- the fetus is unable to make its own IgG to respond to the new environment it is about to encounter;
- the mother provides the fetus with some protection against the spectrum of pathogens to which she has been exposed by transferring her antibodies across the placenta. The maternal antibodies persist for approximately 6 months in the neonate's circulation, which is why diagnosis of certain immunodeficiency diseases can take several months.

The transfer of IgG is not in any way selective: useful, useless and potentially harmful antibodies are all transferred with the same efficiency. This is vividly illustrated in the case of children who suffer **haemolytic anaemia of the newborn**, caused by maternal antibodies against the rhesus D antigen. This potent immunogen is found on erythrocyte membranes. If a mother who does not express this antigen carries a rhesus-positive child as could result, for example, if the father is rhesus D-positive and carries the same ABO blood group as the mother, and the child inherits these genes. A first pregnancy with a rhesus-positive child will generally pass without incident, but some fetal cells leak into the maternal circulation and the mother is exposed to many fetal cells during parturition. These events sensitize her to the rhesus D antigen. If a second pregnancy with a rhesus-positive child occurs, the mother's immune response generates IgG antibodies, which are transferred across the placenta, bind to fetal erythrocytes and recruit complement in the fetal bloodstream to destroy them. The fetus may be rendered severely anaemic or killed. This reaction will occur at all subsequent pregnancies with a rhesus-positive child.

Table 16.2 Summary of the structural features of the five immunoglobulin classes, their presence in particular body fluids and their biological activities

Antibody class	Size (kDa)	Oligomeric structure	Levels in serum (mg ml⁻¹)	Chain types		Distribution	Function
				Heavy	Light		
IgM	950	Pentamer	1	μ	κ or λ	Serum only	Primary response to new antigen
IgG	150	Monomer	12	γ	κ or λ	Serum and tissue fluids	Principal antibody in serum involved in humoral 'arm' of immune response. F_c region involved in effector functions, e.g. complement fixation
IgA	200–500	Monomers or dimers	3	α	κ or λ	Saliva, sweat, tears, other secretions	Immediate defence against invading organisms
IgD	170	Monomers	0.1	δ	κ or λ	Surfaces of B cells	Not known
IgE	190	Monomers	0.0001	ε	κ or λ	Various tissues	Binds to mast cells, releasing histamines during allergic responses. Elevated in individuals with allergies

IgG

The most abundant of the circulating soluble antibodies is IgG, which has the ability to traverse blood vessel walls and enter tissues. It is also important in triggering the activation of the complement pathway, which is instrumental in the destruction of foreign cells by promoting their lysis (Figure 16.9).

IgM

This immunoglobulin, produced in the primary immune response, consists of five molecules (subunits) of the basic four-chain Y-shaped immunoglobulin (Figure 16.9). The subunits are held together by a glycoprotein component called the J-chain, which stabilizes the structure of the polymer. The high molecular mass of IgM restricts it to the bloodstream.

IgG molecules have much higher affinity for a given antigen than IgM, but the IgM antibodies bind very tightly to the antigen because of their **avidity** for the antigen. Avidity is a function of the number of binding sites per molecule, which in the case of IgM is ten. Thus, although the intrinsic affinity of a single combining site of an IgM molecule is low, the rapid dissociation of a single combining site from an antigen is masked by the fact that of the nine other binding sites of the same low affinity on the same IgM molecule at least one is likely to remain bound to the antigen, thereby creating an illusion of high-affinity bonding. The same is true for IgG molecules but to a much lesser degree, as they are only bivalent. IgG binding sites do, however, have extremely high intrinsic affinity for their specific antigen.

IgA

IgA is found as a monomer similar to IgG in serum, and as a dimer in secretions such as sweat, tears and saliva and in mucosal fluids protecting the walls of the bronchi and small intestine. sIgA also has an important role in conferring passive immunity to the newborn infant (see Box 16.4), being found in high amounts in breast milk (colostrum). As either form, IgA interacts with and neutralizes invading microorganisms, preventing them from entering host tissues.

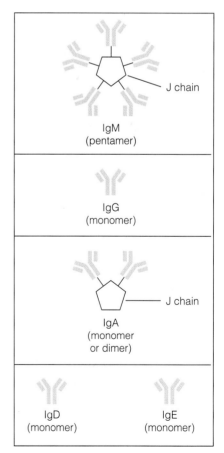

Figure 16.8 Schematic representation of the oligomeric structure of antibodies of the different classes.

Box 16.5   IgE and allergic responses

Normal individuals have a small amount (about 50 ng ml^{-1}) of IgE in their serum, but allergic or atopic people can have levels of IgE up to 10 000-fold higher. On first encountering an allergen (an antigen that can induce an allergic response), a normal IgM primary response occurs and immunological memory with a significant IgE component is established. There are no clinical symptoms at this stage. Most of the IgE is associated with the specific receptor for the F_c region of IgE found on mast cells in the airways and gut mucosa or on basophils in the bloodstream. On a second exposure, the allergen binds to IgE and cross-links adjacent F_c receptors. This acts as a signal to the mast cells, which degranulate and release vasoactive substances, such as **histamine** and **leukotrienes**, into the local tissue. The onset of clinical symptoms, which are due to the vasoactive compounds, is rapid and occurs within a few minutes of encountering an allergen. This type of allergic response is called **anaphylaxis** (or **Type I hypersensitivity**). Common allergens include pollens, house dust mites, animal dander, insect stings and certain foodstuffs.

In the airways, degranulation of mast cells and release of histamine causes local smooth muscle to contract. This is manifest, for example, as the bronchoconstriction leading to the shortness of breath characteristic of hay fever. Insect stings can cause much more damaging responses: massive degranulation of circulating basophils occurs and the agents released cause vascular leakage and reduced blood flow, with potentially fatal consequences. The symptoms can be treated by adrenaline, which has the immediate effect of increasing cardiac output.

Adrenaline and theophylline are commonly used for treating allergies. Immunological treatments, such as desensitization, may also be effective, particularly for the life-threatening insect venoms. The strategy is to immunize the subject with a very small amount of the allergen in the hope of generating IgG antibodies with higher affinity for the allergen than the IgE antibodies already present. When an allergen enters the host as a result of an insect sting IgG antibodies compete with the IgE antibodies, preventing the allergen from cross-linking receptors on basophils and avoiding an anaphylactic attack.

Figure 16.9   (a) Reactions of the classical and alternative pathways of complement fixation. (b) Electron micrograph of an erythrocyte after lysis by complement. The 10-nm diameter pores are polymers of C9, the final component of the cascade. Photograph courtesy of Dr Eckhard Podack.

IgD and IgE

The activities of IgD and IgE, which are both monomers, are less well understood than those of the other immunoglobulins. IgD is present on the surface of B cells, where it functions as a receptor for antigen, but its function in serum is as yet undefined. IgE is involved in responses aimed at the elimination of intestinal parasitic infections and in the development of allergic responses. The F_c region of IgE can attach to a specific receptor on mast cells, a type of connective tissue cell which releases histamine as part of the allergic response (Box 16.5).

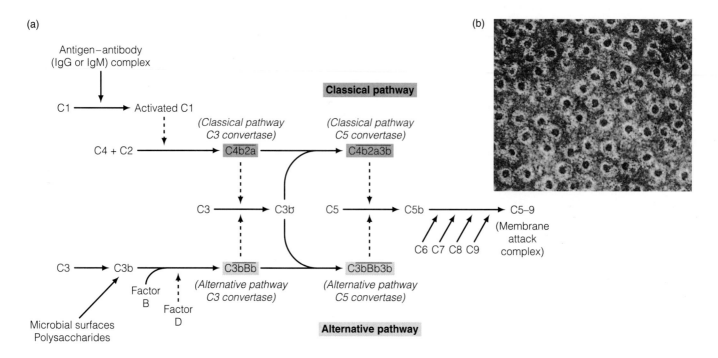

(a)

(b)

Box 16.6   The complement cascade

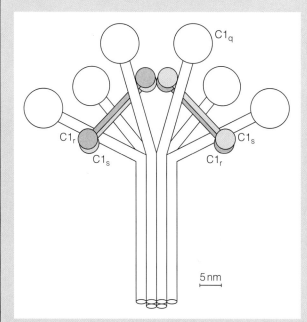

Box Figure 16.6
Schematic representation of the structure of the C1 component of the complement cascade. Six C1 molecules form a cylinder, each strand being composed of triple helices similar to those found in collagen. The globular heads are involved in binding the F_c region of antibodies once bound in immune complexes, thereby activating the C1 and $C1_s$ components.

The C1 component has an unusual organization, as illustrated in Box Figure 16.6. It is composed of a recognition unit (C1q), which forms a hexamer resembling a bunch of flowers in overall design with the stems composed of six triple-stranded α-helices. This helical arrangement of polypeptide is similar to that found in collagen (see Chapter 15). The globular domains of the flower heads contain the recognition sites for the F_c regions of antibodies locked in a multivalent complex with antigen. Within the cavity created by the globular domains and stems of C1q lie two separate polypeptides, C1r and C1s. The binding of C1q to several antigen–antibody conjugates simultaneously causes a conformational change in C1q, which promotes conversion of C1r from a zymogen to a protease by an internal cleavage similar to the conversion of pepsinogen to active pepsin (see Chapter 14). C1r in turn activates C1q by proteolytic processing, leading to sequential cleavage of peptide bonds in C4, C2, C3 and C5. Finally, the C5b fragment of C5 combines with C6, C7 and C8, assembling a complex on the membrane bearing the antigen. This process of **complement fixation** is effected most often and most efficiently by IgM, which is formed early in the primary response.

Chapter 15. Collagen, page 470.
Chapter 14. Proteolysis in the stomach, page 416.

Constant regions of immunoglobulins mediate complement activation

The complement proteins are important in the processes of lysis and destruction of virally infected cells and invading microorganisms. The binding of antibody to antigen on the surface of cells expressing foreign antigens triggers a cascade of proteolytic reactions involving approximately 15 complement proteins in the plasma. This sequence of reactions results in the formation of the **membrane attack complex** on the surface of the cell targeted for destruction. The 'attack' lyses the plasma membrane of the targeted cell.

The induction of the complement cascade by the classical pathway requires the interaction of the initial component of the pathway (C1) with a multivalent antigen–antibody complex, but it cannot be activated by free antibody or a single antibody molecule complexed to its antigen. The **alternative complement pathway** can be triggered directly by bacterial cell wall polysaccharides binding to host cell surface antigens. The two pathways may be seen in Figure 16.9.

The molecular and genetic basis of antibody diversity

Multiple genes code for the variable and hypervariable regions of both heavy and light chains of immunoglobulins

Different regions of the immunoglobulin light and heavy chains are classified as constant, variable or hypervariable, depending on the extent to which their amino acid sequences differ in the different immunoglobulin species. In the light chains, which can be isolated as **Bence–Jones proteins** from the urine of patients with myeloma, the N-terminal half of the polypeptide (108 amino acids) is distinct in each case but the C-terminal half (106 amino acids) has only two types, κ or λ chains.

The γ heavy chain is approximately 440 amino acids in length, with a variable region of about 100 amino acids, roughly the same size as that of light chain. Its constant region thus comprises three-quarters of the polypeptide. Comparison of the primary sequences of the variable regions of many myeloma proteins revealed that specific sections exhibit a much higher degree of variability than the rest of the molecule: these are the hypervariable regions (also called the **complementarity determining regions**). The hypervariable segments, four on each of the heavy chains and three on each of the light chains, are directly involved in antibody recognition and the amino acids in these regions determine the specificity of the antibody (Figure 16.10).

The minimum number of antibodies with distinct specificities that an animal is capable of producing is extremely large, probably greater than 10^8. Part of this diversity stems from the fact that they are encoded by many genes. In mice, for example, about 300 genes code for the variable regions of heavy chains.

Antibody diversity is increased by J (joining) and D (diversity) genes

The V (variable region) genes and C (constant region) genes for heavy and light chains are grouped in separate parts of the chromosome so, to permit the

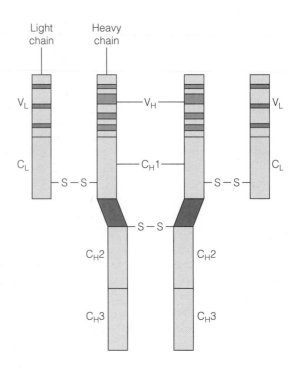

Figure 16.10 Location of hypervariable regions in immunoglobulin structure involved in direct contact with the antigenic determinant. Hypervariable regions (marked in pink) that provide antibody specificity are located in the N-terminal sections of the heavy and light chains where antigens are known to bind. Variable and constant domains on the heavy and light chains are designated V_H, C_H and V_L, C_L, respectively.

expression of intact light or heavy chains during the differentiation of an antibody producing cell, a particular V gene must be translocated to a site adjacent to a C gene by intrachromosomal rearrangement. This is accomplished by means of a set of J (joining) genes, not by direct joining of V and C genes. In the heavy chain and κ light chain genes, five J genes are located in tandem array near the C genes on the appropriate chromosome (Figure 16.11). These five J genes code for most of the final hypervariable region (CDR3) of heavy and light chains (13 amino acids), as the V genes code for only 95 amino acids of the variable region (108 are required in the intact immunoglobulin molecule). The availability of the J genes increases the possible number of combinations of heavy and κ light chains. A V gene is joined to a J gene by recombination (see Chapter 6). This process involves recognition of specific motifs flanking the genes to be joined, scission of the DNA once the V and J coding sequences have been brought close together and DNA ligase activity to seal the joint. All DNA between the V and J genes joined is lost as part of this somatic recombination process (Figure 16.11). In mice, defects in genes that influence recombination are associated with SCID-like syndromes (see Box 16.1).

In the variable regions of the heavy chains (V_H), further diversity is generated by the presence of 15 D (diversity) genes lying between the cluster of V and J genes and encoding the third hypervariable region of the heavy chain. Other mechanisms alter the precision by which V, D, J and C genes are linked in yet another significant source of antibody variation.

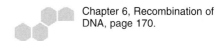

Chapter 6, Recombination of DNA, page 170.

Organization of heavy chain gene segments

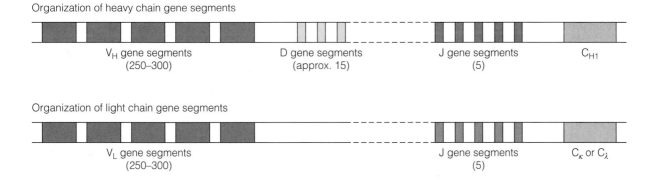

V_H gene segments (250–300)   D gene segments (approx. 15)   J gene segments (5) C_{H1}

Organization of light chain gene segments

V_L gene segments (250–300)   J gene segments (5) $C_κ$ or $C_λ$

Production of different antibody classes involves translocation of genes

Antibody-producing cells manufacture IgM when first stimulated and then switch to production of one of the other classes with an identical or enhanced specificity. In humans the genes encoding the constant region of the heavy chain are clustered on one chromosome (Figure 16.13).

This group of genes is preceded by an array of J_H genes. A complete gene for the heavy chain of IgM, for example, is generated by the translocation of a V_H gene segment to a DJ_H gene segment adjacent to the Cμ gene. When antibody production is switched, for example from IgM to IgA, the complete VDJ_H gene segment is moved to a new location adjacent to the Cα gene. This phenomenon is referred to as **isotype switch**. DNA between the spliced genes and within the individual genes on the chromosome is removed in the **isotype switch recombination event**, which is mediated by an enzyme system different from that responsible for joining the V, D and J genes. The purpose of transferring variable regions that confer antibody specificity to new constant domains is that these constant regions of the immunoglobulin molecule mediate different effector functions.

Figure 16.11 Variable regions of heavy chains and light chains are made from combinations of V, D and J gene segments. The presence of multiple segments is one of the major factors increasing antibody diversity. J genes are located near the constant gene on the chromosome, at a long distance from the V and D genes. The three constant domains and the 'hinge' region of heavy chains are also each encoded by separate genes.

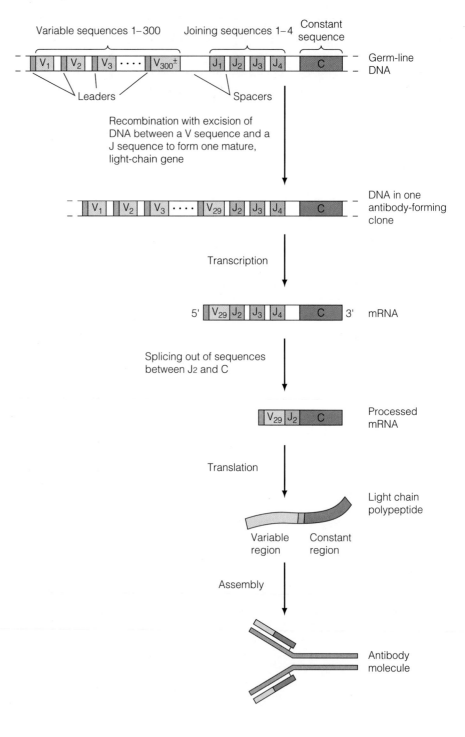

Figure 16.12 Recombination, transcription and translation in the production of an antibody molecule.

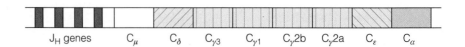

Figure 16.13 Organization of constant-region genes for immunoglobulin heavy chains. There are eight constant region genes for heavy chains, giving rise to the five antibody classes IgM ($C\mu$), IgG ($C\gamma$), IgA($C\alpha$), IgD ($C\delta$) and IgE ($C\varepsilon$) and four subclasses of IgG ($C\gamma1$, $C\gamma2a$, $C\gamma2b$ and $C\gamma3$). Light chains have two constant region genes, giving rise to the κ or λ chains.

The functions of T lymphocytes

The **cellular immune response**, mediated primarily by T lymphocytes, provides defensive mechanisms against parasites and viruses which invade animal cells and against cells that have become cancerous. T lymphocytes (T-cells) have specific transmembrane receptors on their plasma membranes. These receptors, which are similar to secreted immunoglobulins, recognize foreign proteins on the surface of other cells of the body that have become infected or cancerous and this recognition initiates destruction of the infected or cancerous cell.

T-cell responses depend upon 'foreign' peptides binding to major histocompatibility complex proteins

There are two types of major histocompatibility complex that are recognized by T-cells: MHC I and MHC II. MHC I proteins are found in all cell types and bind to peptides synthesized within the cell. MHC II are found only in cells of the immune system (referred to as *antigen-presenting cells*).

When a virus replicates within a host cell, cellular proteases digest some of the newly formed viral proteins. These peptide fragments are taken up by the Golgi complex where they bind to a groove in the surface of a protein of the MHC I class. The peptide-MHC I complex then migrates into the plasma membrane where it can be detected by circulating T lymphocytes. The foreign peptide, bound to MHC, is described as being *presented* to the T lymphocyte. The receptor on the surface of the T lymphocyte recognizes and binds to the presented peptide-MHC complex.

MHC II are specialized to bind to peptides from pathogens which have been taken into the antigen-presenting cell by phagocytosis and degraded by the proteasomal or lysosomal pathways. The degraded fragments bind to the MHC groove and are presented to the T lymphocyte receptor in a manner analogous to that described for MHC I.

The terminology is complicated by the fact that MHCs are referred to as antigens because each individual has slightly different MHC proteins and those from one person provoke an immune response if injected or transplanted into another person. The importance of this is that when tissues are transplanted the immune system of the recipient recognizes the MHC proteins of the donor tissue as foreign and attempts to reject the transplanted tissue.

In humans, MHC antigens are referred to as HLA antigens (Human Leucocyte Antigens). There are three main types of class I antigen: HLA-A, HLA-B and HLA-C. There are up to 50 types of each of these within the human population and each individual expresses the HLA antigen from both chromosomes. Consequently each cell has six types of HLA antigen, each able to accommodate a different peptide in the groove. In the case of Class II, there are again three types called HLA-DR, -DQ and -DP. In assessing material to match transplant and donor material HLA-B and HLA-DR are usually measured since these are the most diverse.

T-cell receptors are similar to secreted antibodies but have distinctive properties.

- There is no secreted form of the T-cell receptor, so the immunological action of the T cell depends on expression of the receptor at the cell surface.
- T-cell receptors do not recognize intact foreign antigens. Foreign molecules are recognized by the T cell only if they are present at the surface of a cell in association with MHC antigens.

- T-cell receptors recognize short peptides derived from the foreign molecule (8–15 amino acids in length), which are found as an extended chain in a groove on the surface of the appropriate MHC antigen.
- T-cell receptors have many structural features in common with immunoglobulins, including variable and constant region domains.

A structural representation of the T-cell receptor is shown in Figure 16.14.

The polypeptides of T-cell receptors exhibit homology with immunoglobulin chains

T-cell receptors are composed of two polypeptide chains, α and β, both of which traverse the plasma membrane. The bulk of each subunit is located on the external surface of the T cell in accord with their role in cell–cell recognition. Each subunit contains several intrachain disulphide bonds and the subunits are covalently linked by a single disulphide bond to form a heterodimer (Figure 16.14).

As with the immunoglobulins, there is a large repertoire of T-cell receptor genes with the ability to react with a multitude of antigenic determinants. The domain organization of the α and β chains is similar to that of the immunoglobulin light and heavy chains, with *N*-terminal V region genes (which contain internal hypervariable regions) and *C*-terminal constant domains. The organization of the genes comprising T-cell receptors is equally complex; multiple copies of V gene segments linked to C genes by several possible D and J genes means that more than 10^{12} T-cell receptors with differing specificities may exist (Figure 16.15).

A second type of T-cell receptor, comprising γ and δ chains, has similar properties to the αβ T-cell receptors. Although the γδ T-cell receptor-bearing T-cell population is minor and of undefined function in humans, the peripheral blood of certain large animals, including cattle, contains up to 50% γδ⁺ T cells.

T-cell receptors interact with foreign peptides presented on the surface of infected cells

The antigen-combining site on the T-cell receptor recognizes a complex epitope comprising parts of the MHC antigen and bound peptide. Specific target cell–T cell

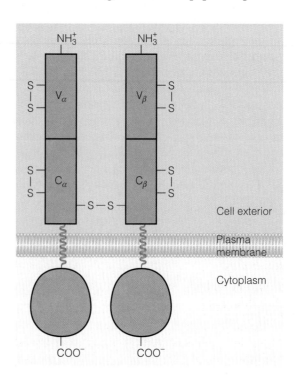

Figure 16.14 Structural representation of the T-cell receptors. The receptor is an α, β heterodimer linked by a disulphide bond. Each polypeptide contains a variable and a constant domain on the external surface, a membrane-spanning α-helix and a small C-terminal domain in the cytoplasm.

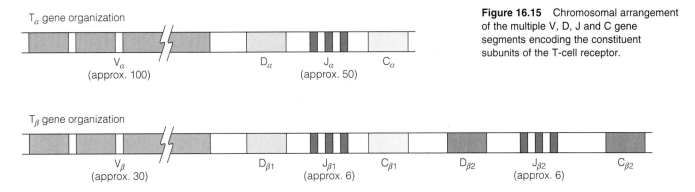

Figure 16.15 Chromosomal arrangement of the multiple V, D, J and C gene segments encoding the constituent subunits of the T-cell receptor.

interactions are stabilized by other plasma membrane proteins, such as CD4 and CD8, found on helper and cytotoxic T cells respectively.

Two principal types of T cells, cytotoxic or killer T cells (CD8⁺) and helper T cells (CD4⁺), exist. The interaction between T-cell receptors on a killer T cell and a cell displaying foreign peptides in combination with the MHC antigen leads to killing of the target cell, often by the release of granules containing **perforin**. This protein polymerizes in the membrane of the target cell and generates pores of 10 nm diameter, eventually destroying the target cell by leakage. Helper (CD4⁺) T cells influence the behaviour of CD8⁺ T cells and B cells. For example, helper T cells bind directly to B cells expressing the appropriate peptide antigen in MHC class II molecules on their surface, and promote their proliferation by directed release of **cytokines** such as **interleukin-2** (IL-2) (see Box 16.7). Thus, specific interactions between B and T cells are important in immune defence mechanisms, linking the humoral and cellular arms of the response (Figure 16.16).

Box 16.7 Cytokines

Cytokines, proteins or glycoproteins secreted by haematopoietic cells, govern the activities of other cells. The cytokines include **interferons, intercrine** molecules, **transforming growth factors, tumour necrosis factors, colony stimulatory factors** and **interleukins**. All cytokines share the following general properties:

- very high biological specific activities (they are active at picomolar concentrations);
- low concentrations in plasma;
- short life;
- pleiotropic effects (a single cytokine can act on several different target cell types);
- action via specific receptors at the surface of target cells.

A typical example of a cytokine is interleukin-2 (IL-2), which is produced by CD4⁺ helper T cells following stimulation with antigen. IL-2 is a glycoprotein, M_r 15–17 000, that acts on T and B cells. IL-2 enhances the growth of CD8⁺ cytotoxic T cells acting in a paracrine manner, and will sustain the growth of activated CD4⁺ T cells: it also acts in an autocrine fashion. The growth of activated B lymphocytes is promoted by IL-2; it can also accelerate antibody secretion in these cells by increasing the rate of synthesis of the J chains necessary to stabilize the IgM pentamer.

The receptor for IL-2 comprises three distinct polypeptides: an α chain of 55 kDa, a β chain of 75 kDa and a γ chain of 64 kDa. The β chain has an intermediate affinity for IL-2 and is present constitutively on lymphocytes, the α chain has a low affinity and must be induced in B and T cells, for example by antigen binding. The αβ heterodimer forms a high-affinity IL-2 receptor complex. The γ chain is required for effective signal transduction; the receptors for a number of other cytokines (including IL-4, IL-7, IL-13 and IL-15) share the IL-2 receptor's γ chain. The importance of the γ chain is emphasized by the finding that mutations in its cytoplasmic domain are associated with X-linked immunodeficiency syndromes (see Box 16.1).

Cytokine therapy has potential in cancer treatment. Most patients with cancers have a population of CD8⁺ T cells directed against the tumour but the sheer mass of tumour may overwhelm this population. One strategy to overcome this is to remove T cells from the patient and culture them *in vitro* with IL-2 to increase the numbers of cytotoxic cells. These cells are then infused back into the patient (with or without a simultaneous infusion of IL-2). Although this treatment with lymphokine-activated killer cells (termed LAK therapy) has had limited success, significant side-effects of nausea, shock and fever have been reported.

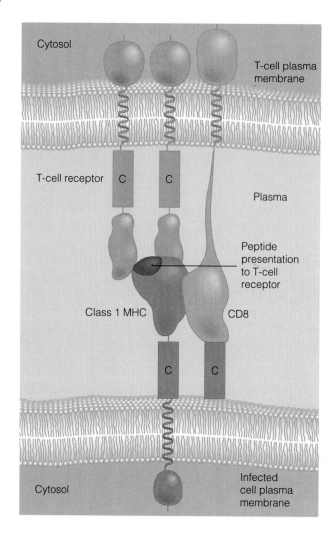

Figure 16.16 Schematic diagram of interactions between target cells and cytotoxic T cells. T-cell recognition involves an interaction with a peptide antigen of appropriate specificity presented on class I (or class II) MHC of the target cell. Interaction also requires the presence of an associated factor, designated CD4 in helper T cells and CD8 in cytotoxic T cells. All these components have related domains that contain an immunoglobulin fold and resemble antibody molecules (see Figure 16.19).

Box 18.7, Influenza A virus: humans, ducks and pigs, page 572.

Separate B-cell and T-cell responses to foreign antigens ensure sensitivity and specificity of the immune system

Membrane-bound or secreted antibodies produced by B lymphocytes recognize foreign antigens, either in solution or displayed on cell surfaces, whereas T-cell mediated responses depend on recognition of foreign peptides presented on cell surfaces in combination with MHCs. An advantage of the two distinct recognition systems is that it increases the overall sensitivity and specificity of detection of foreign proteins. This is important because some invading organisms have the ability to elude parts of the immune surveillance system. For example, the malaria parasite evades the host's defences by coating itself with host proteins from the bloodstream; the influenza virus alters its coat proteins in an attempt to avoid detection (see Chapter 18).

Peptide antigens are presented to T cells by two types of MHC

The MHCs were first recognized by their involvement in tissue transplantation and rejection. Transfer of tissues and organs between individuals, unless identical twins or highly inbred strains of laboratory animals, is beset with problems of rejection. Many individual genes can influence tissue rejection. In humans the most important genes are confined to a region of chromosome 6, the human MHC, referred to as the **human leukocyte antigen** (HLA) locus. Three classes of integral membrane polypeptides are encoded by the genes of this complex.

1. MHC class I proteins are present on almost all cell types, where they present foreign peptides to cytotoxic T cells
2. MHC class II proteins are found exclusively on phagocytes and other cells of the immune system (referred to as antigen-presenting cells – APCs)
3. MHC class III proteins are components of the complement cascade involved in the destruction of invading microorganisms (see Figure 16.10).

The genes encoding class I and II MHC antigens are highly polymorphic within a given species. Humans possess three different class I genes (HLA-A, HLA-B and HLA-C) and there are about 80 different HLA-A alleles within the human population. Each individual has a maximum of two HLA-A alleles, one inherited from each parent. The more alleles available in a population, the less likely a pathogen is to escape immune attack by finding peptides that cannot be presented. A corollary of this is that some HLA genes will afford enhanced protection against particular diseases but others might predispose to disease (see Box 16.8). The best example of the value of MHC polymorphism is seen in animals such as the Syrian hamster or African cheetah, in which species the lack of polymorphism has predisposed them to repeated viral infection.

As with other genes of the immune system, those encoding class I and II MHCs are present in multiple copies and allelic variation within a species is considerable, so that the chances of two unrelated individuals exhibiting the same phenotype are extremely small. This accounts for the observed problems of tissue rejection unless the genotypes of donor and recipient are closely matched.

Foreign peptides presented on MHCs can be derived from intracellular or extracellular sources

Peptides, typically 12–25 amino acids long (average 15), that associate with class II MHC antigens are derived from internalized foreign proteins taken into cells by

---

Box 16.8   HLA and disease

Ankylosing spondylitis is a disease in which damage to the spinal cord causes a pronounced stoop and stiffness of movement in the sufferer. The HLA phenotype of a significant number of sufferers expresses the class I allele HLA-B27. The correlation is sufficiently strong that individuals expressing the HLA-B27 allele are regarded as 90 times more likely to develop the disease than those who do not. Of course, this is only a statistical correlation; some patients do not express HLA-B27 and only a tiny proportion of all HLA-B27+ individuals actually develop the disease. The molecular basis for this effect is not known; the only clue to date is that all peptides that bind to HLA-B27 are nine amino acids in length and all have arginine at position 2. Whether the peptide leading to development of ankylosing spondylitis is of self origin or is derived from an infectious agent is not known.

Insulin-dependent diabetes mellitus can also be correlated with class II HLA alleles and, in particular, with the amino acid present at position 57 of the β chain of the MHC class II molecule. The disease shows a pronounced geographic distribution, Chinese populations having essentially none, the incidence in Scandinavian populations being rather high. The Asian population has an abundance of class II alleles with aspartate at position 57 of the β chain, but this is rare in Scandinavians, in whom serine is the predominant residue. How could this small change explain the predisposition to or, more importantly, protection against the disease? Position 57 of the β chain is located on the side of one of the α-helices of the class II peptide binding grooves, and the side chain points across the groove towards an arginine residue at position 84 on the corresponding α-helix of the α chain. An aspartate at position 57 is able to form a salt link with the arginine on the opposite α-helix, closing the right-hand end of the class II groove and restricting the number of peptides that can bind to this groove; serine can not form such a salt link. Absence of a salt link across the class II peptide-binding groove may be sufficient to permit a peptide that can be recognized by the immune system to enter the groove and give rise to the autoimmune response present in this disease. Individuals with aspartate at position 57 are protected against the disease.

Class 1 major histocompatibility complex

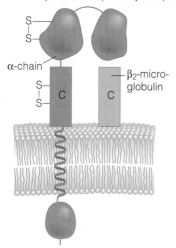

Class 2 major histocompatibility complex

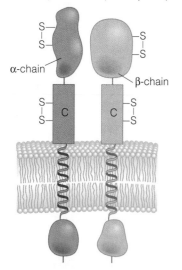

Figure 16.17 Schematic representation of the domain organization of class I and class II MHCs. Class I MHCs are present on nearly all cells and present foreign peptides to cytotoxic T cells. In contrast, class II MHCs are confined to certain types of cells of the immune system (antigen-presenting cells or APCs) and phagocytes, and interact specifically with helper T cells.

normal cellular mechanisms involving receptor-mediated endocytosis, lysosomal degradation and receptor recycling. Smaller peptides (8–10 amino acids) associate with class I MHC antigens and are derived from intracellular sources, from proteins actively synthesized within the cell.

The specificity of an MHC protein for any peptide sequence requires the presence of particular amino acids at specific positions. It is estimated that a single class I molecule can bind any one of 13.7×10^6 different nine-amino acid peptides, provided that certain amino acids are present at certain positions in the peptide. Thus, each MHC recognizes a range of peptides containing conserved sequence elements, termed a **consensus** or **aggretope sequence**. Aggretopes of peptides associating with class I and class II MHC antigens are shown in Table 16.3.

This latitude in the overall sequence specificity of individual MHCs plus the presence of multiple ($>10^4$) variants of these molecules ensures that foreign protein digested within a virally infected cell will produce at least some peptides with aggretope sequences that can be recognized by the host MHCs. A T-cell response can then be elicited against the infected cell.

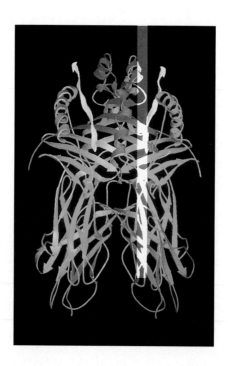

Figure 16.18 Three-dimensional structure of a class II MHC in association with its peptide antigen.

Immunoglobulins, T-cell receptors and histocompatibility antigens are members of an immunoglobulin superfamily

Many proteins belong to multigene families. A **multigene family** is a group of genes encoding proteins of similar sequence and function, such as those encoding different haemoglobins or tissue-specific collagen types. A **superfamily** is more extended, encompassing multigene families and single-copy genes related by sequence and common ancestry but possessing a diverse range of functions.

The individual genes of multigene families have arisen by gene duplication and independent evolution, but their basic function is essentially unaltered. In the immunoglobulin superfamily the immunoglobulin fold, a basic feature of the domain structure of all antibody classes (Figure 16.19), is also evident in the variable regions of the T-cell receptor, class I and class II MHCs, CD4 and CD8

Table 16.3 Aggretopes of peptides associating with class I and II MHC antigens

Class I (HLA-B27)

The HLA-B27 molecule, one of many alleles of the HLA- locus in the human population, binds nonapeptides in an extended conformation. It is of interest because individuals who possess this allele are 90 times more likely to suffer from ankylosing spondylitis than those who do not possess it. However, this is only a disease *association*; not all HLA-B27+ individuals develop the disease and some patients with the disease do not express HLA-B27.

The most notable feature of nonapeptides found bound in HLA-B27 grooves is that all possess arginine at position 2.

		Ile			Ile	Thr		
Arg		Tyr	Lys	Gly	Leu	Ile	Lys	Lys
Lys	ARG	Phe	Phe	Glu	Pro	Val	His	Arg
Gly		Trp		Lys	Val	Arg	Leu	
						Gly		

For each position, only the most frequently occurring residues are shown. From an analysis of the total amino acids noted for any one position, an estimate of the number of possible peptides that can bind to HLA-B27 can be obtained, thus:

$$6 \times 1 \times 8 \times 11 \times 8 \times 7 \times 7 \times 11 \times 6$$

which is approximately 13.7 million possible peptides.

Class II (H-2I-E^d)

Leu				Arg	Arg			Ala			
Val	Glu	X	X	Lys	Lys	X	X	Val	X	X	Lys

(I-E^d refers to the allelic product of the d haplotype encoded at the class II I-E locus in the mouse; the small superscript denotes the allele in the mouse, whereas a number is used in the nomenclature of human MHC antigens.)

Box 16.9 MHC antigens in infection

Class I and II MHC molecules play a key role in guiding the activity of T cells; T cells can recognize a foreign antigen only if peptides derived from the antigen are presented to the T-cell receptor within a peptide binding groove of an MHC molecule. The MHC molecules are themselves antigens because injection of cells from one person into another with different MHC genes leads to production of antibodies to the 'foreign' MHC (and, of course, graft rejection). Such antibodies occur naturally in women who have had multiple pregnancies and are used to form the basis of tissue typing for matching of donor and recipients in organ transplantation.

The importance of MHCs in disease resistance is evident from a group of inherited diseases called **bare lymphocyte syndrome**, in which cells fail to express class I, class II, or both, MHC antigens. These patients have normal numbers of functional B and T cells, but suffer severe fungal and viral infections. The reason for this is that they have no MHC antigens with which to present foreign peptides to their own functional T cells. They are therefore unable to eliminate infected cells and an infection spreads in an essentially uncontrolled manner.

How do class I and II molecules cooperate in protection? The first event in viral infection is entry of the virus into the cell, often via receptor-mediated endocytosis. This places the virion in the endosomal pathway, where its nucleic acids are uncoated for replication and translation and where the coat proteins are cleaved. Peptide fragments of the virion coat may associate with class II molecules in the endosome and be transported to the cell surface, where they are recognized by CD4+ helper cells. The helper cells are activated by this recognition and begin to secrete cytokines, such as IL-2, which help the growth of the activated helper cells and any B cells that have bound other viral particles and are available to secrete antibody. Meanwhile, viral proteins are synthesized in the host cell. Some are broken down and are available within or are transported to the endoplasmic reticulum to associate with class I molecules made there. Loading of a peptide into the class I peptide binding groove allows the complete molecule to travel to the cell membrane, where it is recognized by CD8+ cytotoxic cells. These cells kill the infected cell and, because IL-2 is locally available from activated T cells, the number of CD8+ T cells will increase to ensure that all other infected cells in the area can be eliminated.

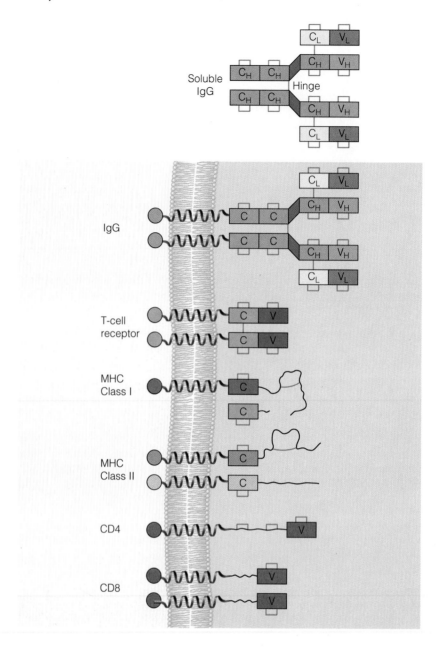

Figure 16.19 Similarities in domain organization of several prominent members of the immunoglobulin superfamily. Many receptors involved in the immune response and cell–cell recognition contain variable and constant regions related in sequence and structure to immunoglobulin domains.

– all molecules that play distinctive, independent roles in immune response mechanisms. However, this structural motif is also found on cell-surface proteins involved in cell–cell recognition processes outwith the immune system – such as N-CAM, a developmentally regulated cell adhesion factor. Although these structurally related proteins have evolved to the extent of performing quite distinctive and independent roles, their individual functions can still be classified under the general heading of cell–cell interactions and recognition phenomena.

The molecular genetic basis of disease

Many diseases have a genetic component, and the study of classical human genetics has contributed greatly to the advance of medicine. In recent years developments in molecular biology (see Part B) have helped in the elucidation of many aspects of genetics and the field of molecular genetics is now of crucial and growing importance in many areas of medicine. In this chapter we examine some molecular genetic aspects of disease and their biochemical connections – both in diseases caused by single-gene defects and in those that are multifactorial.

Cancer is also essentially a 'genetic disease' but not, in most cases, a hereditary one. The **somatic cell theory of cancer**, now widely accepted, explains cancers in terms of mutations in the DNA of somatic cells that cause loss of control of cell division and other changes in their social behaviour. Some of the molecular changes associated with cancer are described. We conclude the chapter with a brief look at 'genetic engineering' and some of its medical applications.

Box 10.5. Phenylketonuria is an inborn error of metabolism which leads to mental retardation, page 512.

Many diseases have a genetic component

Humans suffer from a wide variety of rare inherited disorders. The inherited diseases that are best understood are those caused by defects in single genes, of which phenylketonuria (see Box 10.5) and sickle-cell anaemia are probably the most well known. In these relatively rare conditions a mutation in a gene causes an enzyme or protein to be produced that is partly or completely defective in its function. This molecular defect gives rise to the clinical symptoms of the disease, although the connection between them can be complex. Several thousands of such conditions have been described and continuing analysis of the human genome is allowing more detailed characterization of the defective gene. The inheritance of these genes obeys the laws of classical genetics, and so counselling on the risks can be given to afflicted families.

It has become apparent that other diseases, more common than single-gene defects, are at least partly genetic in origin but that their causation is **polygenic**, involving the products of several different genes. Some of these diseases, such as type I diabetes, are relatively common, and dissection of the molecular genetics of these conditions is a major challenge. Other diseases, coronary heart disease for example, involve environmental as well as genetic factors and are often termed **multifactorial**.

Cancer is also a genetic disease

Cancer in its many forms remains perhaps the greatest challenge to medical science. Three principal approaches (to which biochemistry and molecular genetics are making major contributions) are being taken to reduce the threat posed by cancer. These approaches are

- development of more effective therapies for treatment of tumours;
- elucidation of the fundamental biology of tumorigenesis;
- identification and reduction or avoidance of environmental factors involved in tumorigenesis.

It is anticipated that an understanding of the molecular and cellular biology of cancer will lead to new avenues of treatment and prevention.

Most malignancies are not genetic diseases in the strict sense of being inherited, although there *are* some rare familial cancers. Tumours do, however, have a genetic origin in the sense that tumour cells are mutated descendants of healthy cells that are no longer subject to normal growth regulation and which have become capable of spreading into other tissues. This origin of tumours is referred to as the **somatic cell theory** of cancer. No single 'cancer gene' exists, but defects in one or more of a number of genes can contribute to the onset of cancer. Thus cancer may be regarded as a (generally) noninheritable, multifactorial genetic disease.

Defects in the genes involved in the control of normal cell growth and behaviour are often intimately involved in tumorigenesis. In particular, defects in genes connected with the normal process of apoptosis (programmed cell death (see Box 17.10)) can lead to the defective growth seen in tumour cells. The products of these genes also seem to participate in the mechanisms underlying established anticancer therapies (see below).

Chapter 17. Anticancer drug therapy is useful, but has limitations, page 538.

DNA manipulation and 'genetic engineering' are making great contributions to medicine

The most significant recent technological advance in biology has been the development of the ability to manipulate genes and to determine their sequences. This 'genetic engineering' (see below) has many potential applications in medicine, including

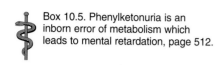
Chapter 17. 'Genetic engineering' and some medical applications, page 539.

- synthesis of human proteins in large amounts for clinical use;
- insertion of human genes into vectors for 'gene therapy' of genetic defects.

Another major development in biology and medicine over the last 20 years is the extensive mapping and sequencing of human and other genomes, giving an 'information explosion' that is advancing our understanding of the fundamental biology underlying medical problems.

Genetics and medicine

A knowledge of genetics is important in understanding many diseases

The science of genetics is concerned with analysis of the molecular basis of variation in living organisms. Medical genetics deals with the human population and the principles underlying the inheritance of specific characteristics or traits, especially those linked to the development of particular diseases. It has long been apparent that each person is a unique blend of characteristics, of appearance and personality, distinguishing that individual from others. Similarities of physical features such as height, hair and eye colour tend to occur within a family (Figure 17.1). The inheritance of such features obeys the rules of 'classical' genetics.

The recognition of DNA as the genetic material and the development of molecular biology (see Introduction to Part B) led to the development of 'molecular' genetics. Human molecular genetics has been given new impetus in recent years as rapid progress in the detection, isolation, sequencing and manipulation of individual genes has permitted detailed analysis of over 2500 human genes. Many types of disease have genetic causes (Table 17.1), and some of these will be discussed later in this chapter.

Table 17.1 Some genetic causes of disease

Chromosomal disorders
Defects in the mitochondrial genome
Multifactorial disorders
Somatic cell disorders
Single-gene defects

 Part B. Information in biological systems and its transfer, page 76.

Box 17.1 Some genetic terms

Alleles
Alternative forms of a gene at the equivalent locus (site) on a given pair of homologous chromosomes.

Aneuploid
Any odd chromosome number which is not an exact multiple of the haploid number.

Autosome
Any chromosome other than the sex chromosomes.

Diploid
The normal chromosome complement of somatic cells.

Dominant
A trait or defect which is evident in heterozygotes.

Genotype
The genetic constitution of the individual or cell.

Haploid
The chromosome number present in the gametes or reproductive cells, i.e. half the diploid number.

Heterozygote
An individual with one normal and one mutant allele at a particular locus (site) on homologous chromosomes.

Homozygote
An individual with a pair of identical alleles at a particular locus on homologous chromosomes.

Karyotype
The characteristic chromosome complement of an individual or cell.

Phenotype
The observable physical characteristics of an individual or organism.

Recessive
A trait or defect which is evident only in individuals who are homozygous for that trait.

Polyploid
An abnormal chromosome number, which is an exact multiple of the haploid complement but exceeds the diploid number.

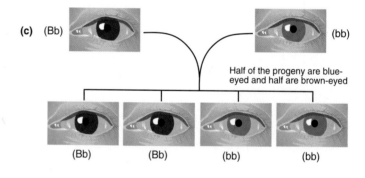

Figure 17.1 Genetic analysis of the inheritance of eye colour. In humans this trait is conferred by a single gene, with that for brown eyes dominating over that for blue eyes (autosomal recessive). Thus, (a) children of brown-eyed (BB) and blue-eyed parents (bb) who are homozygous for these traits will all be brown-eyed with the genotype (Bb). If these children eventually reproduce with individuals of the same genotype (Bb), the recessive trait, blue eyes (bb), will appear on average with a frequency of 25% in this generation. Brown-eyed children will be of mixed genotype, either homozygous (BB, 25%) or heterozygous (Bb, 50%) on average. It will be evident from the above that individuals with the same phenotype, brown eyes, may have different genotypes. These phenotypes can be altered by the influence of other genes. Children of one Bb (brown-eyed) parent and one bb(blue-eyed) parent will be 50% brown-eyed and 50% blue-eyed (c) on average.

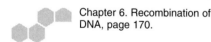

Chapter 6. Recombination of DNA, page 170.

Genes are inherited according to defined laws

The rules governing the inheritance of single traits or characteristics were deduced from Mendel's experiments on garden peas. A modern expression of **Mendel's first law of inheritance** would be that single traits are carried on pairs of genes that segregate into individual germ cells (spermatozoa or ova) during meiosis and are subsequently inherited by the fertilized egg. **Mendel's second law** states that unrelated traits sort independently into the germ cells; such unrelated traits are capable of recombination (see Chapter 6).

Before Mendel's work the genetic traits of parents were thought to blend continuously into those of their offspring. In humans this appeared intuitively to be correct for features such as height or hair colour: it did not, however, account for the inheritance of discontinuous traits such as eye colour or for the pattern of occurrence of disorders such as colour blindness. Mendel's laws offered a rational explanation for the re-emergence of traits in subsequent generations, which may be summarized thus:

Each individual trait in the diploid state (that is, with paired chromosomes) is represented by a pair of alleles (or genes) that segregate at each generation into the reproductive cells (or gametes: sperm and ova), in which the chromosomes are haploid (that is, unpaired). Offspring inherit one allele from each parent.

Genes can be dominant or recessive in their behaviour

Individuals who carry two identical copies of one form of a particular gene are termed **homozygous** for the characteristic determined by that gene: they are **homozygotes**. The two forms of the gene are respectively dominant or recessive and homozygotes show the dominant and recessive phenotypes respectively. **Heterozygotes** carry one copy of the dominant gene and one copy of its recessive counterpart, and typically show the dominant phenotype. The genotypes of the offspring of two heterozygotes will be distributed as follows:

- homozygous for the dominant gene (25% on average)
- heterozygous (50% on average)
- homozygous for the recessive gene (25% on average)

and so only one in four show the recessive phenotype.

Many human genetic disorders that are caused by **single gene defects** (Table 17.2), such as phenylketonuria, are **autosomal** (that is, not sex-linked – see below) recessive conditions. In these cases, heterozygous individuals act as 'carriers'. They do not display any obvious symptoms of the disease but pass the defective gene to their offspring. In some cases, such as sickle-cell anaemia, the heterozygotes are not identical in phenotype to the dominant homozygotes, but display an intermediate biological phenotype (see below). In these cases both forms of the gene are expressed and individuals contain both normal and sickled forms of haemoglobin. This is termed **codominance** and, in a sense, is the opposite of dominance-recessiveness. There is a spectrum across the range of genes between dominance and codominance, depending on the nature of the protein products of the genes (the molecular phenotype).

Single-gene defects in humans were first recognized in amino acid metabolism

The earliest recorded gene defects in humans were those associated with disorders of the degradative pathways of the amino acids, in particular phenylalanine and tyrosine. Defects in individual enzymes catalysing specific steps in the catabolism of these amino acids causes abnormal levels of metabolites to accumulate in the blood and urine of affected individuals, often with serious consequences. The condition of **alkaptonuria** is due to defects in the enzyme homogentisate oxidase, leading to the accumulation of homogentisate from the catabolism of phenylalanine (see Chapter 10). As homogentisate is not toxic the afflicted individual suffers no serious consequences. However, homogentisate accumulates in the urine and quickly turns black as it is oxidized by air – thus explaining the early recognition of this defect. The demonstration by Garrod in 1902 that alkaptonuria is transmitted as a single Mendelian recessive trait was a landmark in medical genetics. This eventually led to the general conclusion that such **inborn errors of metabolism** are caused by single defective genes producing an altered or inactive enzyme or protein.

Some gene defects can have fatal consequences

A second genetic disorder of phenylalanine metabolism, **phenylketonuria**, is one of the most common single-gene defects and, in contrast to alkaptonuria, has extremely serious consequences for the afflicted individual. In phenylketonuria,

Table 17.2   Some diseases caused by single-gene defects

Phenylketonuria (autosomal recessive)
Cystic fibrosis (autosomal recessive)
α_1-Antitrypsin deficiency (autosomal codominant)
Familial hypercholesterolaemia (autosomal dominant)
Haemophilias A and B (X-linked)
Severe combined immunodeficiency disorder (several types, X-linked or autosomal recessive)

Chapter 10. The pathway for catabolism of phenylalanine and tyrosine, page 311.

Box 10.5. Phenylketonuria is an inborn error of metabolism which leads to mental retardation, page 312.

Chapter 10. Melanin, the pigment for skin and hair and eye, is derived from tyrosine, page 312.

Chapter 17. Single-gene defects, page 525.

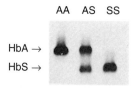

Figure 17.2 Starch gel electrophoresis of haemoglobin from normal patients (AA), those with sickle cell anaemia (SS) and heterozygous carriers (AS).

the conversion of phenylalanine to tyrosine by phenylalanine hydroxylase is blocked, leading to the accumulation of phenylalanine and derived products such as phenylpyruvate (see Box 10.5) in blood, urine and other body fluids. Unless the condition is detected and treated within the first week of life, patients become severely mentally retarded with greatly reduced life expectancies. The basis of the toxicity of metabolites of phenylalanine is not clear at present, although there are many other derangements in aromatic amino acid metabolism associated with imbalances in the level of intermediates. For example, sufferers tend to have lighter hair and skin than normal because synthesis of the pigment melanin is reduced. The first step in melanin production involves additional hydroxylation of the tyrosine ring (see Chapter 10) and this reaction is inhibited by the high levels of phenylalanine in phenylketonurics.

The incidence of phenylketonuria is about 1 in 10 000 neonates. In the developed world, blood samples of all infants are tested at birth for the relevant metabolites. Prenatal diagnosis can also be performed employing DNA probes for the phenylalanine hydroxylase gene. From early infancy, phenylketonurics must follow a diet low in phenylalanine, or irreversible brain damage rapidly occurs. The aim of the dietary programme is to supply just enough phenylalanine for growth and replacement. Heterozygous individuals comprise about 1.5% of the population and are relatively unaffected, except for a raised fasting level of blood phenylalanine. As with alkaptonuria, phenylketonuria is inherited as a Mendelian recessive trait.

In 1959, when phenylketonuria was recognized, only four genetic disorders were known to be the result of defective enzymes or proteins. Currently more than 4500 single-gene defects are known, although in about 85% of cases the gene product has yet to be identified (see below: 'Single-gene defects').

Sickle-cell anaemia is a genetic disorder whose molecular basis is well understood

Sickle-cell anaemia is a common condition among black people of African origin and was the first genetic disorder to be characterized at the molecular level. In 1956 it was recognized that the disease was caused by the presence of defective adult haemoglobin (Figure 17.2). This mutation is caused by a single amino acid replacement; glutamic acid at position 6 of the β-chain is replaced by valine. The hydrophobic nature of valine means that the mutant protein has a marked tendency to form insoluble aggregates, particularly at low oxygen tensions. The aggregated haemoglobin leads to distortion of the erythrocyte structure into the characteristic sickle-cell shape.

Table 17.3 Mutant haemoglobins found in the human population

Mutation	Phenotypic effect
Single amino acid substitutions on surface of haemoglobin A	Generally harmless – haemoglobin S is a striking exception
Substitutions in the vicinity of the haem group, e.g. replacement of proximal or distal His with Tyr	Tend to impair oxygen binding; e.g. iron trapped in its ferric state in His—Tyr replacements
Amino acid substitutions in interior of haemoglobin A	Often distort the three-dimensional structure and produce unstable molecules
Substitutions at subunit interfaces	Usually affect oxygen binding and interfere with allosteric properties
Thalassaemias – defective production of either α or β chains	Abnormal haemoglobin aggregates of α and β chains with impaired oxygen-binding properties
Haemoglobin A_{1c} formed by nonenzymic reaction of glucose with N-termini of β chains	Harmless, but levels of haemoglobin A_{1c} are useful indicators of regulation of blood glucose levels in diabetics

Sickling is an example of a condition which is apparently recessive at the level of the biological phenotype, in that heterozygous carriers are relatively unaffected, suffering only mild anaemia. At the molecular level, however, the sickled form of globin is codominant – that is, heterozygotes contain both normal and sickled forms of haemoglobin in their erythrocytes. It appears that possession of the sickling trait is advantageous in areas where the prevalence of malaria is high. In such regions, heterozygotes form up to 10% of the population, with sickle homozygotes comprising about 0.2%. It seems that the mixture of the two haemoglobins is in some way unfavourable for the development of the malarial parasite.

Study of mutant haemoglobins has increased understanding both of mutations and of protein structure and function

In addition to sickling and the thalassaemias (discussed below), several hundreds of variations are known in human haemoglobin genes (Table 17.3). Many variant haemoglobins appear to have no effect on the individual and were detected by random screening. Identification of these changes in the sequence and three-dimensional structure of haemoglobin has contributed to our understanding of the fine detail of protein structure and function (see Chapter 12).

Chapter 12. Globins have distinctive three-dimensional structures, page 370.

DNA and chromosomes

Each chromosome contains a single molecule of DNA

The genomes of all eukaryotic organisms are divided into a set of **chromosomes**, the number and type being characteristic of the organism – the **karyotype**. Each chromosome contains a single molecule of DNA, a specific sequence of nucleotides comprising a set of genes and intergenic regions. The complete DNA sequences of two of the 16 chromosomes of the yeast *Saccharomyces cerevisiae* (chromosomes III and XI – 315 and 666 kilobase pairs (kbp), respectively) have already been determined, along with large parts of other chromosomes, and the complete sequence of the yeast genome (13 500 kbp) is expected to be complete by 1996. Detailed physical maps of the DNAs of human chromosomes 21 and X have already been made and complete sequences of several human chromosomes should be available by the end of the century (see Box 17.6).

The human genome probably does not contain many more genes than that of the simplest vertebrate

In the chromosomes of yeast more than 70% of the DNA sequence is occupied by **open reading frames** (see Chapter 5), 330 being located on the 660 kbp of chromosome XI. This extrapolates to approximately 7000 genes in the complete yeast genome, compared with an estimated 15 000 in the 100 000 kbp genome of the intensively studied nematode worm *Caenorhabditis elegans*. The simplest known vertebrate, the fish *Fugu rubripes*, has a genome of 400 000 kbp, which is estimated to contain 70 000 genes, very similar to the 'best guess' for the human genome, even though the human genome is some eight times longer than that of *Fugu*. Thus, gene number does not increase in proportion with genome size – the genomes of the higher vertebrates appear longer because of the extra DNA present in introns (see Box 5.3) and in repeated sequences.

Chapter 5. A functional mRNA must contain an open reading frame, page 132.

Box 5.3. The discovery and significance of introns, page 124.

Box 17.2   Repeated DNA sequences

Microsatellites

These are usually less than 1 kbp in size. Examples are the $(AC)_n$ repeats (where $n = 10-60$), of which there are about 50 000 in the human genome. Aberrations of another microsatellite, $(CCG)_n$, play an important role in the pathogenesis of **Fragile X syndrome**, a relatively common (about 1 in 1000) genetic disorder in which, as the name implies, the X chromosome is unusually prone to breakage.

Minisatellites

Minisatellites are about 1–30 kbp in size. Some of these are found at a single location in the human genome, others occur at multiple locations. **Multilocus VNTRs** (for example [AGGGCTGGAGG]$_n$) are targets for cleavage by restriction enzymes used in DNA fingerprinting.

Macrosatellites

Macrosatellites may be up to 1000 kbp long and are clustered in the centromeres and telomeres. They appear to have important functions in chromosomal structure, in the pairing and alignment of chromosomes during mitosis and in genetic recombination events during meiosis and formation of the gametes.

Chapter 6. A group of basic proteins, the histones, package DNA into nucleosomes, page 151.

The human genome contains repeated DNA sequences with structural roles

The human genome probably contains between 50 000 and 100 000 genes that encode proteins (see above). If we assume that the average gene (including introns) is encoded in 10 kbp of DNA, a simple calculation indicates that protein-coding genes constitute 1 000 000 kbp of DNA or, at the very most, one-third of the genome. Much of the remaining DNA is composed of 'moderately repeated' or 'highly repeated' sequences, occurring as hundreds or thousands of copies per genome respectively. Moderately repeated DNA contains few active genes, with the important exceptions of those for ribosomal RNA and for histones. Highly repeated sequences do not appear to contain any genes.

Highly repeated DNA sequences belong to two major types – interspersed sequences and satellite sequences. Interspersed sequences include the *Alu* sequences (so named because they often have a cleavage site for *Alu*I restriction endonuclease), which are dispersed throughout the chromosomes and comprise about 4% of total DNA. They consist of 300 bp repeats of related sequences and are estimated to occur every 5–10 kbp throughout all chromosomes. In forensic testing, the presence of *Alu* sequences is diagnostic of tissues of human or higher primate origin.

Satellite sequences form the basis for DNA fingerprinting (profiling)

The genomes of humans, and indeed most animals, contains various types of **repeated** or **satellite** sequences, stretches of DNA that contain a relatively simple sequence repeated many times. Satellite sequences are composed of **tandem repeats** and belong to the group called variable number of tandem repeats (VNTRs). These multiply-repeated sequences are classified into three groups (Box 17.2).

The presence of some of these repeated sequences provides the basis for DNA or genetic **fingerprinting**. In this technique a tiny sample of human DNA, which may be obtained even from the cells in a mouth washing, is subjected to cleavage by restriction enzymes and electrophoresis to generate a pattern of DNA fragments characteristic of the individual from which the sample was taken. Individual patterns are sufficiently different to allow this technique to be used for forensic purposes. It is also used in paternity cases – at least to say that a particular male could not be the father of the child in question.

Nucleosomes allow packing of DNA into a highly condensed state in chromosomes

Nucleosomes (see Chapter 6) form the **elementary fibre**, 10 nm in diameter, that is the basic unit of structure in the chromosome. The elementary fibre is itself coiled in a helical arrangement by the addition of **histone H1**, forming the **chromatin fibre**, which is 30 nm in diameter. Further compaction of the chromatin fibre in metaphase chromosomes is achieved by attachment to a central acidic protein 'scaffold' at repeated DNA sequences, from which loops of fibre extend to produce the **chromatid strand**, 0.6 μm in diameter. These loops, termed **Laemmli loops** after their discoverer, give a 'hairbrush' appearance to the chromatid strand and each contain about 200 kbp of DNA. The highest level of chromosomal compaction and organization involves the further winding of the chromatid fibre into a helical structure.

During interphase, chromosomal structures are not apparent within the nucleus as DNA is distributed in a less highly condensed form. Two classes of dispersed DNA may be distinguished – **heterochromatin** and **euchromatin**. The DNA is more tightly packed in the former type, which appears to contain few active genes. The euchromatin has a less compact organization and contains most of the actively transcribing genes.

Chromosomes and cells

Normal human cells contain a characteristic complement of chromosomes (karyotype)

All nucleated cells in the human body contain a complete DNA genome, organized into 46 chromosomes, 44 of which comprise the 22 homologous pairs of **autosomes** (the nonsex chromosomes). In addition there are two sex chromosomes, females possessing two copies of the X chromosome, males possessing a single X chromosome and a Y chromosome (a truncated version of the X chromosome). This means that males are **haploid** in those genes of the X chromosome that are absent from the Y. One of the copies of the X chromosome in females is inactivated during development of the embryo. With the exception of the Y chromosome, all chromosomes are X-shaped, with short and long arms, designated p and q respectively. For example, 21q designates a location on the long arm of chromosome 21.

Box 17.3   Human cells can be grown and studied in culture

Cells from most mammalian tissues can be grown *in vitro* (outside the body) in culture medium containing amino acids, vitamins and a balanced solution of salts supplemented by 10% serum. Such **primary cultures** normally survive for a few days, but very occasionally **clones** of cells arise that are capable of growth for many generations. These **established cell lines** are widely used for the detailed investigation of many aspects of cell growth and division. They provide ideal systems for the analysis of viral growth and the effects of toxic substances or new drugs. They have replaced laboratory animals in many experiments.

Cell lines have also been derived from tumour material taken directly from patients. One of the most famous cell lines, still in use in many laboratories, is the **HeLa cell line**, which is derived from a cervical carcinoma taken from Henrietta Lacks who died in Baltimore, USA in 1953. Normal cells in culture divide on average every 16–24 hours until they come into contact with neighbouring cells and form a monolayer by **contact inhibition**. Cancer cells, or cells that have been **transformed** by tumour viruses or carcinogens, have lost this property and often also exhibit altered morphology .

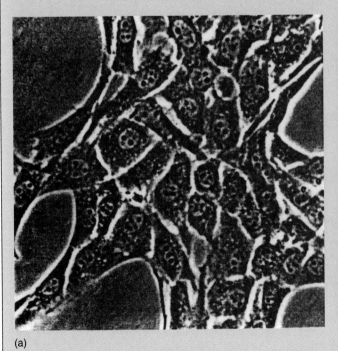
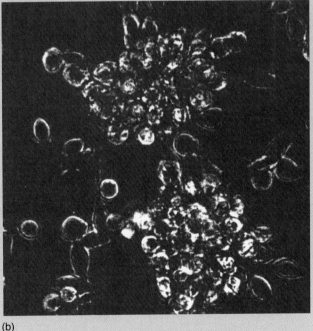

(a)

(b)

Box Figure 17.3 Altered morphology and lack of contact inhibition transformed hamster kidney (BHK 21) cells. In tissue culture, normal cells grow in an orderly manner, exhibiting a distinctive morphology and the phenomenon of contact inhibition whereby cell movement and growth cease once cells come into physical contact (a). After transformation by chemicals or viruses contact inhibition is lost and cells grow in a disorderly array as shown in (b).

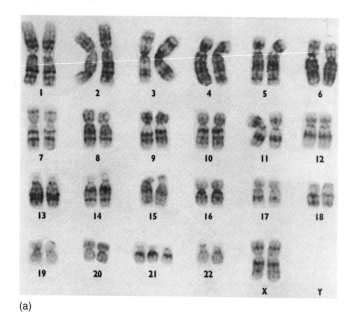

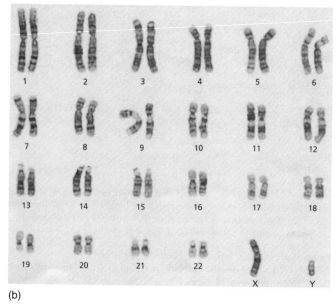

(a)

(b)

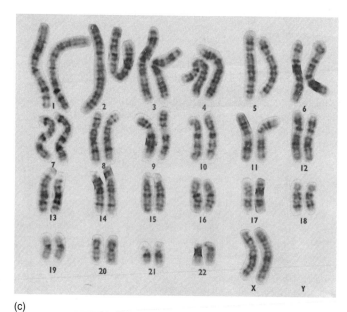

Figure 17.3 Karyotype analysis of normal male and female with Down's syndrome. (a) Down's syndrome female (trisomy 21) (b) Normal male karyotype showing G-banding patterns (c) Normal female karyotype showing G-banding patterns.

(c)

Box 17.4 Karyotype analysis of chromosomes

Peripheral blood lymphocytes are usually used for isolation and analysis of human chromosomes but bone marrow cells, skin fibroblasts, fetal cells from amniotic fluid or biopsy samples from various tissues may also be used. Lymphocytes represent about 1% of the cells in venous blood but are readily separated from erythrocytes by centrifugation. The cells are induced to divide by treatment with phytohaemagglutinin and the dividing cells are cultured for 2–3 days before adding colchicine to accumulate cells in metaphase. After swelling the cells to help spread out the chromosomes, the chromosomes from a single cell are examined to reveal the karyotype (the total number and type of chromosomes) and any major abnormalities in their number or size.

The normal human karyotype is described as 46, XX for females and 46, XY for males. In genetic disorders with an altered chromosome number, this nomenclature is followed by a description of the abnormality (for example, 47, XX, Down's syndrome for a female with an extra copy of chromosome 21) (Figure 17.3).

Chromosomes can be seen dividing only during mitosis

Most cells in tissues such as liver, heart and kidney undergo cell division only occasionally, as a normal part of tissue turnover and renewal. In contrast, stem cells (see Chapter 12) and epithelial cells, such as those of the skin and the lining of the gut, divide frequently because wear and tear demands constant renewal. Highly specialized cell types, particularly neurones, rarely divide throughout adult life.

Mitosis is the process by which the duplicated chromosomes of a cell that is about to divide are segregated into two sets and passed to the two daughter cells by **cytokinesis** (see Figure 3.6). The period between successive mitoses is known as **interphase** and constitutes most of the cell cycle. In an unsynchronized population of dividing cells only 2–3% are undergoing cell division at any one time and so, for analysis of chromosomes, it is necessary to inhibit mitosis and allow cells to accumulate in metaphase. This is usually achieved using colchicine, an alkaloid inhibitor of microtubule formation which prevents chromosomal separation during metaphase. These metaphase chromosomes may be used for karyotypic analysis (Box 17.4).

Chapter 12. The erythrocyte must survive wear and tear, page 365.

Staining reveals details of chromosomal organization

The ultrastructure of specific regions of chromosomes can be defined more precisely by staining with specific dyes such as Giemsa (G-banding) or quinacrine (Q-banding). When stained, chromosomes display a pattern of 300–400 alternating light and dark bands characteristic of each chromosome pair (Figure 17.4). This pattern reflects variations in DNA compaction in different sections of the chromosome; for example, dark bands in Giemsa-stained preparations are AT-rich and contain relatively few actively transcribing genes. Such banding profiles can reveal abnormalities, caused by transfer of DNA between

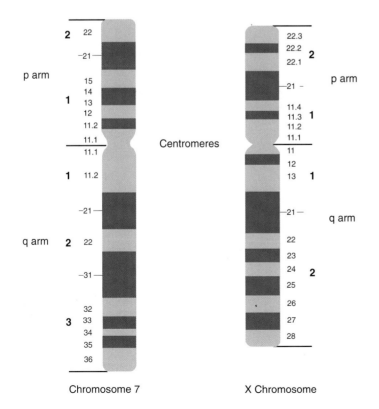

Chromosome 7   Centromeres   X Chromosome

Figure 17.4 Characteristic banding patterns of chromosomes after staining with Giemsa (G-banding). Specific patterns of light and dark bands are observed when metaphase chromsomes are stained with a variety of dyes, of which Giemsa is often used for routine analysis. These alternating bands are believed to reflect differential chromatin condensation in particular regions and are valuable in chromosomal identification, genetic mapping and in detection of specific chromosomal abnormalities. The nomenclature for defining loci on each chromsome based on G-banding patterns is described in the text.

nonhomologous chromosomes or by deletions, in particular regions of chromosomes. Specific regions are identified by a standardized indexing procedure based on G-banding patterns. For example, band 1q42.3 is located on the long arm (q) of chromosome 1, region 4, band 2, sub-band 3. This nomenclature is widely used in the mapping of human DNA and the determination of the locations of human genes.

Chromosomal disorders

Chromosomal disorders involve alterations in chromosome number or gross changes in chromosome structure that are visible under the light microscope, encompassing many hundreds or thousands of genes. The normal human diploid complement of chromosomes was established conclusively as 46 only as recently as 1956. The first recognition of a genetic disease resulting from a chromosomal abnormality, an extra copy of chromosome 21 in Down's syndrome, was in 1959 and by 1970 only 20 different chromosomal disorders had been described. This number has risen to several hundreds as staining techniques for resolving the fine structure of chromosomes have improved, permitting the detection of relatively small changes in specific regions of chromosomes – including a host of apparently normal variations.

Chromosome abnormalities found in early spontaneous abortions are different from those found in neonates

Spontaneously aborted human fetuses show a high incidence of abnormalities of chromosome number. The incidence may be as high as 50% of spontaneous abortions occurring within the first 3–4 weeks of pregnancy, falling to 5% in late spontaneous abortions and stillbirths; only 0.6% of neonates have abnormal karyotypes. Abnormalities observed in neonates are distinct from those detected in early aborted fetuses (Table 17.4) because most chromosome alterations are not

Table 17.4 Differences in patterns of chromosomal abnormalities in early spontaneous abortions and neonates

Chromosomal abnormality	Frequency of occurrence
Early spontaneous abortions	
None	40%
Trisomies (all observed except chromosome 1)	30%
Monosomy (45, X; Turner's syndrome)	10%
Triploidy	10%
Tetraploidy	5%
Miscellaneous	5%
Neonates	
Inversion	1/100 births
Translocation	1/500
Trisomy 13 (Patau syndrome)	1/5000
Trisomy 18 (Edward's syndrome)	1/3000
Trisomy 21 (Down's syndrome)	1/700
47, XXX females	1/1000
47, XXY males (Klinefelter's syndrome)	1/1000
47, XYY females	1/1000
45, X females (Turner's syndrome)	1/10 000

compatible with normal fetal development. Addition of an extra chromosome to the normal pair (**trisomy**) is less deleterious than loss (**monosomy**), probably because monosomic cells have only a single copy of the genes on the monosomic chromosome. Turner's syndrome, monosomy of the X chromosome, is the only deletion of a complete chromosome that results in a viable fetus and survival to adulthood.

Autosomal trisomies, except of chromosome 21, are inimical to life

All possible trisomies, with the exception of those involving chromosome 1, have been detected in fetuses aborted early in pregnancy. In surviving neonates only trisomies of chromosomes 13, 18 and 21 are seen; these are referred to as Patau, Edward's and Down's syndromes, respectively. Down's syndrome is the only common autosomal trisomy, with an incidence of 1 in 700 live births. Patau syndrome and Edward's syndrome are less common, with frequencies of 1 in 5000 and 1 in 10 000, and these individuals rarely survive beyond a few months.

The molecular bases of the effects of trisomy have not yet been elucidated, but are thought to be related to a 'gene dosage' effect of the extra copies of genes on the trisomic chromosome. Chromosome 21 is one of the smallest chromosomes and presumably contains no genes for which an extra copy is inimical to survival into adult life.

Abnormalities of sex chromosomes usually involve extra chromosomes

Abnormalities of the sex chromosomes are relatively common in neonates; early aborted fetuses rarely suffer from defects in these chromosomes. In general, the presence of extra X and Y chromosomes is more readily tolerated than autosomal alterations (Table 17.5).

Individuals with trisomies involving the X and Y chromosomes are generally much less severely affected than those with autosomic trisomies and may be virtually asymptomatic. The two most common defects in males, both occurring with a frequency of 1 in 1000 live births, are the 47, XYY and 47, XXY (Klinefelter's syndrome) karyotypes. In the former case, intelligence may be slightly reduced, patients tend to be taller than average and have more aggressive tendencies but are otherwise normal. In Klinefelter's syndrome, patients have poorly developed sexual characteristics and low testosterone levels, which can be alleviated by hormone

Table 17.5 Sex chromosome abnormalities and their effects

(Abnormal) karyotype	Average frequency	Phenotype
47, XYY males	1/1000 males	No clear symptoms. Often tall, slightly lowered IQ, tendency to aggressive behaviour
47, XXY males (Klinefelter's syndrome)	1/1000 males	Elongated limbs, poorly developed secondary sexual characteristics. Often diagnosed in adult life owing to infertility. Breast cancer rates similar to females
47, XXX females	1/1000 females	Most patients appear normal but are mildly mentally handicapped. The majority are infertile
46, XX males, caused by abnormal transfer of Y chromosome-specific sequences into X chromosome	1/20 000 males	Sterile, with small testes, IQ normal. Diagnosis usually made in adulthood when infertility is being investigated, or at amniocentesis
45, X (Turner's syndrome)	1/5000 females	Short stature. Failure of development of secondary sexual characteristics. IQ and lifespan normal. Sex hormone and growth hormone therapy is beneficial

Box 17.5   Two types of chromosomal translocations

Reciprocal

Exchange of material between any pair of chromosomes, either homologous or nonhomologous. The example involves long arms of chromosomes 10 and 11.

Break points

10   11   10   11

Insertional

In all cases, translocations are balanced and carriers are usually healthy but problems arise in offspring due to abnormal alignment and pairing of chromosomes, leading to loss or duplication of chromosome material in unbalanced offspring.

Break points

1   5   1   5

replacement therapy. However, the defects are usually detected only in adult life, when patients present with fertility problems which by that time are not reversible.

An unusual chromosome abnormality has revealed the site of the gene for 'maleness'

A small number of 'males' (1 in 20 000) have an apparently normal *female* karyotype (46, XX). In these individuals accidental recombination between the X and Y chromosomes seems to have occurred, and the presence of Y chromosome-specific sequences can be detected on one of the X chromosomes. These sequences include regions of the chromosome responsible for the normal development of the male gonads. Analysis of such patients has revealed that male characteristics are attributable to a small region on the short arm of the Y chromosome. The principal gene in this sex-determining region encodes a DNA-binding protein, the testis determining factor (TDF), which is a transcriptional regulator (see Box 5.1) that controls a group of genes involved in testes formation at a crucial stage in embryonic development (Figure 17.5).

Box 5.1. Control of transcription in bacteria and humans, page 117.

Local aberrations of chromosome structure can also occur

More localized changes in the chromosome can also be observed. When such changes are large enough to be detected under the light microscope as alterations in chromosomal size or in G- or Q-banding patterns they are classed as structural aberrations. The limit of detection for small deletions, insertions or translocations is presently 3000–4000 kbp, or 0.1% of the genome. Such changes result from chromosomal breakage (see Box 17.5). When a breakage occurs cellular mechanisms rapidly repair the resultant ends. If two or more breaks happen simultaneously, erroneous rejoining can occur, leading to abnormal combinations of DNA. The background rate of chromosomal breakage is about 1 per 1000 gametes but this rate is markedly enhanced by exposure to ionizing radiation or mutagenic chemicals. Genetic defects in DNA repair, such as that causing xeroderma pigmentosum (see below), can also increase breakage.

Chapter 17. Similar gene defects can arise from different genes in the same pathway, page 524.

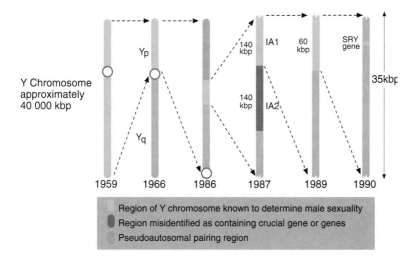

Region of Y chromosome known to determine male sexuality
Region misidentified as containing crucial gene or genes
Pseudoautosomal pairing region

Figure 17.5 Tracking down the sex-determining region of the Y-chromosome from 1959 to 1990. Male sexuality is determined by the presence of a Y-chromosome; however, it is now clear that a small 35 kbp region located on the short arm of the chromosome near the pseudoautosomal region situated at the tip of this arm is sufficient to determine 'maleness'. This arises from analysis of males with an XX genotype in whom it has been found that there has been abnormal translocation of Y chromosome-specific sequences corresponding to this region into the X chromosome. There is evidence that a particular gene in this 35 kbp segment, SRY (sex determining region, Y), encodes a DNA binding protein, the testis determining factor (TDF), which is essential for the expression of a battery of genes which are involved in testes development. As subsequent male sexual differentiation is a consequence of hormonal production by the testis, the function of this gene may be crucial for sex determination. Females with an XY genotype, including one with a frameshift mutation in the SRY gene, have also helped in this search.

Mitochondrial genome disorders

The mitochondrial genome encodes 13 polypeptides

The human mitochondrial genome is a closed, double-stranded circle of 16.5 kbp containing genes for the following (Figure 17.6):

- two ribosomal RNAs form the mitochondrial ribosomes;
- 22 transfer RNAs, used in mitochondrial translation;
- 13 polypeptides, constituents of the mitochondrial inner membrane. These polypeptides are components of the assemblies that catalyse electron transport linked to ATP synthesis. Seven of these proteins belong to the NADH–CoQ reductase complex (Complex I), one to the CoQ–cytochrome *c* reductase complex (Complex III), three to cytochrome *c* oxidase (Complex IV) and two to the ATP synthase complex (see Chapter 7).

Mitochondria use a modified form of the genetic code

In Chapter 5 we emphasized the universal nature of the genetic code but some interesting minor differences in codon usage in the translation of the products encoded by the mitochondrial gene have been identified in the DNA sequence of the human mitochondrial genome. For example, UGA does not function as a stop codon but codes for tryptophan, AGA and AGG act as stop codons and AUA is read as methionine instead of isoleucine. These slight alterations in the genetic code may be attributed to features of mitochondrial tRNAs, which are less discriminating than cytoplasmic tRNAs in distinguishing between the third nucleotide of codons (Table 17.6).

Mitochondrial genetic disorders show maternal inheritance

The limited genetic capacity of the mitochondrial genome, encoding 13 out of a total of several hundreds of mitochondrial proteins, means that this organelle is heavily reliant on polypeptides encoded by the nuclear genome. In humans, where mitochondrial DNA comprises less than 1% of the cellular DNA, many of the mutations affecting mitochondria are encoded in the nucleus and transmitted in normal Mendelian fashion. In contrast, alterations in mitochondrial DNA exhibit a maternal pattern of inheritance, in which all the offspring of an affected mother

Chapter 7. The mitochondrial respiratory chain comprises four multisubunit assemblies of proteins, page 199.

Table 17.6 Distinct features of the mitochondrial genetic code

Codon	Universal code	Mitochondrial code
AGA	Arg	Stop
AGG	Arg	Stop
AUA	Ile	Met
UGA	Stop	Trp

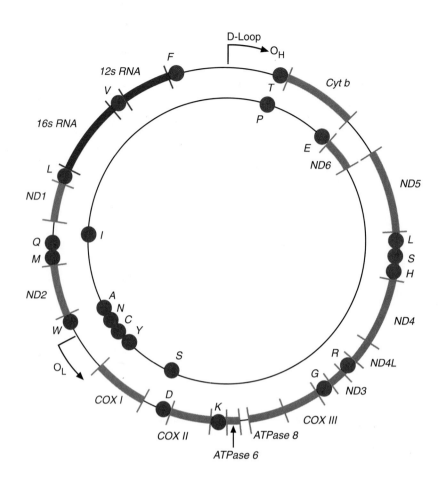

Figure 17.6 Map of the human mitochondrial genome illustrating regions encoding DNAs and mitochondrial polypeptides. Mitochondrial DNA is a closed duplex of 16 569 base pairs illustrated schematically with separate heavy (H, outer) and light (L, inner) strands. Ribosomal and tRNA genes are denoted by their sedimentation coefficients; tRNA genes (circles) are identified by the single letter code corresponding to the appropriate amino acid. Mitochondrial polypeptides are subunits of the NADH–CoQ reductase complex (Complex I, ND genes), CoQ–cytochrome *c* reductase (Complex III, *cyt b* gene), cytochrome *c* oxidase (Complex IV, COX genes), and the ATP synthetase complex (ATP genes) are denoted as indicated. Origins of heavy (O_H and light O_L strand replication are indicated as shown. The D-loop region contains the transcriptional promoters for both H and L strands.

acquire the mutation (the offspring of an affected father are normal). The physical basis for this phenomenon stems from fertilization; the sperm contributes only its nuclear material but the ovum supplies almost all of the cytoplasm of the fertilized egg cell, including all the organelles. Analysis of the mitochondrial genome and its maternal inheritance has been used to investigate the migratory movements of early human ancestors, and supports the idea of an African origin for *Homo sapiens*.

Specific alterations in mitochondrial DNA are implicated in several clinical conditions

Clinical presentation of maternally transmitted mitochondrial defects can occur at any age. Patients are normally classified into three groups, according to symptoms.

1. Relatively mild symptoms, principally muscle weakness (myopathy).
2. Multisystem disease affecting the central nervous system. Patients show myoclonus (uncontrolled jerking movements), ataxia (staggering gait), epilepsy and dementia.
3. Heart muscle problems (cardiomyopathy). These are less common.

In most cases there is no clear correlation between the observed genetic defect and the clinical features. **Heteroplasmy**, in which mitochondrial DNA is a heterogeneous mixture of defective and normal genotypes, is also common. A variety of mutations, including large deletions or duplications or point mutations in mitochondrial tRNA or mRNA genes, is detected in more than 50% of patients with mitochondrial myopathies.

Associations of specific genetic defects with particular phenotypes have also been noted. For example, there is a correlation between ophthalmoplegia (uncontrolled and irregular eye movements) and mitochondrial deletions. The MELAS phenotype (myoclonus, encephalopathy with stroke-like episodes) is associated with a substitution in the mitochondrial tRNA *leu* gene; a similar mutation in the tRNA *lys* gene occurs frequently in the MERRF phenotype (myoclonus, epilepsy with ragged red fibres). A single nucleotide substitution in the ATPase 6 gene occurs in patients with neurogenic disorders, weakness, retinitis pigmentosa and Leigh's disease. Individuals with Leber's hereditary optic neuropathy (the LHON phenotype, characterized by sudden irreversible blindness, principally in young adult males) carry mutations within the *ND1* or 6 genes.

As all of these mutations limit the ability of mitochondria to generate ATP, it is surprising that the phenotypic expression of these mutations is so different – presumably the genetic background of the individual influences the clinical presentation. For example, LHON mainly affects males, and the presence of tissue-specific isoforms of subunits of respiratory chain assemblies and the differing dependence of tissues on mitochondrial metabolism are likely to affect its symptoms. The presence of developmental isoforms of cytochrome *c* oxidase subunits has been implicated in a rare mitochondrial defect causing 'floppy baby syndrome', in which neonates suffer from severe respiratory distress associated with lowered activity of this enzyme. They recover after being maintained in a high oxygen environment for several months.

Multifactorial diseases

Many diseases involve defects in more than one gene, and sometimes also environmental factors

In many human ailments, such as diabetes, heart disease, multiple sclerosis and certain cancers, no simple pattern of inheritance can be discerned. These disorders are examples of **discontinuous multifactorial traits** (in contrast to characteristics such as height, intelligence or blood pressure, which are considered as **continuous multifactorial traits**). The overall response of an individual is determined by a number of different genes at separate loci, which, along with environmental factors, determine the relative susceptibility of a particular individual to a particular disease. Although the genetic contribution to multifactorial traits is now widely accepted, the number of genes involved, the nature of their interactions and the influence of environmental factors are largely unknown. The rapid advances in recombinant DNA technology – including gene isolation, characterization and mapping – give us the hope that considerable progress will be made in the foreseeable future in delineating the molecular details of these complex multifactorial conditions.

Familial connections can indicate a genetic component

In the analysis of multifactorial disease it is important to establish that a genetic component is involved by demonstrating that the incidence of such disorders is higher within affected families than in the general population. The actual frequency of occurrence of discontinuous multifactorial traits is much lower than for single gene defects and usually only close relatives of the affected individual are found to be at slightly higher risk than the general population. Many congenital malformations such as spina bifida and cleft palate fall into this category, as do many of the common diseases of adult life such as diabetes, heart disease and multiple sclerosis.

Box 17.6   The link between genetic make up and response to environmental factors – the P-450 system and hydroxylation

The hydroxylation of foreign substances such as barbiturates and polycyclic aromatic hydrocarbons is an interesting example of how genotype can affect interaction of the body with its environment. The hydroxylation is accomplished by the cytochrome P-450 system (see figure and Box 7.5), located in the endoplasmic reticulum of hepatocytes. Many foreign substances, collectively known as **xenobiotics**, are hydroxylated in the liver (see Chapter 13), generating sites for attachment of glucuronate or sulphate residues. This produces soluble conjugates, which are readily excreted in the urine and thus removed from the body. Unfortunately, this 'detoxifying' hydroxylation is sometimes harmful as it can generate chemically reactive carcinogens from noncarcinogenic substances such as polycyclic hydrocarbons (see Box 6.4). An individual's susceptibility to developing certain kinds of cancer, for example lung cancer, is related not only to the degree of exposure to harmful chemicals but also to the activity of the enzyme systems converting them into carcinogens. This partly explains why not all heavy smokers develop lung cancer.

There is a family of cytochrome P-450 proteins, each with a range of specificities towards the molecules that it hydroxylates. Expression of the genes encoding them is influenced by a number of factors, including the presence of xenobiotics such as the polycyclic hydrocarbons. Cytoplasmic receptor proteins in the cell are able to bind these molecules, then move to the nucleus and activate the P-450 expressing genes, in a manner similar to that of steroid hormone receptors (see Chapter 11).

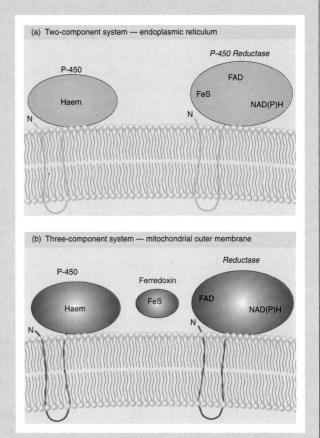

Box Figure 17.6 Two major types of cytochrome P-450 systems found in mammalian cells

Unravelling the complex interactions between the contributing genes, once they have been identified, and establishing the influence of environmental agents (see Box 17.6) is essential to our understanding of the aetiology of these diseases.

Similar gene defects can arise from different genes in the same pathway

The autosomal recessive condition xeroderma pigmentosum (XP) is characterized by unusual sensitivity to the ultraviolet component of sunlight, leading, among other things, to an abnormal predisposition to skin cancers. The XP defect was traced to an inability to repair DNA that has been damaged by ultraviolet light. Seven different subgroups of this condition occur (*XP-A* to *XP-G*) and dissection of the molecular mechanisms of DNA repair has revealed that each group probably represents a defect in a different step of nucleotide excision repair (see Box 6.2). The products of the *XP-A* and *XP-E* genes are involved in the incision process, and those of the *XP-F* and *XP-G* in nucleotide excision. The other gene products catalyse intermediate steps in the repair process, those of *XP-C* in the transcription-independent pathway and *XP-B* and *XP-D* both transcription-dependent and transcription-independent pathways. The overall biological XP phenotype is similar in all cases, indicating that gene defects at different points in a pathway can give rise to similar pathological defects.

Single-gene defects

Single-gene defects are detected only if they are compatible with the viability of the fetus

The concept of single-gene defects was mentioned earlier in this chapter. More than 4500 single gene disorders have been identified, which together affect about 1% of the population. Some of these will be discussed in more detail to illustrate their biochemical aspects. The human genome may contain 70 000 genes, each of which could be defective. In many cases, however, partial or complete loss of function of a gene product would be incompatible with fetal survival or even cell viability. Extremely deleterious mutations of this type are not observed in the human population.

Many single-gene defects produce a mutated gene product with altered properties but the capability to fulfil most of its function. Gene defects producing completely inactive polypeptides are associated with pathways or functions that are not essential for viability. Synthesis of a polypeptide that is altered in some way may be more deleterious than its total absence – an altered peptide may form abnormal interactions with other proteins essential for cellular function, as in cystic fibrosis (see below).

Chapter 17. Cystic fibrosis results from defects in the synthesis of a cyclic AMP-regulated chloride channel, page 525.

There are several types of single-gene disorders

In the following sections we will describe examples of the three main types of disorder arising from a single-gene defect (summarized in Table 17.7). Each type occurs in about 1 in 1000 live births, a rate similar to the rate of occurrence of chromosomal abnormalities.

Table 17.7 The main types of single-gene disorders

Type of defect	Human examples
Autosomal recessive	Cystic fibrosis Thalassaemias
Autosomal codominant	α_1–Antitrypsin deficiency
Autosomal dominant	Familial hypercholesterolaemia
X-linked	Haemophilias

Cystic fibrosis results from defects in the synthesis of a cyclic AMP-regulated chloride channel

Cystic fibrosis (CF) is an inherited defect with symptoms of varying severity, including pancreatic insufficiency and lung disease induced by chronic infections. The lungs and air passages are typically clogged by thick, sticky mucus. The disorder is an autosomal recessive trait for which the gene lies on the long arm of chromosome 7 and encodes a large (1480 amino acids) transmembrane protein that functions as a cAMP-regulated chloride transporter (see Box 17.11). The discovery of the role of defective protein was a triumph for modern molecular biology and provided a basis for an understanding of the symptoms of this disease. Afflicted individuals have high concentrations of Na^+ in body fluids such as sweat and mucus. The lung-clogging mucus is caused by poor osmoregulation resulting from defects in the chloride channel (Figure 17.7).

The disease has a particularly high incidence in northern Europe, affecting 1 in 2000 individuals with a carrier frequency of 1 in 22. Over 70% of patients have a mutation resulting in the deletion of a single phenylalanine residue (Phe 508). This amino acid is located in an adenine nucleotide binding domain and is implicated in the regulation of the chloride channel. Surprisingly, patients who produce no detectable protein often manifest only mild clinical symptoms – confirming the suggestion made above that production of a mutant protein with deleterious secondary effects is more harmful than making no protein at all. Many patients have markedly reduced life expectancy (median age 25), often dying from severe lung problems; the gene for CF is thus a prime candidate for **gene therapy** (see below).

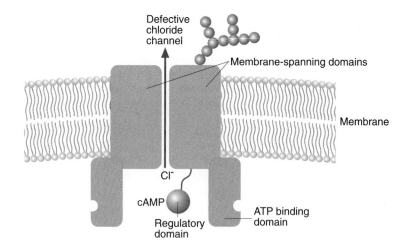

Figure 17.7 The cAMP-regulated chloride transporter that is defective in cystic fibrosis.

Thalassaemias are defects in the synthesis of various globins

Many mutant variants of haemoglobin have been detected in which a single amino acid substitution leads to production of normal levels of a haemoglobin with altered properties (such as sickle-cell, see above). In the thalassaemias, however, a defect in an α-globin (α-thalassaemias) or a β-globin (β-thalassaemias) gene results in reduced levels or absence of specific globins. This defect is produced in one of several ways:

- complete deletion of a globin gene;
- decrease in transcription of a globin gene due to a mutation in the promoter;
- aberrant processing of the globin pre-mRNA transcript;
- production of a grossly abnormal or truncated globin by a frameshift mutation.

Such chains are rapidly degraded.

Although these are potentially very serious mutations, they have a wide geographical distribution that coincides with the prevalence of the malarial parasite. As in the case of the sickle-cell trait, thalassaemic heterozygotes appear to benefit from increased resistance to malarial infection. Asymptomatic individuals can act as carriers of α-thalassaemia. In parts of Italy up to 20% of the population are carriers of this disease. There is also a high incidence in the USA, 25% of Americans of African origin being carriers of α-thalassaemia.

Presence of multiple copies of a gene complicates the pattern of inheritance

Humans possess two α-globin genes on chromosome 16, so there are five possible abnormal genotypes of α-thalassaemia, the normal homozygote being (αα/αα):

- (− −/− −) is incompatible with the viability of the fetus;
- (− −/− α) gives rise to chronic haemolytic anaemia, termed haemoglobin H disease;
- (− α/− α), (− −/αα) or (αα/− −) display no obvious symptoms.

The β-thalassaemias are also inherited as autosomal recessive traits. They are associated with a variety of molecular lesions in the β-globin gene cluster on the short arm of chromosome 11. The situation with the β-thalassaemias is more complex genetically than the α-type because there are several β-globin genes in the cluster.

Deficiency in α₁–antitrypsin causes emphysema and is exacerbated by cigarette smoking

α_1-Antitrypsin is a component of plasma involved in the inhibition of elastase, a proteolytic enzyme secreted by activated neutrophils when they are engulfing bacteria. The importance of this protease inhibitor is apparent from genetic defects that reduce its levels, leading to increased levels of active elastase. Patients with elevated elastase suffer damage to the walls of alveolar cells in the lung, promoted by the digestion of elastic fibres and other connective tissue proteins. The associated emphysema is characterized by acute shortness of breath and inability to exchange blood gases efficiently across the alveolar air sacs. Pulmonary emphysema usually develops during the third or fourth decade of life, and susceptibility is greatly increased (by 70–80%) in cigarette smokers.

The most common mutation results in the replacement of lysine 53 by glutamic acid, a substitution that interferes with the normal secretion of the inhibitor from hepatocytes. Serum α_1-antitrypsin is reduced to 15% of normal in homozygotes, who represent 0.06% of the population; 3.5% of the population are heterozygous carriers. Components of cigarette smoke promote the oxidation of an essential methionine residue (Met 358) to methionine sulphoxide, inactivating the α_1-antitrypsin (Figure 17.8). The effects of smoking are particularly severe in homozygotes, who already have greatly diminished levels of the inhibitor but heterozygotes are also at risk, showing that the disease is inherited as an autosomal codominant trait. The gene for α_1-antitrypsin is located on chromosome 14.

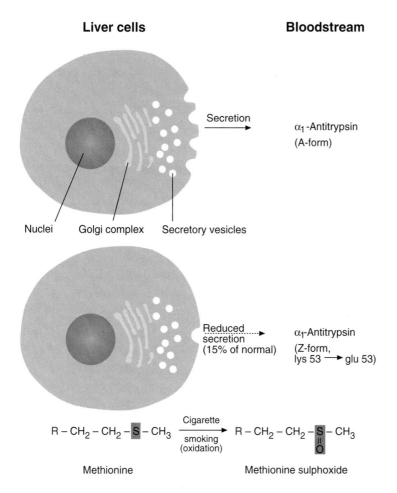

Liver cells **Bloodstream**

Secretion → α_1-Antitrypsin (A-form)

Nuclei   Golgi complex   Secretory vesicles

Reduced secretion (15% of normal) → α_1-Antitrypsin (Z-form, lys 53 → glu 53)

$R-CH_2-CH_2-S-CH_3$ — Cigarette smoking (oxidation) → $R-CH_2-CH_2-\overset{\parallel}{\underset{O}{S}}-CH_3$

Methionine   Methionine sulphoxide

Figure 17.8 Effects of mutation and cigarette smoking on α_1-antitrypsin production by neutrophils. Neutrophils in the bloodstream secrete the protease elastase as part of their immune response. Excess elastase is rapidly neutralized by α_1-antitrypsin, a general protease inhibitor present in plasma. However, the Z-variant is only poorly secreted by liver cells, so that individuals who are homozygous for this trait have only 15% of normal levels. Excess elastase causes damage to lung alveoli and connective tissue, leading to the development of emphysema. This effect is exacerbated in cigarette smokers, particularly those with the Z-variant, as components in smoke promote oxidation of methionine 358 which is essential to the antiprotease action of α_1-antitrypsin.

Hypercholesterolaemia is an autosomal dominant defect of the LDL receptor

Patients with familial hypercholesterolaemia have markedly increased fasting levels of low-density lipoprotein (LDL) and its associated cholesterol. This is due to reduced clearance of LDL from plasma by defective LDL receptors on cell membranes (see Chapter 9). The prognosis for untreated individuals is poor, 50% of affected males are dead from ischaemic heart disease by the age of 60.

Hypercholesterolaemia, inherited as an autosomal dominant trait, is caused by the defects in the LDL receptor gene on the short arm of chromosome 19. Many types of mutation in this gene have been detected in the human population, ranging from deletions and frameshift mutations to premature chain terminations and single base alterations. Prenatal and clinical diagnosis can be made using molecular biological methods such as analysis of intragenic restriction fragment length polymorphisms. Elevated plasma concentrations of LDL and cholesterol are reliable indicators in homozygotes but not in heterozygotes.

Chapter 9. Cholesterol biosynthesis must be regulated to accommodate its many uses, page 290.

Haemophilias A and B are X-linked recessive disorders caused by defects in proteins of the blood clotting cascade

Haemophilias A and B are both X-linked recessive traits and so primarily affect males. They result in defects in blood clotting and spontaneous haemorrhaging into the joints and soft tissues. Although the clinical symptoms of the two forms of the disease are indistinguishable, different factors in the blood clotting cascade – factor VIII in haemophilia A and factor IX in haemophilia B (see Chapter 12) – are affected. The two defective genes are located close to each other on the long arm of the X chromosome. Haemophilia A is the more common defect, accounting for 1 in 5000 male births, compared with 1 in 27 000 for haemophilia B. Intravenous administration of the appropriate clotting factors ensures a near-normal lifespan, although development of antibodies to the infused factor can be a complication, particularly in patients with haemophilia A.

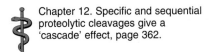

Chapter 12. Specific and sequential proteolytic cleavages give a 'cascade' effect, page 362.

Molecular aspects of cancer

The nature of cancer poses many interesting biological problems

Cancer is one of the major health problems of the present day, especially in the developed world, where public health and other developments have meant that a greater proportion of the population now lives beyond the sixth decade, when the risk of cancer increases. Cancer is not a single disease; it is a group of diseases typified by loss of the normal controls over the division and social behaviour of cells. Tumours may be **benign**, when they remain localized, or **malignant**, when the primary tumour releases cells that are capable of invading other tissues and forming **metastases** or secondary tumours. This second type poses the greatest threat to life and the greatest challenge to medical science.

The 'cancer problem' has many aspects and raises many questions, including the following.

- What are the primary causes of cancer?
- How does a malignant tumour become established?
- Are some individuals more susceptible to cancer than others, and why?
- How can tumour growth be stopped or reversed?
- How can metastasis be prevented?

Many workers in biology and medicine are energetically seeking answers to these questions. The main purpose of this section is to illustrate some of the areas in which biochemistry and molecular biology have contributed to the answers to some of these questions.

Cancer cells have lost the normal controls on proliferation

The generally accepted view is that most, if not all, cancers are both somatic and clonal in origin. That is, the genome of a single cell of an individual undergoes a change which alters the behaviour of that cell so that it loses its normal controls over cell division. If the altered cell multiplies, a clone of daughter cells with similar aberrant properties may appear and grow in size. An event of this sort probably initiates most cancers and is necessary, but not sufficient, for the establishment of malignancy (see below).

Somatic cell mutations are probably responsible for the induction of cancer

Inheritance depends only on mutations that are passed on from generation to generation via the germ cells but mutations or chromosomal rearrangements can also occur in somatic cells during the lifetime of an individual. Mutations can occur during the growth and development of the embryo, other than in precursor gonadal cells, or during adult life in cells of a specific tissue such as lung, kidney or brain. These are referred to as **somatic cell mutations** and are not inherited by any offspring. Such mutations have recently been seen to be involved in the causation of many common cancers. They may also be implicated in the development of autoimmune diseases and in the ageing process.

Cancers are rare in children and young adults, affecting only about 1 in 1000 persons under 25 years old. The incidence increases progressively with age – a quarter of all adults will eventually suffer from some form of cancer. Tumorigenesis is a complex, multistage event which probably develops over a period of years. It is becoming increasingly clear that one or more somatic cell mutations are essential for tumour induction.

Cancer is not generally a hereditary disease, but some rare familial cancers do occur (see below). In families with such conditions, a tumour-inducing mutation may be inherited, giving rise to an increased susceptibility to certain types of cancer. Study of these rare tumours has revealed that mutations in two classes of genes appear to be important, the **proto-oncogenes** and the **tumour suppressor genes**, also known as anti-oncogenes.

Chapter 17. Rb protein, a tumour suppressor, was discovered by study of retinoblastoma, page 535.

Oncogenes and proto-oncogenes

Cancer-causing genes (oncogenes) are found in some retroviruses and some DNA viruses

The discovery of oncogenes was a landmark in the understanding of the underlying causes of cancer and its induction: it also opened up totally new approaches to the regulatory mechanisms governing the proliferation and development of cells and tissues. Genes capable of causing cell transformation were first discovered in retroviruses causing tumours in chickens, mice and monkeys. Retroviruses have a characteristic genomic organization, with three fundamental genetic units (*gag-pol-env*: see Chapter 18). Some retrovirus genomes contain a fourth gene, either in addition to or replacing part of one of these three viral genes. In the second case the virus is defective and is able to replicate only in the presence of a helper virus with the full set of genes (Figure 17.9).

Chapter 18. Retroviral genomes have a distinctive *gag-pol-env* organization that yields the viral proteins, page 576.

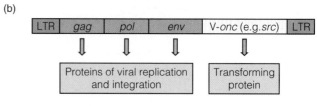

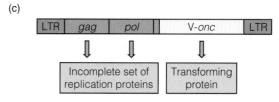

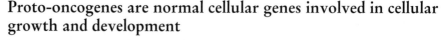

Figure 17.9 Structure of retroviral genomes in the integrated state. (a) A non-oncogenic virus; (b) an oncogenic virus such as Rous sarcoma virus, showing the viral oncogene downstream (rightwards) from the viral replication genes; (c) a defective oncogenic virus, such as Moloney murine sarcoma virus, with the viral oncogene replacing part or all of a gene (*env*) essential to virus replication.

Box 18.5. Classification and general features of human viruses, page 568.

The *src* gene of the Rous sarcoma virus (*sarcoma*) – a retrovirus that induces tumours in chickens – was the first oncogene to be discovered. This gene had been much studied and the discovery that its product was a tyrosine protein kinase and that its sequence had similarity to certain cellular genes, caused much excitement. Many other types of retroviral oncogenes have since been described.

Oncogenes are also found in certain DNA viruses. Examples include the *E1A* gene of adenoviruses, the gene for the large T antigen of polyomaviruses and the *E7* gene of papillomaviruses (see Chapter 18). The significance of viruses in human tumours is discussed below.

Retroviruses cause tumours in two main ways (Figure 17.10).

1. Expression of the viral oncogene after integration of the provirus into the cellular genome perturbs cell regulatory processes.
2. The presence of the provirus in the cellular genome causes a mutation in, or a change in the expression of, a cellular gene (proto-oncogene). This, in turn, perturbs cell regulatory processes.

Proto-oncogenes are normal cellular genes involved in cellular growth and development

The proteins encoded by various retroviral oncogenes are closely related in their sequences to a range of normal cellular genes (proto-oncogenes) that are not themselves oncogenic. Oncogenes are thought to have entered viral genomes as a result of recombination between the host genome and an ancestor of the virus in its DNA form, followed by mutation of the viral oncogene.

A number of proto-oncogenes and their functions in normal cells are shown in Box 17.7. Others have no assigned function as yet, but all known functions are related to the regulation of cell growth and division. The abnormal expression of proto-oncogenes is thought to cause a breakdown in the normal regulatory mechanisms controlling cell proliferation, leading to abnormal patterns of cell growth and behaviour – and eventual tumour growth.

(a) **Integrated provirus expresses a viral oncogene**

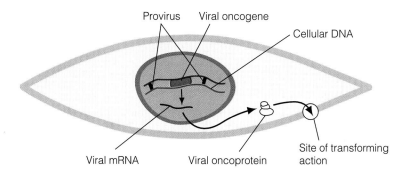

(b) **Integrated provirus causes abnormal expression of a cellular proto-oncogene**

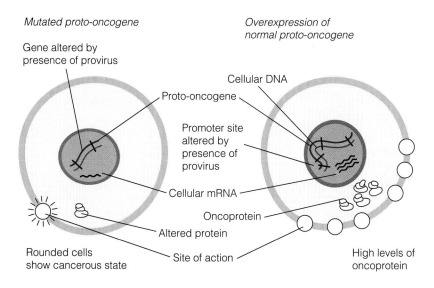

Figure 17.10 Retroviruses can cause tumours in two ways. A proto-oncogene is a normal cellular gene that can be converted into an oncogene, and cause transformation to a cancer cell, by two possible mechanisms. (a) Infection by a retrovirus, which integrates into a chromosomal site next to a proto-oncogene and carries that gene along with its own genome when the virus replicate. (b) Mutation of the cellular proto-oncogene. Mutation of the integrated cellular DNA converts the proto-oncogene to an oncogene, which can cause transformation when this virus infects another cell.

Proto-oncogenes can be abnormally expressed by mutation or chromosomal rearrangement

More than 50 proto-oncogenes have been identified and mapped to specific locations on human chromosomes. Cells from some bladder tumours, lung cancers or melanomas often contain point mutations in one of the *ras* oncogenes; certain mutations occur commonly, for example substitution of glycine 12 by valine. Such mutations can arise by spontaneous mutation or chemical damage to DNA (see Chapter 6).

Some weakly tumorigenic retroviruses do not carry oncogenes in their genome but cause mutations in cellular proto-oncogenes by inserting their DNA provirus at a crucial location near the proto-oncogene. This insertion of retroviral promoters increases the level of expression of the proto-oncogene.

Proto-oncogenes can also be converted to oncogenes by chromosomal rearrangement; an example is chronic myeloid leukaemia. Most patients with this disease have a chromosome 22 that is smaller than normal and in which the *c-abl* proto-oncogene has been transferred from its normal site on the long arm of chromosome 9 to the long arm of chromosome 22, causing production of an aberrant protein. Other types of cancer are also associated with oncogene activation by chromosomal translocations In the case of Burkitt's lymphoma (a B-cell malignancy) translocation between chromosome 8 and one of the three chromosomes containing antibody genes is induced by presence of the Epstein–Barr virus, a herpesvirus.

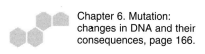

Chapter 6. Mutation: changes in DNA and their consequences, page 166.

Box 17.7 Proto-oncogenes and their functions

The normal cellular functions of the products of proto-oncogenes fall into five categories:

1. **Growth factors** are secreted by cells that stimulate the growth of cells. An example of a growth factor proto-oncogene is *c-sis*, the β-subunit of platelet-derived growth factor (PDGF).
2. **Cell-surface receptors for growth factors**. An example of a proto-oncogene that encodes a growth factor receptor is *c-erb-B*. An oncogene form of *c-erbB* encodes a truncated form of the receptor for epidermal growth factor and exhibits tyrosine kinase activity.
3. **Tyrosine kinases**. An example of such a gene is *c-src*, the cellular homologue of the *src* gene of Rous sarcoma virus. These enzymes participate in intracellular signalling and regulate protein function by phosphorylation.
4. **GTP-binding proteins**. Such genes include members of the *c-ras* family, which are implicated in the activation of several hormone and growth factor-mediated signalling pathways (see Chapter 11).
5. **DNA-binding proteins**. These are transcription factors involved in the regulation of gene expression (see Box 5.1). Examples are *c-fos*, *c-jun*, *c-myb* and *c-myc*.

Chapter 11. Growth factors, page 353.
Box 5.1. Control of transcription in bacteria and humans, page 117.

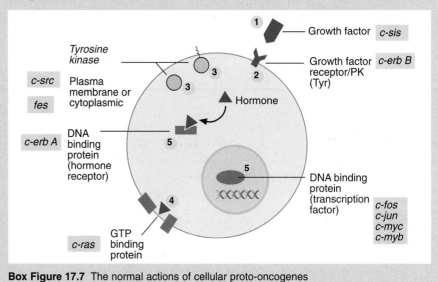

Box Figure 17.7 The normal actions of cellular proto-oncogenes

Initiation of tumorigenesis

The action of oncogenic viruses is only one possible initiating step in the establishment of a malignant tumour. Other initiators include chemical carcinogens, exposure to radiation and hereditary deficiencies in DNA repair (Box 17.8).

Chemicals can participate in both initiation and promotion of tumours

It has long been known that many chemical agents (**carcinogens**) are capable of inducing malignancy. Some of these cause tumours in one particular tissue. The precise mechanisms for these effects are still obscure, although many carcinogens are either mutagens or can be metabolized to mutagens in the body, especially the liver (see Box 6.4).

Chapter 17. Similar gene defects can arise from different genes in the same pathway, page 524.

Box 17.8 Inherited deficiencies in DNA repair are associated with increased cancer risks

In addition to the relatively rare, but well studied, condition of xeroderma pigmentosum (see above and Box 6.2), a connection has been made between a DNA repair gene and colorectal cancer – one of the most common human tumours. **Hereditary nonpolyposis colon cancer** comprises about 10% of all colonic cancers and cells derived from these tumours show a high rate of replication errors. The gene responsible for this defect (*hMSH2*, at chromosomal location 2p16) is similar in sequence to bacterial genes involved in mismatch repair (a cellular repair mechanism that corrects errors made during DNA replication). Mismatch repair is especially important in correcting errors in repeated sequences such as the $(AC)_n$ microsatellites (see Box 17.2), where excision repair is ineffective. Deficiency in mismatch repair causes expansion of the microsatellite leading to chromosomal instability. Individuals with an altered *hMSH2* gene have a 70% chance of developing colon cancer. Women with this altered gene also have a 50% chance of developing uterine cancer.

The recognition of chemical carcinogens and their removal from the environment is an extremely important public health measure. One of the most important groups of chemical carcinogens is the agents found in cigarette smoke – particularly the polycyclic hydrocarbons such as benzpyrene. The carcinogenicity of some of these compounds has long been known – from the days when small boys, who were sent naked up sooty Victorian chimneys, suffered a high incidence of scrotal tumours. The mechanism of action of these compounds has been established only recently (see Box 6.4).

Chemical carcinogens serve to **initiate** the cancerous state by causing mutations that change the behaviour of the affected cell, allowing it to lose the usual regulatory controls on growth. Other chemicals that are not themselves mutagenic or carcinogenic have been shown to increase the carcinogenicity of substances that are – these are the **tumour promoters** (or cocarcinogens) (Figure 17.11). A classic example of these is the group of phorbol esters, for example, tetradecanoyl phorbol acetate, that probably act by stimulating cell division.

(a)

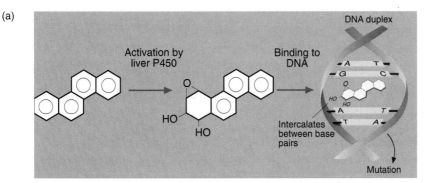

(b)

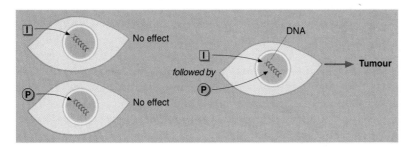

Figure 17.11 Tumour initiators and promoters. Tumor initiators are generally mutagenic; for example, polycyclic hydrocarbons (a) (see Box 6.4). Tumorigenesis by an initiator alone may need prolonged or repeated exposure. Tumor promoters (b) are neither mutagenic nor carcinogenic. Application of a promoter (P) after (but not before) an initiator (I) gives increased carcinogenesis. Promoters may act by stimulating cell division.

Radiation damages DNA and induces tumours

Radiation, both ionizing (X-rays or γ-rays) and nonionizing (ultraviolet light) have long been known to cause tumours. This is almost certainly related to their ability to damage DNA. The simple view, that these agents damage the DNA and cause mutations leading to the initiation of tumour formation, is probably correct in some (or even many) cases. The situation is almost certainly more complex than this, however, because cells are capable of repairing damage to their DNA. It is likely that some tumours arise from the operation of these repair processes rather than directly from damage to the DNA. Programmed cell death, or apoptosis, is part of the normal cellular response to severe damage by chemicals or radiation and has important links to cancer (see Box 17.10).

Cancer caused by radiation is wholly environmental in origin and would be largely preventable if proper precautions were observed, for example, in the use of X-rays, especially in children and women of childbearing age. The recent concern over skin cancers induced by exposure of fair-skinned individuals to intense sunlight (and its associated ultraviolet radiation) is another case in point.

Viruses and human cancers

There is a long-standing connection, well-established in birds, mice and cats, between certain viruses and certain tumours. This is more difficult to prove conclusively in humans because of the ethical limitations on performing definitive experiments. The original idea in this field was that part of the viral genome, encoding a tumour-causing protein, would transform the cell. This is readily demonstrated in cells grown in culture infected by many DNA viruses and retroviruses.

Human retroviruses and cancer

Retroviruses do not appear to be a significant cause of human cancers but the human T-cell lymphotropic viruses (HTLV-I and HTLV-II) have been clearly linked to certain chronic leukaemias with distinct geographical distributions. The HTLVs have been overshadowed in the public eye by concern over HIV and AIDS (see Chapter 18), especially because HIV was originally called HTLV-III. It should be noted that HIV is *not* a tumorigenic virus. Hepatitis B virus, a distant relative of the retroviruses, is linked to primary liver cancer.

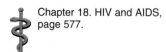

Chapter 18. HIV and AIDS, page 577.

Human DNA viruses and cancer

There have been many attempts to establish links between certain viruses and human tumours, but proving direct causal connections has proved difficult. For example, for many years workers believed that genital herpes (caused by herpes simplex, type 2 (HSV-2)) caused cervical carcinoma in women. One of the main reasons for this belief was the ability to detect RNA complementary to HSV-2 DNA in samples from individuals with cervical carcinoma. Later work showed that these women carried a whole range of pathogens as well as HSV-2 and, although there is a connection between HSV-2 and cervical carcinoma, it is *not* a direct causal one. More recent work has connected several types of human papillomavirus with cervical carcinoma, but whether this 'causal connection' will stand the test of further investigation remains to be seen.

The link between another herpesvirus (Epstein–Barr virus: EBV) and cancer is intriguing. Very strong connections have been made between EBV and Burkitt's lymphoma and nasopharyngeal carcinoma. Both these tumours have limited geographical distributions (they are found mainly in Africa and China, respectively) but the virus is found world-wide and does not appear to cause these cancers elsewhere. This implies that EBV does play a role in the causation of these cancers, but is not sufficient on its own. The Chinese nasopharyngeal carcinoma may have a dietary component in the form of a chemical present in dried fish.

The links between viruses and cancer may be less common than has been claimed. It is very likely that there is a connection between some viruses and some, if not most, types of cancer. It may be, however, that a virus is important or even essential in the overall establishment of a malignancy (see below), but is not sufficient for the whole process. It is significant that only viruses with a 'DNA phase' (the DNA viruses, the retroviruses and their relatives) have been implicated in cancers – no 'true' RNA virus has yet been linked in this way. This is, in itself, indirect support for both a somatic origin for cancer and for the connection with viruses.

Tumour suppressor genes (anti-oncogenes)

Several steps lead to the development of a malignant tumour

The initiation phase and the action of tumour promoters have been already mentioned, but, at least for colorectal cancer, cells pass through several states between the initiation of the process and the development of a malignant, metastasizing tumour. Several distinct stages of colorectal tumour progression have been recognized, the earliest of which are benign, and the process may take 10–30 years (Figure 17.12).

Cells contain tumour suppressor genes

The role of cellular proto-oncogenes in the development of tumours has been discussed above. Passage to the next stage of a sequence such as that shown in Figure 17.12 typically involves escaping the attentions of the product of another type of gene – the **tumour suppressor genes** or anti-oncogenes. These genes were identified from the study of rare familial cancers such as retinoblastoma (see below), in which mutation of a particular tumour gene gives a predisposition towards cancer in afflicted individuals. About eight such genes have so far been identified, but it is thought that there may be as many as 50. Most colorectal cancers have been found to contain 7–10 mutations in 4–5 genes, both proto-oncogenes and tumour suppressors.

Rb protein, a tumour suppressor, was discovered by study of retinoblastoma

The first tumour suppressor gene was discovered by study of a rare inherited form of cancer, **retinoblastoma**, the most common malignant eye tumour in children. Of those who develop tumours in both eyes 20–30% inherit the condition as an autosomal dominant trait associated with the absence of expression of a gene on the long arm of chromosome 13. The protein encoded by the normal form of this gene, Rb protein, acts as a tumour suppressor, probably by stopping cells from progressing from G_1 to S phase in the cell cycle (see Figure 3.6). The Rb protein achieves this by inhibiting transcription factors that switch on genes for proteins, such as DNA polymerase, which are required during S phase.

Protein p53 has been called the 'guardian of the genome'

The cellular protein p53 (named from its apparent molecular mass of 53 000) has been intensively studied because it is involved in the control of DNA replication in cells infected by oncogenic DNA viruses. Viral **oncoproteins** such as the large T antigen of SV40, protein E7 of papillomaviruses and protein E1A of adenoviruses bind p53 and allow these viruses to escape the normal controls of DNA replication.

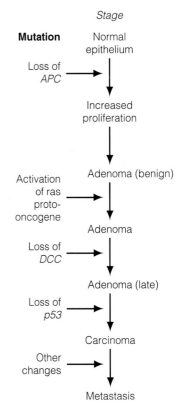

Tumour suppressor genes
APC = Adenomatous polyposis coli
(chromosome 5q 21)
DCC = Deleted in colon cancer
(chromosome 18q 21)
p53 = Protein 53
(chromosome 17p 13)

Figure 17.12 Mutations leading to colorectal cancer.

Box 5.1. Control of transcription in bacteria and humans, page 117.

Chapter 4. X-ray crystallography can be used to study parts of proteins, page 100.

Box 17.9   Protein p53, mutations and cancer

Functions of p53
- Controls a cell cycle checkpoint
- Induces cell cycle arrest in response to DNA damage
- Controls entry into apoptosis (see Box 17.10)
- Acts as a transcriptional activator in the above by binding to specific DNA sequences ($[PuPuPuC(A/T)(A/T)GPyPyPy]_z$)
- Acts as a tetrameric form.

Domain structure
Protein p53 has three principal domains: a floppy, unstructured *N*-terminal region that activates transcription by binding to TFIID (see Chapter 5); a core region that binds to DNA-recognition sequences; a *C*-terminal region involved in forming the active tetramer and in directing the protein to the nucleus (Box Figure).

Analysis of DNA-binding domain three-dimensional structure
The core and *C*-terminal regions can be isolated as discrete fragments because of their resistance to proteolysis. The structure of the latter has been solved by NMR (see Chapter 4), but the core has been crystallized as a complex bound to DNA and its three-dimensional structure determined.

The core region has a structure quite different from those of other known DNA-binding proteins. It consists of a loop–sheet–helix made from conserved regions II and V and two large loops (regions III and IV).

p53 Mutations in human cancers
When the sequences of p53 genes from the tumours of 2000 patients were determined, it was found that the mutations were almost all in the core region, mostly in the conserved regions. Six mutations were especially common, five of them in arginines, two of which bind directly to DNA. This demonstrates a very direct link between amino acid sequence, protein structure and causation of cancer by defective p53.

Attempts are being made to design drugs that can reverse the effects of these common mutations and so prevent cancer in patients with defective p53.

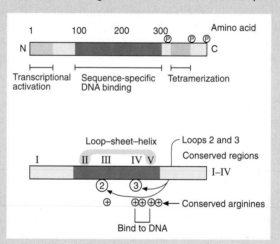

Box Figure 17.9 Protein p53.

Protein p53 does not appear to be expressed in normal, undamaged cells but more than half of human tumours have been found to react with antibody to p53 and the p53 gene is inactivated by mutation in about 75% of human colorectal tumours (Box 17.9). In the Li-Fraumeni syndrome, which has a high familial incidence of cancer, a single amino acid is changed in p53.

The suggestion has been made that p53 acts as 'guardian of the genome' and that it is expressed only after DNA has been damaged, halting DNA synthesis until the defective DNA has been repaired. Many common tumours may progress to malignancy because a mutation renders p53 unable to switch off DNA replication after damage to DNA (Figure 17.12). Thus, instead of entering apoptosis (Box 17.10), the cell starts uncontrolled proliferation. Defects of p53 and Rb are also found in many breast and lung tumours.

Box 17.10   Cell proliferation or cell death?

Apoptosis, programmed cell death, is a normal element in cellular behaviour

We are so conditioned to think of life as a 'good thing' and death as a 'bad thing' that we have only recently recognized that, at the cellular level, cell death may be every bit as important as cell proliferation – at least for multicellular organisms. Apoptosis (programmed cell death) is important in many aspects of proper cell function within the body. Apoptosis differs from necrosis (death caused by heat, anoxia or other severe damage) by following a distinct series of events and by being initiated by a complex network of checks and balances that respond to a variety of signals, such as damage to the cell's DNA. Apoptosis probably plays a role in many aspects of body function, such as the involution of the post-lactating breast gland (Box Figure).

Loss of ability to enter apoptosis may be crucial in the initiation of tumours

There is a growing body of evidence to support the idea that an important reason why cells of malignant tumours do not show the social behaviour of normal cells is their ability to ignore the signals that would send a normal cell into the apoptotic state. The present picture is exceedingly complex, but both Rb and p53 proteins play central roles in deciding whether a cell enters apoptosis; Rb by inhibiting and p53 by inducing the process.

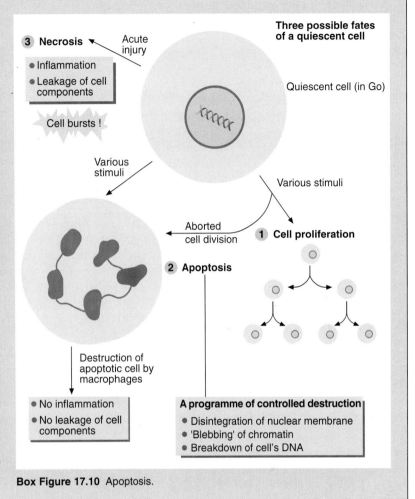

Box Figure 17.10 Apoptosis.

Tumours that evade tumour suppressors may be eliminated by cell-mediated immune responses

The body almost certainly carries out continuous immune surveillance of its cells by the cell-mediated response (see Chapter 16), in addition to the action of tumour suppressors. Most potential cancers are probably recognized and destroyed by this system and only a tiny minority 'slip though the net'. If this is indeed the case, the long-term reduction of cancer levels will be assisted by any factors that keep an individual's immune system effective throughout life.

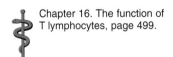

Chapter 16. The function of T lymphocytes, page 499.

Control of cancer

A simple 'cure' for cancer is very unlikely

Although the idea of a 'cure' for cancer has much popular appeal, it is probably an unrealistic goal. Cancer, unlike a bacterial or viral infection, is essentially not of external origin (though it probably often has an external trigger); it lies in the very genome of the human cell. It is therefore unlikely that some magic drug or group of drugs will be able to cure all types of cancer. It is much more realistic to think in terms of control of cancer using a broadly based approach involving various considerations such as

- elimination of environmental carcinogens;
- identification and monitoring of high-risk groups and individuals;
- earlier and/or less invasive diagnosis;
- more effective and/or less traumatic drug therapies.

Measures that aid prevention (such as the first two points in the above list), although intellectually and politically less dramatic than the development of 'wonder drugs', are almost certainly more cost-effective and beneficial to the population at large. Biochemistry and molecular biology will continue to play leading roles in the pursuit of these strategies.

Anticancer drug therapy is useful, but has limitations

When a patient is diagnosed as having a malignant tumour, the prognosis for the particular condition must be evaluated and the nature of any intervention decided. Unfortunately, for many of the common malignancies the prognosis is not good and it is too late for effective treatment. In some cases surgery and/or radiotherapy are appropriate, in others anti-cancer chemotherapy may be the only available option or may be used as an adjunct to other treatments.

Almost all anti-cancer chemotherapy is crude and nonspecific because it uses cytotoxic drugs that prevent cells (including the tumour cells) dividing. Fortunately, the cells of most tissues in the adult are not multiplying, but the undesirable effects of chemotherapy mainly stem from the effects of the drugs on hair, skin, gut lining and blood-forming cells (which are actively dividing). Nevertheless, by careful dosage and monitoring and using a combination of drugs the patient's life may be considerably extended in some cases.

A range of drugs has been used in anti-cancer chemotherapy (Table 17.8) and two important examples (fluorouracil and methotrexate) stop cell division by interfering with the synthesis of thymidine nucleotides (see Chapter 10). It has recently been discovered that induction of the cellular process of apoptosis (Box 17.10), rather than simple inhibition of cell proliferation, appears to be important in the killing of tumour cells by drugs of this type.

A drug therapy that is effective in preventing metastases, especially from common tumours such as colorectal, breast and lung cancer, would be particularly valuable. It would be especially beneficial if combined with more effective early diagnosis and monitoring of tumour development. One problem has been the occurrence of multidrug resistance (Box 17.11).

Some individuals are probably at high risk of cancer because of their genetic make up

Although a strong link has been established between cigarette smoking and lung cancer, not all heavy smokers develop lung cancer while some nonsmokers do. This implies that individuals differ in susceptibility to chemical carcinogens. Similar considerations may apply to viral initiation of tumours. More susceptible individuals may differ from the bulk of the population in possessing, for example:

- higher levels of metabolic activation of carcinogens in the liver (see Box 17.6);
- defective DNA repair (see Box 6.2);
- increased susceptibility of cells to tumour promoters;
- a defect in their proto-oncogenes and/or tumour suppressor genes.

These are **constitutional** predispositions; that is, they occur in the 'normal' cells of afflicted individuals and are not simply mutations in any tumours that

Table 17.8 Anticancer agents

5-Fluorouracil
Methotrexate
Cyclophosphamide
Vinblastine*
Doxorubicin*
Actinomycin D*
Mitomycin C*

*Involved in multidrug resistance (Box 17.11)

Chapter 10. Production of dTTP requires a special set of reactions, page 320.

Box 17.11   Cancer multidrug resistance and the ABC transporter protein superfamily

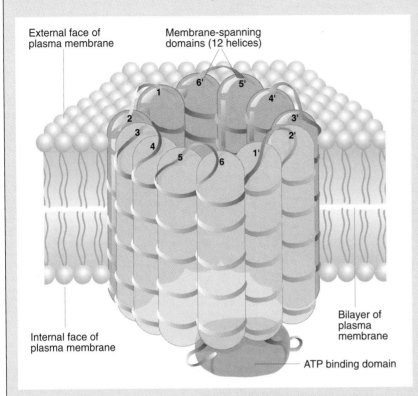

Box Figure 17.11 Structure of proteins of the ABC transporter superfamily.

Structure of ABC-type transporter proteins:

The structure has been predicted from amino acid sequences to possess:

- N-terminus
- six transmembrane helices
- ATP-binding domain
- six transmembrane helices
- ATP-binding domain
- C-terminus

Member of ABC-transporter protein 'superfamily'

A number of proteins have similar amino acid sequences and are said to belong to the ABC-transporter 'superfamily'. They are predicted to have very similar three-dimensional structures in the cell membrane. Members of the 'superfamily' include:

- Multidrug resistance (MDR) protein
- Cystic fibrosis transmembrane regulator (see Figure 17.7)
- Chloroquine resistance in malaria

Multidrug resistance protein

The MDR protein is a normal component of healthy human cells and is capable of pumping a wide range of molecules out of cells. These molecules are typically rather nonpolar in nature and include a number of drugs used in anti-cancer chemotherapy (see Table 17.8). In cancerous cells, MDR protein may be over-expressed and the resultant high levels of protein can lead to the tumour becoming simultaneously resistant to a range of anti-cancer drugs.

---

occur (as in the case of p53, above). Identification of such genes and the individuals who possess them will be of value in indicating the need for more frequent checks for cancer, avoidance of environmental carcinogens or other appropriate precautions.

'Genetic engineering' and some medical applications

What is 'genetic engineering'?

'Genetic engineering' is a popular term that covers an essentially practical field. It is, however, a term that has many theoretical implications and which can be used to answer fundamental questions about every type of organism. In essence, it is an assortment of techniques (Table 17.9) that allows manipulation of DNA, indirectly also of RNA and proteins, study of the organization of genomes and the structure and expression of genes.

Table 17.9 Some important techniques used in genetic engineering

- Sequence-specific cutting of DNA, using restriction endonucleases (Box 17.12)
- Separation of DNA fragments by electrophoresis
- Physical mapping of DNA genomes using restriction fragments
- Joining of DNA fragments using DNA ligase
- Manipulation and cloning of DNA using bacterial plasmids and viruses as vectors (Figure 17.13)
- Amplification of DNA by the polymerase chain reaction (PCR)
- Determination of DNA sequences (Box 17.13)
- Making DNA copies (cDNA) of RNA molecules and cDNA 'libraries'
- Expression of genes *in vitro* to make proteins
- Precisely controlled mutation of genes (site-directed mutagenesis)

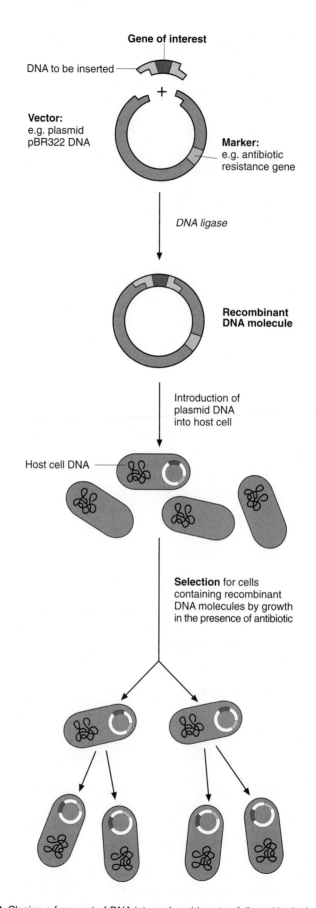

Figure 17.13 Cloning a fragment of DNA into a plasmid vector, followed by its introduction into bacteria.

How can genetic engineering be applied to human medicine?

The possibilities are virtually limitless within the constraints of time, money and ingenuity. In medicine, ethical and moral considerations dictate that not everything that *can* be done *should* be done. Some applications are outlined in the following sections.

The ability to cleave DNA specifically was the big breakthrough

The great technical breakthrough came in 1970, with the isolation of restriction endonucleases from bacterial cells. The function of these enzymes appears to be to protect the bacterial genome from corruption by foreign DNA. They cut double-stranded DNA at specific sequences, usually six base pairs long, and so cleave a given DNA molecule into a distinct set of fragments. Many hundreds of restriction endonucleases have now been discovered, and most are commercially available. The action of these enzymes usually makes staggered cuts in the two strands, generating 'sticky ends', which facilitate the rejoining of DNA fragments by DNA ligase. Before the discovery of these enzymes the manipulation and sequencing of DNA had seemed virtually impossible; now it is routine and the tasks involved are becoming increasingly automated.

DNA molecules and genomes can be physically mapped

If a pure DNA molecule, such as that of a plasmid or virus, is cut with a single restriction endonuclease, a precise set of DNA fragments is generated and can be analysed by gel electrophoresis. By analysing the patterns of fragments obtained using a range of restriction endonucleases DNA genomes may be mapped. This approach has been applied successfully to many viruses and is being extended to bacteria, yeast cells and other model eukaryotes – culminating in the human genome project (see Box 17.16).

The cloning of DNA is an essential step

The other major routine type of work in genetic engineering is that of DNA or gene **cloning**, in which unique fragments of DNA are manipulated into **vectors** (DNA plasmids or viruses of bacteria, usually *Escherichia coli*). These are used to **amplify** DNA (producing workable amounts of pure DNA from very small initial samples) and to select clones for further analysis. Genomic 'libraries' of random DNA fragments can be prepared, from which individual genes can be selected and cloned. Amplification of DNA can be achieved without a vector by the polymerase chain reaction (PCR), performed using heating cycles to denature DNA and a heat-resistant DNA polymerase that can survive the heat. This highly automated technique allows minute amounts of DNA, such as those obtainable from traces of forensic material, to be amplified and mapped.

Although RNA can not be manipulated in the same way as DNA, it is easy to copy RNA sequences into complementary (cDNA) sequences using the retroviral enzyme reverse transcriptase (see Chapter 18). In this way mRNAs or the RNAs of RNA viruses and retroviruses can be manipulated and analysed. 'Libraries' of cDNA made from mRNAs of different tissues can be used to determine the particular genes expressed in various tissues of the body. Thus, a 'liver cDNA library' or a 'muscle cDNA library' may be used to study tissue-specific proteins. The use of cDNA probes is very useful in detecting proteins that are made in very small amounts or are difficult to detect directly.

Box 17.12 Restriction endonucleases

The action of a restriction endonuclease (*Eco*RI) cleaves DNA containing self-complementary sequence GAATTC at the arrows, leaving sticky ends:

5'—G↓AATTC—3'
3'—CTTAA↑G—5'

Some other commonly used enzymes are:

*Bam*H1   G↓GATCC
*Hin*dIII   A↓AGCTT
*Hpa*I   GTT↓AAC
*Sma*I   CCC↓GGG

The last two enzymes make *blunt*, not sticky, ends.

Chapter 18. Retroviruses have a life cycle in which the viral genome alternates between RNA and DNA forms, page 574.

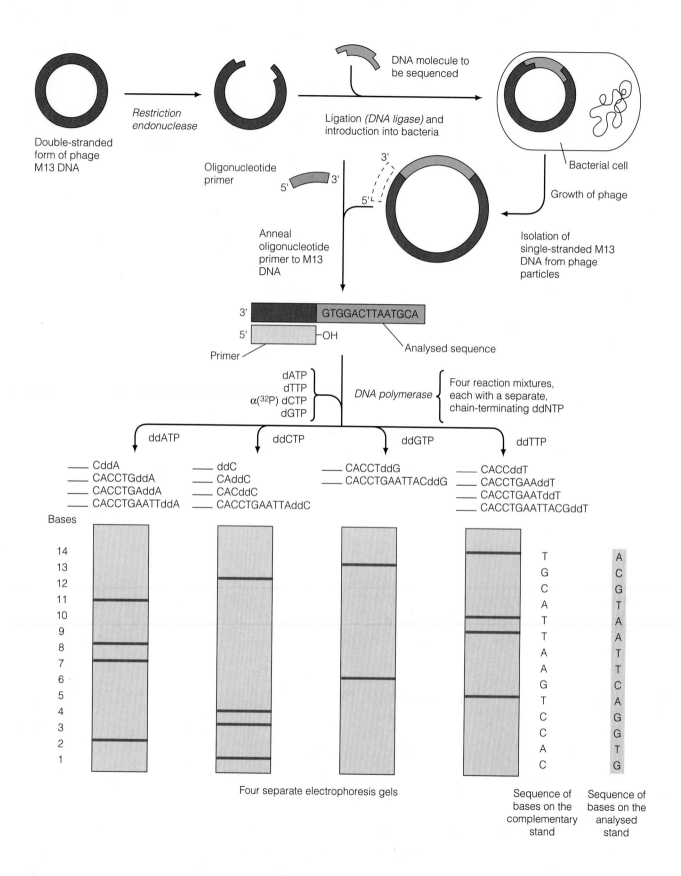

Figure 17.14 Cloning into M13 and sequencing by the Sanger method.

The ability to determine the sequence of DNA has led to an 'information explosion'

The great importance of DNA sequencing (Box 17.13) is the access it gives to the biological information present in the genomes of organisms (see Part B). Before 1980, the only substantial source of sequence information was that derived from the determination of the amino acid sequences of highly purified proteins. Now it is possible to deduce the sequence of a protein from DNA sequence data, sometimes even before the protein has itself been identified. It is impossible to exaggerate the significance of this breakthrough for the medical and biological sciences. With the sequencing of human and other genomes proceeding rapidly, huge amounts of sequence information will appear over the next decade, revealing many genes and proteins about which we are presently ignorant.

Box 17.13   DNA sequencing methodology

The process of DNA sequencing requires the section of DNA to have been cloned and inserted into a sequencing vector, typically phage M13. The phage DNA containing the sequence is then copied using a DNA polymerase and dideoxy-ribonucleoside triphosphates (ddNTPs). These are analogues of the normal deoxy-ribonucleoside triphosphates (dNTPs) used by DNA polymerase, but lack the 3-hydroxyl and thus act as chain terminators. Adding each ddNTP (ddATP, ddCTP, ddGTP and ddTTP) in turn to one of a set of four reaction mixtures with the normal dNTPs leads to a set of labelled DNA products, which may be identified on gel electrophoresis; one band representing each position at which an A (or C or G or T) was present in the sequence. The complete sequence of the DNA insert may thus be read directly from the gel on which the four labelled DNA reaction mixtures were separated. These procedures have also been automated, allowing the determination of many tens of thousands of nucleotides per day.

Interpreting DNA sequence data

What can be done with a DNA sequence once it has been obtained? First, it is important to realize that the use of computers with appropriate software packages and access to sequence databases is essential in interpretation of DNA sequencing data. This is because so much information is generated that it can not be handled manually. Moreover, new sequences must always be compared with existing ones, because a 'new' gene may have already been discovered by someone else.

Many computer programs and sets of programs facilitate the manipulation of sequence data in a wide variety of ways and compare new data with those already in sequence databases. One of the most important uses of computers in handling sequence data is in the identification of open reading frames (ORFs) (see Chapter 5). Computer programs are able to identify all possible ORFs in a given sequence, but it is for the experimenter to judge (and also to prove experimentally) the one probably used in the cell. Criteria for a 'real' ORF include

Figure 5.12. An open reading frame in an mRNA molecule, page 133.

- the presence of consensus promoter sequences upstream from the start codon; and
- the presence of a polyadenylation signal downstream from the stop codon.

Attention should also be paid to the possibility of introns, which can be recognized by the presence of consensus splice sites. These, and other relevant matters, are discussed in more detail in Chapter 5.

Cloned human genes can be used to make human proteins *in vitro*

What can be done with a gene once it has been cloned? At the experimental and theoretical levels the possibilities are limited only by the imagination of the experimenter, but at the clinical level there is one very important application. If a gene is put into a suitable cellular environment it may be expressed to very high levels – making large amounts of the protein encoded by the gene in a controlled culture vessel.

This potential is extremely important in the production of, for example, human proteins such as hormones, which are present in very low concentrations or are species-specific in their function – isolates from large nonhumans such as cows are of no use to humans. Once the gene for such a protein has been isolated it can be inserted into a suitably designed **expression vector**, such as a bacterial plasmid containing suitable promoters and designed to give a high level of gene expression. It may be possible to express sufficiently large amounts of the protein to allow its purification in commercially useful quantities. Human growth hormone was the first hormone to be produced successfully by this method. The hormone produced by this method is routinely injected into children whose retarded growth is due to its lack, replacing material purified from the pituitaries of human cadavers. Similar success has been achieved with a number of other proteins, including insulin.

There can be problems in the production of cloned proteins, especially if the gene for the protein of interest contains introns or the protein is a glycoprotein,

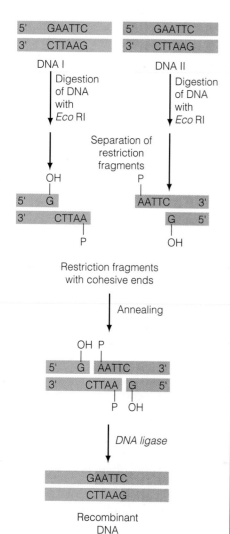

Figure 17.15 Creation of recombinant DNA molecules *in vitro*.

Box 17.14 *In vitro* expression of human genes

The problem

Some human diseases can be treated by administration of the appropriate protein, for example insulin for type 1 diabetes. When the protein is species specific (for example, growth hormone) it can not be obtained from animal sources.

The solution

The gene for the protein can be isolated, inserted into a suitable vector (Figure 17.13) in which the gene is transcribed and translated to give the human protein. This can be isolated and purified from the bacterial culture (Box Figure).

Addition of a signal sequence to the protein causes the bacterium to secrete the protein, simplifying the subsequent purification.

Advantages
- Reproducibility, and large-scale production is possible
- Material is free of human pathogens – before 1985, growth hormone was isolated from human cadavers and was contaminated with the prion causing Creutzfeld–Jakob disease. Since 1985 growth hormone has been made by *in vitro* methods.

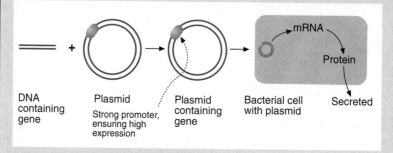

Box Figure 17.14 *In vitro* expression of human genes.

because bacterial cells are unable to carry out splicing or mammalian-type glycosylation. These problems can be overcome by

- removing the intron from the gene before inserting it into the vector; or
- using yeast cells, which can carry out splicing and glycosylation, for expressing the protein.

Genes can be finely engineered by site-directed mutagenesis

The deliberately induced mutation of genes has played a vital role in the elucidation of both classical and molecular genetic problems, but such mutations are essentially random. Gene cloning and associated technologies now permit mutagenesis to be carried out in a precise fashion, changing any nucleotide within a gene. Such site-directed mutagenesis has a wide range of applications, both experimental and practical.

In elucidating the mechanism of action of a protein or enzyme it is possible to use site-directed mutagenesis to mutate individual amino acids in a protein, then express the mutant protein and study the effects of the mutation on the function of the protein. Changing a single crucial amino acid could render a protein completely inactive or make it temperature sensitive. This has become a very powerful tool in the study of protein function. Other mutations can make a protein temperature resistant, which have been of commercial importance – for example, heat-resistant lipases have been produced for use in washing powders.

Site-directed mutagenesis may also be applied to noncoding sequences, such as those present in promoter regions controlling gene expression or in origins of DNA replication. Mutations in either of these types of sequence alter their properties; their use is increasing our understanding of the transcription and replication of DNA.

Gene therapy is a potential cure for some genetic disorders

Many human inherited disorders are due to a defect in the gene for a single enzyme or protein. Fortunately these are all relatively rare conditions, but many are extremely distressing for the individuals concerned. Some, such as phenylketonuria, respond to simple treatment, but most do not or the treatment may be only partially successful, leaving other problems and even a reduced lifespan, as happens in juvenile onset diabetes mellitus. The obvious answer, and the only reliable lasting cure, is to supply the individual with sufficient copies of the non-defective gene.

Such **gene therapy** has been advocated since single gene defects were first recognized, but it has come near to being a practical reality only recently. Clearly the first problem is to isolate the necessary gene in a suitable form, but this is unlikely to present much of a problem in modern medicine.

Practical difficulties of gene therapy include

- Delivery of the gene to the target tissue(s) – those that would normally express the healthy gene. This requires a suitable vector.
- Retention of the gene within the target cells in a stable form; either integrated into the genomic DNA or as part of a stable plasmid.
- Expression of the gene in the cell in a continuing fashion; many genes introduced into cells grown in culture are 'switched off' after some time.

All of these are major technical obstacles and vary considerably from one case to another, requiring expensive 'customization' for treatment of each genetic defect rather than a standardized procedure that could be applied to many different defects.

Box 18.9. Two developments in antiviral chemotherapy, page 581.

Chapter 3. Purified membrane proteins may be inserted into artificial membranes (liposomes), page 57.

Box 17.15 Gene therapy for inherited defects

The principle
The administration of functional copies of a gene to individuals in whom the gene is absent or defective.

The problems
- To deliver the functional gene to the appropriate cells and tissues – the gene defect may affect several different tissues.
- To ensure expression of the gene.
- To ensure retention of the gene within the cells.

Possible solutions

Viruses as vectors
A weakened form of virus (adeno- herpes-, retro – see Box 18.9) containing the gene is used to infect cells with defective genes (Box Figure 1).

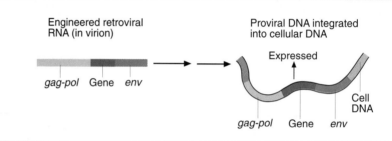

Box Figure 17.15.1 Integration of engineered retroviral genome carrying gene for therapy.

   Advantages: The gene can be integrated into the cell DNA (retroviruses); viruses can target tissues (e.g. adenoviruses target the lungs, herpesviruses the neural tissues).
   Problems: The body's antiviral responses, and 'switching off' of the targeted gene.

Direct delivery of DNA
Liposomes: Lipid vesicles (see Chapter 3) can be prepared that enclose DNA molecules containing the functional gene (Box Figure 2). Aerosols of such vesicles containing a healthy cystic fibrosis gene have been administered to the lungs of sufferers of the disease, with some success.
   'Shotgun': Incredibly, DNA molecules shot into cells at high velocity can express their genes. Could this be a development for the future?

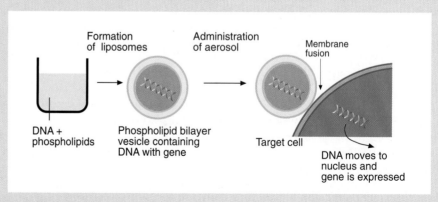

Box Figure 17.15.2 Delivery of gene using liposomes.

Viruses can be used as vectors for the delivery of genes to tissues

Viruses have been the favoured vectors for delivering genes, which has so far been performed only in cultured cells or in animal disease models. Some of the problems of using virus vectors include the following.

- The need for the virus to be 'safe', in both the short and the long term.
- The ability to target the appropriate tissue. Many viruses have a built-in advantage, because they have natural tissue **tropisms** (see Chapter 18), but this means that more than one virus vector might be necessary for a disease that affects a number of different tissues. An example is the deficiency of β-glucuronidase (see Chapter 3), correction of which would require expression in seven or eight different tissues.
- The continued and appropriate expression of the gene in the target tissue. Retroviruses have been favoured as experimental vectors because they are able to introduce the gene directly into the cellular genome, but successful introduction of the gene has often been followed by its being 'switched off'.

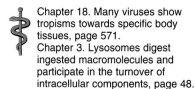

Chapter 18. Many viruses show tropisms towards specific body tissues, page 571.
Chapter 3. Lysosomes digest ingested macromolecules and participate in the turnover of intracellular components, page 48.

Besides these technical problems, and the great expense of developing therapies for very small populations of individuals, major ethical and moral questions are involved in the use of gene therapy – even if it were to become available.

Box 17.16   The human genome project

As the ability to map genomes and to isolate and sequence their constituent genes became easier and more routine, large DNA viruses with genomes exceeding 10^5 nucleotide pairs were completely sequenced. Extensive progress has been made in sequencing the genomes of the bacterium *Escherichia coli*, the yeast *Saccharomyces cerevisiae* and many individual human genes have also been sequenced; it was therefore natural that a long-term project should be undertaken to sequence the human genome completely. This is a task of great magnitude (3×10^9 base pairs, compared with 3×10^6 for *E. coli*), which will cost billions of dollars and necessitate collaboration between many laboratories. There was much debate between those who felt it was a waste of scarce resources, which could be better spent, and those who felt it was a 'Big Science' project with many valuable spin-offs, analogous to landing humans on the Moon. The project is going ahead, and is expected to be complete by 2005.

Gene technology raises ethical and moral considerations

Many of the medically related developments in 'genetic engineering' have already raised ethical and moral questions, which are likely to increase with time. These currently fall into two broad areas – 'information' and 'technical development'. As information on individuals becomes available from gene technologies, what use will be made of the information and who should have access to it? How confidential should it be? Should insurance companies be able to discriminate against individuals on the basis of their gene sequences? If technical developments make something possible, does that make it ethically and morally acceptable? Over all this are the questions Who is going to decide? How can such decisions be enforced?

Certain individuals or groups with sufficient money may proceed regardless of whether something is accepted as morally or ethically wrong.

The whole area of *in vitro* fertilization of human ova is controversial, but when combined with the enormous possibilities opened up by gene manipulation and sequencing the potential ethical and moral problems and dilemmas are multiplied. Satisfactory resolution of these issues will require the active participation of many groups: medical and legal professionals, scientists, politicians, religious leaders and the general public.

CHAPTER 18

Invaders of the body

This chapter considers biochemical aspects of infectious diseases, especially those caused by bacteria and viruses.

Bacteria may be beneficial to humans as well as posing a health risk. Those that are pathogenic for humans typically cause disease by the production of toxins, either endogenous or secreted. An antibiotic is a substance produced by one microorganism that inhibits the growth of another. A few dozen known antibiotics are clinically useful in the control of bacterial disease; others are useful inhibitors in experimental biology. The use of antibiotics as antibacterial agents illustrates the concepts of selective toxicity and of 'targets' for the action of antibacterial drugs, notably in the inhibition of bacterial cell wall synthesis.

Viruses are noncellular agents that are able to multiply only within a suitable host cell. Viruses exist outside a host cell as virions (or virus particles). Different families of viruses have distinctive virion morphologies that enclose viral nucleic acids composed of either DNA or RNA. The viral nucleic acid is the genome of the virus, containing the viral genes and determining both the detailed nature of the replication programme of the virus and its interaction with its host.

The relationship between viruses and disease is often complex and, for most viruses, remains to be dissected at the molecular level. Some important molecular aspects of viral replication are discussed in this chapter, as is as the treatment of viral disease by immuno- and chemotherapies.

The molecular basis of parasitic diseases is still poorly understood, but recent developments have revealed that parasites have molecular biological mechanisms not found in humans. For example, some parasites use *trans*-splicing or RNA editing to evade the host's immune response.

The nature of infectious disease

Infectious diseases of humans are caused mainly by pathogenic bacteria, viruses or parasites

The work of Pasteur, Koch and others in the 19th century established that infectious diseases were caused, not by 'vapours' or 'evil spirits', but by microscopic organisms (microorganisms or microbes). Their work laid the foundations of microbiology, the study of microorganisms in general, with its human and veterinary branches concerned with the microorganisms causing disease in humans and animals. The study of the biochemistry and molecular biology of microorganisms forms a large part of microbiology and this chapter concerns itself with some medical aspects of these areas.

The healthy body presents few routes for the entry of pathogens

Chapter 16. The innate immune response is a collection of barriers to infection, page 484.

The body's first line of defence against infectious pathogens is to prevent them gaining access. This is achieved by the skin, which forms a virtually impenetrable barrier. The areas that are most exposed to contact with infectious agents, such as the eyes, ears, nose, mouth, lungs and gut lining, are bathed and protected by secretions including tears, waxes and mucus (see Chapter 16). Some pathogenic microorganisms are transmitted by **vectors**, such as biting insects, which allow them to penetrate the skin's barrier.

The immune response is usually the best defence against infectious disease

Chapter 16. The basis for the specific or adaptive immune response resides in the lymphocyte, page 485.

Perhaps the most important of the body's defences against infection is the immune system (see Chapter 16). The body's resistance to infection may be artificially enhanced by **immunization** using preparations of bacteria or viruses that have been killed or weakened. Cell-mediated immune responses, especially the action of the T lymphocytes, are important in the destruction of cells that have been infected by virus.

Microbial pathogens have mechanisms for evading the body's defences

Pathogenic microorganisms may be divided into those that appear to be specifically adapted to the human host and those that are merely fortuitous or opportunistic invaders. The long-term survival of the former type depends on its ability to resist or evade the immune system by either

- becoming effectively 'invisible' to immune recognition by masking the antigenic properties of its surface molecules; or
- mutating sufficiently rapidly to remain 'one step ahead' of the immune response.

Rapid mutation is a common behaviour of microbial pathogens; many bacteria and viruses exist as mixtures of strains, which may be closely related in a biological sense but may have quite different pathogenic effects. Two populations are involved in a host/pathogen interaction – the population exposed to the pathogen and the population of the pathogenic microorganism itself – and there is a dynamic relationship between the two. There are important connections between infectious diseases of humans and those of animals, especially domestic animals (see Box 18.7).

The young, old and sick are generally most vulnerable to infectious agents

Although this seems an obvious point, it is extremely important. Many microorganisms are relatively innocuous to healthy adults and young people but pose a much more serious risk to infants, the elderly and the infirm of any age. This is partly due to 'lifestyle' – small children put all sorts of items into their mouths – but is mainly due to deficiencies of the immune system. The very young have not been exposed to many pathogens and so have no immunological 'memory' (see Box 16.4). The immune system in the infirm may be defective or already heavily burdened. Older people are mainly at risk only if they are also infirm, because they generally possess an extensive repertoire of antimicrobial antibodies.

 Box 16.4. IgG: passive protection and damage, page 492.

Hygiene and public health measures have eradicated many bacterial diseases

Pioneering work in the 19th century alerted society to the dangers of bacterial infections and showed how many of the problems could be solved by simple measures – such as cleanliness, hand-washing and care in food preparation. Other measures, such as the provision of clean drinking water and the safe disposal of human excreta by efficient sewage systems, required expensive public works. The implementation of such measures was rapidly followed by dramatic improvement in the health of the populations of the industrialized countries. Conversely, countries that have been unable to introduce these changes still have high levels of infectious disease. Even in developed countries, when water and sewage systems are damaged by war, floods or earthquakes, the human population can soon fall prey to typhoid, cholera and other old scourges.

Many bacterial infections can be controlled by antimicrobial chemotherapy

Although public measures controlled many infectious bacterial diseases; others, notably tuberculosis, remained a major cause of mortality and morbidity until drug therapies were developed – one of the major achievements of 20th century science and medicine. Antimicrobial chemotherapy can be extremely effective for treating diseases for which a successful immunization regimen is not yet available or if time is needed to allow the immune system to develop a response. The use of antibiotics has virtually eliminated some bacteria as troublesome pathogens and has lessened the risk associated with others (see below).

Development of effective antiviral drugs has proved difficult

Viral diseases have proved much more refractory to treatment, and only limited successes have been reported, notably the treatment of herpesvirus infections with acyclovir (see Box 18.8).

Bacterial infections

Bacteria seem to be capable of colonizing almost any environment and of metabolizing almost any type of molecule as a food source. Indeed, the metabolic capabilities of some bacteria are used to dispose of some chemical pollutants.

Many bacteria are relatively harmless or even beneficial to humans

Humans, like all animals, carry a substantial bacterial flora, both externally on their skin and internally in the digestive system. The 'normal' bacterial flora of humans, especially that of the lower intestine, is beneficial in certain nutritional respects, for example in supplying certain vitamins and essential amino acids (see Chapter 14). However, even 'innocuous' bacteria can cause problems, for example perforation of the bowel wall allows gut bacteria to enter the bloodstream, causing septicaemia. Bacteria thrive under different circumstances; anaerobic types such as species of *Clostridium* (agents of diseases such as tetanus and gas gangrene) are generally a problem only when tissues become anoxic, as happens when the circulation to the extremities is impaired. This illustrates the general principle that bacteria are mainly pathogenic through being 'in the wrong place at the wrong time'.

Because of their association with human diseases, bacteria generally have a 'bad press', but they have a number of positive aspects for human society (Box 18.1).

Chapter 14. Biotin, page 449.

Box 18.1 Some beneficial aspects of bacteria

Ruminant animals
Plants are the prime captors of solar energy, and the whole biosphere ultimately depends on them. Among mammals, only ruminants and other herbivores are able to digest the cellulose of plants – this requires the presence of symbiotic bacteria, fungi and protozoa in a fermentation organ such as the rumen.

Nitrogen fixation
Plants that can fix atmospheric nitrogen do so because of their symbiotic relationship with bacteria such as *Rhizobium* species.

The environment
Toxic substances in the environment, many made as a result of human activity, are often degraded only because of the actions of bacteria and other microorganisms. Without these organisms such poisons would persist for many years.

Antibiotic production
Most antibiotics are produced in huge cultures of selected bacteria or fungi and the antibiotics purified from the culture medium.

Genetic engineering
The techniques used in gene manipulation and popularly called genetic engineering depend absolutely on the use of bacteria, their plasmids, viruses and enzymes, notably the restriction endonucleases (see Box 17.12).

Box 17.12. Restriction endonucleases, page 541.

Relatively few bacteria are pathogenic to humans

Considering the number of types of bacteria (there are probably well over a million different species), relatively few are pathogenic to humans. Some of the important bacterial pathogens of humans are listed in Table 18.1.

Pathogenic bacteria are often harmful because of the toxins they produce

Individual types of bacteria are pathogenic to humans or animals either directly by exerting a toxic effect on some aspect of body function or indirectly by

Table 18.1 Important bacterial pathogens of humans

Disease	Bacterium	Toxin
Cholera	*Vibrio cholerae*	Enterotoxin
Diphtheria	*Corynebacterium diphtheriae*	Diphtheria toxin
Dysentery	*Shigella dysenteriae*	Neurotoxin
Food poisoning	*Salmonella* species	Lipopolysaccharide endotoxin
	Staphylococcus aureus	Enterotoxin
Pneumonia	*Streptococcus pneumoniae*	Haemolysin
Scarlet fever	*Streptococcus pyogenes*	Streptolysins
Tetanus	*Clostridium tetani*	Neurotoxin
Tuberculosis	*Mycobacterium tuberculosis*	(Invades lung tissue)
Typhoid	*Salmonella typhi*	Lipopolysaccharide endotoxin
Whooping cough	*Bordetella pertussis*	Pertussis toxin

provoking host mechanisms (inflammation, immune response) that give rise to the pathogenic effects of the bacterial infection (for example, staphylococcal infections of wounds). Pathogenicity may be due to a combination of these two effects.

The direct toxicity of bacteria is caused by two main types of toxins (Table 18.1):

- **exotoxins** – the bacterial cell secretes a protein or peptide with specific toxic effects;
- **endotoxins** – a surface component of the bacterium itself is toxic.

Exotoxins are secreted by some pathogenic bacteria

Many of the most harmful bacteria exert their effects by secreting exotoxins, which circulate in the infected host and cause disease symptoms. The effect of exotoxins does not depend on the bacteria that excrete them. The deadly form of food poisoning (botulism) is caused by the exotoxin released by *Clostridium botulinum* in improperly canned foodstuffs. *Vibrio cholerae*, the agent of cholera, causes an infection of the lower intestine characterized by severe diarrhoea that is frequently fatal because of dehydration. The symptoms caused by different exotoxins are quite distinct and characteristic of the pathogen, in contrast to the rather nonspecific symptoms (aching and feverishness) generally caused by bacteria carrying endotoxins. The molecular targets of some exotoxins are shown in Table 18.2.

Table 18.2 Some bacterial exotoxins and their modes of action

Toxin	Bacterium	Mode of action
Cholera	*Vibrio cholerae*	Causes ADP-ribosylation of the α-subunit of G_s, inhibiting its GTPase activity and causing permanent activation of adenylate cyclase (Chapter 11).
Pertussis	*Bordetella pertussis*	Causes ADP-ribosylation of G_i, blocking the inhibition of adenylate cyclase by G_i.
Diphtheria	*Corynebacterium diphtheriae*	Causes ADP-ribosylation and inactivates protein synthesis elongation factor eEF2, stopping protein synthesis (Figure 5.16).

Chapter 11. Hormonal responses mediated through adenylate cyclase and the production of cAMP, page 338.
Figure 5.16. Signal recognition particles in the direction of ribosomes to the rough endoplasmic reticulum, page 139.

Some bacterial cells are themselves toxic and act as endotoxins

Many Gram-negative bacteria of the gut, especially members of the genus *Salmonella*, exert their pathogenic effects through components of their outer membranes. These bacteria commonly cause food poisoning, but *Salmonella typhi* causes the much more serious typhoid fever. The outermost layers of the cell surfaces of salmonellae are formed by the 'O-specific' side chains of the lipopolysaccharide, long chains of many different sugars in repeating units that are specific to the bacterial strain. This is in contrast to the underlying 'core' layer of oligosaccharide, which is common to many bacteria. The core region is attached to several fatty acid chains that bind it to the outer membrane of the bacterial cell (see Figure 18.1). This high degree of strain variability of the outermost surface makes it difficult for the immune system to respond to the many hundreds of strains of salmonellae that have been recorded.

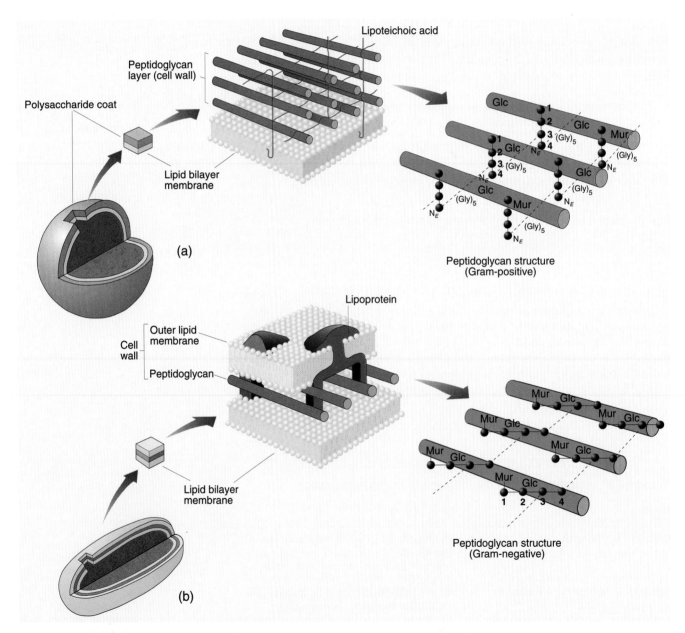

Figure 18.1 The structures of the cell walls of Gram-positive and Gram-negative bacteria (a) Gram positive: *Bacillus subtilis*; (b) Gram negative: *Escherichia coli*.

Bacterial cell walls are of two main types

The properties of salmonellae emphasize the importance of the bacterial cell wall in interactions with animal cells and their immune systems. Two major groups of bacteria are distinguished by their response to the Gram stain. Both types have a lipid bilayer 'plasma membrane' (see Chapter 3) enclosed within a layer of peptidoglycan but the nature of their outer surfaces differ (Figure 18.1).

- Gram-positive bacteria have a thick layer of peptidoglycan anchored to the cell membrane by lipoteichoic acid.
- Gram-negative bacteria have a thin layer of peptidoglycan, different from that of the Gram-positive strains and enclosed by an outer lipid membrane in which lipopolysaccharide is embedded.

The synthesis of the bacterial cell wall is an important target for antibacterial agents (see Box 18.2)

Chapter 3. The plasma membrane marks the boundary of the cell, page 40.

The control of bacterial infections

Some bacterial infections can be controlled by immunization

Immunization (see Chapter 16) is employed against a number of serious bacterial diseases of humans, including anthrax, bubonic plague, streptococcal pneumonia, tetanus and tuberculosis. In the case of tetanus a harmless chemically polymerized form of the tetanus toxin protein, tetanus toxoid, is used as vaccine because the toxin itself is one of the most toxic substances known. To combat *Mycobacterium tuberculosis*, the well known BCG (bacille Calmette–Guérin) vaccine is prepared from an attenuated (weakened) strain of the bacterium.

Although useful for certain bacterial diseases, immunization is essentially a preventive measure and is of limited use for acute bacterial infections, which generally require administration of antibacterial drugs.

Chapter 16. The primary immune response is characterized by IgM production and the secondary response by IgG production, page 486.

Effective antibacterial agents show 'selective toxicity'

Following the recognition of the nature of bacteria, Paul Ehrlich sought a 'magic bullet' for the treatment of syphilis. This was the birth of the concept later formulated more fully by Adrian Albert as **selective toxicity**: the principle of a drug being toxic towards the pathogenic microbe but relatively harmless to the host.

The effectiveness of a drug may be measured by a parameter such as the concentration required to reduce the rate of bacterial growth by 50% or to kill 50% of a bacterial population. The toxicity of a drug towards the host may be measured by the concentration required to cause undesirable side-effects in 50% of the treated population. The ratio of the concentrations for these two parameters is a measure of the **therapeutic index** (effectiveness/toxicity) of the drug, a low value being the hallmark of a successful drug.

Ehrlich's 'magic bullet' had to have a 'target', and to be 'magic' the 'bullet' needed to seek out a specific 'target' in the bacterium. The concept of a target is still valid and has been applied to the less tractable problems of antiviral chemotherapy (see below).

Box 18.8. Potential targets for antiviral drugs, page 580.

Bacteria compete with each other in the wild by producing antibiotics

In everyday speech, the term antibiotic effectively means 'a drug with antibacterial activity', but the biological definition of an antibiotic is 'a substance produced

Table 18.3   Antibiotics that are useful experimentally rather than clinically

Process	Chapter reference	Antibiotic examples
Translation	Chapter 5	Cycloheximide, puromycin
Transcription	Chapter 5	Actinomycin D, α-amanitin
Electron transport	Chapter 7	Antimycin A
Ion transport	Chapter 3	Monensin, valinomycin

by one organism that inhibits the growth of another'. Many bacteria, and some other microorganisms, produce such substances, which presumably give them a competitive edge 'in the wild'. Over 2000 antibiotics have been characterized, but only about 50 have clinical value as antibacterial agents (see Table 18.4). This is because most lack selective toxicity; that is, they have toxicity towards humans equal to or greater than that to other bacteria.

Although not clinically useful as antibacterial agents, some of these toxic antibiotics are useful experimental tools, especially in the analysis of complex biological systems. Their highly specific properties have aided the analysis of various processes (Table 18.3).

Some antibiotics are clinically useful in the control of pathogenic bacteria

Most antibiotics are produced by fungi of the group *Aspergillales* and by bacteria such as the *Actinomyces* and *Streptomyces*. Several of those with clinical applications are listed in Table 18.4; a major industry has developed for the large-scale commercial production of these compounds. Some antibiotics, especially many penicillins, are described as 'semi-synthetic' because the basic molecule is produced biologically and is then modified chemically to produce molecules with more desirable properties such as a broader spectrum of action or reduced development of resistance. The sites of action of many antibiotics are typically in the inhibition of synthesis of bacterial cell walls, proteins or RNA (Table 18.4).

Other antibacterial agents are produced chemically

The sulphonamides and nalidixic acid are clinically useful antibacterial agents but are not 'true' antibiotics – they were developed in chemical laboratories. The sulphonamides interfere with bacterial production of folic acid (see Chapter 10); nalidixic acid inhibits bacterial topoisomerase (see Chapter 6).

Chapter 10. Folic acid and 'one-carbon' metabolism, page 323.
Chapter 6. Molecular events at the replication fork are centred on DNA polymerase, page 156.

Table 18.4   Antibiotics with useful antibacterial properties

Antibiotic type	Site of action	Molecular action
Penicillin	Cell wall synthesis	Inhibits transpeptidase
Cephalosporin	Cell wall synthesis	Inhibits cross-linking
Vancomycin	Cell wall synthesis	Polysaccharide synthesis
Aminoglycoside (e.g. streptomycin)	Protein synthesis	Causes misreading of codons
Macrolide (e.g. erythromycin)	Protein synthesis	Inhibits ribosome translocation
Tetracycline	Protein synthesis	Inhibits binding of aminoacyl-tRNAs
Chloramphenicol	Protein synthesis	Inhibits peptidyl transferase
Rifamycin	Transcription	Inhibits RNA polymerase

Antibacterial agents act on 'targets' in the bacterium that are different in, or absent from, human cells

In order to consider how the concept of 'selective toxicity' (see above) might be applied to the problem of designing drugs that kill or inhibit the growth of bacteria without harming human cells, it is necessary to consider the major differences between the human and bacterial cells.

Box 18.2 Penicillins interfere with bacterial cell wall synthesis

The peptidoglycan of bacterial cell walls consists of polysaccharide chains linked by short peptides (see Figure 18.1). A crucial step in the synthesis of this peptidoglycan is the action of peptidoglycan transpeptidase, which cross-links polysaccharide chains via the pentaglycine peptide unit with loss of the terminal D-alanine (Box Figure 1). The activity of transpeptidase is inhibited by the penicillins (Box Figure 2) because the configuration of the β-lactam ring resembles that of D-alanine, allowing penicillin to bind to the active site (Box Figure 3). Lack of transpeptidase activity results in failure to cross-link the peptidoglycan and loss of strength of the bacterial cell wall. This in turn causes the production of aberrant and unstable forms of bacterial cells during bacterial growth.

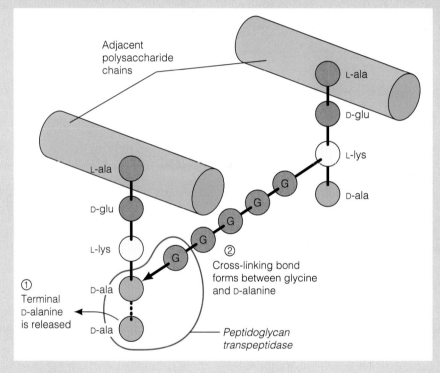

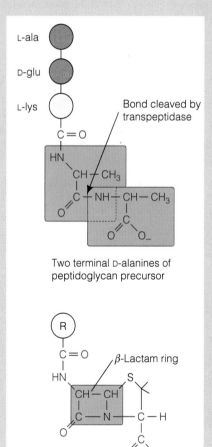

Box Figure 18.2.1 The action of peptidoglycan transpeptidase.

Two terminal D-alanines of peptidoglycan precursor

Box Figure 18.2.3 Transpeptidase is inhibited by the β-lactam ring of penicillin.

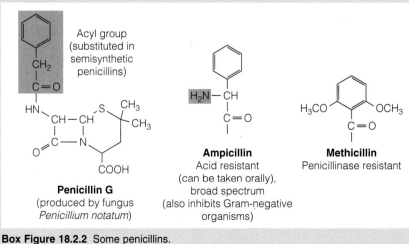

Box Figure 18.2.2 Some penicillins.

The bacterial cell wall

The cell wall is the most obvious difference between the two cell types because all bacteria possess a cell wall but human cells do not (see Figure 18.1). The importance of the cell wall to the bacterium lies in its great physical strength, which is vital for the free-living lifestyle of most bacteria. Bacterial cells without a cell wall (protoplasts) can be produced in the laboratory, but they are extremely fragile and could not survive in the wild. Impairment of bacterial cell wall synthesis is an obvious 'target' for the design of antibacterial agents. The penicillins have been used very effectively to exploit this target (Box 18.2).

Cell membranes

Both bacterial and human cells are bounded by a membrane but there are substantial differences between the plasma membrane of the human cell and the bacterial cell membrane. Whereas in eukaryotic cells various functions are distributed among the several membranes of the cell – for example, electron transport on the mitochondrial inner membrane and protein synthesis on membrane-bound ribosomes of the rough endoplasmic reticulum – the bacterial cell membrane is a multipurpose structure that carries out all of these functions.

Box 18.3   The modes of action of some antibiotics that inhibit bacterial protein synthesis

Streptomycin is one of the aminoglycoside group of antibiotics that interfere with protein synthesis by binding to the 30S subunit of bacterial ribosomes, thus causing misreading of codons.

Erythromycin is one of a group of antibiotics (the macrolides) that stop protein synthesis by inhibiting ribosome translocation.

The **tetracyclines** are a group of related antibiotics that block protein synthesis by inhibiting the binding of aminoacyl-tRNAs to the 50S subunit of bacterial ribosomes.

Chloramphenicol blocks protein synthesis by inhibiting peptidyl transferase.

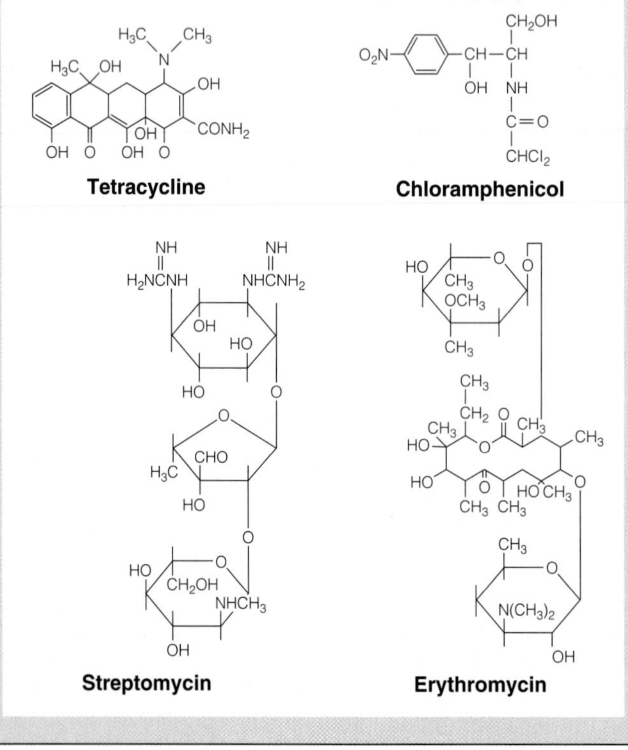

Tetracycline

Chloramphenicol

Streptomycin

Erythromycin

Box Figure 18.3
The structures of streptomycin, erythromycin, tetracycline and chloramphenicol.

Protein synthesis

Although the general principles of protein synthesis are similar in bacterial and human cells, the functional and structural properties of the ribosomes differ in important ways. Structurally, bacterial ribosomes are smaller than those of eukaryotes (70S compared with 80S) because they contain shorter RNAs and fewer proteins. Functionally, bacterial ribosomes, unlike their eukaryotic counterparts, are capable of 'starting' and 'stopping' translation in the middle of an mRNA; that is, bacterial mRNAs may be polycistronic and able to encode more than one polypeptide chain. Antibiotics that inhibit bacterial protein synthesis are effective antibacterial drugs (Box 18.3) because they block bacterial translation but not the translation on the 80S ribosomes of human cells. Some of the side-effects of the prolonged use of some of these drugs (for example the tetracyclines) are due to inhibition of protein synthesis on the ribosomes of the mitochondria of human cells (see Chapter 17). These resemble bacterial ribosomes in a number of respects.

Chapter 17. The mitochondrial genome encodes 13 polypeptides, page 521.

Nucleic acid synthesis

As with protein synthesis, the broad principles (described in Chapters 5 and 6) are similar in the two cell types, but the details differ in important aspects (Table 18.5) that provide potential targets for chemotherapy.

Bacteria can become resistant to antibacterial agents

Soon after clinicians began to use antibiotics, it was found that bacteria that were initially sensitive to a particular antibiotic ceased to be so – they had acquired **resistance**. Resistance to antibiotics has become a widespread and troublesome phenomenon. Examples of mechanisms of resistance are listed in Table 18.6. The genes

Table 18.5 Differences in nucleic acid synthesis between bacterial and human cells

Target	Bacterial cells	Human cells
Gene expression	One RNA polymerase	Three RNA polymerases
	Direct translation of mRNA	Translation in cytoplasm
	No mRNA processing	Capping and tailing of mRNA
	No splicing of RNA	Splicing out of introns
DNA replication and cell division	Single origin	Multiple origins
	Occurs continuously	Occurs only during S phase
	No organized nucleoprotein	DNA occurs in nucleosomes
	Segregation by membrane	Mitotic apparatus

Table 18.6 Some mechanisms of bacterial resistance to antibiotics

Production of an enzyme that inactivates the antibiotic
- The hydrolytic action of penicillinase, opening the essential β-lactam ring of penicillins
- Acetylation of chloramphenicol by chloramphenicol transacetylase. This enzyme is widely used as a marker on bacterial plasmids used in genetic engineering (see Chapter 17).

Reduction in permeability of the bacterial cell membrane
- One of the mechanisms of resistance to tetracyclines depends on a gene that causes a change in membrane permeability

Mutation of the target in the bacterium
- Mutation in a protein of the 30S subunits allows the bacterial ribosome to become insensitive to inhibition by streptomycin. In contrast to the genes for the types of resistance mentioned above, which are typically located on plasmid genomes, this mutation occurs on the main bacterial genome

Chapter 17. 'Genetic engineering' and some medical applications, page 539.

responsible for the resistance can reside either on the main bacterial chromosome or on a **plasmid** (see Box 18.4). They can move between bacterial and plasmid genomes because they are located on **transposons**, a type of mobile genetic element.

Acquisition of a plasmid by a bacterium may confer resistance to several types of antibiotics in a single step

Some plasmids carry a number of genes, each of which, when expressed as its protein, confers resistance to a different antibiotic. A bacterium that acquires such a plasmid immediately becomes resistant to these antibiotics – some plasmids carry genes conferring resistance to as many as eight different antibiotics. Plasmids were first discovered as agents of **multidrug resistance**.

The nature of viruses

Viruses are also agents of infectious disease

In the early years of the 20th century, some infectious microbes were found to pass through the unglazed porcelain filters used to remove bacteria from aqueous suspensions and were termed 'filterable viruses'. By the middle of the century, it was recognized that, although some (for example, the psittacosis agent) were simply very small bacteria, most of them belonged to another class of microbial agent – the viruses. Viruses are totally different from bacteria, being noncellular in nature and with a distinctive mode of replication, dependent on a suitable cellular host.

The study of clinical virology is extremely important in the diagnosis and epidemiology of viral disease but a deeper understanding of viruses and their nature has come from analysis of their molecular nature, replication and genomes. These studies are firmly rooted in the areas of biochemistry, genetics and molecular biology.

Viruses are noncellular and require a host cell for replication

Virus particles (**virions**) are inert: they do not carry out any metabolic activities or make any molecules. Only when they come into contact with a suitable host cell (one with which a virus can interact so as to bring about its own multiplication or replication) do they become active. Viruses infect every type of cellular organism – bacteria, plants, animals, birds, insects, fungi – but individual viruses show a high degree of specificity towards a host. Many viruses that infect humans are specific for humans, but some are almost certainly viruses of other animal species that have been transmitted to humans. These include some rare, but extremely pathogenic, viruses, such as Marburg and Ebola viruses (Filoviridae) and Lassa fever virus (Arenaviridae) (see Table 18.5), as well as more common viruses such as influenza A virus (see Box 18.7). Viruses are not the only type of noncellular infectious agent, but the others – viroids, plasmids and prions – are fundamentally different from viruses (see Box 18.4).

A typical virus life cycle includes a 'virion phase' and a 'replicative phase'

We can distinguish two phases in the life of a typical virus (Figure 18.2)

1. A **replicative** phase, in which the virus enters its host cell and goes through a precisely timed programme of events leading to the production of many progeny virions from a single infecting particle. This is sometimes known as growth by 'burst'.

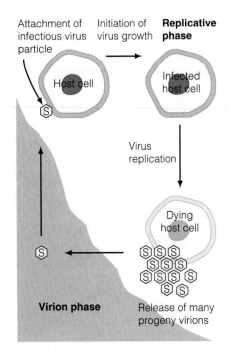

Attachment of infectious virus particle

Initiation of virus growth

Replicative phase

Host cell

Infected host cell

Virus replication

Dying host cell

Virus replication

Release of many progeny virions

Virion phase

Figure 18.2 The two phases of a viral life cycle.

Box 18.4   Other noncellular infectious agents: viroids, plasmids and prions

Viruses are not the only noncellular agents able to move from cell to cell, but they are the only ones with both a genome and a viable extracellular existence.

Plasmids resemble viruses in a number of respects (indeed, some viruses behave like plasmids), but the genomes of true plasmids lack genes for making a virion and must be passed directly from cell to cell. The plasmid–cell relationship appears to be one of symbiosis, as the plasmids usually confer some advantage on the host (such as genes for resistance to antibiotics) in return for the use of its 'facilities'. The experimental use of plasmids is central to the methodology of 'genetic engineering' (see Chapter 17).

Viroids are superficially similar to viruses in their size and infectivity and have generally been studied as part of 'virology'; they are, however, fundamentally different. Viroids are 'self-replicating' RNA molecules that infect plants, but are not genomes, because they contain no genes and make no proteins. The hepatitis delta agent of humans is a highly unusual and defective 'virus', part of its genome resembling viroid RNA (see Box 18.6).

Prions are a distinctive type of entity in that they appear to consist solely of a modified form of a protein encoded by a cellular gene. This protein is able to modify existing protein molecules when it enters a suitable nerve cell. In contrast to viruses, prions contain no nucleic acid, either DNA or RNA. Prions cause severe and irreversible neurodegenerative disease in humans (Jacob–Creutzfeld syndrome) and animals (scrapie of sheep; bovine spongiform encephalopathy of cows – known as BSE or 'mad cow disease'). Prions are highly resistant to most commonly used sterilizing procedures

Chapter 17. 'Genetic engineering' and some medical applications, page 539.

2. A **virion** phase, a period of variable duration. In this phase the virion leaves the cell in which it was made and remains inert until it encounters another host cell and begins another replicative cycle. The length of time that the virus can survive outside a host depends on the nature of the virus and the conditions it encounters. Many viruses are not very stable outside their host and so need fairly direct contact between hosts.

The replicative type of interaction between virus and host is often called **lytic growth**, because it usually results in the death of the host cell. Other types of interactions do occur between a virus and its host; these can be clinically important (see below).

Viruses do not always kill their host cell, page 566.

The infectious unit of a virus is an inert particle, or virion

Viruses are typically studied by growing them in cultured cells rather than in whole animals, and the infectious unit is taken as the amount of virus preparation that will produce a localized **plaque** of dying cells on the cell sheet. Productive infection commonly results in the production of several thousand infectious progeny virions from a single plaque. In a 'natural' infection, these would be capable of spreading the infection to adjacent host cells and possibly throughout the body.

The particles of different viruses range in size, from about 20 nm to about 300 nm, as measured under the electron microscope. Each type of virus has a distinctive **virion morphology**, which can be visualized by high-resolution electron microscopy and is an important factor in identifying viruses. Various types of virion are shown in Figure 18.3.

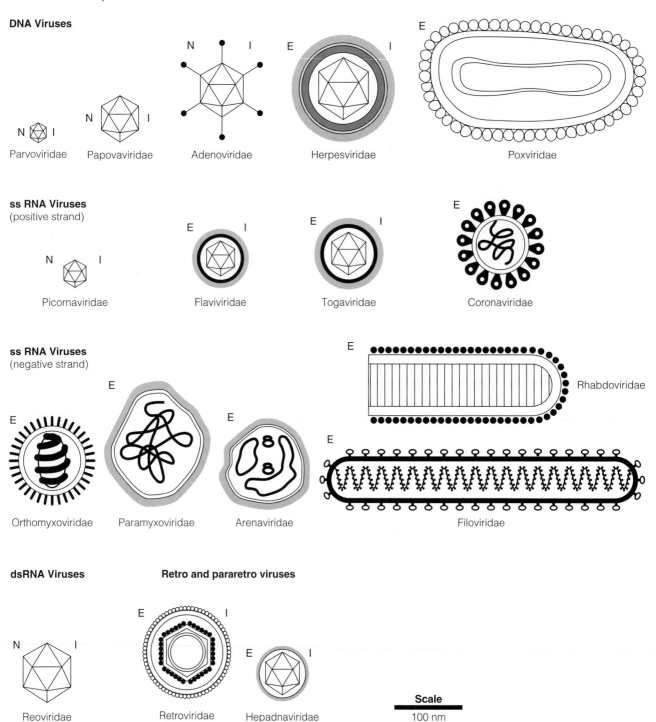

Figure 18.3 Some types of virus particles that infect humans. N, naked; E, enveloped; I, icosahedral.

Virions contain protein and nucleic acid and have a defined three-dimensional structure

All types of virion contain protein and nucleic acid, the protein forming a 'shell' or 'coat' around the viral nucleic acid – the genome of the virus. Virions are typically composed of large numbers of a few species of protein molecule, and the characteristic virion morphologies are due to the geometry of the proteins in the virus particle. The three-dimensional structures of several of the smaller viruses have been determined, and this has increased our understanding of the precise interactions between protein molecules (Figure 18.4).

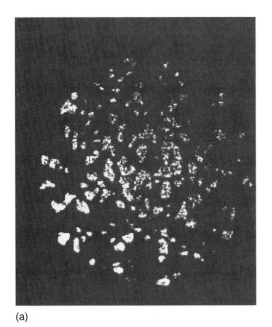

(a)

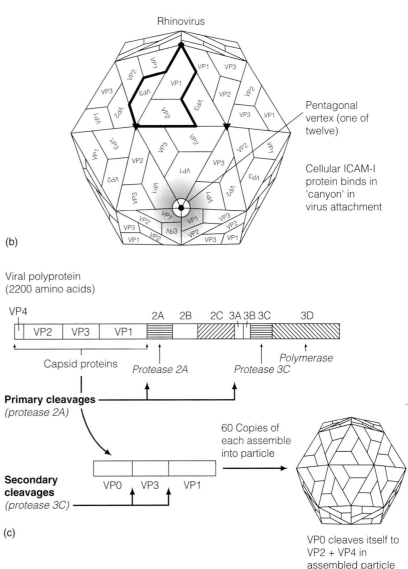

(b)

(c)

Figure 18.4 The three-dimensional structure of human rhinovirus and the cleavage of the viral polyprotein. (a) Cryoelectron micrograph of human rhinovirus. (b) Schematic view of rhinovirus showing capsid proteins VP1, VP2 and VP3. (c) Viral polyprotein and its cleavage. Rhinoviruses belong to the Picornaviridae, all of which have similar virion structures and cleavage patterns.

Virions of many types of viruses are enclosed by a membranous envelope

Although many types of viruses consist of particles composed only of protein and nucleic acid ('naked' virions), many others also contain lipid and display a membranous 'envelope' structure under the electron microscope (Figure 18.3). This envelope is acquired from cellular membranes, either the plasma membrane or one of the intracellular membranes of the infected cell (see Chapter 3), depending on the type of virus, during virion assembly. The membrane of the envelope is modified by the insertion of viral proteins, usually glycoproteins.

Chapter 3. The endoplasmic reticulum and Golgi complex are membrane-bound compartments, page 46.

The genome of a virus consists of a set of genes and control sequences

One of the first discoveries of 'modern' virology, once viruses had been isolated, purified and analysed, was that virions contained DNA or RNA but never both. There is a fundamental distinction between the 'DNA viruses' and the 'RNA viruses' (see below). The genomes of all viruses are composed of a number of genes,

Table 18.7   Examples of viral genomes

Virus family*	Typical length of genome	No. of viral genes†	Figure(s)	
Papova	5 kbp	5	⊚	⎫
Adeno	40 kbp	40		⎬ ds DNA
Herpes	160 kbp	75		⎭
Picorna	8 kb	10		⎫ (+) RNA
Corona	25 kb	6		⎭
Orthomyxo	14 kb	10		⎫ (−) RNA
Paramyxo	14 kb	7–9		⎭
Retro	9.4 kb	16		⎫ Retroid
Hepadna	3.3 kbp	5	⊚	⎭

*See Box 18.5
†Comparison of the precise number of viral 'genes'
is difficult because of the different modes of replication

Scale: ——— 10 kb (Single strand)
═══ 10 kbp (Double strand)

ranging from 5 to 200 (Table 18.7), that direct the synthesis of a corresponding number of viral proteins during viral replication. Typically at least 90% of a viral genome is occupied by protein-coding sequences and most of the remainder consists of sequences that control the expression of viral genes or direct the assembly of the virion. The genome contains the information that directs the programme of replication leading to the production of progeny virions. Viruses that have alternative interactions with their host, such as latency or persistence (see below), probably possess genes that allow them to act in a noncytolytic fashion.

RNA virus genomes are generally small (3–30 kb); some, such as influenza (see Box 18.7) comprise several RNA molecules. DNA viruses cover a tenfold greater range, the genomes of the largest DNA virus being almost as big as those of the smallest bacteria (see Table 18.7).

There are three main types of virus: DNA viruses, RNA viruses and retroviruses

The 'true' DNA viruses possess DNA genomes and pass their inheritance from DNA to DNA, resembling cellular organisms in this respect. The 'true' RNA viruses possess RNA genomes and pass their inheritance from RNA to RNA. This feature is unique to the RNA viruses but may reflect an ancient situation in the postulated 'RNA world' of about 4000 million years ago (see Box 6.5).

Discovery of the retroviruses and analysis of their lifestyle unexpectedly revealed a third type of viral inheritance. In this, the viral genome alternates between RNA and DNA forms during the replicative cycle: retroviruses contain

 Box 6.5. Was RNA the first genetic material in the origin of life?, page 170.

RNA in their virion, but pass through a DNA stage (see below). Other viruses – notably hepatitis B virus of humans – that have DNA in their virion have since been found to resemble retroviruses in passing through an RNA phase during their replicative cycle. Hepatitis B virus also resembles the retroviruses in its genomic organization (see Table 18.7 and Box 18.5). These relatives of retroviruses are often termed 'retroid viruses' or 'pararetroviruses'.

Viruses are classified into a number of families

On the basis of properties such as type of nucleic acid, virion morphology and biological and immunological behaviour, animal viruses have been classified into a number of families. Members of each family (with a name characterized by the suffix -**viridae**) share distinctive genomic organizations, replicative mechanisms and virion structures and are subdivided into genera (suffix -**virus**). Some families, such as the Herpesviridae and Retroviridae, are subdivided into subfamilies (suffix -**virinae**). Box 18.9 discusses some important groups of viruses with members that infect humans.

Viruses need a number of host cell functions

A virus needs its host cell to provide:

- energy, in the form of ATP and other nucleoside triphosphates;
- small molecule precursors, such as amino acids and nucleotides;
- the machinery of protein synthesis;
- membrane sites for processes such as the assembly of virions.

We may think of the host cell as providing a range of subcellular and molecular 'niches', which different viruses exploit in different ways to replicate.

Virus replication consists of a programmed sequence of events

The details of replication vary greatly between individual viruses but all follow a broadly similar sequence of events within the host cell during the replicative cycle (Table 18.8).

Attachment and uptake of virus need cellular processes

Many viruses attach to their host cell by a virion protein binding to a specific cell-surface receptor. For example the cell, for some purpose of its own, carries a receptor which binds human rhinovirus (see Figures 18.4 and 18.5a), but the virus has 'learned' to parasitize this function in order to gain entry to the cell. Once the virus particle has attached it becomes a **passive** participant during uptake by endocytosis or other processes (see Chapter 3). Viral replication is then initiated by 'uncoating' – partial or complete disassembly of the virus particle (Figure 18.5b).

Viruses replicate in different cellular compartments

A major distinction between the DNA viruses is the site of replication – nuclear or cytoplasmic (Box 18.5). Most RNA viruses replicate in the cytoplasm, an exception being influenza, which replicates in the nucleus. Different enveloped viruses assemble and mature at various intracellular sites, including the plasma membrane, the endoplasmic reticulum and the Golgi complex.

Table 18.8 Steps in a typical virus replicative cycle

Attachment to the surface of the host cell
Uptake into the host cell
Uncoating of the virion
Gene expression to make viral proteins
Genomic replication to make progeny viral nucleic acids
Assembly of virions
Maturation and release of virions

 Box 3.1. Endocytosis and exocytosis, page 49.

(a) Attachment of virions to surface of host cell

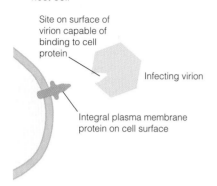

Site on surface of virion capable of binding to cell protein

Infecting virion

Integral plasma membrane protein on cell surface

(b) Uptake of virion by host cell

There are different mechanisms of uptake of virions but several, including Influenza virus, are taken up by **receptor-mediated endocytosis**.

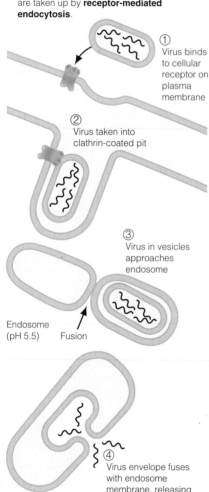

① Virus binds to cellular receptor on plasma membrane

② Virus taken into clathrin-coated pit

③ Virus in vesicles approaches endosome

Endosome (pH 5.5)   Fusion

④ Virus envelope fuses with endosome membrane, releasing viral nucleic acid into cell

Figure 18.5 Attachment and uptake of virions at the cell surface.

DNA Viruses

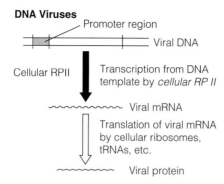

Promoter region

Viral DNA

Cellular RPII

Transcription from DNA template by *cellular RP II*

Viral mRNA

Translation of viral mRNA by cellular ribosomes, tRNAs, etc.

Viral protein

RNA Viruses: negative strand
(viral RNA is complement of mRNA)

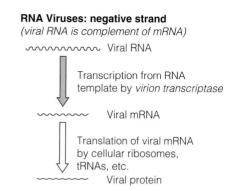

Viral RNA

Transcription from RNA template by *virion transcriptase*

Viral mRNA

Translation of viral mRNA by cellular ribosomes, tRNAs, etc.

Viral protein

RNA Viruses: positive strand
(viral RNA is mRNA)

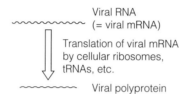

Viral RNA (= viral mRNA)

Translation of viral mRNA by cellular ribosomes, tRNAs, etc.

Viral polyprotein

Retroviruses

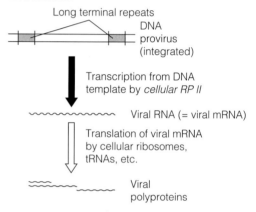

Long terminal repeats

DNA provirus (integrated)

Transcription from DNA template by *cellular RP II*

Viral RNA (= viral mRNA)

Translation of viral mRNA by cellular ribosomes, tRNAs, etc.

Viral polyproteins

Figure 18.6 Viral mRNAs in viral replication.

Viruses do not always kill their host cell

The process of virus replication usually kills the host cell (cytolytic growth), resulting in many of the pathogenic effects of the diseases caused by virus infection. Some viruses are also capable of **noncytolytic interactions** with their host.

The molecular mechanisms of these noncytolytic interactions are not yet known in such great detail as those of cytolytic growth but are being investigated because they are vital for a proper understanding of viral disease.

Latency

Some viruses enter a state in which infectious virus can no longer be detected, but resides in latent form within the host cells. At a later stage some stimulus reactivates the virus and infectious particles are once more produced. Most herpesviruses show this phenomenon – as evidenced in the recurring 'cold sores' of herpes simplex virus and the reappearance of childhood chickenpox in later life as shingles (varicella-zoster virus).

Persistence

Some viruses do not kill their host cell, but enter a persistent state in which low levels of virus are produced. This can occur with paramyxoviruses, such as measles virus.

Cell transformation

A number of DNA viruses, and most retroviruses, are able to **transform** host cells; that is, alter their social behaviour so that they behave more like cancer cells.

Replication of different viruses show many common features

Viral mRNAs play a central role in viral replication

Viral mRNAs play a central role in the replication of any virus and direct the production of viral proteins. Different viruses have different mechanisms for production of their mRNAs that are central to the replicative cycles and are governed by the nature of the viral genome (Figure 18.6). Translation of viral mRNAs requires the host cell's protein-synthetic machinery.

Single-stranded RNA viruses are divided into two distinct groups:

1. Positive strand viruses (+RNA viruses), whose RNA is translated directly into viral proteins after uncoating. Viral RNA acts as mRNA and is typically translated into a single large protein or polyprotein. Examples include rhinoviruses and poliovirus.

2. Negative strand viruses (–RNA viruses), whose RNA must be transcribed into its complementary viral mRNA before translation can occur. Transcription is achieved by an RNA transcriptase present in the virion. Examples include influenza and rabies viruses.

Viruses encode both structural and nonstructural viral proteins

Viral proteins found within mature virions are termed **structural** proteins; other, **nonstructural** proteins are found only in the infected cell and function in various aspects of the replicative process. Structural proteins are often made late in the replicative cycle and usually in much larger amounts than the nonstructural proteins, which often function in a catalytic manner, for example the action of RNA replicases of picornaviruses in replication of the RNA viral genome.

Replication of a virus genome is essential for virus replication

If infection by a single virion (with its genome) is to lead to the production of many thousands of virions, then virus replication must involve the synthesis of many thousands of copies of the viral genome. These associate with a corresponding number of viral structural proteins and progeny virions are assembled.

DNA Viruses

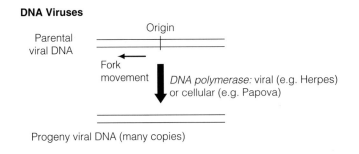

RNA Viruses: positive strand

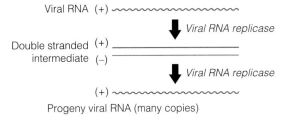

RNA Viruses: Negative strand

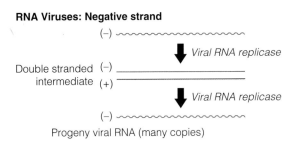

Retroviruses

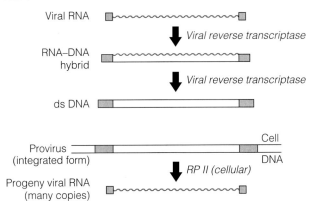

Figure 18.7 Genomic replication in different virus types.

Box 18.5 Classification and general features of human viruses

Please note that this is not an exhaustive or rigorous list, but simply presents some of the most important families of viruses with members causing human disease. See also Figures 18.2, 18.4 and 18.5 and Table 18.8.

DNA viruses

Papovaviridae
Naked icosahedral virions enclose circular dsDNA of 5–7 kbp. Some viruses (for example, SV40) are widely studied as an experimental system, especially for mammalian DNA replication; they encode a multifunctional **large T** (tumour) **antigen** that, along with cellular proteins, directs viral DNA replication. Human papillomaviruses cause warts and are implicated in cervical cancer. Papovaviruses possess oncogenes and are able to transform cells in culture.

Adenoviridae
Naked, icosahedral virions with distinctive spikes enclose a linear dsDNA of about 40 kbp. Viral transcripts are very extensively spliced to produce roughly 40 viral proteins, including a DNA polymerase. Over 40 human adenoviruses are known to cause a range of diseases, including upper respiratory infections and diarrhoea; a few are capable of transforming cells in culture because they possess oncogenes.

Herpesviridae
Enveloped virions contain an icosahedral capsid enclosing linear dsDNA of 120–240 kbp that contains a core of about 40 genes common to all herpesviruses and 30 or more other genes specific to the individual virus. Several genes encode viral enzymes, including DNA polymerase and thymidine kinase (see Box 18.8). The seven known human herpesviruses include herpes simplex virus (Table 18.10), varicella-zoster virus and Epstein–Barr virus (Table 18.11).

Other families of DNA viruses
Parvoviridae: small viruses whose DNA is single-stranded. Human member: B19.
Poxviridae: large, complex virions enclose dsDNA of about 200 kbp. In contrast to the nuclear location of the replication of other DNA viruses, poxviruses replicate in the cytoplasm and encode an RNA polymerase resembling cellular RNA pol II (see Chapter 5). Smallpox, the most significant human member, was officially eradicated from the world in 1977.

Positive strand RNA viruses

Picornaviridae
Naked, icosahedral virions enclose a ssRNA of approximately 8 kb that is translated into a polyprotein of about 3000 amino acids. This is cleaved into ten viral proteins, including three virion proteins (see Figure 18.3) and an RNA replicase. Members infecting humans include the human rhinoviruses (see Table 18.9), poliovirus and hepatitis A virus (see Box 18.6)

Flaviviridae
Enveloped icosahedral virions enclose a ssRNA of roughly 10 kb. Members affecting humans include agents of mainly tropical diseases, including yellow fever. Hepatitis C virus (see Box 18.6) is distantly related to these viruses.

Togaviridae
Enveloped icosahedral virions enclose a ssRNA of approximately 11 kb. Gene expression involves production of a subgenomic RNA that is translated into the virion proteins. Human-affecting members are mainly agents of tropical diseases; rubella (which causes 'German measles') is distantly related to these viruses.

Coronaviridae
Large enveloped virions with a distinctive 'crown' shape and enclosing an unusually large ssRNA of 25 kb. Gene expression involves an unusual mechanism of viral mRNA production that produces a 'nested set' of viral RNAs of various lengths, but with a common 3'-end. Members are responsible for about 30% of all 'colds' in humans.

Negative strand RNA viruses

Orthomyxoviridae
These are the influenza viruses (see Box 18.6) with their eight-segment genome and nuclear replication (almost unique in RNA viruses). Influenza B generally causes milder infections than influenza A; influenza C can be almost asymptomatic and contains only seven RNA segments.

Paramyxoviridae

Enveloped virions with unsegmented helical nucleocapsid (RNA about 14 kb) and virion RNA transcriptase. Members cause many important human diseases, including mumps, measles and viral pneumonia of infants (respiratory syncytial virus).

Double-stranded RNA viruses

Reoviridae

Distinctive double-shelled icosahedral virions enclose (typically) ten segments of dsRNA (totalling about 25 kbp) and contain enzymes for the production of mature viral mRNAs. Members of the genus rotavirus (11 segments totalling 81 kbp) cause diarrhoea and are a major human pathogen, killing some three million children every year.

Other RNA viruses

Rhabdoviridae

Members of this group have many similarities to paramyxoviridae. The most significant member is rabies virus.

Arenaviridae

These are bisegmented RNA viruses that contain ribosomes in their virions. They cause very severe illness (for example Lassa fever) and are probably not strictly human viruses, but viruses of typically unidentified wild mammals that occasionally stray into the human population.

Filoviridae

Unsegmented RNA viruses with a distinctive virion shape; they resemble the paramyxoviruses and rhabdoviruses in genome organization. Like the arenaviruses, they are probably also animal viruses that only occasionally infect humans, but cause extremely virulent haemorrhagic disease (Marburg and Ebola viruses).

Retroviruses and pararetroviruses

These differ from the 'true' DNA and RNA viruses listed above, but both alternate their genome between RNA and DNA forms. They also share a general similarity in genome organization. The only pararetrovirus of significance in human disease is hepatitis B virus.

Retroviridae

Enveloped virions enclose a nucleocapsid containing two identical molecules of ssRNA and reverse transcriptase. Reverse transcriptase copies the RNA into dsDNA and the viral DNA 'provirus' is incorporated into cellular DNA. Transcription of the provirus yields viral RNA that goes to form new virions and is also translated to form viral proteins. The viral genome (8–11 kb) has a characteristic *gag-pol-env* organization.

Oncovirinae are the 'classical' retroviruses that generally transform rather than kill their host cell. They vary in their ability to cause tumours and many are defective; that is, they are able to replicate only in the presence of another, competent retrovirus. Many contain a fourth gene (an oncogene) that replaces part of the *gag-pol-env* structure, rendering the virus defective. The only known human members are the human T-cell lymphotropic viruses I and II, which cause a type of adult leukaemia.

Lentivirinae do not cause tumours, but kill their host cell. In addition to the *gag-pol-env* set, they carry up to six 'extra' genes that are probably responsible for the complex interactions these viruses make with their host. The only known human members are the human immunodeficiency viruses I and II, associated with AIDS.

Spumavirinae show similarities to the other two subfamilies, but are unusual because they appear to cause no disease in their hosts (which include humans).

Hepadnaviridae

Small virions enclose a small, circular dsDNA containing a 'gap' in one strand that is filled and joined early in infection to give closed circular dsDNA that is transcribed to form viral RNAs. Full-length viral RNA acts as a 'pregenome' that associates with viral proteins and is copied into viral DNA by the viral polymerase. The only member affecting humans is hepatitis B virus (see Box 18.6 and Table 18.11).

Single-stranded viral nucleic acids replicate via double-stranded forms

Genomic replication of those viruses with single-stranded genomes, either RNA or DNA, must pass through a double-stranded intermediate because nucleic acid molecules can be copied only into complementary strands (see Introduction to Part B). All RNA viruses copy RNA to RNA and pass through double-stranded RNA intermediates. The various types of viral genomic replication are shown in Figure 18.7.

Virus replication requires both viral and cellular functions

Most viruses employ a combination of viral and cellular mechanisms to achieve their replication: the larger the viral genome, the more viral functions the virus is able to specify. For example, small DNA viruses (for example, the large T antigen of polyomaviruses) modify cellular DNA replication to their own ends by involving a single viral protein. In contrast, large DNA viruses such as the herpesviruses probably encode a complete set of DNA replication proteins.

Assembly and maturation of virions is poorly understood

Once the viral genome has been replicated and the viral structural proteins synthesized, virus particles may be assembled. The details of this assembly are not yet fully understood and differ substantially between viruses. Virion assembly generally involves

- protein–protein interactions, between viral structural proteins; and
- protein–nucleic acid interactions, between structural proteins and viral nucleic acid.

Viral nucleic acids contain **encapsidation signals** that direct the packaging of viral nucleic acid. The initial product of viral assembly typically undergoes maturation events that yield the mature virion. These events involve specific proteolytic cleavages of viral proteins, as occurs in the rhinoviruses (Figure 18.3c); this cleavage probably allows conformational changes in the virion proteins, causing them to adopt a more stable state in the mature virion.

Viruses and disease

The relationship between viruses and the diseases they cause is complex and still only poorly understood – elucidation of the details remains an outstanding challenge for molecular virologists. The following sections illustrate some of the complexities.

Different viruses may cause similar diseases in the same host

Symptoms described as 'upper respiratory infections' or 'common colds' may be caused by members of five different families of virus (Table 18.9). The symptoms caused by these viruses are so similar that accurate diagnosis may be made only by detailed clinical virological analysis, especially using immunological methods. The number of immunologically distinct strains (serotypes) of a particular virus in circulation within the human population varies. The frequent recurrence of 'colds' is partly due to the large number (more than 100) of serotypes of rhinovirus – individuals continue to encounter previously unseen strains against which they have little immunity.

Table 18.9 Viruses that cause 'upper respiratory infections' or 'common colds'

(+)RNA viruses
Picornaviridae (genus Rhinovirus): The 'common cold virus' causing about 30% of 'colds' contains over 100 human serotypes
Picornaviridae (genus Enterovirus): Some human types, for example some Coxsackie viruses, cause 'colds'
Coronaviridae: Human coronaviruses (two serotypes) cause up to 30% of 'colds'

(–)RNA viruses
Orthomyxoviridae: Influenza viruses A, B or C may all cause 'cold' symptoms
Paramyxoviridae (genus Paramyxovirus): Human parainfluenza viruses (four serotypes) also cause 'croup' in infants
Paramyxoviridae (genus Pneumovirus): Human respiratory syncytial virus causes 'cold' symptoms in adults, but is a major cause of pneumonia in children under 12 months of age

DNA viruses
Adenoviridae: Human adenoviruses of subgroups B, C and E can give 'cold' symptoms

A single type of virus may cause several different diseases in the same host

A virus may be able to cause more than one disease. An example is the human DNA virus, herpes simplex virus (HSV), which is capable of causing several distinct disease states (Table 18.10). Several of these involve latency – the oral lesions enter a latent state in the trigeminal nerve ganglia, genital lesions lie latent in the sacral ganglia. Another example of variable pathogenicity occurs with poliovirus, which generally causes a rather mild febrile condition but occasionally gives rise to the much-feared paralysing poliomyelitis.

Many viruses show tropisms towards specific body tissues

Many virus infections do not act in a generalized fashion, but affect specific parts of the body – for example, varicella-zoster virus (neurotropic) affects nervous tissue, the various hepatitis viruses (hepatotropic) affect the liver (see Box 18.6) and Epstein–Barr virus (lymphotropic) affects lymphoid tissue in glandular fever. A virus may pass through different tissues of the body, from a primary site, often the

Table 18.10 Diseases caused by herpes simplex viruses (HSV)

Very common	Oral lesions (HSV type 1 predominates)
	Genital lesions (HSV type 2 predominates)
Less common	Ocular conjunctivitis and keratitis
Rare	Herpetic encephalitis

Box 18.6 Viral hepatitis

Viruses are probably the major cause of hepatitis, a serious and often life-threatening disease. A number of new hepatitis viruses have been discovered, the principal features of which are described below.

Hepatitis A virus (HAV)
This is an RNA virus distantly related to other members of the family Picornaviridae. It has an unusually stable virion that contains ss(+)RNA of 8 kb and generally causes a mild or inapparent infection. Virulent strains of this virus do occur.

Hepatitis B virus (HBV)
This is the member of the family Hepadnaviridae (members of the 'retroid' or pararetroviruses) that affects humans. Virions of HBV are very stable, contain circular, partially single-stranded dsDNA of 3.3 kbp and usually cause very serious disease, in both acute and chronic forms. Extremely high levels of virus can be found in the blood of affected individuals, including asymptomatic 'carriers'. This explains the danger of infected blood to healthcare workers and transfusion recipients and the need to check that blood products are HBV-negative. Individuals who have been infected with HBV have an increased risk of liver cancer.

Hepatitis D virus (δ agent)
This is a defective RNA virus that can multiply only in the presence of HBV as a 'helper' virus. It causes both acute and chronic hepatitis but does not resemble any other known type of virus. Part of its RNA sequence is similar to the viroids of plants (Box 18.4).

Non-A, non-B hepatitis
The above viruses have been intensively studied for some time and, although other types of viral hepatitis were known, little progress was made by conventional virological procedures. Genetic manipulation technology, including cloning of the viral RNAs, has led to the rapid discovery and detailed characterization of hepatitis C and E viruses, neither of which is closely related to any known virus. Other hepatitis viruses probably remain to be discovered.

Hepatitis C virus (HCV)
This RNA virus is a major cause of parenteral hepatitis; that is, caused by injection. Its study in the laboratory is difficult , but has rapidly become a major field of study following the sequencing of its viral RNA, a ss(+)RNA of 9.4 kb. There appears to be a number of related viruses of this type worldwide. They form a distinct group of viruses, but are probably distantly related to the Flaviviridae.

Hepatitis E virus (HEV)
This RNA virus has a ss(+)RNA of 7.5 kb and is an important cause of enteric hepatitis. It is not closely related to any other virus, although its sequence distantly resembles that of rubella, the agent of 'German measles'.

Box 18.7   Influenza A virus: humans, ducks and pigs

Many human viruses pose a 'once in a lifetime' threat to an individual. Typical of these are the childhood diseases, in which a single exposure generally confers lifelong immunity. Other viruses recur – for example, the rhinoviruses occur in over 100 serotypes so we regularly encounter strains against which we have little or no immunity.

In the case of influenza viruses, and particularly influenza A, which generally inflicts the most severe symptoms, repeated reinfection is due to continual change in the antigenic nature of the surface of influenza virions circulating in the human population. The change takes two distinct forms – **antigenic drift** and **antigenic shift**. In antigenic drift the genome of the virus undergoes continual, but limited mutation that keeps the virus 'one step ahead' of human immune responses. Antigenic shift involves a much more radical change in the viral genome. It is much rarer than antigenic drift and gives rise to the sudden appearance of a substantially different virus. Shift is the cause of worldwide epidemics (pandemics) of infection that are exacerbated by the rapid spread of virus by modern human travel patterns.

The antigenicity of influenza virus depends principally on the two major glycoproteins of the virion envelope (Box Figure 1): haemagglutinin (HA) and neuraminidase (NA). The three-dimensional structures of the external domains of these proteins are known in detail. **Haemagglutinin** is the major envelope protein, a trimer involved in fusion of the viral envelope with the cell membrane and the major target for neutralizing antibodies. It is a polypeptide of 63 kDa encoded by RNA segment 4. **Neuraminidase** is a tetrameric 'spike' projecting from the surface of the virion and removes sialic acid residues from glycoproteins. It is probably involved both in uptake of virus into the cell and release of progeny virions after virus replication. It is a polypeptide of 50 kDa encoded by RNA segment 6.

Several serotypes are known for both HA and NA, and a strain of influenza A virus can be classified immunologically according to the types present on the surface of the virion – for example, H1N1 contains haemagglutinin type 1 and neuraminidase type 1. Influenza A/H1N1 was in circulation before 1957 and reappeared after 1976. Influenza A/H3N2 appeared after 1968; it is known as 'Hong Kong.flu'.

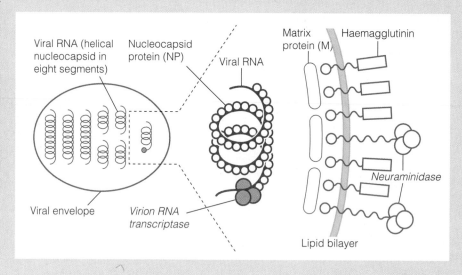

Box Figure 18.7.1 (a) Structure of the influenza virion.

mucosa of the upper respiratory tract, into the blood, giving a viraemia, and then on to the major site of pathogenesis. A major, and largely unsolved, problem is to understand the molecular basis of these tissue tropisms. This presumably depends on the function of specific viral gene products and their interactions with different cell types within the body.

Some viruses infect other animals in addition to humans

The major driving force in the study of viruses has always been their role as human pathogens but there is also keen interest in the viruses of domestic animals. There are important connections between human and veterinary virology, as some viruses are able to cross species boundaries. This interest has been heightened by the great volume of virus genomic sequence data now available, which has emphasized the similarities between viruses. These connections have been further enhanced by the concern over HIV (see below) and its possible origin from a monkey or primate virus. Recent discoveries with influenza A virus have highlighted links between human and animal viruses (Box 18.7).

A possible explanation for the origin of antigenic shift and influenza pandemics follows (see Box Figure 2).

- The influenza genome is composed of eight segments of RNA.
- Mixed infection of a cell by two different influenza viruses gives the possibility of reassortment of segments, producing a new type of influenza A virus by the acquisition of new genome segments encoding HA and/or NA.
- Influenza A is found in three species of domestic animals – horses, pigs and ducks – and in wild ducks and seagulls.
- Influenza A virus is probably an avian virus that has spread into the other species.
- New strains of human influenza A that cause pandemics have probably arisen by rare segment reassortment between a human and an avian influenza A virus.
- The mechanism of the reassortment is unclear because avian viruses do not infect humans and vice versa. Pigs have been implicated as 'mixing vessels' for segment reassortment following infection by an avian virus. The 'new' pig virus containing avian viral RNA segments the spreads to humans.
- Most pandemic strains have arisen in China where large numbers of pigs and ducks are bred in close proximity to each other and to humans.
- Sequence comparison suggests that influenza A may be a relatively new virus of humans and is not yet completely adapted to its new host. Antigenic drift may reflect the process of adaptation, antigenic shift being an undesirable complication that may have been brought about by human activities.

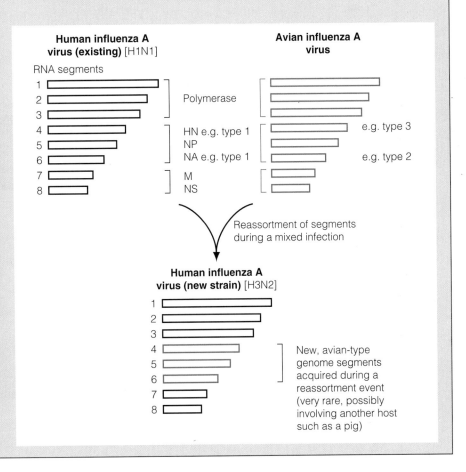

Box Figure 18.7.2 Reassortment of genome segments and antigenic shift.

Some viruses are associated with the induction of certain cancers

There has been a long and continuing interest in the connection between viruses and human cancer. Many associations between certain viruses and certain cancers have been demonstrated, the strongest of which are shown in Table 18.11 but for ethical reasons it is extremely difficult to establish a direct causal connection. As the multistep nature of the causation of cancer is becoming more clear (see Chapter 17), it is apparent that there are points in the process at which a virus could participate without necessarily being sufficient for the induction of cancer: Epstein–Barr virus, for example, causes specific chromosomal translocations (see Chapter 17) linked to the induction of tumours.

Chapter 17. Several steps lead to the development of a malignant tumour, page 535.
Chapter 17. Human DNA viruses and cancer, page 534.

The molecular mechanisms of viral disease are still not well understood

Although great strides have been made in dissecting the nature of viral genomes and the details of viral replication, the progress in understanding the mechanisms

Table 18.11   Connections between viruses and human cancers

Virus	Type of cancer
Human papillomavirus (especially type 16 and other related viruses)	Cervical carcinoma
Human T-cell lymphotropic virus (HTLV-1, HTLV-2)	Adult leukaemias
Hepatitis B virus	Primary liver carcinoma
Epstein–Barr virus	Burkitt's lymphoma, nasopharyngeal carcinoma

of the causation of disease by viruses has been much slower. This is because the advances made so far have depended mainly on growth of viruses in cell culture – systems that can not be used to study the behaviour of viruses in whole animals. The modern techniques of genetic manipulation may be able to solve these problems. The solutions may lie in studying the many viral genes for which functions have not yet been assigned, and many of which are not required for growth in cell culture.

Retroviruses, HIV and AIDS

The distinctive nature of retroviruses was first recognized in 1970. For most of the 1970s retroviruses were of great interest because of their ability to cause cancers in birds and mice. Interest waned somewhat as it was realized that they were not a significant cause of cancer in humans but then increased dramatically with the appearance of the **acquired immunodeficiency syndrome** (AIDS) and the recognition of the associated **human immunodeficiency virus** (HIV) as a retrovirus.

Retroviruses have a life cycle in which the viral genome alternates between RNA and DNA forms

Retroviruses contain two identical molecules of ssRNA in their distinctive enveloped virions (Figure 18.8) and they were originally referred to as **RNA**

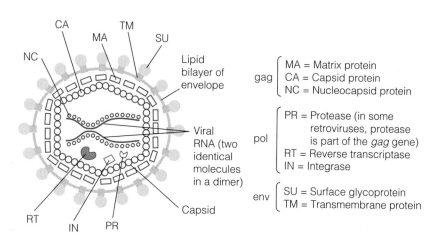

Figure 18.8 The virion of a typical retrovirus showing its constituent proteins.

tumour viruses. Although some DNA viruses are able to transform cells and are associated with cancers, this is not so for the 'true' RNA viruses. RNA tumour viruses were recognized as unusual among RNA-containing viruses because they require DNA synthesis during their replication. The significance of these features was eventually recognized when it was shown that retrovirus virions contain an unusual multifunctional enzyme, reverse transcriptase (RT), capable of copying RNA into DNA.

The role of RT in the retrovirus life cycle is to convert the RNA form of the genome in virions through an RNA–DNA hybrid into a dsDNA that can be integrated into cellular DNA through its long terminal repeats (LTRs) (Figure 18.9 and 18.10). This integrated form is often called the **provirus** and its presence may transform the infected cell – leading to the production of cancer cells by the oncogenic retroviruses. The action of RT is only one-half of the life cycle, however, because the viral genome must return to its RNA form to allow production of new virions. This is catalysed by the cellular enzyme RNA polymerase II, which recognizes promoter sequences in the LTR region and makes an RNA copy of the proviral DNA. Some of this RNA is used in the production of new virions, the remainder is processed, spliced and translated (like a cellular pre-mRNA) to form viral proteins (see next section).

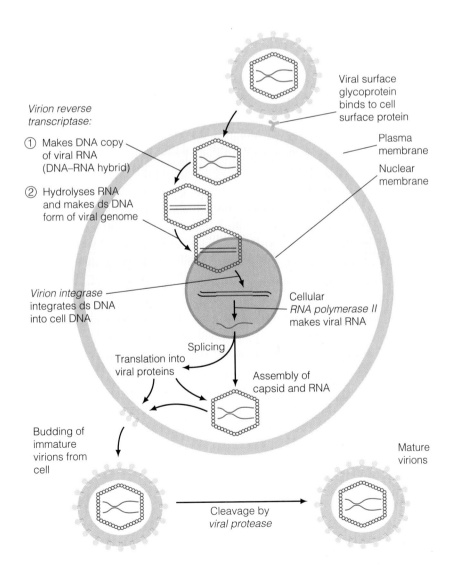

Figure 18.9 Generalized retrovirus replicative cycle.

Retroviral genomes have a distinctive *gag-pol-env* organization that yields the viral proteins

All retroviral genomes contain the same basic organization of three 'gene units' – *gag, pol* and *env* – although many contain variants of this pattern (Figure 18.10 and see below). Each of these units encodes two or more proteins – *gag* encodes the proteins of the capsid, *pol* the various enzyme activities of the virus (DNA polymerase, RNase H, integrase and proteinase) and *env* the two proteins of the envelope (Figure 18.8). Translation of full-length viral transcripts gives the gag polyprotein about 1/20 chance of 'reading through' the stop codon to produce the gag–pol polyprotein. This ensures that the viral enzymes are produced in appropriately smaller amounts than the capsid proteins. The *env* mRNAs are produced from full-length viral RNA by splicing and translated to give the env polyprotein. All three polyproteins are subjected to specific proteolysis to generate the final viral polypeptides (see Figure 18.8). The gag and gag–pol polyproteins (in a ratio of 20:1) are assembled into **pre-virions**, where cleavage occurs during maturation of virions.

Retroviruses mutate rapidly

When retroviruses replicate they show a rate of mutation higher than that of most other viruses. In contrast, while they are in their proviral state their rate of mutation is broadly similar to that of a typical cellular gene, implying that the high mutation rate is associated with the replicative cycle of retroviruses. The high mutation rate occurs because retroviral genomic replication involves two enzymes – viral reverse transcriptase and cellular RNA polymerase II (Figure 18.9) – that lack proofreading functions and have low inherent fidelities of copying.

Cancer-causing retroviruses carry an extra gene that resembles certain cellular genes: the oncogenes

Some retroviruses contain a fourth gene in addition to the *gag-pol-env* organization. It often replaces part of one of these genes, rendering the virus defective or incapable of replicating except in the presence of a replication-competent retrovirus (Figure 17.9). These 'extra' genes are homologues of cellular genes, typically involved in the control of cell division, and are called **oncogenes** or cancer-causing genes (see Chapter 17).

Surprise and excitement was caused in the scientific world when the genomes of humans and animals were found to contain sequences resembling those of retroviruses. These are thought to represent defective retroviral proviruses that are unable to express themselves. They are probably 'fossils' of retroviruses that once flourished in the ancestral human population.

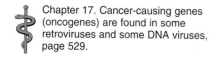

Chapter 17. Cancer-causing genes (oncogenes) are found in some retroviruses and some DNA viruses, page 529.

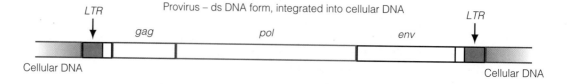

Provirus – ds DNA form, integrated into cellular DNA

Viral RNA – parts of LTR at 5' and 3' ends, polyadenylated

Figure 18.10 The genome structure of a typical retrovirus, shown in both provirus and viral RNA forms.

Lentiviruses are retroviruses that kill their host cell and can contain 'extra' genes

Whereas most retroviruses (subfamily Oncovirinae) transform their host cell but do not kill it, members of one subfamily – the Lentivirinae, mainly viruses that infect domestic animals – are cytolytic (*lenti* = 'slow' – of growth). The sequence of HIV was seen to be closely related to known lentiviruses. The connection between HIV and AIDS has prompted a great deal of research on HIV and the lentiviruses of other animals.

Lentivirus genomes contain 'extra' genes not found in other retroviruses. Of these, *tat* activates viral transcription, *rev* controls viral RNA processing and transport. The functions of *vif, nef, vpu* and *vpr* are not fully understood; they do not appear to be necessary for growth of HIV in cell culture and probably function in as yet unknown aspects of the biology of the virus in the whole body. These 'extra' genes are thought to be crucial in the complex interactions between HIV and its human host that have made the biology of AIDS so difficult to unravel.

HIV infects T lymphocytes by binding to the CD4 receptor

The attachment of viruses to their host cell is thought to occur via specific receptors (see above), and one of the first breakthroughs in this area was when the specificity of HIV for T lymphocytes was shown to be due to the virion surface glycoprotein gp120 binding to the CD4 receptor. This is not, however, the only way that HIV interacts with human cells because it also infects cells in the brain that do not express the CD4 receptor.

HIV mutates in infected individuals and becomes drug resistant

In common with all retroviruses HIV is highly mutable, especially for the gp120 surface glycoprotein that binds to the CD4 receptor. This rapid change undoubtedly makes it more difficult for the immune response to clear the virus from the body. The drug azidothymidine (AZT) (see Box 18.8) is given to AIDS sufferers over many months; HIV isolated at various times shows some mutants with an altered reverse transcriptase that has become resistant to the drug. The mutations conferring resistance always occur at the same few sites on the protein.

HIV and AIDS

The connection between HIV and AIDS is complex and controversial and it is not appropriate to discuss this in detail in a book of this type. Suffice it to say that

- primary infection of an individual by HIV ('becoming HIV positive') is followed by a long and variable period before the onset of full-blown AIDS;
- development of AIDS in HIV-seropositive individuals probably requires further trigger events. The nature of these is not yet clear, but it has been shown that HIV replicates and kills CD4+ T cells in the period before AIDS develops; loss of T cells is balanced by new production;
- the onset of AIDS is preceded by a dramatic fall in the level of circulating T lymphocytes, presumably due to a failure to replace losses; this fall continues to the point at which chance infections by normally nonfatal agents such as the parasite *Pneumocystis carinii* or viruses such as cytomegalovirus (a herpesvirus present in most of the adult human population) lead to death of the afflicted individual.

The control of viral diseases

Most virus infections are limited by a healthy body

Most viruses that are endemic in a population present low health risk to most of the healthy individuals and generally cause relatively mild and short-lived ailments. Within any population, however, there are 'high-risk' groups including infants (especially neonates), the elderly (especially when frail and/or ill) and immuno-compromised individuals. Some viruses are significantly pathogenic only towards such groups and most viruses afflict these groups far more severely than the general population. It is towards these high-risk groups that most medical intervention is addressed.

Perhaps the main reason for this situation is the role of the immune system in determining the outcome of virus–host interactions (see Box 16.9). It is rare for any virus to establish an infection in a well immunized individual and most virus infections are limited by the immune response in an otherwise healthy individual who has not previously been exposed to the virus.

Box 16.9. MHC antigens in infection, page 505.

Some viruses strike only once, other virus infections recur

Some viruses, especially those causing the so-called 'childhood diseases' (such as measles, mumps and chickenpox), run a typically relatively innocuous course. In so doing they stimulate the production of lifelong immunity.

Other viruses are capable of infecting an individual repeatedly, either because they occur in a large number of serotypes (for example, the rhinoviruses) or because the virus population changes significantly with time (for example, influenza A) (see Boxes 18.6 and 18.7).

A third group of viruses establishes a primary infection with resulting production of immunity, but then progress to a lifestyle that evades the attentions of the immune system by mechanisms such as latency (for example, the herpesviruses) or persistence (the paramyxoviruses).

A further group, fortunately rare, is either highly pathogenic in a time too short to allow mobilization of the immune response or are weakly immunogenic but do not provoke an adequate response (examples include rabies, arenaviruses and filoviruses). It seems likely that these viruses usually infect wild mammals but have been accidentally introduced into humans, for example, by animal bite or insect vector.

Antiviral vaccines can be made using killed or attenuated viruses

Box 16.3. Polio virus vaccines, page 488.

The best defence against viral infection is immunization with a vaccine that provokes an effective immune response, causes minimal pathogenicity and provides lasting immunity (Box 16.3). Active immunization after infection with the virus is generally less useful but may still be beneficial. The administration of γ-globulin against the virus or passive immunization may be the only suitable therapy for some highly pathogenic viruses.

Preparation of safe and effective vaccines can be extremely difficult and expensive, but the reward is usually worth the effort. Two strategies – the use of killed and attenuated viruses as vaccines – are classically used; these have been supplemented by a range of newer methods.

Killed vaccines

A preparation of virus is inactivated by a suitable method (heat, formaldehyde) to a level which is safe, but which does not seriously impair the antigenicity of the virus. An example of this is the Salk poliovirus vaccine, which was formaldehyde-inactivated. However, use of this vaccine had a low but definite risk because of the presence of a low level of active virus.

Attenuated vaccines

These are live viruses, which actually cause an infection in the individual and thus stimulate immunity much nearer that of a 'natural' infection than killed vaccines. The virus strain used has been attenuated (made less pathogenic), for example by growth and multiple passage through another species. An example of this is the Sabin vaccine for poliovirus, which is still used. Its use has virtually eliminated poliomyelitis as a major health problem in developed countries.

Modern approaches to vaccine production

Although the traditional approaches have had many successes, it has not proved possible to apply them effectively to many virus pathogens. More recent approaches include the following.

- Subunit vaccines: protein subunits isolated from the virus particle stimulate immunity but are not infectious.
- Genetically engineered vaccines: a viral gene from a pathogenic virus is cloned and expressed *in vitro* (see Box 17.14) and the resultant protein used as a vaccine. Alternatively, the viral gene is introduced into a 'safe' viral vector that can be used to infect an individual harmlessly while provoking an immune response to the protein from the pathogenic virus.

 Box 17.14. *In vitro* expression of human genes, page 544.

Viral replication mechanisms provide potential targets for antiviral drugs

Although vaccination has proved highly effective and will probably always be the method of choice where available, effective antiviral drugs would be advantageous in the treatment of a significant number of viruses. As much of the molecular biology underlying the replication of most types of viruses is now understood, it is possible to identify steps in viral replicative cycles that are potential 'targets' for antiviral drugs (Box 18.8).

Why have antiviral drugs been so difficult to develop?

In comparison with the dramatic successes achieved in the field of antibacterial drugs, developments in the antiviral field have been modest. It could be argued that the 'credit' for the antibacterials really belongs to nature, because most are naturally occurring antibiotics or their derivatives. Unfortunately, nature does not seem to have developed correspondingly effective antiviral substances – or we have not yet managed to discover them.

A number of factors and difficulties may be identified.

1. Most viral diseases are self-limiting and relatively trivial in the bulk of the population (see above).
2. Because most viruses do not cause life-threatening diseases, any drug used to treat them must be of low toxicity and without, for example, any carcinogenic or teratogenic effect.
3. By the time many viral infections have become apparent, it may be too late for even an effective drug to dramatically alter the course of the disease.
4. Because of the varied nature of viruses, the spectrum of action of a successful antiviral agent is likely to be limited.
5. The costs of drug development are high, so the disease must have a sufficiently lucrative potential market for manufacturers to continue with development.

Box 18.8   Potential targets for antiviral drugs

Attachment
Drug binds to virion, blocking attachment to cellular receptor. An example is rhinovirus (Box Figure 1).

Uptake and uncoating
Drug binds to virion protein, blocking the change of pH and protein conformational change needed for uncoating of virion. For example, influenza virus (amantidine).

Inhibition of virion enzymes
Retroviruses and negative-strand RNA viruses contain virion transcriptases different from enzymes of the uninfected cell and required for virus replication.

It should be possible to design specific inhibitors of these viral enzymes, but this has not yet been achieved. The nearest is probably azidothymidine (AZT), used in the treatment of HIV and AIDS.

AZT is phosphorylated by three cellular enzymes, thymidine, NMP and NDP kinases, to AZT triphosphate (Box Figure 2). This is used as a substrate by the virion reverse transcriptase and stops production of the DNA form of the virus.

Inhibition of viral replication proteins
The replication proteins of viruses are prime targets for antiviral drugs.

Assembly and maturation of virions
Replication of retroviruses requires a viral proteinase for cleavage of viral polyproteins in the assembled virion (Box Figure 3).

Inhibitors of this enzyme have been developed and have antiviral action that complements AZT, but resistance is still a problem.

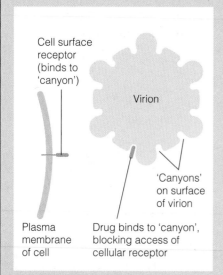

Box Figure 18.8.1 Drug blocking access of receptor to virion.

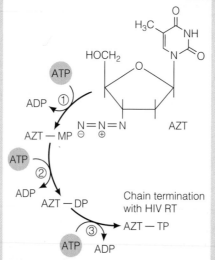

Box Figure 18.8.2 Phosphorylation of AZT

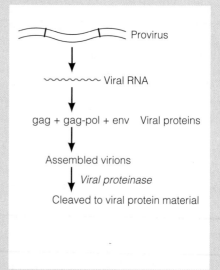

Box Figure 18.8.3 Replication of retroviruses requires a viral proteinase.

In spite of these difficulties, there are cases for which antiviral agents could be useful:

- viruses for which an effective vaccine is not available;
- viruses with alternative, immune-evading lifestyles (latency, persistence);
- applications for 'high risk' groups, especially the immunocompromised.

Some success has been achieved and knowledge-based rational drug design holds promise for the future (Box 18.9).

Box 18.9   Two developments in antiviral chemotherapy

Acyclovir and herpes simplex virus: exploitation of viral enzymes

Acyclovir is a synthetic acyclic analogue of deoxyriboguanosine lacking the 2 and 3 carbon atoms but retaining a hydroxymethyl group corresponding to the 5 carbon (Box Figure 1).

Acyclovir effectively inhibits the replication of herpes simplex virus (HSV) and, to a lesser extent, varicella-zoster virus, but has very low toxicity for humans. This is because, although it is taken up into cells, it is not phosphorylated by any cellular enzymes (like most nucleoside analogues, it must be converted to its triphosphate form to be effective).

In HSV-infected cells, however, two enzymes encoded by viral genes – thymidine kinase and DNA polymerase – behave differently from their cellular counterparts. HSV thymidine kinase is an enzyme with a much lower specificity than the cellular form and phosphorylates a number of nucleosides, including acyclovir. The monophosphate produced is further phosphorylated to a triphosphate by cellular enzymes (see Chapter 10). Acyclovir triphosphate is a substrate for HSV DNA polymerase and is incorporated into viral DNA, where it acts as a chain terminator, thus stopping viral DNA replication. (Box Figure 2). AZT acts in a broadly similar fashion (see Box 18.8).

The application of rational drug design to the control of influenza

Although influenza is not life threatening for most individuals, it kills more people than AIDS and is a major cause of work loss and inconvenience for the general population. A safe and effective anti-influenza drug would find a large market. The one existing drug – rimantidine – is of limited effectiveness.

A recent approach to developing an anti-influenza drug has been 'knowledge-based' – the known three-dimensional structure of influenza proteins has been carefully examined in order to design drug molecules that interfere with their functions in viral replication.

The best target so far has been the viral neuraminidase (NA) that hydrolyses sialic acid residues from the surface of the cell after influenza virus has bound to its receptor. This probably functions both at the beginning and the end of viral replication. Using computer modelling of the active site of NA and of the binding of sialic acid analogues to the site researchers have designed analogues of NA (Box Figure 3) that inhibit the growth of influenza virus in cell culture, are not toxic and have shown promise in early clinical trials.

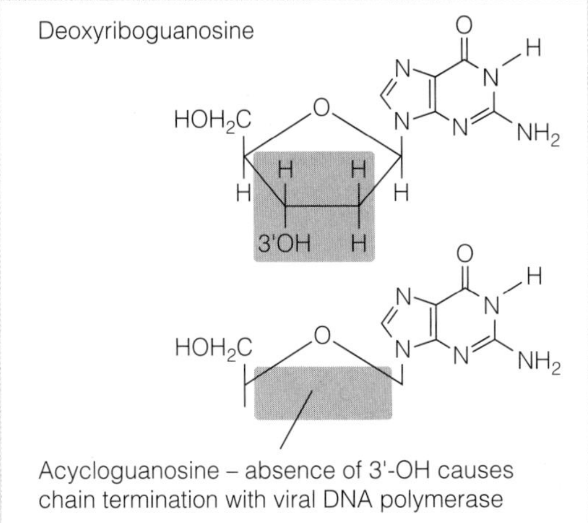

Box Figure 18.9.1 Structure of acyclovir

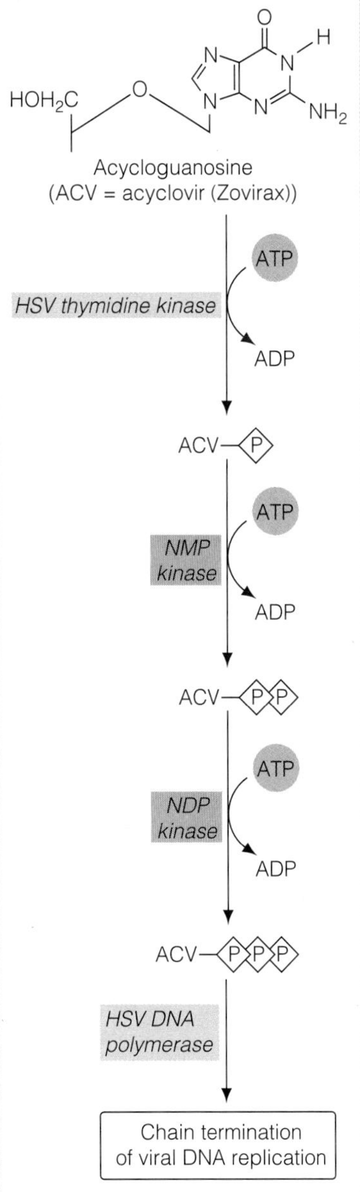

Box Figure 18.9.2 Action of acyclovir

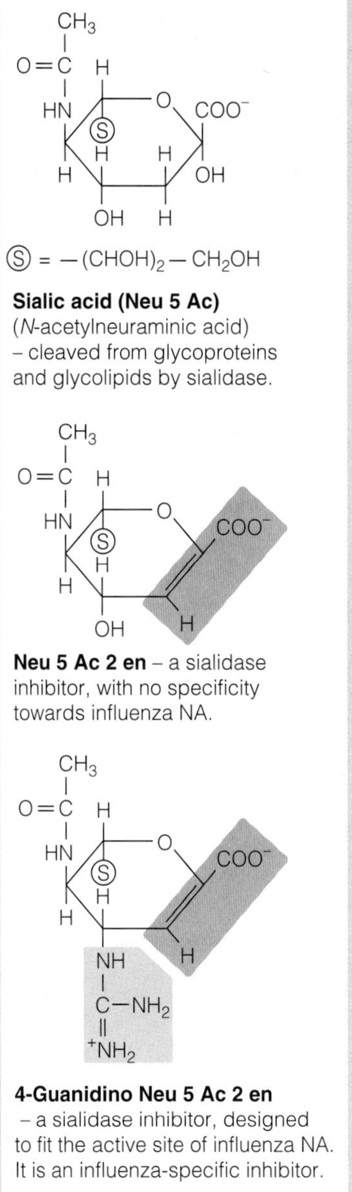

Box Figure 18.9.3 Sialic acid analogues as viral neuraminidase inhibitors

Parasitic diseases

Some simple eukaryotic organisms, both unicellular and multicellular, can cause human disease

So far, we have considered only bacteria and viruses as major human pathogens but a significant number of lower eukaryotes also cause disease in humans. These organisms are collectively described as **parasites** and often have complicated lifestyles involving other hosts and vectors (Table 18.12). Many of these parasites infect large proportions of the world's population, especially in the developing world. Some, such as trypanosomes and malaria parasites, cause many deaths; others, such as schistosomes, are thought to weaken hundreds of millions of people by the burden imposed on the body by the presence of the parasite. The immunological burden in particular probably renders individuals more prone to other infections – this may be part of the explanation of the different response to HIV and AIDS seen in developed and developing countries. Parasites typically persist within the body for years, an ability that depends on their becoming 'invisible' or resistant to the host's immune responses. Their ability to do this makes them particularly difficult to combat.

Some parasites have unusual strategies for gene expression and control

Until recently our knowledge of genome organization and gene expression in simple eukaryotic parasites has been only rudimentary. However, understanding in this area is advancing rapidly and studies, especially of trypanosomes (for example *Trypanosoma brucei*), have revealed processes very different from those in the bacteria or the more complex higher eukaryotes. It is to be hoped that by understanding the basic molecular biology of parasites effective new anti-parasitic therapies will be possible.

The genomes of lower eukaryotes contain multicistronic transcriptional units

It has long been thought that eukaryotic mRNAs are monocistronic, in contrast to bacteria whose mRNAs are often polycistronic (encoding several polypeptides in tandem). This is still thought to be the case for human and other mammalian mRNAs but at least a quarter of all the genes of simple eukaryotes are organized into **clusters**. The tandem arrays of closely spaced genes are expressed as **polycistronic** transcripts. The mature mRNAs, however, are monocistronic because they are all processed by *trans*-splicing.

Table 18.12   Lower eukaryotes causing parasitic diseases of humans

Protozoa	*Giardia lamblia*
	Trichomonas vaginalis
	Trypanosoma species (sleeping sickness)
	Plasmodium species (malaria)
	Pneumocystis carinii (AIDS-related pneumonia)
Nematodes	*Ascaris lumbricoides*
	Trichinella spiralis
Trematodes	*Schistosoma* species

Trans-splicing is common in lower eukaryotes

The splicing that occurs during the processing of pre-mRNA is termed *cis*-splicing because it occurs within a single molecule of RNA. Splicing that joins exons from two *different* pre-mRNA molecules is called *trans*-splicing and was recently discovered in trypanosomes, in which it seems to be an obligatory step in the processing of all pre-mRNA species. It involves the participation of spliced leader (SL) RNA, a 5'-capped molecule that is spliced to all trypanosome mRNAs. Polyadenylation of trypanosome mRNAs occurs via *trans*-splicing, not the AAUAAA signal.

In contrast to trypanosomes, the RNAs of which lack conventional introns and carry out only *trans*-splicing, many other parasites possess both *cis*- and *trans*-splicing mechanisms.

RNA editing in lower eukaryotes produces diversity of gene expression

Another novel mechanism of gene expression has been discovered in lower eukaryotes, including trypanosomes – 'editing' of the RNA of mitochondrial transcripts. This process introduces uracil residues into the 5'-region or throughout the RNA, with the result that open reading frames are created within 'cryptogenes' (genes that lack an open reading frame). An example of a protein expressed in this way is subunit III of cytochrome oxidase.

This process is directed by special RNAs (gRNAs) in ribonucleoprotein complexes containing gRNAs, mRNA and proteins. RNA editing may be a primitive feature, perhaps from the 'RNA world' (see Box 6.5), that has survived only in some simple eukaryotes.

 Box 6.5. Was RNA the first genetic material in the origin of life?, page 170.

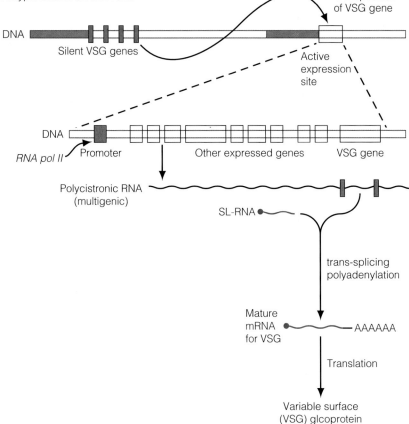

Figure 18.11 VSG 'gene switching' in trypanosomes allows change in the surface of the parasite.

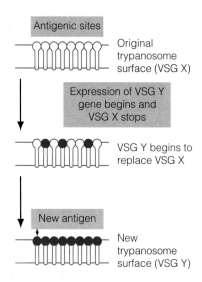

583

Most control of gene expression in lower eukaryotes appears to occur at the post-transcriptional level

Two features distinguish the genomic organization of lower eukaryotes from that of the higher ones: lower eukaryotes possess tandem arrays of closely spaced genes and apparently have no promoter sequences on individual genes (see Figure 18.11). This suggests that, in contrast to the situation in higher eukaryotes, most control of gene expression in lower eukaryotes appears to occur at the post-transcriptional level. Control takes the form of *trans*-splicing and its associated mechanism of polyadenylation.

Trypanosomes are able to rapidly vary their surface antigens

The surface coat of trypanosomes consists of a dense layer of a single protein species, the **variant surface glycoprotein** (VSG). This elongated molecule is very tightly packed on the surface of the parasite, so that only its very tip is exposed at the surface (Figure 18.11). Small changes in the tip region are sufficient to achieve a dramatic change in the antigenicity of the parasite, requiring the host to respond.

About a thousand VSG genes are known in the trypanosome genome, which gives the parasite an enormous potential variability in its surface. The variable expression is achieved by 'switching' from one VSG gene to another, either by replacing the gene at an active expression site (see Figure 18.11) or by switching expression to another site. The active expression sites are located near telomeres and contain a number of associated genes. The VSG gene is replaced by transposing a copy of the 'silent' VSG gene to be expressed into the polycistronic transcription unit (Figure 18.11). The nature of all the VSG proteins is sufficiently similar that the newly expressed VSG fits snugly into the surface coat next to the molecules from the previously expressed gene.

A-bands, muscle, 457
Abetalipoproteinaemia, Box 14.5 447
ABO blood groups, 23, 369
Absorption, 412
Acetaldehyde dehydrogenase, 253
Acetyl CoA, 244, 302, 404, 436
  citric acid cycle, 213
  oxidation, 213
  pyruvate metabolism, 250
Acetyl CoA carboxylase, 275
  glucagon inhibition, 348
Acetylcholine, 333, 335, 403-4
Acetylcholine receptor, 404
Acetylcholinesterase, 404
N-Acetylglutamate, 302
N-Acetylglutamate synthetase, 302
Acetylsalicylic acid, see aspirin
Acid-base balance, 391
Acinar cells
  pancreas, 394
Aconitase, 214-5
Acquired immunodeficiency syndrome (AIDS), 574
ACTH, 339, 351, Box 11.10 353, 383
Actin, 44, 84, 368, 457, 461
Action potential, 402
Activation reaction, 14
Active transport, 63
Acyclovir (Zovirax)
  anti-HSV drug, Box 18.9 580
Acyl carrier protein, 277
Acyl CoA synthase, 265
Adenine nucleotide translocator, 210
Adenosine deaminase, immune responses, Box 10.6 320
Adenosine triphosphate
  energy content, 183
  see also structure, 183
S-Adenosyl methionine, 309, 343
  carnitine synthesis, 314
Adenylate cyclase, 70, 335-9, 346, 348, Box 11.9 350, 351, 383, 396, 403,
  glycogen metabolism, 227-9
Adenylate kinase, 300, 385
  smooth muscle, 464
ADH, see antidiuretic hormone
Adipocytes, see adipose tissue
Adipose tissue, 262, 431-2, 443
  brown, 381
  white, 380
ADP, pyruvate metabolism, 250
Adrenal cortex, 351
Adrenal medulla, 342
Adrenaline, 224, 227-9, 332, 339, 342, Box 11.3 343, 350, 383, 404
  smooth muscle activation, 464
α-Adrenergic receptors, 339
β-Adrenergic receptors, 335, Box 11.3 343
Adrenocorticotrophic hormone, see ACTH

Aggretope sequence consensus, 504
AIDS in haemophiliacs, infected factor VIII, 363
ALA synthetase, Box 5.1 117-119
Alanine, 299
  collagen, 472
Alanine cycle, 251-2, 435
Alanine:glyoxylate AT, Box 5.4 137
Albumin,
  fatty acid transport, 383
  bilirubin binding, 328
Alcohol dehydrogenase, 253
Aldolase, 232-3
Aldose, 16
Aldosterone, 351, 393
Aldosterone synthase, Box 11.10 353
Alkaline phosphatase, 479
Alkaptonuria, Box 10.4 311, 511
All-trans-retinal, 407
Allergic response, Box 16.5 494
Allopurinol, xanthine oxidase inhibition, 323
Allosteric
  behaviour, 87
  effectors, 92
  enzyme, 91, 194
  regulation of glycogen synthase, 226
α-Amanitin, 114
Ames test, 168, Box 6.3 169
Amino acid
  biosynthesis, 306-8
  branched-chain, 442
  conserved in proteins, 97
  essential, 428-9
  in proteins, 94
  nonessential, 428-9
  not found in protein, 31
  structure, 29-30
  zwitterion properties, 29
Amino acid sequence
  biosynthesis, 306-8
  deduced from DNA sequence, 97, Box 4.7 98
  determination, Box 4.6 97
  glucogenic, 310
  ketogenic, 310
  nonessential, 307
Aminoacyl tRNA, 14
Aminoacyl-tRNA synthetase, 129-30
γ-Aminobutyric acid, 299, 335, 344, 404-5, 405
Aminopeptidases
  collagen, 474
  digestion, 416
Aminotransferase, 242, 294-301, 307, 442
Ammonium ions, 298-302
AMP deaminase, 300, 385
Amphibolic pathway, 259
Amylase, 412, 417
  antibacterial, in saliva, 484

Amylopectin, 417, 425
Amylose, 417, 425
Anabolism, 6, 185
Anaemia
  folic acid, 449
  haemolytic, Box 10.7 326
  iron-deficiency, 453
  vitamin B12, 450
Analbuminaemia, Box 9.7 287, 360
Anaphylaxis, Box 16.5 494
Anaplerotic reactions, 248
Androgens, 353
Angiotensin, 339, 393
Angiotensin converting enzyme, 393
Angiotensin II, Box 11.10 353
Angiotensinogen, 393
Anion exchange protein
  bicarbonate/chloride, 368
  CO₂ metabolism, 372
Anisotropic, 457
Ankyrin, 368
Anti-oncogene, 535-37
Antibacterial agent
  chemical, 556
  resistance to, 559
  targets for, 557
Antibacterial target
  cell membrane, 558
  cell wall, 558
  nucleic acid synthesis, 559
  protein synthesis, 559
Antibiotic
  clinically useful, 556
  competition between bacteria, 555
  experimental tools, 556
Antibody, 485
  classes
  isotype switch, 497
  diversity, molecular basis, 496-98
Anticancer drug therapy, 160, 538
Anticodon, 129
Antidiuretic hormone, 393
Antigen, 485
Antigenic determinant, see epitope
Antihistamines, 344
Antimicrobial chemotherapy, 551
Antioxidants
  vitamin A, 446
  vitamin C, 45
  vitamin E, 447
Antithrombin
  binding of thrombin, 364
α₁-Antitrypsin, 527
Antiviral drug, 579-81
  targets for, Box 18.8 580
  therapy, 551
Antiviral vaccine
  attenuated virus, 578
  genetically engineered, 578
  killed virus, 578
  virus subunit, 578

Apoferritin, 453
Apolipoprotein, 287
Apoptosis, 162, Box 6.2 166-65, 354, Box 17.10 537
  in tumour initiation, Box 17.10 537
  of immature B cells, 489
Arachidonic acid, 283, 349, 423
Archaebacteria, see archea
Archea, Box 6.6 171
  evolution, Box 5.2 120
Arenaviruses
  poor immunogenicity, 578
Arginine , Box 11.2 334
  diet, 429
Argininosuccinase, 304
Argininosuccinate, 303
Argininosuccinate synthase, 303
Arthritis, Box 11.8 349, 350
Ascorbic acid, 312
  collagen synthesis, 473
  see also vitamin C
Asparagine, 307
Asparagine synthase, 308
Aspartate, 303, 307
  calcium binding proteins, 466
Aspartate aminotransferase, 306
Aspartate proteases, Box 14.2 416
Aspirin, 350
Asthma, 351
Atherosclerosis, 289-91
  calcium, 478
Atoms and molecules, 12
ATP, 8
  adenylate kinase, 191
  common intermediate , 190
  energy status, 190-2
  export from mitochondria, 210
  hydrolysis, 465
  structural features, 191
ATP synthase, 199, 208
ATPases, families of, 67
Atractyloside, 207
Atrial natriuretic peptide, 393
Autocrine factor, 40
Autocrine signalling, 333-4
Autoimmune disease, 399
Autoimmunity
  survival of antiself B cells, 490
Azidothymidine (AZT)
  anti-HIV drug, Box 18.8 580
  resistance to, 577

B cell, 485
  antibody production, 502
  different antigen recognition from T cell, 502
B cell clone
  antibody formation, 487
B lymphocyte, see B cell
Bacteria, 10, Box 6.6 171
  evolution, Box 5.2 120
Bacteria
  beneficial aspects, Box 18.1 552
  Gram negative, 555
  Gram positive, 555

585

Bacteria (*cont.*)
  in antibiotic production, Box 18.1 552
  in genetic engineering, Box 18.1 552
  in ruminant animals, Box 18.1 552
  infections, 551–7
  pathogenic toxins, 552
Basal lamina, 467
Base pair, 79, 148
  tautomeric forms, 159
Base pairing, 34
  codon-anticodon, 130
Basement membrane
  kidney, 390, 467
Bee sting, 350
Bence-Jones protein, 496
Benzodiazepines, 344
Beriberi, 448
Bicarbonate, 368
Bile acids, *see* bile salts
Bile pigments, digestive tract, 419
Bile salts, 418–22
  conjugation, 419
Bilirubin, 327, 375
Bilirubin diglucuronide, 375
Bilirubin-UDP-glucuronyl transferase, 328
Biliverdin, 328, 375
Binding protein, 82, Table 4.2 85, 85–88
  binding site, 86
Biotin, 449
  pyruvate carboxylase, 246
1,3–Bisphosphoglycerate, 233
2,3–Bisphosphoglycerate, 368
  binding to haemoglobin, 371
Bisphosphoglycerate isomerase, 376
Blood clotting, 356, 360–64
  extrinsic pathway, 362
  intrinsic pathway, 362
Blood glucose, 241
Blood pressure, 393
Blood vessels, 354
Bohr effect, 372
Bone, 451, 478
Bordetella pertussis, Box 11.1 337
Bradykinin, 350
Brain, 438
Branching enzyme, 225
Brittle hair syndrome, Box 6.2 166–65
Bruton's agammaglobulinaemia, Box 16.1, 487
Building block, 13
Burkitt's lymphoma
  chromosome translocation, 531
  Epstein-Barr virus, 531

C-peptide, 398
Ca^{2+}-ATPase
  absence from erythrocyte, 369
  sarcoplasmic reticulum, 57, 466
Caffeine, 393
Calcitonin, 347, 452
Calcitriol, *see* vitamin D
Calcium, 338, 340, 346, 399, 403
  bone, 478
  glycogen metabolism, 227
Calcium binding
  troponin, 463
Calcium binding proteins
  calsequestrin, 466
  high affinity , 466
  see also calmodulin

Calcium channels, 403
Calcium ion concentration
  muscle, 466
Calmodulin, Box 11.2 334, 340
  glycogen metabolism, 228
  smooth muscle, 463
Calsequestrin, 466
cAMP, *see* cyclic AMP
cAMP phosphodiesterase, 340
cAMP-activated kinase, 339
cAMP-dependent protein kinase, 464
cAMP-phosphodiesterase, 339
Cancer, 10
  genetic defect, 508
  molecular aspects, 528–39
  mutation and, 162
  somatic cell mutations, 529
  somatic cell theory, 508
5'-Capping of pre-mRNA, 123
Captopril, 393
Carbaminohaemoglobin, 372
Carbamoyl phosphate, 301–3
Carbamoyl phosphate synthetase, 107, 301–3, 316
Carbohydrate
  building blocks, 15
  conformations, 19
  energy source, 15
  membrane components, 15
  structural component, 15
Carbon dioxide
  binding to haemoglobin, 371
Carbon monoxide, 205
Carbonic anhydrase, 368, 372
  kidney, 392
Carbonic anhydrase, 88
Carboxyatractyloside, 210
Carboxypeptidases
  digestion, 416
  procollagen, 474
Carcinogens, 161, Box 6.4 16
  are mutagens, 168
  chemical, 532–33
  metabolic activation, 538
Cardiolipin, 26
Carnitine, 265
  exchange carrier, 266
  synthesis, 314
  transferase I, 265
β-Carotene, 445
Cartilage
  collagen, 471
Cascade amplification, 198
Catabolism, 6, 185–6
Catecholamine O-methyltransferase, 405
cDNA, *see* complementary DNA
CDP-diacylglycerol, 281
Cell, 40–51
  death , *see* apoptosis, necrosis
  division, 127
  organization, 40
Cell adhesion factor
  N-CAM, 506
Cell culture, Box 17.3 515
Cell cycle, S phase, 152
Cell wall (bacterial), 555
  target for penicillin, Box 18.2 557
Cellulase, 425
Cellulose, 425
Central Dogma, 78, 110
Ceramide, 283
Cerebrosides, 283
cGMP, *see* cyclic GMP

Chaperonin, protein assembly, 142
Chemiosmotic theory, Box 7.4 204, 213
Chenodeoxycholic acid, 419
Chloramphenicol, Box 18.3 553
Chloride, 368
Cholecystokinin, 414
Cholera toxin, Box 11.1.337
Cholesterase, 418
Cholesterol, 351, 406, 419
  biosynthesis, 284–5, 290–1
  functions, 28, 284
  membrane permeability, Box 5.5 140
  regulation, 291
  structure, 27
  synthesis, 143
  uptake, 290
  vitamin D, 446
Cholic acid, 419
Choline acetyltransferase, 404
Chondroitin sulphate, 469
Chromatin, 46, 151, 514
Chromium, 451
Chromosome, 46, 146, 149
  disorders, 517
  DNA in, 513
  nomenclature, 517
  nucleosomes in, 151
  staining patterns, 517
Chromosome number
  abnormal, 517
Chronic granulomatous, Box 16.1 487
Chylomicrons, 289, 381
Chymotrypsin, 460
Chymotrypsinogen, 417
Cigarette tar, carcinogens in, Box 6.4 169
Citrate lyase, 252
Citrate synthase, 214
Citrate transporter, 252–3
Citrate-malate transporter, 275
Citric acid cycle, 213–20
  amphibolic role, 219
  anaplerotic reactions, 219
  energy production , 217
  interactions with urea cycle, 305
  link to glycolysis, 220
  regulation, 220
  reversibility, 218
Citrulline, 303
Clearing factor lipase, 381
Clonal selection theory, 488
Clostridium histolyticum, 475
Cobalamin, *see* vitamin B12
Cobalt, 451
Cockagne syndrome, Box 6.2 164–65
Codominance, 511
Codon, 131
Coenzyme A, *see* pantothenic acid
Coenzymes
  vitamin derivatives, 443
Colipase, 418
Collagen, 84, 467–77
  hydroxylated amino acids in, 94
Collagenases
  bacterial, 475
  mammalian, 475
Compartmentation, 194
Complement, 484
  alternative pathway, Box 16.6 495
  cascade, Box 16.6 495
  classical pathway, Box 16.6 495
  Ig constant region, Box 16.6 495

Complementary DNA (cDNA)
  copying from RNA, 541
Conjugation
  glucuronates, 389
  hydroxylation, 389
  sulphates, 389
Connective tissue
  dense, 468–76
  loose, 467–9
Connexin, in gap junctions, 60
Consensus sequence
  signal sequences, 108
  start codon, 132
Cooperative binding , 197
Copper, 343, 451, 454
  -dependent elastin cross-linking, 477
  -dependent lysyl oxidase, 475
Cori cycle, 251, 436
Cori's disease, Box 8.2 230
Cornea
  collagen, 471
Corticoid binding protein, 352
Corticosterone, 351
Cortisol, 351
Covalent haem binding, 370
Covalent modification, 195
Creatine kinase, 384
Creatine phosphate, 384, 406
Creatinine, Box 13.1 385
Cross-bridges
  muscle, 459
Cross-linking
  collagen, 476
Cyanide, 205
Cyclic AMP, 227–9, 338, 393
Cyclic AMP phosphodiesterase, 393
Cyclic GMP, Box 11.2 334, 338, 407
Cyclic GMP phosphodiesterase, 407
Cyclo-oxygenase, 350
Cysteine, 308
  collagen, 472
Cysteine protease, Box 14.2 416
Cystic fibrosis
  cAMP-regulated chloride channel, 525
Cytochrome *c*, structure, 203
Cytochrome oxidase, 365, 454
Cytochrome *P*-450, Box 6.4 169, Box 7.5 206, 351, 365
  hydroxylation by, Box 17.6 524
Cytochromes, haem rings, 326
Cytokine, 500, Box 16.7 501
  anticancer therapy, Box 16.7 501
Cytoskeleton , 44, 49, Table 3.2, 50
  matrix, 403

DAG, *see* diacylglycerol
Debranching enzyme, 226
Dehydrogenase, two substrates, 89
Denaturation of proteins, Box 4.8 99
Deoxyribonucleotides in DNA replication, 160
Depolarization, 402
Dermatan sulphate, 469
Desaturase, 454
Desaturation, fatty acids, 279
Detoxification reactions, 389
Dexamethasone, Box 11.10 353
Di-iodotyrosine, 344

Diabetes
  insulin-dependent, 399
  insulin-independent, 399
  maturity onset of young, 399
  renal complications, 390
Diabetes mellitus, 68
Diacylglycerol, 284, 338, 339
Diacylglycerol lipase, 382
Diet, 294
DiGeorge syndrome, Box 16.1 487
Digestion, 412
Dihydrofolate reductase, 322
  thymidine synthesis, 322
Dihydroxyacetone phosphate, 233
1,25–Dihydroxycholecalciferol ,
  see vitamin D
Dihydroxyphenylalanine, 342
Dimethylallyl pyrophosphate, 286
2,4–Dinitrophenol, 207
Dipeptidases, 412
Disaccharidases, 412, 417
Disulphide bond, 83, Box 4.1 83
Diuretics, 393
DNA, 6
  Alu sequences, 514
  antisense strand, 113
  cloning into vectors, 541
  delivery to cells by liposomes,
    Box 17.15 546
  fingerprinting (profiling), 514
  in chromosomes, 513
  'junk', 110
  physical mapping, 541
  recombination, 146, 163, 170
  repeated sequences, 514, Box
    17.2 514
  sequencing, 149, 543, Box 17.13
    543
  stability, 160
  stacking, 150
  structure, 76, 147
  template strand, 113, 154
DNA binding protein, 148
  proto-oncogene, Box 17.7 532
  single strand, 157
DNA damage
  by external agents, 161
  deamination, 161
  depurination, 161
  spontaneous, 160
DNA glycosylase, 163, Box 6.2
  166–65
DNA helicase, 156
DNA ligase, 157
DNA polymerase, 79, 154–59
  fidelity, 154, 158
  in DNA repair, Box 6.2 166–65
  proofreading, 154
DNA polymerase α, 156
DNA polymerase δ, 156
  3'-5' exonuclease activity, 156
DNA primase, 157
  in DNA polymerase a, 156
DNA repair, 160–66
  defective, 538
  deficiency and cancer risk, Box
    17.8 533
  enzymes, 162
  excision, 163
  incision, 163
  xeroderma pigmentosum, Box
    6.2 166–65
DNA replication, 146, 151–160
  lagging strand, 157
  leading strand, 157
  origin of, 152–54
  replication fork, 152

DNA sequence
  interpretation of, 543
DNA structure
  antiparallel, 112
  complementary, 112
DNA topoisomerase, 156
DNA virus, 564, Box 18.5 568–9
  double-stranded, Box 18.5
    568–9
  oncogenes in, 530
DNA-binding domain
  protein p53, 100
DNP, *see* 2,4–dinitrophenol
Domain
  in immunoglobulins, 490
  in protein, 84, 86, 101–105, 367
Dopa, *see* dihydroxyphenylalanine
Dopa decarboxylase, 343
Dopamine, 342, Box 11.4 343, 404
Double helix, 146
Down's syndrome, 517
  trisomy, 517
Down-regulation receptors, 335
Drug design, knowledge-based,
  94, 580

E. coli DNA pol I, 155
  Klenow fragment, 155, Fig. 6.6
Ebola virus (filovirus)
  high pathogenicity, 559
EDRF, *see* endothelial derived
  relaxing factor
Ehler-Danlos syndrome, 475
Eicosanoids, 262, 332, 348
Eicosapentaenoic acid, 349
Elastase
  inhibition by α₁–antitrypsin 17,
    527
Elastin, 477
Electrically active tissues, 401
Electron transfer
  coenzymes, 188
  mechanisms, 186–8
  complexes, 199–206
Electron transport chain, 206
  inhibitors, 206
  linkage to proton gradient, 206
Electrostatic interactions, in
  protein structure, Box 4.10
  102
Emulsification
  digestion, 418
Endocrine gland, secretion of
  hormones, 40
Endocrine signalling, 333–4
Endocytosis, 44, 67, Box 3.1 49, 67
  insulin, 348
  receptor mediated, Box 3.1 49,
    143
  uptake of viruses, 565
Endonuclease, incision, 163
Endopeptidases, 416
Endoplasmic reticulum, 44
  rough, 47
  smooth, 47
Endosome, 48
Endosymbiotic theory, 26,
  Box 7.2 200
Endothelial derived relaxing
  factor, Box 11.2 334
Endotoxins, 554
Energy generation, glycolysis, 231
Energy investment, glycolysis,
  231–43
Energy transfer metabolism, 182
Enhancer, 115

Enolase, 234
Enoyl CoA hydratase, 267
Entactin, 390, 467
Enteroglucagon, 414
Enterohepatic circulation, 419–21
Enterokinase, 416
Enzyme, 88–94
  activation energy, 89
  active (catalytic) site, 89
  allosteric, 87
  basis of catalysis, 90
  covalent modification, 91
  denaturation, 91
  diagnosis of disease, 94
  inhibition, Box 4.4 93
  measurement of activity, Box
    4.3 92
  stereospecificity, 91
  types of reaction, 91
Enzyme induction, 195–6
  amino acid catabolism, 309
  tyrosine metabolism, 311
  urea cycle, 306
  vitamin D synthesis, 446
Epidermal growth factor, 353
  receptor, 72
Epinephrine, *see* adrenaline
Epithelial cell, 40
  apical surface, 40
  basolateral surface, 40
  intestinal, 395
  zonula adherens, 41
Epitope, 489
Epstein-Barr virus
  and human tumours, 534
  lymphotropic, 571
Equilibrium constant, Box 7.1 189
Erythrocyte, 356, 364–374
  breakdown, 328
  cytoskeleton, 366
  loss of nucleus, 44
  membrane, 366–69
  membrane proteins, 367
Erythromycin, Box 18.3 553
Erythropoiesis, 356, 365
Essential amino acids, 306
Essential fatty acids, 24, 279, 350
Ethanol metabolism, 253
N-Ethylmaleimide, 207
Eubacteria, *see* bacteria
Eucarya, Box 6.6 171
  see also eukaryote
Euchromatin, 514
Eukaryote, 113, Box 5.2 120
  parasitic, 582
Eukaryotic cell, 42
  internal organization , 42
  organelles, 44–51
Eukaryotic membrane,
  composition, 26
Evening primrose oils, Box 11.8
  349
Evolution, mitochondria, Box 7.2
  200
Excitory transmitters, 403
Exocytosis, Box 3.1 49
  transmitters, 401
3'-5' Exonuclease, proofreading,
  159
Exopeptidases, 416
Exotoxins, 553
Extracellular fluid, composition, 37
Extracellular matrix, 58, 467–70
Eye, rod cells, 406

Facilitated diffusion, 63

Factor VII, 362
Factor VIII, 362
  absence in haemophilia, 363
Factor VIII, in haemophilia A, 528
Factor IX, in haemophilia B, 528
Factor XII, 362
Factor XIII, transglutaminase, 364
FAD, *see* flavin adenine
  dinucleotide
Farnesyl pyrophosphate, 286
Fatty acid
  activation, 265
  binding protein, 421
  cellular functions, 262
  membrane component, 23
  mobilization, 263–4
  oxidation, 262–271
  polyunsaturated, 423–5
  saturated, 24, 423
  structure, 23
  synthesis, 274–9
  unsaturated, 24
Fatty acid oxidation
  energy production, 270
  genetic defects, Box 9.2 268
Fatty acid synthase, 275–6
Fatty acyl adenylate, 265
Fatty acyl CoA synthetase, 265
Fatty acyl groups on membrane
  proteins, 55
Fatty acyl transferase, 280
Fe, *see* iron
Ferritin, 453
Ferrous ions
  collagen synthesis, 473
  haem rings, 326
Fetal haemoglobin, 372
Fibre, dietary, Box 14.3 424
Fibrillin, 475
Fibrin, 84
Fibrinogen, 357, 362
Fibronectin receptor, 58
Filamin, 469
Filoviruses
  poor immunogenicity
  see Ebola, 578
Flavin adenine dinucleotide
  energy transfer, 188
  structure, 188
5–Fluorouracil, 160
Fluorouracil
  anticancer drug, 538
Fodrin, 469
Folic acid, 319, 323, 449
Follicle stimulating hormone, 351
Frameshift, 133
Free energy, 190
Fructokinase, 427
Fructose-1,6–bisphosphatase, 248
Fructose-1,6–bisphosphate, 231
Fructose-1–phosphate, 224
Fructose-2,6–bisphosphatase, 239
Fructose-2,6–bisphosphate, 239
Fructose-6–phosphate, 224
  pentose phosphate pathway,
    259
Fumarase, 217, 306
Functional group, 12
Furanose ring, 18

G-proteins, 227, 335–9, 383, 403,
  408
G6P dehydrogenase deficiency in
  haemolytic anaemia, 377
GABA, *see* γ-aminobutyric acid
GABA deaminase, 299

Index

Galactokinase, 196, 427
Galactose, 425
Galactosyl-1–phosphate uridylyl transferase, 428
Gall bladder, 419
Gangliosides, 284
GAP dehydrogenase, 368
Gap junction, 60
Gastric inhibitory peptide, 414
Gastric secretions, 350
Gastrin, 395, 414
Gene, 10, 110, 149
    dominance/recessiveness, 511
    insulin, 398
Gene expression, 110, 333
    cholesterol, 142–44
    hormone synthesis, 353
Gene therapy, 509
    of inherited defects, 545–47, Box 17.15 546
Genetic code, 78, 79, 130–36, Fig. 5.10
Genetic engineering, 80, 508–9, 539–47
Genome, 110, 146, 149
    yeast, as model for human, 513
Geranyl pyrophosphate, 286
Germ cells, 510
Globin
    deficient in thalassaemia, 526
    evolution, 365
    exons, 373
    gene expression, 373
    haem binding, 365
    structure, 370
Globular protein, 83, 84
α-Globulins, 357
β-Globulins, 358
γ-Globulins, 358
Glomerulus
    kidney, 390
Glucagon, 224, 227, 240, 335, 339, 347–8, 383, 397, 425, 434
    activation, 348
Glucocorticoids, 351
Glucokinase, 223, 388, 399
Gluconeogenesis, 236, 244–53, 442
    alcohol metabolism, 254
    control, Box 5.1 117–119
    energy consumption, 251
Glucose
    kidney reabsoption, 391
    metabolism, 222
    oxidation, energy yield, 244
    tolerance test, 391
    use by erythrocyte, 376
Glucose transporters, 55, 222–3, 347
    family of, 68–69, Table 3.6, 69
    GLUT 1, 367
    transmembrane segments, 69, 222, 226, 230, 249
Glucose-6–phosphatase, 386, 388
Glucose-6–phosphate transporter, Box 8.2 230
Glucose/fatty acid cycle, 437
Glucosuria, 391
Glucuronic acid
    hormone conjugation, 353
GLUT, see glucose transporters
Glutamate, 294–301, 307, 335, 404
    calcium binding proteins, 466
Glutamate dehydrogenase, 299, 301–2
    kidney, 392

Glutaminase
    kidney, 392
Glutamine, 294–301, 307
Glutamine synthase, 307
γ-Glutamyl cycle, 417
γ-Glutamyl transferase, 314
Glutathione, 83, 451
    amino acid transport, 312
    detoxification of superoxide, 378
    free radicals, 312
Glutathione peroxidase, 378, 451
Glutathione reductase, 378
Glyceraldehyde-3–phosphate, 233
    pentose phosphate pathway, 259
Glyceraldehyde-3–phosphate dehydrogenase, Box 5.1 117–119, 233, 236, 264, 377
Glycerol, 437
    gluconeogenesis, 252
Glycerol kinase, 264
Glycerol-3–phosphate, 382
α-Glycerolphosphate shuttle, 258
Glycine, 308, 404
    collagen, 472
Glycocholate, 419–21
Glycogen, 426
    metabolism, 222–30
    glycogen debranching enzyme, Box 8.2 230
Glycogen phosphorylase, 226–30
Glycogen phosphorylase kinase, 341
Glycogen storage disease, Box 8.2 230
Glycogen synthase, 224–5, 227–9
    glucagon inhibition, 348
Glycogenesis, 224, 434
Glycogenin, 225
Glycogenolysis, 225, 434
Glycolysis, 231–41
    control, Box 5.1 117–119, 238
    energy yield, 235
    free energy changes, 237
Glycophorin, 54, 367
Glycoprotein, 94
    structure, 21
Glycosaminoglycans, see mucopolysaccharides
Glycosyl transferases
    ABO blood groups, 369
Glycosylation, 394, 400
    collagen, 473
Goitre, Box 11.6 345
Golgi complex, 44, 47
    protein targeting, 138
Gonads, 351
Gramicidin, 207
Graves' disease, Box 11.5 345
gRNA
    in RNA editing, 583
Ground substance, 467
Growth factor, proto-oncogene, Box 17.7 532
Growth factor receptor, proto-oncogene, Box 17.7 532
Growth factors, 332, 335, 353
GTP-binding protein, proto-oncogene, Box 17.7 532
GTPase, 336
Guanylate cyclase, 407, Box 11.2 334

H⁺, binding to haemoglobin, 371
Haem, synthesis, 374
Haem oxygenase, 375

Haemoglobin, 356, 369–75
    allosteric properties, 92, 371
    degradation, 374–75
    haem rings, 326
    human mutations, 513
    in erythrocyte, 365–66
    in sickle cell anaemia, 512
    synthesis, 374
Haemolytic anaemia of newborn, Box 16.4, 492
Haemophilia, 528
Haemopoetin, 356
Hapten, 485
Haptoglobin, 375
Haworth projection, 18
HDL, 287–8
Heat shock protein, 141
    hsp 70, 141
Helices
    transmembrane, 404
α-Helix in proteins, 101, Box 4.11 104
Heparan sulphate, 390
    connective tissue, 467
Hepatitis B virus, 565
Hepatitis delta virus, similarity to viroids, Box 18.4 561
Hepatitis virus
    types A, B, C, D and E, Box 18.6 571
Hereditary nonpolyposis, Box 17.8 533
Herpes simplex virus, 534
    Acyclovir therapy, Box 18.9 580
    several diseases, 571
Herpes virus, 570
Heterochromatin, 514
Heterozygote, 511
Hexokinase, 223, 232, 386
    regulation, 239
Hexose, 17
High affinity calcium binding protein, 466
High energy molecules, 182
Histamine, 344
Histidine decarboxylase, 344
Histidine, in diet, 429
Histone, 151, 514
    basic properties, Box 4.5 96
HIV, 534
    binding to CD4 receptor, 577
    lentivirus, 577
    relationship to AIDS, 577
    resistance to AZT, 577
    specificity for T-lymphocytes 18, 577
    tat and rev genes, 577
HIV reverse transcriptase, mutation of, 108
HLA, 499, 502
HLA class I
    ankylosing spondylitis, Box 16.8 503
HLA class II
    insulin-dependent diabetes, Box 16.8 503
Homo-γ-linoleic acid, 349
Homocystinuria, Box 10.3 307
Homogentisate, 311
Homogentisate dioxygenase, 311
Homologous sequences, Box 6.6 171
Homozygote, 511
Hormonal control, 197
Hormone, 9, 40, 332
    receptor, 86

Hormone-sensitive lipase, 263, 437
HTLV and human leukaemias, 534
Human genome
    compared with simpler organisms, 513
Human Genome Project, 97, 110, 150, Box 17.16 547
Human growth hormone
    in vitro expression, 544
Human immunodeficiency virus (HIV), 574
Human leucocyte antigen (HLA), see MHC
Human papillomavirus and cervical carcinoma, 534
Human rhinovirus, structure and assembly, Figure 18.4 563
Hyaluronic acid, 469–70
Hyaluronidase, Box 15.1 469
Hydrochloric acid
    production in stomach, Box 14.1 413
Hydrogen bonds in protein structure, Box 4.10 102, Box 4.11 104
Hydrogen peroxide, 365
Hydrophobic interactions in protein structure, Box 4.10 102
3–Hydroxy acyl CoA dehydrogenase, 267
Hydroxyapatite, 454, 478
Hydroxylation reactions, collagen, 473
Hydroxylysine, collagen, 472
3–Hydroxy,3–methylglutaryl CoA reductase, 285
Hydroxyproline, collagen, 472
Hygiene, control of microbial pathogens, 551
Hypercalcaemia, 452
Hypercholesterolaemia
    an inherited disorder, 528
    LDL receptor, Box 9.8 291
Hyperglycaemia, 400
Hyperinsulinaemia, 443
Hyperpolarization, 407
Hypertension, Box 11.10 353, 393
Hypocalcaemia, 452
Hypoglycaemia, 400
    alcoholic, 254
Hypothermia, alcoholic, 254
Hypothyroidism, Box 11.7 346
Hypoxanthine, 323

I-bands, muscle, 457
Ibuprofen, 350
IDL, 287–8
Immune response, 484
    adaptive, 484
    cell-mediated, 550
    cellular, 485
    cellular, 499
    humoral, 485
    innate, 484
    memory, 484
    to microbial infection, 550
Immunization
    against bacteria, 555
    against viruses, 578
Immunodeficiency (ID), congenital, Box 16.1 487
Immunogen, 485

Immunoglobulin, 9, 485
  antigen recognition site, 491
  classes, 492
  constant region, 490, 495
  domains in, 102
  flexibility, 491
  F_{ab} region, 491
  F_c region, 491
  hypervariable region, 490, 496
  multiple domains, 490
  structure, 490–95
  superfamily, 504
  variable region, 490
Immunoglobulin A (IgA), 493
Immunoglobulin A, secretory
  (sIgA), Box 16.3 488
Immunoglobulin D (IgD), 494
Immunoglobulin E (IgE), 494
Immunoglobulin fold, 504
Immunoglobulin G (IgG), 493
  secondary response, 485
Immunoglobulin gene
  C gene, 497
  D gene, 497
  J gene, 497
  V gene, 497
Immunoglobulin M (IgM), 493
  primary response, 485
IMP aminotransferase, 377
In vitro expression of human
  genes, Box 17.14 544
Induction, Box 5.1 117–119
  enzymes, *see* enzyme induction
Influenza A virus, 572, Box 18.7
  572–3
  antigenic drift, Box 18.7 572–3
  antigenic shift, Box 18.7 572–3
  changing types, 578
  haemagglutinin, Box 18.7 572–3
  neuraminidase, Box 18.7 572–3
  neuraminidase inhibitors, Box
    18.9 580
  RNA segments, Box 18.7 572–3
Influenza virus, 559
  immune evasion, 502
  protein structures, 100
Information, 77, 80
Inherited disorders, 508
Inhibitor-1, 228
Inhibitory transmitters, 403
Inorganic constituent, 13
Inorganic ions, 6
Inositol-1,4,5–trisphosphate, 284,
  339
Insulin, 335, 347, 382, 397–8, 425,
  443
  secretion, 399
  receptors, 399
  response elements, Box 5.1
    117–119
Integral protein of membranes,
  106
Interleukin-2, Box 16.7 501
Intermediate filament, 50
Intrinsic factor, 421
Introns, Box 5.3 124
  within ORFs, 543
Iodine, 344, 451
IP_3, *see* inositol-
  1,4,5–trisphosphate
Iron, 451–2
  metabolism, Box 5.1 117–119
  response elements, Box 5.1
    117–119
Iron-sulphur centres, Box 7.3 202
Islets of Langerhans, 339, 347

Isocitrate dehydrogenase, 215
Isoelectric point of proteins, Box
  4.5 96
Isoenzyme, 107
Isomaltase, 417
Isopentenyl pyrophosphate, 285
Isoprenoid groups on membrane
  proteins, 55
Isotropic, 457

Jacob-Creutzfeld syndrome, prion
  agent, Box 18.4 561
Jaundice, Box 10.7 326
Juxtaglomerular apparatus, 393

Kallikrein, 362
Karyotype, 46, 513
  analysis of chromosomes, Box
    17.4 516
  of chromosomes, 515
Keratan sulphate, 469
Keratin, 445
Keto acids, 295–7
  branched-chain, 442
Ketogenesis, 389, 437
α-Ketoglutarate, 299–301
α-Ketoglutarate dehydrogenase,
  217
Ketone bodies, 400, 406, 436–8
  as nutrients, 273
  energy source, 274
  synthesis, 271–2
Ketose, 16
Kidney
  metabolism, 390
Kininogen, 362
Klinefelter's syndrome, 519
Krebs cycle, *see* citric acid cycle

lac operon, 115
Lactase, 418
Lactate, 242
  production by erythrocyte, 376
Lactate dehydrogenase, 236, 242,
  254
  isoenzymes, 107, Box 13.2 387
Lactoferrin, 454
Lactose, 424
  intolerance, 427
  structure, 19
Laemmli loops
  in chromosome, 514
Laminin, 390, 467
Lanosterol, 286
LDL, 287–8
LDL receptor, 292
  in hypercholesterolaemia, 528
Lecithin:cholesterol
  acyltransferase, 290
Lentivirus
  cytolytic retroviruses, 576
Leucocytes, 351, 356
Leukotrienes, 348, 350
Li-Fraumeni syndrome
  protein p53 mutation, 536
Ligandin, 328
N-Linked oligosaccharide, 139
Linoleic acid, 423
Linolenic acid, 423
Lipid
  membrane, 14
  triacylglycerol, 14
  bilayer, 7
  transport, 285–9
Lipo-oxygenases, 351
Lipoate, 243

Lipofuscin, 447
Lipopolysaccharide
  Salmonella endotoxin, 554
Lipoprotein classes, 287
Lipoprotein lipase, 381
Liposome, 57
  in gene therapy, Box 17.15 546
Liver, 431–2
  conjugation reactions, Box 10.7
    326
  metabolism, 388
Long chain fatty acids, synthesis,
  278
Low density lipoprotein, 143
Luteinizing hormone, 351
Lymphocytes
  source of chromosomes, Box
    17.4 516
Lysine
  collagen, 472
Lysophosphatides, 280
Lysosomes, 44, 48, 440–1
  protein targeting, 139
Lysozyme, bacteriostatic, in tears,
  484
Lysyl hydroxylase, 473
Lysyl oxidase, 454, 475

M-line
  muscle, 458
Macromolecule, 4, 6
Macrophages, Box 11.2 334
Magnesium, 451–2
Major histocompatibility complex,
  see MHC
Malaria parasite
  immune evasion, 502
Malate dehydrogenase, 217, 248
Malate-α-ketoglutarate
  transporter, 248
Malate-aspartate shuttle, 255–6
Malic enzyme, 277
Malonyl CoA, 275
Maltose, structure, 19
Marfan's syndrome, 475
McArdle's disease, Box 8.2 230
Meiosis, 46
Melanin, 312
Membrane, 7, 51–70
  composition, 51
  dynamic behaviour, 57–60
  electron microscopy, 52
  fluidity, 59
  fluidity, effect of cholesterol,
    60
  functions, 51
Membrane lipid
  'flip-flop', 59
  lateral diffusion, 57
Membrane protein, 53–57, 84
  anchor, 84
  diffusion, 57
  integral, 53–55, 84
  lipid anchor, 55
  membrane-spanning domain,
    84
  peripheral, 53
  transmembrane, 54–55
Membrane transport, 38, 63–70
  mechanisms, 63
Mendel's laws, 510
Menke's disease, 454
Messenger RNA (mRNA), 76, 113
  5'-cap, 134
  monocistronic, 113, 121
  polycistronic, 113, 121

Metabolic activation of
  carcinogens, 169
Metabolic pathways, general
  properties, 184–5
Metabolic regulation, 194
Methaemoglobin, 377
  increased by primaquine, 378
Methaemoglobin reductase, 378
Methionine, 308
Methotrexate, 160, 321
Methyl groups, reactive, 309
MHC,
  bare lymphocyte syndrome,
    Box 16.9 505
  binding of peptide fragments,
    499
  peptide recognition, 504
  polymorphic genes, 503
MHC I, 499, 503
MHC II, 499, 503
MHC III, 503
Micelles, digestion, 419
Michaelis-Menten kinetics, 91,
  Box 4.3 92
Microfilament, 50
Microtubule, 50
Microvilli, intestinal, 412
Mineralocorticoid, 351
Mismatch repair, 163, Box 6.2
  166–65
Mismatch repair, deficiency and
  cancer risk, Box 17.8 533
Mitochondrial complexes
  composition/ properties, 201
  diffusion in bilayer, 201
Mitochondrial genome, 521
Mitochondrial membrane
  transporters, 213–20
Mitochondrial structure, Box 7.4
  204
Mitochondrion, 44, 47–8
  contact sites, 63
  cristae, 47
  genetic defects, 521
  genome, 48
  inner membrane, 47
  intermembrane space, 47
  maternal inheritance, 521
  matrix, 47
  outer membrane, 47
  use of modified genetic code,
    521
Mitosis, 46, 517
Molecular chaperone, 112, 135,
  142
  in protein folding, 107
Molecular phylogeny, 78, 97, Box
  6.6 171
Molydenum, 451
  xanthine oxidase, 323
Monacylglycerol lipase, 382
Mono-iodotyrosine, 344
Monoaminoxidase, 405
Mucin, 22
Mucopolysaccharide, 469–70
Multifactorial diseases, 523
Multigene family, 504
Multiple myeloma, 489
Multiple sclerosis, 406
Muscle cells, 456
  red , 385
  white , 385
Muscle, metabolism, 384, 433
Mutagen, 161, 168
Mutagenesis
  site-directed, 545

Mutation, 162, 166–69
  chain-terminating, 168
  conservative, 108, 167
  deletion, 167
  effect on proteins, 108, 167–68
  frameshift, 108, 167, 168
  insertion, 167
  in microbial pathogens, 550
  radical, 108, 168
  silent, 167
  substitution, 108, 167–68
Myasthenia gravis, Box 13.6 405
Myelin, 406
Myocytes, *see* muscle cells
Myoglobin, 370, 386
  haem rings, 326
Myosin, 84, 458
Myosin light chain kinase, 463
Myosin light chains, 460, 464
Myxodema, Box 11.7 346

Na^+ channels, 407–8
Na^+-pump, 402
Na^+/glucose antiporter, 68
Na^+/K^+ ATPase, 64, 298, 367, 354, 391, 402
  inhibition by digitalis, 65
  mechanism of action, 65
  muscle, 456
NAD(H), *see* nicotinamide adenine dinucleotide
NADPH, 241
Nalidixic acid
  topoisomerase inhibitor, 556
NDP kinases, 316
Necrosis, 162
NEM, *see* N-ethyl maleimide
Nerve growth factor, 353
Neuraminidase inhibitors
  influenza therapy, Box 18.9 580
Neurotransmitters, 403
Niacin, *see* nicotinic acid
Nicotinamide adenine dinucleotide
  energy transfer, 188
  structure, 188
Nicotinamide adenine dinucleotide phosphate, 188
Nicotinic acid, 448
Nitric oxide, Box 11.2 334
Nitrogen, 294
Nitrogen balance, 430
Nitrogen equilibrium, 301, 428, 430
NMP kinases, 316
NMR of proteins, 100
Nonessential amino acids, 306
Noradrenaline, 342, 383, 404
Norepinephrine, *see* noradrenaline
Nuclear localization signal, 141, 144
Nuclear pore complex, 46, 63, 125
Nucleic acid
  information in sequences, 77
  sequence, 78
Nucleolus, 44, 46, 114, 127
Nucleoside diphosphate kinase, 216
Nucleoside kinases, 316
Nucleoside modification, 122
Nucleoside triphosphate, 79
Nucleosome, 151, 514
  on newly-replicated DNA, 160
Nucleotide
  coenzyme, 36
  metabolism, 315

  reoxidation, 254
  structure, 32
  sugar, 14
  terminology, Box 2.3 35
  triphosphate, 14
Nucleus, 44–46
  envelope, 46
Nutrient stores, 184

Obesity, 443
  diabetes, 400
Odd-numbered fatty acids
  oxidation, 267–8
Oestradiol, 353
Oestrogens, 353
Oleic acid
  antibacterial, in sweat, 484
Oligomeric structure
  of haemoglobin, 106
  of protein p53, 106
Oligopeptide, 32
Oncogene, 529
  in some retroviral genomes, 576
Open reading frame (ORF), 130–36, Box 6.1 150, 513
  identification of, 543
Opsin, 406
Organelle membranes, 44
Organic constituent, 13
Origin of life, 170
Origin of replication, mutation in, 166
Ornithine, 303
Ornithine transcarbamoylase, 303
Orthophosphate, mitochondrial concentration, 211
Osteoblasts, 346, 478
Osteocalcin, 478
Osteoclasts, 346, 478
Osteocytes, 478
Osteogenesis imperfecta, 475
Osteopontin, Box 15.4 478
Osteoporosis, Box 10.3 307, Box 15.5 479
Oxaloacetate, 308
α-Oxidation, 271
β-Oxidation pathway, 266–271
ω-Oxidation, 271
Oxidative phosphorylation, 199–213
Oxygen, 364–65
  binding to haemoglobin, 371–72

Pancreas
  endocrine function, 394
  exocrine function, 394
Pancreatic lipase, 418
Pancreozymin, 395, 414
Pantothenic acid, 449
Paracrine factor, 40
Paracrine signalling, 333–4
Pararetrovirus, 565, Box 18.5 568–9
Parasite
  and disease, 582
  gene expression, 582
Parathyroid hormone, 346, 452
Passive protection, Box 16.4 492
Pathways, reversibility, 197
Pellagra, 448
Penicillin
  allergic reaction, Box 16.2 487
  immunogenic properties, Box 16.2 487
Pentose, 17

Pentose phosphate pathway, 241, 258
  in erythrocyte, 377
Pepsin, 344, 412, 416
Pepsinogen, 416
Peptic ulcers, 344
Peptide bond, 31, 94–95
Peptide hormones, 346
Peptides, binding to MHCs, 499
Peptidoglycan transpeptidase
  inhibition by penicillin, Box 18.2 557
Peroxidation
  cells, 447
  foods, 447
  see also antioxidants
Peroxisome, 44, 49, Box 5.4 137
Phagocytes, 484
Phenylacetate, Box 10.5 312
Phenylalanine hydroxylase, 311
  in phenylketonuria, 511
Phenylalanine metabolism, 311
Phenylketonuria, Box 10.5 312, 508, 511
Phenyllactate, Box 10.5 312
Phenylpyruvate, Box 10.5 312
Phorbol esters, 340
Phosphatases, 316
Phosphate, 346, 454
Phosphate ester, 34
Phosphate transporter, 210
Phosphatidyl-4,5–bisphosphate, 339
Phosphatidylcholine, 282
Phosphatidylethanolamine, 282
Phosphatidylglycerol, 281
Phosphatidylinositol, 281
Phosphatidylinositol-4,5–bisphosphate, 282
Phosphatidylserine, 282
Phosphoanhydride bonds, energy content, 192
Phosphodiester bond, 37
Phosphoenolpyruvate, 234
Phosphoenolpyruvate carboxykinase, 248
Phosphofructokinase, 232
  citrate regulation, 239
  regulation, 239
Phosphofructokinase-2, 239
Phosphoglucoisomerase, 232
Phosphoglucomutase, 226
6–Phosphogluconate, 258
2–Phosphoglycerate, 234
3–Phosphoglycerate, 234
Phosphoglycerate kinase, 233–4
Phosphoglycerate mutase, 234
Phospholipase A2, 350
Phospholipase C, 70, 336, 339, 396, 404
Phospholipases, 284, 418
Phospholipid, 24
  bilayer, 28, 52
  composition, 25
  synthesis, 280
Phosphopantetheine, 276
Phosphoprotein, 94
Phosphoribosylpyrophosphate, 316
Phosphoryl transfer potential, 231
Phosphorylation
  enzymes, 333
Phosphorylation
  glycogen metabolism, 227
  pyruvate dehydrogenase, 243
Phytanic acid, 271

PIP_2, *see* phosphatidylinositol-4,5–bisphosphate
Placenta, 351
Plasma albumin, 359–60
  bilirubin binding, 359
  fatty acid binding, 359
  metal ion binding, 360
  nonpolar drug binding, 359
  osmoregulation, 359
  steroid hormone binding, 359
Plasma cell, 489
Plasma membrane, 40
  from Golgi complex, 40
  muscle, 456
  phagocytosis, 40
  pinocytosis, 40
  protein targeting, 138
Plasma proteins, 357–60
  binding properties, 359
  electrophoresis, 357, Fig. 12.1
  glycoprotein nature, 358
  ligands, Table 12.4
  synthesis in liver, 358
Plasmalogens, 283
Plasmid, Box 18.4 561
  drug resistance gene, 559
  in genetic engineering, Box 18.4 561
  multidrug resistance, 559
Plasmin, 364
Plasminogen
  drug activation, 364
Platelet-activating factor, 361
Platelet-derived growth factor, 353
Platelet-derived growth factor receptor, 72
Platelets, 350, 356
  role in blood clotting, 360
Polio virus, 571
  vaccines, Box 16.3 488
Poly A polymerase, 123
Polyadenylation, 123
  of pre-mRNA, 120
Polyadenylation
  in parasites, 583
  signal near ORFs, 543
Polycyclic hydrocarbon, mutagenic, Box 6.4 169
Polymerase, template-directed, 78, 79
Polymerization, 6
Polyomavirus, large T antigen, 570
Polypeptide
  amino terminus, Box 4.5 96
  carboxy terminus, Box 4.5 96
  see also protein
Polysaccharide, 6, 15, 77
Polysome, 51, 128
Polyunsaturated fatty acids, 424
Porphyrin metabolism, 326
Post-translational modification of membrane proteins, 55
Postsynaptic cell, 401
Postsynthetic modification of protein, 84, 107
Pre-mRNA
  processing, 123–27
  tracking, 127
Pre-rRNA, 113
  cleavage, 122
Pre-tRNA
  self-cleavage, 130
Pregnenalone, 351
Presynaptic cell, 401
Primary hyperoxaluria, Box 5.4 137

Primary structure of proteins, 94–98
Prion, Box 18.4 561
Procarboxypeptidase, 417
Processing of RNA, 111, 112, 121–27
Procollagen, 472
Proelastase, 417
Progenote, Box 5.2 120
Progesterone, 351
Prohormones, 346
Prokaryotes, 113
    evolution, 113
Prokaryotic cell, 43
Proline, 307
    collagen, 472
Prolyl hydroxylase, 473
Prolyl isomerase, 135
Promoter sequence, 115
    mutation in, 166
Promoter sequences
    presence near ORFs, 543
Propionyl CoA, metabolism, Box 9.3 270
Prostaglandins, 348
Protein, 6, 76
    3D structure, 86
    amino acid sequence, Box 4.5 96
    catalytic, 82
    charge, Box 4.5 96
    cytoskeletal, 84
    dispensable, 441
    fibrous, 83
    globular, 83
    homologous, 97
    indispensable, 441
    intracellular, 82
    molecular mass, Box 4.5 96
    oligomeric structure, 105
    sequence, 78
    structural, 82
Protein folding, 101–102, 135
    heat-shock proteins, 141
    molecular chaperones, 107
Protein kinase C, 339–40
Protein p53, 97, 80, 154
    anti-oncogene, 535–36, Box 17.9 536
    DNA binding by, Box 17.9 536
    domain structure, Box 17.9 536
    domains in, 86
    human colorectal cancer, 536
    Li-Fraumeni syndrome, Box 17.9 536
    mutations in human cancer, Box 17.9 536
    structure, 100
Protein superfamily, 108, 504
Protein synthesis
    inhibition by antibiotics, Box 18.3 553
Protein turnover, 430, 440
Prothrombin, 362
Proto-oncogene, 530, Box 17.7 532
Proton pump
    lysosomal, 440
Protoporphyrin IX, 374
PRPP, see phosphoribosyl pyrophosphate
PRPP synthase, 316
Psoriasis, Box 11.8 349
PTH, see parathyroid hormone
Public health
    control of microbial pathogens, 551

Pulmonary surfactant, Box 9.5 281
Purine, 315
    biosynthesis, 316
    hydrogen bonding, 34
    structure, 33
    pyranose ring, 18
Pyridoxal phosphate, 297–8, Box 10.3 307, 325, 343
Pyridoxine, see vitamin B6
Pyrimidine, 315
    hydrogen bonding, 34
    structure, 33
Pyrophosphate, 35
Pyruvate carboxylase, 220, 246
    acetyl CoA activation, 250
    ADP inhibition, 250
Pyruvate carboxylase, 436
Pyruvate dehydrogenase, 436
    acetyl CoA inhibition, 250
Pyruvate dehydrogenase complex, 213, 242–4
Pyruvate kinase, 234, 436
    acetyl CoA inhibition, 250
    ADP inhibition, 250
    allosteric regulation, 240
    isoenzymes, 240
    phosphorylation, 240
    regulation, 240
Pyruvate, metabolism, 241

Quaternary structure, 86
    of proteins, 105–106

Rabies virus, poor immunogenicity, 578
Radiation, DNA damage and tumour induction, 534
ras oncogene
    mutations in cancers, 531
Rb protein
    anti-oncogene, 535
Receptor
    cell surface, 70–75
    channel linked, 335
    enzyme linked, 335
    G-protein linked, 70, 335
    growth factor, 72
    hormone, 70
    insulin, 72
    intrinsic tyrosine kinase, 72
Recombination, 510
    antibody classes, 497
    of immunoglobulin genes, 497
Red blood cell, see erythrocyte
Reductive biosynthesis, role of NADP, 188
Refsum's disease, 271
Renal glomerulus, 467
Renal osteodystrophy, Box 14.7 452
Renal threshold, glucose, 391
Renaturation of proteins, Box 4.8 99
Renin, 393
Repolarization, 402
Repression, 116
Respiratory chain, 199–213
Respiratory control, 210
Respiratory distress syndrome, 282
Resting potential, 402
Restriction endonucleases, 541
11-cis-Retinal , 406
Retinal isomerase, 407
Retinol, 445
Retinol binding protein, 445

Retrovirus, 564, 574–77, Box 18.5 568–9
    gag-pol-env genome, 575
    long terminal repeat, 575
    oncogenes in, 529
    presence of oncogenes, 576
    provirus, 575
    rapid mutation, 576
Reverse transcriptase
    in retrovirus replication, 575
    synthesis of cDNA, 541
Rhesus antigen, Box 16.4 492
Rhinovirus, 570
    many serotypes, 578
Rhodopsin, 406
Riboflavin, see vitamin B2
Ribose
    methylation, 122
    nucleotide synthesis, 315
Ribose-5–phosphate, 241, 258
Ribosomal RNA (rRNA), 113
    evolution, Box 5.2 120, 123
Ribosome, 44, 51, 112, 134–39
    bacterial, 128
    mitochondrial, 128
    peptidyl transferase, 112
    signal recognition particle, 138
    subunit, 127
    subunit export, 125
Ribosome cycle, 128
Ribozyme, 89, 130
Ribulose-5–phosphate, 258
Rickets, Box 15.5 479
Rigor complexes, 465
RNA, 6
    as original genetic material, Box 6.5 170
RNA editing
    in parasites, 583
RNA polymerase, 79, 112
    initiation, 114
    three forms of, 107
RNA polymerase I, 127
RNA polymerase II, Box 5.1 117–119, 123
    TF II in DNA repair, Box 6.2 166–65
RNA polymerase II
    in retrovirus replication, 575
RNA polymerase III, 130
RNA virus, 564
    negative strand, Box 18.5 568–9
    positive strand, Box 18.5 568–9
'RNA world', 130, 564, 583
RNase H, 157
Robison's Booster theory, 479
Rod cells, 406
Rough endoplasmic reticulum, 112
    ribosomes, 128, 136
    sterol sensor, 143

Salmonella, endotoxin variability, 554
SAM, see S-adenosyl methionine
Sarcolemma, 456
Sarcomere, muscle, 457–9
Sarcoplasmic reticulum, 456, 464
Schiffs base, 297
Scurvy, Box 14.6 450
Second messengers, 338
Secondary structure of proteins, 100–101, Box 4.11 104
Secretagogues, digestive hormones, 395
Secretin, 395, 414

Secretory protein, 47
    protein targeting, 138
Selective toxicity, 555
Selenium, 451–2
Sequence, information in, 543
Sequence of macromolecules, 78
Serine, 308
    phosphorylation, 339
Serine deaminase, 297
Serine hydroxymethylase, 308
Serine proteases, Box 14.2 416
Serotonin, 344
Serum albumin, see plasma albumin
Severe combined immunodeficiency, Box 10.6 320, Box 16.1 487
Sex chromosome, abnormal, 519
Sex hormone binding globulin, 352
β-Sheet in proteins, 101, Box 4.11 104
Sickle cell anaemia, 508
    defective haemoglobin, 512
Signal recognition particle, 138, Box 5.5 140
Signal sequence, 136–141
    hormones, 346
Signal transduction, 335, 338
Signals, 6
    hormones, 332
Single gene defect, 524–28
    autosomal nature, 511
Skin, collagen, 471
Sliding filament theory, 458
Smooth muscle, Box 11.2 334, 351, 463
Snake venom, Box 9.6 284
SnRNA, 125
Somatostatin, 347, 397
Spectrin, 368
Sphingolipid, 24
Sphingomyelin, 283
Sphingosine, 283
β-Spiral, elastin, 477
Spliceosome, 125
Splicing
    of HIV mRNAs, 576
    of pre-mRNA, 125
    trans-, in parasites, 582
Squalene, 286
SRE-binding protein, 143
Standard free energy, Box 7.1 189
    phosphorylated compounds, 193
Starch, 417, 425
    amylopectin, 19
    amylose, 19
Start codon, 133
Starvation, 437–9, 441–2
Steroid hormone, 332, 341–2, 351
Sterol regulatory element, 143
Stomach pH
    antibacterial action, 484
Stop codon, 120, 132
Streptomycin, Box 18.3 553
Substrate, 88
Substrate level phosphorylation, 231, 234
Succinate dehydrogenase, 216
Succinyl-CoA synthetase, 216
Succinyl-CoA:3–oxoacylCoA transferase, 389
Sucrase, 418, 428
Sucrose, 424
    structure, 19

Index

Sudden infant death , Box 9.1 267
Sugar, *see* polysaccharide
Sunflower seed oils, Box 11.8 349
Superhelix, collagen, 470
Superoxide dismutase, 378, 454
Superoxide radical, 365, 377
Supersecondary structure in
    proteins, 105
Suxamethonium, 404
SV40 virus, large T antigen, 154
Synapses, 401
Synaptic cleft, *see* synapses
Synaptic gap, *see* synapses

T cell, 485
    cytotoxic (CD8$^+$), 500
    functions, 499–506
    helper (CD4$^+$), 489, 500
T cell receptor, 499
    similarity to Ig, 500
T lymphocyte, 485
    see also T cell,
T-tubule, *see* transverse tubule
Targeting
    defects, Box 5.4 137
    of immunoglobulin, 139
    of protein, 82, 85, 107, 112
    of protein to mitochondria, 139
    of protein to nucleus, 141
    stop-transfer sequence, 139
'TATA Box ', 115, Box 5.1,
    117–119
Taurine, 404
Taurocholate, 419–21
Temperature regulation, *see*
    thermogenesis
Template, 79
    copying, Box 6.5 170
Tendon, collagen, 471
Teratogen, 161
Tertiary structure, 86
    of proteins, 100–105
Testis determining factor,
    transcriptional regulator, 520
Testosterone, 353
Tetra-iodothyronine, 344
Tetracyclines, Box 18.3 553
Tetrahydrobiopterin, Box 10.4 311
Tetrahydrofolate, 325
Thalassaemia, 526
Theobromine, 393
Theophylline, 393
Therapeutic index, of drugs, 555
Thermodynamics, Box 7.1 189
    phosphoglycerate kinase, 236
Thermogenesis, 383, 443
Thiamine, *see* vitamin B1
    deficiency, 243
Thiamine pyrophosphate, 448
Thick filaments, 460
Thin filaments, 457
Thioredoxin, 320
Three-dimensional structure of
    proteins, 98–106
Threonine, phosphorylation, 339
Threonine deaminase, 297
Thrombin, action on fibrinogen,
    362–64

Thromboxanes, 348
Thymidine kinase in S phase, 160
Thymidine nucleotides, inhibition
    of synthesis, 538
Thymidylate synthetase, 320
Thyroglobulin, 344–5
Thyroid gland, 344
Thyroid hormones, 332, 341–2, 344
Thyroid stimulating hormone,
    344
Thyroid-binding globulin, 345
Thyrotoxicosis, Box 11.5 345
Thyrotropin, *see* TSH
Tight junction, 61
Titratable acidity, 393
Tocopherols, *see* vitamin E
trans-Golgi network, 47
Transaldolase reactions, 258
Transcription, 111, 112–21, 149
    bacterial, Box 5.1 117–119
    basal, 116
    control, Box 5.1 117–119
    error rate, 116
    factor, 114, Box 5.1 117–119,
        144
    hormone response, 116
    multicistronic, in parasites, 582
    start site, 113
    termination, 120
Transcription factor, 166
    in DNA repair, 163
Transducin, 336, 408
Transfer RNA (tRNA), 79, 113,
    129–30,
    3'-CCA terminus, 129
    stem-loops, 129
Transferrin, 446, 453
Transferrin receptor, Box 5.1
    117–119
Transketolase reactions, 258
Translation, 111, 127–36,
    accessory proteins, 128
    cap independent, 136
    elongation, 135
    initiation, 134
    termination, 135
    use by viruses, 136
Transmembrane pump, 64
Transmitters, 332
Transport, 6
Transporter proteins, endoplasmic
    reticulum, 249
Transposon
    drug resistance gene, 559
Transverse tubules, 456
Tri-iodothyronine, 344
Triacylglycerol, 24
    synthesis, 279
    fuel storage molecules, 262
    metabolism, 380
    re-esterification, 382
Triacylglycerol lipase, 382
Tricarboxylic acid cycle , *see* citric
    acid cycle
Triose phosphate isomerase, 233
Trisomy, gene dosage, 518
Tropomyosin, 462
Troponin, 462

Trypanosome, 582
    'gene switching', 584
    surface antigen (VSG), 584
Trypsin, 416
Trypsin inhibitor, 417
Trypsinogen, 417
TSH, 345
Tubulin, 44
Tumour initiator, 533
Tumour promoters, 340, 533
Tumour suppressor gene,
    see anti-oncogene
b-Turn in proteins, 101, Box 4.11
    104
Tyrosinase, 312
Tyrosine hydroxylase, 342
Tyrosine kinase, 335, 354
    proto-oncogene, Box 17.7 532
Tyrosine metabolism, 311
Tyrosine phosphatase, 337

Ubiquitin, 440
UDP-glucose, 224–5
UDP-glucose pyrophosphorylase,
    225
Ulcers, Box 11.9 350
Uncoupling, mitochondria, 209–10
Unsaturated fatty acids , 267–9
'Upstream', 115
Urea, 298
Urea cycle, 301–7
    enzyme deficiencies, Box 10.2 304
Uric acid, 323
Uridine diphosphate, *see* UDP
Urobilin, 328
Urobilinogen, 328

Vaccination, 484
Van der Waals forces, Box 4.10 102
Varicella-zoster virus
    neurotropic, 571
Vasodilation, Box 11.2 334
Vasopressin, 339, 393
Vector
    expression, 544
    for cloning of DNA, 541
    for microbial pathogens, 550
    viruses as, 547, Box 17.15 546
Vibrio cholerae, Box 11.1 337
Viral genome, 563
    DNA, 564
    replication, 567
    RNA, 564
    single-stranded, 570
Viral infections, diabetes, 399
Viral mRNA
    central role, 567
Viral protein
    non-structural, 567
    structural, 567
Virion, 559
    assembly and maturation, 570
    encapsidation signal, 570
    enveloped, 563
    morphology, 561
    naked, 563
    uncoating, 565
Viroid, Box 18.4 561

Viruses, 10, 136, 559–81
    and disease, 570–74
    and human cancer, 534
    cell transformation, 566
    classification, Box 18.5 568–9
    families, 565, Box 18.5 568–9
    human cancer, Table 18.11 574
    latency, 566
    lytic growth, 561
    non-cellular nature, 559
    persistence, 566
    tissue tropisms, 571
    upper respiratory infection, 570
Virus binding, cell surface
    receptor, 565
Virus life cycle
    replicative phase, 559
    virion phase, 559
Virus particle, *see* virion
Virus replication
    host cell functions, 565
    programmed events, 565
Vitamin A, 406, 445
Vitamin B$_1$, 448
Vitamin B$_2$, 448
Vitamin B$_6$, 297, 448
Vitamin B$_{12}$, 326, 449
Vitamin C, 450
    see also ascorbic acid
Vitamin D, 346, 446, 452
    absorption, 419
    bone formation, 478
Vitamin E, 447
Vitamin K, 447
Vitamins
    fat soluble, 445
    water soluble, 443
VLDL, 287–8
Von Gierke's disease, Box 8.2 230
Von Willebrand factor, 361, Box
    12.1 361

Weak forces, *see* weak interactions
    ligand interaction, 86
Weak interactions
    in haemoglobin, 370
    in protein structure, Box 4.10 102
Whooping cough, Box 11.1 337
Wilson's disease, 454
'Wobble', 129

X-ray diffraction, 370
    double helical model of DNA,
        Box 4.9 100
    in biology, Box 4.9 100
    of proteins, 98–100
Xanthine, 210, 323
Xanthine oxidase, 323, 365
Xenobiotics, 389
Xeroderma pigmentosum, Box 6.2
    166–65
    defective DNA repair, 524
    human genes in, Box 6.2 166–65
Xylulose-5–phosphate, 258
Z-lines, 457
Zinc, 451
Zymogens, 394
    digestive enzymes, 415